T0368017

EQUILIBRIUM AND NON-EQUILIBRIUM STATISTICAL THERMODYNAMICS

This book gives a self-contained exposition at graduate level of topics that are generally considered fundamental in modern equilibrium and non-equilibrium statistical thermodynamics.

The text follows a balanced approach between the macroscopic (thermodynamic) and microscopic (statistical) points of view. The first half of the book deals with equilibrium thermodynamics and statistical mechanics. In addition to standard subjects, such as the canonical and grand canonical ensembles and quantum statistics, the reader will find a detailed account of broken symmetries, critical phenomena and the renormalization group, as well as an introduction to numerical methods, with a discussion of the main Monte Carlo algorithms illustrated by numerous problems. The second half of the book is devoted to non-equilibrium phenomena, first following a macroscopic approach, with hydrodynamics as an important example. Kinetic theory receives a thorough treatment through the analysis of the Boltzmann–Lorentz model and of the Boltzmann equation. The book concludes with general non-equilibrium methods such as linear response, projection method and the Langevin and Fokker–Planck equations, including numerical simulations. One notable feature of the book is the large number of problems. Simple applications are given in 71 exercises, while the student will find more elaborate challenges in 47 problems, some of which may be used as mini-projects.

This advanced textbook will be of interest to graduate students and researchers in physics.

MICHEL LE BELLAC graduated from the Ecole Normale Supérieure and obtained a Ph.D. in Physics at the Université Paris-Orsay in 1965. He was appointed Professor of Physics in Nice in 1967. He also spent three years at the Theory Division at CERN. He has contributed to various aspects of the theory of elementary particles and recently has been working on the theory of the quark–gluon plasma. He has written several textbooks in English and in French.

FABRICE MORTESSAGNE obtained a Ph.D. in high-energy physics at the Université Denis Diderot of Paris in 1995, and then was appointed Maître de Conférences at the Université de Nice–Sophia Antipolis. He has developed semi-classical approximations of wave propagation in chaotic systems and was one of the initiators of the 'Wave Propagation in Complex Media' research group. In

1998 he extended his theoretical research activities with wave chaos experiments in chaotic optical fibres and microwave billiards.

G. GEORGE BATROUNI obtained a Ph.D. in theoretical particle physics at the University of California at Berkeley in 1983 and then took a postdoctoral fellowship at Cornell University. In 1986 he joined Boston University and later the Lawrence Livermore National Laboratory. He became professor at the Université de Nice–Sophia Antipolis in 1996. He was awarded the Onsager Medal in 2004 by the Norwegian University of Science and Technology. He has made important contributions in the development of numerical simulation methods for quantum field theories and many body problems, and in the study of quantum phase transitions and mesoscopic models of fracture.

EQUILIBRIUM AND NON-EQUILIBRIUM STATISTICAL THERMODYNAMICS

MICHEL LE BELLAC, FABRICE MORTESSAGNE
AND G. GEORGE BATROUNI

CAMBRIDGE
UNIVERSITY PRESS

University Printing House, Cambridge CB2 8BS, United Kingdom

Cambridge University Press is part of the University of Cambridge.

It furthers the University's mission by disseminating knowledge in the pursuit of education, learning and research at the highest international levels of excellence.

www.cambridge.org
Information on this title: www.cambridge.org/9780521821438

First published 2004
Reprinted 2006

A catalogue record for this publication is available from the British Library

Library of Congress Cataloguing in Publication data

Le Bellac, Michel.
Equilibrium and non-equilibrium statistical thermodynamics/Michel Le Bellac, Fabrice Mortessagne, G. George Batrouni.
p. cm.
Includes bibliographical references and index.
ISBN 0 521 82143 6
1. Thermodynamic equilibrium. 2. Irreversible processes. 3. Statistical thermodynamics.
I. Mortessagne, Fabrice, 1966– II. Batrouni, G. George (Ghassan George), 1956– III. Title.
QC318.T47L43 2004
536′.7–dc22 2003055894

ISBN 978-0-521-82143-8 Hardback

Contents

Preface

This book attempts to give at a graduate level a self-contained, thorough and pedagogic exposition of the topics that, we believe, are most fundamental in modern statistical thermodynamics. It follows a balanced approach between the macroscopic (thermodynamic) and microscopic (statistical) points of view.

The first half of the book covers equilibrium phenomena. We start with a thermodynamic approach in the first chapter, in the spirit of Callen, and we introduce the concepts of equilibrium statistical mechanics in the second chapter, deriving the Boltzmann–Gibbs distribution in the canonical and grand canonical ensembles. Numerous applications are given in the third chapter, in cases where the effects of quantum statistics can be neglected: ideal and non-ideal classical gases, magnetism, equipartition theorem, diatomic molecules and first order phase transitions. The fourth chapter deals with continuous phase transitions. We give detailed accounts of symmetry breaking, discrete and continuous, of mean field theory and of the renormalization group and we illustrate the theoretical concepts with many concrete examples. Chapter 5 is devoted to quantum statistics and to the discussion of many physical examples: Fermi gas, black body radiation, phonons and Bose–Einstein condensation including gaseous atomic condensates.

Chapter 6 offers an introduction to macroscopic non-equilibrium phenomena. We carefully define the notion of local equilibrium and the transport coefficients together with their symmetry properties (Onsager). Hydrodynamics of simple fluids is used as an illustration. Chapter 7 is an introduction to numerical methods, in which we describe in some detail the main Monte Carlo algorithms. The student will find interesting challenges in a large number of problems in which numerical simulations are applied to important classical and quantum models such as the Ising, XY and clock (vector Potts) models, as well as lattice models of superfluidity.

Kinetic theory receives a thorough treatment in Chapter 8 through the analysis of the Boltzmann–Lorentz model and of the Boltzmann equation. The book ends with general non-equilibrium methods such as linear response, the projection method, the fluctuation-dissipation theorem and the Langevin and Fokker–Planck equations, including numerical simulations.

We believe that one of this book's assets is its large number of exercises and problems. Exercises pose more or less straightforward applications and are meant to test the student's understanding of the main text. Problems are more challenging and some of them, especially those of Chapter 7, may be used by the instructor as mini-research projects. Solutions of a selection of problems are available on the web site.

Statistical mechanics is nowadays such a broad field that it is impossible to review in its entirety in a single volume, and we had to omit some subjects to maintain the book within reasonable limits or because of lack of competence in specialized topics. The most serious omissions are probably those of the new methods using chaos in non-equilibrium phenomena and the statistical mechanics of spin glasses and related subjects. Fortunately, we can refer the reader to excellent books: those by Dorfman [33] and Gaspard [47] in the first case and that of Fisher and Hertz [42] in the second.

The book grew from a translation of a French version by two of us (MLB and FM), *Thermodynamique Statistique*, but it differs markedly from the original. The text has been thoroughly revised and we have added three long chapters: 4 (Critical phenomena), 7 (Numerical simulations) and 9 (Topics in non-equilibrium statistical mechanics), as well as a section on the calculation of transport coefficients in the Boltzmann equation.

1

Thermostatics

The goal of this first chapter is to give a presentation of thermodynamics, due to H. Callen, which will allow us to make the most direct connection with the statistical approach of the following chapter. Instead of introducing entropy by starting with the second law, for example with the Kelvin statement 'there exists no transformation whose sole effect is to extract a quantity of heat from a reservoir and convert it entirely to work', Callen assumes, in principle, the existence of an entropy function and its fundamental property: the principle of maximum entropy. Such a presentation leads to a concise discussion of the foundations of thermodynamics (at the cost of some abstraction) and has the advantage of allowing direct comparison with the statistical entropy that we shall introduce in Chapter 2. Clearly, it is not possible in one chapter to give an exhaustive account of thermodynamics; the reader is, instead, referred to classic books on the subject for further details.

1.1 Thermodynamic equilibrium

1.1.1 Microscopic and macroscopic descriptions

The aim of statistical thermodynamics is to describe the behaviour of macroscopic systems containing of the order of $N \approx 10^{23}$ particles.[1] An example of such a macroscopic system is a mole of gas in a container under standard conditions of temperature and pressure.[2] This gas has 6×10^{23} molecules[3] in incessant motion, continually colliding with each other and with the walls of the container. To a first approximation, which will be justified in Chapter 2, we may consider these molecules as classical objects. One can, therefore, ask the usual question of classical mechanics: given the initial positions and velocities (or momenta) of the

[1] With some precautions, one can apply thermodynamics to mesoscopic systems, i.e. intermediate between micro- and macroscopic, for example system size of the order of 1 μm.

[2] The reader will allow us to talk about temperature and pressure even though these concepts will not be defined until later. For the moment intuitive notions of these concepts are sufficient.

[3] In the case of a gas, we use the term 'molecules' instead of the generic term 'particles'.

molecules at $t = 0$, what will the subsequent evolution of this gas be as a function of time? Let us, for example, imagine that the initial density is non-uniform and ask how the gas will evolve to re-establish equilibrium where the density is uniform. Knowing the forces among the molecules and between molecules and walls, it should be possible to solve Newton's equations and follow the temporal evolution of the positions, $\vec{r}_i(t)$, $i = 1, \ldots, N$, and momenta, $\vec{p}_i(t)$, as functions of the positions and momenta at $t = 0$. We could, therefore, deduce from the trajectories the evolution of the density $n(\vec{r}, t)$. Even though such a strategy is possible in principle, it is easy to see that it is bound to fail: if we simply wanted to print the initial coordinates, at the rate of one coordinate per microsecond, the time needed would be of the order of the age of the universe! As for the numerical solution of the equations of motion, it is far beyond the capabilities of the fastest computers we can imagine, even in the distant future. This kind of calculation, called molecular dynamics, can currently be performed for a maximum of a few million particles.

The quantum problem is even more hopeless: the solution of the Schrödinger equation is several orders of magnitude more complex than that of the corresponding classical problem. We keep in mind, however, that our system is, at least in principle, susceptible to a microscopic description: positions and momenta of particles in classical mechanics, their wave function in the quantum case. If this information is available, we will say that a system has been attributed a *microscopic configuration* or *microstate*. In fact, this microscopic description is too detailed. For example, if we are interested, as above, in the temporal evolution of the density of the gas, $n(\vec{r}, t)$, we have to define this density by considering a small volume, ΔV, around the point $\vec{r}$, and count (at least in principle!) the average number of gas molecules in this volume during a time interval, Δt, centred at t. Even though ΔV is microscopic, say of the order of 1 μm on a side, the average number of molecules will be of the order of 10^7. We are only interested in the average number of molecules in ΔV, not in the individual motion of each molecule. In a macroscopic description, we need to make spatial and temporal averages over length and time scales that are much larger than typical microscopic scales. Length and time scales of 1 μm and 1 μs are to be compared with characteristic microscopic scales of 0.1 nm and 1 fs for an atom. In this averaging process, only a small number of combinations of microscopic coordinates will play a rôle, and not each of these coordinates individually. For example, we have seen that to calculate the density, $n(\vec{r}, t)$, we have to count all molecules found at time t in the volume ΔV around point $\vec{r}$, or, mathematically,

$$n(\vec{r}, t) = \frac{1}{\Delta V} \int_{\Delta V} \mathrm{d}^3 r \sum_{i=1}^{N} \delta\left(\vec{r} - \vec{r}_i(t)\right) \tag{1.1}$$

This equation selects a particular combination of positions, $\vec{r}_i(t)$, and gives an example of what we will call *macroscopic variables*. Another example of a combination of microscopic coordinates yielding a macroscopic variable will be given below for the energy (Equation (1.2)).

Since the microscopic approach leads to a dead end, we change descriptions, and take as fundamental quantities global macroscopic variables related to the sample: number of molecules, energy, electric or magnetic dipole moment, etc. Macroscopic variables, or more precisely, their densities (density of molecules, of energy etc.) define a *macrostate*. The evolution of macroscopic variables is governed by deterministic equations: Newton's equations for elastic objects, Euler's equations for fluids, Maxwell's equations for electric or magnetic media, etc. However, this purely mechanical description is insufficient since a macrostate is compatible with a very large number of different microstates. Therefore, we cannot forget the microscopic degrees of freedom that have been eliminated by averaging. For these microscopic degrees of freedom, which, for the moment, we have ignored in the macroscopic approach, we will use a probabilistic description; this will in turn lead to the concept of entropy, which is needed to complete our macroscopic picture. This probabilistic approach will be discussed in Chapter 2. Contrary to other macroscopic variables, *entropy is not a combination of microscopic variables*: it plays a singular rôle compared to other macroscopic quantities.

In the remainder of this chapter, we will limit ourselves to a thermodynamic description and only consider macroscopic variables and the entropy.

1.1.2 Walls

A particularly important macroscopic variable in thermodynamics is the energy, which can take many forms. In a mechanical system with only conservative forces (derivable from a potential), the mechanical energy, which is the sum of the kinetic and potential energies, is conserved. A mechanical system protected from all external influences, i.e. *isolated*, finds its energy conserved, in other words independent of time. Mathematically, the energy can be written as

$$\boxed{E = \sum_{i=1}^{N} \frac{\vec{p}_i^{\,2}}{2m} + \frac{1}{2} \sum_{i \neq j} U(\vec{r}_i - \vec{r}_j)} \tag{1.2}$$

To simplify the writing, we have assumed in this equation that the particles are identical and of mass m; $\vec{p}_i$ is the momentum of particle i, $\vec{r}_i$ its position, and U the potential energy of two molecules. We also assumed the molecules to have no internal structure. Equation (1.2) also gives the expression for the classical or quantum Hamiltonian, H, of the isolated system. In the quantum case, $\vec{p}_i$ and

$\vec{r}_i$ are the canonically conjugate momentum and position operators for particle i. When the system is not isolated, we know that we can transfer mechanical energy to it: when we compress a spring, it acquires additional energy, which it stores in the form of elastic potential energy. During the compression process, the point where the force is applied moves with the consequence that energy is given to the system in the form of *work*. Similarly, we supply energy to a gas by compressing it with the help of a piston. In both cases, the *external parameters*[4] of the system, length of the spring in one case, volume of the gas in the other, are modified in a known way. However, we know, experimentally, that we can transfer energy to an object in many other ways. Any handyman knows that we can transfer energy to a drill bit by drilling a hole in concrete. The bit heats up due to friction, and, according to a popular but thermodynamically incorrect statement (Footnote 10), some of the mechanical energy supplied by the drill motor is 'changed to *heat*'. We can obtain the same result by leaving the drill bit in the sun on a hot summer day, which corresponds to 'transforming electromagnetic energy into heat', or by immersing it in boiling water, i.e. by using thermal contact. In the latter case, there is no visible modification of external variables (see Section 2.5.1) either of the bit or of the water. Only the microscopic degrees of freedom are involved in the exchange of energy. The heating of the bit corresponds to bigger vibrations of its atoms around their equilibrium positions, the concomitant cooling of the water corresponds to a reduction of the average speed of its molecules.[5] *Energy transfer in the form of heat is characterized by the fact that neither the external parameters of the system, nor the configuration of the external medium, are modified.* This heat transfer can be effected by conduction (the drill bit in contact with water), or by radiation (between the sun and the bit).

In summary, a system can receive energy either in the form of work, or in the form of heat. The energy supplied in the form of work is, at least in principle, measurable from mechanical considerations because work is supplied by a macroscopic mechanical device whose parameters (masses, applied forces, etc.) are, supposedly, perfectly known.[6] Work is obtained by causing a change, either of the external parameters, or the configuration of the external medium, or both. However, the amount of energy received by an object is not known with precision from the principle of conservation of total energy unless we are able to eliminate energy exchange in the form of heat. This can be accomplished by isolating the system using a heat insulating wall, or an *adiabatic wall*; on the other hand, a *diathermic wall*

[4] External parameters are those quantities that are under the direct control of the experimentalist: volume, external electric or magnetic fields, etc.

[5] To simplify the discussion, we neglect complications due to the potential energy of the molecules.

[6] From this point of view, the energy supplied by an electric device will be considered as work since it can be determined by electric measurements performed with a voltmeter or an ammeter.

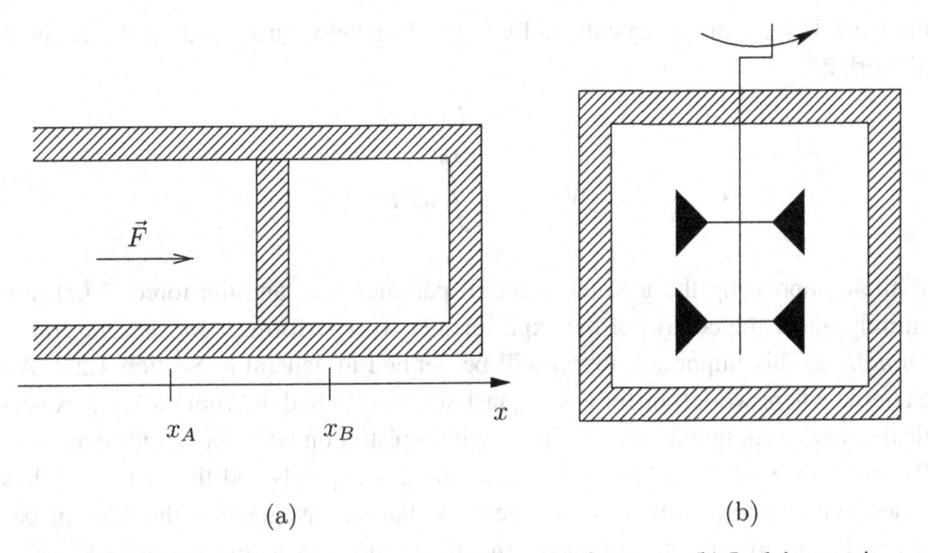

Figure 1.1 Two ways to supply work. (a) Compressing a gas, (b) Joule's experiment.

allows heat transfer. Inversely, we eliminate energy transfer in the form of work by using a *rigid wall* not penetrated by any mechanical device.[7] The possibility of a thermodynamic description is founded on the existence, at least in theory (since walls are never perfectly adiabatic or rigid!), of walls capable of controlling the different forms of energy transfer.

An *isolated system* is a system that cannot exchange energy with its surroundings in any form whatsoever: it is isolated from the external world by walls that are perfectly adiabatic, rigid, impermeable to molecules and shielded electrically and magnetically.

1.1.3 Work, heat, internal energy

We now develop more quantitatively the concepts defined above. First we need to make an essential distinction between two different ways of supplying work. To fix ideas, let us consider work done on a gas by compressing it with the help of a piston displaced between positions x_A and x_B (Figure 1.1(a)), the position of the piston being given by its abscissa x. If at every instant the position of the piston and $F(x)$, the component of the applied force parallel to Ox, are perfectly controlled,

[7] Rigorously speaking, for a mechanical system one also should eliminate energy transfer by other processes such as electric or magnetic.

the work $W_{A\to B}$ can be calculated by integrating between x_A and x_B the element of work $đW$

$$đW = F(x)\,\mathrm{d}x$$
$$W_{A\to B} = \int_{x_A}^{x_B} \mathrm{d}x\, F(x) \tag{1.3}$$

While compressing the gas, the external parameter, x, and the force, $F(x)$, are entirely under the control of the experimenter who performs a *quasi-static transformation*. This important notion will be defined in general in Section 1.2.2. An example where work transfer is *not* quasi-static is studied in Problem 1.7.1. A vertical cylinder containing gas is closed with a piston on which a weight is placed. When this weight is suddenly removed, the gas expands and the piston reaches a new equilibrium position after a few oscillations. Neglecting the friction between the piston and the cylinder, the work supplied to the gas is $-P_{\text{ext}}\Delta V$, where P_{ext} is the external pressure and ΔV the change in volume. This change in volume was not controlled during the expansion of the gas. Another example of non-quasi-static transfer of work is illustrated in Figure 1.1(b). The system is isolated from the exterior by an adiabatic wall, but a motor turns vanes in the fluid, which heats up due to viscosity. The energy supplied in the form of work can be calculated from the characteristics of the motor.[8] These two examples of non-quasi-static work transfer appear very different but do have a point in common. In the first example, the final temperature is higher than it would have been had the change been quasi-static, and, as is the case in the second example, viscous forces are responsible for the increase in temperature. The example of the heated drill bit, given earlier, is another illustration of work done non-quasi-statically.

We now examine the energy transfer between states of a system. Let A and B be two possible arbitrary states. The energy of each of these states is, in principle, a well-defined quantity, for example by Equation (1.2). In thermodynamics, this is called the *internal energy* and will be denoted by E. We know that only energy differences have physical meaning, and that *a priori* the interesting quantity is $E_B - E_A$. Our goal is to demonstrate that this energy difference is accessible experimentally. Note that all transferred energies, be they in the form of work or heat, are algebraic quantities that can be positive or negative.

Taking E_A as the reference energy, we will be able to determine E_B if it is possible to go from A to B by supplying the system only with work, positive or

[8] A more modern version of this experiment, which dates back to Joule, consists of putting in the fluid a known resistance across which we apply a known potential difference: the amount of electrical energy 'transformed into heat' is thus known.

negative, since this work, furnished by a mechanical device, is measurable. To determine whether such a transformation is possible, we start from the following empirical observation. It is possible to go either from state A to state B, or from state B to state A by a process whose sole effect is to supply work to the system. However, under this condition, only one of the two transformations is allowed. We justify this statement as follows. If states A and B have the same volume, and if $E_B > E_A$, a mechanism similar to that in Figure 1.1(b) allows us to go from A to B, which will be impossible if $E_B < E_A$.[9] If the volumes of A and B are different, we can use an adiabatic expansion or compression, $A \to A'$, which brings the system to the desired volume, $V_{A'} = V_B$, with an energy $E_{A'}$. If $E_{A'} < E_B$, work can be done to arrive at the final state with energy E_B. To summarize, we can determine, either $E_B - E_A$ by a transformation $A \to B$, or $E_A - E_B$ by a transformation $B \to A$, by supplying only measurable work to the system.

If the transformation $A \to B$ is now performed in an arbitrary manner, in other words it involves an exchange of both work and heat, we can control the work, $W_{A\to B}$, which is determined by macroscopic mechanical parameters. The energy difference, $(E_B - E_A)$, has previously been determined, and we thus obtain the amount of heat, $Q_{A\to B}$, supplied in this process

$$\boxed{Q_{A\to B} = (E_B - E_A) - W_{A\to B}} \tag{1.4}$$

This equation, which simply expresses conservation of energy, constitutes the 'first law of thermodynamics'. It is often written in the differential form

$$\boxed{đQ = \mathrm{d}E - đW} \tag{1.5}$$

Unlike the increase in the internal energy, $E_{A\to B} = E_B - E_A$, the work, $W_{A\to B}$ and the amount of heat, $Q_{A\to B}$, are not determined by the initial and final states: they depend on the transformation itself.[10] This is why, unlike the differential $\mathrm{d}E$, the infinitesimal quantities $đQ$ and $đW$ are not differentials. We can understand this intuitively by making an analogy with mechanics. If a force, $\vec{F}$, is such that $\vec{\nabla} \times \vec{F} \neq 0$, the work it does between points A and B

$$W_{A\to B} = \int_A^B \vec{F} \cdot \mathrm{d}\vec{l}$$

does not depend only on the points A and B. It also depends on the path taken, and there is no function whose differential gives the infinitesimal work.

[9] Anticipating what is to follow, $E_B > E_A$ means that the temperature, T_B, of B is greater than T_A of A: it is impossible to cool down a volume simply by exchanging work.

[10] We cannot, therefore, ascribe to a system a work or heat content. The concepts of heat and work expose two different forms of *energy exchange* between two systems: heat and work are 'energy in transit'.

1.1.4 Definition of thermal equilibrium

For simplicity, we limit ourselves to the ideal case of a system that is shielded from electric or magnetic influences, leaving these more complex cases to the references. This restriction is not a limitation of the theory, but merely a convenience to simplify the discussion. Suppose that our system is isolated in a completely arbitrary initial state, for example with a spatially dependent density. *Experience tells us that if we wait long enough, the system will evolve to an equilibrium state, that is a state which depends neither on time nor on the past history of the system.* The equilibrium state is entirely characterized by macroscopic variables and external parameters describing the system: the volume, V, the energy, E, and the numbers, $N^{(1)}, \ldots, N^{(r)}$, of molecules of type $1, \ldots, r$.

The time that characterizes the approach to equilibrium is called the *relaxation time*. Relaxation times can be as short as a few microseconds, and as long as several millennia. It is, therefore, not obvious in practice to decide whether or not we have attained an equilibrium state. In numerous cases, we only reach a state of metastable equilibrium whose average lifetime can be extremely long. Such a state only appears to be independent of time, and in fact also depends on its past history. A very familiar example is hysteresis: if we magnetize an initially unmagnetized sample of magnetic material by applying a magnetic field, the magnetization does not disappear when the field is removed. The evolution of the magnetization, M, as a function of the applied magnetic field, B, describes a hysteresis cycle as shown in Figure 1.2. On this figure, the dashed line represents the case where the initial magnetization is zero. The magnetic state, therefore, depends on its past history even though we have obtained a magnet that is apparently in a stable state. However, we may, for example, reach a state whose magnetization is opposite to the magnetic

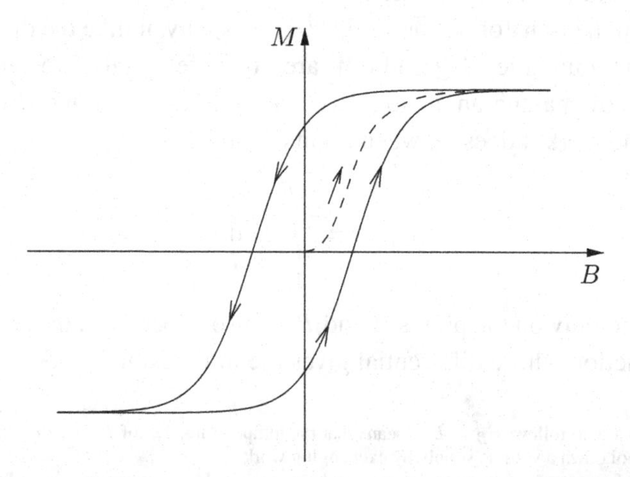

Figure 1.2 Hysteresis cycle for a magnetic system in an external field.

field, a state that is clearly metastable. Another common case is the existence of a variety of metastable crystals: for example, graphite is the stable crystalline form of carbon under standard conditions of pressure and temperature, and diamond is metastable. There are many other examples: glasses, alloys, memory materials, etc. In such cases, if we mistakenly assume that the system is at equilibrium, we may arrive at conclusions that contradict experiment.

While keeping in mind the difficulty related to extremely long relaxation times, *we take as our first postulate the existence of equilibrium states*: an isolated system will attain, after sufficiently long time, an equilibrium state that is independent of its past history and characterized by its own intrinsic properties such as volume V, energy E and numbers $N^{(i)}$, of molecules of different types. In what follows, we shall often limit ourselves to a single type of molecule, N in number, and to homogeneous equilibrium states whose properties, for example the density, are uniform. The quantities E, V, and N are said to be *extensive*: if we merge into one system two identical subsystems at equilibrium, the energy, volume and number of molecules are doubled.

1.2 Postulate of maximum entropy

1.2.1 Internal constraints

As we have emphasized in the introduction, our presentation of thermodynamics postulates the existence of an entropy function. To define it correctly, it is necessary to introduce the notion of *internal constraint*, of which we shall give a simple example. Consider an isolated system that we have divided in two subsystems (1) and (2), separated by a piston (Figure 1.3). As always, we assume that the contributions of the walls (or the piston) to the energy, E, to the volume, V, and to the total number of molecules, N, are negligible since they arise from surface effects. Consequently, E, V, and N represent the sums of energies, volumes and

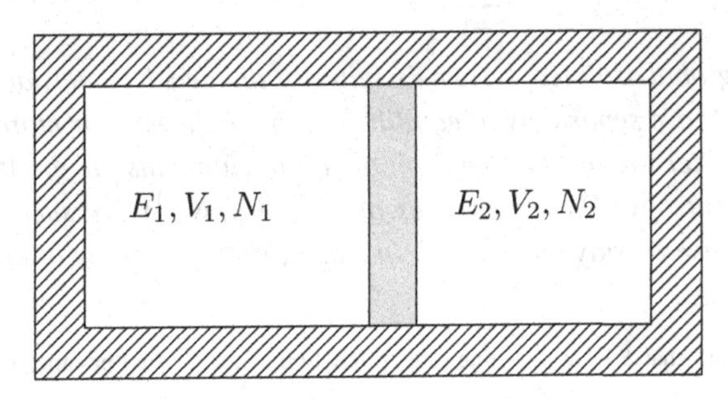

Figure 1.3 An isolated system which is divided into two 'subsystems'.

numbers of particles in the two subsystems

$$E = E_1 + E_2 \qquad V = V_1 + V_2 \qquad N = N_1 + N_2 \tag{1.6}$$

The piston creates the following internal constraints:

- If it is fixed, it creates a constraint in preventing the free flow of energy from one subsystem to the other in the form of work.
- If it is adiabatic, it creates a constraint in preventing the free flow of energy from one subsystem to the other in the form of heat.
- If it is impermeable to molecules, it creates a constraint in preventing the flow of molecules from one subsystem to the other.

We lift a constraint by rendering the piston mobile, diathermic or permeable. We can, of course, lift more than one constraint at a time.

Let us start with the following initial situation: the piston is fixed, adiabatic, impermeable, and both subsystems are separately at equilibrium. We lift one (or several) of the constraints and we await the establishment of a new equilibrium state. We can then pose the following question: what can we say about this new equilibrium state? We shall see that the answer to this fundamental question is given by the principle of maximum entropy.

1.2.2 Principle of maximum entropy

We make the following postulates, which are equivalent to the usual statement of the 'second law of thermodynamics':

(*i*) *For any system at equilibrium, there exists a positive differentiable entropy function* $S(E, V, N^{(1)}, \ldots, N^{(r)})$.[11] *As a general rule, this function is an increasing function of* E *for fixed* V *and* $N^{(i)}$.[12]

(*ii*) *For a system made of* M *subsystems,* S *is additive, or extensive: the total entropy* S_{tot} *is the sum of the entropies of the subsystems,*

$$S_{\text{tot}} = \sum_{m=1}^{M} S(E_m, V_m, N_m^{(1)}, \ldots, N_m^{(r)}) \tag{1.7}$$

(*iii*) *Suppose the global isolated system is initially divided by internal constraints into subsystems that are separately at equilibrium: if we lift one (or more) constraint, the final entropy, after the re-establishment of equilibrium must be greater than or equal to the initial entropy. The new values of* $(E_m, V_m, N_m^{(i)})$ *are such that the entropy can only increase or stay unchanged. In summary: the entropy of an isolated system cannot decrease.*

[11] The reader will remark that S is, at the same time, a function of an external parameter V, and macroscopic variables E and N.

[12] See Problem 3.8.2 for an exception.

We emphasize that *a priori* entropy is only defined for a system at equilibrium. When a system is very far from equilibrium, we cannot in general define a unique entropy. However, when a system, not globally at equilibrium, can be divided into subsystems that are almost at equilibrium with their neighbours with whom they interact weakly, we can, once again, define the global entropy of the system with the help of Equation (1.7). As we shall see in more detail in Chapter 6, this permits us to define the entropy of a system that is not in global but only in local equilibrium. This case is very important in practice: for example, all of hydrodynamics rests on the notion of local equilibrium.

We now introduce the notion of *quasi-static processes*, already evoked in Section 1.1.2. A transformation $A \to B$ is said to be quasi-static if the system under consideration stays infinitesimally close to a state of equilibrium. A quasi-static transformation is necessarily infinitely slow since we need to wait after each step for equilibrium to be re-established. Clearly, a quasi-static transformation is an idealized case: a real transformation cannot be truly quasi-static. Along with walls that are perfectly adiabatic or perfectly rigid, quasi-static transformations are theoretical tools for formulating precise reasonings in thermodynamics. In practice, we consider a transformation quasi-static if it takes place over a time much larger than that needed to re-establish equilibrium. In the case of a quasi-static transformation, we can define the entropy at each instant because the system is infinitesimally close to equilibrium. This entropy, $S_{\text{tot}}(t)$, will be *a priori* time dependent and, for an isolated system, will be an increasing function of time.

An essential consequence of the principles mentioned above is that the entropy is a concave function of its arguments. Recall that a function, $f(x)$, is concave if for any x_1 and x_2 (Figure 1.4) we have

$$f\left(\frac{x_1 + x_2}{2}\right) \geq \frac{f(x_1) + f(x_2)}{2} \tag{1.8}$$

If a concave function is twice differentiable, its second derivative is either negative or zero: $f''(x) \leq 0$. For a convex function, $f''(x) \geq 0$.

Concavity of the entropy: *For a homogeneous system, the entropy is a concave function of the extensive variables* (E, V, N).

In effect, suppose that S is locally a strictly convex function of, for example, the energy in an interval around the value E; we would then have

$$2S(E) < S(E - \Delta E) + S(E + \Delta E)$$

From additivity of the entropy, $2S(E) = S(2E)$: application of a constraint on the energy that renders the system inhomogeneous would allow entropy to increase. The maximum entropy principle would then lead to an inhomogeneous equilibrium

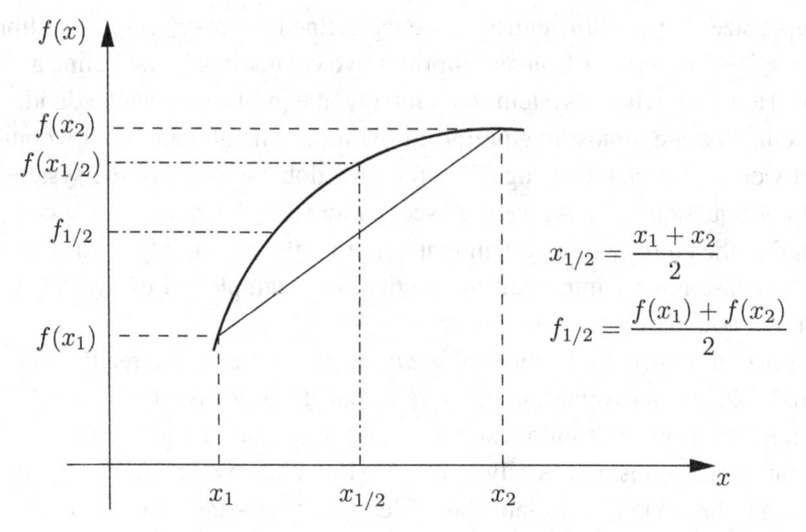

Figure 1.4 Properties of a concave function.

state, in contradiction with our starting hypothesis. Such behaviour is the signature of a phase transition (see Section 3.5.2). In our reasoning we have used a strict inequality, the case where the inequality is not strict will be examined soon. The generalization to several variables will be given in 1.4.1.

1.2.3 Intensive variables: temperature, pressure, chemical potential

We will now define the temperature, T, the pressure, P, and the chemical potential, μ, starting with the entropy, $S(E, V, N^{(1)}, \ldots, N^{(r)})$. We shall later show that these definitions correspond to intuitive notions.

Temperature T
$$\left.\frac{\partial S}{\partial E}\right|_{V,N^{(i)}} = \frac{1}{T} \tag{1.9}$$

Pressure P
$$\left.\frac{\partial S}{\partial V}\right|_{E,N^{(i)}} = \frac{P}{T} \tag{1.10}$$

Chemical potential $\mu^{(i)}$
$$\left.\frac{\partial S}{\partial N^{(i)}}\right|_{E,V,N^{(j\neq i)}} = -\frac{\mu^{(i)}}{T} \tag{1.11}$$

The variables T, P and μ^i are called *intensive* variables: for example, if we multiply by two all the extensive variables, the temperature, pressure and chemical potential remain unchanged, which is clear from their definitions. On the other hand, the extensive variables, E, V and $N^{(i)}$ are doubled if we double the size of the system while keeping T, P and μ^i constant.

Temperature: thermal equilibrium

Let us first show that our definition of temperature allows us to arrive at the notion of thermal equilibrium. We use, again, the system depicted in Figure 1.3, where the piston is initially fixed, adiabatic and impermeable to molecules, and the two compartments (subsystems) are separately at equilibrium. In order to lighten the notations, and while assuming the wall to be impermeable to molecules, it will be convenient to define the entropy functions, $S_1(E_1, V_1)$ and $S_2(E_2, V_2)$, of the two subsystems by

$$S_1(E_1, V_1) = S\big(E_1, V_1, N_1^{(1)}, \ldots, N_1^{(r)}\big)$$
$$S_2(E_2, V_2) = S\big(E_2, V_2, N_2^{(1)}, \ldots, N_2^{(r)}\big)$$

Now make the piston diathermic: a new equilibrium will be established corresponding to a maximum of the entropy. All infinitesimal variations around this new equilibrium must obey the extremum condition $\mathrm{d}S = 0$, where S is the total entropy

$$\mathrm{d}S = \left.\frac{\partial S_1}{\partial E_1}\right|_{V_1} \mathrm{d}E_1 + \left.\frac{\partial S_2}{\partial E_2}\right|_{V_2} \mathrm{d}E_2 = 0 \tag{1.12}$$

The energy conservation condition, $E_1 + E_2 = E =$const, implies $\mathrm{d}E_1 = -\mathrm{d}E_2$, and, keeping in mind the definition (1.9), we conclude that the final temperatures, T_1 and T_2, are equal, $T_1 = T_2 = T$. When energy is allowed to flow freely in the form of heat, the final equilibrium corresponds to the temperatures of the two subsystems being equal, that is to say thermal equilibrium.

We now show that energy flows from the hot to the cold compartment. Figure 1.5 shows $\partial S/\partial E$ for the two subsystems (i.e. $1/T_1$ and $1/T_2$) as a function of E. Because $S(E)$ is concave, the two curves, $\partial S_1/\partial E_1$ and $\partial S_2/\partial E_2$ are decreasing functions of E_1 and E_2 respectively. We have assumed for the initial temperatures that $1/T_1' > 1/T_2'$, i.e. $T_1' < T_2'$. In other words, compartment (1) is initially colder than compartment (2). Suppose that the final energy, E_1, of (1) is smaller than its initial energy ($E_1 < E_1'$), and thus $T_1 < T_1'$. From energy conservation, we have $E_2 > E_2'$ and, therefore, $T_2 > T_2'$. It is then impossible to have equal temperatures and the final state does not satisfy the maximum entropy principle. Equality of temperatures is possible only if $E_1 > E_1'$ and $E_2 < E_2'$, that is only when energy flows from the hotter to the colder compartment. This property conforms to our intuitive idea of temperature and heat.

We have so far assumed that $\partial^2 S/\partial E^2$ is strictly negative, i.e. that $\partial S/\partial E$ is strictly decreasing. It can happen that $\partial^2 S/\partial E^2 = 0$ over an interval in E, in which case $S(E)$ has a linear part (Figure 1.6). This means that an energy transfer does

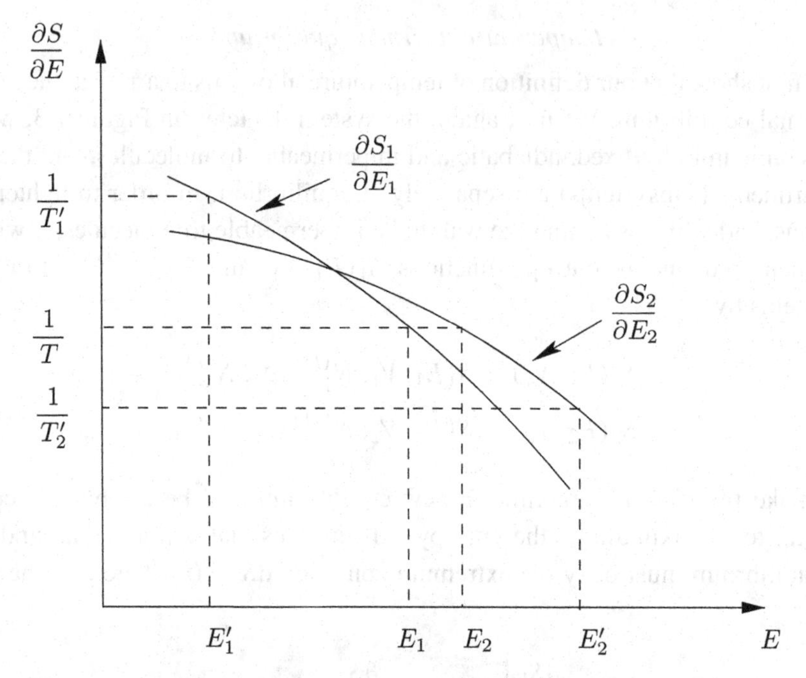

Figure 1.5 Graphic illustration of energy flow.

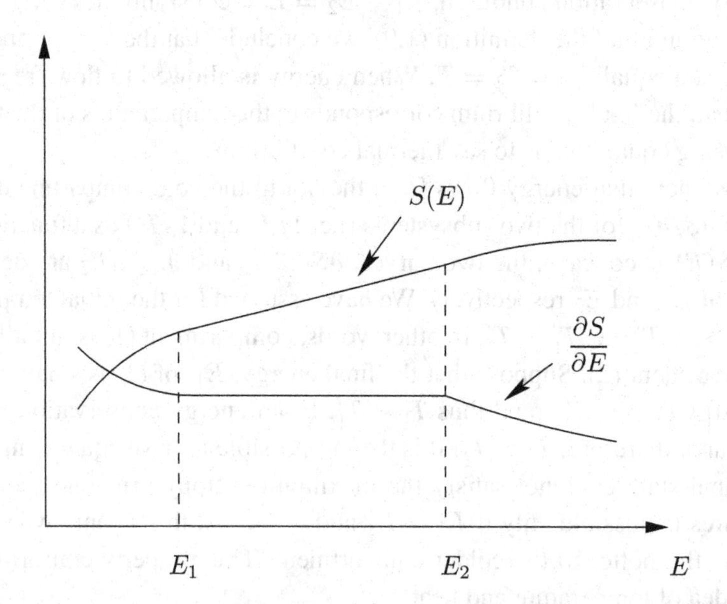

Figure 1.6 The entropy and its derivative in the presence of a phase transition.

not change the temperature. Such a situation arises at a phase transition: when we supply heat to a mixture of ice and water, we first melt the ice without changing its temperature. In such a case, the system is no longer homogeneous: in order to characterize it, one has to give, in addition to the temperature, the fraction of the system in each of the phases.

Pressure: mechanical equilibrium

Once again, we start with the situation shown in Figure 1.3, but now we make the piston both diathermic and mobile. Once equilibrium is reached, the entropy variation, $\mathrm{d}S$, should vanish for all small fluctuations of the energy and volume

$$\mathrm{d}S = \left(\left.\frac{\partial S_1}{\partial E_1}\right|_{V_1} \mathrm{d}E_1 + \left.\frac{\partial S_2}{\partial E_2}\right|_{V_2} \mathrm{d}E_2 \right) + \left(\left.\frac{\partial S_1}{\partial V_1}\right|_{E_1} \mathrm{d}V_1 + \left.\frac{\partial S_2}{\partial V_2}\right|_{E_2} \mathrm{d}V_2 \right) = 0$$

The conservation conditions of energy, $E_1 + E_2 = E =$ const, and volume, $V_1 + V_2 = V =$ const, along with the definitions (1.9) and (1.10) imply

$$\mathrm{d}S = \left(\frac{1}{T_1} - \frac{1}{T_2} \right) \mathrm{d}E_1 + \left(\frac{P_1}{T_1} - \frac{P_2}{T_2} \right) \mathrm{d}V_1 = 0 \tag{1.13}$$

Since the energy and volume fluctuations are independent to first order in $(\mathrm{d}E, \mathrm{d}V)$ (see Footnote 13), the extremum condition for the entropy, $\mathrm{d}S = 0$, implies thermal equilibrium, $T_1 = T_2 = T$, and mechanical equilibrium, $P_1 = P_2 = P$. The connection between $(\partial S/\partial V)_E$ and the usual concept of pressure will be made very soon.

The reader will remark that the case where the piston is mobile and adiabatic is not symmetric with that when the piston is diathermic and fixed: a variation in the volume causes a transfer of energy, whereas a fixed piston prevents the exchange of volume. If the piston is made mobile but not diathermic, it will oscillate indefinitely in the absence of friction. If we take into account the friction of the piston against the cylinder and the viscosities of the two fluids, we obtain a final stationary situation where the temperatures of the two compartments are *a priori* different and are determined by the viscosities and friction.

Returning to Equation (1.13) and assuming that we are in the neighbourhood of an equilibrium with $T_1 = T_2 = T$ and P_1 slightly larger than P_2, the variation, $\mathrm{d}V_1$, has to be such that the entropy increases ($\mathrm{d}S \geq 0$), which implies that $\mathrm{d}V_1 > 0$. In other words, the volume of the compartment that was at higher initial pressure increases.[13]

[13] The alert reader will remark that the final equilibrium temperature will be slightly different from T. However, the change in temperature, calculated explicitly for a van der Waals gas in Problem 1.7.2, is of the order of $(\mathrm{d}V_1)^2$ and so is $\mathrm{d}E_1$. On the other hand, for finite variations, energy and volume are not independent parameters.

Chemical potential: equilibrium of particle (or molecule) flux

Finally, the piston is made diathermic, fixed, and permeable to molecules of type (i). Returning to the general notation $S(E_m, V_m, N_m^{(1)}, \ldots, N_m^{(r)})$, we again use the principle of maximum entropy

$$\mathrm{d}S = \left(\left.\frac{\partial S}{\partial E_1}\right|_{V_1, N_1^{(i)}} \mathrm{d}E_1 + \left.\frac{\partial S}{\partial E_2}\right|_{V_2, N_2^{(i)}} \mathrm{d}E_2 \right) + \left(\left.\frac{\partial S}{\partial N_1^{(i)}}\right|_{E_1, V_1, N_1^{(j \neq i)}} \mathrm{d}N_1^{(i)} + \left.\frac{\partial S}{\partial N_2^{(i)}}\right|_{E_2, V_2, N_2^{(j \neq i)}} \mathrm{d}N_2^{(i)} \right) = 0$$

The conservation laws for the energy and the number of molecules of type (i) give $\mathrm{d}E_1 = -\mathrm{d}E_2$ and $\mathrm{d}N_1^{(i)} = -\mathrm{d}N_2^{(i)}$, and using the definitions (1.9) and (1.11) we get

$$\mathrm{d}S = \left(\frac{1}{T_1} - \frac{1}{T_2} \right) \mathrm{d}E_1 - \left(\frac{\mu_1^{(i)}}{T_1} - \frac{\mu_2^{(i)}}{T_2} \right) \mathrm{d}N_1^{(i)} = 0 \tag{1.14}$$

The entropy being stationary implies that at equilibrium the chemical potentials are equal: the transfer of molecules between the two subsystems stops at equilibrium. The term 'chemical potential' is used for historical reasons, but this terminology is unfortunate because we see clearly in our example that no chemical reaction is involved in its definition. It is easy to generalize this to all types of molecules present. The condition for equilibrium is[14]

$$\mu_1^{(1)} = \mu_2^{(1)}, \ldots, \mu_1^{(i)} = \mu_2^{(i)}, \ldots, \mu_1^{(r)} = \mu_2^{(r)}$$

Reasoning along the same lines as for the pressure allows us to show that the number of particles decreases in the compartment where the chemical potential was the larger.

Equation of state

Assuming $N^{(i)}$ to be constant, the definitions (1.9) and (1.10) of T and P define the functions f_T and f_P of E and V

$$\frac{1}{T} = \left.\frac{\partial S}{\partial E}\right|_V = f_T(E, V) \qquad \frac{P}{T} = \left.\frac{\partial S}{\partial V}\right|_E = f_P(E, V)$$

Since f_T is a strictly decreasing function of E for a homogeneous system, we can determine E as a function of T at fixed V, $E = g(T, V)$, and substituting in the

[14] We assume here that no chemical reactions take place when the subsystems are put in contact. This case will be studied in Chapter 3.

equation for the pressure gives

$$P = T\, f_P(g(T, V), V) = h(T, V) \tag{1.15}$$

The relation $P = h(T, V)$ is the *equation of state* for a homogeneous substance. The best known equation of state is that of the *ideal gas* written here in three useful forms

$$\boxed{P = \frac{RT}{V} = \frac{\mathcal{N}kT}{V} = nkT} \tag{1.16}$$

This equation is written for one mole of gas, R is the ideal gas constant, $\mathcal{N}$ is Avogadro's number, $k = R/\mathcal{N}$ is the Boltzmann constant and $n = \mathcal{N}/V$ is the density of molecules.

1.2.4 Quasi-static and reversible processes

Putting together the definitions (1.9) to (1.11) and considering only one kind of molecules, we can write

$$\mathrm{d}S = \frac{1}{T}\,\mathrm{d}E + \frac{P}{T}\,\mathrm{d}V - \frac{\mu}{T}\,\mathrm{d}N \tag{1.17}$$

or

$$\boxed{T\,\mathrm{d}S = \mathrm{d}E + P\,\mathrm{d}V - \mu\,\mathrm{d}N} \tag{1.18}$$

which is the '$T\,\mathrm{d}S$ equation in the entropy variable', referring to the fact that the basic function is $S(E, V, N)$. But since S is an increasing function of E, the correspondence between them is one-to-one, which allows us to find an energy function $E(S, V, N)$. The $T\,\mathrm{d}S$ equation can now be written in the equivalent form

$$\boxed{\mathrm{d}E = T\,\mathrm{d}S - P\,\mathrm{d}V + \mu\,\mathrm{d}N} \tag{1.19}$$

This is the '$T\,\mathrm{d}S$ equation in the energy variable'. When $\mathrm{d}V = 0$ (fixed piston) and $\mathrm{d}N = 0$ (no molecule transfer), the energy transferred is in the form of heat. We can then make the identification: $\mathrm{d}E = T\,\mathrm{d}S = đQ$.

At this point, it is important to make a crucial remark: we have emphasized that the existence of the entropy function supposes a situation of equilibrium, or infinitesimally close to it, and that the $T\,\mathrm{d}S$ equations are *a priori* valid only for quasi-static processes. In particular, the equation $đQ = T\,\mathrm{d}S$ is also valid only for a quasi-static process. Similarly, let us re-consider the expression for mechanical work given by Equation (1.3): if the area of the piston is A, the force is given by $F = P_{\mathrm{ext}}A$ where P_{ext} is the external pressure applied by the gas (if the force is due to an external pressure). In a quasi-static process, the external pressure, P_{ext}, is infinitesimally close to the internal pressure, P. Consequently,

$$đ W = F\,\mathrm{d}x = P_{\mathrm{ext}} A\,\mathrm{d}x = -P\,\mathrm{d}V \qquad (1.20)$$

The minus sign is due to the fact that we supply work by compressing the gas: $\mathrm{d}x > 0$ corresponds to $\mathrm{d}V < 0$ (Figure 1.1(a)). Note that the first two equalities giving $đ W$ are valid in general, but that the last is valid only for a quasi-static process since the pressure, P, of a gas is only defined at equilibrium (see Problem 1.7.1). By comparing with (1.5) we see that the quantity $(-\partial E/\partial V)_{S,N}$ is indeed the pressure, as is $T(\partial S/\partial V)_{E,N}$.

We emphasize, yet again, the essential point that (for N fixed) conservation of energy or the 'first law', $\mathrm{d}E = đ Q + đ W$, is always valid. The equation $\mathrm{d}E = T\,\mathrm{d}S + đ W$, which relies on the notion of entropy and, therefore, the 'second law', is valid only for quasi-static processes. For such processes, $đ W$ can be expressed as a function of variables that are internal to the system with no reference to the characteristics of the external medium. A good starting point to solve a thermodynamic problem is to write the $T\,\mathrm{d}S$ equation adapted to the specific problem under consideration.

We finally introduce the notion of *reversible transformation*: by definition, a quasi-static transformation is reversible if it takes place at constant total entropy. In general, lifting one or more internal constraints in an isolated system increases the entropy, $S_{\mathrm{final}} > S_{\mathrm{initial}}$. Re-imposing these constraints cannot return the system to its initial state. The transformation is, therefore, *irreversible*. For a strictly isolated system, evolution takes place spontaneously in a certain direction, with the entropy increasing until equilibrium is reached. On the other hand, for a reversible transformation, we can return the system to its initial state by choosing appropriately the manner in which we change the internal constraints. This explains the term 'reversible transformation': it can be effected in both directions. By means of external intervention, a small modification of the constraints fixes the sense of the reversible transformation. For example, we can choose the sense of the displacement of the piston by increasing slightly the pressure (or adding energy[15]) in one of the compartments of Figure 1.3. We note that a transformation that is, at the same time, quasi-static and takes place without any energy exchange in the form of heat, is reversible. In fact, since the transformation is quasi-static, all energy exchange in the form of heat can be written as $đ Q = T\,\mathrm{d}S$, and since $đ Q = 0$, this means that $\mathrm{d}S = 0$. We have already seen that all transformations of an isolated system produce an entropy change $\Delta S \geq 0$. On the other hand, it can happen that parts of an isolated system experience a decrease in entropy, as long as this is compensated by an increase in S in other parts of the system.

[15] We remark that, if a system at a temperature $T - \mathrm{d}T$ and with specific heat C absorbs, in the form of heat, a quantity of energy $đ Q = C\,\mathrm{d}T$ from a reservoir at T, the entropy change is proportional to $(\mathrm{d}T)^2$.

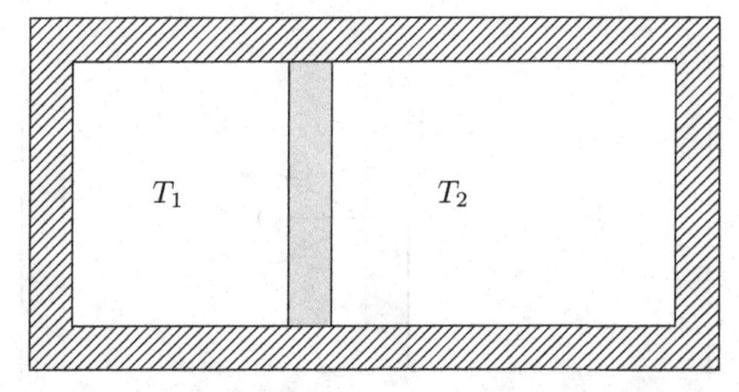

Figure 1.7 Example of a quasi-static but non-reversible process: two compartments separated by an almost adiabatic partition.

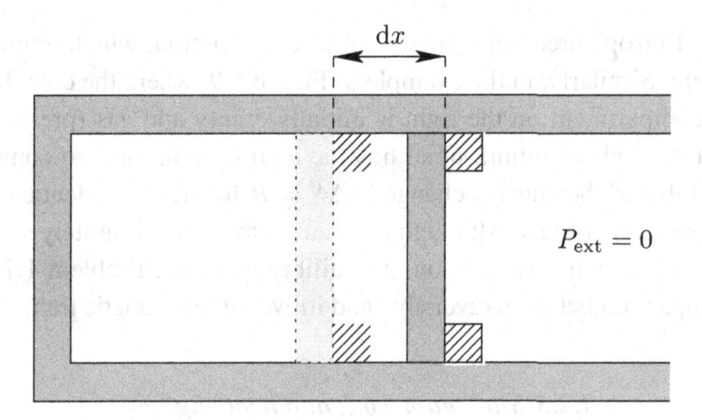

Figure 1.8 Example of an infinitesimal but not quasi-static process.

It is useful to give an example of a quasi-static transformation that is *not* reversible. Consider Figure 1.7: the two compartments are at different temperatures, but the wall is almost adiabatic (heat flow is very slow) and each compartment is infinitesimally close to equilibrium. The entropy, S, is well defined but increases with time (Section 2.6). As we shall see in Chapter 6, the wall is in a state of local equilibrium because the temperature is position dependent but well defined at all points and entropy production takes place in this wall.

It is also useful to keep in mind that an infinitesimal transformation is not necessarily quasi-static. In the example of Figure 1.8, the container and the piston are assumed to be adiabatic: the piston is first released and moves an infinitesimal distance, $\mathrm{d}x$, before being blocked. Assuming the gas is ideal, the entropy change is $\mathrm{d}S = R\,\mathrm{d}V/V$ for one mole of gas (see Problem 1.7.1). The transformation is adiabatic, and were it quasi-static it would also be reversible, which is not the case since $\mathrm{d}S \geq 0$. The sudden movement of the piston creates turbulence in its neighbourhood with the consequence that the transformation no longer is near

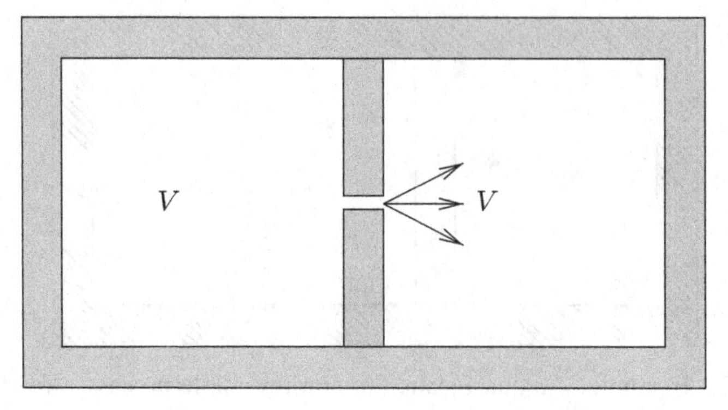

Figure 1.9 An infinitely slow process which is not quasi-static.

equilibrium. Entropy creation comes from viscous friction, which returns the gas to equilibrium. Similarly, in the example of Figure 1.9, where the container is adiabatic, the compartment on the right is initially empty and gas (presumed ideal) leaks into it through an infinitesimal hole: at equilibrium, the two compartments are equally full and the entropy change is $\Delta S = R \ln 2$ for two identical compartments and one mole of gas. Although the transformation is infinitely slow, it does not take place through a succession of equilibrium states. Problem 1.7.1 gives a detailed comparison between reversible and irreversible adiabatic transformations.

1.2.5 Maximum work and heat engines

In the usual presentation of thermodynamics, a central rôle is played by the heat engine. A device, $\mathcal{X}$, connected to two heat sources[16] at temperatures T_1 and T_2, $T_1 > T_2$, supplies work to the outside (Figure 1.10(a)). We establish first a preliminary result: *the theorem of maximum work*. Suppose that $\mathcal{X}$ receives a quantity of energy,[17] Q, from a heat source, S, at temperature T. We assume that S is big enough for Q to be infinitesimal, compared to its total energy, so that its temperature, T, stays essentially unchanged (Figure 1.10(b)). The transformation experienced by the source is quasi-static and its entropy change is $-Q/T$: such a source is called a *heat reservoir*. The system, $\mathcal{X}$, supplies an amount of work, $\mathbb{W}$, to the outside world, symbolized by a spring in Figure 1.10(b). This spring is purely a mechanical object that does not contribute to the entropy. The combination $[\mathcal{X} + \mathcal{S}]$ is thermally isolated, but the wall separating $\mathcal{X}$ from $\mathcal{S}$ is rendered diathermic for a time period during which $\mathcal{X}$ and $\mathcal{S}$ come in thermal contact. This thermal contact and the work done will cause $\mathcal{X}$ to go from one equilibrium state to

[16] By heat source we mean a source that can exchange energy only in the form of heat.
[17] We should be careful with the signs: the source receives a quantity of heat $-Q$, the system receives an amount of work $W = -\mathbb{W}$.

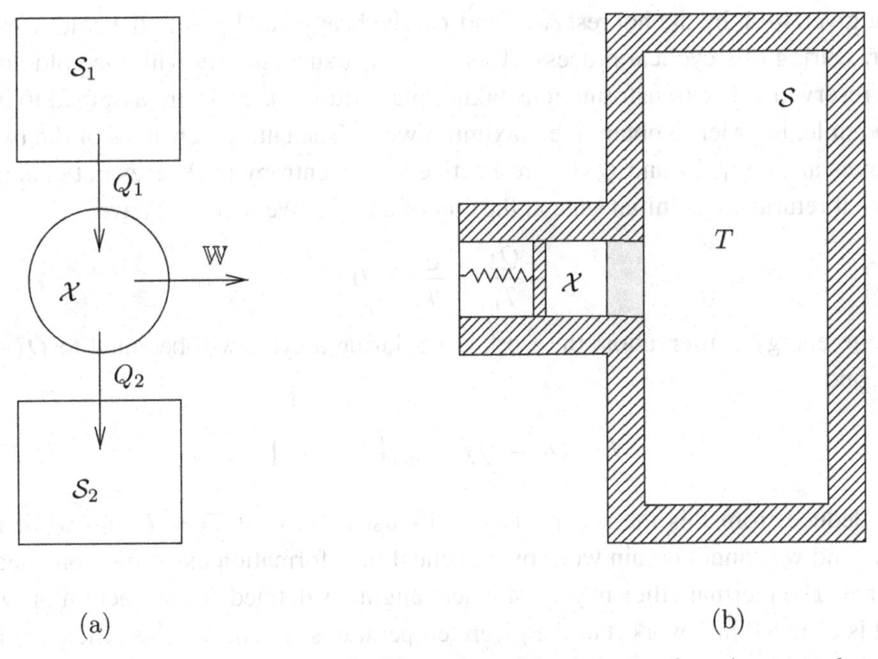

Figure 1.10 (a) Heat engine, (b) illustration of the theorem of maximum work.

another, and with this transformation its energy will change by ΔE and its entropy by ΔS. Conservation of energy gives the following expression for the *supplied* work $\mathbb{W}$

$$\mathbb{W} = -\Delta E + Q$$

The work, $\mathbb{W}$, is maximum if Q is maximum. But, the maximum entropy principle applied to the isolated system $[\mathcal{X} + \mathcal{S}]$ implies

$$\Delta S_{\text{tot}} = \Delta S - \frac{Q}{T} \geq 0$$

From this the theorem of maximum work follows

$$\boxed{\mathbb{W} \leq T\Delta S - \Delta E} \tag{1.21}$$

Maximum work is obtained from a transformation when it is reversible and $\Delta S_{\text{tot}} = 0$.

The device $\mathcal{X}$ functions in a cycle, that is to say, it returns periodically to the same state. Let $\mathbb{W}$ be the work supplied[18] by $\mathcal{X}$ during one such cycle, Q_1 the

[18] We will often use the term 'work (or heat) supplied' as shorthand for energy supplied in the form of work (or heat).

heat given to $\mathcal{X}$ by the hot reservoir and Q_2 the heat given by $\mathcal{X}$ to the cold reservoir. During this cyclical process, $\mathcal{X}$ is in contact successively with the cold and hot reservoirs. The transformations taking place during a cycle are assumed to be reversible, in order to obtain the maximum work. The entropy changes of the two sources are $-Q_1/T_1$ and Q_2/T_2 respectively. The entropy of $\mathcal{X}$ does not change since it returns to its initial state at the end of a cycle. We therefore have

$$-\frac{Q_1}{T_1} + \frac{Q_2}{T_2} = 0$$

Due to energy conservation, the work done during a cycle will be equal to $Q_1 - Q_2$, and

$$\mathbb{W} = Q_1 - Q_2 = Q_1\left(1 - \frac{T_2}{T_1}\right) \tag{1.22}$$

We deduce from this the second law in its usual form: if $T_1 = T_2$, no work is done and we cannot obtain work by a cyclical transformation using only one heat source. The thermal efficiency, η, of a heat engine is defined as the fraction of Q_1 that is changed into work. For the given temperatures, T_1 and T_2, this efficiency is maximum for an engine functioning reversibly and gives, from Equation (1.22),

$$\boxed{\eta = 1 - \frac{T_2}{T_1}} \tag{1.23}$$

Entropy is defined up to a multiplicative constant: replacing S by λS, where λ is the same constant for all systems, preserves the fundamental extremum properties. With this transformation, $T \to T/\lambda$ as given by the definition (1.9). We fix λ by demanding that the temperature, in kelvins (K), of the triple point of water be equal to 273.16. Since energy is measured in joules (J), entropy is measured in $\mathrm{J\,K^{-1}}$. The temperature, $\mathcal{T}$, measured on the Celsius scale is defined by $\mathcal{T} = T - 273.15\,\mathrm{K}$.

1.3 Thermodynamic potentials

1.3.1 Thermodynamic potentials and Massieu functions

Assuming for the moment N to be constant, the energy, E, is then a function of S and V. In thermodynamics, it is very often convenient to be able to change variables and use, for example, the temperature and pressure rather than the entropy and volume. These variable changes are implemented with the help of thermodynamic potentials or Massieu functions that are, respectively, the Legendre transforms (Section A.1) of the energy and entropy. We start with the energy and show how to go from a function of entropy to one of temperature. From Equation (1.19) we have

$$\left.\frac{\partial E}{\partial S}\right|_V = T$$

The Legendre transform, F, of E with respect to S is given by

$$\boxed{F(T, V) = E - TS \qquad \left.\frac{\partial F}{\partial T}\right|_V = -S, \quad \left.\frac{\partial F}{\partial V}\right|_T = -P} \tag{1.24}$$

The function F is called the *free energy*, and going from E to F has allowed us to use the temperature instead of the entropy. The two other thermodynamic potentials are the *enthalpy* $\overline{H}$, which is the Legendre transform of the energy with respect to the volume,

$$\boxed{\overline{H}(S, P) = E + PV \qquad \left.\frac{\partial \overline{H}}{\partial S}\right|_P = T, \quad \left.\frac{\partial \overline{H}}{\partial P}\right|_S = V} \tag{1.25}$$

and the *Gibbs potential*, which is obtained by a double Legendre transform

$$\boxed{G(T, P) = E - TS + PV \qquad \left.\frac{\partial G}{\partial T}\right|_P = -S, \quad \left.\frac{\partial G}{\partial P}\right|_T = V} \tag{1.26}$$

The *Maxwell relations* are an important consequence of the existence of these functions. In fact, the equality

$$\frac{\partial^2 F}{\partial T\, \partial V} = \frac{\partial^2 F}{\partial V\, \partial T}$$

combined with Equation (1.24), implies

$$\left.\frac{\partial S}{\partial V}\right|_T = \left.\frac{\partial P}{\partial T}\right|_V \tag{1.27}$$

whereas the same reasoning applied to the Gibbs potential gives

$$\left.\frac{\partial S}{\partial P}\right|_T = -\left.\frac{\partial V}{\partial T}\right|_P \tag{1.28}$$

Instead of taking the Legendre transforms of the energy, we can take those of the entropy and thus obtain the Massieu functions. For example, from Equation (1.17) we have

$$\left.\frac{\partial S}{\partial E}\right|_V = \frac{1}{T}$$

and the Massieu function Φ_1 is obtained by taking the Legendre transform with respect to the energy

$$\begin{aligned} &\Phi_1\left(\frac{1}{T}, V\right) = S - \frac{E}{T} = -\frac{1}{T} F \\ &\left.\frac{\partial \Phi_1}{\partial 1/T}\right|_V = -E \qquad \left.\frac{\partial \Phi_1}{\partial V}\right|_{1/T} = \frac{P}{T} \end{aligned} \tag{1.29}$$

The reader can easily construct the two other Massieu functions (Exercise (1.6.1)).

For historical reasons in thermodynamics, the variable T is more commonly used than the variable $1/T$ and the energy representation (1.19) is preferred to that of the entropy (1.17), and F to Φ_1. However, since entropy is of a different nature from the mechanical variables E and V, the entropy representation would be the more natural. We shall also see that the natural variable in statistical mechanics is $1/T$, that Φ_1 is the logarithm of the partition function, and that the entropy representation imposes itself in out-of-equilibrium situations.

1.3.2 Specific heats

Suppose we give a system a quantity of heat, $đQ$, in a quasi-static process, while keeping fixed one or more thermodynamic variables, y. If $\mathrm{d}T$ is the increase in temperature, the *specific heat* (or heat capacity), C_y, at fixed y, is the ratio

$$C_y = \left.\frac{đQ}{\mathrm{d}T}\right|_y = T\left.\frac{\partial S}{\partial T}\right|_y \tag{1.30}$$

The substitution of $T\,\mathrm{d}S$ for $đQ$ is justified because we have assumed a quasi-static process. The classic cases are the specific heat at constant volume, C_V,

$$\boxed{C_V = \left.\frac{đQ}{\mathrm{d}T}\right|_V = T\left.\frac{\partial S}{\partial T}\right|_V = \left.\frac{\partial E}{\partial T}\right|_V} \tag{1.31}$$

and the specific heat at constant pressure, C_P,

$$\boxed{C_P = \left.\frac{đQ}{\mathrm{d}T}\right|_P = T\left.\frac{\partial S}{\partial T}\right|_P = \left.\frac{\partial \overline{H}}{\partial T}\right|_P} \tag{1.32}$$

It is the enthalpy, $\overline{H}$, and not the energy, that appears in this last equation!

We now present a classic calculation leading to a useful relation between the specific heats at constant volume and pressure. It is helpful to define the following three coefficients which we define as intensive variables.

Expansion coefficient at constant pressure:

$$\alpha = \frac{1}{V}\frac{\partial V}{\partial T}\bigg|_P \tag{1.33}$$

Coefficient of isothermal compressibility:

$$\kappa_T = -\frac{1}{V}\frac{\partial V}{\partial P}\bigg|_T \tag{1.34}$$

Coefficient of adiabatic compressibility (or at constant entropy):

$$\kappa_S = -\frac{1}{V}\frac{\partial V}{\partial P}\bigg|_S \tag{1.35}$$

We need the following relation between partial derivatives. Consider a function of two variables, $z(x, y)$, and its differential

$$\mathrm{d}z = \frac{\partial z}{\partial x}\bigg|_y \mathrm{d}x + \frac{\partial z}{\partial y}\bigg|_x \mathrm{d}y$$

If we now restrict ourselves to a surface $z = $const, i.e. $\mathrm{d}z = 0$, we get

$$\frac{\partial z}{\partial x}\bigg|_y \mathrm{d}x = -\frac{\partial z}{\partial y}\bigg|_x \mathrm{d}y$$

or, in a form that is easy to write by circular permutation of the three variables (x, y, z),

$$\boxed{\frac{\partial x}{\partial y}\bigg|_z \frac{\partial y}{\partial z}\bigg|_x \frac{\partial z}{\partial x}\bigg|_y = -1} \tag{1.36}$$

Applying this relation to the variables (T, P, V)

$$\frac{\partial T}{\partial P}\bigg|_V \frac{\partial P}{\partial V}\bigg|_T \frac{\partial V}{\partial T}\bigg|_P = -1 \tag{1.37}$$

We start with $T\,\mathrm{d}S$ expressed in terms of the variables (T, P)

$$T\,\mathrm{d}S = C_P\,\mathrm{d}T + T\frac{\partial S}{\partial P}\bigg|_T \mathrm{d}P = C_P\,\mathrm{d}T - T\frac{\partial V}{\partial T}\bigg|_P \mathrm{d}P = C_P\mathrm{d}T - TV\alpha\,\mathrm{d}P$$

where we have first used the Maxwell relation (1.28) and then the definition of α (1.33). From this we find, by going from variables (T, P) to (T, V),

$$T\frac{\partial S}{\partial T}\bigg|_V = C_P - TV\alpha\frac{\partial P}{\partial T}\bigg|_V = C_P - TV\alpha\frac{\alpha}{\kappa_T}$$

The second equality has been obtained by evaluating $(\partial P/\partial T)_V$ with the help of (1.37). The final result is the classic relation for $C_P - C_V$

$$\boxed{C_P - C_V = \frac{TV\alpha^2}{\kappa_T}} \tag{1.38}$$

Another relation is (Exercise 1.6.3)

$$\boxed{\frac{C_P}{C_V} = \frac{\kappa_T}{\kappa_S}} \tag{1.39}$$

1.3.3 Gibbs–Duhem relation

We now consider a situation where the number of particles, N, can change. If we scale all the extensive variables by a factor λ

$$E \to \lambda E \qquad V \to \lambda V \qquad N \to \lambda N$$

the entropy, being also extensive, will scale as $S \to \lambda S$, and consequently

$$\lambda S = S(\lambda E, \lambda V, \lambda N)$$

By differentiating with respect to λ, using (1.17), and then taking $\lambda = 1$, we obtain

$$S = \frac{E}{T} + \frac{PV}{T} - \frac{\mu N}{T}$$

which allows us to identify μN as the Gibbs potential, G (1.26)

$$\mu N = E - TS + PV = G \tag{1.40}$$

The differential of the Gibbs potential can be written in two ways by using (1.26) or (1.40)

$$\mathrm{d}G = -S\,\mathrm{d}T + V\,\mathrm{d}P + \mu\,\mathrm{d}N = \mu\,\mathrm{d}N + N\,\mathrm{d}\mu$$

from which is obtained the 'Gibbs–Duhem' relation

$$\boxed{N\,\mathrm{d}\mu + S\,\mathrm{d}T - V\,\mathrm{d}P = 0} \tag{1.41}$$

We apply this to the isothermal case, $\mathrm{d}T = 0$. With the particle density defined by $n = N/V$, the Gibbs–Duhem relation yields

$$\left.\frac{\partial P}{\partial \mu}\right|_T = n$$

We now calculate $(\partial\mu/\partial n)_T$ with the help of definition (1.34) for the isothermal compressibility, κ_T, and with $v = V/N = 1/n$

$$\frac{1}{\kappa_T} = -v\left.\frac{\partial P}{\partial v}\right|_T = n\left.\frac{\partial P}{\partial n}\right|_T = n\left.\frac{\partial P}{\partial \mu}\right|_T\left.\frac{\partial \mu}{\partial n}\right|_T = n^2\left.\frac{\partial \mu}{\partial n}\right|_T$$

and therefore

$$\left.\frac{\partial \mu}{\partial n}\right|_T = \frac{1}{n^2\kappa_T} \tag{1.42}$$

1.4 Stability conditions

1.4.1 Concavity of entropy and convexity of energy

We now generalize to several variables the concavity analysis of the entropy that we did in 1.2.2 for one variable. To simplify the notation, we limit ourselves to two variables, the energy E and volume V, but the generalization to more variables is straightforward. Let $S(2E, 2V)$ be the entropy of a homogeneous, isolated system at equilibrium with energy $2E$ and volume $2V$. We now divide the system into two subsystems with energies $(E \pm \Delta E)$ and volumes $(V \pm \Delta V)$, and we suppose that the entropy is locally convex,

$$S(E + \Delta E, V + \Delta V) + S(E - \Delta E, V - \Delta V) > 2S(E, V) = S(2E, 2V)$$

In this case, applying an internal constraint would render the system inhomogeneous (unless $\Delta E/E = \Delta V/V$) while allowing the entropy to increase. The principle of maximum entropy would then lead to an inhomogeneous equilibrium state in contradiction with the initial assumption. We obtain the condition that the entropy must be concave

$$S(E + \Delta E, V + \Delta V) + S(E - \Delta E, V - \Delta V) \le 2S(E, V) \tag{1.43}$$

This inequality can also be written in the form

$$\boxed{(\Delta S)_{(E,V)} \le 0} \tag{1.44}$$

which can be interpreted as follows. At fixed E and V, internal constraints can only decrease the entropy. Taking the Taylor expansion of $S(E \pm \Delta E, V \pm \Delta V)$ for small ΔE and ΔV, we get

$$S(E \pm \Delta E, V \pm \Delta V) \simeq S(E, V) \pm \Delta E\frac{\partial S}{\partial E} \pm \Delta V\frac{\partial S}{\partial V} + \frac{1}{2}(\Delta E)^2\frac{\partial^2 S}{\partial E^2} + \frac{1}{2}(\Delta V)^2\frac{\partial^2 S}{\partial V^2} + \Delta E\,\Delta V\frac{\partial^2 S}{\partial E\,\partial V}$$

Substituting this expansion in Equation (1.43) we obtain the inequality

$$(\Delta E)^2 \frac{\partial^2 S}{\partial E^2} + (\Delta V)^2 \frac{\partial^2 S}{\partial V^2} + 2\Delta E\, \Delta V \frac{\partial^2 S}{\partial E\, \partial V} \le 0 \tag{1.45}$$

The concavity condition on the entropy can be transformed into a convexity condition on the energy. Consider an isolated system with energy E and volume V, which we divide into two subsystems at equilibrium

$$E = E_1 + E_2 \qquad V = V_1 + V_2$$

We now apply an internal constraint where

$$E_1 \to E_1 + \Delta E \qquad E_2 \to E_2 - \Delta E$$

and

$$V_1 \to V_1 + \Delta V \qquad V_2 \to V_2 - \Delta V$$

The principle of maximum entropy gives

$$S(E_1 + \Delta E, V_1 + \Delta V) + S(E_2 - \Delta E, V_2 - \Delta V) \le S(E, V)$$

But S is an increasing function of the energy. There exists, therefore, an energy $\tilde{E} \le E$ such that

$$S(E_1 + \Delta E, V_1 + \Delta V) + S(E_2 - \Delta E, V_2 - \Delta V) = S(\tilde{E}, V)$$

At constant entropy, internal constraints can only increase the energy: $E_1 + E_2 = E \ge \tilde{E}$. Therefore, the analogue of Equation (1.44) for the energy is

$$\boxed{(\Delta E)_{(S,V)} \ge 0} \tag{1.46}$$

which says that the energy is a convex function of S and V. Another demonstration of this result is proposed in Exercise 1.6.2.

1.4.2 Stability conditions and their consequences

The convexity condition on the energy leads to an equation analogous to (1.45). Only the sense of the inequality changes when we go from concavity of the entropy to convexity of the energy

$$(\Delta S)^2 \frac{\partial^2 E}{\partial S^2} + (\Delta V)^2 \frac{\partial^2 E}{\partial V^2} + 2\Delta S \Delta V \frac{\partial^2 E}{\partial S\, \partial V} \ge 0 \tag{1.47}$$

This condition can be conveniently expressed in matrix form by introducing the symmetric 2×2 matrix $\mathcal{E}$ whose elements are second derivatives of E

$$\mathcal{E} = \begin{pmatrix} \frac{\partial^2 E}{\partial S^2} & \frac{\partial^2 E}{\partial S \partial V} \\ \frac{\partial^2 E}{\partial V \partial S} & \frac{\partial^2 E}{\partial V^2} \end{pmatrix} = \begin{pmatrix} E''_{SS} & E''_{SV} \\ E''_{VS} & E''_{VV} \end{pmatrix} \tag{1.48}$$

Introducing the two-component vector, $x = (\Delta S, \Delta V)$, and its transpose, x^T, Equation (1.47) becomes $x^T \mathcal{E} x \geq 0$, which means that the matrix $\mathcal{E}$ must be *positive*. By definition, a symmetric[19] (and therefore diagonalizable) $N \times N$ matrix, A_{ij}, is said to be positive if for any vector x with components x_i, $i = 1, \ldots, N$

$$x^T A\, x = \sum_{i,j=1}^{N} x_i A_{ij} x_j \geq 0 \tag{1.49}$$

The positivity condition for a matrix can be expressed in terms of its eigenvalues. A necessary and sufficient condition for a matrix to be positive is that all its eigenvalues, λ_i, be positive, $\lambda_i \geq 0$. For a positive definite matrix, this inequality, and those in Equation (1.49), become strict inequalities. In fact, we can diagonalize a symmetric matrix by $A = R^T \Lambda R$, where Λ is a diagonal matrix and R an orthogonal matrix $R^T = R^{-1}$. The positivity condition (1.49) becomes

$$y^T \Lambda y = \sum_{i=1}^{N} \lambda_i\, y_i^2 \geq 0$$

where the N-component vector y is given by $y = Rx$. Clearly, this implies that $\lambda_i \geq 0$. We return now to the case of a 2×2 matrix,

$$\begin{pmatrix} a & b \\ b & c \end{pmatrix} \tag{1.50}$$

An elementary calculation gives the positivity condition. For the two eigenvalues to be positive, we must have $a + c \geq 0$ and $ac - b^2 \geq 0$, which means that $a \geq 0$ and $c \geq 0$ separately. In the case of a negative matrix we have $a \leq 0$ and $c \leq 0$ whereas $ac - b^2 \geq 0$.

The analysis of the stability conditions is most conveniently done with the help of the thermodynamic potentials. The Legendre transform changes concavity to convexity (Section A.1), and consequently we have

$\overline{H}$	convex in E	concave in P
F	concave in T	convex in V
G	concave in T	concave in P

[19] For real matrices one needs to impose the symmetry condition, but in a complex vector space, a positive matrix is automatically Hermitian and thus diagonalizable.

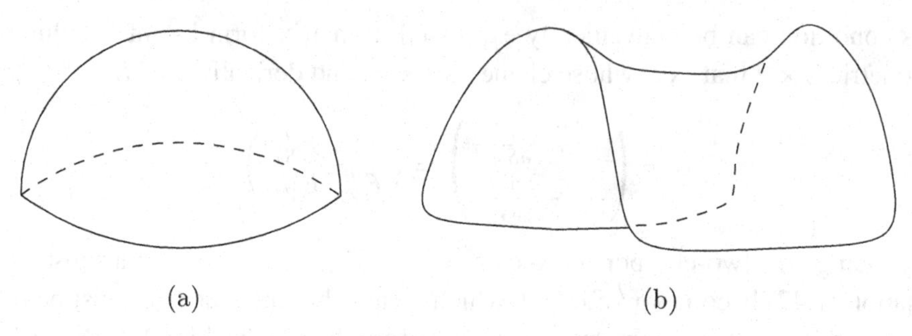

Figure 1.11 Surfaces with (a) positive and (b) negative curvatures.

Geometrically, the curvature K of a surface, $z = f(x, y)$, is given by

$$K = \frac{f''_{xx} f''_{yy} - (f''_{xy})^2}{(1 + f'^2_x + f'^2_y)^2} \tag{1.51}$$

E, S, and G are therefore positively curved functions of their respective variables (sphere-like shape) whereas H and F are negatively curved functions (saddle-like shape), see Figure 1.11.

The stability conditions are obtained most simply by studying the Gibbs potential. The matrix constructed from the second derivatives, G''_{TT}, G''_{TP}, G''_{PP} must be negative since G is a concave function[20] of T and P. We therefore have

$$G''_{TT} = -\left.\frac{\partial S}{\partial T}\right|_P = -\frac{C_P}{T} \le 0 \Rightarrow C_P \ge 0$$

and

$$G''_{PP} = \left.\frac{\partial V}{\partial P}\right|_T = -V\kappa_T \le 0 \Rightarrow \kappa_T \ge 0$$

where κ_T is the isothermal compressibility defined in (1.34). The last condition is

$$G''_{TT} G''_{PP} - (G''_{PT})^2 \ge 0$$

Using $G''_{PT} = (\partial V/\partial T)_P = \alpha V$, where α is the constant pressure expansion coefficient defined in (1.33), the above condition can be written as

$$C_P - \frac{\alpha^2 V T}{\kappa_T} \ge 0$$

Equation (1.38) leads to the conclusion that $C_V \ge 0$. To summarize, the two stability conditions are

$$\boxed{C_V \ge 0 \qquad \kappa_T \ge 0} \tag{1.52}$$

[20] In Section A.1.2 we show that G is indeed a concave function of both variables T and P.

With Equations (1.38) and (1.39) in mind, the stability conditions (1.52) yield $C_P \geq C_V \geq 0$ and $\kappa_T \geq \kappa_S \geq 0$. The condition $C_V \geq 0$ can be obtained directly from the concavity of the entropy, but the condition on κ_T is more difficult to show directly from S. Although one might think intuitively that the expansion coefficient, α, must be positive (an increase in temperature normally leads to expansion, not contraction), we should note that the stability conditions do not impose any restrictions on this coefficient. In fact, it can happen that α becomes negative!

1.5 Third law of thermodynamics

1.5.1 Statement of the third law

The 'third law' is fundamentally related to low temperatures. Before stating it, we give orders of magnitude of temperatures reached by various techniques[21]

Pumped helium-4	1 K
Mixture helium-3–helium-4	10 mK
Helium-3 compression (Problem 5.7.6)	2 mK
Electronic spin demagnetization (Problem 1.7.7)	3 mK
Nuclear spin demagnetization	10 μK
Bose–Einstein condensates of atomic gases (Problem 5.7.5)	1 nK

Laser cooling also produces nK temperatures, but there is no thermal equilibrium here, only an effective temperature. The temperature of an atomic Bose–Einstein condensate corresponds to metastable equilibrium, moreover these condensates cannot be used to cool other systems.

The third law of thermodynamics allows us to fix the entropy at zero temperature: 'The entropy tends to zero as the temperature vanishes.' More precisely, for a system of volume V and entropy $S(V)$[22]

$$\boxed{\lim_{T\to 0}\lim_{V\to\infty}\frac{1}{V}S(V) = 0} \tag{1.53}$$

The first limit in Equation (1.53) corresponds to the *thermodynamic limit*, the intensive quantities remain finite.

Current techniques allow cooling down to the mK scale. At that temperature there remains the residual entropy due to the nuclear spins, which vanishes only

[21] The method of cooling by nuclear spin demagnetization can reach temperatures of 0.3 nK, but this is the temperature of the spin lattice, which is very different in nature from the usual temperature, it can even be negative. See Problem 3.8.2.

[22] There are, however, model systems which retain a non-vanishing entropy at zero temperature. It has been known for a long time that the antiferromagnetic Ising model on a two-dimensional triangular lattice has an entropy of $0.338k$ per site at $T = 0\,\mathrm{K}$ [120]. This zero-point entropy does not persist for the quantum Heisenberg model on the same lattice. However, we find again a zero-point entropy on the Kagome lattice [93].

below 1 μK, with the notable exception of helium-3 (see Problem 5.7.6). If S_0 is the entropy per unit volume due to nuclear spins, and if we do not go much below 1 mK, we can replace (1.53) with an effective statement of the third law

$$\lim_{T\to 0}\lim_{V\to\infty}\frac{1}{V}S(V)=S_0 \tag{1.54}$$

S_0 is a reference entropy that is independent of chemical composition, pressure, solid, liquid or gaseous state, crystalline form, etc. because nuclear spins are insensitive to these parameters. The third law has its origins in quantum physics, and we shall have several occasions to verify its validity in Chapter 5 from calculations in quantum statistical mechanics.

1.5.2 Application to metastable states

Consider a system that can exist in two crystalline forms (a) and (b), one stable and the other metastable but with such a long lifetime that we can apply to it the usual equations. The form (a) is stable for $T < T_c$ and form (b) for $T > T_c$ where T_c is, therefore, the temperature of the phase transition. This phase transition takes place with a latent heat L, with

$$\frac{L}{T_c}=S^{(b)}(T_c)-S^{(a)}(T_c) \tag{1.55}$$

where $S^{(a)}(T)$ and $S^{(b)}(T)$ are the entropies of (a) and (b) at T. By varying the temperature along a quasi-static path where the variables y are held constant, we can calculate $S^{(a)}(T)$ and $S^{(b)}(T)$ in terms of $C_y^{(a)}(T)$ and $C_y^{(b)}(T)$, the specific heats at fixed y defined in (1.31)

$$S^{(a)}(T_c)=S_0+\int_0^{T_c}\mathrm{d}T\,\frac{C_y^{(a)}(T)}{T} \tag{1.56}$$

and

$$S^{(b)}(T_c)=S_0+\int_0^{T_c}\mathrm{d}T\,\frac{C_y^{(b)}(T)}{T} \tag{1.57}$$

The crucial point is that S_0, being independent of the crystalline form, is the same in the two equations. We can therefore determine the difference $S^{(b)}(T_c)-S^{(a)}(T_c)$ either by measuring the latent heat (1.55), or by measuring the specific heat and taking the difference of (1.56) and (1.57) thus eliminating S_0. The fact that the two results coincide is a verification of the third law. A classic case is that of grey and white tin. Grey tin is a semiconductor, stable for $T < T_c = 292\,\mathrm{K}$, white

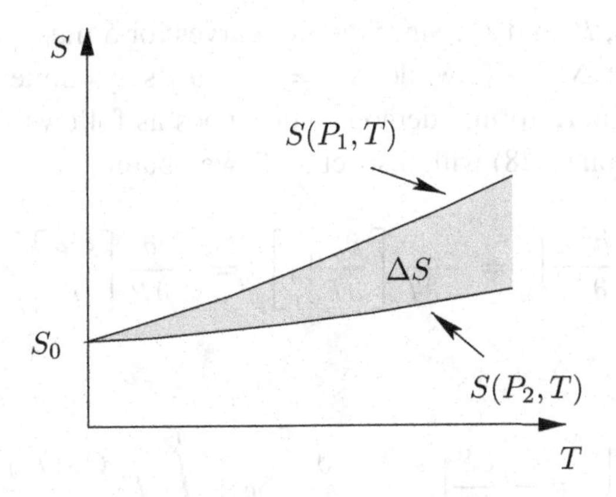

Figure 1.12 Behaviour of the entropy near $T = 0$ for two different pressures.

tin is a metal, stable for $T > T_c$. Experimental measurement of $S^{(b)}(T_c) - S^{(a)}(T_c)$ using (1.55) gives 7.7 J K^{-1} whereas the difference between (1.56) and (1.57) gives 7.2 J K^{-1}. Taking into account experimental uncertainty, the agreement is satisfactory.

1.5.3 Low temperature behaviour of specific heats

The third law constrains the behaviour of the specific heats as $T \to 0$. We start with the expression for $S(T, P)$ or $S(T, V)$

$$S(T; P, V) - S_0 = \int_0^T \mathrm{d}T' \, \frac{C_{P,V}(T')}{T'} \tag{1.58}$$

and follow a quasi-static path at constant pressure or volume. Since the integral in (1.58) has to converge for $T' \to 0$ (otherwise the entropy would be infinite), C_P and C_V must vanish[23] as $T \to 0$

$$T \to 0 \qquad C_P, C_V \to 0 \tag{1.59}$$

Furthermore, the expansion coefficient at constant pressure α (1.33) as well as $(\partial P/\partial T)_V$ also vanish

$$\alpha = \frac{1}{V}\frac{\partial V}{\partial T}\bigg|_P \to 0 \qquad \frac{\partial P}{\partial T}\bigg|_V \to 0 \tag{1.60}$$

We demonstrate, as an example, the first of relations (1.60). We start with the Maxwell relation (1.28) and we plot on Figure 1.12 S as a function of T for two

[23] This argument only assumes that the entropy is finite at zero temperature.

different pressures, P_1 and P_2. Since the two curves for S must pass through S_0 for $T=0$, we see that $\Delta S \to 0$, while $\Delta P = P_2 - P_1$ stays finite and consequently $\Delta S/\Delta P \to 0$. A more formal demonstration goes as follows. By differentiating the Maxwell relation (1.28) with respect to T, we obtain

$$\left.\frac{\partial^2 V}{\partial T^2}\right|_P = -\frac{\partial}{\partial T}\left[\left.\frac{\partial S}{\partial P}\right|_T\right]_P = -\frac{\partial}{\partial P}\left[\frac{C_P}{T}\right]_T$$

On the other hand

$$\left.\frac{\partial V}{\partial T}\right|_P = -\left.\frac{\partial S}{\partial P}\right|_T = -\frac{\partial}{\partial P}\left[S_0 + \int_0^T \mathrm{d}T' \, \frac{C_P(T')}{T'}\right]_T$$

The important point is that S_0 is a constant independent of P which leads to

$$\left.\frac{\partial V}{\partial T}\right|_P = -\int_0^T \mathrm{d}T' \, \frac{1}{T'}\left.\frac{\partial C_P}{\partial P}\right|_T = \int_0^T \mathrm{d}T' \left.\frac{\partial^2 V}{\partial T'^2}\right|_P = \left.\frac{\partial V}{\partial T}\right|_P - \left.\frac{\partial V}{\partial T}\right|_P^{T=0}$$

The demonstration of the result for $\partial P/\partial T_V$ follows the same lines, by using the Maxwell relation (1.27).

We can obtain an additional result for $C_P - C_V$ by making an assumption on the behaviour of C_P. We assume that C_P follows a power law in T

$$T \to 0 \qquad C_P \simeq T^x f(P)$$

with $x > 0$ and $f(P) > 0$. We then obtain

$$\alpha V = \left.\frac{\partial V}{\partial T}\right|_P = -\int_0^T \mathrm{d}T' \, T'^{x-1} f'(P) = -\frac{1}{x} f'(P) T^x$$

It follows that $\alpha V/C_P$ is a function only of P for $T \to 0$

$$\lim_{T\to 0} \frac{\alpha V}{C_P} = -\frac{f'(P)}{x f(P)} = g(P)$$

Using (1.38), and keeping in mind that κ_T stays finite for $T \to 0$, we see that $(C_P - C_V)$ decreases more rapidly than C_P and C_V separately

$$T \to 0 \qquad \frac{C_P - C_V}{C_P} \sim \alpha T$$

1.6 Exercises

1.6.1 Massieu functions

1. Construct the three Massieu functions

$$\Phi_1\left(\frac{1}{T}, V, N\right) \qquad \Phi_2\left(\frac{1}{T}, \frac{P}{T}, N\right) \qquad \Phi_3\left(\frac{1}{T}, V, \frac{\mu}{T}\right)$$

2. Applications

Consider a system where the energy is a function only of the temperature T. Show that the volume is a function only of P/T, $V = V(P/T)$, and verify the immediate consequence of this (see also Problem 1.7.1)

$$\left.\frac{\partial P}{\partial T}\right|_V = \frac{P}{T} \tag{1.61}$$

Show that the difference between the specific heat at constant pressure and constant volume depends only on P/T

$$C_P - C_V = -\left(\frac{P}{T}\right)^2 V'\left(\frac{P}{T}\right)$$

1.6.2 Internal variable in equilibrium

We consider a system at fixed volume and particle number. The system is characterized either by its total energy E or its total entropy S, as well as one internal variable y. For example in the system in Figure 1.3, y may be the position of the partition, or the distribution of the energy between the two compartments. At equilibrium and for a given E, the value of y is fixed by the condition of maximum entropy

$$\left.\frac{\partial S}{\partial y}\right|_E = 0 \qquad \left.\frac{\partial^2 S}{\partial y^2}\right|_E \leq 0 \tag{1.62}$$

We have seen in Section 1.4.1 that the condition of maximum entropy at fixed E becomes a condition of minimum energy at fixed entropy. We give another demonstration of this result by showing that (1.62) implies

$$\left.\frac{\partial E}{\partial y}\right|_S = 0 \qquad \left.\frac{\partial^2 E}{\partial y^2}\right|_S \geq 0 \tag{1.63}$$

which is equivalent to (1.47).

1. Show first that

$$\left.\frac{\partial S}{\partial y}\right|_E = -\left.\frac{\partial E}{\partial y}\right|_S \left.\frac{\partial S}{\partial E}\right|_y \tag{1.64}$$

and then establish (1.63).

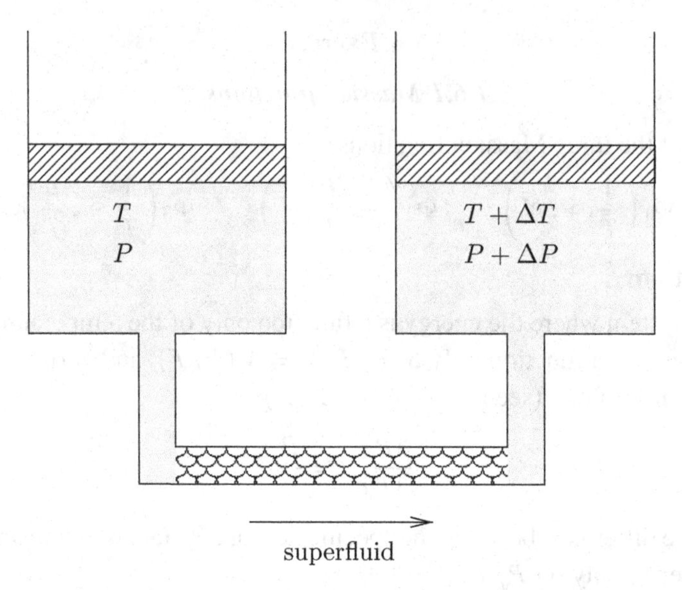

Figure 1.13 Superfluid flow through a porous plug.

2. Return to Figure 1.3 and suppose that the two compartments have different temperatures and pressures, and that the piston, while still adiabatic, is made mobile. Show that the condition of maximum entropy does not allow us to fix an equilibrium position.

3. Below 2 K, helium-4 may be considered as a mixture of a fluid of zero entropy flowing with zero viscosity through very small openings, and a normal fluid (see Problem 8.6.6 for more detail). In the setup shown in Figure 1.13, the tube connecting the two containers of fixed volumes allows only the superfluid to pass. With the help of the minimum energy condition, show that the chemical potentials are equal in the two containers, but that the pressures and temperatures may be different. If ΔP and ΔT are the pressure and temperature differences between the two containers, show that

$$\frac{\Delta P}{\Delta T} = \frac{S}{V}$$

This result illustrates the ability of the superfluid to balance a pressure difference with a temperature difference, provided the normal fluid is unable to move.

1.6.3 Relations between thermodynamic coefficients

Demonstrate the following relation

$$\frac{\kappa_S}{\kappa_T} = \frac{C_V}{C_P} \tag{1.65}$$

1.6.4 Contact between two systems

Consider two identical copper blocks, A and B, of mass m, specific heat C and respective temperatures T_A and T_B with $T_A \leq T_B$. We assume that C does not depend on the temperature and that the volumes of the blocks are constant. We put A and B in contact while maintaining the ensemble thermally insulated.

1. What is the change in the internal energy of each of the blocks? What is the total energy change? What is the final temperature?

2. What is the entropy change of each of the blocks? What is the total entropy change?

3. We construct a heat engine which uses the two blocks as heat sources. What is the maximum work that can be obtained? What is the final temperature of the two blocks in this case?

1.6.5 Stability conditions

1. Establish the following relations at constant N

$$\left.\frac{\partial \mu}{\partial V}\right|_T = \frac{V}{N}\left.\frac{\partial P}{\partial V}\right|_T$$

$$\left.\frac{\partial T}{\partial V}\right|_S = -\frac{T}{C_V}\left.\frac{\partial P}{\partial T}\right|_V$$

2. An experimentalist claims to have found a material with the following properties

$$(i)\ \left.\frac{\partial P}{\partial V}\right|_T < 0 \qquad (ii)\ \left.\frac{\partial P}{\partial T}\right|_V > 0 \qquad (iii)\ \left.\frac{\partial \mu}{\partial V}\right|_T < 0 \qquad (iv)\ \left.\frac{\partial T}{\partial V}\right|_S > 0$$

Which of the relations above is compatible with the stability conditions? Are all these relations compatible with each other?

1.6.6 Equation of state for a fluid

A fluid has the following properties

(*i*) At a given fixed temperature T_0, the work $\mathbb{W}$ furnished to the outside world during isothermal expansion from volume V_0 to V is

$$\mathbb{W} = RT_0 \ln \frac{V}{V_0}$$

(*ii*) Its entropy is given by (a is a positive constant)

$$S(T, V) = R\left(\frac{V_0}{V}\right)\left(\frac{T}{T_0}\right)^a$$

What is the equation of state of this fluid?

1.7 Problems

1.7.1 Reversible and irreversible free expansions of an ideal gas

The most general way to define an ideal gas is as follows

(*i*) The internal energy E is independent of the volume at constant temperature.
(*ii*) PV is a function of only the temperature: $PV = \phi(T)$.

1. Show that $\phi(T)$ is a linear function of T: $\phi(T) = aT$.

2. In what follows we consider a mole of gas and assume that the specific heat is independent of temperature

$$C_V = \frac{l}{2}R \qquad l = 3, 5, \ldots$$

We recall that $C_P - C_V = R$ and we write $\gamma = C_P/C_V$. Starting at an initial temperature T_0 and initial volume V_0, calculate the entropy $S(V, T)$ as a function of R, l and $S(V_0, T_0)$. Repeat but now taking P_0 and T_0 as initial conditions.

3. Using the fact that the entropy is constant for a reversible adiabatic expansion, show that $PV^\gamma =$ const or $TP^{(1-\gamma)/\gamma} =$ const.

4. We subject a mole of gas, initially in a state $(P_\mathrm{i}, T_\mathrm{i}, V_\mathrm{i})$, to a reversible adiabatic expansion taking it to the state $(P_\mathrm{f}, T_\mathrm{f}, V_\mathrm{f})$ with $P_\mathrm{f} < P_\mathrm{i}$. Calculate T_f and show that the work done on the gas is $W = C_V(T_\mathrm{f} - T_\mathrm{i})$.

5. A mole of gas is compressed by a piston on which is placed a mass m such that the initial pressure is

$$P_\mathrm{i} = P_\mathrm{f} + \frac{mg}{A}$$

where P_f is atmospheric pressure and A is the area of the piston (Figure 1.14). The mass is instantaneously removed, we assume the gas to be perfectly isolated thermally and the piston to be frictionless. What is the final temperature T_f' as a function of T_i, $P_\mathrm{i}/P_\mathrm{f}$ and γ? Plot representative curves for $T_\mathrm{f}/T_\mathrm{i}$ and $T_\mathrm{f}'/T_\mathrm{i}$ as a function of $P_\mathrm{f}/P_\mathrm{i}$ for $0 \le P_\mathrm{f}/P_\mathrm{i} \le 1$.

6. Could we have predicted without calculation that $T_\mathrm{f}' > T_\mathrm{f}$? Which expansion, that of Question 4 or Question 5, furnishes more work to the outside world? In the experiment of Question 5 we put back the mass on the piston. The final temperature and volume are T_f'' and V_f''. Compare T_f'' and V_f'' with T_i and V_i.

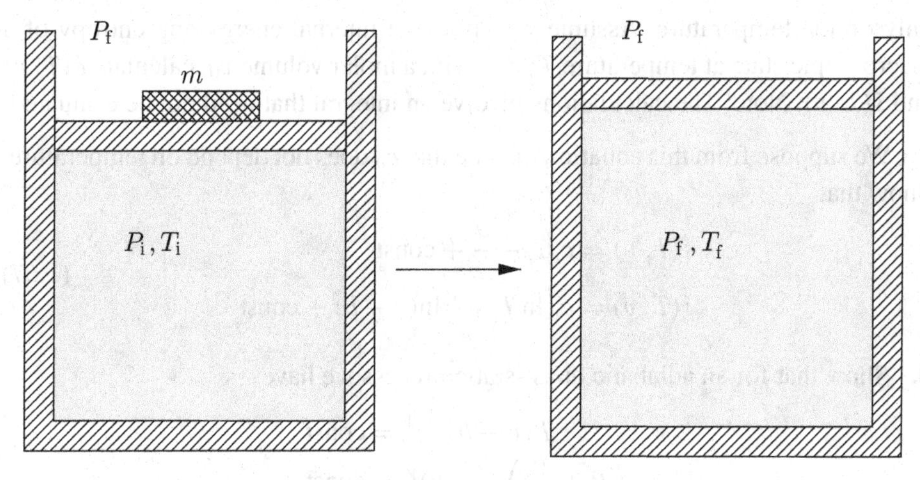

Figure 1.14 Compression and expansion of a gas.

7. In the experiment of Figure 1.8, the (ideal) gas is at a pressure P and temperature T, and the container is in vacuum. By freeing the piston, we allow the gas to expand by $\mathrm{d}V$. What is the final temperature? What is the entropy increase? Deduce that although the transformation is infinitesimal it is not quasi-static.

8. In Figure 1.9, the ideal gas is initially in the left compartment, which is assumed adiabatic. A very tiny hole is made in the central partition. What is the final temperature? What is the entropy increase? Show that although this transformation is infinitely slow, it is not quasi-static.

1.7.2 van der Waals equation of state

Consider a gas whose equation of state is given by

$$\left(P + \frac{a}{v^2}\right)(v - b) = kT \tag{1.66}$$

where v is the specific volume, $v = V/N$. This is the equation of state of a 'van der Waals gas'. It is an empirical equation that includes some of the residual interactions between molecules (attractive at long range, repulsive at short range) that were ignored in the ideal gas approximation.[24]

1. Give a physical interpretation of the terms a/v^2 and b. In which limit does the van der Waals equation become the ideal gas equation of state?

2. Let $\check{\epsilon}(T, v)$ and $\check{s}(T, v)$ be respectively the internal energy and entropy per gas molecule. Show that the constant volume specific heat per molecule, c_v, depends

[24] We emphasize that the van der Waals equation is in no way an exact equation.

only on the temperature. Assume we know the internal energy and entropy of a macroscopic state at temperature T_0 and with a molar volume v_0, calculate $\check{\epsilon}(T, v)$ and $\check{s}(T, v)$. Note: The calculations involve an integral that need not be evaluated.

3. We suppose from this equation of state that c_v does not depend on temperature. Show that

$$\begin{aligned}\check{\epsilon}(T, v) &= c_v T - \frac{a}{v} + \text{const} \\ \check{s}(T, v) &= c_v \ln T + k \ln(v - b) + \text{const}\end{aligned} \tag{1.67}$$

4. Show that for an adiabatic quasi-static process we have

$$T(v - b)^{\gamma - 1} = \text{const}$$
$$\left(P + \frac{a}{v^2}\right)(v - b)^{\gamma} = \text{const}$$

where $\gamma = \dfrac{c_v + k}{c_v}$.

5. Determine the sign of the temperature change of this gas when it undergoes an adiabatic expansion in vacuum as in Problem 1.7.1 Question 8. We assume that the initial (final) state is at T_1 and v_1 (T_2 and v_2).

6. Consider the example of Figure 1.3 and assume each compartment to contain half a mole of the van der Waals gas. Also assume that the two compartments are at the same temperature and that the initial volumes are V_1 and $2V - V_1$. What is the final temperature if the diathermal piston is free to move?

1.7.3 Equation of state for a solid

An often used phenomenological equation of state for solids is

$$E = A\mathrm{e}^{b(V - V_0)^2} S^{4/3} \mathrm{e}^{S/3R} = f(V)g(S) \tag{1.68}$$

where A, b and V_0 are positive constants and R is the ideal gas constant; E is the internal energy, S the entropy and V the volume.

1. Show that this equation of state satisfies the third law of thermodynamics.

2. Show that the constant volume specific heat, C_V, is proportional to T^3 at low temperature (Debye law), and that it approaches $3R$ at high temperature (Dulong–Petit law).

3. Calculate the pressure P. What is the physical interpretation of V_0? How does the thermal expansion coefficient at constant pressure behave for $P = 0$? Is this behaviour reasonable?

4. Show that

$$\left.\frac{\partial S}{\partial V}\right|_T = -\frac{f'(V)g'(S)}{f(V)g''(S)}$$

and verify that the thermal expansion coefficient at constant pressure may be written as

$$\left.\frac{\partial V}{\partial T}\right|_P = C_V \frac{f'(V)g''(S)}{(f'(V))^2(g'(S))^2 - f(V)f''(V)g(S)g''(S)} \quad (1.69)$$

1.7.4 Specific heats of a rod

Consider a rod whose relaxed length is L_0. Assume that the tension τ in this rod when stretched to a length L is given by

$$\tau = aT^2(L - L_0)$$

where L_0 and a do not depend on the temperature. We define the constant length specific heat of the rod by

$$C_L \,\widehat{=}\, T \left.\frac{\partial S}{\partial T}\right|_L$$

where S is its entropy. In addition we know C_L as a function of temperature when the rod is relaxed

$$C_L(L_0, T) = bT$$

where b is a constant.

1. Write the expressions for the differentials $\mathrm{d}E$ and $\mathrm{d}S$ of the rod internal energy and entropy in the variables (T, L).

2. Calculate the specific heat $C_L(L, T)$ for arbitrary length L.

3. Let the temperature of the rod at length L_0 be T_0. Calculate its entropy $S(L, T)$ in terms of $S(L_0, T_0)$.

4. The rod is in an initial state characterized by L_i and T_i. By applying an external force on it, its length is taken adiabatically and reversibly to L_f with $L_\mathrm{f} > L_\mathrm{i}$. Give the expression for the final temperature T_f and verify that the lengthening of the rod is accompanied by cooling.

5. Calculate the specific heat at constant tension $C_\tau(L, T)$. How does the rod behave when it is heated at constant tension?

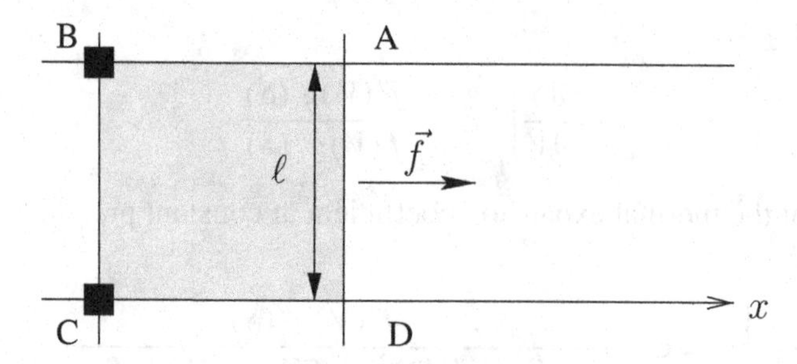

Figure 1.15 Soap film on a frame with a movable side.

1.7.5 Surface tension of a soap film

A soap film is held by the four sides of a rectangle ABCD (Figure 1.15). The wire AD, length ℓ, can be moved while kept parallel to BC allowing the film to be stretched. The modulus of the force $\vec{f}$, applied in the positive x direction, needed to keep AD fixed is $\sigma\ell$, where σ is the surface tension.

1. Write the $T\,\mathrm{d}S$ equation for this problem.

2. We define the specific heat at constant length as usual

$$C_x = T\,\frac{\partial S}{\partial T}\bigg|_x$$

Express the partial derivatives of the energy E in the variables (T, x) in terms of C_x, T and the partial derivatives of the surface tension σ.

3. For a wide range of temperatures around $T = 300$ K, the surface tension of soapy water varies linearly with T

$$\sigma = \sigma_0(1 - a(T - T_0)) \tag{1.70}$$

where σ_0, a, T_0 are constants, $\sigma_0 = 8 \times 10^{-2}\,\mathrm{N\,m^{-1}}$, $a = 1.5 \times 10^{-3}\,\mathrm{K^{-1}}$, $T_0 = 273.16\,\mathrm{K}$.

We stretch the film by $\mathrm{d}x$ *quasi-statically* and at *constant temperature*. Calculate the concomitant infinitesimal increase in the internal energy $\mathrm{d}E$. Verify that energy, in the form of heat, must be given to the film in order to maintain its temperature constant.

4. Verify that when σ obeys (1.70), C_x is independent of the length of the film.

5. We may define the 'isothermal shrinkage' coefficient by

$$\kappa_T = -\frac{\ell}{\mathcal{A}}\,\frac{\partial x}{\partial f}\bigg|_T$$

where $\mathcal{A}$ is the area of the film. What do you obtain for κ_T by studying the behaviour of $(\partial f/\partial x)_T$?

6. We assume that in the range of temperatures under consideration C_x remains constant. We stretch the film *quasi-statically* and *adiabatically* by dx. Calculate the resulting temperature increase dT that accompanies this stretching for an initial temperature $T = 300$ K. Assume that a reasonable order of magnitude estimate for C_x is obtained by calculating it from the bulk specific heat of the water forming the soap film. The film thickness is $e = 10^{-3}$ cm, $x = 2$ cm and we take for the bulk specific heat $C = 4185\,\mathrm{J\,kg^{-1}\,K^{-1}}$.

1.7.6 Joule–Thomson process

During a Joule expansion, a gas initially confined in a fixed volume escapes freely without exchanging heat with its environment. In this problem we consider a Joule–Thomson expansion where a gas initially in a compartment of volume V_1, temperature T_1 and pressure P_1, expands into a container of volume V_2 under a constant pressure P_2 ($P_2 < P_1$). The two compartments are thermally insulated and are connected by a porous plug which allows us to maintain the pressure difference. We assume the gas flux to be stationary. This adiabatic (but not quasi-static) expansion brings the temperature of the gas in the second compartment to a value T_2. We shall determine the sign of the temperature change.

1. Show that this expansion takes place at constant enthalpy $\overline{H}$.

2. Can we cool down an ideal gas with the Joule–Thomson expansion?

3. The quantity $\mu_{\mathrm{JT}} \,\hat{=}\, (\partial T/\partial P)_{\overline{H}}$ is called the Joule–Thompson coefficient. Express μ_{JT} in terms of V, T, C_P and α. Verify that the value μ_{JT} takes in the case of an ideal gas is consistent with your answer to Question 2.

4. Figure 1.16 shows isenthalpy curves ($\overline{H} = \text{const}$) in the (T, P) plane for gaseous nitrogen. Comment on this figure. What does the dashed line represent? What is the maximum temperature beyond which nitrogen cannot be cooled with Joule–Thomson expansion?

1.7.7 Adiabatic demagnetization of a paramagnetic salt

When a paramagnetic solid is subjected to a uniform external magnetic field, it acquires a non-zero average macroscopic magnetization $\mathcal{M}$. An infinitesimal increase dB in the applied field B produces a change $đW = -\mathcal{M}\,dB$ in the internal

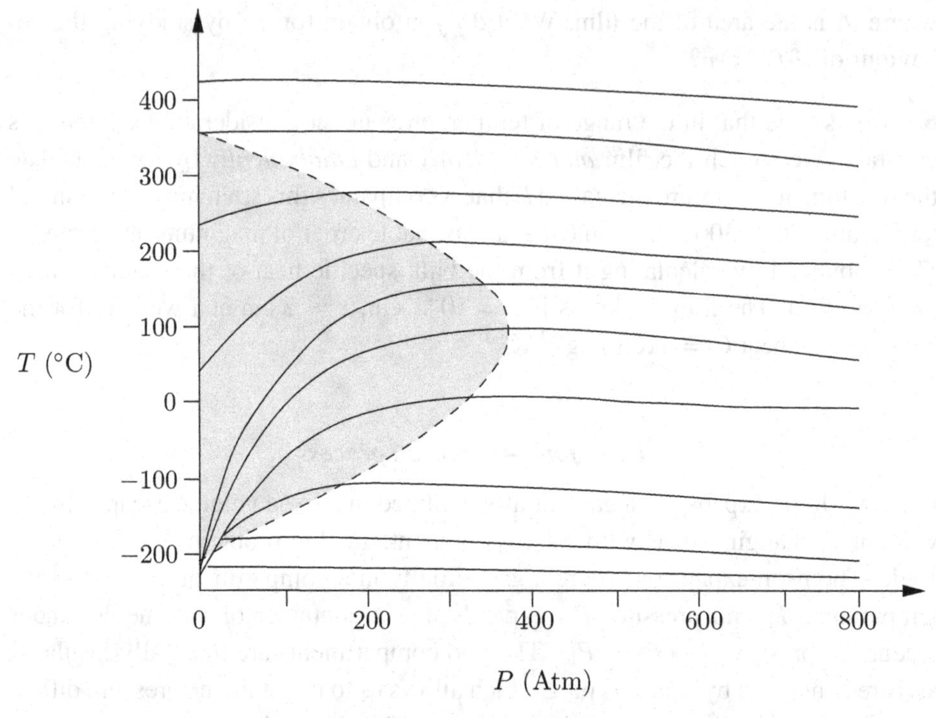

Figure 1.16 Isenthalpy curves in the (T, P) plane.

energy in the form of work. We define $\mathcal{M}$ from its relation to the magnetic susceptibility per unit volume $\chi(T, B)$

$$\chi = \frac{1}{V}\frac{\partial \mathcal{M}}{\partial B}$$

where V is the volume of the solid. We will assume that χ is given by

$$\chi = \frac{1}{V}\frac{\mathcal{M}}{B}$$

The adiabatic demagnetization of a paramagnetic crystal (also called paramagnetic salt) was one of the first methods used to attain sub-kelvin temperatures. The principle is as follows. A paramagnetic material, initially in thermal equilibrium with a heat bath at temperature T_{i}, is subjected to a magnetic field B_{i}. Although work is exchanged during this process, the heat bath maintains the crystal at its initial temperature. The solid is then thermally insulated and the magnetic field is reduced adiabatically and quasi-statically to a value B_{f} (usually $B_{\mathrm{f}} = 0$). The final temperature is lower than the initial one.

1. Give, without calculation, the analogous mechanical equivalent (gas, piston, etc.) of this cooling process.

2. In order to characterize the temperature change accompanying the demagnetization, we introduce the coefficient $\mu_D = (\partial T/\partial B)_S$. Show that

$$\mu_\mathrm{D} = -\frac{VTB}{C_B}\left.\frac{\partial \chi}{\partial T}\right|_B$$

where $C_B(T, B)$ is the specific heat at constant magnetic field.

3. Show that in order to calculate $C_B(T, B)$, and therefore μ_D, it is sufficient to know $\chi(T, B)$ and $C_B(T, 0)$.

4. In the temperature range (which depends on the crystal) where adiabatic demagnetization is used, we may assume that the magnetic susceptibility is given by the Curie law $\chi = a/T$ and that the heat capacity in zero field is given by $C_B(T, 0) = Vb/T^2$, where a and b are constants. Determine the final temperature T_f, after adiabatic demagnetization, in terms of T_i, B_i and B_f.

5. Calculate T_f under the following conditions: $T_\mathrm{i} = 2\,\mathrm{K}$, $B_\mathrm{i} = 0.71\,\mathrm{T}$, $B_\mathrm{f} = 0\,\mathrm{T}$, $a = 78.7\,\mathrm{J\,K\,T^{-2}}$ and $b = 2.65\,\mathrm{J\,K}$.

6. By assuming the atoms in the solid to be independent, an argument based on typical energy scales in the problem leads to the conclusion that $S = S(B/T)$. With such a hypothesis, what is the minimum temperature attainable by adiabatic demagnetization? What can you conclude from this result concerning the constant b?

1.8 Further reading

The discussion in Sections 1.1 and 1.2 is inspired by the presentation of thermodynamics in Callen's book [24] to which we refer the reader for useful comments. See in particular Chapters 1 to 8. The book by Balian [5], Chapters 5 and 6, also gives insight into this presentation. In addition there is a large number of books on thermodynamics: a classic is Zemansky [124]. Chapters 2 to 5 of Reif [109] or Chapters 1 to 5 of Schroeder [114] are strongly recommended; the latter show a great variety of applications. The reader can find discussions of long range forces in Balian [5] (Section 6.6.5), Reif [109] (Chapter 11) or Mandl [87] (Chapter 1). A demonstration of relation (1.51), which gives the curvature of a surface, is given for example in [34] Chapter 2. The main technical principles of cooling are discussed by Lounasmaa [82]; see also a more recent article by the same author [83]. The zero point entropy is discussed by Wannier [120], Mila [93] and see also Ramirez *et al.* [106].

2

Statistical entropy and Boltzmann distribution

Our main objective in this chapter is to derive the expression for the entropy starting with a microscopic description. We shall first recall the density operator description of a statistical mixture in quantum mechanics. We shall then introduce the concept of statistical entropy (information entropy) of a probability distribution and the closely related entropy of a quantum statistical mixture. This will allow us to construct the density operator at equilibrium, i.e. the Boltzmann distribution, and to obtain from it the corresponding statistical entropy. By comparing with the results of Chapter 1, we shall be able to identify, at equilibrium, the statistical entropy of the Boltzmann distribution with the thermodynamic entropy. We conclude the chapter with a discussion of irreversibility.

2.1 Quantum description

2.1.1 Time evolution in quantum mechanics

In quantum mechanics, the most complete information on the state of a system at a given time $t = t_0$ is given by the eigenvalues, $a_1, \ldots, a_N$, of a set of Hermitian operators[1] $\mathsf{A}_1, \ldots, \mathsf{A}_N$, acting on a Hilbert space, representing a set of dynamical variables and commuting among each other ('a complete set of commuting observables'). This state is represented mathematically by a state vector $|\psi(t_0)\rangle = |a_1, \ldots, a_N\rangle$ in the Hilbert space which satisfies $\mathsf{A}_i|\psi(t_0)\rangle = a_i|\psi(t_0)\rangle$. The time evolution of this state is governed by the Schrödinger equation,

$$i\hbar\frac{\mathrm{d}}{\mathrm{d}t}|\psi(t)\rangle = H(t)|\psi(t)\rangle \tag{2.1}$$

where $H(t)$ is the Hamiltonian. For an isolated system, H is independent of time, but if the system interacts with the external world, for example a time dependent

[1] With the exception of the Hamiltonian, operators representing dynamic variables will be written in sans serif typeface.

electric field, the Hamiltonian can depend explicitly on time. The time evolution given by Equation (2.1) can be represented equivalently by using the time evolution operator $U(t, t_0)$

$$|\psi(t)\rangle = U(t, t_0)|\psi(t_0)\rangle \tag{2.2}$$

which satisfies

$$\mathrm{i}\hbar \frac{\mathrm{d}}{\mathrm{d}t} U(t, t_0) = H(t)U(t, t_0) \tag{2.3}$$

This evolution operator is unitary: $U^\dagger(t, t_0) = U^{-1}(t, t_0)$, it satisfies the group property

$$U(t, t_0) = U(t, t_1)U(t_1, t_0)$$

and, if H is time independent,

$$U(t, t_0) = \exp(-\mathrm{i}H(t - t_0)/\hbar) \tag{2.4}$$

What we have just described is called 'the Schrödinger picture'. Another description is possible, this is 'the Heisenberg picture' where the state vector is independent of time

$$|\psi_{\mathrm{H}}\rangle = U^{-1}(t, t_0)|\psi(t)\rangle = |\psi(t_0)\rangle \tag{2.5}$$

while the dynamic variables do depend on time according to

$$\mathsf{A}_{\mathrm{H}}(t) = U^{-1}(t, t_0)\mathsf{A}(t)\, U(t, t_0) \tag{2.6}$$

For the sake of generality in Equation (2.6), we have shown the explicit time dependence of the operator A, but usually the dynamic variables are time independent in the Schrödinger picture. If, with the help of Equation (2.6), we define the Hamiltonian and $(\partial \mathsf{A}(t)/\partial t)_{\mathrm{H}}$ in the Heisenberg picture, we obtain

$$H_{\mathrm{H}}(t) = U^{-1}(t, t_0)H(t)\, U(t, t_0) \qquad \left(\frac{\partial \mathsf{A}(t)}{\partial t}\right)_{\mathrm{H}} = U^{-1}(t, t_0)\, \frac{\partial \mathsf{A}(t)}{\partial t}\, U(t, t_0)$$

The differential form of (2.6) is

$$\mathrm{i}\hbar \frac{\mathrm{d}\mathsf{A}_H}{\mathrm{d}t} = [\mathsf{A}_{\mathrm{H}}(t), H_{\mathrm{H}}(t)] + \mathrm{i}\hbar \left(\frac{\partial \mathsf{A}(t)}{\partial t}\right)_{\mathrm{H}} \tag{2.7}$$

where the last term comes from a possible explicit dependence on time of the operator in the Schrödinger picture. In the frequently encountered case where H does not depend on time in the Schrödinger picture, we simply have $H_{\mathrm{H}}(t) = H$, also time independent. The expectation values of dynamic variables or, more generally, their matrix elements that are the measurable quantities,[2] are of course independent of the point of view adopted

$$\langle\varphi(t)|\mathsf{A}(t)|\psi(t)\rangle = \langle\varphi_{\mathrm{H}}|\mathsf{A}_{\mathrm{H}}(t)|\psi_{\mathrm{H}}\rangle \tag{2.8}$$

Actually, the correspondence between the physical state and the Hilbert space vector is not one-to-one: $|\psi\rangle$ and $e^{i\alpha}|\psi\rangle$, where $e^{i\alpha}$ is a phase factor, represent the same physical state. Instead of using $|\psi\rangle$ to describe the physical state, we can do that by defining the projection operator $D = |\psi\rangle\langle\psi|$ which is independent of the phase factor. The expectation values are then given by (cf. (A.15))

$$\langle\mathsf{A}\rangle = \langle\psi|\mathsf{A}|\psi\rangle = \mathrm{Tr}\,\mathsf{A}D \tag{2.9}$$

2.1.2 The density operators and their time evolution

The above description of a quantum state presupposes complete knowledge of the system: the system is said to be in a *pure state*. This is not the general case. Very often, and this is the case in statistical mechanics, we only know the probabilities of finding the system in a given state. Consider again the pure case described by the projection operator $D = |\psi\rangle\langle\psi|$. If $\{|i\rangle\}$ is a basis in the Hilbert space, we can use it to decompose the vector $|\psi\rangle$

$$|\psi\rangle = \sum_i c_i\,|i\rangle$$

and therefore

$$D = \sum_{i,j} c_i c_j^* |i\rangle\langle j| \qquad D_{kl} = \langle k|D|l\rangle = c_k\,c_l^* \tag{2.10}$$

Suppose now that we only know the probability p_n of finding the system in the state $|\psi_n\rangle$ with

$$|\psi_n\rangle = \sum_i c_i^{(n)}\,|i\rangle$$

The vectors $|\psi_n\rangle$ are normalized, $|\langle\psi_n|\psi_n\rangle|^2 = 1$, but not necessarily orthogonal. The probabilities, p_n, satisfy the necessary conditions: $p_n \geq 0$ and $\sum_n p_n = 1$. We then say that the system is in a *statistical mixture*, and we define the *density operator* D by[3]

$$\boxed{D = \sum_n |\psi_n\rangle\, p_n\, \langle\psi_n|} \tag{2.11}$$

[2] In fact what is directly measurable are the squares of the absolute values of the matrix elements.
[3] Exercise 2.7.1 familiarizes the reader with density operators.

This definition generalizes that given previously for a pure state and allows one to compute expectation values of dynamical variables as in (2.9)

$$\boxed{\langle \mathsf{A} \rangle = \sum_n p_n \langle \psi_n | \mathsf{A} | \psi_n \rangle = \mathrm{Tr}\, \mathsf{A} D} \tag{2.12}$$

The matrix elements of D are $D_{kl} = \sum_n p_n c_k^{(n)} c_l^{(n)*}$, and the basic properties of D, which can be derived easily from the definition (2.11), are

(*i*) D is Hermitian: $D = D^\dagger$
(*ii*) D has unit trace: $\mathrm{Tr}\, D = \sum_n p_n = 1$
(*iii*) D is a positive operator: for any vector $|\varphi\rangle$, $\langle \varphi | D | \varphi \rangle = \sum_n p_n |\langle \varphi | \psi_n \rangle|^2 \geq 0$
(*iv*) The necessary and sufficient condition for D to describe a pure state is $D^2 = D$: in this case, D is a projection operator.

Let us now consider time evolution. If at time t_0 the density operator is given by

$$D(t_0) = \sum_n |\psi_n(t_0)\rangle\, p_n\, \langle \psi_n(t_0)|$$

Equation (2.2) shows that $D(t)$ is given by

$$D(t) = U(t, t_0) D(t_0) U^{-1}(t, t_0) \tag{2.13}$$

or in differential form

$$\boxed{\mathrm{i}\hbar \frac{\mathrm{d}D}{\mathrm{d}t} = [H(t), D]} \tag{2.14}$$

If we compare (2.6) and (2.13) we see that the order of the operators U and U^{-1} is different. Equivalently, in the differential form (2.7) the sign of the commutator on the right hand side is opposite to that in Equation (2.14). Naturally, we can use the density operator formalism in the Heisenberg picture. The density operator is independent of time, $D(t_0) = D_{\mathrm{H}}$, whereas the dynamic variables depend on time according to (2.6). We obtain the expectation value of a variable A at time t using the two equivalent expressions which generalize (2.8)

$$\boxed{\langle \mathsf{A} \rangle (t) = \mathrm{Tr}\, \mathsf{A}(t) D(t) = \mathrm{Tr}\, \mathsf{A}_{\mathrm{H}}(t) D_{\mathrm{H}}} \tag{2.15}$$

We can now make more explicit the discussions in Section 1.1. If we had perfect knowledge of the state of a macroscopic system, we could attribute to it a state vector: we would then say that there is a microstate corresponding to this system. As we discussed in that section, it is impossible to have such complete information and, consequently, we have to settle for information on macroscopic quantities, which only give us expectation values of a small number of dynamic variables.

It is this macroscopic information which defines a macrostate. *A huge number of microstates is compatible with this macrostate*, and the best we can hope for is to know the probability that the system is in one microstate or another. The system will, therefore, be described by a density operator and the entire problem is now to determine its form. This will be the object of Section 2.4.

2.1.3 Quantum phase space

In what follows it will be crucial to know how to count energy levels. We start with a simple case: the counting of energy levels in a one-dimensional box of length L, with the quantum particle of mass m confined to the interval $[0, L]$. Let the energy be ε. The Schrödinger equation is

$$-\frac{\hbar^2}{2m}\frac{\mathrm{d}^2}{\mathrm{d}x^2}\psi(x) = \varepsilon\,\psi(x) \tag{2.16}$$

with the boundary conditions: $\psi(0) = \psi(L) = 0$. The solutions are labelled by an integer n

$$\psi_n(x) = A\sin(k_n x) \qquad k_n = \frac{\pi n}{L} \tag{2.17}$$

Note that we can choose n to be strictly positive, $n \geq 1$, because the substitution $n \to -n$ simply changes the sign of the wave function, which is equivalent to multiplying by a phase factor. As discussed above, this does not lead to a different physical state. Instead of the boundary conditions $\psi(0) = \psi(L) = 0$, it is generally more convenient to take *periodic boundary conditions*: $\psi(x) = \psi(x + L)$, which gives the solution to Equation (2.16) in the form

$$\psi_n(x) = \frac{1}{\sqrt{L}}\,\mathrm{e}^{\mathrm{i}k_n x} \qquad k_n = \frac{2\pi n}{L} \tag{2.18}$$

Here n takes both positive and negative integer values: $n \in \mathbb{Z}$. Since we are interested in very large values of n, the counting stays the same. In fact, the number of values of n is doubled but the spacing between the k_n is twice as big. It is very important to note that since L is macroscopic, the levels are very closely spaced. We can therefore make a continuum approximation for the counting. To do this, we introduce the *density of states* $\rho(k)$:[4] $\rho(k)\Delta k$ is the number of states in the interval $[k, k + \Delta k]$. To Δk corresponds an interval Δn in n ($\Delta n \gg 1$) given by

$$\Delta n = \frac{L}{2\pi}\,\Delta k = \rho(k)\Delta k \tag{2.19}$$

[4] In order to avoid excessive proliferation of notation, we have chosen to represent consistently the density of states by ρ. Clearly this notation includes different functional dependence on the nature of the argument.

The number of states per unit k is then given by $\rho(k) = L/2\pi$. This result can be generalized easily to three dimensions. For a particle confined to a rectangular box of sides (L_x, L_y, L_z) with $\vec{k} = (k_x, k_y, k_z)$, the components of $\vec{k}$ are given by

$$\vec{k} = \left(\frac{2\pi n_x}{L_x}, \frac{2\pi n_y}{L_y}, \frac{2\pi n_z}{L_z} \right)$$

where (n_x, n_y, n_z) are integers and

$$\rho(\vec{k}) = \frac{L_x}{2\pi} \frac{L_y}{2\pi} \frac{L_z}{2\pi} = \frac{V}{(2\pi)^3}$$

where V is the volume of the box. The number of states in $\mathrm{d}^3 k$ is then[5]

$$\boxed{\rho(\vec{k})\, \mathrm{d}^3 k = \frac{V}{(2\pi)^3}\, \mathrm{d}^3 k} \tag{2.20}$$

Instead of the wave vector, $\vec{k}$, we can use the momentum $\vec{p} = \hbar \vec{k}$. The number of states in $\mathrm{d}^3 p$ is

$$\boxed{\rho(\vec{p})\, \mathrm{d}^3 p = \frac{V}{h^3}\, \mathrm{d}^3 p} \tag{2.21}$$

Note that in obtaining (2.20) and (2.21) we have not made any non-relativistic approximations. These equations are valid in all kinematic regimes.

We have just determined the density of states in wave vector or momentum. Now we return to our initial problem, namely the determination of the *density of energy levels*, i.e. the number of levels per unit energy. Let $\Phi(\varepsilon)$ be the number of levels whose energy is less than or equal to ε. Using the non-relativistic dispersion relation, $p = \sqrt{2m\varepsilon}$, and the fact that in the continuum approximation sums are replaced by integrals, we get

$$\Phi(\varepsilon) = \int_{p \le \sqrt{2m\varepsilon}} \rho(\vec{p})\, \mathrm{d}^3 p = \frac{V}{h^3} \int_{p \le \sqrt{2m\varepsilon}} \mathrm{d}^3 p = \frac{4\pi V}{h^3} \int_0^{\sqrt{2m\varepsilon}} \mathrm{d}p\, p^2$$

where the last expression was obtained by going to spherical coordinates. The integral is easily evaluated

$$\Phi(\varepsilon) = \frac{4\pi V}{3h^3} (2m\varepsilon)^{3/2} = \frac{V}{6\pi^2 \hbar^3} (2m\varepsilon)^{3/2} \tag{2.22}$$

[5] This result is also valid for a box of arbitrary shape. The corrections are of the order of $(kL)^{-1}$ where L is a characteristic length of the box. The first correction is a surface term, just like the difference between the fixed, (2.17), and periodic, (2.18), boundary conditions. These surface effects are negligible in the thermodynamic limit.

The level density, $\rho(\varepsilon)$, is simply the derivative of $\Phi(\varepsilon)$

$$\boxed{\rho(\varepsilon) = \Phi'(\varepsilon) = \frac{Vm}{2\pi^2\hbar^3}(2m\varepsilon)^{1/2}} \tag{2.23}$$

In two dimensions, for a particle confined to a rectangle of area S, the level density is independent of ε (see Exercise 2.7.2)

$$\boxed{\rho(\varepsilon) = \frac{Sm}{2\pi\hbar^2}} \tag{2.24}$$

2.1.4 (P, V, E) relation for a mono-atomic ideal gas

We can establish a very useful relation between the pressure P, the volume V and the internal energy E of an ideal mono-atomic gas at equilibrium, i.e. a gas where the only molecular degrees of freedom are translational, and where the internal energy is the sum of the kinetic energies of the molecules. The origin of this relation is purely mechanical, which is why it can be established independently of all entropic considerations. It is valid for classic as well as quantum ideal gases. We start with the non-relativistic case where, from (2.17), an energy level ε_r has the form

$$\varepsilon_r = \frac{\hbar^2\vec{k}^2}{2m} = \frac{\hbar^2\pi^2}{2m}\left(\frac{n_x^2}{L_x^2} + \frac{n_y^2}{L_y^2} + \frac{n_z^2}{L_z^2}\right) \tag{2.25}$$

The isotropy of the equilibrium state requires that when the average is taken over the microstates, the three momentum components all have the same mean value

$$\frac{\langle n_x^2\rangle}{L_x^2} = \frac{\langle n_y^2\rangle}{L_y^2} = \frac{\langle n_z^2\rangle}{L_z^2} = \frac{1}{3}\frac{2m}{\pi^2\hbar^2}\langle\varepsilon_r\rangle \tag{2.26}$$

By differentiating ε_r with respect to one of the sides of the box, L_x for example, we get

$$\frac{\partial\varepsilon_r}{\partial L_x} = -\frac{2}{L_x}\frac{\hbar^2\pi^2}{2m}\frac{n_x^2}{L_x^2}$$

Taking the average gives[6]

$$\frac{\partial\langle\varepsilon_r\rangle}{\partial L_x} = -\frac{2}{3L_x}\langle\varepsilon_r\rangle \tag{2.27}$$

[6] See Equation (2.83): in the case considered here, only the variation in the Hamiltonian contributes to the variation of the average of the energy.

Suppose now that we vary L_x adiabatically and quasi-statically, the change in the internal energy is equal to the supplied work

$$\mathrm{d}E = \text{đ}W = -PL_yL_z\,\mathrm{d}L_x$$

The internal energy is $E = N\langle\varepsilon_r\rangle$, where N is the number of molecules. Using (2.27) we get

$$\frac{\partial E}{\partial L_x} = -\frac{2}{3L_x}E$$

from which we obtain[7]

$$\boxed{PV = \frac{2}{3}E \text{ (non-relativistic case)}} \qquad (2.28)$$

Similarly, we can treat the ultra-relativistic case (c is the speed of light)

$$\varepsilon_r = \hbar c|\vec{k}| = \hbar c\pi\left(\frac{n_x^2}{L_x^2}+\frac{n_y^2}{L_y^2}+\frac{n_z^2}{L_z^2}\right)^{1/2}$$

The derivative with respect to L_x becomes

$$\frac{\partial\varepsilon_r}{\partial L_x} = -\frac{\hbar c\pi n_x^2}{L_x^3}\left(\frac{n_x^2}{L_x^2}+\frac{n_y^2}{L_y^2}+\frac{n_z^2}{L_z^2}\right)^{-1/2}$$

Isotropy gives

$$\frac{n_x^2}{L_x^2} \to \frac{1}{3}\left(\frac{n_x^2}{L_x^2}+\frac{n_y^2}{L_y^2}+\frac{n_z^2}{L_z^2}\right)$$

and

$$\frac{\partial\langle\varepsilon_r\rangle}{\partial L_x} = -\frac{\hbar c\pi}{L_x}\frac{1}{3}\left\langle\left(\frac{n_x^2}{L_x^2}+\frac{n_y^2}{L_y^2}+\frac{n_z^2}{L_z^2}\right)^{1/2}\right\rangle = -\frac{1}{3L_x}\langle\varepsilon_r\rangle$$

The pre-factor 2/3 of the non-relativistic case is replaced by 1/3

$$\boxed{PV = \frac{1}{3}E \text{ (ultra-relativistic case)}} \qquad (2.29)$$

In the intermediate case where the molecular speeds are neither very small nor very large compared to c, there is no simple relation between (P, V, E). We know only that $E/3 \leq PV \leq 2E/3$.

[7] Kinetic theory gives a shorter derivation of Equation (2.28). The advantage of the present demonstration is that no classical notions were used.

2.2 Classical description

2.2.1 Liouville's theorem

Recall that in analytical mechanics one often describes a system by its Hamiltonian $H(q_i, p_i)$, which depends on $2N$ coordinates q_i and p_i, $i = (1, \ldots, N)$ where N is the number of *degrees of freedom*. While it is possible for H to depend explicitly on time, we know that for an isolated system the Hamiltonian is independent of time. In the simple case of a system with one degree of freedom, and therefore two coordinates (q, p), a common case is

$$H = \frac{p^2}{2m} + U(q) \tag{2.30}$$

where $U(q)$ is the potential energy. Hamilton's equations are

$$\frac{\partial H}{\partial p_i} = \dot{q}_i \qquad \frac{\partial H}{\partial q_i} = -\dot{p}_i \tag{2.31}$$

with $\dot{q}_i = \mathrm{d}q_i/\mathrm{d}t$ and $\dot{p}_i = \mathrm{d}p_i/\mathrm{d}t$. The set of coordinates (q_i, p_i) constitutes the *phase space*. A trajectory in phase space is determined from the initial conditions $q_i(0)$ and $p_i(0)$. Such a trajectory cannot self-intersect. If there were an intersection, then choosing the intersection point as an initial condition would not uniquely determine the trajectory. *Defining a microstate in classical mechanics corresponds to knowing all the coordinates* $(q_i(t), p_i(t))$ *at an arbitrary time* t.

An important result in phase space dynamics is Liouville's theorem. We shall establish it for one degree of freedom leaving the (straightforward) generalization to the reader. Consider two times, t and $t + \mathrm{d}t$, and defining $q = q(t)$ and $q' = q(t + \mathrm{d}t)$, $p = p(t)$ and $p' = p(t + \mathrm{d}t)$, we have

$$q' = q + \dot{q}\,\mathrm{d}t = q + \frac{\partial H}{\partial p}\,\mathrm{d}t$$

$$p' = p + \dot{p}\,\mathrm{d}t = p - \frac{\partial H}{\partial q}\,\mathrm{d}t$$

The Jacobian, $\partial(q', p')/\partial(q, p)$, is given by

$$\frac{\partial(q', p')}{\partial(q, p)} = \begin{vmatrix} 1 + \frac{\partial^2 H}{\partial p\,\partial q}\,\mathrm{d}t & +\frac{\partial^2 H}{\partial p^2}\,\mathrm{d}t \\ -\frac{\partial^2 H}{\partial q^2}\,\mathrm{d}t & 1 - \frac{\partial^2 H}{\partial p\,\partial q}\,\mathrm{d}t \end{vmatrix} = 1 + \mathcal{O}(\mathrm{d}t)^2 \tag{2.32}$$

This equation implies that the Jacobian of the transformation $(q(0), p(0)) \to (q(t), p(t))$ satisfies

$$\frac{\mathrm{d}J(t)}{\mathrm{d}t} = 0 \tag{2.33}$$

and therefore $J(t) = 1$. This is Liouville's theorem, or the theorem of conservation of area in phase space

$$\mathrm{d}q(0)\,\mathrm{d}p(0) = \mathrm{d}q(t)\,\mathrm{d}p(t)$$

In concise notation

$$\boxed{\mathrm{d}q\,\mathrm{d}p = \mathrm{d}q'\,\mathrm{d}p'} \tag{2.34}$$

For a system of N particles in three dimensions, the coordinates $\vec{q}_i(t) \equiv \vec{r}_i(t)$ and the momenta $\vec{p}_i(t)$ are three-dimensional vectors and therefore there are $3N$ degrees of freedom. Liouville's theorem then becomes

$$\mathrm{d}^3r_1 \dots \mathrm{d}^3r_N\,\mathrm{d}^3p_1 \dots \mathrm{d}^3p_N = \mathrm{d}^3r'_1 \dots \mathrm{d}^3r'_N\,\mathrm{d}^3p'_1 \dots \mathrm{d}^3p'_N \tag{2.35}$$

This invariance suggests taking as integration measure in phase space

$$\mathrm{d}\Gamma = \mathcal{C}\prod_{i=1}^{N}\mathrm{d}^3p_i\,\mathrm{d}^3r_i \tag{2.36}$$

The constant $\mathcal{C}$ is *a priori* arbitrary in classical mechanics. It will be fixed later by resorting to quantum mechanics.

2.2.2 Density in phase space

Consider our classical system at time $t = 0$ with positions $\vec{r}_i = \vec{r}_i(t = 0)$ and momenta $\vec{p}_i = \vec{p}_i(t = 0)$. A microstate at $t = 0$ is defined by the set of coordinates $\{\vec{r}_i, \vec{p}_i\}$. Given the macroscopic constraints, the probability of observing a specific microstate at $t = 0$ is fixed by a probability density $D(\vec{r}_i, \vec{p}_i, t = 0) = D_0(\vec{r}_i, \vec{p}_i)$. This quantity is a probability density in phase space which is of course positive and normalized by

$$\int \mathrm{d}\Gamma\, D_0(\vec{r}_i, \vec{p}_i) = \mathcal{C}\int \prod_{i=1}^{N}\mathrm{d}^3p_i\,\mathrm{d}^3r_i\, D_0(\vec{r}_i, \vec{p}_i) = 1 \tag{2.37}$$

The probability density D is the classical analogue of the density operator D of the previous section (we have chosen the same notation on purpose) and the normalization condition (2.37) is the analogue of $\mathrm{Tr}\, D = 1$. We now consider time evolution using the collective notation $x = \{\vec{p}_1, \dots, \vec{p}_N; \vec{r}_1, \dots, \vec{r}_N\}$. When going from $t = 0$ to t, the coordinates evolve according to

$$\vec{p}_i \to \vec{p}_i(t) \qquad \vec{r}_i \to \vec{r}_i(t)$$

where $\vec{p}_i(t), \vec{r}_i(t)$ are obtained by solving Hamilton's equations (2.31). With the notation defined above, we have

$$x \to y = \varphi_t(x) \tag{2.38}$$

Conversely, we can write $x = \varphi_t^{-1}(y) = \varphi_{-t}(y)$ and from Liouville's theorem, $\mathrm{d}x = \mathrm{d}y$, where $\mathrm{d}x$ is the measure (2.36). Let $\mathcal{A}$ be a classical dynamical variable,[8] in other words a function of coordinates, without explicit time dependence

$$\mathcal{A}(t=0) = \mathcal{A}(\vec{p}_i, \vec{r}_i) = \mathcal{A}(x) \qquad \mathcal{A}(t) = \mathcal{A}(\vec{p}_i(t), \vec{r}_i(t)) = \mathcal{A}(\varphi_t(x))$$

At $t = 0$, the mean value of $\mathcal{A}$ is obtained from a weighted average using the probability density $D_0(x)$

$$\langle \mathcal{A} \rangle (t=0) = \int \mathrm{d}x \, D_0(x) \mathcal{A}(x) \tag{2.39}$$

whereas at time t

$$\langle \mathcal{A} \rangle (t) = \int \mathrm{d}x \, D_0(x) \mathcal{A}(\varphi_t(x)) \tag{2.40}$$

Equation (2.40) is analogous to the expectation value in the Heisenberg picture: D is independent of time while the dynamical variables do depend on time. This is the usual picture in classical mechanics. The analogue of the Schrödinger picture is obtained by changing variables in (2.40), $x \to \varphi_t(x) = y$, and using Liouville's theorem

$$\langle \mathcal{A} \rangle (t) = \int \mathrm{d}y \, D_0(\varphi_{-t}(y)) \mathcal{A}(y) = \int \mathrm{d}x \, D(x(t), t) \mathcal{A}(x) \tag{2.41}$$

where the time dependent probability density, $D(x(t), t)$, is

$$D(x(t), t) = D_0(\varphi_{-t}(x)) \tag{2.42}$$

In (2.41), the phase space density depends on time whereas the dynamical variables do not. $D(x(t), t)$ is the classical analogue of the density operator in the Schrödinger picture.

Finally, we give the evolution law of the density $D(x(t), t)$ by using the conservation of probability. Considering, for simplicity, a single degree of freedom $x = (q, p)$, this conservation is expressed as follows

$$D(q(t+\mathrm{d}t), p(t+\mathrm{d}t), t+\mathrm{d}t) \, \mathrm{d}q' \, \mathrm{d}p' = D(q(t), p(t), t) \, \mathrm{d}q \, \mathrm{d}p$$

[8] In order to avoid confusion, we denote classical dynamic variables with calligraphic letters ($\mathcal{A}$). The dynamic variables of the corresponding quantum problem are written as A.

Table 2.1 *Classical and quantum descriptions.*

	Quantum	Classical
Microstate	$\lvert\psi_n(t)\rangle$	$p_i(t), q_i(t)$
Macrostate	$D(t)$	$D(p,q;t)$
Probability law	$\mathrm{Tr}D = \sum_n p_n = 1$	$\mathcal{C}\int \mathrm{d}p\,\mathrm{d}q\, D(p,q;t) = 1$
Time evolution	$\dfrac{\mathrm{d}D}{\mathrm{d}t} = \dfrac{1}{i\hbar}[H, D]$	$\dfrac{\partial D}{\partial t} = -\{H, D\}$
Average values	$\mathrm{Tr}[\mathsf{A}D(t)]$	$\mathcal{C}\int \mathrm{d}p\,\mathrm{d}q\, \mathcal{A}(p,q;t)D(p,q;t)$

with $q' = q(t+\mathrm{d}t)$, $p' = p(t+\mathrm{d}t)$. Using Liouville's theorem, $\mathrm{d}q\,\mathrm{d}p = \mathrm{d}q'\,\mathrm{d}p'$, and expanding in a Taylor series to first order in $\mathrm{d}t$, we obtain

$$\frac{\partial D}{\partial q}\dot{q} + \frac{\partial D}{\partial p}\dot{p} + \frac{\partial D}{\partial t} = 0$$

We note that Liouville's theorem and probability conservation imply that the total derivative of D with respect to time vanishes: D is constant along a trajectory in phase space. On the other hand, the partial time derivative of D is taken at a fixed point in phase space. This generalizes to a system of N particles

$$\sum_{i,\alpha}\left(\frac{\partial D}{\partial q_{i,\alpha}}\dot{q}_{i,\alpha} + \frac{\partial D}{\partial p_{i,\alpha}}\dot{p}_{i,\alpha}\right) + \frac{\partial D}{\partial t} = \{H, D\} + \frac{\partial D}{\partial t} = 0 \qquad (2.43)$$

where α labels the components of the vectors $\vec{q}_i(t)$ and $\vec{p}_i(t)$: $\alpha = (x, y, z)$. $\{\mathcal{A}, \mathcal{B}\}$ is the Poisson bracket[9] for the variables $\mathcal{A}$ and $\mathcal{B}$

$$\{\mathcal{A}, \mathcal{B}\} = \sum_{i,\alpha}\left(\frac{\partial \mathcal{A}}{\partial p_{i,\alpha}}\frac{\partial \mathcal{B}}{\partial q_{i,\alpha}} - \frac{\partial \mathcal{A}}{\partial q_{i,\alpha}}\frac{\partial \mathcal{B}}{\partial p_{i,\alpha}}\right) \qquad (2.44)$$

We mention that for a dynamical variable which can depend explicitly on time, the evolution law becomes

$$\frac{\mathrm{d}\mathcal{A}}{\mathrm{d}t} = \{H, \mathcal{A}\} + \frac{\partial \mathcal{A}}{\partial t} \qquad (2.45)$$

The difference in sign of the Poisson bracket between (2.43) and (2.45) is the analogue of the sign difference of the commutator between (2.7) and (2.14), that is the sign difference between, on the one hand, the evolution of the density operator in the Schrödinger picture, and, on the other hand, the evolution of a dynamical variable in the Heisenberg picture. We also remark that $\partial D/\partial t = 0$ if D is a function only of H.

Table 2.1 summarizes the quantum and classical descriptions.

[9] The reader should be aware that some authors use a sign convention opposite to that used in Equation (2.44).

2.3 Statistical entropy

2.3.1 Entropy of a probability distribution

The last concept we need to introduce before arriving at the equilibrium distribution is that of *statistical entropy* (or *information entropy*). We will first define the concept of entropy of a probability distribution and then the closely related statistical entropy of a quantum state (or, more precisely, that of a statistical quantum mixture). Let $\{e_m\}$ be a set of possible events with $m = 1, \ldots, M$ and the probability of such an event e_m be P_m with

$$P_m \geq 0 \qquad \sum_{m=1}^{M} P_m = 1$$

By definition, the *entropy of the probability distribution* $\mathcal{P}$ defined by the P_ms is given by

$$\boxed{S[\mathcal{P}] = -\sum_{m=1}^{M} P_m \ln P_m} \tag{2.46}$$

Note that if $P_m = 0$, event e_m does not contribute to $S[\mathcal{P}]$ since $\lim_{x\to 0} x \ln x = 0$. The limiting cases are as follows.

(*i*) The probability law gives complete information on the succession of events: If one of the events has unit probability, for example $P_1 = 1$, $P_m = 0$, $m \geq 2$, then $S[\mathcal{P}] = 0$.

(*ii*) The probability law gives information which is as incomplete as possible on the succession of events: all events have probabilities $P_m = 1/M$ for any m, then $S[\mathcal{P}] = \ln M$.

We can then say: *the less the information in the probability law, the larger the entropy of* $\mathcal{P}$. In other words, $S[\mathcal{P}]$ describes the lack of information due to the probabilistic nature of events. We now present some properties of $S[\mathcal{P}]$.

- $0 \leq S[\mathcal{P}] \leq \ln M$

To demonstrate this property, we look for the extrema of $S[\mathcal{P}]$ subject to the constraint $\sum_m P_m = 1$ by using the method of Lagrange multipliers (Section A.2). We define $\tilde{S}[\mathcal{P}] = S[\mathcal{P}] - \lambda(\sum_m P_m - 1)$, where λ is a Lagrange multiplier. The extrema of $\tilde{S}[\mathcal{P}]$ are obtained from

$$\frac{\partial \tilde{S}}{\partial P_m} = -[\ln P_m + 1 + \lambda] = 0$$

Therefore, all the P_m are equal to $1/M$, which gives for the extremum of S the value $\ln M$. But since we know at least one smaller value of S (zero) we conclude that the extremum of S with the value $\ln M$ is a maximum. This maximum is unique and obtained when all the probabilities are equal.

- Additivity of the entropy of a probability distribution

Let $\{e'_{m'}\}$ be a second set of possible events statistically independent of $\{e_m\}$. The probability of observing the pair $\{e_m, e'_{m'}\}$ is then $P_{m,m'} = P_m P'_{m'}$. Let $\mathcal{P} \otimes \mathcal{P}'$ be the joint probability,[10] a simple calculation then gives

$$S[\mathcal{P} \otimes \mathcal{P}'] = S[\mathcal{P}] + S[\mathcal{P}'] \tag{2.47}$$

If events $\{e_m\}$ and $\{e'_{m'}\}$ are not independent, we can show that in general

$$S[\mathcal{P} \otimes \mathcal{P}'] \leq S[\mathcal{P}] + S[\mathcal{P}'] \tag{2.48}$$

the equality being valid only if $P_{m,m'} = P_m P'_{m'}$ (Exercise 2.7.5). Qualitatively, the interpretation of the inequality is that the information on the correlations contained in $P_{m,m'}$ is missing on the right hand side of Equation (2.48) with the consequence that the entropy is larger. By way of example, the reader can examine the case of perfect correlations ($M = M'$)

$$P_{m,m'} = \frac{1}{M}\,\delta_{mm'}$$

and show that in this case $S[P_{m,m'}] = S[P_m] = S[P'_{m'}]$.

2.3.2 Statistical entropy of a mixed quantum state

The definition of the statistical entropy of a mixed quantum state is modeled after that of the entropy of a probability distribution. Let D be the density operator of a statistical mixture (2.11)

$$D = \sum_n |\psi_n\rangle\, p_n\, \langle\psi_n| \tag{2.49}$$

The vectors $|\psi_n\rangle$ are not necessarily orthogonal, but it is always possible to diagonalize D in some basis $|m\rangle$[11]

$$D = \sum_m |m\rangle\, P_m\, \langle m| \tag{2.50}$$

[10] This notation is used in analogy with that used in the following paragraph.

[11] We stress the need to diagonalize D since, in general, $\mathrm{Tr}\, D \ln D \neq \sum_n p_n \ln p_n$. It is easy to convince oneself of that by considering a 2×2 density matrix.

We thus define the statistical entropy $S_{\rm st}[D]$ of the mixture described by D

$$\boxed{S_{\rm st}[D] = -k \sum_m P_m \ln P_m = -k {\rm Tr}\, D \ln D} \tag{2.51}$$

where k is the Boltzmann constant, a multiplicative factor introduced in order to obtain, at equilibrium, the thermodynamic entropy (see Section 2.5.2). Note that the statistical entropy of a pure case vanishes: $S_{\rm st}[D] = 0$.

The additivity property of statistical entropy is obtained by examining a system made of two non-interacting subsystems. These two systems can be separated by a wall, but they also can be in the same space as for example in the case of lattice spin systems where, to leading approximation, the spins are decoupled from the lattice vibrations (Problem 3.8.2). Let $\mathcal{H}^{(a)}$ be the Hilbert space of states of the first system and $\mathcal{H}^{(\alpha)}$ that of the second. The total Hilbert space is the tensor product $\mathcal{H} = \mathcal{H}^{(a)} \otimes \mathcal{H}^{(\alpha)}$. Since the two subsystems do not interact, the total density operator is the tensor product (cf. (A.17)) of the separate density operators $D^{(a)}$ and $D^{(\alpha)}$. The Latin (Greek) letters denote the elements of the density operators acting in $\mathcal{H}^{(a)}$ $(\mathcal{H}^{(\alpha)})$.

$$D = D^{(a)} \otimes D^{(\alpha)} \quad \text{where} \quad D_{a\alpha;b\beta} = D^{(a)}_{ab} D^{(\alpha)}_{\alpha\beta} \tag{2.52}$$

The additivity of the entropy is exhibited by putting $D^{(a)}$ and $D^{(\alpha)}$ in diagonal form

$$D_{a\alpha;b\beta} = D^{(a)}_{aa} D^{(\alpha)}_{\alpha\alpha} \delta_{ab} \delta_{\alpha\beta}$$

which yields

$${\rm Tr}\, D \ln D = \sum_{a,\alpha} D^{(a)}_{aa} D^{(\alpha)}_{\alpha\alpha} \left(\ln D^{(a)}_{aa} + \ln D^{(\alpha)}_{\alpha\alpha} \right) = {\rm Tr}\, D^{(a)} \ln D^{(a)} + {\rm Tr}\, D^{(\alpha)} \ln D^{(\alpha)}$$

This, therefore, gives

$$S_{\rm st}\big[D\big] = S_{\rm st}\big[D^{(a)}\big] + S_{\rm st}\big[D^{(\alpha)}\big] \tag{2.53}$$

When the two subsystems are not independent, we define the density operator of one subsystem by taking the trace with respect to the other, or *partial trace* (cf. (A.18)).[12] For example

$$D^{(a)} = {\rm Tr}_{\alpha} D \quad \text{where} \quad D^{(a)}_{ab} = \sum_{\alpha} D_{a\alpha;b\alpha} \tag{2.54}$$

[12] Taking the partial trace is analogous to the operation of calculating in probability theory the distribution of one variable when given the distribution for two variables: $D^{(a)}(x) = \int {\rm d}y\, D^{(a\alpha)}(x, y)$. A density operator corresponding to a tensor product (2.52) is the analogue of a probability distribution of two statistically independent variables: $D^{(a\alpha)}(x, y) = D^{(a)}(x) D^{(\alpha)}(y)$.

In fact, if A is an operator which acts only on $\mathcal{H}^{(a)}$: $\mathsf{A} = \mathsf{A}^{(a)} \otimes \mathbb{I}^{(\alpha)}$, then

$$\langle \mathsf{A} \rangle = \mathrm{Tr}\, \mathsf{A} D = \sum_{a,\alpha,b,\beta} A^{(a)}_{ab} \delta_{\alpha\beta} D_{b\beta;a\alpha} = \sum_{ab} A^{(a)}_{ab} \sum_{\alpha} D_{b\alpha;a\alpha} = \mathrm{Tr}\, \mathsf{A}^{(a)} D^{(a)}$$

which shows that $D^{(a)}$ is indeed the density operator of subsystem (a), also called the reduced density operator. This definition is of course consistent with (2.52) in the case of independent subsystems and (see Exercise 2.7.5)

$$S_{\mathrm{st}}[D] \leq S_{\mathrm{st}}\big[D^{(a)}\big] + S_{\mathrm{st}}\big[D^{(\alpha)}\big] \tag{2.55}$$

which is the analogue of (2.48) for the entropy of a probability distribution.

We now show an inequality which will be crucial for what is to follow. Let X and Y be two positive Hermitian operators.[13] They then satisfy a convexity inequality[14]

$$\boxed{\mathrm{Tr}\, X \ln Y - \mathrm{Tr}\, X \ln X \leq \mathrm{Tr}\, Y - \mathrm{Tr}\, X} \tag{2.56}$$

the equality being satisfied only for $X = Y$. To show this inequality, we note that X and Y can be diagonalized in the bases $|m\rangle$ for X and $|q\rangle$ for Y, with positive eigenvalues X_m and Y_q: $X_m > 0$ and $Y_q > 0$. The left hand side of the inequality (2.56) takes the following form by writing explicitly the traces in the basis $|m\rangle$

$$\sum_{m,q} X_m \langle m|q\rangle \ln Y_q \langle q|m\rangle - \sum_m X_m \ln X_m$$

By using $\sum_q |\langle m|q\rangle|^2 = 1$, the above expression becomes

$$\sum_{m,q} |\langle m|q\rangle|^2 X_m \ln \frac{Y_q}{X_m}$$

The concavity of the logarithm function yields $\ln x \leq x - 1$, with the equality holding only when $x = 1$. By taking $x = Y_q/X_m$ we find

$$\sum_{m,q} |\langle m|q\rangle|^2 X_m \ln \frac{Y_q}{X_m} \leq \sum_{m,q} |\langle m|q\rangle|^2 (Y_q - X_m) = \mathrm{Tr} Y - \mathrm{Tr} X$$

The equality

$$|\langle m|q\rangle|^2 X_m \ln \frac{Y_q}{X_m} = |\langle m|q\rangle|^2 (Y_q - X_m)$$

[13] The following demonstration assumes the two operators to be strictly positive. The results apply by continuity to two positive operators.

[14] Equation (2.56) is a typical convexity inequality as its validity relies on the convexity properties of the logarithm.

cannot apply for a pair (m, q) unless $X_m = Y_q$ or if $|\langle m|q\rangle|^2 = 0$. Consequently, the equality in (2.56) implies

$$|\langle m|q\rangle|^2 (Y_q - X_m) = 0 \ \forall (m, q)$$

and thus

$$\sum_{m,q} |m\rangle\langle m|q\rangle(Y_q - X_m)\langle q| = 0$$

or in other words $Y = X$.

2.3.3 *Time evolution of the statistical entropy*

We would now like to determine the time evolution of the statistical entropy, which necessitates calculating the time derivative of $-k\text{Tr}D \ln D$. The calculation is not so obvious due to the non-commutativity of the operators. To calculate the time derivative of a function $f(\mathsf{A})$ of a time-dependent operator A, we expand the function in a Taylor series

$$f(\mathsf{A}) = \sum_{n=0}^{\infty} \frac{1}{n!} f^{(n)}(0)\mathsf{A}^n$$

The derivative of A^n is

$$\frac{\mathrm{d}}{\mathrm{d}t}\mathsf{A}^n = \frac{\mathrm{d}\mathsf{A}}{\mathrm{d}t}\mathsf{A}^{n-1} + \mathsf{A}\frac{\mathrm{d}\mathsf{A}}{\mathrm{d}t}\mathsf{A}^{n-2} + \cdots + \mathsf{A}^{n-1}\frac{\mathrm{d}\mathsf{A}}{\mathrm{d}t}$$

The different terms cannot be regrouped to give $n\mathsf{A}^{n-1}\mathrm{d}\mathsf{A}/\mathrm{d}t$, unlike the case of the derivative of an ordinary function, because in general $\mathrm{d}\mathsf{A}/\mathrm{d}t$ does not commute with A. However, if we take the trace of this result we can perform cyclical permutations (see (A.12)), then

$$\text{Tr}\left(\frac{\mathrm{d}}{\mathrm{d}t}\mathsf{A}^n\right) = n\text{Tr}\left(\mathsf{A}^{n-1}\frac{\mathrm{d}\mathsf{A}}{\mathrm{d}t}\right)$$

which gives, by re-summing the Taylor series,

$$\text{Tr}\left(\frac{\mathrm{d}}{\mathrm{d}t}f(\mathsf{A})\right) = \text{Tr}\left(f'(\mathsf{A})\frac{\mathrm{d}\mathsf{A}}{\mathrm{d}t}\right) \tag{2.57}$$

Another useful way of writing (2.57) is

$$\mathrm{d}(\text{Tr}\, f(\mathsf{A})) = \text{Tr}(f'(\mathsf{A})\mathrm{d}\mathsf{A}) \tag{2.58}$$

We apply this to the case of time evolution of the statistical entropy. Using $(x \ln x)' = \ln x + 1$ we obtain

$$\frac{\mathrm{d}S_{\mathrm{st}}[D]}{\mathrm{d}t} = -k\mathrm{Tr}\ \ln D \frac{\mathrm{d}D}{\mathrm{d}t} \tag{2.59}$$

where we have also used $\mathrm{Tr}\,\mathrm{d}D/\mathrm{d}t = 0$ which follows immediately from $\mathrm{Tr}\, D = 1$. For Hamiltonian evolution, including the case where the Hamiltonian depends explicitly on time, Equation (2.14) yields

$$\frac{\mathrm{d}S_{\mathrm{st}}[D]}{\mathrm{d}t} = -\frac{k}{\mathrm{i}\hbar}\mathrm{Tr}\,(\ln D[H, D]) = +\frac{k}{\mathrm{i}\hbar}\mathrm{Tr}\,(H[\ln D, D]) = 0 \tag{2.60}$$

For Hamiltonian evolution, the statistical entropy is conserved!

2.4 Boltzmann distribution

2.4.1 Postulate of maximum of statistical entropy

Recall the outline briefly discussed in Section 1.1.1. We assume knowledge of a macrostate, i.e. a certain number of macroscopic variables, with which is compatible a huge number of microstates $|m\rangle$. We shall now address the fundamental question: *what is the probability law* $\{P_m\}$ *of the microstates which determines the density operator D in (2.50) subject to the macroscopic constraints*, i.e. constraints imposed by the available information on the macrostate? Intuitively, we wish to choose the least biased probability law possible, in other words, a law that contains only the available information and nothing more. When no information is available, the answer is simple: $P_m = 1/M$ where M is the number of accessible states, this corresponds to the maximum possible statistical entropy $k \ln M$. We generalize this to the case where we have partial information. This partial information, provided by the macrostate, can be of two types.

- We have at our disposal data of which we are certain, for example we know that the energy lies in the interval $[E, E + \Delta E]$ where E is a known energy (not an average) with uncertainty ΔE, which is very small at the macroscopic scale ($\Delta E \ll E$), but very large at the microscopic one. In other words, ΔE is very large compared to the average level spacing or, more quantitatively, if $\rho(E)$ is the density of energy levels, $\Delta E \gg 1/\rho(E)$. In such a situation, we say that we have *microcanonical* information which constrains the Hilbert space of states. In the above example, the only acceptable states $|r\rangle$ are the eigenvectors of H ($H|r\rangle = E_r|r\rangle$) such that

$$E \leq E_r \leq E + \Delta E \tag{2.61}$$

If there are no other restrictions, the unbiased probability law will be, once again, $P_r = 1/M$ where $M = \rho(E)\Delta E$ is the number of states satisfying (2.61). All accessible states are equally probable. The density operator is then[15]

$$D = \sum_r |r\rangle \frac{1}{M} \langle r| \qquad E \leq E_r \leq E + \Delta E \tag{2.62}$$

which corresponds to $S_{\text{st}} = k \ln M$. Despite the simple form of the density operator in (2.62), the microcanonical ensemble is not very convenient in practice. It is better to impose statistical constraints that, in addition, are more natural.

- We have data of statistical nature: average values, correlations, etc. For example, if we know that the average energy E is fixed, we have the constraint on D: $\langle H \rangle = \text{Tr}\, DH = E$. In order to express these constraints conveniently, we introduce the following notation: the average values of operators A_i representing dynamic variables will be written as A_i

$$A_i \equiv \langle \mathsf{A}_i \rangle = \text{Tr}\, D\mathsf{A}_i \tag{2.63}$$

It is shown in information theory that the non-biased probability law is that which gives the maximum entropy. It is, therefore, reasonable to take as the fundamental postulate of *equilibrium* statistical mechanics the postulate of maximum statistical entropy.

Postulate of maximum statistical entropy

Among all the density operators consistent with the macroscopic constraints, we must choose the density operator D that gives the maximum statistical entropy $S_{\text{st}}[D]$. At equilibrium, a macrostate will therefore be represented by this density operator.

In more intuitive terms, we choose the most disordered macrostate consistent with the available information. The density operator D is the one which contains no information beyond what is necessary to satisfy the macroscopic constraints.

2.4.2 Equilibrium distribution

Henceforth, we will consider the canonical situation defined by the constraints (2.63). The *canonical ensemble*[16] corresponds to situations where one is given the Hamiltonian operator H and the average energy $E = \text{Tr}DH$. If, in addition, one is given the particle number operator N and its average value $\langle \mathsf{N} \rangle = \text{Tr}D\mathsf{N}$, one has the *grand canonical ensemble*. These are the two most common cases. In order to

[15] It is very important to note that in (2.62) the index r labels the set of quantum numbers identifying a given state $|r\rangle$ of eigenvalue E_r. The frequently occurring level degeneracies are taken into account in (2.62) because the sum over r includes all the states.

[16] The 'theory of ensembles', developed by Gibbs, has played a major historical rôle in the founding of statistical mechanics. We will maintain the terminology without attempting to give details that are not needed for what follows.

find the maximum statistical entropy subject to constraints, we will naturally use Lagrange multipliers (see Section A.2) with $-\lambda_i$ the multiplier associated with A_i.[17] We should also satisfy the constraint $\mathrm{Tr}D = 1$. We therefore define

$$\frac{1}{k}\tilde{S}_{\mathrm{st}}[D] = -\mathrm{Tr}\, D \ln D + \sum_i \lambda_i\, (\mathrm{Tr}\, D\mathsf{A}_i - A_i) - \lambda_0(\mathrm{Tr}\, D - 1)$$

To determine the extrema of $\tilde{S}_{\mathrm{st}}[D]$, we use (2.58) to differentiate the above equation thus obtaining

$$\mathrm{Tr}\Big[\mathrm{d}D(\ln D + 1 - \sum_i \lambda_i \mathsf{A}_i + \lambda_0)\Big] = 0$$

The term in braces should vanish for all $\mathrm{d}D$ which gives the solution

$$\ln D = -\Big(1 + \lambda_0 - \sum_i \lambda_i \mathsf{A}_i\Big)$$

The term $(\lambda_0 + 1)$ should be chosen so that $\mathrm{Tr}D = 1$. We call the density operator thus obtained D_{B} (B for Boltzmann)

$$\boxed{D_{\mathrm{B}} = \frac{1}{Z}\exp\Big(\sum_i \lambda_i \mathsf{A}_i\Big) \qquad Z = \mathrm{Tr}\exp\Big(\sum_i \lambda_i \mathsf{A}_i\Big)} \tag{2.64}$$

and $\mathrm{Tr}D_{\mathrm{B}} = 1$ by construction. These equations define *the Boltzmann distribution* (or Boltzmann–Gibbs), and Z is called the *partition function*. The statistical entropy of the Boltzmann distribution, or the *Boltzmann entropy*,[18] is

$$\begin{aligned} S_{\mathrm{B}} = S_{\mathrm{st}}[D_{\mathrm{B}}] &= -k(\mathrm{Tr}\, D_{\mathrm{B}} \ln D_{\mathrm{B}}) \\ &= k \ln Z - k\sum_i \lambda_i A_i \end{aligned} \tag{2.65}$$

This entropy is an extremum by construction; we now show that it is a maximum. Let D be a density operator satisfying the constraints $\mathrm{Tr}\, D\mathsf{A}_i = A_i$; by using (2.56) with the substitutions $Y \to D_{\mathrm{B}},\ X \to D$, we obtain

$$\begin{aligned} -\mathrm{Tr}\, D \ln D \le -\mathrm{Tr}\, D \ln D_{\mathrm{B}} &= \ln Z - \sum_i \lambda_i \mathrm{Tr}\, D\mathsf{A}_i \\ &\le \ln Z - \sum_i \lambda_i A_i \end{aligned}$$

[17] By introducing a minus sign, which may appear ill-advised here, we avoid writing many more in later expressions.

[18] Unfortunately, the terminology varies with authors. For some authors, the expression 'Boltzmann entropy' is reserved for the entropy defined in (8.110).

and consequently

$$S_{\text{st}}[D] \leq S_{\text{st}}[D_{\text{B}}] = S_{\text{B}}$$

The Boltzmann distribution, which is the equilibrium distribution, is the one that gives the *maximum statistical entropy* subject to the constraints (2.63). This argument also shows that if the extremum problem admits a solution,[19] this solution is unique since, according to (2.56), the entropies cannot be equal unless $D \equiv D_{\text{B}}$.

2.4.3 Legendre transformation

The average values of the operators A_i can be related to the derivatives of the partition function with respect to the Lagrange multipliers λ_i. Using relation (2.58) we get

$$\frac{1}{Z}\frac{\partial}{\partial \lambda_j}\operatorname{Tr}\exp\Big[\sum_i \lambda_i \mathsf{A}_i\Big] = \frac{1}{Z}\operatorname{Tr}\Big(\mathsf{A}_j \exp\Big[\sum_i \lambda_i \mathsf{A}_i\Big]\Big)$$
$$= \operatorname{Tr}(D_{\text{B}}\mathsf{A}_j) = A_j$$

in other words,

$$\boxed{A_j = \frac{\partial}{\partial \lambda_j}\ln Z[\lambda_i]} \qquad (2.66)$$

Equation (2.66) emphasizes that the partition function Z must be considered a function of the Lagrange multipliers λ_i. Equation (2.65) for the Boltzmann entropy can be written as

$$\frac{1}{k}S_{\text{B}} = \ln Z - \sum_i \lambda_i \frac{\partial \ln Z}{\partial \lambda_i} \qquad (2.67)$$

This shows that S_{B}/k is the Legendre transform of $\ln Z$, which can be expressed as a function of the A_i. We therefore have[20]

$$\mathrm{d}\ln Z = \sum_i A_i\, \mathrm{d}\lambda_i \qquad (2.68a)$$

$$\mathrm{d}S_{\text{B}} = -k\sum_i \lambda_i\, \mathrm{d}A_i \qquad (2.68b)$$

In practice, the calculation of the partition function is the fundamental calculation in statistical mechanics from which we deduce all the thermodynamic properties. As a general rule, it is not necessary to resort to the explicit form of the density operator.

[19] It can happen that there is no solution, for example if E is chosen below the ground state energy.
[20] The next section will allow us to show that Equation (2.68b) leads to a $T\,\mathrm{d}S$ equation.

We now show concavity/convexity properties of $\ln Z$ and $S_{\rm B}$. We give a simplified proof by assuming that all the operators A_i inter-commute. The general proof is left to Exercise 2.7.8. If all the operators A_i inter-commute, we can easily take two derivatives with respect to λ_i

$$\frac{\partial^2 Z}{\partial\lambda_i\partial\lambda_j} = \mathrm{Tr}\left[\mathsf{A}_i\mathsf{A}_j \exp\left(\sum_k \lambda_k \mathsf{A}_k\right)\right] = Z\langle \mathsf{A}_i\mathsf{A}_j\rangle \tag{2.69}$$

which gives

$$\frac{1}{Z}\frac{\partial^2 Z}{\partial\lambda_i\partial\lambda_j} - \left(\frac{1}{Z}\frac{\partial Z}{\partial\lambda_i}\right)\left(\frac{1}{Z}\frac{\partial Z}{\partial\lambda_j}\right) = \frac{\partial^2 \ln Z}{\partial\lambda_i\partial\lambda_j} = \frac{\partial A_i}{\partial\lambda_j}$$

This leads to

$$\frac{\partial^2 \ln Z}{\partial\lambda_i\,\partial\lambda_j} = \frac{\partial A_i}{\partial\lambda_j} = \langle(\mathsf{A}_i - A_i)(\mathsf{A}_j - A_j)\rangle = \langle \mathsf{A}_i\mathsf{A}_j\rangle - A_i A_j \tag{2.70}$$

This equation is called the *fluctuation-response theorem* because $\langle(\mathsf{A}_i - A_i)(\mathsf{A}_j - A_j)\rangle$ measures *fluctuations* relative to the average values A_i and A_j (the two indexes can be equal), whereas $\partial A_i/\partial\lambda_j$ is a *response* since this quantity describes the response of the average value A_i to a variation of the Lagrange multiplier λ_j.

The above results also allow us to show a very important positivity result. Defining the operator B as $\mathsf{B} = \sum_k a_k(\mathsf{A}_k - A_k)$, where the a_k are real numbers, we use (2.70) to obtain

$$\sum_{i,j} a_i a_j \frac{\partial^2 \ln Z}{\partial\lambda_i\,\partial\lambda_j} = \langle \mathsf{B}^2\rangle \geq 0 \tag{2.71}$$

The matrix $C_{ij} = \partial^2 \ln Z/\partial\lambda_i\partial\lambda_j$ is then positive, which implies that $\ln Z$ is a convex function of the λ_i, and that its Legendre transform $S_{\rm B}$ is a concave function of the A_i, exactly like the thermodynamic entropy S defined in Chapter 1. The general form of the fluctuation-response theorem when the operators A_i do not inter-commute is to be found in Exercise 2.7.8, where we also show that the concavity properties remain valid.

2.4.4 Canonical and grand canonical ensembles

The canonical and grand canonical ensembles will be treated in detail in the next chapter. We introduce them here to illustrate concretely the previous discussion. The canonical ensemble corresponds to exact knowledge of the volume, V, the number of molecules, N, as well as the average value of the Hamiltonian E (a statistical quantity). We therefore have a single operator $\mathsf{A}_1 = H$, and conventionally, the Lagrange multiplier is written as $\lambda_1 = -\beta$.[21] We shall soon see that $\beta = 1/kT$. The partition function is

$$\boxed{Z = \mathrm{Tr}\, \mathrm{e}^{-\beta H}} \tag{2.72}$$

whereas the density operator[22] becomes

$$D_{\mathrm{B}} = \frac{1}{Z} \mathrm{e}^{-\beta H} = \frac{1}{Z} \sum_r |r\rangle \mathrm{e}^{-\beta E_r} \langle r| \tag{2.73}$$

with $H|r\rangle = E_r|r\rangle$. When the system under consideration is a sum of two independent subsystems, the Hamiltonian is a sum of two commuting Hamiltonians[23]

$$H = H^{(a)} + H^{(\alpha)} \qquad [H^{(a)}, H^{(\alpha)}] = 0$$

In addition, as we have seen in (2.52), the total density operator of the system is given by the tensor product of the density operators of the two subsystems. We can easily calculate the partition function in the factorized basis $|l\rangle = |a\rangle \otimes |\alpha\rangle$ ($|m\rangle = |b\rangle \otimes |\beta\rangle$) where H is diagonal

$$\langle l|H|m\rangle = \left(H^{(a)}_{ab} + H^{(\alpha)}_{\alpha\beta}\right) \delta_{ab}\delta_{\alpha\beta}$$

$$\langle l|\mathrm{e}^{-\beta H}|m\rangle = \mathrm{e}^{-\beta H^{(a)}_{ab}}\, \mathrm{e}^{-\beta H^{(\alpha)}_{\alpha\beta}}\, \delta_{ab}\delta_{\alpha\beta}$$

and therefore

$$Z = \mathrm{Tr}\, \mathrm{e}^{-\beta H} = \left(\sum_a \mathrm{e}^{-\beta H^{(a)}_{aa}}\right)\left(\sum_\alpha \mathrm{e}^{-\beta H^{(\alpha)}_{\alpha\alpha}}\right) = Z^{(a)} Z^{(\alpha)} \tag{2.74}$$

In the grand canonical ensemble, the two operators whose average values are fixed are the Hamiltonian, $\mathsf{A}_1 = H$ and the number of particles, $\mathsf{A}_2 = \mathsf{N}$. The corresponding Lagrange multipliers are $\lambda_1 = -\beta$ and $\lambda_2 = \alpha$. We shall show in the following section that α is related to the chemical potential: $\alpha = \mu/kT$.

The *grand partition function* $\mathcal{Q}$ is

$$\boxed{\mathcal{Q} = \mathrm{Tr}\, \exp(-\beta H + \alpha \mathsf{N})} \tag{2.75}$$

and the density operator

$$\boxed{D_{\mathrm{B}} = \frac{1}{\mathcal{Q}} \exp(-\beta H + \alpha \mathsf{N})} \tag{2.76}$$

[21] Attention to the sign!

[22] To avoid cumbersome notations, we do not distinguish with an additional index the Boltzmann density operator expressed in one ensemble or another.

[23] Rigorously, one should write $H^{(a)} \otimes \mathbb{I}^{(\alpha)}$ and $\mathbb{I}^{(a)} \otimes H^{(\alpha)}$ where $\mathbb{I}^{(a)}$ ($\mathbb{I}^{(\alpha)}$) is the identity operator acting in the Hilbert space $\mathcal{H}^{(a)}$ ($\mathcal{H}^{(\alpha)}$).

2.5 Thermodynamics revisited

2.5.1 Heat and work: first law

We now revisit the concepts of thermodynamics and identify the statistical quantities of the preceding section with the thermodynamic quantities of Chapter 1. We start with the first law and consider a system $\mathcal{A}$ (Figure 2.1) that can exchange work with an external medium and heat with a reservoir $\mathcal{R}$. The system and the reservoir are together thermally isolated from the outside world. Here, as in Chapter 1, energy exchanges in the form of work can be calculated in terms of external parameters, x_i, which are controllable (at least in principle) by the experimenter and whose time dependence guides that of the Hamiltonian: $H(t) = H(x_i(t))$. We therefore only consider quasi-static processes, non-quasi-static processes such as Joule's experiment cannot be described by a Hamiltonian. We further assume that the wall between system and reservoir is impermeable to molecules. At equilibrium, the system is therefore described by the canonical ensemble with Hamiltonian

$$H_{\text{tot}} = H + H_{\mathcal{R}} + \mathcal{V} \tag{2.77}$$

where H is the Hamiltonian of the system, $H_{\mathcal{R}}$ that of the reservoir and $\mathcal{V}$ describes the interaction between the system and the reservoir. Since the interaction $\mathcal{V}$ is a surface effect, its contribution to the energy and entropy is negligible, but $\mathcal{V}$ is clearly essential for the heat exchange between the system and the reservoir to take place. If $\mathcal{V} = 0$, the system and the reservoir are independent.

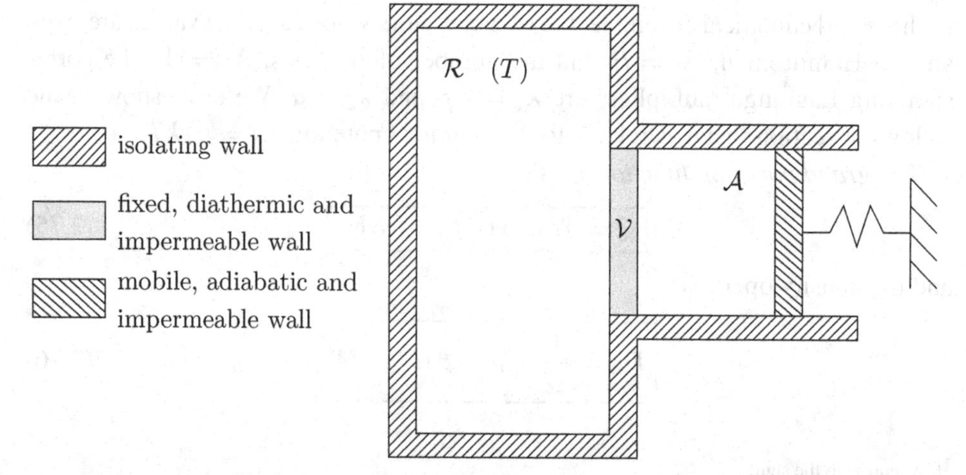

Figure 2.1 Exchange of heat and work. The spring represents exchange of work with the outside world.

Let D_{tot} be the total density operator for the system/reservoir combination. From (2.14) and (2.77), the evolution equation for D_{tot} is

$$i\hbar \frac{\mathrm{d}D_{\text{tot}}}{\mathrm{d}t} = [H, D_{\text{tot}}] + [H_{\mathcal{R}}, D_{\text{tot}}] + [\mathcal{V}, D_{\text{tot}}] \tag{2.78}$$

The density operator of the system is obtained by taking the partial trace of D_{tot} with respect to the reservoir variables (cf. (A.18)): $D = \text{Tr}_{\mathcal{R}} D_{\text{tot}}$. Let us take the trace over the reservoir variables in Equation (2.78). Using (A.20) we can show

$$\text{Tr}_{\mathcal{R}}[H_{\mathcal{R}}, D_{\text{tot}}] = 0$$

and the evolution equation of D then becomes

$$i\hbar \frac{\mathrm{d}D}{\mathrm{d}t} = [H, D] + \text{Tr}_{\mathcal{R}}[\mathcal{V}, D_{\text{tot}}] \tag{2.79}$$

This equation is not closed: $\mathrm{d}D/\mathrm{d}t$ does not depend only on the system variables and consequently the evolution is non-Hamiltonian. In other words, we cannot find a Hamiltonian such that the evolution of D be governed by Equation (2.14).

We now determine the evolution equation of the average energy E by differentiating, with respect to time, the definition $E = \text{Tr}DH$ (Tr here is the trace over the system variables)

$$\frac{\mathrm{d}E}{\mathrm{d}t} = \text{Tr}\left(H \frac{\mathrm{d}D}{\mathrm{d}t}\right) + \text{Tr}\left(D \frac{\mathrm{d}H}{\mathrm{d}t}\right) \tag{2.80}$$

The first term on the right hand side of (2.80) depends on the thermal contact with the reservoir. In fact

$$\begin{aligned} \text{Tr}\left(H \frac{\mathrm{d}D}{\mathrm{d}t}\right) &= \frac{1}{i\hbar} \text{Tr}\,(H[H, D]) + \frac{1}{i\hbar} \text{Tr}\,\text{Tr}_{\mathcal{R}}(H[\mathcal{V}, D_{\text{tot}}]) \\ &= \frac{1}{i\hbar} \text{Tr}_{\text{tot}}\,(D_{\text{tot}}[H, \mathcal{V}]) \end{aligned} \tag{2.81}$$

This term describes the energy exchange in the form of heat because it depends on the coupling $\mathcal{V}$ between the system and the reservoir ($\text{Tr}_{\text{tot}} \equiv \text{Tr}\text{Tr}_{\mathcal{R}}$). The second term in (2.80) describes the energy exchange in the form of work. It can be written in terms of the controllable external parameters $x_i(t)$

$$\text{Tr}\left(D \frac{\mathrm{d}H}{\mathrm{d}t}\right) = \sum_i \text{Tr}\left(D \left.\frac{\partial H}{\partial x_i}\right|_{x_{j \neq i}}\right) \frac{\mathrm{d}x_i}{\mathrm{d}t} \tag{2.82}$$

Therefore, Equation (2.80) provides the statistical mechanical interpretation of the first law of thermodynamics

$$\begin{aligned} \mathrm{d}E = \mathrm{d}(\mathrm{Tr}\,(DH)) &= \mathrm{Tr}\,(H\,\mathrm{d}D) + \mathrm{Tr}(D\,\mathrm{d}H) \\ &= đQ + đW \end{aligned} \tag{2.83}$$

We identify the exchanges of heat, $đQ$, and work, $đW$, as follows

$$đQ = \mathrm{Tr}\,(H\,\mathrm{d}D) \tag{2.84a}$$

$$đW = \mathrm{Tr}\,(D\,\mathrm{d}H) \tag{2.84b}$$

Equation (2.84a) provides the interpretation of the energy exchange in the form of heat as a modification of D and, consequently, of the probabilities of energy level occupations. This relation remains valid for any infinitesimal transformation, and even applies to finite transformations if H does not change: $Q = \mathrm{Tr}(H\Delta D)$. On the other hand, Equation (2.84b), which identifies work as a modification of the Hamiltonian, is only valid for quasi-static processes.

2.5.2 Entropy and temperature: second law

We now give a physical interpretation of the Lagrange multiplier β in the canonical ensemble by showing that $1/\beta$ defines a temperature scale. Consider two independent systems (a) and (α), in the canonical ensemble, whose average energies and Lagrange multipliers are (E_a, β_a) and (E_α, β_α). If we put these two systems in thermal contact, the average energy of the total system will be $E = E_a + E_\alpha$. To this average energy will correspond a single, unique Lagrange multiplier, β, because the only constraint now present concerns the total energy and not the individual energies of the two subsystems: at thermal equilibrium, the Lagrange multipliers of the two subsystems take the same value. In the canonical ensemble where $\lambda_1 = -\beta$, Equation (2.66) becomes

$$E = -\frac{\partial \ln Z}{\partial \beta}$$

Since $\ln Z$ is a convex function of β (cf. (2.71)) this equation means that the energy is a decreasing function of β. The same reasoning as in Section 1.2.3 allows us to show that the equality of the two β values can only happen if the energy flows from the subsystem with the smaller β to that with the larger value, and that the final equilibrium value of β lies between β_a and β_α. In other words, $1/\beta$ defines a temperature scale. Two systems in thermal contact will reach the same equilibrium temperature, and during this thermal contact, energy flows from the hotter to the cooler subsystem.

It remains to be shown that β is proportional to $1/T$. We do this by comparing the Boltzmann and thermodynamic entropies. During a quasi-static process, the density operator is infinitesimally close to the equilibrium Boltzmann density operator (2.73). For a variation $D_B \to D_B + \mathrm{d}D_B$ around equilibrium, we obtain the change $\mathrm{d}S_B$ of the Boltzmann entropy by using $\mathrm{Tr}\,\mathrm{d}D_B = 0$ and Equation (2.58)

$$\mathrm{d}S_B = -k\mathrm{Tr}\,(\mathrm{d}D_B(\ln D_B + 1)) = k\mathrm{Tr}\,(\mathrm{d}D_B[\ln Z + \beta H]) = k\beta\mathrm{Tr}\,(H\,\mathrm{d}D_B)$$

Therefore, using (2.84a) we have

$$\mathrm{d}S_B = k\beta\,đ\,Q \tag{2.85}$$

But, for a quasi-static process[24]

$$\mathrm{d}S = \frac{đ\,Q}{T}$$

where T is the equilibrium thermodynamic temperature of the system. Combining the two preceding equations allows us to relate $\mathrm{d}S_B$ to $\mathrm{d}S$

$$\mathrm{d}S_B = k\beta T\,\mathrm{d}S$$

and to derive

$$\frac{\partial S_B}{\partial \beta} = k\beta T\frac{\partial S}{\partial \beta}$$

as well as

$$\frac{\partial S_B}{\partial x_i} = k\beta T\frac{\partial S}{\partial x_i}$$

where the x_i are the external parameters. Then, using the fact that T does not depend on the external parameters, we obtain

$$\begin{aligned}\frac{\partial^2 S_B}{\partial\beta\partial x_i} &= k\beta T\frac{\partial^2 S}{\partial\beta\partial x_i}\\ &= k\beta T\frac{\partial^2 S}{\partial\beta\partial x_i} + \frac{\partial(k\beta T)}{\partial\beta}\frac{\partial S}{\partial x_i}\end{aligned}$$

which shows that $k\beta T$ is a constant. We can always choose a system of units where $k\beta T = 1$ and thus conclude that $S_B - S$ is a constant. In fact this constant is zero. This can be seen by recalling that S vanishes at $T = 0$ because of the third law, while S_B vanishes because a quantum system at zero temperature is in its ground state whose statistical entropy is zero.[25] To summarize, *the Boltzmann entropy,*

[24] We keep the convention of Chapter 1 by using S for the thermodynamic entropy.

[25] Unless one encounters an exceptional ground state degeneracy such as mentioned in Footnote 22, Chapter 1.

which measures the microscopic disorder compatible with the macroscopic constraints, becomes identical to thermodynamic entropy.

We note that $\beta > 0$ or, equivalently, $T \geq 0$; in other words and as a general rule, the entropy is an increasing function of energy. In fact, the Hamiltonian of a quantum system is unbounded and $\mathrm{Tr}\exp(-\beta H)$ can only be defined if $\beta > 0$. The only exception possible is that of a quantum system whose Hilbert space is finite, for example a system of spins. In such a case, we can observe negative (spin) temperatures (see Problem 3.8.2).

2.5.3 Entropy of mixing

In the previous paragraph, the relation found between the thermodynamic entropy and the Boltzmann entropy (and therefore the statistical entropy) may lead one to believe that entropy is always related to heat. We use the simple example of the ideal gas in the canonical ensemble to emphasize that this is not necessarily the case.

We first calculate the partition function Z_N of a classical mono-atomic ideal gas of N molecules.[26] The Hamiltonian H of the gas is the sum of individual Hamiltonians, H_i,

$$H = \sum_{i=1}^{N} H_i = \sum_{i=1}^{N} \frac{\vec{\mathsf{P}}_i^{\,2}}{2m} \tag{2.86}$$

where $\vec{\mathsf{P}}_i$ is the momentum operator acting on the Hilbert space of particle (i).[27] We have already seen in (2.74) that if the Hamiltonian is the sum of two independent Hamiltonians, the partition function is then the product of the two partition functions corresponding to each Hamiltonian separately. An immediate generalization gives $Z_N = \zeta^N$, where ζ is the partition function for a single molecule. Such a simple result is clearly only valid because the molecules are non-interacting. It is easy to calculate ζ if we express the trace in (2.72) in the basis of eigenstates $|\vec{p}\,\rangle$ of $\vec{\mathsf{P}}$,

$$\zeta = \sum_{\vec{p}} \langle \vec{p}\,|\exp\left(-\beta\frac{\vec{\mathsf{P}}^{\,2}}{2m}\right)|\vec{p}\,\rangle = \frac{V}{h^3}\int \mathrm{d}^3 p \, \exp\left(-\frac{\beta\vec{p}^{\,2}}{2m}\right) = V\left(\frac{2\pi m}{\beta h^2}\right)^{3/2}$$

where we use (2.21) to obtain the second equality and (A.37) for the last one. We

[26] Although the gas is 'classical', the particles of the gas will be treated quantum mechanically: it is the treatment of the statistical properties which will remain classical. This subtlety will be explained more precisely in Chapter 5.

[27] We use the same simplification of notation discussed in Footnote 23.

therefore obtain for the partition function of an ideal gas,

$$Z_N = V^N \left(\frac{2\pi m}{\beta h^2}\right)^{3N/2}$$

This result needs to be modified to take into account the fact that the molecules are identical. It seems natural to divide by $1/N!$ because a permutation of two identical molecules does not modify the configuration of the gas.[28] However, the factor $1/N!$ cannot be justified rigorously in classical mechanics because it is always possible to label the molecules without changing their properties. For example, we can suppose they are like little numbered billiard balls. A physical justification of the $1/N!$ factor, which we give in Chapter 5, can be obtained by considering the classical gas as the limit of a quantum gas obeying either the Bose–Einstein or Fermi–Dirac statistics, and in both cases we do indeed obtain the factor of $1/N!$. The common classical limit of the Bose–Einstein and the Fermi–Dirac statistics will be called the Maxwell–Boltzmann statistics. The final result for Z_N is then

$$\boxed{Z_N = \frac{V^N}{N!}\left(\frac{2\pi m}{\beta h^2}\right)^{3N/2}} \tag{2.87}$$

We discuss, in passing, the semi-classical approach to phase space. By replacing the momentum operators with classical vectors in Equation (2.86), we could have calculated Z_N using an (almost) entirely classical approach

$$Z_N^{\text{class}} = C\int \prod_{i=1}^{N} \mathrm{d}^3 r_i\, \mathrm{d}^3 p_i \exp\left(-\sum_i \frac{\vec{p}_i^{\,2}}{2m}\right)$$

where we have used Equation (2.36) for the classical phase space measure and where the $\vec{r}_i$ and $\vec{p}_i$ are classical variables ($\int \mathrm{d}^3 r = V$). But we still have the constant C which classical mechanics cannot determine. In fact we can see, by comparing with Equation (2.87), that this constant contains quantum effects

$$C = \frac{1}{N!h^{3N}} \tag{2.88}$$

Thus, the 'classical' expression for the correctly normalized phase space is

$$\boxed{\mathrm{d}\Gamma = \frac{1}{N!}\prod_{i=1}^{N} \frac{\mathrm{d}^3 r_i \mathrm{d}^3 p_i}{h^3}} \tag{2.89}$$

With this, we conclude our discussion of the semi-classical approach.

[28] See, however, the discussion of Equation (3.13).

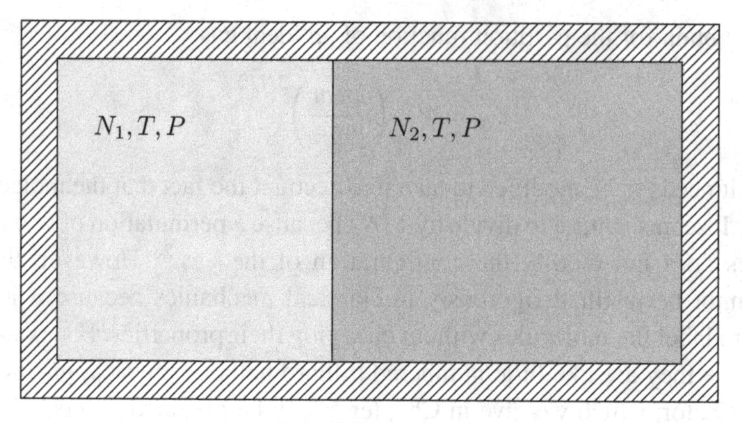

Figure 2.2 Mixing two gases.

Knowledge of the partition function allows us to obtain directly the entropy of a gas, this being the Legendre transform (2.67) of $\ln Z_N$ with $\lambda_1 = -\beta$,

$$S = k \ln Z_N - k\beta \left.\frac{\partial \ln Z_N}{\partial \beta}\right|_V \tag{2.90a}$$

$$= kN \left[\ln \frac{V}{N} + \frac{3}{2} \ln \left(\frac{2\pi mkT}{h^2} \right) + \frac{5}{2} \right] \tag{2.90b}$$

where, to obtain (2.90b), we use the Stirling approximation: $\ln N! \approx N \ln N - N$. It is important to remark that the normalization $1/N!$ in (2.87) is indispensable for the extensivity of the entropy: without this term, we would obtain $\ln V$ instead of $\ln(V/N)$ in Equation (2.90b), and the entropy would not be extensive.

Consider now two different, classical, ideal gases initially in the two compartments (volumes V_1 and V_2, $V = V_1 + V_2$) of a rigid adiabatic container (Figure 2.2). They have the same pressure P and temperature T, the numbers of molecules of each compartment are N_1 and N_2 ($N = N_1 + N_2$) and we assume there are no chemical reactions. We open the separation between the two compartments without introducing energy. The initial partition function $Z_N^{(\text{in})}$ is

$$\begin{aligned} Z_N^{(\text{in})}(T, V) &= Z_{N_1}(T, V_1) Z_{N_2}(T, V_2) \\ &= \frac{V_1^{N_1}}{N_1!} \left(\frac{2\pi m_1 kT}{h^2} \right)^{3N_1/2} \frac{V_2^{N_2}}{N_2!} \left(\frac{2\pi m_2 kT}{h^2} \right)^{3N_2/2} \end{aligned} \tag{2.91}$$

whereas the final partition function $Z_N^{(\text{fin})}$ is

$$\begin{aligned} Z_N^{(\text{fin})}(T, V) &= Z_{N_1}(T, V) Z_{N_2}(T, V) \\ &= \frac{V^{N_1}}{N_1!} \left(\frac{2\pi m_1 kT}{h^2} \right)^{3N_1/2} \frac{V^{N_2}}{N_2!} \left(\frac{2\pi m_2 kT}{h^2} \right)^{3N_2/2} \end{aligned} \tag{2.92}$$

The second term in Equation (2.90a) gives the same contribution to the entropies, $S_{\rm in}$ and $S_{\rm fin}$.[29] Then, the entropy difference due to the mixing operation is given by

$$S_{\rm fin} - S_{\rm in} = k \ln \frac{Z_N^{(\rm fin)}}{Z_N^{(\rm in)}} = k\left[N_1 \ln \frac{V}{V_1} + N_2 \ln \frac{V}{V_2}\right] > 0 \qquad (2.93)$$

Mixing two different gases is accompanied by an increase in entropy. It is clear that the process is irreversible: once the mixing is done, the two gases will not return spontaneously to their initial compartments. The final state is more disordered than the initial one, and the entropy corresponding to this increase in disorder is called the *entropy of mixing*.

If the two gases are identical, mixing them does not lead to an increase in entropy since the initial and final states are the same, and if we re-install the partition, we will regain the initial state. We can show this easily, and the demonstration emphasizes the importance of the $1/N!$ factor in (2.87). For two identical gases, the expression for $Z_N^{(\rm in)}$ is obtained by putting $m_1 = m_2 = m$ in Equation (2.91) and noting that $V_{1,2} = (N_{1,2}/N)V$. The final partition function is that of an ideal gas of N identical molecules

$$Z_N^{(\rm fin)} = \frac{V^N}{N!}\left(\frac{2\pi mkT}{h^2}\right)^{3N/2} \neq \frac{V^N}{N_1!N_2!}\left(\frac{2\pi mkT}{h^2}\right)^{3N/2}$$

where we have emphasized the difference with the expression obtained by putting $m_1 = m_2 = m$ in Equation (2.92). We verify that indeed the entropy does not change

$$S_{\rm fin} - S_{\rm in} = k\left[N_1 \ln \frac{N_1 V}{N V_1} + N_2 \ln \frac{N_2 V}{N V_2}\right] = 0$$

2.5.4 Pressure and chemical potential

We now evaluate the energy exchanged in the form of work during a quasi-static process. The equilibrium distribution is that of Boltzmann and, therefore, we use in Equation (2.82) the density operator (2.73). Equation (2.84b) then becomes[30]

$$đ W = {\rm Tr}[D_{\rm B} {\rm d}H] = {\rm Tr}\left[D_{\rm B} \sum_i \frac{\partial H}{\partial x_i}{\rm d}x_i\right] = \sum_i X_i {\rm d}x_i \qquad (2.94)$$

[29] The reader will remark that only the term $V^N/N!$ in (2.87) enters our reasoning.

[30] Note that we have simplified the notation:

$$\frac{\partial H}{\partial x_i} \equiv \left.\frac{\partial H}{\partial x_i}\right|_{x_{j\neq i}}$$

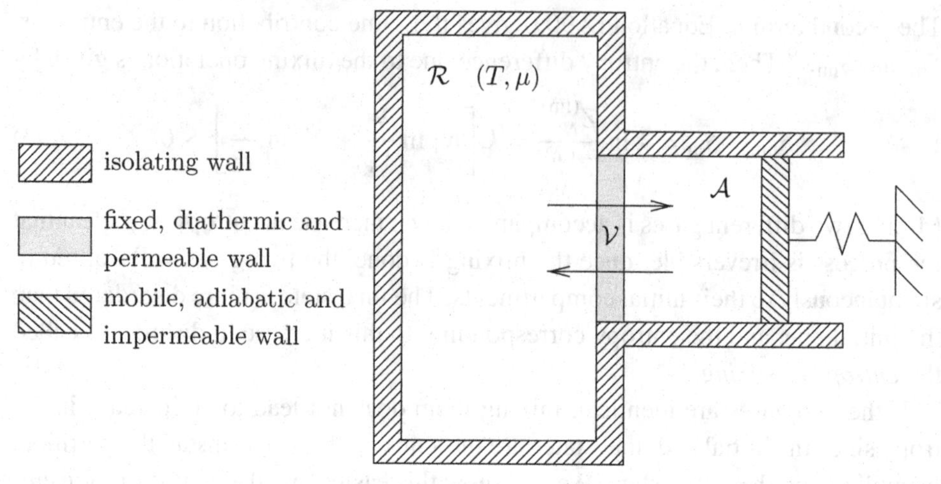

Figure 2.3 Grand canonical situation.

where X_j is the *conjugate variable*, calculated at equilibrium, of the controllable external variable x_j

$$X_j = \left\langle \frac{\partial H}{\partial x_j} \right\rangle = \frac{1}{Z} \mathrm{Tr}\left[\frac{\partial H}{\partial x_j} \mathrm{e}^{-\beta H} \right] = -\frac{1}{\beta} \left. \frac{\partial \ln Z}{\partial x_j} \right|_{\beta, x_{i \neq j}} \tag{2.95}$$

In particular, the pressure P is the conjugate variable of $-V$

$$\boxed{P = \frac{1}{\beta} \left. \frac{\partial \ln Z}{\partial V} \right|_{\beta, N}} \tag{2.96}$$

Consider now a system like that in Figure 2.1 but where the coupling $\mathcal{V}$ allows the exchange of particles between the reservoir and the system (Figure 2.3). Only the volume is known exactly; the number of particles, like the energy, is known only on average, $N = \mathrm{Tr}(D_{\mathrm{B}}\mathsf{N})$. The change in the Boltzmann entropy is, once again, given by

$$\mathrm{d}S_{\mathrm{B}} = -k\mathrm{Tr}\,(\mathrm{d}D_{\mathrm{B}}(\ln D_{\mathrm{B}} + 1))$$

But D_{B} is now the density operator in the grand canonical ensemble (2.76):

$$\mathrm{d}S_{\mathrm{B}} = k\beta \mathrm{Tr}\,(H\,\mathrm{d}D_{\mathrm{B}}) - k\alpha \mathrm{Tr}\,(\mathsf{N}\,\mathrm{d}D_{\mathrm{B}}) = k\beta\,\mathrm{d}E - k\alpha\,\mathrm{d}N$$

In order to identify the Boltzmann entropy with the thermodynamic entropy we must choose, by comparison with Equation (1.18),

$$\boxed{\beta = \frac{1}{kT} \qquad \alpha = \frac{\mu}{kT}} \tag{2.97}$$

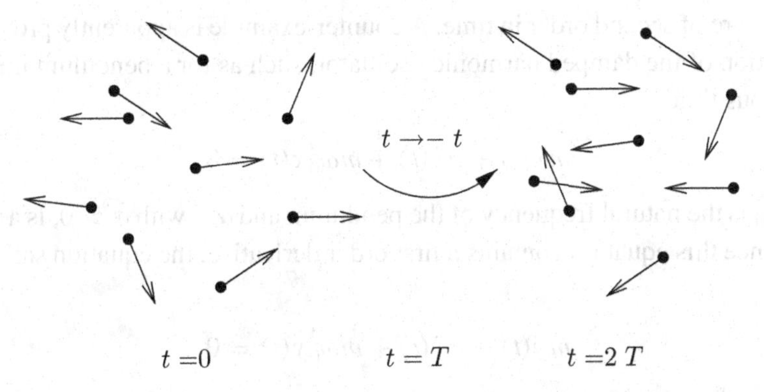

Figure 2.4 Time reversal invariance.

The grand canonical ensemble[31] accords the chemical potential (more precisely μ/T) a status comparable with that of the temperature (actually $1/T$).

2.6 Irreversibility and the growth of entropy

2.6.1 Microscopic reversibility and macroscopic irreversibility

The equations of classical mechanics are invariant under time reversal. Intuitively, this says that if a film of a sequence of events is run backward, what is seen on the screen appears to be physically possible.[32] For example, if it were possible to film the collisions of molecules in a volume containing only a few of them, it would be impossible upon viewing the film to tell if it was being run forward or backward. Mathematically, the equations of classical mechanics are invariant under the substitution $t \to -t$.[33] Imagine, for example, that we follow the motion of the molecules of a gas from $t = 0$ to $t = T$, and at time T we reverse all velocities. Then, at time $t = 2T$, all the molecules would be exactly where they were at $t = 0$ but with their velocities reversed: $\vec{r}_i(2T) = \vec{r}_i(0)$ and $\vec{v}_i(2T) = -\vec{v}_i(0)$ (Figure 2.4). This is called *micro-reversibility*. We give a simple example of this: the motion of a particle of mass m in a one-dimensional force field $F(x)$ and governed by the Newton equation $m\ddot{x}(t) = F(x(t))$. Putting $y(t) = x(-t)$ gives

$$m\ddot{y}(t) = m\ddot{x}(-t) = F(x(-t)) = F(y(t)) \tag{2.98}$$

We see that $y(t)$ indeed satisfies Newton's equations, in other words the motion described by $y(t)$ is physically possible. The key to this result is that Newton's

[31] Thermodynamics in the grand canonical ensemble will be treated in Chapter 3.

[32] In the same way, physics is invariant under parity transformation if an experiment viewed in a mirror also appears possible. There are very small violations of time reversal symmetry observed in the $K^0 - \overline{K}^0$ meson system, but that does not concern us here.

[33] For the Schrödinger equation, which is of first order in t, the time reversal operator is anti-unitary.

equations are of second order in time. A counter-example is apparently provided by the equation of the damped harmonic oscillator, such as for a pendulum immersed in a viscous fluid

$$m\ddot{x}(t) + \alpha\dot{x}(t) + m\omega_0^2 x(t) = 0 \tag{2.99}$$

where ω_0 is the natural frequency of the pendulum and $\alpha\dot{x}$, with $\alpha > 0$, is a viscous force. Since this equation contains a first order derivative, the equation satisfied by $y(t)$

$$m\ddot{y}(t) - \alpha\dot{y}(t) + m\omega_0^2 y(t) = 0$$

does not correspond to physically allowed motion since the pendulum would be accelerated by the viscosity! If, as in the previous discussion, we reverse the velocity at time T, the pendulum does not regain its initial position at time $2T$ because it continues to be slowed down by the viscosity. However, this example does not really contradict micro-reversibility. In fact, the viscous force is an effective force representing the interaction of the pendulum with an enormous number of degrees of freedom of the fluid as we shall see in Section 9.3. Micro-reversibility is only valid if we write the equations of motion of *all* the degrees of freedom. One can show (see Section 2.7.3) that the irreversible equations of motion in classical physics (diffusion equations, Navier–Stokes equations in hydrodynamics, etc.) rely on the existence of two very different time scales: a microscopic time scale τ^*, and a much longer macroscopic time scale τ, with $\tau \gg \tau^*$. In the case of the damped pendulum, τ^* is the average time between two collisions of the pendulum with molecules of the fluid, and τ is the macroscopic time $\tau = m/\alpha$.

Everyday phenomena do not appear to satisfy micro-reversibility as can be seen easily by running backward a film depicting daily events! For example, a viscous force always brakes, never accelerates. However, energy conservation alone does not preclude that the fluid molecules conspire to accelerate the pendulum instead of braking it. In fact, it is the second law that forbids this from happening, otherwise we could obtain work from a single source of heat. Everyday life is irreversible and corresponds to an inescapable increase in entropy. The natural tendency is to transform order into disorder and not the reverse. It remains to explain the 'paradox of irreversibility': how is it possible that equations which are reversible at the microscopic scale lead to phenomena which are irreversible at the macroscopic scale? To elucidate this paradox, we shall consider more deeply the concept of entropy.[34] The following discussion (like the preceding one) is presented in the context of

[34] The discussion to be presented here is based on ideas first discussed by Boltzmann and accepted by a large majority of physicists. However, a number of authors contest this point of view and instead seek the origin of irreversibility in singular evolution of probability distributions due to chaotic dynamics. The interested reader can consult the references cited at the end of the chapter.

classical mechanics, even though we use the notations of quantum mechanics, due to their simplicity.

2.6.2 Physical basis of irreversibility

So far, we have not used the microcanonical ensemble, but an intuitive discussion of irreversibility is a little more convenient in this framework. We shall return to the canonical ensembles in the following section. The (semi-)classical approximation for the number of states whose energy E_r is between E and $E + \Delta E$ is given by (2.89) for a system of N particles

$$\Omega(E) = \frac{1}{N!h^{3N}} \int_{E \le E_r \le E+\Delta E} \prod_{i=1}^{N} \mathrm{d}^3 r_i \, \mathrm{d}^3 p_i \tag{2.100}$$

The explicit expression for the phase space volume $\Omega(E)$ is given in Exercise 2.7.9 for N non-interacting particles. At equilibrium, the entropy of the system is written as $S(E) = k \ln \Omega(E)$ since all the states which satisfy the energy constraint are equally probable. We reconsider the example of irreversible expansion of a gas which we have already discussed in Figure 1.9 of Section 1.2.4. A gas is initially in the left compartment of the container, at $t = 0$ the partition is pierced allowing the gas to leak into the right compartment eventually filling the entire container. Let Ω_{in} and Ω_{fin} be respectively the initial and final phase space volumes. For a dilute gas, the ratio $\Omega_{\text{fin}}/\Omega_{\text{in}}$ is equal to the ratio of the volumes raised to the power N: $(V_{\text{fin}}/V_{\text{in}})^N = 2^N$ if the two compartments have equal volumes (see Exercise 2.7.9). This last result is in fact obvious because in the absence of interactions the integrals over $\vec{r}_i$ and $\vec{p}_i$ are decoupled and the value of the space integral is V^N. Among all the microscopic configurations (microstates) accessible in the final state, only a very small fraction, of the order of 2^{-N}, are such that all the molecules are in the left compartment. Therefore, the probability that the gas regains its initial configuration is negligible, of the order of 2^{-N}. Irreversibility appears, then, as the evolution from the less probable to the more probable and results from our ability to construct states with low entropy, for example, by compressing the gas into the left compartment with the aid of a piston. In this process, the total entropy of the universe increases whereas that of the gas decreases. The asymmetry in time reversal comes from the fact that just after the lifting of a constraint, the volume of phase space actually occupied is very small compared with the whole accessible volume.

Several arguments have been presented to contest this point of view since it was first put forth by Boltzmann. The first such argument is that of *Poincaré recurrences* (Zermelo's paradox): for an isolated system, if one waits long enough, all trajectories in phase space will end up passing arbitrarily close to their starting points. In the preceding example, all the molecules will eventually end up regrouping in the left compartment, which would result in a spontaneous decrease in the entropy. Poincaré recurrences are easy to observe in systems of few degrees of freedom, but for a macroscopic system the time needed for a Poincaré recurrence increases like e^N. In rigorous arguments, we take the thermodynamic limit in order to avoid such recurrences; for a macroscopic system the estimated time needed for a Poincaré recurrence is several times the age of the universe. In addition, even in the case of a finite system, the fact that the system is never perfectly isolated would inevitably perturb the observation of such recurrences.[35]

Loschmidt's paradox is based on micro-reversibility. Considering the same preceding example, if we reverse all velocities of the molecules at time T, then at time $2T$ they should all be back in the left compartment, which would result in spontaneous reduction in the entropy. Such an experiment can be done in computer simulations using molecular dynamics where in fact we see that the molecules will regain the left compartment. However, very small numerical errors in the positions and velocities at time T are enough to reduce and eventually eliminate this effect since phase space trajectories are extremely unstable with respect to initial conditions. In fact, the dynamics are chaotic, i.e. two trajectories with very close initial conditions will move apart very rapidly. To regain by going back in time a configuration whose probability is 2^{-N}, one has to aim extremely accurately.[36] In addition, one can never avoid the perturbations due to the imperfect isolation of the system. Consequently, the analogue of the numerical experiment is not possible, even approximately, for a real gas.

In summary, the statistical mechanical explanation of irreversibility is due to the conjunction of the following factors:

- The use of a probabilistic argument which relates the probability of a macrostate to the occupied phase space volume.
- The existence of an enormous number ($\sim 10^{23}$) of degrees of freedom.
- The existence of an initial condition that leads to a macrostate whose probability becomes negligible after lifting a constraint due to the enormous number of degrees of freedom.
- The existence of chaotic dynamics, which is not important qualitatively, but which plays a quantitative rôle in the manner in which equilibrium is approached.

[35] However, and contrary to what is found sometimes in the literature, the origin of irreversibility is not at all related to the fact that the system may not be perfectly isolated.

[36] It is, however, possible to perform so-called 'echo' experiments, for example spin or plasma echo experiments. These experiments rely on the existence of metastable states and a non-chaotic dynamics, which allow a 'hidden order' to reappear. This leads to an apparent decrease in entropy for an observer unaware of this hidden order.

2.6.3 Loss of information and the growth of entropy

We now return to the canonical ensemble. Our objective is to generalize the Boltzmann distribution (2.64) to a time dependent situation in order to follow the temporal evolution of the entropy. We have already identified the statistical entropy $S_{\rm st}$ with the thermodynamic entropy S at equilibrium. This identification is lost away from equilibrium. The statistical entropy is always defined by $-k\,\mathrm{Tr}\,D\ln D$, whether or not the density operator D corresponds to an equilibrium distribution. However, the thermodynamic entropy S is defined only at equilibrium or very close to it. We saw in (2.60) that statistical entropy is independent of time for an isolated system and, more generally, for Hamiltonian evolution, which might appear to contradict the principle of increase of entropy for an isolated system. To resolve this apparent paradox, we remark first that under the same conditions, the time evolution of the density operator $D(t)$ is perfectly well determined by (2.14) with the knowledge of the Hamiltonian $H(t)$ and the initial conditions $D(t=t_0)$. This allows us to know, for all times, the average values $A_i(t)$ of the observables A_i which enter in the definition of the thermodynamic entropy,

$$A_i(t) = \mathrm{Tr}\,D(t)\mathsf{A}_i \tag{2.101}$$

As usual, we have assumed that the observables A_i are independent of time in the Schrödinger picture. Let us now examine the evolution between two equilibrium states corresponding to times $t = t^{(\rm in)}$ and $t = t^{(\rm fin)}$. We can take as example the case of thermal contact of Section 1.2.2, Figure 1.7, by rendering the partition diathermic without supplying work to the system. The Hamiltonian depends on time because the molecules of the two subsystems, initially not interacting, are interacting in the final state. We could model this time dependent interaction by introducing in the Hamiltonian a term that depends explicitly on time

$$\mathcal{V}(t) = \theta(t)(1 - \mathrm{e}^{-t/\tau}) \sum_{a,\alpha} U_{a\alpha}(\vec{r}_a - \vec{r}_\alpha)$$

In this equation, θ is the Heaviside function, U the potential energy, τ a time that is very long compared to a characteristic microscopic time (e.g. the time between two successive collisions of the same molecule) and $\vec{r}_a$ and $\vec{r}_\alpha$ are the positions of two molecules taken in the two different subsystems. The evolution is Hamiltonian and, taking (2.60) into account, the statistical entropy does not change

$$S_{\rm st}^{(\rm fin)} = S_{\rm st}^{(\rm in)}$$

The initial thermodynamic entropy is equal to the initial statistical entropy by construction

$$S^{(\rm in)} \equiv S_{\rm B}^{(\rm in)} = S_{\rm st}^{(\rm in)} = -k\,\mathrm{Tr}\,D_{\rm B}^{(\rm in)} \ln D_{\rm B}^{(\rm in)}$$

but the final thermodynamic entropy cannot be the same because it has to take into account new constraints defined by the new equilibrium state

$$A_i^{(\text{fin})} = \text{Tr}\left(D^{(\text{fin})}\mathsf{A}_i\right) = \text{Tr}\left(D_{\text{B}}^{(\text{fin})}\mathsf{A}_i\right)$$

By definition, $D^{(\text{fin})}$ and $D_{\text{B}}^{(\text{fin})}$ must give the same average values, and by construction, the final Boltzmann density operator, $D_{\text{B}}^{(\text{fin})}$, maximizes the entropy subject to the constraints imposed on the averages of the observables. Consequently, we have

$$S^{(\text{in})} \equiv S_{\text{B}}^{(\text{in})} = S_{\text{st}}^{(\text{in})} = S_{\text{st}}^{(\text{fin})} \leq S_{\text{B}}^{(\text{fin})} \equiv S^{(\text{fin})} \tag{2.102}$$

The statistical entropy does not change, but the thermodynamic entropy increases in agreement with the postulate of entropy increase in Chapter 1. *Behind this simplistic reasoning hides the hypothesis that the entropy is determined by the averages A_i*. For a simple system, these averages are the energy and particle densities, which are of course uniform in the final equilibrium state. At the final equilibrium state, the new averages determine a new set of probability densities, P_m, which can then be used to construct a new statistical entropy (Equation (2.51)). This new statistical entropy is equal to the new thermodynamic (Boltzmann) entropy and different from $S_{\text{st}}^{(\text{in})}$. Having said this, we also emphasize that the result (2.102) depends only on the Hamiltonian character of the evolution. The initial and final energies can well be different, the only constraint is that there be no heat transfer.

In the preceding discussion we did not address the question of what happens between two equilibrium states. In general, and unlike statistical entropy, the thermodynamic entropy is not defined when the system is out of equilibrium in transition between two equilibrium states. However, knowing the average values $A_i(t)$ permits generalizing (2.64) as follows

$$D_{\text{B}}(t) = \frac{1}{Z(t)} \exp\left(\sum_i \lambda_i(t)\mathsf{A}_i\right) \qquad Z(t) = \text{Tr}\exp\left(\sum_i \lambda_i(t)\mathsf{A}_i\right) \tag{2.103}$$

with time dependent Lagrange multipliers and partition function. The density operator $D_{\text{B}}(t)$ is constructed by maximizing $S_{\text{B}}(t) = -k\text{Tr}[D_{\text{B}}(t)\ln D_{\text{B}}(t)]$ subject to the constraints

$$\text{Tr}\left(D_{\text{B}}(t)\mathsf{A}_i\right) = A_i(t)$$

Our discussion can be illustrated with the following example. Consider a container divided into two compartments each of which contains a liquid, one red and one yellow. The vertical partition parallel to the yOz plane is rendered porous, without creating any turbulence, and the two liquids are allowed to mix slowly by diffusion (we assume no chemical reactions). At every instant the densities, $n_{\text{R}}(x,t)$ and

$n_Y(x,t)$, of the red and yellow liquids play the rôle of the average values $A_i(t)$, where the index i stands for R (or Y) and x. The corresponding operators are $\mathsf{n}_R(x)$ and $\mathsf{n}_Y(x)$ with

$$n_{R,Y}(x,t) = \mathrm{Tr}\,[D(t)\mathsf{n}_{R,Y}(x)]$$

The mixing operation is manifestly irreversible: one cannot expect the resulting orange liquid suddenly to separate into the original red and yellow liquids. Whereas the statistical entropy remains constant, since the evolution is Hamiltonian, the Boltzmann entropy increases.

We note that $D_B(t)$ does not satisfy the evolution equation (2.14). The operators A_i that appear in the arguments leading to (2.103) are called *relevant operators*, the $A_i(t)$ *relevant variables* and the Boltzmann entropy $S_B(t)$ is also called *relevant entropy*. If at time $t = t_0$, the density operator $D(t_0)$ is the same as $D_B(t_0)$, i.e. if all the available information at t_0 is contained in the average values $A_i(t_0)$ of the relevant variables at t_0, then an argument identical to the preceding one shows that $S_B(t_1) \geq S_B(t_0)$ for $t_1 > t_0$. However, nothing guarantees that $S_B(t)$ be uniformly increasing in the interval $[t_0, t_1]$. One can, nonetheless, show that $\mathrm{d}S_B(t)/\mathrm{d}t \geq 0$ on condition that the average values $A_i(t)$ obey a system of autonomous (or Markovian) differential equations[37] of the form $\dot{A}_i(t) = f(A_j(t))$. In general, one expects the time derivatives $\dot{A}_i(t)$ to contain memory effects[38] (see Section 9.3). $\dot{A}_i(t)$ depends not only on the values of A_j at time t, but also on all the values taken in the interval $[t_0, t]$ which expresses the indirect influence of the dynamics of irrelevant variables. Under these conditions, a piece of information available at a time preceding t may have repercussions that lead to a momentary reduction of the Boltzmann entropy. There remains the question of the optimal choice of relevant variables. It is recommended to include in the description a sufficient number of variables, in particular all those that obey conservation laws. This will give the best hope of obtaining a system of autonomous differential equations. We illustrate this discussion with the case of a simple fluid in local equilibrium (Chapter 8). The relevant variables are, then, the densities of particles $n(\vec{r},t)$, energy $\epsilon(\vec{r},t)$ and momentum $\vec{g}(\vec{r},t)$ whose evolution is governed by the autonomous partial differential

[37] In general a system of partial differential equations that depends on the gradients of the average values.

[38] For one variable, a simple example of a (linear) equation with memory is

$$\dot{A}(t) = -\int_{t_0}^{t} \mathrm{d}t'\, K(t-t')A(t')$$

The memory kernel $K(t)$ depends on the dynamics of the irrelevant variables. The short memory approximation assumes that $K(t)$ is non-zero only in the interval $0 \leq t \leq \tau^*$, where τ^* is very small compared to the characteristic evolution time τ of the A_i. This approximation gives an autonomous (or Markovian) differential equation $\dot{A}(t) = -cA(t)$, with $c = \int_0^\infty \mathrm{d}t\, K(t)$. What allows us to obtain such an autonomous equation is the separation of the microscopic and macroscopic time scales $\tau^* \ll \tau$. See Section 9.3 for more details on memory effects.

equations of hydrodynamics (Section 6.3). For a dilute gas not at local equilibrium the correct description is that of Boltzmann (Section 8.3) which includes as a relevant variable the distribution function $f(\vec{r}, \vec{p}, t)$.[39] This Boltzmann description is more detailed than the hydrodynamic one since the densities are given by integrals of $f(\vec{r}, \vec{p}, t)$ over $\vec{p}$

$$n(\vec{r}, t) = \int \mathrm{d}^3 p\, f(\vec{r}, \vec{p}, t) \qquad \epsilon(\vec{r}, t) = \int \mathrm{d}^3 p\, \frac{p^2}{2m} f(\vec{r}, \vec{p}, t)$$
$$\vec{g}(\vec{r}, t) = \int \mathrm{d}^3 p\, \vec{p}\, f(\vec{r}, \vec{p}, t)$$

and also because the Boltzmann equation is an autonomous equation for f, whereas away from local equilibrium the densities suffer from memory effects. For a non-dilute gas away from local equilibrium the Boltzmann description does introduce memory effects (see Section 8.3.3 for more details).

In summary, one should understand the difference between S_{st}, which is constant for Hamiltonian evolution because $D(t)$ contains the same information as the operator D^{in}, and S_{B} which, in general, increases with time. With Hamiltonian evolution, we could imagine following the evolution from time $t = 0$ to time $t = T$ and then reversing the direction of time to return to the initial situation (however with $\vec{p} \to -\vec{p}$) at $t = 2T$, thus taking advantage of micro-reversibility and complete knowledge of the state at time T. The information contained in $D_{\mathrm{B}}(T)$ is different, it relies on knowing the average values $A_i(T)$ of a few macroscopic variables. We thus lose an enormous amount of information. To regain the initial situation, not only do we need to know all the average values but also all the correlations. Between $t = 0$ and $t = T$ there has been a leakage of available information toward unobservable degrees of freedom, which manifests itself as an increase in $S_{\mathrm{B}}(t)$. The information at our disposal at $t = T$ is not enough to return to the initial state.

2.7 Exercises

2.7.1 Density operator for spin-1/2

We consider a two-level quantum system, which is the simplest non-trivial case since the Hilbert space is two dimensional. We shall study the density operator in this space. There are many applications of this example: the polarization of a massive spin-1/2 particle, photon polarization, two-level atoms, etc. Since a very common application of this system is that of spin-1/2, we shall consider this example in order to fix ideas and terminology. We choose two basis vectors of state

[39] $f(\vec{r}, \vec{p}, t)\mathrm{d}^3 r\, \mathrm{d}^3 p$ is the number of particles at time t contained in the phase space volume $\mathrm{d}^3 r \mathrm{d}^3 p$.

space, $|+\rangle$ and $|-\rangle$, which can be for example the eigenvectors of the z-component of spin. In this basis, the density operator is represented by the 2×2 density matrix D.

1. The density matrix has the general form of a Hermitian 2×2 matrix of unit trace

$$D = \begin{pmatrix} a & c \\ c^* & 1-a \end{pmatrix} \tag{2.104}$$

where a is a real number and c is complex. Show that the positivity of D (and thus of its eigenvalues) adds the following constraint on the matrix elements:

$$0 \le a(1-a) - |c|^2 \le \frac{1}{4} \tag{2.105}$$

Show that the necessary and sufficient condition for D to describe a pure case is $a(1-a) = |c|^2$. Calculate a and c for the density matrix that describes the normalized state vector $|\psi\rangle = \alpha|+\rangle + \beta|-\rangle$ with $|\alpha|^2 + |\beta|^2 = 1$, and verify that $|\psi\rangle$ represents a pure state.

2. Show that we may write D in terms of a vector $\vec{b}$ (called the Bloch vector)

$$D = \frac{1}{2}\begin{pmatrix} 1+b_z & b_x - \mathrm{i}b_y \\ b_x + \mathrm{i}b_y & 1-b_z \end{pmatrix} = \frac{1}{2}\left(1 + \vec{b}\cdot\vec{\sigma}\right) \tag{2.106}$$

as long as $|\vec{b}|^2 \le 1$. We introduce the vector operator $\vec{\sigma}$ whose components are the Pauli matrices σ_i

$$\sigma_x = \begin{pmatrix} 0 & 1 \\ 1 & 0 \end{pmatrix} \qquad \sigma_y = \begin{pmatrix} 0 & -\mathrm{i} \\ \mathrm{i} & 0 \end{pmatrix} \qquad \sigma_z = \begin{pmatrix} 1 & 0 \\ 0 & -1 \end{pmatrix}$$

Show that the pure case corresponds to $|\vec{b}|^2 = 1$. The pure case $|\vec{b}| = 1$ is also called totally polarized, the case $\vec{b} = 0$ not polarized or with zero polarization. To interpret physically the vector $\vec{b}$, we calculate the average value of the spin (measured in units of $\hbar$) $\vec{S} = \frac{1}{2}\vec{\sigma}$. By using

$$\sigma_i \sigma_j = \delta_{ij} + \mathrm{i}\varepsilon_{ijk}\sigma_k \tag{2.107}$$

show that

$$\mathrm{Tr}\left(D\frac{1}{2}\sigma_i\right) = \frac{1}{2}b_i \tag{2.108}$$

$\frac{1}{2}\vec{b}$ is then the average value of the spin.

3. When the spin is placed in a constant magnetic field $\vec{B}$, the Hamiltonian is $H = -\gamma\,\vec{S}\cdot\vec{B}$, where γ is the gyromagnetic ratio. For an electron of charge q and mass m, we have $\gamma \simeq q\hbar/m$. Assuming that $\vec{B}$ is parallel to the Oz axis, $\vec{B} = (0, 0, B)$, use Equation (2.14) to calculate the equation of motion for D and show that the vector $\vec{b}$ precesses around $\vec{B}$ with an angular frequency to be determined. Compare with the case of a classical magnetic moment (A.24).

4. Consider now two spin-1/2 objects coupled in a singlet state of zero total angular momentum[40]

$$\frac{1}{\sqrt{2}}(|+-> - |-+>) \tag{2.109}$$

Calculate the density operator of the first spin by taking the partial trace over the second spin. What type of density matrix do you obtain? Show that the density matrix of the total system cannot be a tensor product.

2.7.2 Density of states and the dimension of space

1. Consider a spinless particle of mass m constrained to move in a region of volume L^d in d-dimensional space with $d = 1, 2$ or 3. For each value of d calculate the density of states in the following three cases:

(*i*) non-relativistic particle, $\varepsilon = p^2/2m$;
(*ii*) ultra-relativistic particle, $\varepsilon = pc$;
(*iii*) relativistic particle of total energy $\mathcal{E} = \sqrt{p^2c^2 + m^2c^4}$.

2. In the non-relativistic case for $d = 2$, draw the curve $\Phi(p)$ that is the number of states whose momentum modulus is less than or equal to p. With the help of this curve, show that the density of states does not depend on the boundary conditions imposed on the wave function: periodic boundary conditions $p_i = n_i(2\pi\hbar/L_i)$ ($n_i \in \mathbb{Z}$), fixed boundary condition $p_i = n_i(\pi\hbar/L_i)$ ($n_i \in \mathbb{N}$).

2.7.3 Liouville theorem and continuity equation

We shall show here that, due to Liouville's theorem, Equation (2.43) is in fact a continuity equation in phase space of the type studied in detail in Chapter 6. For simplicity, we focus on two variables p and q, the generalization to N variables being obvious. The velocity in phase space is a two-component vector $\vec{v}$: $v_p = \dot{p}$, $v_q = \dot{q}$, and probability conservation may be expressed in terms of the continuity equation in phase space (see (6.6))

$$\frac{\partial D}{\partial t} + \vec{\nabla} \cdot (D\vec{v}) = 0 \tag{2.110}$$

[40] Such states, called 'entangled states', play a crucial rôle in discussions of fundamental problems of quantum mechanics such as the Einstein–Podolsky–Rosen (EPR) effect.

where, for a two-dimensional vector $\vec{A}$ of components (A_p, A_q), we have

$$\vec{\nabla} \cdot \vec{A} = \frac{\partial A_p}{\partial p} + \frac{\partial A_q}{\partial q}$$

Show that

$$\frac{\partial \dot{p}}{\partial p} + \frac{\partial \dot{q}}{\partial q} = 0 \tag{2.111}$$

from which you may deduce

$$\frac{\partial D}{\partial t} + \frac{\partial D}{\partial p}\dot{p} + \frac{\partial D}{\partial q}\dot{q} = \frac{\partial D}{\partial t} + \{H, D\} = 0 \tag{2.112}$$

2.7.4 Loaded dice and statistical entropy

Consider a die with the six faces numbered from 1 to 6. The average obtained after a large number of throws is 4 instead of the 3.5 expected from a fair die. If we have no other information, what is the probability of throwing each of the faces of the die?

2.7.5 Entropy of a composite system

Consider a system made of two interacting parts (a) and (α), described by a density operator D acting in the Hilbert space of states $\mathcal{H}^{(a)} \otimes \mathcal{H}^{(\alpha)}$. We define with the help of partial traces (2.54) the density operators $D^{(a)}$ and $D^{(\alpha)}$ of the two subsystems (a) and (α). Show that

$$\ln\left(D^{(a)} \otimes D^{(\alpha)}\right) = \ln\left(D^{(a)} \otimes \mathbb{I}^{(\alpha)}\right) + \ln\left(\mathbb{I}^{(a)} \otimes D^{(\alpha)}\right) \tag{2.113}$$

where $\mathbb{I}^{(a)}$ and $\mathbb{I}^{(\alpha)}$ are the identity operators acting in the spaces $\mathcal{H}^{(a)}$ and $\mathcal{H}^{(\alpha)}$. Show also that

$$\mathrm{Tr}\, D \ln\left(D^{(a)} \otimes \mathbb{I}^{(\alpha)}\right) = \mathrm{Tr}_a\, D^{(a)} \ln D^{(a)} \tag{2.114}$$

and establish the inequality (2.55). Hint: Introduce the density operator $D' = D^{(a)} \otimes D^{(\alpha)}$ and use (2.56). Give the interpretation of the inequality (2.55).

2.7.6 Heat exchanges between system and reservoir

Equation (2.81) was obtained from the description of the heat exchange between system and reservoir in the setup shown in Figure 2.1. Show that it may also be put in the form

$$\frac{đQ}{\mathrm{d}t} = -\frac{1}{\mathrm{i}\hbar}\,\mathrm{Tr}_{\mathrm{tot}}\left(D_{\mathrm{tot}}[H_R, \mathcal{V}]\right) \tag{2.115}$$

Discuss this relation.

2.7.7 Galilean transformation

Consider a system of mass M at equilibrium and moving at a velocity u with respect to the laboratory frame R. For simplicity we consider the one-dimensional case. We consider an ensemble where the average values of the Hamiltonian H and momentum P are fixed. Show that the density operator is the same in the frame R as in the frame R' where the system is at rest

$$D = \frac{1}{Z}\,\mathrm{e}^{-\beta H + \lambda \mathsf{P}} = \frac{1}{Z'}\,\mathrm{e}^{-\beta' H'} \tag{2.116}$$

The primed quantities are measured in R' and λ is the Lagrange multiplier conjugate to P.

By using (2.116) show that: (*i*) $\beta = \beta'$, (*ii*) $\beta u = \lambda$, (*iii*) entropy is unchanged, (*iv*) the average value of the momentum is given by

$$\langle \mathsf{P} \rangle = \frac{\partial \ln Z}{\partial \lambda} = Mu \tag{2.117}$$

and calculate the energy E' in terms of E.

2.7.8 Fluctuation-response theorem

1. *Preliminary operator identities.* The exponential of an operator A is defined, in the same way as for a number, by the Taylor series

$$\mathrm{e}^{\mathsf{A}} = \sum_0^\infty \frac{1}{n!}\mathsf{A}^n$$

However, care must be taken because identities valid for numbers are no longer valid for operators. For example, in general for two operators A and B we have

$$\mathrm{e}^{\mathsf{A}}\mathrm{e}^{\mathsf{B}} \neq \mathrm{e}^{\mathsf{B}}\mathrm{e}^{\mathsf{A}} \qquad \mathrm{e}^{\mathsf{A}+\mathsf{B}} \neq \mathrm{e}^{\mathsf{A}}\mathrm{e}^{\mathsf{B}}$$

A sufficient condition to re-establish the equality of both sides is $[\mathsf{A}, \mathsf{B}] = 0$. To evaluate $\exp(\mathsf{A}+\mathsf{B})$, we will consider the operator

$$K(x) = \mathrm{e}^{x(\mathsf{A}+\mathsf{B})}\,\mathrm{e}^{-x\mathsf{A}}$$

Show that

$$\frac{\mathrm{d}K(x)}{\mathrm{d}x} = \mathrm{e}^{x(\mathsf{A}+\mathsf{B})}\mathsf{B}\,\mathrm{e}^{-x\mathsf{A}}$$

Verify therefore that

$$\mathrm{e}^{\mathsf{A}+\mathsf{B}} = \mathrm{e}^{\mathsf{A}} + \int_0^1 \mathrm{d}x\, \mathrm{e}^{x(\mathsf{A}+\mathsf{B})}\,\mathsf{B}\,\mathrm{e}^{(1-x)\mathsf{A}} \tag{2.118}$$

Application: If $\lambda \to 0$, show that

$$\mathrm{e}^{\mathsf{A}+\lambda\mathsf{B}} = \mathrm{e}^{\mathsf{A}} + \lambda \int_0^1 \mathrm{d}x\, \mathrm{e}^{x\mathsf{A}}\,\mathsf{B}\,\mathrm{e}^{(1-x)\mathsf{A}} + \mathcal{O}(\lambda^2) \tag{2.119}$$

Express $\mathrm{Tr}[\exp(\mathsf{A}+\lambda\mathsf{B})]$ in the same approximation. Note that, in general, we *cannot* write

$$\mathrm{e}^{\mathsf{A}+\lambda\mathsf{B}} = \mathrm{e}^{\mathsf{A}}\,\mathrm{e}^{\lambda\mathsf{B}} \simeq \mathrm{e}^{\mathsf{A}}(1+\lambda\mathsf{B})$$

but that taking the trace of this incorrect equation yields the right answer! Using (2.118), and for an operator A which depends on a parameter λ, show that

$$\frac{\mathrm{d}}{\mathrm{d}\lambda}\mathrm{e}^{\mathsf{A}(\lambda)} = \int_0^1 \mathrm{d}x\, \mathrm{e}^{x\mathsf{A}(\lambda)}\,\frac{\mathrm{d}\mathsf{A}}{\mathrm{d}\lambda}\,\mathrm{e}^{(1-x)\mathsf{A}(\lambda)} \tag{2.120}$$

from which you establish (2.57) for $\mathrm{d}(\mathrm{Tr}[\exp\mathsf{A}(\lambda)]/\mathrm{d}\lambda)$.

2. Let D be a density operator in the general form (2.64)

$$D = \frac{1}{Z}\exp\left(\sum_k \lambda_k \mathsf{A}_k\right)$$

Show that, with the notation $A_i = \langle \mathsf{A}_i \rangle$, the fluctuation-response theorem may be written in general as

$$C_{ij} = \frac{\partial A_i}{\partial \lambda_j} = \frac{\partial^2 \ln Z}{\partial\lambda_i \partial\lambda_j} = \int_0^1 \mathrm{d}x\, \mathrm{Tr}\Big[(\mathsf{A}_i - A_i)D^x(\mathsf{A}_j - A_j)D^{(1-x)}\Big] \tag{2.121}$$

Deduce from (2.121) that C_{ij} is a positive matrix, i.e.

$$\sum_{ij} a_i a_j C_{ij} \geq 0 \qquad \forall a_i$$

Consequence: $\ln Z$ is a convex function of the λ_i.

2.7.9 Phase space volume for N free particles

Calculate the number of accessible levels $\Omega(E)$, Equation (2.100), for a system of N particles of mass m in volume V and whose energy is between E and $E + \Delta E$. To start, calculate the accessible number of states $\Phi(E)$ when the particles' total energy is less than or equal to E. Use a dimensional argument to show that $\Phi(E)$ can be written in the form

$$\Phi(E) = \frac{V^N}{N!h^N}\left(\frac{2mE}{N}\right)^{3N/2} C(N) \tag{2.122}$$

and determine the dimensionless constant $C(N)$. Hint: As a preliminary result, calculate the following integral in polar coordinates

$$I_n = \int \mathrm{d}x_1 \ldots \mathrm{d}x_n \mathrm{e}^{-(x_1^2+\cdots+x_N^2)}$$

Use this formula to determine the surface area of the sphere S_n with unit radius in n dimensions

$$S_n = \frac{2\pi^{n/2}}{\Gamma(n/2)} \tag{2.123}$$

2.7.10 Entropy of mixing and osmotic pressure

In this exercise we study the properties of a dilute solution.[41] N_B molecules of type B (the solute) are in solution in N_A molecules of type A (the solvent) with $N_A \gg N_B \gg 1$. This being a dilute solution, we ignore the interactions between solute molecules but not between solute and solvent molecules.

1. Let δS be the entropy change upon adding a B molecule to the initially pure solvent. Show that

$$\delta S = k \ln N_A + \text{ terms independent of } N_A$$

and that if we add N_B molecules we have

$$\delta S = kN_B \ln \frac{\mathrm{e}N_A}{N_B} + \text{ terms independent of } N_A$$

2. In this part we write the form of the total Gibbs potential. If we add a B molecule to the initially pure solvent, the change δG in G at *fixed temperature and pressure* is

$$\delta G = \delta E - T\delta S + P\delta V$$

[41] This exercise is inspired by Schroeder [114].

Show that δE and $P\delta V$ do not depend on N_A and that we can write

$$\delta G = f(T, P) - kT \ln N_A$$

Justify the following form for the Gibbs potential of 'solvent + solute'

$$G = N_A \mu_0(T, P) + N_B f(T, P) - N_B kT \ln \frac{\mathrm{e} N_A}{N_B} \tag{2.124}$$

where $\mu_0(T, P)$ is the Gibbs potential of the pure solvent. Verify that G is indeed an extensive quantity under the change of scale $N_A \to \lambda N_A$, $N_B \to \lambda N_B$. Deduce from (2.124) the chemical potentials μ_A and μ_B of the solvent and solute.

3. Consider the following setup: a semi-permeable membrane, permeable to A but not to B molecules, divides a container into two compartments. The left compartment contains pure solvent, the right one contains solution, and the temperature is uniform. Assuming the difference in pressure is small between the left compartment, at P_0, and the right one, at P_1, show that

$$P_1 - P_0 \simeq \frac{N_B kT}{V}$$

where V is the volume of the right compartment. Hint: Expand in a Taylor series to first order in $(P_1 - P_0)$. This pressure difference is the *osmotic pressure*.

2.8 Further reading

A discussion of the foundations of quantum mechanics and further details on the density operator are found in Cohen-Tannoudji *et al.* [30] (Chapter III) and in Messiah [89] (Chapter VII). For the Hamiltonian formulation of classical mechanics, see Goldstein [48] (Chapters VII and VIII) and Landau and Lifschitz [69] (Chapter VII). The discussion in Sections 2.4 and 2.5 follows that in Balian [5] (Chapters 2 to 5) which contain further useful details. The reader interested in questions of irreversibility should consult the two papers by Lebowitz [74, 75], Bricmont [20] or Schwabl [115] (Chapter 10). For a less technical approach, see the books by Feynman [39] (Chapter 5) or Penrose [100] (Chapter 7). Excellent discussion of this problem may also be found in Kreuzer [67] (Chapter 11) and numerical simulations in Levesque and Verlet [79]. The spin echo experiments are discussed in Balian [5] (Chapter 15). See also Ma [85] (Chapter 24). An opposing viewpoint concerning irreversibility is discussed in Prigogine [104]. As a more advanced complement to Section 2.6.3, we recommend the articles by Rau and Müller [107] and Balian [6].

3

Canonical and grand canonical ensembles: applications

This chapter is devoted to some important applications of the canonical and grand canonical formalisms. We shall concentrate on situations where the effects of quantum statistics, to be studied in Chapter 5, may be considered as negligible. The use of the formalism is quite straightforward when one may neglect the interactions between the elementary constituents of the system. Two very important examples are the ideal gas (non-interacting molecules) and paramagnetism (non-interacting spins), which are the main subjects of Section 3.1. In the following section, we show that at high temperatures one may often use a semi-classical approximation, which leads to simplifications in the formalism. Important examples are the derivation of the Maxwell velocity distribution for an interacting classical gas and of the equipartition theorem. These results are used in Section 3.3 to discuss the behaviour of the specific heat of diatomic molecules. In Section 3.4 we address the case where interactions between particles cannot be neglected, for example in a liquid, and we introduce the concept of pair correlation function. We show how pressure and energy may be related to this function, and we describe briefly how it can be measured experimentally. Section 3.5 shows the fundamental rôle played by the chemical potential in the description of phase transitions and of chemical reactions. Finally Section 3.6 is devoted to a detailed exposition of the grand canonical formalism, including a discussion of fluctuations and of the equivalence with the canonical ensemble.

3.1 Simple examples in the canonical ensemble

3.1.1 Mean values and fluctuations

In the preceding chapter, we defined the canonical ensemble as the equilibrium distribution (2.64) with only one relevant operator, the Hamiltonian, and consequently only one Lagrange multiplier $\lambda_1 = -\beta$ ($\beta = 1/kT$). Recall the main results of

Sections 2.4.4 and 2.5.4. The density operator D_{B} and the partition function are given by

$$D_{\mathrm{B}} = \frac{1}{Z}\mathrm{e}^{-\beta H} = \frac{1}{Z}\sum_r |r\rangle \mathrm{e}^{-\beta E_r}\langle r| \tag{3.1a}$$

$$Z = \mathrm{Tr}\,\mathrm{e}^{-\beta H} = \sum_r \mathrm{e}^{-\beta E_r} \tag{3.1b}$$

$$Z = \int \mathrm{d}E\,\rho(E)\,\mathrm{e}^{-\beta E} \tag{3.1c}$$

with $\beta = (kT)^{-1}$, $H|r\rangle = E_r|r\rangle$, $\rho(E)$ is the level density and we draw the reader's attention to the very important Footnote 15 in Chapter 2.

In the canonical ensemble, we know with precision the values of a set of external parameters[1] $x = \{x_i\}$, controllable from outside the system under consideration, for example the volume, the electric or magnetic field components, etc. We now undertake to explain precisely the rôle they play. The Hamiltonian depends on these external parameters: $H = H(x)$, but only its average value, $E(\beta, x)$ is known. According to (2.66), and writing explicitly the external parameters, $E(\beta, x)$ can be written as

$$E(\beta, x) = -\left.\frac{\partial \ln Z}{\partial \beta}\right|_x \tag{3.2}$$

For a quasi-static process, and for each external parameter x_i, we can obtain, with the help of the equilibrium distribution (3.1) and Equation (2.95), the average value of the corresponding conjugate variable (or 'conjugate force') X_i

$$X_i = \left\langle \frac{\partial H}{\partial x_i} \right\rangle = -\frac{1}{\beta}\left.\frac{\partial \ln Z}{\partial x_i}\right|_{\beta, x_{j\neq i}} \tag{3.3}$$

We remark that in calculating macroscopic quantities at thermodynamic equilibrium, $\ln Z$ plays the central rôle, and not Z. The logarithm of the partition function is in fact the 'good' thermodynamic potential for a system described by the canonical distribution: $\ln Z$ is an extensive quantity, which is not true for Z. Equating, in this chapter, the Boltzmann entropy with the thermodynamic entropy (cf. Section 2.5.2) we can write Equation (2.67) as

$$\ln Z = \frac{S}{k} - \beta E \tag{3.4}$$

[1] We reserve the term 'external parameter' for quantities that appear in Equation (2.94) for the work. The number of particles, N, which might appear at first sight to be an external parameter in the canonical ensemble, does not satisfy our definition.

This expression allows us to complete (2.68b) by adding the contributions of the conjugate forces. Using Equations (3.2) and (3.3) we obtain the entropy differential

$$
\begin{aligned}
\frac{\mathrm{d}S}{k} &= \mathrm{d}\ln Z + \mathrm{d}(\beta E) = -E\,\mathrm{d}\beta - \beta\sum_i X_i\,\mathrm{d}x_i + \beta\,\mathrm{d}E + E\mathrm{d}\beta \\
&= \beta\,\mathrm{d}E - \beta\sum_i X_i\,\mathrm{d}x_i
\end{aligned}
$$

which establishes the general form of the '$T\mathrm{d}S$ equation' in the canonical ensemble[2]

$$
\boxed{T\mathrm{d}S = \mathrm{d}E - \sum_i X_i\,\mathrm{d}x_i} \tag{3.5}
$$

When we use the variable T instead of β, the relevant thermodynamic potential is represented by the free energy $F(T, x)$

$$
\boxed{F = -\frac{1}{\beta}\ln Z \qquad \mathrm{d}F = -S\,\mathrm{d}T + \sum_i X_i\,\mathrm{d}x_i} \tag{3.6}
$$

The conjugate forces are obtained by taking the derivatives of the free energy with respect to external parameters. As an example, we obtain the pressure by considering $-V$ as an external parameter

$$
P = \frac{1}{\beta}\left.\frac{\partial\ln Z}{\partial V}\right|_{\beta,N} = -\left.\frac{\partial F}{\partial V}\right|_{T,N} \tag{3.7}
$$

Although N is not considered an external parameter, we can clearly calculate the chemical potential in the canonical ensemble by using its thermodynamic definition (1.11) and Equation (3.6)

$$
\mu = \left.\frac{\partial F}{\partial N}\right|_{T,V} = -\frac{1}{\beta}\left.\frac{\partial\ln Z}{\partial N}\right|_{\beta,V} \tag{3.8}
$$

To conclude, we examine fluctuations in the canonical ensemble. The fluctuation-response theorem can be written here as

$$
\left\langle H^2\right\rangle - E^2 = \left\langle (H-E)^2\right\rangle = \left.\frac{\partial^2\ln Z}{\partial\beta^2}\right|_x \tag{3.9}
$$

[2] This equation can be also obtained from (3.4), which shows that $\ln Z$ is the Legendre transform of the entropy with respect to the energy: $\ln Z$ is a Massieu function.

$$
\ln Z \equiv \Phi_1'(\beta, x) = \frac{1}{k}\Phi_1(1/T, x) \qquad \mathrm{d}\ln Z = -E\,\mathrm{d}\beta - \beta\sum_i X_i\,\mathrm{d}x_i
$$

where Φ_1 is the Massieu function defined in (1.29); also see Exercise 1.6.1.

By using (3.2) and also taking $\Delta H = H - E$, Equation (3.9) becomes

$$\left\langle (\Delta H)^2 \right\rangle = - \left. \frac{\partial E}{\partial \beta} \right|_x = kT^2 C_x \geq 0 \tag{3.10}$$

where C_x is the specific heat at constant x (generally C_V). This demonstrates the positivity of C which is required for stability (cf. (1.58)).

We see in (3.2) and (3.9) the rôle played by $\ln Z$ in generating, by differentiation, the moments of physical quantities. However, we caution against a hasty generalization[3]

$$\left\langle (\Delta H)^n \right\rangle \neq \left. \frac{\partial^n \ln Z}{\partial \beta^n} \right|_x \qquad \text{for } n > 3$$

3.1.2 Partition function and thermodynamics of an ideal gas

The canonical ensemble provides us the occasion to re-examine the concept of an ideal gas, introduced in Chapter 2, and to discuss its fundamental hypotheses. As we saw in Section 2.1.3, the quantum state of a molecule in a mono-atomic gas is determined by three integer quantum numbers (n_x, n_y, n_z) that give the components of the wave vector $\vec{k}$: $k_\alpha = (2\pi n_\alpha)/L_\alpha$. In the ideal gas approximation, the energy (entirely kinetic) of a molecule is given by $\varepsilon \sim E/N$. The *average occupation number*[4] $\langle n_r \rangle$ of a level r, i.e. the average number of molecules occupying the energy level ε_r, is of the order of $N/\Phi(\varepsilon)$, where $\Phi(\varepsilon)$ is the number of energy states below ε calculated in (2.22). In other words, to a first approximation, we assume that the molecules are equally divided among all the levels whose energy is less than ε. By using (2.22) and neglecting numerical factors of order unity, we then have

$$\langle n_r \rangle \sim \frac{N}{\Phi(\varepsilon)} \sim \frac{V}{d^3} \frac{h^3}{V} \frac{1}{(2m\varepsilon)^{3/2}}$$

where $d \sim (V/N)^{1/3} = n^{-1/3}$ is the average distance between molecules. We note that $h/(2m\varepsilon)^{1/2}$ is the de Broglie wavelength of a molecule of energy ε. In anticipation of a result that we shall show very soon, $\varepsilon \sim kT$, we define the *thermal*

[3] In fact $\ln Z$ is the generating function of *cumulants* of H (see for example Ma [85] Chapter 12)

$$\left\langle H^n \right\rangle_c = \frac{\partial^n \ln Z}{\partial \beta^n}$$

Only for $n = 2$ and $n = 3$ do we have $\left\langle (\Delta H)^n \right\rangle = \left\langle H^n \right\rangle_c$.

[4] Mathematically: $\langle n_r \rangle = \mathrm{Tr}\, D \mathsf{n}_r$, where n_r is the particle number operator in state r.

wavelength λ by

$$\boxed{\lambda = \frac{h}{\sqrt{2\pi mkT}} = \sqrt{\frac{\beta h^2}{2\pi m}}} \tag{3.11}$$

where the numerical factor has been chosen in order to simplify subsequent expressions. We can then write

$$\langle n_r \rangle \sim \left(\frac{\lambda}{d}\right)^3 \tag{3.12}$$

When $T \to \infty$, $\lambda \to 0$ and $\langle n_r \rangle \to 0$ at constant density, the probability that a state be doubly occupied is negligible, and so *a priori* are the quantum effects. For example, if the molecules are fermions, the Pauli exclusion principle will have negligible effect under these conditions. The wavelength, λ, is a measure of the 'quantum size' of the molecules, and the fact that $\lambda \ll d$ means that the wave packets of different molecules do not overlap. We may thus consider the molecules individually as in classical mechanics. So, the classical approximation will be valid when $\lambda \ll d$. We shall see that in general, in statistical mechanics, a high enough temperature (compared to a given scale) will yield the conditions necessary for this approximation to be valid. In this limit, the partition function for an ideal gas takes the form (2.87)[5]

$$\boxed{Z_N = \frac{V^N}{N!}\left(\frac{2\pi m}{\beta h^2}\right)^{3N/2} = \frac{1}{N!}\left(\frac{V}{\lambda^3}\right)^N} \tag{3.13}$$

[5] The second expression in (3.13) permits a probabilistic interpretation of the partition function. We can consider V/λ^3 as the number of 'boxes', M, in volume V; M^N is then the number of distinct ways to distribute N distinguishable 'marbles' in M labeled boxes with an arbitrary number of marbles per box. We also remark that

$$\langle n_r \rangle \sim \frac{N}{(V/\lambda^3)} = \frac{N}{M}$$

where N/M is indeed the average number of marbles per box. We shall justify in Chapter 5 that classical statistical mechanics is valid in the limit $\langle n_r \rangle \ll 1$. In this limit, we can justify the factor $1/N!$ by making the marbles indistinguishable. By considering the two cases to which Chapter 5 will give physical meaning, the combinatorics is written as

(i) an arbitrary number of marbles per box ($\equiv$ Bose–Einstein statistics)

$$\frac{(M+N-1)!}{N!(M-1)!} \xrightarrow{N/M \ll 1} \frac{M^N}{N!}$$

(ii) at most one marble per box ($\equiv$ Fermi–Dirac statistics)

$$\frac{M!}{N!(M-N)!} \xrightarrow{N/M \ll 1} \frac{M^N}{N!}$$

It is straightforward to calculate thermodynamic functions starting with Z_N, or F, if we use the Stirling approximation

$$\ln N! \simeq N \ln N - N = N \ln \frac{N}{\mathrm{e}} \qquad N \gg 1$$

By introducing the specific volume, $v = V/N$, $\ln Z_N$ becomes

$$\begin{aligned} \ln Z_N &= N\left[\ln \frac{v}{\lambda^3} + 1\right] \\ &= N\left[\ln \frac{V}{N} + \frac{3}{2}\ln\left(\frac{2\pi m}{\beta h^2}\right) + 1\right] \end{aligned} \tag{3.14}$$

The average energy is obtained by using (3.2)

$$E = \frac{3N}{2\beta} = \frac{3}{2}NkT \tag{3.15}$$

It is worth noting that the specific heat at constant volume, $C_V = (3/2)Nk$, does not obey the third law. This should not be surprising within the framework of the classical approximation we used here. The free energy $F(T, V, N)$ is given by

$$\begin{aligned} F(T, V, N) &= -NkT\left[\ln \frac{v}{\lambda^3} + 1\right] \\ &= -NkT\left[\ln \frac{V}{N} + \frac{3}{2}\ln\left(\frac{2\pi mkT}{h^2}\right) + 1\right] \end{aligned} \tag{3.16}$$

from which we obtain the entropy of the ideal gas

$$S = -\left.\frac{\partial F}{\partial T}\right|_{V,N} = kN\left[\ln \frac{V}{N} + \frac{3}{2}\ln\left(\frac{2\pi mkT}{h^2}\right) + \frac{5}{2}\right] \tag{3.17}$$

This result agrees with that obtained in (2.90b) using the Legendre transform (3.4). The expression for the pressure follows easily

$$P = -\left.\frac{\partial F}{\partial V}\right|_{T} = \frac{N}{V}kT = nkT$$

where $n = N/V = 1/v$ is the density. This is the 'ideal gas law'. With this expression for the pressure along with Equation (3.15), we verify Equation (2.28) $PV = (2/3)E$. Furthermore, we note that the arbitrary constant $S(V_0, T_0)$ of Problem 1.7.1, which is indeterminate in thermodynamics, is now fixed. We give also the expression for the chemical potential of an ideal gas

$$\mu = \left.\frac{\partial F}{\partial N}\right|_{T,V} = -kT\ln\left(\frac{V}{N}\left(\frac{2\pi mkT}{h^2}\right)^{3/2}\right) = -kT\ln\left(\frac{v}{\lambda^3}\right) = kT\ln(n\lambda^3) \tag{3.18}$$

The chemical potential is negative if $v > \lambda^3$. In fact, this condition is always satisfied because the classical approximation that we have made assumes $v \gg \lambda^3$. We can understand this negative value of μ by starting from $\mu/T = -(\partial S/\partial N)_{E,V}$, which, for large N, is approximately equal to $\mu/T = -(S(N+1) - S(N))$ at constant E and V. If we add a molecule with zero kinetic energy to the system, neither the energy nor the volume of the system will change, but the number of accessible states will increase;[6] therefore $S(N+1) > S(N)$ and μ is thus negative. When we take into consideration the potential energy due to molecular interactions, the preceding argument is no longer valid because adding a molecule, even if it has zero kinetic energy, will change the energy of the gas. Under these conditions, the chemical potential can become positive, which in fact happens to the ideal Fermi gas but for different reasons (Chapter 5).

Finally, we give a few typical orders of magnitude. The average kinetic energy of a molecule is obtained from the average energy (3.15)

$$\frac{1}{2}m\left\langle v^2\right\rangle = \frac{E}{N} = \frac{3}{2}kT$$

We can then estimate the mean square velocity $\sqrt{\langle v^2\rangle}$ of a molecule, which we express in units of the velocity of the lightest molecule H_2

$$\sqrt{\langle v^2\rangle} = \sqrt{\frac{3kT}{m}} = \sqrt{\frac{m_{H_2}}{m}}\sqrt{\frac{3kT}{2m_p}} \tag{3.19}$$

where we used $m_{H_2}/m_p \approx 2$ with m_p the proton mass. For a temperature of $T = 300\,\text{K}$, the numerical value is $\sqrt{\langle v^2\rangle} \approx 1.93 \times 10^3$ m s^{-1} for hydrogen, and for other molecules it suffices to divide by $\sqrt{m/m_{H_2}}$. For example, for oxygen, with $m_{O_2} \approx 16 m_{H_2}$, we find a speed of 500 m s^{-1}. Under standard temperature and pressure, the molecular volume is 22.4 litres, and the density is $n \approx 2.7 \times 10^{25}$ molecule/m^3 which gives the mean inter-molecular distance $d \sim n^{-1/3} \sim 3 \times 10^{-9}$m. The thermal wavelength of hydrogen at $T = 300\,\text{K}$ is 2×10^{-10} m and it is four times smaller for oxygen: the condition $\lambda \ll d$ is well satisfied.

3.1.3 Paramagnetism

Another very important, and in some cases simple, illustration of the canonical formalism is the study of magnetism. We can relate the macroscopic magnetic properties to the collective behaviour of the microscopic magnetic moments. Some

[6] An estimate of the number of accessible states is $\Omega_N(E) \approx (\Phi(E/N))^N$ where Φ is given by (2.22). See Exercise 2.7.9 for an exact expression.

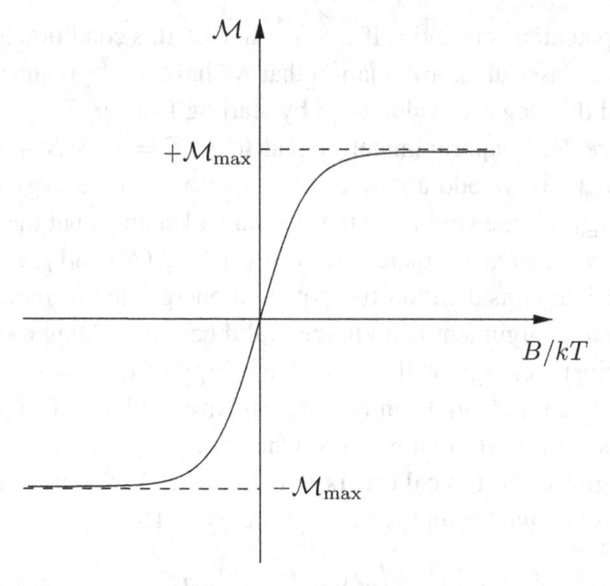

Figure 3.1 Magnetization curve for a paramagnetic material.

materials that do not exhibit spontaneous macroscopic magnetization can be magnetized by subjecting them to an external magnetic field (Figure 3.1). When the macroscopic magnetic moment thus created is aligned with the magnetic field, and when the magnetization does not persist upon removal of the external field, we are dealing with *paramagnetism*. We only describe here the *paramagnetic crystal* model, which is an idealization of a real crystal where we ignore all interactions among the atoms and all degrees of freedom not related to the angular momentum of the atom (or ion). The magnetic field $\vec{B}$ is the external parameter of the problem. Under these conditions, the energy is due entirely to the interaction of the individual moments with the external field. The total Hamiltonian of a system of N magnetic moments is the sum of the individual energies[7]

$$H = \sum_{i=1}^{N} (-\vec{\mu}_i \cdot \vec{B}) \equiv \sum_{i=1}^{N} h_i \tag{3.20}$$

Each atom carries a magnetic moment directly proportional to its angular momentum, $\vec{J}$, measured in units of $\hbar$

$$\vec{\mu} = \gamma \vec{J} \tag{3.21}$$

where γ is the *gyromagnetic factor*. The relationship between magnetic moment $\vec{\mu}$ and angular momentum is well known in classical physics. The orbital rotation of

[7] We use μ for the magnetic moment unless there is confusion with the chemical potential in which case the magnetic moment is denoted by $\tilde{\mu}$.

a charge q, subject to a central force, induces a magnetic moment $\vec{\mu} = (q\hbar/2m)\vec{J}$ ($\gamma_{\text{class}} = q\hbar/2m$), where m is the mass of the orbiting particle.[8] However, classical mechanics does not provide a correct description of magnetism in matter. One must also take into account the magnetism associated with the intrinsic angular momentum of the particles, also called the spin of the particles, the main rôle being played by the electron magnetic moment. The electron has spin-1/2, and the absolute value of its gyromagnetic factor is about twice the *Bohr magneton* $\mu_B = e\hbar/2m_e$, where $-e$ ($e > 0$) and m_e are the charge and mass of the electron.[9]

For illustration, we will take a model system of spin-1/2 where the components of the spin operator $\vec{S}$ (or $\vec{J}$) are proportional to the Pauli matrices σ_i

$$\vec{S} = \frac{1}{2}\vec{\sigma}$$

Even with the simplicity of this model, the calculation of the partition function (3.1b) of the paramagnetic crystal

$$Z = \mathrm{Tr}\,\mathrm{e}^{-\beta H} = \sum_{\{|l\rangle\}} \langle l|\mathrm{e}^{-\beta H}|l\rangle \tag{3.22}$$

may appear to be less than obvious due to the very large number (2^N) of possible configurations. But Hamiltonian (3.20) is the sum of independent and identical Hamiltonians, in which case, as we have shown in (2.74) and also used to establish (2.87) for the ideal gas, the partition function factorizes[10]

$$Z = \zeta^N \qquad \zeta = \mathrm{Tr}\,\mathrm{e}^{-\beta h} \tag{3.23}$$

where ζ is the partition function for a single spin. While the trace of an operator is independent of the basis used, it is simpler to evaluate this trace in the basis of the eigenvectors of the operator, especially if we want the trace of its exponential. It is particularly easy to do this here by choosing the z-axis parallel to the magnetic field, ($\vec{B} = (0, 0, B)$). Then the eigenstates of an individual Hamiltonian are the eigenstates of σ_z, i.e. the states 'spin up', $|+\rangle$, and 'spin down', $|-\rangle$

$$h|\pm\rangle = \mp\mu B|\pm\rangle$$

[8] Clearly, the appearance of $\hbar$ in γ_{class} is due to our choice of units for $\vec{J}$.

[9] There is also a magnetic moment due to the nuclear spin whose order of magnitude is given by the nuclear magneton $\mu_N = e\hbar/2m_p$ where m_p is the proton mass: $\mu_B/\mu_N = m_p/m_e \approx 2000$. Its effects are therefore negligible as far as magnetism in matter is concerned. On the other hand, it is the basis of nuclear magnetic resonance (NMR).

[10] There is no factor $1/N!$ as in the case of the ideal gas because the spins are localized to the site of the crystal lattice and can thus be distinguished. If the partition functions are not identical (different B_i or μ_i), $Z = \prod_{i=1}^N \zeta_i$ generalizes (3.23).

In the basis $\{|+\rangle, |-\rangle\}$, the matrix representing the operator $\mathrm{e}^{-\beta h}$ is diagonal

$$\mathrm{e}^{-\beta h} = \begin{pmatrix} \mathrm{e}^{\beta\mu B} & 0 \\ 0 & \mathrm{e}^{-\beta\mu B} \end{pmatrix}$$

The partition function for a spin magnetic moment is then

$$\zeta = \mathrm{e}^{\beta\mu B} + \mathrm{e}^{-\beta\mu B} = 2\cosh(\beta\mu B) \tag{3.24}$$

Knowing Z, we can use Equation (3.3) to calculate the z-component of the average magnetization in the material, which is, in fact, the conjugate variable of the external field (actually $(-B)$)

$$\mathcal{M} = \frac{1}{\beta}\frac{\partial \ln Z}{\partial B}\bigg|_{\beta} = N\mu\tanh(\beta\mu B)$$

The factorization of the partition function (3.23) leads to a simple physical interpretation of the macroscopic magnetization. From Equation (3.24) for the individual partition function, we easily calculate the probability p_+ of finding the spin parallel to the magnetic field (state $|+\rangle$) as well as the probability p_- to find it antiparallel to the field (state $|-\rangle$)

$$p_{\pm} = \frac{1}{\zeta}\mathrm{e}^{\pm\beta\mu B} \tag{3.25}$$

The average value of a single magnetic moment along z follows immediately[11]

$$\langle\mu\rangle = \mu(p_+ - p_-) = \mu\tanh(\beta\mu B) = \mu\tanh\left(\frac{\mu B}{kT}\right) \tag{3.26}$$

The average magnetization of the sample is, then, simply the sum of the average individual moments

$$\mathcal{M} = N\langle\mu\rangle = N\mu\tanh\left(\frac{\mu B}{kT}\right) \tag{3.27}$$

Since the hyperbolic tangent is an increasing function of its argument, we see that the magnetization is a decreasing function of the temperature. When the temperature tends to zero, the spins will increasingly align themselves with the magnetic field, but when the temperature increases, thermal fluctuations tend to destroy this alignment. In fact we are witnessing here an 'energy–entropy competition'. As we will see in Section 3.5, a system at temperature T will try to minimize its free

[11] Clearly, $\langle\mu\rangle$ corresponds to the force conjugate to $(-B)$ for a one-spin system. Equation (3.26) can also be obtained from

$$\langle\mu\rangle = \frac{1}{\beta}\frac{\partial \ln \zeta}{\partial B}\bigg|_{\beta}$$

energy $F = E - TS$. At low temperature, it is advantageous to reduce the energy but at high temperature, advantage is gained by increasing the entropy. When the temperature is high enough so that $\mu B \ll kT$, we obtain the Curie law by using the approximation $\tanh x \approx x$, valid for $|x| \to 0$

$$\mathcal{M} = \frac{N\mu^2 B}{kT} \tag{3.28}$$

Then, at high temperature, the magnetic susceptibility $\chi = \partial\mathcal{M}/\partial B$ takes the value $\mu^2 N/kT$. The susceptibility at zero magnetic field is given, for any temperature, by

$$\chi_0 = \lim_{B\to 0} \frac{\partial\mathcal{M}}{\partial B} = \frac{N\mu^2}{kT}$$

We have so far only discussed electronic paramagnetism. But, in spite of the small value of the nuclear magneton, nuclear paramagnetism also plays an important rôle. As a widespread application, we mention nuclear magnetic resonance and its offshoots such as medical imaging by magnetic resonance,[12] MRI (Magnetic Resonance Imaging). The central rôle is played by the magnetic moment of the proton which equals 1.41×10^{-26} J/T. In a field of 10 T at normal temperature, $p_+ - p_- \approx \mu B/kT$ is approximately equal to 3×10^{-5}. If we reduce the temperature to that of liquid helium (about 2 K) the orientation effect of the spins becomes much more pronounced: $p_+ - p_- \approx 5 \times 10^{-3}$ (but of course we cannot cool the patients to such convenient temperatures to perform medical imaging!). The interesting effects are obtained by flipping the magnetization, which is done by applying a magnetic field oscillating at the resonance frequency $2\mu B/h$, i.e. 420 MHz for a field of 10 T.

3.1.4 Ferromagnetism and the Ising model

Paramagnetism is a collective exogenous effect: the atomic magnetic moments are totally independent and the magnetization is due solely to the external magnetic field. Certainly, then, a paramagnetic material will not exhibit a magnetization cycle like the one shown in Figure 1.2. That figure shows, on the one hand, that the magnetization is non-zero even when the external magnetic field vanishes and, on the other hand, that it evolves with the magnetic field and traces a hysteresis cycle. This behaviour cannot be explained at all with a Hamiltonian of the form (3.20). Rather, this behaviour is characteristic of a *ferromagnetic* substance. Below a certain critical temperature T_C, called the *Curie temperature* (about 10^3 K for iron),

[12] One should say of course 'medical imaging by nuclear magnetic resonance', but the word *nuclear* has been dropped so as not to alarm the public even though this technique has nothing to do with radioactivity. An interesting example of political correctness and marketing savvy.

magnetic substances undergo a second order phase transition from a disordered paramagnetic phase to an ordered ferromagnetic phase characterized notably by the presence of *spontaneous magnetization*[13] where the magnetization can be non-zero even in the absence of a magnetic field. As for paramagnetism, this magnetization is a result of the alignment of the individual electronic magnetic moments. However, for spontaneous magnetization, this alignment is entirely of an endogenous origin. There is an interaction between the individual electronic magnetic moments which, below the Curie temperature, succeeds in aligning them along the same direction thus forming magnetized domains of a few micrometres. The origin of this interaction is not the direct coupling between the magnetic moments themselves: a simple estimate of the orders of magnitude shows that the Curie temperature would not exceed 1 K in that case. In fact the interaction comes from a combination of the electrostatic repulsion between neighbouring electrons and the antisymmetry of the total wave function. The magnetism of spin-1/2 systems offers a simple example of this 'exchange interaction'.[14] We first consider the spin wave function, which is given by either one of the three states of the spin triplet, which are symmetric under exchange of two spins, or the spin singlet state, which is antisymmetric.[15] The total wave function has to be antisymmetric under the exchange of two electrons, therefore the spatial wave function must be

(*i*) antisymmetric when the spins are in a triplet state,
(*ii*) symmetric when the spins are in the singlet state.

Ferromagnetism is associated with the triplet, whereas the singlet leads to zero spontaneous magnetization.[16] Nonetheless, the singlet state can lead to an ordered configuration. There exist *antiferromagnetic* materials, which, at a critical temperature called the Néel temperature, undergo a phase transition from a disordered phase to an ordered one corresponding to long range order of alternating spins.

It remains to determine how the selection of one state or another is made. In case (*i*), the Pauli exclusion principle forbids the electrons to be close to each other, consequently the positive Coulomb potential energy is less important in this

[13] It can happen that we measure zero spontaneous magnetization in a sample even in its ferromagnetic phase. This is due to cancellation effects between the magnetized domains which form as a result of the phase transition.

[14] This example can be directly applied to magnetism in insulating materials, such as the oxides. Magnetism in conducting materials is more complicated because of the band structure.

[15] The triplet states are given by: $|++\rangle, |--\rangle, (|+-\rangle + |-+\rangle)/\sqrt{2}$, and the singlet by: $(|+-\rangle - |-+\rangle)/\sqrt{2}$.

[16] A helpful picture that is often used is to consider the magnetic moments in the singlet state as antiparallel vectors whereas they are parallel in the triplet. This is not obvious for the triplet state $(|+-\rangle + |-+\rangle)/\sqrt{2}$, but we can convince ourselves, by writing the rotation matrix of spin 1, that there is a direction $\vec{u}$, such that the triplet can be written as a linear combination of the states $|++\rangle_{\vec{u}}$ and $|--\rangle_{\vec{u}}$ of the operator $\vec{S}_{\text{tot}} \cdot \vec{u}$. Clearly, no rotation permits writing the singlet state in such a form.

case than in case (*ii*). On the other hand, the kinetic energy here is more important because the momentum is proportional to the gradient of the wave function. A competition between the potential and kinetic energies is thus established. If the former wins, case (*i*) is chosen, which leads to ferromagnetism, otherwise case (*ii*) is chosen leading to antiferromagnetism.

To model, in the simplest way possible, the exchange interactions between two electrons on lattice sites i and j of a crystal, we take the following interaction Hamiltonian

$$h_{ij} = -J_{ij}\vec{\sigma}_i \cdot \vec{\sigma}_j$$

where the parameter J_{ij} is taken positive to describe a ferromagnetic interaction and negative for the antiferromagnetic case. By assuming that the interaction strength decreases rapidly enough with distance so that only interactions between nearest neighbours are important, we obtain the Heisenberg Hamiltonian in the presence of an external magnetic field $\vec{B}$

$$H = -\mu \sum_{i=1}^{N} \vec{\sigma}_i \cdot \vec{B} - J \sum_{\langle i,j \rangle} \vec{\sigma}_i \cdot \vec{\sigma}_j \tag{3.29}$$

where $\langle i, j \rangle$ indicates a sum over only the nearest neighbours. We recognize the first term as the Hamiltonian for paramagnetism. In spite of the apparent simplicity of this Hamiltonian, we do not know how to calculate the partition function exactly except in one dimension.

As a solvable approximation of the Heisenberg Hamiltonian, the *Ising model* has played a big rôle in the study of magnetism.[17] It provoked great advances in the study of the para–ferromagnetic phase transition and contributed to a much deeper understanding of critical phenomena. Our goal here is not an exhaustive presentation of the Ising model, an enormous task due to the huge number of applications it has found even beyond the study of critical phenomena. Our more modest goal is to use an Ising-like model to study the ferromagnetism of spin-1/2, and to give an example of a simple calculation of the partition function when the Hamiltonian is not a simple sum of independent Hamiltonians. Further properties of the Ising model will be examined in Section 4.1.

One of the difficulties of the Heisenberg Hamiltonian is the presence of products of Pauli matrices that do not commute, another is the vector nature of the variables. It is therefore much more convenient to take a model of classical spins, or 'Ising spins'. An *Ising spin*, S, is a variable that can take the values ± 1. The classical

[17] The Ising model (actually proposed by Lenz) was solved in one dimension by Ising in 1925 (see Reference [10]). It was necessary then to wait till 1942 for Onsager [97] to give the first exact calculation of the partition function in two dimensions. At the present time, no exact calculation is available in three dimensions.

Ising Hamiltonian corresponding to (3.29) is[18]

$$H = -\mu B \sum_{i=1}^{N} S_i - J \sum_{\langle i,j \rangle} S_i S_j \tag{3.30}$$

The constant, J, is the effective coupling between the spins, the range of this interaction is given by the distance between i and j in $\langle i, j\rangle$. When $J = 0$, Equation (3.30) becomes the Ising version of the Hamiltonian (3.20) which models paramagnetism. The partition function is then

$$Z = \sum_{S_1=\pm 1} \cdots \sum_{S_N=\pm 1} \exp\left(\beta J \sum_{\langle i,j \rangle} S_i S_j + \beta\mu B \sum_i S_i\right) \tag{3.31}$$

and contains 2^N terms, one for each possible configuration of the N spins. The trace here becomes a sum over all 2^N possible spin configurations. We can formally express the average total magnetization as

$$\begin{aligned} \mathcal{M} &= \mu \left\langle \sum_i S_i \right\rangle \\ &= \frac{1}{Z} \sum_{S_1=\pm 1} \cdots \sum_{S_N=\pm 1} \left(\sum_i S_i\right) \exp\left(\beta J \sum_i S_i S_{i+1} + \beta\mu B \sum_i S_i\right) \\ &= \frac{1}{\beta} \left.\frac{\partial \ln Z}{\partial B}\right|_{\beta} \end{aligned} \tag{3.32}$$

Once again, we find the partition function playing the rôle of a generating function.[19]

In spite of the great simplification due to the classical approximation, the partition function Z with the Hamiltonian (3.30) can be calculated exactly only in very specific cases: in one dimension, and in two dimensions but with $B = 0$. It is however straightforward to simulate it numerically (see Chapter 7). We shall evaluate the partition function (3.31) in the one-dimensional case with periodic boundary conditions. We thus have a closed geometry shown in Figure 3.2. With these boundary conditions, the partition function can be evaluated exactly using the transfer matrix method which we now outline.

[18] While it is important for the direction of the magnetic field to be uniform in the Ising model, its value can change from site to site. We shall treat this more general situation in Section 4.2.2.

[19] Note that Equation (3.32) has the same form as (2.66). In fact we can treat ferromagnetism by considering the Hamiltonian in zero magnetic field, $H = -J\sum_i S_i S_{i+1}$, and introducing the 'magnetic moment' operator M ($\langle \mathsf{M} \rangle = \mathcal{M}$) as a new relevant operator. The partition function is given by Equation (2.64)

$$Z = \mathrm{Tr}\, \exp\left(\lambda_1 H + \lambda_2 \mathsf{M}\right)$$

We can easily show, by considering a quasi-static process, that $\lambda_1 = -\beta$ and $\lambda_2 = \beta\mu B$.

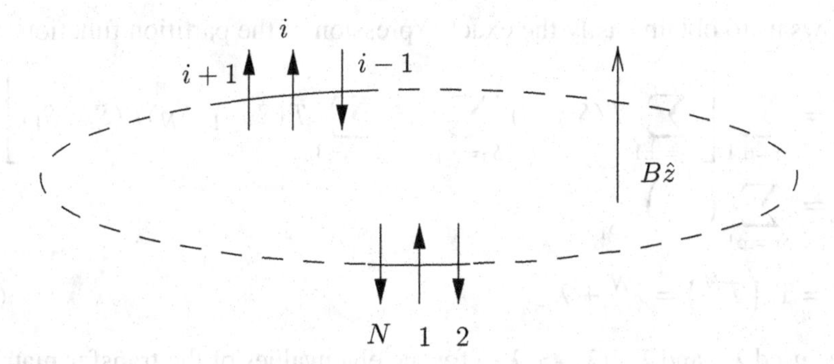

Figure 3.2 One-dimensional system of N Ising spins with periodic boundary conditions.

We first write (3.31) in the form

$$Z = \sum_{S_1=\pm1} \cdots \sum_{S_N=\pm1} \prod_{i=1}^{N} \exp\left(K S_i S_{i+1} + \frac{K'}{2}(S_i + S_{i+1}) \right)$$
$$= \sum_{S_1=\pm1} \cdots \sum_{S_N=\pm1} \prod_{i=1}^{N} T(S_i, S_{i+1}) \tag{3.33}$$

with

$$T(S_i, S_j) = \exp\left(K S_i S_j + \frac{K'}{2}(S_i + S_j) \right) \quad \text{and} \quad K = \beta J, \quad K' = \beta \mu B$$

$T(S_i, S_j)$ is a quantity that can take four values determined by the four possible combinations of the two variables S_i and S_j. It can therefore be considered as an element of a 2×2 matrix called the *transfer matrix* that is defined by $\mathcal{T}_{S_i S_j} = T(S_i, S_j)$, i.e.

$$\mathcal{T} = \begin{pmatrix} T(+1,+1) & T(+1,-1) \\ T(-1,+1) & T(-1,-1) \end{pmatrix} = \begin{pmatrix} \mathrm{e}^{(K+K')} & \mathrm{e}^{-K} \\ \mathrm{e}^{-K} & \mathrm{e}^{(K-K')} \end{pmatrix}$$

The advantage of this form is in the following equality in which one recognizes a matrix multiplication rule

$$\sum_{S_j=\pm1} T(S_i, S_j) T(S_j, S_k) = \left(\mathcal{T}^2 \right)_{S_i S_k}$$

and allows us to obtain easily the exact expression of the partition function[20]

$$\begin{aligned}
Z &= \sum_{S_1=\pm 1}\left[\sum_{S_2=\pm 1} T(S_1, S_2)\sum_{S_3=\pm 1}\cdots\sum_{S_N=\pm 1} T(S_{N-1}, S_N)T(S_N, S_1)\right] \\
&= \sum_{S_1=\pm 1}\left(\mathcal{T}^N\right)_{S_1 S_1} \\
&= \mathrm{Tr}\left(\mathcal{T}^N\right) = \lambda_+^N + \lambda_-^N
\end{aligned} \tag{3.34}$$

We have used λ_+ and λ_- ($\lambda_+ > \lambda_-$) for the eigenvalues of the transfer matrix

$$\lambda_\pm = \mathrm{e}^K\left[\cosh K' \pm \sqrt{\sinh^2 K' + \mathrm{e}^{-4K}}\right]$$

In the limit $N \gg 1$ we have

$$Z = \lambda_+^N$$

In practice we will use this second expression for Z. Once the partition function is known, we may find expressions for the interesting physical quantities. We can verify that for $J = 0$, Equation (3.32) gives the previous result (3.27) for the magnetization of a paramagnetic substance. We can also evaluate the free energy

$$F = -\frac{N}{\beta}\ln\lambda_+ \tag{3.35}$$

and verify that it is extensive. In the simple case with $B = 0$, the free energy becomes

$$F_0(T, N) = -NkT\ln\left(2\cosh\frac{J}{kT}\right)$$

We return now to the concept of spontaneous magnetization and its origin. As mentioned earlier, spontaneous magnetization is the result of an energy–entropy competition. The minimization of the free energy $F = E - TS$ requires the best compromise between minimal energy E (all spins aligned) and maximum entropy S (random spins). When the temperature is very low, the entropy plays a smaller rôle, and it is more efficient to minimize E in order to minimize F. Consequently, the spins can become ordered giving a spontaneous magnetization. As the temperature increases, the maximization of the entropy and, therefore, disordered spins are favoured. Thus, thermodynamically, the existence of spontaneous magnetization

[20] We propose another method to calculate Z for the simpler case with $B = 0$ in Problem 3.8.1.

is easily understood. However, this reasoning faces a serious difficulty. Examining Equations (3.30) and (3.32) with zero external field, we see clearly that the Hamiltonian is invariant under the symmetry $S_i \to -S_i$, which leads immediately to the absence of magnetization in zero external field. To apply the above energy–entropy argument, we need to break the symmetry explicitly, for example, by applying an infinitesimal external field.[21] This forces us to take the thermodynamic limit of the magnetization, more precisely the intensive quantity $\langle m \rangle = \mathcal{M}/N$, which is the magnetization per spin, in the presence of a tiny (positive) external magnetic field, before taking the limit of zero field

$$\langle m \rangle = \lim_{B\to 0^+} \lim_{N\to\infty} \frac{1}{N}\mathcal{M} \tag{3.36}$$

The order of the limits is crucial to obtain $\langle m \rangle \neq 0$, i.e. the appearance of spontaneous magnetization.[22] If we reverse the order of the limits in (3.36) we obtain $\langle m \rangle = 0$. This situation is typical of *spontaneous symmetry breaking*. In its ferromagnetic phase, the configuration taken by the system has a lower degree of symmetry than the Hamiltonian. The limiting process (3.36) destroys the symmetry $S_i \to -S_i$. The magnetization per spin, $\langle m \rangle$, which vanishes in the symmetric phase but acquires a non-zero value in the phase with broken symmetry, is called an *order parameter*. Order parameters and symmetry breaking will be discussed further in Section 4.1.

As far as spontaneous magnetization is concerned, the one-dimensional model is special. To see this, it is sufficient to examine the limit (3.36). A lengthy but easy calculation gives

$$\langle m \rangle = \lim_{B\to 0^+} \lim_{N\to\infty} kT \left.\frac{\partial \ln \lambda_+}{\partial B}\right|_\beta = 0$$

There is no spontaneous magnetization in one-dimension! The absence of symmetry breaking can be understood easily: in one dimension, fluctuations travel the whole system no matter how big it is. There is no sufficiently strong energy constraint (see Problem 3.8.1 for an illustration), and, except for $T = 0$, it is impossible to establish an ordered phase.

[21] We can also explicitly break the symmetry by aligning all the boundary spins in the same direction and then taking the thermodynamic limit.

[22] Obviously

$$\lim_{B\to 0^-} \lim_{N\to\infty} \frac{1}{N}\mathcal{M} = -\langle m \rangle$$

3.1.5 Thermodynamic limit

Consider a given extensive quantity[23] $A(T, V, N)$ and assume

$$\boxed{\lim_{\substack{N,V\to\infty \\ n \text{ finite}}} \frac{1}{V} A(T, V, N) = a(n, T)} \tag{3.37}$$

where $n = N/V$ is the density. We say that the thermodynamic limit of A exists if a has a finite value. In this limit, the physical properties of the system are independent of its size.

In the simple case of an ideal gas, the limit (3.37) of the average energy (3.15) gives a quantity proportional to nkT which is finite for finite n. The energy of an ideal gas is an extensive quantity. Let us now introduce gravity into the problem. The energy acquires a gravitational potential energy term

$$E_G \propto -G\frac{(\rho V)^2}{V^{1/3}} \sim -\rho^2 V^{5/3}$$

where G is the gravitational constant and $\rho = mn$ is the mass per unit volume. When the size of the system increases at constant ρ (or n) we have

$$\frac{E_G}{V} \sim -V^{2/3} \xrightarrow[V\to\infty]{} -\infty$$

We see then that the thermodynamic limit no longer exists.[24] The dependence on $V^{2/3}$ is a manifestation of the long range nature of the gravitational potential ($\propto 1/r$). The thermodynamic limit is intimately related to the short range nature of interactions. For a system bound by Coulomb interactions, which have infinite range, the existence of the thermodynamic limit is ensured by the presence of charges of opposite signs and screening effects,[25] which is not the case for gravitation.

We illustrate the above discussion with the help of the two-dimensional Ising model in zero field.[26] Consider a square system of $2N \times 2N = 4N^2$ Ising spins (Figure 3.3(a)). To begin with, we take the unphysical case of identical interactions

[23] The restriction to the variables T, V, N is only for the clarity of presentation. A may depend on any number of variables, extensive or intensive. However, a can be a function only of intensive variables including the ratios of extensive variables.

[24] Nonetheless, stars are stable. This is because the gravitational collapse is balanced by pressure if the mass is not too great. However, we do not observe a thermodynamic limit: for example, neutron stars (Exercise 5.6.2) are such that $E \propto M^{-1/3}$. A similar observation applies to white dwarfs (Problem 5.7.3).

[25] More precisely, the existence of a thermodynamic limit was proven by Dyson and Lenard [36] under the following three assumptions: (*i*) the system is globally neutral, (*ii*) the kinetic energy is given by quantum mechanics, and (*iii*) at least one of the kinds of charges (positive or negative) is carried by a fermion.

[26] Note that for the Ising model the rôle of volume is played by the number of spins.

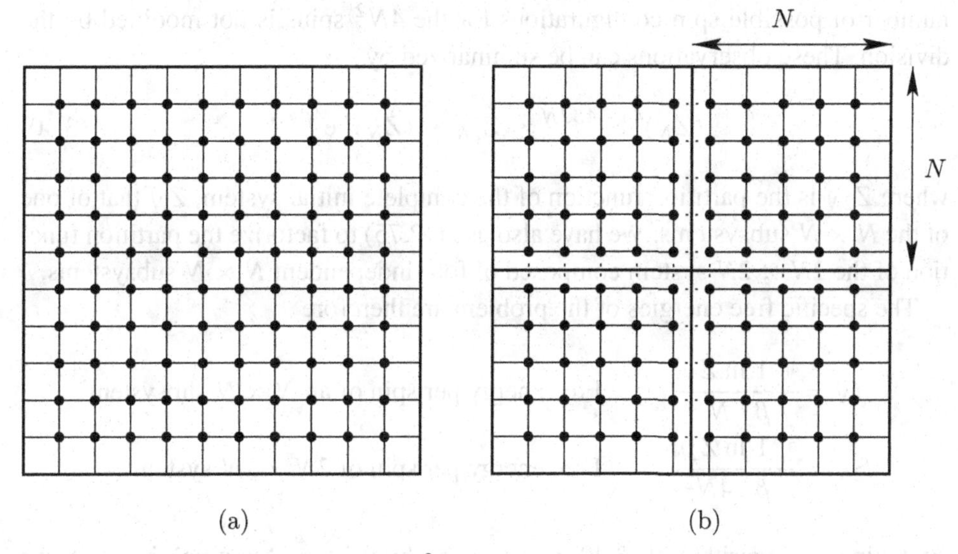

Figure 3.3 (a) Ising system of $4N^2$ spins, and (b) the same system divided into four subsystems of N^2 spins each.

between all the spins, i.e. infinite range interactions. The Hamiltonian becomes

$$H = -J \sum_{i \neq j} S_i S_j \tag{3.38}$$

At $T = 0$, all the spins are aligned and $E \sim N^4$. The limit (3.37) then gives

$$\lim_{N \to \infty} \frac{E}{N^2} \sim N^2 \to \infty$$

which shows that the thermodynamic limit does not exist for the Hamiltonian (3.38). Nonetheless, this model is sometimes used as a model for mean field but with an interaction (J/N^2), which leads to a good thermodynamic limit. We can show that, with this renormalized interaction, the model is exactly solvable and gives the mean field results (Section 4.2).

Now consider the Ising model with interactions only between nearest neighbour spins, and with free boundary conditions

$$H = -J \sum_{\langle i,j \rangle} S_i S_j \tag{3.39}$$

Divide the $4N^2$ spins into four identical subsystems (Figure 3.3(b)), which we assume to be independent. By neglecting the interactions of subsystems at their boundaries, we make an error in the energy ranging between $-4JN$ (boundary spins all parallel) and $4JN$ (boundary spins anti-parallel). On the other hand, the number of possible spin configurations for the $4N^2$ spins is not modified by the division. These observations can be summarized by

$$(Z_N)^4 \, e^{-4\beta JN} \leq Z_{2N} \leq (Z_N)^4 \, e^{4\beta JN} \tag{3.40}$$

where Z_{2N} is the partition function of the complete initial system, Z_N that of one of the $N \times N$ subsystems. We have also used (2.75) to factorize the partition function of the $2N \times 2N$ system composed of four independent $N \times N$ subsystems.

The specific free energies of the problem are therefore

$$f_N = -\frac{1}{\beta}\frac{\ln Z_N}{N^2} \qquad \text{Free energy per spin of an } N \times N \text{ subsystem}$$

$$f_{2N} = -\frac{1}{\beta}\frac{\ln Z_{2N}}{4N^2} \qquad \text{Free energy per spin of } 2N \times 2N \text{ system}$$

By taking the logarithm of (3.40) and dividing by the total 'volume', i.e. $4N^2$, we obtain

$$f_N - \frac{J}{N} \leq f_{2N} \leq f_N + \frac{J}{N}$$

which we write as

$$|f_{2N} - f_N| \leq \text{const} \times \frac{N}{N^2} \sim \frac{L}{\Sigma} \tag{3.41}$$

where $L = N$ is the perimeter of a subsystem and $\Sigma = N^2$ its area. Equation (3.41) shows, on the one hand, that the difference in physical properties between a subsystem and the complete system is due to boundary effects, and, on the other hand, that these boundary effects disappear when $N \to \infty$. In this limit we can use, with impunity, either the whole system or a part thereof. Equation (3.41) shows that the specific free energies form a Cauchy sequence. They converge to a finite limit, demonstrating again the existence of the thermodynamic limit for the Ising Hamiltonian (3.39). The reader can easily generalize (3.41) to three dimensions by considering a cube of volume $V = 8N^3$ divided into eight identical cubes. The term L/Σ in (3.41) becomes Σ/V.

A similar, but more detailed and complicated argument, allowed Lee and Yang to demonstrate that the free energy of a gas or a fluid is extensive. The crucial point in their derivation is the fact that interactions between molecules are short ranged. In conclusion, the existence of a thermodynamic limit is highly non-trivial and has only relatively recently been demonstrated for a Coulomb gas by Dyson and Lenard [36], see Footnote 25.

3.2 Classical statistical mechanics

3.2.1 Classical limit

It often happens that classical mechanics, which is simpler than quantum mechanics, is sufficient to study a particular physical problem. Such a simplification can also happen in statistical mechanics where classical statistical mechanics may be a good enough approximation. However, one should approach this classical limit with care. We have seen that even in the simple case of an ideal gas, Planck's constant was crucial in fixing the normalization of the entropy and that the identity of the particles (not a classical notion) played a fundamental rôle in ensuring its extensivity. The same is true for the correct expression of the chemical potential. On the other hand, other quantities such as the energy or pressure do not depend on factors of h^{3N} or $N!$ which appear in the partition function (2.87). Except for certain simple cases like Langevin paramagnetism, there is no classical limit, in the strict sense, of statistical mechanics but rather a semi-classical formulation.

Our starting point is the semi-classical equation (2.89) giving the trace for N structureless particles[27]

$$\mathrm{Tr} \xrightarrow[\text{limit}]{\text{classical}} \frac{1}{N!}\int \frac{\prod_{i=1}^{N} \mathrm{d}^3 p_i\, \mathrm{d}^3 r_i}{h^{3N}} \equiv \frac{1}{N! h^{3N}} \int \mathrm{d}\boldsymbol{p} \mathrm{d}\boldsymbol{q} \tag{3.42}$$

The integration measure, $\prod_i \mathrm{d}^3 p_i \mathrm{d}^3 r_i / N! h^{3N}$, can be interpreted in the quantum phase space as discussed around Equation (2.88).

Once the trace is defined, all the results we established in the course of the formal construction of the Boltzmann distribution are immediately usable. In particular, $\ln Z$ maintains its rôle as a generating function and, if the classical limit of a quantum dynamical variable A exists

$$\mathsf{A} \xrightarrow[\text{limit}]{\text{classical}} \mathcal{A}(\boldsymbol{p}, \boldsymbol{q})$$

then its average value A can be written as

$$\begin{aligned} A = \frac{1}{Z}\,\mathrm{Tr}\,[\mathsf{A}\mathrm{e}^{-\beta \mathsf{H}}] &\xrightarrow[\text{limit}]{\text{classical}} \frac{1}{Z}\int \frac{\mathrm{d}\boldsymbol{p}\,\mathrm{d}\boldsymbol{q}}{N! h^{3N}} \mathcal{A}(\boldsymbol{p}, \boldsymbol{q})\mathrm{e}^{-\beta H(\boldsymbol{p},\boldsymbol{q})} \\ &= \frac{\int \mathrm{d}\boldsymbol{p}\,\mathrm{d}\boldsymbol{q}\, \mathcal{A}(\boldsymbol{p}, \boldsymbol{q})\mathrm{e}^{-\beta H(\boldsymbol{p},\boldsymbol{q})}}{\int \mathrm{d}\boldsymbol{p}\,\mathrm{d}\boldsymbol{q}\, \mathrm{e}^{-\beta H(\boldsymbol{p},\boldsymbol{q})}} \end{aligned} \tag{3.43}$$

The last line of (3.43) shows that in calculating average values, the quantum origin is no longer apparent: neither h nor the combinatoric factor $N!$ appear in the result.

[27] We use the following concise notation: $\boldsymbol{q} = (\vec{r}_1, \ldots, \vec{r}_N)$, $\boldsymbol{p} = (\vec{p}_1, \ldots, \vec{p}_N)$, $\mathrm{d}\boldsymbol{q} = \mathrm{d}^3\vec{r}_1 \ldots \mathrm{d}^3\vec{r}_N$ and $\mathrm{d}\boldsymbol{p} = \mathrm{d}^3\vec{p}_1 \ldots \mathrm{d}^3\vec{p}_N$. We consider three-dimensional vectors, but the generalization of the results of Sections 3.2.1 and 3.2.2 to any dimensionality is straightforward.

We have already shown, and we shall verify in more detail in Chapter 5, that the classical limit of statistical thermodynamics corresponds to a situation where the average occupation number of a level is small. For gases, the thermal wavelength gives a measure of the importance of quantum effects with the classical limit given by $\lambda/d \to 0$.[28] Equation (3.12) relates these two views of the classical limit.

3.2.2 Maxwell distribution

Consider the frequently encountered situation where the Hamiltonian separates into kinetic and potential energy terms

$$H(\boldsymbol{p},\boldsymbol{q}) = H_K(\boldsymbol{p}) + H_U(\boldsymbol{q}) = \sum_{i=1}^{N} \frac{\vec{p}_i^{\,2}}{2m} + \frac{1}{2}\sum_{i\neq j} U(|\vec{r}_i - \vec{r}_j|) \tag{3.44}$$

In the classical approximation, the term $\exp(-\beta H)$ is a real number called the *Boltzmann weight*. The probability density for a configuration $(\mathbf{p},\mathbf{q})$ is then given by the normalized Boltzmann weight $Z^{-1}\exp(-\beta H(\mathbf{p},\mathbf{q}))$. Since the Hamiltonian (3.44) is the sum of two independent terms, the classical partition function factorizes according to (2.74)

$$Z = Z_K Z_U \tag{3.45}$$

It should be emphasized that this factorization is possible only in classical mechanics where the positions and momenta are independent variables. In quantum mechanics, coordinates and momenta are conjugate variables represented by non-commuting operators and consequently it is much more complicated to treat a gas of interacting particles. Z_U is constructed with the potential term in the Hamiltonian

$$Z_U = \frac{1}{V^N}\int \mathrm{d}\boldsymbol{q}\, \mathrm{e}^{-\beta H_U(\boldsymbol{q})} \tag{3.46}$$

On the other hand, Z_K represents the contribution of the 'kinetic' (translational) degrees of freedom to the partition function. By introducing the thermal wavelength (3.11), Z_K can be written as in (3.13)

$$Z_K = \frac{V^N}{N!}\int \frac{\mathrm{d}\boldsymbol{p}}{h^{3N}}\, \mathrm{e}^{-\beta H_K(\boldsymbol{p})} = \frac{1}{N!}\left(\frac{V}{\lambda^3}\right)^N \tag{3.47}$$

[28] This remark allows us to interpret correctly the entrenched and unfortunate condensed notation for the classical limit: $\hbar \to 0$. It should be understood to signal the regime where the dimensionless parameter that measures quantum effects becomes negligible.

We recognize in (3.47) the partition function of a classical ideal gas, which is expected since the Hamiltonian of an ideal gas is given by (3.44) but with $U \equiv 0$. For an ideal gas $Z_U = 1$. Once again we note that V/λ^3 is the partition function for a single classical free particle in a box of volume V. Because the momentum and position degrees of freedom are decoupled for a Hamiltonian of the form (3.44), Equation (3.47) shows that to calculate the average value of a function $g(\vec{p}_i)$ that depends only on the momentum of particle i, it is not necessary to know the momenta of the remaining $(N-1)$ particles.

$$\begin{aligned}\langle g(\vec{p}_i)\rangle &= \frac{\int \mathrm{d}\boldsymbol{p}\,\mathrm{d}\boldsymbol{q}\; g(\vec{p}_i)\mathrm{e}^{-\beta H(\boldsymbol{p},\boldsymbol{q})}}{\int \mathrm{d}\boldsymbol{p}\,\mathrm{d}\boldsymbol{q}\;\mathrm{e}^{-\beta H(\boldsymbol{p},\boldsymbol{q})}} = \frac{\int \mathrm{d}\boldsymbol{q}\;\mathrm{e}^{-\beta H_U(\boldsymbol{q})}\int \mathrm{d}\boldsymbol{p}\; g(\vec{p}_i)\mathrm{e}^{-\beta H_K(\boldsymbol{p})}}{\int \mathrm{d}\boldsymbol{q}\;\mathrm{e}^{-\beta H_U(\boldsymbol{q})}\int \mathrm{d}\boldsymbol{p}\;\mathrm{e}^{-\beta H_K(\boldsymbol{p})}} \\ &= \frac{\int \mathrm{d}^3 p_i\; g(\vec{p}_i)\exp\left[-\frac{\beta}{2m}\vec{p}_i^{\,2}\right]}{\int \mathrm{d}^3 p_i\;\exp\left[-\frac{\beta}{2m}\vec{p}_i^{\,2}\right]}\end{aligned}$$

Since the particles are identical, the final result does not depend on the particle considered

$$\langle g(\vec{p}_i)\rangle \equiv \langle g(\vec{p})\rangle = \frac{1}{(2\pi mkT)^{3/2}}\int \mathrm{d}^3 p\; g(\vec{p})\exp\left[-\frac{\vec{p}^{\,2}}{2mkT}\right] = \int \mathrm{d}^3 p\; g(\vec{p})\bar{\varphi}(\vec{p}) \tag{3.48}$$

The last equation in (3.48) defines the *Maxwell distribution*[29] for the momenta, $\bar{\varphi}(\vec{p})$, or the velocities, $\varphi(\vec{v})$

$$\boxed{\bar{\varphi}(\vec{p}) \,\widehat{=}\, \frac{1}{(2\pi mkT)^{3/2}}\exp\left[-\frac{\vec{p}^{\,2}}{2mkT}\right] = \frac{1}{m^3}\left(\frac{m}{2\pi kT}\right)^{3/2}\exp\left[-\frac{m\vec{v}^{\,2}}{2kT}\right] \,\widehat{=}\, \frac{1}{m^3}\varphi(\vec{v})} \tag{3.49}$$

$\bar{\varphi}(\vec{p})\mathrm{d}^3 p = \varphi(\vec{v})\mathrm{d}^3 v$ is the probability of finding a particle in the volume $\mathrm{d}^3 p$ around $\vec{p}$ (or $\mathrm{d}^3 v$ around $\vec{v}$). We emphasize that the Maxwell distribution is general and always valid, as long as quantum effects are negligible, no matter what correlations are introduced by H_U.

The distributions (3.49) factorize into the products of three Gaussian distributions for the three independent spatial directions. They are centred at zero with variance mkT for $\bar{\varphi}(\vec{p})$ or kT/m for $\varphi(\vec{v})$

$$\left\langle p_\alpha^2\right\rangle = mkT \qquad \left\langle v_\alpha^2\right\rangle = \frac{kT}{m} \tag{3.50}$$

[29] It is clear, for obvious chronological reasons, that the 'post-Boltzmannian' reasoning we follow here, and which represents the Maxwell distribution as a simple consequence of the canonical equilibrium distribution, is not the reasoning followed by Maxwell who developed his theory in 1860.

We immediately conclude that the average kinetic energy per particle is

$$\left\langle \frac{1}{2} m \vec{v}^{\,2} \right\rangle = 3 \left\langle \frac{1}{2} m v_\alpha^2 \right\rangle = \frac{3}{2} kT \tag{3.51}$$

The average kinetic energy of the system, i.e. the average total energy for an ideal gas, is obtained by a sum of the N identical contributions (3.51), in agreement with (3.15)

$$E = \frac{3}{2} NkT \tag{3.52}$$

We shall see in the following section that these last two results are special cases of the equipartition theorem.

From the average of the kinetic energy (3.51), we can define a typical thermal speed $\sqrt{\langle v^2 \rangle}$. We say 'a typical' and not 'the typical' speed because there are other 'typical' speeds. To understand the origin of this, consider the distribution of the absolute values of the velocities: it is not a Gaussian!

$$\begin{aligned} \phi(v)\mathrm{d}v &= \int \mathrm{d}\Omega\, v^2\, \mathrm{d}v\, \varphi(\vec{v}) \\ &= 4\pi \left(\frac{m}{2\pi kT} \right)^{3/2} v^2 \exp\left[-\frac{m\vec{v}^{\,2}}{2kT} \right] \mathrm{d}v \end{aligned} \tag{3.53}$$

This distribution law is called a *Maxwellian* and is plotted in Figure 3.4. In addition to the typical speed obtained from the equipartition theorem, we also show in this

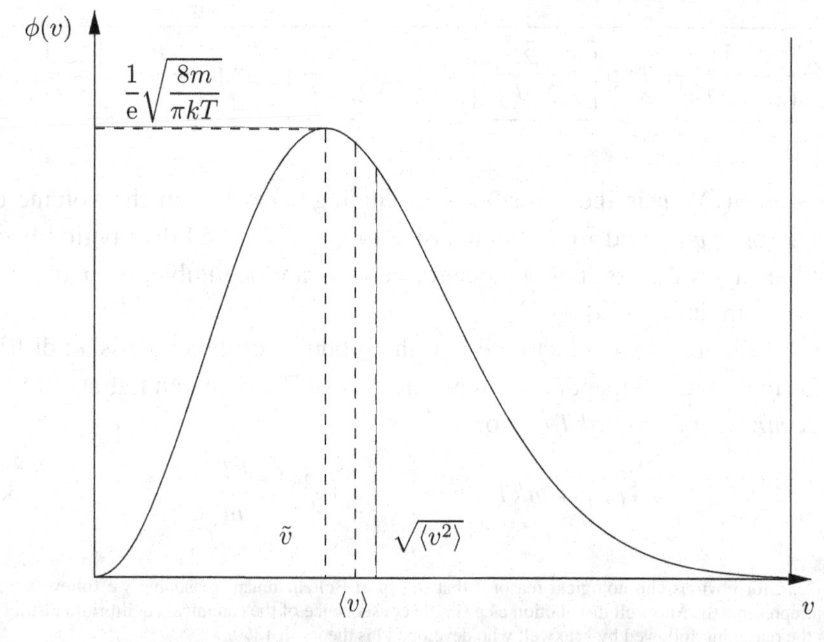

Figure 3.4 Distribution of the velocity modulus.

figure the most probable speed $\tilde{v}$ which corresponds to the maximum of the probability density, and the average speed $\langle v \rangle$. These three speeds are given by

$$\tilde{v} = \sqrt{\frac{2kT}{m}} \tag{3.54a}$$

$$\langle v \rangle = \sqrt{\frac{8kT}{\pi m}} = 1.13 \times \tilde{v} \tag{3.54b}$$

$$\sqrt{\langle v^2 \rangle} = \sqrt{\frac{3kT}{m}} = 1.22 \times \tilde{v} \tag{3.54c}$$

The small differences in the numerical values show that for an estimate of orders of magnitude, we can equally well use one or another.

3.2.3 Equipartition theorem

The result in Equation (3.51), and the large number of its experimental verifications, have greatly contributed to the emergence of the statistical picture of thermodynamics. It is an illustration of a major theorem of classical statistical mechanics called the *equipartition theorem*. The validity and application of this theorem extend far beyond the simple framework of physical situations described by a Hamiltonian of the form (3.44). We shall establish it for a Hamiltonian of the following general form

$$H = \sum_{i,j=1}^{\nu} x_i a_{ij} x_j + \widetilde{H} \tag{3.55}$$

where x_i represents a component of a coordinate or momentum vector. The first term on the right hand side is the 'quadratic part' of the Hamiltonian where a_{ij} is a constant element of the $\nu \times \nu$ symmetric and strictly positive matrix a.[30] The integer parameter ν, $\nu \leq 2M$, where M is the number of degrees of freedom, counts the number of 'quadratic coordinates' of the system. The other part of the Hamiltonian, $\widetilde{H}$, is the non-quadratic part, and regroups all the contributions that cannot be written under that form. It depends *only* on the $(2M - \nu)$ coordinates other than the x_i.

Let us calculate the average value of the quadratic part of the Hamiltonian

$$\left\langle \sum_{i,j=1}^{\nu} x_i a_{ij} x_j \right\rangle = \sum_{i,j=1}^{\nu} a_{ij} \langle x_i x_j \rangle$$

[30] The a_{ij} can be functions of x_k for $k > \nu$ (see (3.60)). We leave this generalization of the theorem as an exercise for the reader.

All average values can be calculated with the total Hamiltonian by using (3.43), but here the contribution of $\tilde{H}$ drops out. This is precisely the simplification that has been used to establish (3.48). The evaluation of the average value $\langle x_i x_j \rangle$ involves only Gaussian integrals, a result which we give in (A.45), with $A_{ij} = 2\beta a_{ij}$

$$\langle x_i x_j \rangle = \frac{kT}{2} (a^{-1})_{ij}$$

We thus establish the equipartition theorem which stipulates that each quadratic degree of freedom contribute $kT/2$ to the average value of the Hamiltonian

$$\langle H \rangle = \nu \frac{kT}{2} + \langle \tilde{H} \rangle \tag{3.56}$$

To illustrate this theorem we take the simple but physically important case of a particle in a one-dimensional harmonic potential

$$H(p, x) = \frac{p^2}{2m} + \frac{1}{2} \kappa x^2$$

The theorem (3.55) gives

$$\left\langle \frac{p^2}{2m} \right\rangle = \left\langle \frac{1}{2} \kappa x^2 \right\rangle = \frac{1}{2} kT \tag{3.57}$$

We remark that the first equality in (3.57) is the virial theorem of classical mechanics: the equality, on average, of the potential and kinetic energies. For our Hamiltonian that has two quadratic coordinates we have

$$E = kT \qquad C_V = k \tag{3.58}$$

We see with this example one of the important consequences of the equipartition theorem: knowing the number ν of quadratic coordinates immediately allows us to calculate their contribution to the specific heat, $C_V = \nu k$. We shall see in the next section how including the quantum effects at the microscopic scale imposes limits on the equipartition theorem. In particular, we shall be led to define the notion of 'frozen degrees of freedom'.

To conclude, note that in the classical limit, fluctuations of the energy, just like its average value, are determined by kT. In fact, for a purely quadratic Hamiltonian, the fluctuation-response theorem (3.10) becomes, with the help of (A.46),

$$\left\langle (\Delta H)^2 \right\rangle = \frac{\nu}{2} k^2 T^2 \tag{3.59}$$

which, according to (3.10), is in agreement with $C_V = \nu k/2$.

3.2.4 Specific heat of a diatomic ideal gas

We now apply the equipartition theorem to an ideal gas of N diatomic molecules. Since the assumptions behind the ideal gas model ensure that the partition function factorizes, it only remains to identify correctly the various degrees of freedom of a molecule. In addition to translational motion, which brings in three degrees of freedom, a diatomic molecule can have rotational and vibrational motion which we take to be independent in first approximation. To model the rotation we assume that the molecule is made of two atoms, masses m_1 and m_2, forming a dumb-bell that can turn around an axis of rotation perpendicular to the line joining the two nuclei, and whose moment of inertia is I (Figure 3.5). Using the angles defined in Figure 3.5, the rotation Lagrangian for a single molecule can be written as

$$\mathcal{L}_{\rm rot} = \frac{1}{2} I\,(\dot{\theta}^2 + \sin^2\theta\,\dot{\varphi}^2)$$

which leads to the Hamiltonian

$$h_{\rm rot} = \frac{1}{2I}\left(p_\theta^2 + \frac{1}{\sin^2\theta}\,p_\varphi^2\right) \tag{3.60}$$

We have introduced the generalized momenta p_θ and p_φ which are defined by

$$p_\theta = \frac{\partial \mathcal{L}_{\rm rot}}{\partial \dot{\theta}} = I\dot{\theta} \qquad p_\varphi = \frac{\partial \mathcal{L}_{\rm rot}}{\partial \dot{\varphi}} = I\sin^2\theta\,\dot{\varphi}$$

According to (3.60), we must associate two degrees of freedom and two quadratic coordinates ($\nu = 2$ in (3.56)) with rotational motion.

The (small) vibrational motion of a molecule can be interpreted as the harmonic oscillations of an effective particle whose mass is the reduced mass,

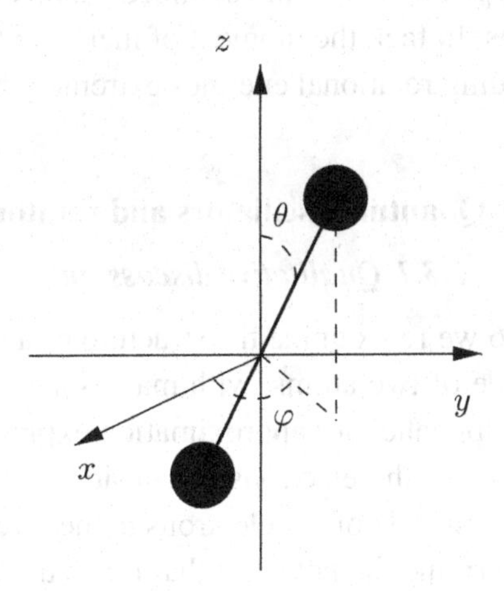

Figure 3.5 A diatomic molecule.

$\mu = m_1 m_2/(m_1 + m_2)$, and whose normal mode frequency, ω_{vib}, depends on the molecule. The corresponding Hamiltonian can be written as

$$h_{\text{vib}} = \frac{p^2}{2\mu} + \frac{1}{2}\mu\,\omega_{\text{vib}}^2 q^2 \tag{3.61}$$

The system has one degree of freedom and two quadratic coordinates.

We can now count the number of quadratic coordinates for an ideal diatomic gas and apply the equipartition theorem to calculate its specific heat

$$\left.\begin{array}{ll} E_{\text{tr}} = \frac{3}{2}NkT & C_V^{\text{tr}} = \frac{3}{2}Nk \\ E_{\text{rot}} = NkT & C_V^{\text{rot}} = Nk \\ E_{\text{vib}} = NkT & C_V^{\text{vib}} = Nk \end{array}\right\} \Rightarrow C_V = \frac{7}{2}Nk \tag{3.62}$$

Figure 3.6 shows the generic behaviour of the specific heat of an ideal diatomic gas. The equipartition theorem appears to be satisfied only at 'high temperature'. In this problem with no reference scales, the ambiguity of the notion of high temperature reflects limitations of the classical description. As the temperature decreases, we observe big deviations from the classical predictions. Whereas the onset of the deviations is determined by the temperature, their values are not. Furthermore, these deviations are well beyond what might be explained by anharmonic effects. However, the plateaux shown in the figure, at $5/2Nk$ and $3/2Nk$, correspond to values that can be explained by the classical approach on condition that we 'forget' certain types of motion and their degrees of freedom. The next section will justify this piece-wise validity of the equipartition theorem by taking into account quantum effects and in particular the discrete nature of the energy levels associated with different kinds of motion. It is this notion of degrees of freedom that justifies taking only rotations perpendicular to the symmetry axis of the molecule and not rotations around that axis. In fact, the moment of inertia in the second case is very small and the corresponding rotational energies extremely high.

3.3 Quantum oscillators and rotators

3.3.1 Qualitative discussion

To understand Figure 3.6 we re-examine the structure of a diatomic molecule. We consider a molecule made of two atoms, with masses m_1 and m_2 (atomic masses A_1 and A_2). The Born–Oppenheimer approximation exploits the big difference in mass between the nuclei and the electrons to consider the former immobile and take into account only the motion of the electrons in the reference frame where the nuclei are at rest. To determine the potential that binds the atoms into a molecule, one follows the evolution of the electronic wave functions, as a function of nuclear

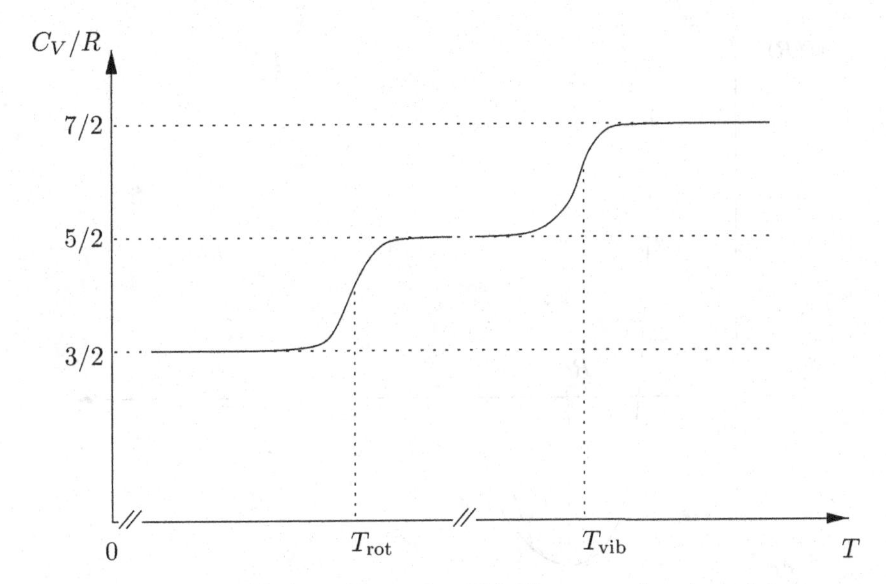

Figure 3.6 The specific heat of a mole of ideal gas as a function of the temperature.

separation. For a large majority of molecules in their ground state, the projection of the total orbital angular momentum along the symmetry axis is zero. The only control parameter is the distance R between the two nuclei. For each value of R, we determine the electronic wave function, which in turn determines the intensity of the interaction between the two nuclei. In this way we determine, step by step, the potential energy $U(R)$ of the molecule in its ground state. The typical qualitative form of $U(R)$ is shown in Figure 3.7.

The potential energy $U(R)$ has the form of a Lennard-Jones potential similar to what is used to describe interaction between molecules. However, the characteristic scales, R_0 and E_0, are very different in the two systems, so this should not lead to any confusion. The exact form of the potential depends a great deal on the molecule under consideration. For order-of-magnitude estimates, we can take as the position of the minimum the radius of the Bohr atom ($R_0 \simeq a_0 \simeq 0.053\,\text{nm}$) and for the depth of the well the Rydberg ($|E_0| \simeq \text{Ry} \simeq 13.6\,\text{eV}$). With R_0 and the reduced mass we can calculate the moment of inertia

$$I \simeq \mu a_0^2$$

To estimate the order of magnitude of a typical vibrational energy, we first state the obvious: a diatomic molecule can only exist if the energies associated with its internal degrees of freedom are much smaller than its dissociation energy. The vibrational levels are therefore at the bottom of the potential well, and, as such, are

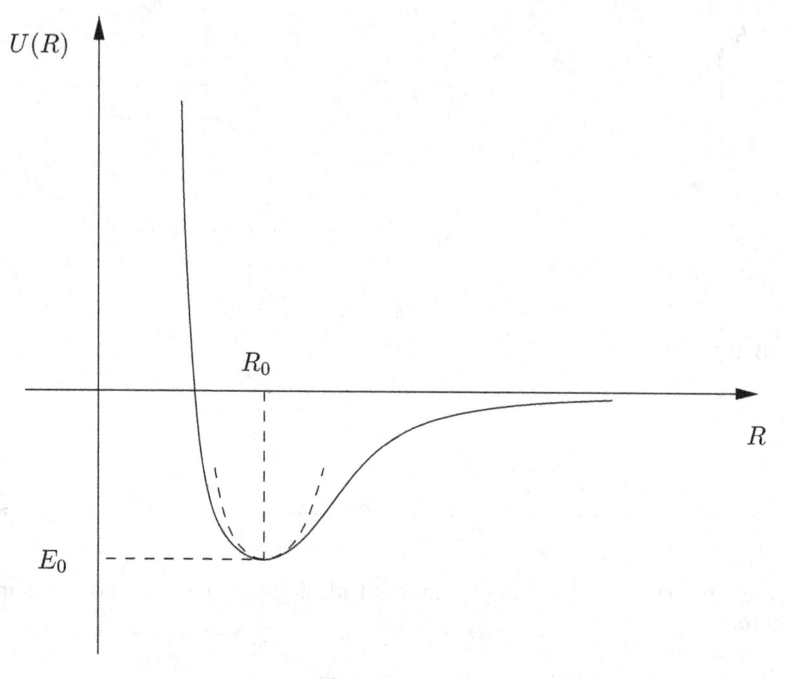

Figure 3.7 The general form of the molecular potential.

very well described by its parabolic approximation (see Figure 3.7)

$$U(R) \simeq U(R_0) + \frac{1}{2}\kappa(R - R_0)^2$$

with

$$\kappa = \left.\frac{\mathrm{d}^2 U}{\mathrm{d}R^2}\right|_{R=R_0} \simeq \frac{\mathrm{Ry}}{a_0^2}$$

We are now ready to calculate the characteristic rotational and vibrational energies, and the relevant temperature scales

$$\bar{\varepsilon}_{\mathrm{vib}} = \hbar\omega_{\mathrm{vib}} = \hbar\sqrt{\frac{\kappa}{\mu}} \simeq \frac{1}{\sqrt{\alpha}}\sqrt{\frac{m_{\mathrm{e}}}{m_{\mathrm{p}}}}\mathrm{Ry} \quad T_{\mathrm{vib}} = \bar{\varepsilon}_{\mathrm{vib}}/k \simeq \frac{1}{\sqrt{\alpha}} \times 3700\,\mathrm{K} \quad (3.63\mathrm{a})$$

$$\bar{\varepsilon}_{\mathrm{rot}} = \frac{\hbar^2}{2I} \simeq \frac{1}{\alpha}\frac{m_{\mathrm{e}}}{m_{\mathrm{p}}}\mathrm{Ry} \quad T_{\mathrm{rot}} = \bar{\varepsilon}_{\mathrm{rot}}/k \simeq \frac{1}{\alpha} \times 85\,\mathrm{K} \quad (3.63\mathrm{b})$$

where α represents the reduced mass in units of the proton mass

$$\mu = \frac{m_1 m_2}{m_1 + m_2} \simeq \frac{A_1 A_2}{A_1 + A_2} m_p = \alpha m_p$$

Let us examine again Figure 3.6. At high temperature, i.e. $T_{\rm vib,rot} \ll T$, all degrees of freedom are excited, and since the number of accessible states is huge, the classical approximation is valid. Recall that for a classical particle the average energy and the amplitude of fluctuations are determined by kT. When the temperature, which is the only source of energy for the molecule, goes below the temperature associated with a particular motion, that motion can no longer be excited. We thus speak of 'frozen' degrees of freedom where the only occupied level is the ground state, a situation very far from the regime of validity of the classical approximation. In Figure 3.6 we first observe the freezing of the vibrational and then the rotational degrees of freedom. We note that for many gases it is impossible to observe the freezing of the rotational degrees of freedom because the required temperature is below the liquefaction temperature of the gas itself, a situation where the ideal gas model is obviously no longer valid for a liquid. It might appear surprising that the specific heat does not vanish at zero temperature as required by the third law. However, the model presented here is not complete since it does not take into account the effects of quantum statistics.

3.3.2 Partition function of a diatomic molecule

To make precise the notion of frozen degrees of freedom, we calculate explicitly the vibrational and rotational partition functions for a diatomic molecule.[31] It is sufficient to calculate the individual partition functions because the different kinds of degrees of freedom are independent and, consequently, the partition function for an ideal gas of diatomic molecules can be written as

$$Z_N = \frac{1}{N!}\left(\zeta(\beta)\zeta_{\rm vib}\zeta_{\rm rot}\right)^N \tag{3.64}$$

where $\zeta(\beta) = V/\lambda^3$ is the translational partition function (see (3.47)) for one molecule. The evaluation of $\zeta_{\rm vib}$ and $\zeta_{\rm rot}$ is the object of this section.

With the harmonic approximation of the potential well, the vibrational energy levels are those of the quantum simple harmonic oscillator

$$\varepsilon_n^{\rm vib} = -u_0 + \left(n + \frac{1}{2}\right)\hbar\omega_{\rm vib} \tag{3.65}$$

[31] We will limit ourselves to molecules composed of two different molecules (heteronuclear). For homonuclear molecules we have the additional requirement of making the total wave function antisymmetric under the exchange of the two nuclei. Problem 3.8.3 treats this situation with the example of the hydrogen molecule.

with $u_0 = -E_0 \simeq$ Ry. The evaluation of the partition function is therefore reduced to summing a geometric series

$$\zeta_{\text{vib}} = \sum_{n=0}^{\infty} \mathrm{e}^{-\beta\varepsilon_n^{\text{vib}}} = \exp[\beta u_0] \frac{\exp[-\beta(\hbar\omega_{\text{vib}}/2)]}{1 - \exp[-\beta\hbar\omega_{\text{vib}}]} \tag{3.66}$$

The average vibrational energy is given by

$$\varepsilon_{\text{vib}} = -\frac{\partial \ln \zeta_{\text{vib}}}{\partial \beta} = -u_0 + \left(\frac{1}{\exp[\beta\hbar\omega_{\text{vib}}] - 1} + \frac{1}{2} \right) \hbar\omega_{\text{vib}} \tag{3.67}$$

Comparing (3.65) and (3.67) we find the average thermally excited energy level of the harmonic oscillator $\langle n \rangle = (\exp[\beta\hbar\omega_{\text{vib}}] - 1)^{-1}$. We will return to the physical interpretation of this quantity in Section 5.4.

The specific heat[32] is then given by

$$C_V = k \left(\frac{T_{\text{vib}}}{T} \right)^2 \frac{\mathrm{e}^{T_{\text{vib}}/T}}{\left(\mathrm{e}^{T_{\text{vib}}/T} - 1\right)^2} \simeq \begin{cases} k\,(T_{\text{vib}}/T)^2\, \mathrm{e}^{-T_{\text{vib}}/T} \to 0 & \text{for } T \ll T_{\text{vib}} \\ k & \text{for } T \gg T_{\text{vib}} \end{cases} \tag{3.68}$$

When $T \gg T_{\text{vib}}$, i.e. when the available energy, kT, is large compared to the level spacing in the harmonic oscillator, $\hbar\omega_{\text{vib}}$, the classical approximation is valid and we regain the prediction of the equipartition theorem. At the other end, $T \ll T_{\text{vib}}$, we verify that this equation is in agreement with the picture of frozen degrees of freedom, and the specific heat vanishes.[33]

The rotational states are described by spherical harmonics, $Y_{jm}(\theta, \varphi)$, which are simultaneous eigenstates of J^2 (eigenvalues $j(j+1)$) and J_z (eigenvalues m). In the Born–Oppenheimer model, the rotational levels are independent of m, and are consequently $(2j+1)$-fold degenerate,

$$\varepsilon_{j,m}^{\text{rot}} = j(j+1) \frac{\hbar^2}{2I} = j(j+1)\, kT_{\text{rot}} \tag{3.69}$$

[32] The specific heat of a quantum oscillator is of great importance in the history of science because it corresponds to that of a solid (up to an overall numerical factor) in the model proposed by Einstein (Exercise 3.7.2). By successfully describing the vanishing of the specific heat of solids at very low temperatures, observed experimentally but not explained by classical physics (the Dulong–Petit law, based on the classical equipartition theorem, predicts a constant specific heat), it reinforced the power of the newly discovered quantum theory. However, Einstein's model does not describe vibrations in solids satisfactorily: the reader is referred to Section 5.4 for a more satisfactory model due to Debye. Einstein's model proposes $\omega_k =$ const in (5.47), which is equivalent to assuming that all atoms vibrate at the same frequency.

[33] The exponential decrease of the specific heat in Einstein's model is to be compared with the power law T^3 found in (5.64) and which is observed experimentally.

The rotational partition function, then, takes the following form

$$\zeta_{\text{rot}} = \sum_{j=0}^{+\infty} \sum_{m=-j}^{+j} \mathrm{e}^{-\beta \varepsilon_{j,m}^{\text{rot}}} = \sum_{j=0}^{+\infty} (2j+1) \exp\left[-j(j+1) \frac{T_{\text{rot}}}{T} \right] \tag{3.70}$$

For most molecules (see (3.63b)) $T_{\text{rot}} \lesssim 10$ K, there is therefore a large range of temperatures such that $T_{\text{rot}}/T \ll 1$. In this regime, the discrete structure of the rotational energy levels is totally washed out, $kT \gg \bar{\varepsilon}_{\text{rot}}$ and we can adopt the classical description. This allows us to replace the discrete sum in (3.70) with an integral

$$\zeta_{\text{rot}} \simeq \int_0^{+\infty} \mathrm{d}x \, (2x+1) \exp\left[-x(x+1) \frac{T_{\text{rot}}}{T} \right] \simeq \frac{T}{T_{\text{rot}}} \tag{3.71}$$

This expression for the partition function was obtained in the classical approximation. It is easy to verify that the average energy and the specific heat obtained from it are in agreement with the equipartition theorem.

$$\varepsilon_{\text{rot}} = -\frac{\partial}{\partial \beta} \ln \zeta_{\text{rot}} = kT \qquad \text{and} \qquad C_V = \frac{\partial \varepsilon_{\text{rot}}}{\partial T} = k$$

For temperatures below a few tens of kelvins, or even for higher temperature but for light molecules such as H_2 or HD, the classical approximation is no longer valid. We therefore enter the non-classical regime where $T_{\text{rot}}/T \gg 1$ and for which the rotation of molecules is not favoured energetically. Under these conditions, only the first few terms of the sum (3.70) will contribute significantly to the partition function

$$\zeta_{\text{rot}} = 1 + 3\mathrm{e}^{-2T_{\text{rot}}/T} + 5\mathrm{e}^{-6T_{\text{rot}}/T} + \cdots \simeq 1$$

The exponential behaviour of the specific heat, for $T \to 0$, observed in (3.68) for the harmonic oscillator and which also holds for rotational motion, *is characteristic of an energy gap, ΔE, between the ground state and the first excited state*

$$C \sim \exp\left(-\frac{\Delta E}{kT} \right) \tag{3.72}$$

3.4 From ideal gases to liquids

The thermodynamics of a gas at low density and high temperature, where interactions between atoms are negligible, can be accurately calculated using the ideal gas model whose partition function is easy to evaluate. We shall also see, in Chapter 5, that at lower temperatures where quantum effects can no longer be ignored, we can construct a quantum ideal gas model. Somewhat surprisingly, these two models can often be generalized to density and temperature regimes characteristic of solids where correlations among constituent particles are very strong. In fact, in favourable cases, the statistical thermodynamics of solids reduces to ideal gas models of collective excitations that give rise in the quantum case to 'quasi-particles'. These quasi-particles, which are often independent to first approximation, are the quanta of collective propagation modes. For example, the interactions between the electrons and the lattice potential in a metal lead to the introduction of the concept of conduction electrons. These are effective, approximately independent quasi-particles whose properties, such as mass, are different from those of free electrons. Their statistical properties, obtained from a quantum ideal gas model, explain the low temperature specific heat in metals (see Section 5.2). In the same way, the specific heat of insulating solids is explained by considering a quantum ideal gas of 'phonons' which are quasi-particles corresponding to the vibration modes of the crystal lattice (see Section 5.4).

What is the situation for liquids? To begin with, it is interesting and instructive to ask what is the basic property of the liquid state. This is an important question because we quickly realize that it is impossible to bring out a basic difference in the behaviour of liquids and gases. Only the presence of an interface between the two states allows us to distinguish immediately between them. This ambiguity is clear when we examine a typical phase diagram such as shown in Figure 3.8. We see from these diagrams that it is possible to go around the critical point, thus passing continuously from the liquid to the gaseous phase, which is not true for the liquid–solid and gas–solid transitions. We also encounter great difficulty in trying to establish a difference between a liquid and an amorphous substance. Briefly, we can say that any substance that is neither a gas nor a crystalline solid is a

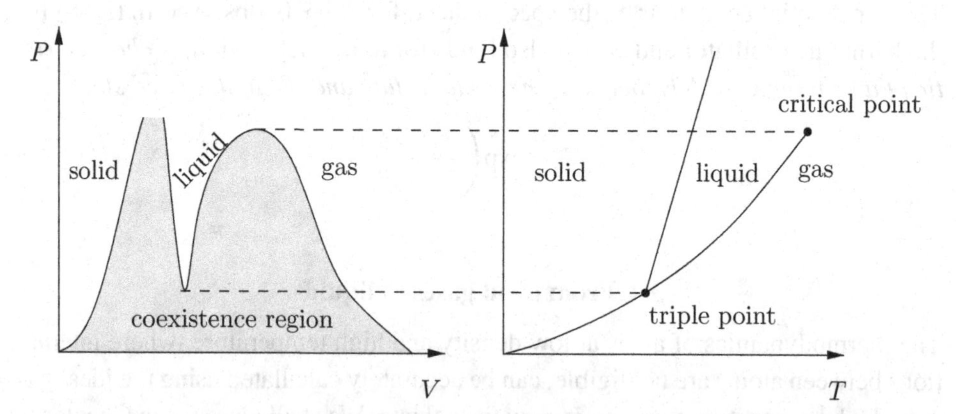

Figure 3.8 The $P-V$ and $P-T$ phase diagrams of a representative material.

liquid. In an ideal gas the molecules are totally independent while in a solid there is *long range order*. In a liquid, no such long range order is present which provides a qualitative difference with solids. However, there are *short range correlations*, which are much stronger in a liquid than in a gas. In the following paragraphs, we will demonstrate how these correlations determine the thermodynamics of liquids and dense gases.

3.4.1 Pair correlation function

We work in the framework of the classical approximation and take a Hamiltonian of the form (3.44). The partition function factorizes according to (3.45) into an 'ideal gas' part

$$Z_K = \frac{1}{N!}\left(\frac{V}{\lambda^3}\right)^N$$

and a potential energy term which controls the dependence on spatial coordinates

$$Z_U = \frac{1}{V^N}\int \prod_{i=1}^{N} \mathrm{d}^3 r_i \, \mathrm{e}^{-\beta H_U} \qquad\qquad H_U = \frac{1}{2}\sum_{i\neq j} U(\vec{r}_i, \vec{r}_j)$$

The integration measure $\prod \mathrm{d}^3 r_i$ represents a volume element in the $3N$ dimensional space formed by the N position vectors. The statistical weight of this volume element depends on its spatial position. The joint probability to find particle 1 in element $\mathrm{d}^3 r_1$ at $\vec{r}_1$, particle 2 in $\mathrm{d}^3 r_2$ at $\vec{r}_2$ ... particle N in $\mathrm{d}^3 r_N$ at $\vec{r}_N$ is given by the normalized Boltzmann weight

$$\mathcal{P}_N(\vec{r}_1, \ldots, \vec{r}_N)\frac{\mathrm{d}^3 r_1 \ldots \mathrm{d}^3 r_N}{V^N} = \frac{\mathrm{e}^{-\beta H_U(\vec{r}_1,\ldots,\vec{r}_N)}}{Z_U}\frac{\mathrm{d}^3 r_1 \ldots \mathrm{d}^3 r_N}{V^N} \tag{3.73}$$

The probability, $\mathcal{P}_1(\vec{r}_i)\mathrm{d}^3 r_i$, of finding particle i in the neighbourhood of $\vec{r}_i$ for any position of the other $N-1$ particles is obtained by integrating over all coordinates $j \neq i$

$$\mathcal{P}_1(\vec{r}_i)\mathrm{d}^3 r_i = \frac{\mathrm{d}^3 r_i}{V}\int_V \cdots \int_V \frac{\prod_{j\neq i}\mathrm{d}^3 r_j}{V^{N-1}}\mathcal{P}_N(\vec{r}_1, \ldots, \vec{r}_N) \tag{3.74}$$

We see that probability distributions are not very convenient since, for example, $\mathcal{P}_1(\vec{r}_i) \sim 1/V \ll 1$. A more direct, but totally equivalent approach, is to choose a region of space and count the average number of molecules it contains.

In this approach we deal with local densities. The simplest among them is the number density, i.e. the number of particles per unit volume

$$\boxed{n(\vec{r}) = \frac{1}{Z_U}\int_V \cdots \int_V \frac{\mathrm{d}^3 r_1 \ldots \mathrm{d}^3 r_N}{V^N} \sum_{i=1}^{N} \delta(\vec{r}-\vec{r}_i)\mathrm{e}^{-\beta H_U} = \left\langle \sum_{i=1}^{N} \delta(\vec{r}-\vec{r}_i) \right\rangle} \tag{3.75}$$

The quantity

$$n(\vec{r})\,\mathrm{d}^3 r = \left\langle \sum_{i=1}^{N} \delta(\vec{r}-\vec{r}_i) \right\rangle \mathrm{d}^3 r$$

is the average number of particles in a volume $\mathrm{d}^3 r$ at $\vec{r}$. Clearly, n is directly related to $\mathcal{P}_1$:[34] $n(\vec{r}) = N\mathcal{P}_1(\vec{r})$.

Similarly we define the *pair density* as follows

$$\boxed{n_2(\vec{r},\vec{r}\,') \widehat{=} \left\langle \sum_{i=1}^{N} \sum_{j\neq i} \delta(\vec{r}-\vec{r}_i)\delta(\vec{r}\,'-\vec{r}_j) \right\rangle} \tag{3.76}$$

where $n_2(\vec{r},\vec{r}\,')\,\mathrm{d}^3 r\,\mathrm{d}^3 r'$ is the joint average number of pairs of particles at $\vec{r}$ and $\vec{r}\,'$ and the average value $\langle \bullet \rangle$ is defined in (3.75). We can easily show that this definition of the pair density satisfies the following expected result

$$\int_V \mathrm{d}^3 r'\, n_2(\vec{r},\vec{r}\,') = (N-1)n(\vec{r})$$

Indeed a given particle has $(N-1)$ chances of pairing up. When the density is uniform, which is the case in the absence of an external potential, we have $n(\vec{r}) \equiv n = N/V$. Then, for $N \gg 1$,

$$\frac{1}{V}\int_V \mathrm{d}^3 r'\, n_2(\vec{r},\vec{r}\,') \simeq n^2$$

We introduce in this case the *pair correlation function*, $g(\vec{r},\vec{r}\,')$, which measures the degree of correlation among the particles[35]

$$n_2(\vec{r},\vec{r}\,') = n^2 g(\vec{r},\vec{r}\,') \tag{3.77}$$

[34] $n(\vec{r}) = \sum_{i=1}^{N} \int_V \mathrm{d}^3 r_i\, \delta(\vec{r}-\vec{r}_i)\mathcal{P}_1(\vec{r}_i)$, and for one type of particle all the probabilities are identical.

[35] If the density is not uniform, we define $g(\vec{r},\vec{r}\,')$ by

$$g(\vec{r},\vec{r}\,') = \frac{V}{(N-1)n(\vec{r})}\, n_2(\vec{r},\vec{r}\,')$$

Note that $g(\vec{r}, \vec{r}\,')\, \mathrm{d}^3 r' / V$ can be interpreted as the conditional probability of finding a particle in volume $\mathrm{d}^3 r'$ at $\vec{r}\,'$ knowing there is a particle at $\vec{r}$. By definition, $g(\vec{r}, \vec{r}\,') = 1$ for an ideal gas where gas particles are totally independent (no correlations). In the general case, g takes the asymptotic value $g = 1$, even in a dense gas, because as the separation between particles increases, their correlation disappears

$$\lim_{|\vec{r} - \vec{r}\,'| \to \infty} g(\vec{r}, \vec{r}\,') = 1 \tag{3.78}$$

Very often, the fluid properties are uniform and isotropic. This is the case when the potential and the correlation function depend neither on position nor on the relative orientation of the two particles: $U(\vec{r}_i, \vec{r}_j) = U(|\vec{r}_i - \vec{r}_j|)$ and $g(\vec{r}, \vec{r}\,') = g(|\vec{r} - \vec{r}\,'|) \equiv g(r)$.

The pair correlation function is a very important quantity because, on the one hand, it can be measured experimentally (Section 3.4.2) and, on the other hand, it is related to important thermodynamic quantities like energy and pressure (Section 3.4.3). We can easily get an idea of the general qualitative behaviour of $g(r)$. Figure 3.9(a) shows $g(r)$ in the two extreme cases of ideal gas and crystalline solid with lattice spacing a and where r is taken along a lattice axis. Ideally, the pair correlation function of a solid is a series of equally spaced Dirac δ-functions, but thermal fluctuations bring about the observed widening of the peaks. Note that

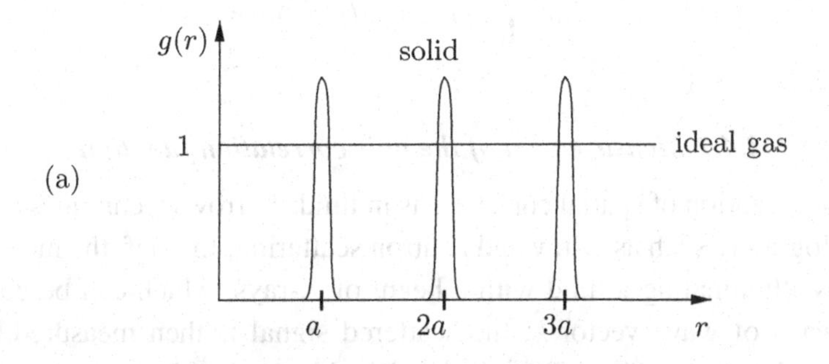

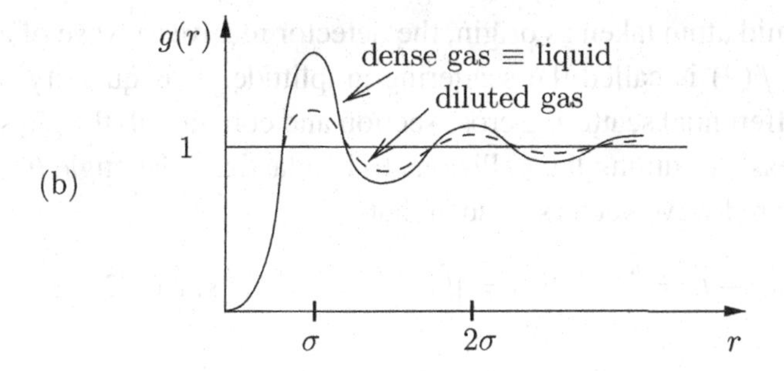

Figure 3.9 Pair correlation function. (a) An ideal gas and a solid, (b) a real gas.

$g(r) = 0$ implies complete anticorrelation, i.e. no particle at a distance r of the reference particle. Figure 3.9(b) shows what we expect for a fluid. We can form a qualitative idea of the correlations with the help of the hard sphere model, where a particle is represented by a hard sphere of diameter σ. This diameter determines the oscillation period of $g(r)$ in the same way that the lattice spacing determines the period for a solid. The average, $g = 1$, means there are no correlations and thus corresponds to the dilute gas case. The amplitude of the oscillations, which decreases as $r \to \infty$ according to (3.78), increases with the density as the gas moves farther away from the ideal gas case. Beyond the range of the hardcore repulsion, i.e. for $r > \sigma$, the correlation function cannot vanish, contrary to what happens in a solid: we cannot have a fluid full of gaps!

To conclude, we establish a result we shall use later. Consider some function $h(\vec{r}_i, \vec{r}_j)$ and calculate its average value

$$
\begin{aligned}
\left\langle \sum_{i \neq j} h(\vec{r}_i, \vec{r}_j) \right\rangle &= \frac{1}{Z_U} \int_V \frac{\mathrm{d}^3 r_1 \ldots \mathrm{d}^3 r_N}{V^N} \sum_{i \neq j} h(\vec{r}_i, \vec{r}_j) \mathrm{e}^{-\beta H_U} \\
&= \frac{1}{Z_U} \int_V \mathrm{d}^3 r \, \mathrm{d}^3 r' \int_V \frac{\mathrm{d}^3 r_1 \ldots \mathrm{d}^3 r_N}{V^N} h(\vec{r}, \vec{r}\,') \\
&\quad \times \sum_{i \neq j} \delta(\vec{r} - \vec{r}_i) \delta(\vec{r}\,' - \vec{r}_j) \mathrm{e}^{-\beta H_U} \\
&= n^2 \int_V \mathrm{d}^3 r \, \mathrm{d}^3 r' \, g(\vec{r}, \vec{r}\,') h(\vec{r}, \vec{r}\,')
\end{aligned}
\tag{3.79}
$$

3.4.2 Measurement of the pair correlation function

The characterization of spatial correlations in fluids borrows techniques developed in crystallography such as X-ray and neutron scattering. In brief, the measurement is done by illuminating a fluid with a beam of X-rays, which can be considered a plane wave of wave vector $\vec{k}$; the scattered signal is then measured far from the fluid and as a function of θ, the angle with the incident beam (see Figure 3.10). From a fluid atom taken as origin, the detector receives a wave of amplitude $f(\theta)/R$ where $f(\theta)$ is called the scattering amplitude. This quantity is directly related to the differential scattering cross section and contains all the physics of the scattering process. Assuming the collisions to be elastic,[36] the angle θ, therefore, selects the scattered wave vectors $\vec{k}\,'$ such that

$$
\vec{k}\,' - \vec{k} = \vec{q} \qquad |\vec{k}\,'| = |\vec{k}| \qquad \vec{q}^{\,2} = 4k^2 \sin^2(\theta/2)
$$

[36] Inelastic scattering allows the measurement of the dynamic structure factor. See Section 9.1.4.

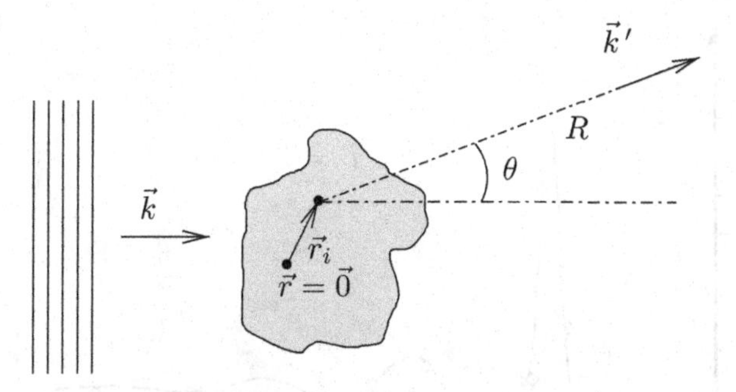

Figure 3.10 X-ray scattering experiment.

The amplitude of a wave scattered by an atom at $\vec{r}_i$ has a phase difference of $(\vec{k}\,' - \vec{k})\cdot\vec{r}_i$ relative to the wave scattered by the atom taken as the origin. This means that the scattering amplitude from this atom reads

$$\frac{f(\theta)}{R}\mathrm{e}^{\mathrm{i}(\vec{k}\,'-\vec{k}).\vec{r}_i} = \frac{f(\theta)}{R}\mathrm{e}^{\mathrm{i}\vec{q}.\vec{r}_i}$$

Since the interaction is the same for all scattering centres, and since we do not know the scatterer, the total scattering amplitude is a sum of amplitudes

$$\mathcal{A}(\theta) = \frac{f(\theta)}{R}\sum_i \mathrm{e}^{\mathrm{i}\vec{q}.\vec{r}_i} \tag{3.80}$$

A priori, Equation (3.80) contains no information on the statistical fluctuations in the fluid. However, detectors always have an integration time such that the measured intensity is not simply the modulus squared of the (instantaneous) amplitude. Instead, it is given by its (thermodynamic) average value

$$\begin{aligned} I(\theta) = \left\langle |\mathcal{A}(\theta)|^2 \right\rangle &= \frac{|f(\theta)|^2}{R^2}\left\langle \sum_{i,j} \mathrm{e}^{\mathrm{i}\vec{q}.(\vec{r}_i-\vec{r}_j)} \right\rangle \\ &= \frac{|f(\theta)|^2}{R^2}\left(N + \left\langle \sum_{i\neq j} \mathrm{e}^{\mathrm{i}\vec{q}.(\vec{r}_i-\vec{r}_j)} \right\rangle \right) \end{aligned} \tag{3.81}$$

We now introduce another quantity borrowed from solid state physics, the *structure factor* $S(\vec{q})$

$$I(\theta) = \frac{|f(\theta)|^2}{R^2} N S(\vec{q})$$

Substituting $h(\vec{r}_i, \vec{r}_j) = \mathrm{e}^{\mathrm{i}\vec{q}.(\vec{r}_i-\vec{r}_j)}$ in (3.79) and using (3.81) we get

$$N S(\vec{q}) = N + n^2 \int_V \mathrm{d}^3 r\, \mathrm{d}^3 r'\, g(\vec{r}, \vec{r}\,')\mathrm{e}^{\mathrm{i}\vec{q}.(\vec{r}-\vec{r}\,')}$$

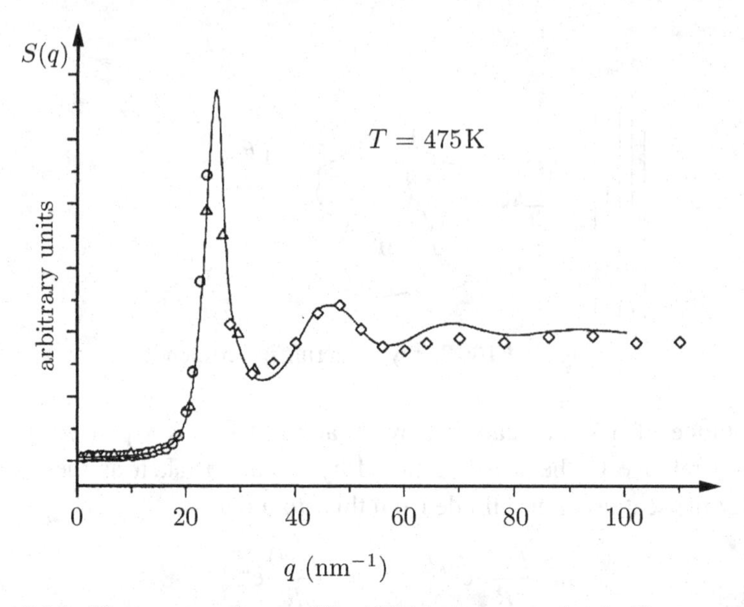

Figure 3.11 The structure factor of liquid lithium measured at the European Synchrotron Radiation Facility (ESRF) at Grenoble, France.

If the fluid is uniform, $g(\vec{r}, \vec{r}\,')$ is a function only of the separation $(\vec{r} - \vec{r}\,')$. We can then integrate over all positions of the reference point which yields a factor of V,

$$S(\vec{q}) = 1 + n \int_V \mathrm{d}^3 r\, g(\vec{r}) \mathrm{e}^{\mathrm{i}\vec{q}\cdot\vec{r}} \tag{3.82}$$

If, in addition, the fluid is isotropic, we have

$$\boxed{S(q) = 1 + \frac{4\pi n}{q} \int_0^{+\infty} \mathrm{d}r\, r\, g(r) \sin(qr)} \tag{3.83}$$

To within a δ-function, the structure factor (Figure 3.11) is proportional to the Fourier transform of the pair correlation function (review the notion of reciprocal lattice in solid state physics).

3.4.3 Pressure and energy

Since the structure factor is measured experimentally, we know the pair correlation function. It remains for us to express thermodynamic quantities in terms of g. We begin with the average energy ($h(\vec{r}_i, \vec{r}_j) = U(\vec{r}_i, \vec{r}_j)$ in (3.79))

$$E = \frac{3}{2}NkT + \frac{1}{2}\left\langle \sum_{i \neq j} U(\vec{r}_i - \vec{r}_j) \right\rangle$$
$$= \frac{3}{2}NkT + \frac{n^2}{2}\int_V \mathrm{d}^3r\,\mathrm{d}^3r'\,U(\vec{r}, \vec{r}\,')g(\vec{r}, \vec{r}\,') \tag{3.84}$$

For a translation invariant fluid, (3.84) becomes

$$\boxed{\frac{E}{N} = \frac{3}{2}kT + \frac{n}{2}\int_V \mathrm{d}^3r\,U(\vec{r})g(\vec{r})} \tag{3.85}$$

We now calculate the pressure starting with (3.7)

$$P = \frac{NkT}{V} + kT\,\frac{\partial}{\partial V}\ln Z_U\bigg|_T \tag{3.86}$$

We note that Z_U depends on the volume only because of the spatial integration. We take $V = L^3$ and consider the case of a uniform isotropic fluid. Since only relative distances matter, we can rescale $\vec{r}_i$: $\vec{r}_i = L\vec{s}_i$, and so $|\vec{r}_i - \vec{r}_j| = |\vec{s}_i - \vec{s}_j|L$. We then have in the Hamiltonian $H_U = 1/2\sum U(|\vec{s}_i - \vec{s}_j|L)$, and furthermore

$$\frac{\partial}{\partial V} \to \frac{L}{3V}\frac{\partial}{\partial L}$$

from which we obtain

$$\frac{1}{Z_U}\frac{\partial Z_U}{\partial V} = \frac{1}{Z_U}\frac{L}{3V}\frac{\partial}{\partial L}\left\{\int_V \frac{\mathrm{d}^3r_1\ldots\mathrm{d}^3r_N}{V^N}\mathrm{e}^{-\beta H_U}\right\}$$
$$= -\frac{1}{Z_U}\frac{L\beta}{6V}\int_{0\le|\vec{s}_i|\le 1}\prod_{i=1}^{N}\mathrm{d}\vec{s}_i\sum_{i\neq j}|\vec{s}_i - \vec{s}_j|U'(|\vec{s}_i - \vec{s}_j|L)\mathrm{e}^{-\beta H_U}$$
$$= -\frac{1}{Z_U}\frac{\beta}{6V}\int_V \frac{\mathrm{d}^3r_1\ldots\mathrm{d}^3r_N}{V^N}\sum_{i\neq j}|\vec{r}_i - \vec{r}_j|U'(|\vec{r}_i - \vec{r}_j|)\mathrm{e}^{-\beta H_U}$$

By putting $h(\vec{r}_i, \vec{r}_j) = |\vec{r}_i - \vec{r}_j|U'(|\vec{r}_i - \vec{r}_j|)$ in (3.79), we finally get for the pressure

$$\boxed{P = nkT - \frac{1}{6}n^2\int_V \mathrm{d}^3r\,g(r)rU'(r)} \tag{3.87}$$

3.5 Chemical potential

3.5.1 Basic formulae

We have already given in Section 1.2.3 the definition (1.11) of the chemical potential μ_i, for particles of type i, at constant energy and volume[37]

$$\mu_i = -T \left(\frac{\partial S}{\partial N_i} \right)_{E,V,N_j \neq N_i} \tag{3.88}$$

Two other definitions use the free energy F (see (1.24))

$$\mu_i = \left(\frac{\partial F}{\partial N_i} \right)_{T,V,N_j \neq N_i} \tag{3.89}$$

or the Gibbs potential G (see (1.26))

$$\mu_i = \left(\frac{\partial G}{\partial N_i} \right)_{T,P,N_j \neq N_i} \tag{3.90}$$

We emphasize that the variables that are kept constant are different for the different cases. Equation (3.90) yields another demonstration of the Gibbs–Duhem relation (1.41). At constant temperature and pressure, we multiply the number of particles by $(1+\varepsilon)$ with $\varepsilon \ll 1$: $N_i \to N_i(1+\varepsilon)$ or $\mathrm{d}N_i = \varepsilon N_i$. Since the Gibbs potential is extensive, this operation gives $G \to G(1+\varepsilon)$ and we therefore have

$$\mathrm{d}G = \varepsilon G = \sum_i \mu_i (\varepsilon N_i) = \varepsilon \sum_i \mu_i N_i$$

This is the generalization of (1.40) to several particle species[38]

$$\boxed{G = \sum_i \mu_i N_i} \tag{3.91}$$

By taking the differential of (3.91) and using (1.26) and (3.90) we obtain

$$\mathrm{d}G = \sum_i (\mu_i \,\mathrm{d}N_i + N_i \,\mathrm{d}\mu_i) = -S\,\mathrm{d}T + V\,\mathrm{d}P + \sum_i \mu_i \,\mathrm{d}N_i$$

from which we get the Gibbs–Duhem relation

$$\boxed{S\,\mathrm{d}T - V\,\mathrm{d}P + \sum_i N_i \,\mathrm{d}\mu_i = 0} \tag{3.92}$$

[37] We do not introduce subsystems here so we can replace the notation $N^{(i)}$ by N_i.

[38] The μ_i are not identical to what we would observe for independent subsystems at the same pressure and temperature (see Problem 2.7.10) except in the case of a mixture of ideal gases.

The introduction of the chemical potential allows us to distinguish between two components of the quantity $T\,\mathrm{d}S$: the *conduction* and the *convection* components.[39] Consider a system at equilibrium, with a thermostat at temperature T, kept at pressure P and able to exchange particles with a reservoir at chemical potential μ. We take only one kind of particle in order to simplify the discussion. Let $\check{s} = S/N$ and $\check{\epsilon} = E/N$ be the entropy and energy per particle. We can write

$$T\,\mathrm{d}S = T\,\mathrm{d}(N\check{s}) = TN\,\mathrm{d}\check{s} + T\check{s}\,\mathrm{d}N \tag{3.93}$$

The first term in (3.93) is the entropy change due to a change in the entropy per particle, in other words this is the conduction term. On the other hand, the second term is the entropy change due to a change in the number of particles, i.e. the convection term. Going back to the equation for $\mathrm{d}E$ at constant volume, it is reasonable to regroup all the terms proportional to $\mathrm{d}N$

$$\mathrm{d}E = TN\,\mathrm{d}\check{s} + (T\check{s} + \mu)\,\mathrm{d}N = TN\,\mathrm{d}\check{s} + \check{h}\,\mathrm{d}N \tag{3.94}$$

where $\check{h} = \overline{H}/N = \check{\epsilon} + Pv$ is the enthalpy per particle. We have used (1.25), the definition of the enthalpy, and (1.40) to write $\mu + T\check{s} = \check{\epsilon} + Pv$ with $v = V/N$. The term $\check{h}\,\mathrm{d}N$ has a simple physical interpretation: if $\mathrm{d}N$ particles are transported into a given volume by convection, the energy increases by $\check{\epsilon}\,\mathrm{d}N$. However, to return to the initial volume, it is necessary to compress the volume by $v\,\mathrm{d}N$, and we need to add into the energy balance the work $Pv\,\mathrm{d}N$ done by the pressure. An example is the Joule–Thomson expansion (Problem 1.7.6). Another example comes from hydrodynamics where one uses in general the Eulerian description in a fixed volume. We then define the energy density $\epsilon = E/V = \check{\epsilon}n$ and we write the separation of the conduction and convection terms in the form

$$\mathrm{d}\epsilon = Tn\,\mathrm{d}\check{s} + \check{h}\,\mathrm{d}n \tag{3.95}$$

Equation (3.95) explains the rôle played by the enthalpy in fluid mechanics. In a process which conserves entropy, the energy current is given by $(k + h)\vec{u}$ rather than $(k + \epsilon)\vec{u}$ where k is the kinetic energy density, $h = \overline{H}/V$ the enthalpy density and $\vec{u}$ the velocity of the fluid (see Problem 6.5.2).

3.5.2 Coexistence of phases

In this section we are concerned with equilibria between two phases. The chemical potentials μ_1 and μ_2 describe the same kind of particle but in two different phases, e.g. a vapour phase (1) and a liquid phase (2). The equilibrium condition between

[39] Consider a fluid in which there is a temperature gradient. If the fluid is kept motionless, thermal equilibrium, which corresponds to a uniform temperature, is reached by conduction. To make the temperature uniform, it is more efficient to mix (stir) the fluid, in other words, to establish convection currents.

the two phases (1.14) means the temperatures are equal and so are the chemical potentials ($\mu_1 = \mu_2$) and consequently the pressures. This equality defines in the (T, P) plane a curve $\mu_1(T, P) = \mu_2(T, P)$ that is the coexistence curve between the two phases. The chemical potentials being equal means, from Equation (3.91), that the Gibbs potentials are equal for the same number of particles in both phases: $G_1(T, P) = G_2(T, P)$. When three phases are in equilibrium, the equality of the chemical potentials

$$\mu_1(T, P) = \mu_2(T, P) = \mu_3(T, P)$$

defines an isolated point in the (T, P) plane called the triple point (Figure 3.8).

We obtain Clapeyron's equation by differentiating along the coexistence curve the relation $\mu_1(T, P) = \mu_2(T, P)$

$$-\check{s}_1 \, \mathrm{d}T + v_1 \, \mathrm{d}P = -\check{s}_2 \, \mathrm{d}T + v_2 \, \mathrm{d}P$$

where $\check{s}_i$ and v_i are the entropy per particle and the specific volume in phase i. This gives

$$\left.\frac{\mathrm{d}P}{\mathrm{d}T}\right|_{\mathrm{coex}} = \frac{\check{s}_1 - \check{s}_2}{v_1 - v_2} \tag{3.96}$$

The *latent heat* per particle for the transition (2) → (1) is $\ell = T(\check{s}_1 - \check{s}_2)$. We can therefore rewrite (3.96) as

$$\boxed{\left.\frac{\mathrm{d}P}{\mathrm{d}T}\right|_{\mathrm{coex}} = \frac{\ell}{(v_1 - v_2)T}} \tag{3.97}$$

For a liquid (2)–vapour (1) transition, the entropy per particle and the specific volume of the liquid are smaller than those for the vapour, and so $\mathrm{d}P/\mathrm{d}T > 0$. In the case of a solid (2)–liquid (1) transition, the solid is generally more ordered than the liquid ($\check{s}_2 < \check{s}_1$) and its specific volume smaller ($v_2 < v_1$) which again leads to $\mathrm{d}P/\mathrm{d}T > 0$. The two exceptions are the water–ice transition where the specific volume of ice is larger than that of water ($v_2 > v_1$), and the case of liquid helium-3 for $T < 0.3$ K where the entropy of the liquid is smaller than that of the solid ($\check{s}_1 < \check{s}_2$) due to the nuclear spins (Problem 5.7.6). In both these cases, $\mathrm{d}P/\mathrm{d}T < 0$ and for helium-3 we also have $\ell < 0$.

3.5.3 Equilibrium condition at constant pressure

Before proceeding with the study of those transitions, we establish a result that gives the condition for equilibrium of a system $\mathcal{A}$ with a reservoir $\mathcal{R}$ whose temperature T_0 and pressure P_0 are fixed (Figure 3.12). The reservoir is always at

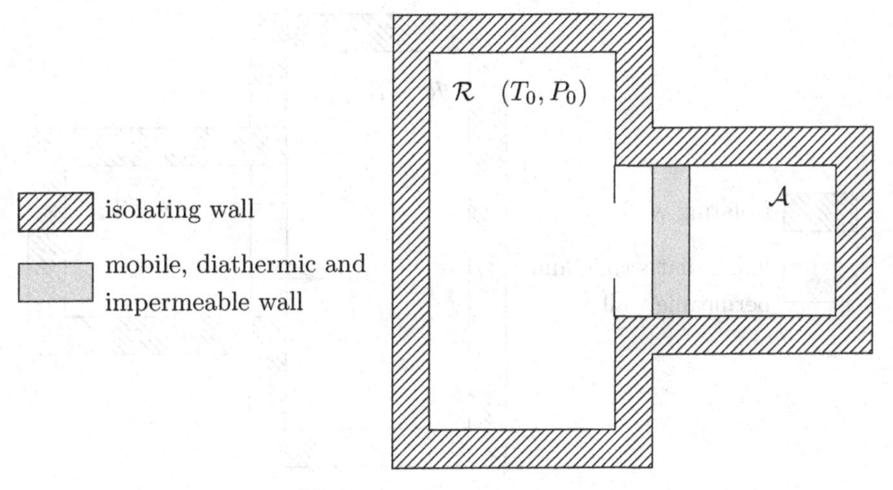

Figure 3.12 Equilibrium at constant temperature and pressure.

equilibrium (its temperature and pressure being well defined and constant) regardless of the arbitrary state of the system $\mathcal{A}$.[40] The combination $(\mathcal{A}+\mathcal{R})$ is isolated from the outside world. The contact between reservoir and system is assumed to be mobile. With the obvious notation where ΔX denotes the change in a quantity X, which will have a subscript zero if it is a reservoir quantity, and recalling that the entropy of an isolated system can only increase, we have

$$\Delta S_{\text{tot}} = \Delta S + \Delta S_0 \geq 0$$

If $\mathcal{A}$ receives a quantity of heat Q, the reservoir receives $-Q$ and $\Delta S_0 = -Q/T_0$. In the same way, if $\mathcal{A}$ gives the reservoir an amount of work $\mathbb{W} = -P_0 \Delta V_0$, then by conservation of total volume we have $\mathbb{W} = P_0 \Delta V$ and therefore $Q = \Delta E + P_0 \Delta V$. The inequality for the change in the total entropy becomes

$$\Delta S_{\text{tot}} = \Delta S - \frac{Q}{T_0} = \frac{1}{T_0}\left(T_0\, \Delta S - (\Delta E + P_0\, \Delta V)\right) \geq 0 \tag{3.98}$$

By defining the function $G_0 = E - T_0 S + P_0 V$, which reduces to the Gibbs potential when the system and reservoir are in equilibrium ($T = T_0$, $P = P_0$), condition (3.98) becomes

$$\boxed{T_0,\, P_0 \text{ fixed}: \quad \Delta G_0 \leq 0} \tag{3.99}$$

If the piston is fixed, then with $F_0 = E - T_0 S$ the above condition becomes

$$\boxed{T_0,\, V \text{ fixed}: \quad \Delta F_0 \leq 0} \tag{3.100}$$

[40] More precisely, we assume the initial and final states to be in local equilibrium in order for the entropy of the system $\mathcal{A}$ to be well defined. However, as the system is being taken from initial to final states through contact with the reservoir, it can pass through intermediate states that are not in local equilibrium.

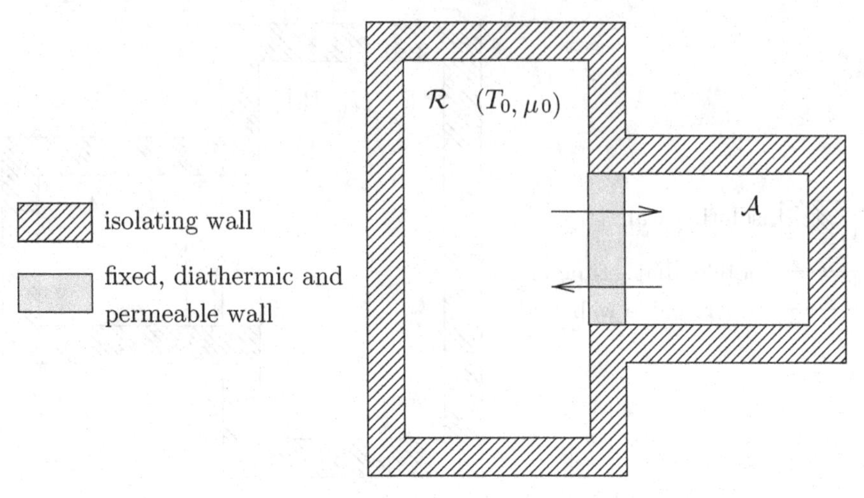

Figure 3.13 Equilibrium at constant temperature and chemical potential.

which we often express by saying that if the temperature and volume are fixed, the system will try to minimize its free energy. The quantity F_0 gives the maximum work that can be supplied by $\mathcal{A}$. In fact, subject to the condition $\Delta S_0 \geq -\Delta S$, the amount of work received by $\mathcal{A}$ is

$$W = \Delta E - Q = \Delta E + T_0 \, \Delta S_0$$

and it satisfies

$$W \geq \Delta E - T_0 \, \Delta S = \Delta F_0$$

and so the work furnished is limited by $-\Delta F_0$

$$\mathbb{W} \leq -\Delta F_0 \tag{3.101}$$

This is the same result we found in (1.21).

3.5.4 Equilibrium and stability conditions at constant μ

For the study of phase transitions, it is most convenient to use a variant of the preceding results, by determining the stability condition for a system $\mathcal{A}$ in contact with a reservoir $\mathcal{R}$ whose temperature T_0 and chemical potential μ_0 (and therefore its pressure P_0) are fixed. Indeed, this situation is well suited for the description of phase transitions where particles can go from one phase to the other. Either phase can be considered as the reservoir. The system and the reservoir are put in thermal contact via a porous wall which allows the exchange of particles while the volume of $\mathcal{A}$ stays constant. As above, the entropy of an isolated system can only increase, and so

$$\Delta S_{\rm tot} = \Delta S + \Delta S_0 \geq 0$$

Since the reservoir stays at equilibrium we have $\Delta S_0 = Q_0/T_0$ and so

$$T_0\,\Delta S_0 = \Delta E_0 - \mu_0\,\Delta N_0 = -\Delta E + \mu_0\,\Delta N$$

where we have used the conservation of energy and number of particles. We therefore obtain

$$\Delta S_{\rm tot} = \frac{1}{T_0}(T_0\,\Delta S - \Delta E + \mu_0\,\Delta N) = -\frac{1}{T_0}\,\Delta(E - T_0 S - \mu_0 N) \geq 0$$

This result shows that the system is trying to minimize $F_0 - \mu_0 N$

$$\boxed{T_0,\ \mu_0 \text{ fixed}:\quad \Delta(F_0 - \mu_0 N) \leq 0} \tag{3.102}$$

It is again important to remark that what appear in Equation (3.102) are the temperature and chemical potential of the reservoir. At equilibrium

$$F_0 - \mu_0 N \to F - \mu N = E - TS - \mu N = -PV \tag{3.103}$$

The quantity $F - \mu N = -PV$ is called the *grand potential*, $\mathcal{J}$, and is a function of (T, V, μ). Its differential is

$$\mathrm{d}\mathcal{J} = -S\,\mathrm{d}T - P\,\mathrm{d}V - N\,\mathrm{d}\mu = -V\,\mathrm{d}P - P\,\mathrm{d}V \tag{3.104}$$

We return now to a system at equilibrium and with free energy F. The free energy is extensive, $F(T, V) = Vf(T, n)$, and since the temperature does not play any rôle in what follows we simply write $f(T, n) = f(n)$. From the following identities

$$\left.\frac{\partial n}{\partial N}\right|_V = \frac{1}{V} \qquad \left.\frac{\partial n}{\partial V}\right|_N = -\frac{n}{V} \tag{3.105}$$

we obtain

$$\mu = \left.\frac{\partial F}{\partial N}\right|_V = f'(n) \tag{3.106}$$

$$P = -\left.\frac{\partial F}{\partial V}\right|_N = -f + nf'(n) \tag{3.107}$$

These relations can be used for another demonstration of (1.42). From (3.107) we get $(\partial P/\partial n)_T = nf''(n)$, which we use to calculate the coefficient of isothermal expansion, κ_T (1.34)

$$\frac{1}{\kappa_T} = -v\left.\frac{\partial P}{\partial v}\right|_T = n\left.\frac{\partial P}{\partial n}\right|_T = n^2 f''(n) = n^2\left.\frac{\partial \mu}{\partial n}\right|_T \tag{3.108}$$

This equation shows that f is a convex function of n ($f''(n) \geq 0$) because a condition for stability is $\kappa_T \geq 0$: when $\partial P/\partial n \propto f''(n)$ becomes negative, this signals a phase transition. The pressure P is an increasing function of n on both the high density side (e.g. liquid phase) and the low density side (e.g. gaseous phase). The pressure becomes a decreasing function in the interval $\overline{n}_1 \leq n \leq \overline{n}_2$, which represents an instability zone. The densities $\overline{n}_1$ and $\overline{n}_2$ are inflexion points of the function f (Figure 3.14) called *spinodal points*. If the chemical potential μ_0 is imposed by a reservoir, then since, on the one hand, $(f - \mu_0 n)$ must be the same in both phases[41] and, on the other hand, we minimize $(f - \mu_0 n)$ according to (3.102), we conclude that $f' = \mu_0$. These two conditions allow us to construct the common tangent (see Figure 3.14(b)), which determines the densities n_1 and n_2 in phases (1) and (2)

$$\frac{f_2 - f_1}{n_2 - n_1} = \mu_0 \tag{3.109}$$

These same densities can also be obtained from the Maxwell construction (Figure 3.14(c))

$$\int_{n_1}^{n_2} \mathrm{d}n\,(\mu - \mu_0) = \int_{n_1}^{n_2} \mathrm{d}n(f'(n) - \mu_0) = [f - \mu_0 n]_1^2 = 0 \tag{3.110}$$

This equation means that the shaded areas in Figure 3.14(c) are equal. The zones defined by $n_1 \leq n \leq \bar{n}_1$ and $\bar{n}_2 \leq n \leq n_2$ are zones of metastability.

3.5.5 Chemical reactions

The chemical potential is an essential tool for studying chemical reactions. We limit ourselves to a few simple examples by considering only reactions in the gaseous phase. Let $B_1, B_2, \ldots, B_m$ be the m types of molecules involved in the reaction, for example $B_1 = H_2$, $B_2 = O_2$ and $B_3 = H_2O$ for the reaction

$$2H_2 + O_2 \rightleftarrows 2H_2O \tag{3.111}$$

It is convenient to rewrite (3.111) in the symbolic form

$$-2H_2 - O_2 + 2H_2O = 0 \tag{3.112}$$

[41] Let x be the fraction (in mass) of molecules in phase 1, $(1 - x)$ the fraction in phase 2, $f - \mu_0 n = x[f_1 - \mu_0 n_1] + (1 - x)[f_2 - \mu_0 n_2]$ should be stationary in the variable x.

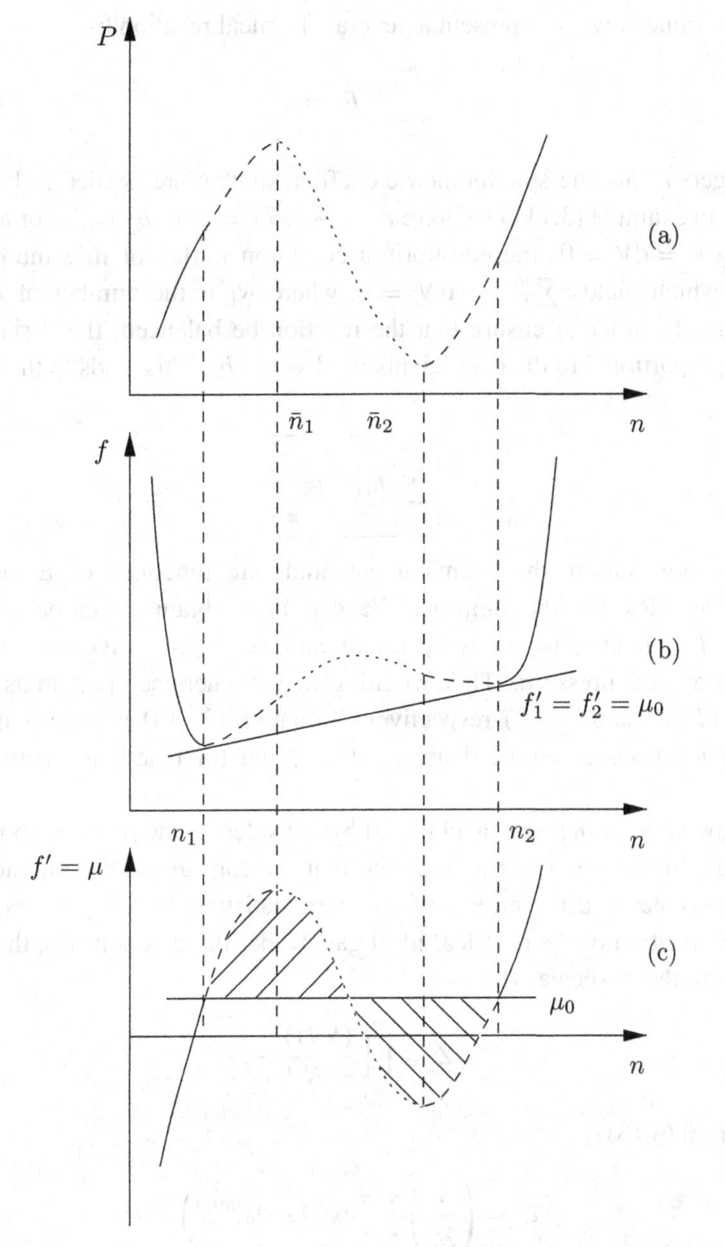

Figure 3.14 (a) The stability curve $P(n)$: the dotted line shows the unstable region, the dashed line the metastable one. The spinodal points $\overline{n}_1$ and $\overline{n}_2$ correspond to the extrema of P (or f'). (b) Determination of the densities n_1 and n_2 of the phases 1 and 2 using the common tangent construction. The spinodal points are the inflexion points of f. (c) Determination of n_1 and n_2 using the Maxwell construction.

and in the same way we represent a general chemical reaction by

$$\sum_{i=1}^{m} b_i B_i = 0 \tag{3.113}$$

The integers b_i are the stoichiometric coefficients that are needed to balance the reaction. In example (3.112) we have $b_1 = -2,\ b_2 = -1,\ b_3 = 2$. For an isolated system ($\mathrm{d}E = \mathrm{d}V = 0$) the equilibrium condition is that of maximum entropy, $\mathrm{d}S = 0$, which yields $\sum_{i=1}^{m} \mu_i \,\mathrm{d}N_i = 0$, where N_i is the number of molecules of type B_i. In order to ensure that the reaction be balanced, the variations $\mathrm{d}N_i$ must be proportional to the coefficients b_i: $\mathrm{d}N_i = \varepsilon b_i$. This leads to the important result

$$\boxed{\sum_{i=1}^{m} b_i \mu_i = 0} \tag{3.114}$$

In this demonstration, the chemical potentials are functions of E and V, the natural variables for the entropy. We can also obtain Equation (3.114) by writing $\mathrm{d}F = 0$ at constant temperature and volume, or $\mathrm{d}G = 0$ at constant temperature and pressure. This would give the chemical potentials as functions of (T, V) and (T, P) respectively. Equation (3.114) is very general, it is valid for reactions in gases, dense or dilute, and for reactions taking place in solution.

The *law of mass action*[42] is obtained by considering the reaction to take place in the gaseous phase, and by assuming that we can ignore the interactions between molecules and the effects of quantum statistics. In other words, consider the approximation of the classical ideal gas. Under these conditions, the partition function for the m species is

$$Z = \prod_{i=1}^{m} \frac{(V\zeta_i)^{N_i}}{N_i!} \tag{3.115}$$

where, from (3.13)

$$\zeta_i = \left(\frac{1}{\lambda_i^3}\right) \sum_s \exp\left(-\beta \varepsilon_s^{\mathrm{int}(i)}\right) \tag{3.116}$$

In this equation, λ_i is the thermal wavelength (3.11) of molecules of type i, the sum over s runs over all internal molecular degrees of freedom such as vibration and rotation, and $\varepsilon_s^{\mathrm{int}(i)}$ are the energies of the corresponding levels. The chemical

[42] Yet another mysterious terminology!

potentials μ_i are then given by

$$\mu_i = \left.\frac{\partial F}{\partial N_i}\right|_{V,T,N_j \neq N_i} = -kT \ln \frac{V\zeta_i}{N_i} \tag{3.117}$$

This gives the equilibrium condition

$$\sum_{i=1}^{N} b_i \ln \frac{\zeta_i}{n_i} = 0$$

which, upon exponentiation, becomes

$$\boxed{n_1^{b_1} \ldots n_m^{b_m} = \zeta_1^{b_1} \ldots \zeta_m^{b_m} = K(T)} \tag{3.118}$$

since the ζ_i are functions only of T. Equation (3.118) expresses the law of mass action for the concentrations $n_1, \ldots, n_m$: the product of concentrations n_i each raised to the power b_i is equal to a function $K(T)$ which depends only on the temperature.

As an application, we demonstrate the Saha law that gives the dissociation percentage of atomic hydrogen H into free electrons e and protons p (e + p $\rightleftarrows$ H) as a function of the temperature. The previously defined quantities ζ are given as follows for the electrons, protons and atomic hydrogen

$$\zeta_{\mathrm{e}} = \frac{2}{\lambda_{\mathrm{e}}^3} = 2\left(\frac{2\pi m_{\mathrm{e}}}{\beta h^2}\right)^{3/2} \tag{3.119}$$

$$\zeta_{\mathrm{p}} = \frac{2}{\lambda_{\mathrm{p}}^3} = 2\left(\frac{2\pi m_{\mathrm{p}}}{\beta h^2}\right)^{3/2} \tag{3.120}$$

$$\zeta_{\mathrm{H}} = 4\left(\frac{2\pi m_{\mathrm{H}}}{\beta h^2}\right)^{3/2} \sum_n \exp(-\beta E_n) \tag{3.121}$$

The factors 2 for the electron and proton and 4 for the hydrogen come from the spin degeneracies. The sum over n in (3.121) runs over the hydrogen energy levels in accordance with (3.116). The ground state of the hydrogen atom has $E_0 = -13.6$ eV, which corresponds to a temperature of 1.6×10^5 K. If we assume $T \ll 10^5$ K, the excited levels can be ignored because their weights are negligible in (3.121). In addition we can take the masses of the proton and atomic hydrogen to be the same, $m_p \simeq m_H$, which gives

$$\zeta_H \simeq 4\left(\frac{2\pi m_p}{\beta h^2}\right)^{3/2} \exp(\beta|E_0|)$$

Using (3.118), we obtain the Saha law

$$\frac{n_e n_p}{n_H} \simeq \left(\frac{2\pi m_e}{\beta h^2}\right)^{3/2} \exp(-\beta|E_0|) \tag{3.122}$$

If the only source of electrons and protons is the dissociation of neutral hydrogen, then $n_e = n_p$ and (3.122) becomes

$$n_e = \left(\frac{2\pi m_e}{\beta h^2}\right)^{3/4} \exp(-\beta|E_0|/2) n_H^{1/2} \tag{3.123}$$

This equation illustrates the energy–entropy competition involved in the minimization of the free energy $E - TS$. At low temperature, it is advantageous to minimize E thus favouring the formation of atomic hydrogen; at high temperature, it is more advantageous to increase entropy thus dissociating the atoms.

3.6 Grand canonical ensemble

3.6.1 Grand partition function

Following the general framework outlined in Chapter 2, we now examine in more detail the grand canonical ensemble. In other words, we study a system that can exchange both energy and particles with a reservoir. Therefore, this fixes the average values $E = \langle H \rangle$ and $\overline{N} = \langle \mathsf{N} \rangle$ of the energy and particle number.[43] We now have to change notation from that used in Chapter 2. Since N designates a well-defined integer number of particles, the eigenvalue of N, it is necessary to use a different notation, $\overline{N}$, for the average value. Following the construction of the Boltzmann distribution in Chapter 2, the density operator in the grand canonical ensemble is given in terms of the two Lagrange multipliers, α and β, by

$$\boxed{D = \frac{1}{\mathcal{Q}} \exp(-\beta H + \alpha \mathsf{N}) \qquad \mathcal{Q} = \mathrm{Tr}\, \exp(-\beta H + \alpha \mathsf{N})} \tag{3.124}$$

$\mathcal{Q}$ is called the *grand partition function*. The density operator D acts in a Hilbert space $\mathcal{H}$ that is the direct sum[44] of Hilbert spaces $\mathcal{H}^{(N)}$ each of which describes a

[43] The correct mathematical description of the grand canonical ensemble requires, in principle, a formalism which allows the number of particles to change, which is not the case for elementary quantum mechanics. Nonetheless, such a formalism exists and is (improperly) called ‘second quantization’. We will construct this for a simple case in Section 5.4.1. We do not need this formalism here, we only need to know that it exists.

[44] One should not confuse the direct sum of spaces with their tensor product. For example, for two spaces of dimensions N_1 and N_2, the dimension of the direct sum space is $N_1 + N_2$, whereas the dimension of the tensor product space is $N_1 N_2$. The trace of a direct sum operator is $\mathrm{Tr}\,\mathsf{A} = \mathrm{Tr}\,\mathsf{A}^{(1)} + \mathrm{Tr}\,\mathsf{A}^{(2)}$ while that for a tensor product operator is $\mathrm{Tr}\,\mathsf{A} = \mathrm{Tr}\,\mathsf{A}^{(1)}\mathrm{Tr}\,\mathsf{A}^{(2)}$.

fixed number, N, of particles

$$\mathcal{H} = \bigoplus_N \mathcal{H}^{(N)}$$

In the 'second quantization' formalism, this space is called a Fock space.

In non-relativistic quantum mechanics, the number of particles is conserved: $[H, \mathsf{N}] = 0$, the Hamiltonian commutes with the number operator, which is not the case in relativistic quantum mechanics. If we consider, for example, electrons and positrons, we find that the number of these particles is not conserved individually; what is conserved is the difference in the numbers of electrons and positrons due to conservation of electric charge. In the Boltzmann distribution, the Lagrange multiplier α (or the directly related chemical potential μ) would then multiply the difference, $(\mathsf{N}_- - \mathsf{N}_+)$, between the number operators for the electrons and positrons. Generally, the chemical potential is associated with a conservation law. In elementary quantum mechanics, to which we will now limit ourselves, this is the conservation of the number of particles.

We now write down the explicit form of the grand canonical partition function. Since $[H, \mathsf{N}] = 0$, we can diagonalize simultaneously these two operators. Let $|N, r\rangle$ be a basis labeled by the eigenvalue, N, of the number operator, N, and r which labels the energies E_r, which are functions of N and external parameters x_i

$$\begin{aligned} H|N, r\rangle &= E_r(N, x_i)|N, r\rangle \\ \mathsf{N}|N, r\rangle &= N|N, r\rangle \end{aligned} \tag{3.125}$$

In the N-particle Hilbert space $\mathcal{H}^{(N)}$ we have

$$\langle N, r|\mathrm{e}^{-\beta H + \alpha \mathsf{N}}|N, r\rangle = \mathrm{e}^{\alpha N}\, \mathrm{e}^{-\beta E_r(N, x_i)}$$

which gives the following trace in $\mathcal{H}^{(N)}$

$$\mathrm{Tr}_N \mathrm{e}^{-\beta H + \alpha \mathsf{N}} = \mathrm{e}^{\alpha N} \sum_r \mathrm{e}^{-\beta E_r(N, x_i)}$$

The grand partition function, which is a function of α, β and x_i, is obtained by summing over N

$$\mathcal{Q}(\alpha, \beta, x_i) = \sum_{N=0}^{\infty} \mathrm{e}^{\alpha N} \sum_r \mathrm{e}^{-\beta E_r(N, x_i)} \tag{3.126}$$

The sum over r in (3.126) is simply the canonical partition function. We can then write (3.126) in the form

$$\mathcal{Q}(\alpha, \beta, x_i) = \sum_{N=0}^{\infty} e^{\alpha N} Z_N(\beta, x_i) \tag{3.127}$$

By proceeding at fixed external parameters, we have expressed in Chapter 2 Equation (2.97) the Lagrange multipliers, α and β, in terms of the temperature and chemical potential

$$\beta = \frac{1}{kT} \qquad \alpha = \frac{\mu}{kT} \tag{3.128}$$

From Equations (2.66) and (2.95), we obtain the average values of the energy, the particle number and the conjugate forces X_i

$$E = -\left.\frac{\partial \ln \mathcal{Q}}{\partial \beta}\right|_{\alpha, x_i} \qquad \overline{N} = \left.\frac{\partial \ln \mathcal{Q}}{\partial \alpha}\right|_{\beta, x_i} \qquad X_i = -\frac{1}{\beta}\left.\frac{\partial \ln \mathcal{Q}}{\partial x_i}\right|_{\alpha, \beta, x_{j \neq i}} \tag{3.129}$$

In particular, for $x_i = -V$ (to which we will limit ourselves in what follows) the pressure is given by

$$P = \frac{1}{\beta}\left.\frac{\partial \ln \mathcal{Q}}{\partial V}\right|_{\alpha, \beta} \tag{3.130}$$

The entropy is the Legendre transform of the partition function (cf. (2.67))

$$\frac{1}{k} S = \ln \mathcal{Q} + \beta E - \alpha \overline{N} \tag{3.131}$$

which gives the entropy differential

$$\begin{aligned} \frac{1}{k}\, \mathrm{d}S &= \mathrm{d}\ln \mathcal{Q} + \mathrm{d}(\beta E) - \mathrm{d}(\alpha \overline{N}) \\ &= \beta\, \mathrm{d}E + \beta P\, \mathrm{d}V - \alpha\, \mathrm{d}\overline{N} \end{aligned} \tag{3.132}$$

This is nothing but Equation (1.18)

$$T\, \mathrm{d}S = \mathrm{d}E + P\, \mathrm{d}V - \mu\, \mathrm{d}\overline{N}$$

Finally, we remark that (3.131) gives a relation between the grand potential, $\mathcal{J}$, and the logarithm of the grand partition function. Using μ and T instead of α and β, we get

$$\boxed{\mathcal{J}(T, V, \mu) = -kT \ln \mathcal{Q} = -PV} \tag{3.133}$$

This equation is the counterpart of Equation (3.6) for the canonical ensemble. The grand potential, $\mathcal{J}$, plays in the grand canonical ensemble the same rôle played by the free energy, F, in the canonical ensemble. Often, instead of the chemical potential, we use the *fugacity* z

$$z = \mathrm{e}^{\beta\mu} = \mathrm{e}^{\alpha} \tag{3.134}$$

which permits us to write (3.127) in the form that we will use preferentially in what follows

$$\boxed{\mathcal{Q}(z, \beta, V) = \sum_{N=0}^{\infty} z^N Z_N(\beta, V)} \tag{3.135}$$

The average values then become[45]

$$\boxed{E = -\left.\frac{\partial \ln \mathcal{Q}}{\partial \beta}\right|_{z,V} \qquad \overline{N} = z\left.\frac{\partial \ln \mathcal{Q}}{\partial z}\right|_{\beta,V} \qquad P = \frac{1}{\beta}\left.\frac{\partial \ln \mathcal{Q}}{\partial V}\right|_{z,\beta}} \tag{3.136}$$

3.6.2 Mono-atomic ideal gas

As a simple application, we calculate the grand partition function for an ideal mono-atomic gas. Using Equations (3.13) and (3.135) we obtain

$$\mathcal{Q} = \sum_{N=0}^{\infty} \frac{z^N}{N!}\left(\frac{V}{\lambda^3}\right)^N = \exp\left(\frac{zV}{\lambda^3}\right) \tag{3.137}$$

With Equations (3.133) and (3.136) this leads to

$$\overline{N} = z\frac{\partial}{\partial z}\left(\frac{zV}{\lambda^3}\right) = \frac{zV}{\lambda^3} \quad \text{and} \quad \left(\frac{zV}{\lambda^3}\right) = \beta P V \tag{3.138}$$

which is the ideal gas law

$$PV = \overline{N}kT$$

This simple example illustrates the strategy to be followed in the grand canonical ensemble. With μ or z given, we calculate the average value $\overline{N}$ using (3.136), which then gives μ or z in terms of $\overline{N}$. This value is then put into Equation (3.133) of the grand potential, which gives the equation of state, i.e. PV as a function of T, V and $\overline{N}$.

[45] Care should be taken because a derivative at fixed μ is not the same as at fixed z or α: $(\partial \ln \mathcal{Q}/\partial\beta)_{\mu,V} = -E + \mu\overline{N}$.

In the ideal gas case, the relation between the density and fugacity is particularly simple. From (3.138) we get

$$z = \beta P \lambda^3 = n\lambda^3 \sim \left(\frac{\lambda}{d}\right)^3 \tag{3.139}$$

where $n = \overline{N}/V$ is the density and $d \sim n^{-1/3}$ the average distance between two molecules. Of course, Equations (3.139) and (3.18) are identical, which means that the classical ideal gas approximation corresponds to the condition $z \ll 1$. Finally, we note that the distribution function, $f(\vec{r}, \vec{p})$, normalized to

$$\int \mathrm{d}^3 p\, f(\vec{r}, \vec{p}) = n \tag{3.140}$$

is given by

$$f(\vec{r}, \vec{p}) = \frac{z}{h^3} \exp\left(-\frac{\beta p^2}{2m}\right) = \frac{1}{h^3} \exp\left(\alpha - \frac{\beta p^2}{2m}\right) \tag{3.141}$$

This relation will be useful in Chapter 8.

3.6.3 Thermodynamics and fluctuations

It remains to show that thermodynamics in the grand canonical ensemble is identical to that in the canonical ensemble for large systems. Physically, this equivalence comes from the fact that fluctuations in the number of particles, $\Delta N = \langle(\mathsf{N} - \overline{N})^2\rangle^{1/2}$, are of the order of $\sqrt{\overline{N}}$, which is negligible in the limit $\overline{N} \to \infty$: $\Delta N/\overline{N} \to 0$. Fixing the value N is equivalent to fixing the average value $\overline{N}$ when the variance is negligible. In order to give a precise argument, we study the distribution $P(N)$ of the variable N. According to (3.127), we have

$$P(N) \propto \mathrm{e}^{\beta\mu N} \mathrm{e}^{-\beta F_N} = P'(N)$$

where $F_N = -kT \ln Z_N$ is the free energy in the canonical ensemble. It is convenient to use the logarithm of $P'(N)$ rather than $P'(N)$ itself. Let $\widetilde{N}$ be the most probable value of N that then satisfies

$$\left.\frac{\partial \ln P'(N)}{\partial N}\right|_{N=\widetilde{N}} = \beta\mu - \beta \left.\frac{\partial F_N}{\partial N}\right|_{N=\widetilde{N}} = \beta(\mu - \mu_{\text{can}}(\widetilde{N})) = 0$$

Therefore, $\widetilde{N}$ is such that the chemical potential μ of the reservoir is the same as $\mu_{\text{can}}(\widetilde{N})$, the chemical potential in the canonical ensemble. Using (3.108) we get the second derivative of $\ln P'(N)$

$$\frac{\partial^2 \ln P'(N)}{\partial N^2} = -\beta \frac{\partial^2 F_N}{\partial N^2} = -\frac{\beta}{V} f''(n) = -\frac{\beta v}{\widetilde{N}\kappa_T}$$

where $v = V/\widetilde{N}$ is the specific volume and κ_T the isothermal compressibility (1.34). Writing the Taylor series of $\ln P'(N)$ in the neighbourhood of $N = \widetilde{N}$, we have

$$\ln P'(N) \simeq \ln P'(\widetilde{N}) - \frac{\beta v}{2\widetilde{N}\kappa_T}(N - \widetilde{N})^2 \tag{3.142}$$

This equation shows that $P'(N)$ is approximately a Gaussian centred at $N = \widetilde{N}$ and whose variance is

$$(\Delta N)^2 = \frac{\kappa_T \overline{N}}{\beta v} = \frac{\overline{N}}{\beta v^2(-\partial P/\partial v)_T} \tag{3.143}$$

Since $\Delta N/\overline{N} \to 0$, we can take the average value to be equal to the most probable value: $\overline{N} = \widetilde{N}$. For the same reason, the numerical value of the Gaussian approximation of $P'(N)$ is negligible for $N \leq 0$, and so we can replace the fluctuation interval of N, $[0, +\infty]$, by the expanded interval $[-\infty, +\infty]$. If κ_T is finite, i.e. if $(\partial P/\partial v)_T \neq 0$, the particle number distribution is very sharply peaked around its average value $\overline{N}$. With this assumption for κ_T, we can calculate $\mathcal{Q}$ using (3.127) by replacing the sum over N by an integral

$$\begin{aligned}\mathcal{Q} &= \sum_{N=0}^{\infty} P'(N) \simeq P'(\overline{N}) \int_{-\infty}^{\infty} \mathrm{d}N \exp\left[-\frac{(N - \overline{N})^2}{2(\Delta N)^2}\right] \\ &= \left(\frac{2\pi \overline{N}\kappa_T}{\beta v}\right)^{1/2} P'(\overline{N})\end{aligned} \tag{3.144}$$

In the calculation of $\ln \mathcal{Q}$, the first term of (3.144) is of the order of $\ln \overline{N}$ and is therefore negligible compared to the second term which is of the order $\overline{N}$

$$\begin{aligned}\ln \mathcal{Q} = \beta PV|_{\text{g.can}} &\simeq \ln P'(\overline{N}) = \beta(\mu \overline{N} - F_{\overline{N}}) \\ &= \beta(\mu_{\text{can}}\overline{N} - F_{\overline{N}}) = \beta(G - F) = \beta PV|_{\text{can}}\end{aligned} \tag{3.145}$$

As long as $(\partial P/\partial v)_T \neq 0$ and $\overline{N} \to \infty$, the canonical and grand canonical ensembles give the same value for PV, and thus the same value for the grand potential and the same thermodynamics. A somewhat longer analysis leads to the same conclusion when $(\partial P/\partial v)_T = 0$, i.e. for a phase transition. Close to a critical point, where $(\partial P/\partial v)_T \simeq 0$, density fluctuations become very large leading to the phenomenon of critical opalescence.

The preceding calculation and Equation (3.10) lead to the conclusion that for homogeneous substances, the stability conditions (1.52) follow from statistical mechanics

$$C_V \propto \langle (H - E)^2 \rangle \geq 0 \qquad \kappa_T \propto \langle (\mathsf{N} - \overline{N})^2 \rangle \geq 0 \tag{3.146}$$

The specific heat is related to the energy fluctuations and the isothermal compressibility to the particle number fluctuations.

3.7 Exercises

3.7.1 Density of states

Consider a dilute gas in contact with a heat reservoir at temperature T ($\beta = 1/kT$) and let $\mathcal{P}(\varepsilon)\,\mathrm{d}\varepsilon$ be the probability that a molecule has an energy between ε and $\varepsilon + \mathrm{d}\varepsilon$. Why is it not correct to write

$$\mathcal{P}(\varepsilon)\,\mathrm{d}\varepsilon \propto \mathrm{e}^{-\beta\varepsilon}\,\mathrm{d}\varepsilon \tag{3.147}$$

3.7.2 Equation of state for the Einstein model of a solid

A crystalline solid contains N atoms that can vibrate around their equilibrium lattice sites. As a first approximation, we may view the solid as $3N$ independent, identical but distinguishable harmonic oscillators. In the Einstein model, all these oscillators have the same frequency ω.

1. Calculate the partition function $\zeta(\beta)$ for a harmonic oscillator. Use ζ to calculate the partition function Z_N of the solid. Derive the expression for the internal energy E and verify that the specific heat has the form (3.68).

2. We introduce the Grüneisen constant γ by assuming that the frequency ω is a power law of the volume V of the solid,

$$\frac{\partial \ln \omega}{\partial \ln V} \simeq -\gamma$$

Show that the pressure is given by

$$P = \gamma \frac{E}{V}$$

and that

$$\gamma = \frac{\alpha V}{\kappa_T C_V} \tag{3.148}$$

where α and κ_T are the thermodynamic coefficients defined in (1.33) and (1.34).

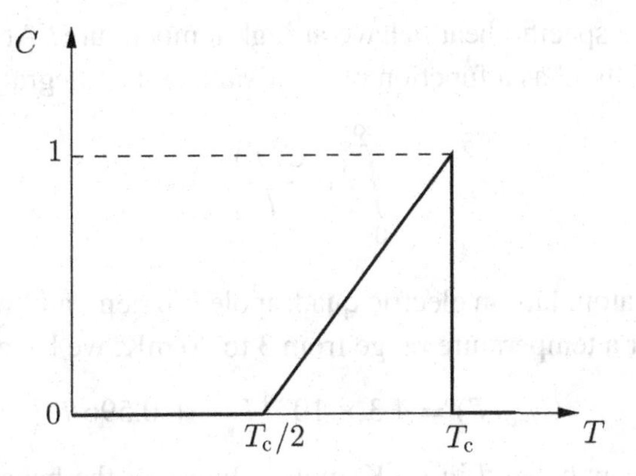

Figure 3.15 Approximate behaviour of the specific heat of a ferromagnet as a function of temperature.

3.7.3 *Specific heat of a ferromagnetic crystal*

A rough approximation of the specific heat C due to atomic spins in a ferromagnetic crystal is (Figure 3.15)

$$C = \begin{cases} NkA\left(2\dfrac{T}{T_C} - 1\right) & \text{for } T_C/2 \leq T \leq T_C \\ 0 & \text{elsewhere} \end{cases}$$

T_C is the Curie temperature, N the number of atoms and A a constant. The atoms of the system are modeled by N Ising spins placed on the lattice sites of the crystal. Estimate A.

3.7.4 *Nuclear specific heat of a metal*

Consider a metal whose atoms have nuclear spin 1 which can be in one of three quantum states: $m = -1$, $m = 0$ or $m = +1$. Each atom also carries an electric quadrupole moment that interacts with the internal electric field of the crystal. We ignore spin–spin interactions. Because of the interactions between the electric field and the quadrupoles, the states $m = \pm 1$ have the same energy, higher by an amount ε than the energy of the $m = 0$ state which we take *as the reference state*.

1. For a mole of these atoms, calculate the average energy E and the specific heat C due to the nuclear quadrupoles.

2. How does the specific heat behave at high temperature? At low temperature? Draw schematically C as a function of $\beta\varepsilon$. Evaluate the integral

$$\int_0^\infty \mathrm{d}T \, \frac{C(T)}{T}$$

3. The gallium atom has an electric quadrupole moment, and we can measure its specific heat. For a temperature range from 3 to 20 mK, we have experimentally[46]

$$C_{\text{exp}}(T) = 4.3 \times 10^{-4} T^{-2} + 0.596\, T$$

where $C_{\text{exp}}(T)$ is expressed in mJ K/ mol^{-1}. Interpret the behaviour of $C_{\text{exp}}(T)$. Estimate the value of ε/k in K and of ε in eV.

3.7.5 Solid and liquid vapour pressures

Consider a container of volume V in which are placed N atoms. These atoms are in two coexisting phases: solid and gas. The pressure P and temperature T of the gas can be perfectly controlled, and the volume occupied by the solid is very small compared with that of the gas. We call u_0 the ionization energy per atom ($u_0 > 0$) of the solid: this is the energy needed to remove an atom from the solid and take it to infinity at zero temperature. Energy and atoms may be exchanged between gas and solid thus allowing equilibrium to be established. We shall calculate the saturated vapour pressure of the solid. We recall that the chemical potential $\mu_{\text{IG}}(T, V)$ of a mono-atomic ideal gas with no internal degrees of freedom is given by (3.18).

1. Let μ_{s} and μ_{g} be the chemical potentials of this substance in its solid and gaseous phases respectively. What precaution should be taken to make sure the two chemical potentials are equal at equilibrium?

2. We assume the solid is incompressible with specific volume v_{s} and specific heat per atom $c(T)$. Calculate the Gibbs potential of the solid and show that its chemical potential may be written as

$$\mu_{\text{s}}(T) = v_{\text{s}} P - u_0 - \int_0^T \mathrm{d}T' \left(\frac{T}{T'} - 1 \right) c(T') \tag{3.149}$$

Write an implicit equation for the vapour pressure $P(T)$ of the solid at equilibrium with its gaseous phase. Give the explicit expression for $P(T)$ in the approximation where v_{s} is negligible compared with the specific volume of the gas.

[46] The experimental conditions are such that the superconductivity normally present for gallium at such low temperatures is suppressed: gallium behaves as a normal metal.

3. We now establish the form of $P(T)$ when we have equilibrium between a gas and a liquid. To this end we use a simple microscopic model for the liquid phase while keeping the mono-atomic ideal gas approximation for the gas. Each atom of the liquid phase occupies a volume v_0, the liquid is incompressible and has a total volume Nv_0 for N atoms. The atoms are independent each with a potential energy $-u_0$, $u_0 > 0$. Clearly, this is a very rough model for a liquid. Calculate the partition function of the liquid. Deduce the expressions for the Gibbs potential G_l and the chemical potential μ_l of the liquid as a function of pressure P

$$\mu_l = Pv_0 - u_0 - kT \ln\left(\frac{ev_0}{\lambda^3}\right)$$

Why is it not correct to use $\mu_l = \partial F_N/\partial N|_V$, where F_N is the free energy? Applying the condition that the chemical potentials of the liquid and gaseous phases are equal, derive an implicit equation for the vapour pressure $P(T)$ at temperature T. Assuming that $Pv_0/(kT) \ll 1$, show that we have the following explicit expression

$$P(T) = \frac{kT}{ev_0}\,e^{-u_0/(kT)} \tag{3.150}$$

Is the condition $Pv_0/(kT) \ll 1$ reasonable physically? For the remainder of the problem we shall assume this condition to hold.

4. Calculate the entropy per atom for the gas (σ_g) and for the liquid (σ_l). Find the expression for the latent heat of vaporization per atom, ℓ. Is the Clapeyron relation

$$\frac{dP}{dT} = \frac{\ell}{T(v_g - v_l)} = \frac{\ell}{T(v - v_0)}$$

satisfied? Hint: (*i*) $v_0 \ll v$ and (*ii*) show that

$$\ln\frac{v}{v_0} = 1 + \frac{u_0}{kT}$$

In the case of water boiling at atmospheric pressure (10^5 Pa), calculate $\mu_0/(kT)$

(*i*) from the latent heat of vaporization of 40.7 kJ/mole,
(*ii*) from the equation of the vaporization curve (3.150).

Compare the two results.

3.7.6 Electron trapping in a solid

Consider a solid with A identical and independent sites each of which is capable of trapping at most one electron. We use for the magnetic moment of the electrons an Ising spin model: the electron spin magnetic moment can only take the two values $\vec{\mu} = \pm\tilde{\mu}\hat{z}$ corresponding to the two states (±). In the presence of an external magnetic field $\vec{B} = B\hat{z}$, the energy of a trapped electron depends on the orientation of the magnetic moment

$$\varepsilon_{\pm} = -u_0 \mp \tilde{\mu} B$$

1. Consider first the case where the number of trapped electrons is fixed at N ($N \leq A$). Calculate $Z_N(T, A)$, the partition function of the system of N trapped electrons. Calculate E, the average energy of the system, and $\overline{N}_+$, the average number of electrons with a magnetic moment in the state (+).

2. Calculate the chemical potential μ of the trapped electrons. Give the expression for N in terms of μ.

3. We now consider the grand canonical case. We assume the solid is in equilibrium with a gas of electrons that we consider to be ideal and classical. This imposes a chemical potential μ. Calculate the grand partition function of the system $\mathcal{Q}(T, A, \mu)$

(*i*) by using the partition function Z_N explicitly,
(*ii*) by first calculating the grand partition function of a single trap $\xi(T, A, \mu)$.

4. Use ξ to calculate the probabilities p_+ and p_- for a trap to be occupied by an electron with a magnetic moment parallel and antiparallel to $\vec{B}$. Find the expressions for $\overline{N}_+$, $\overline{N}_-$ and $\overline{N}$. Verify that the expressions for E obtained in the canonical and grand canonical ensembles are identical if we take $N = \overline{N}$.

3.8 Problems

3.8.1 One-dimensional Ising model

A one-dimensional lattice consists of $N + 1$ equally spaced Ising spins (Figure 3.16) coupled by nearest-neighbour exchange interactions. The Hamiltonian of the system is

$$H = -J \sum_{j=1}^{N} S_j S_{j+1}$$

where J is a positive constant.

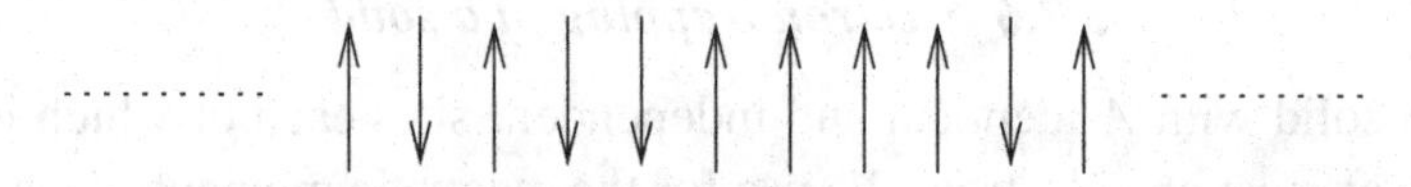

Figure 3.16 One-dimensional lattice of Ising spins.

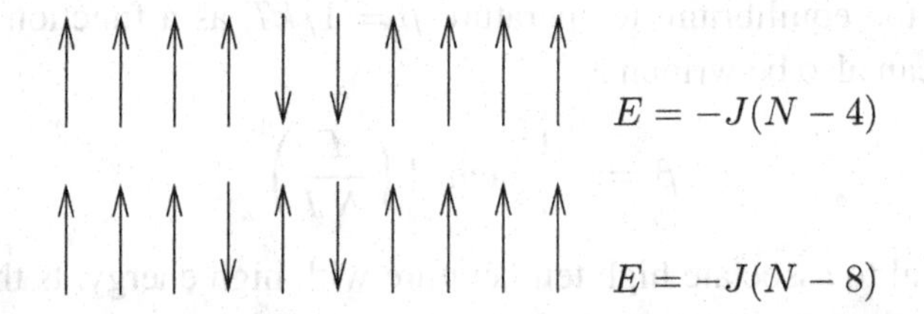

Figure 3.17 Two configurations with the same number of flipped spins but with different energies.

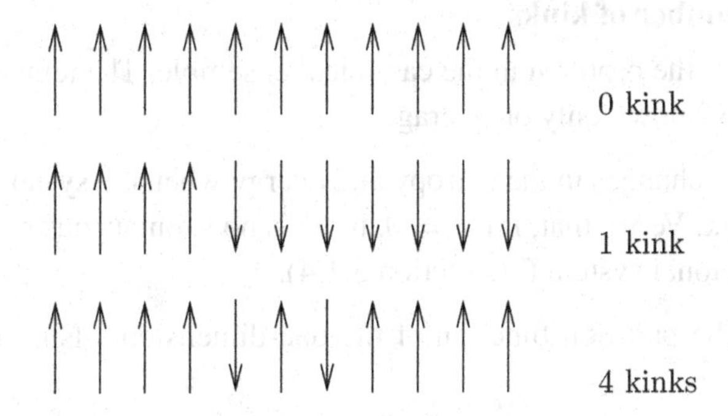

Figure 3.18 Number of kinks in three different configurations.

A Preliminaries

We can see from Figure 3.17 that, in this problem, flipping two neighbouring spins and flipping two distant spins generates configurations with different energies. The energy of a configuration is determined by the number of nearest-neighbour pairs of antiparallel spins, called 'kinks'. To obtain an elementary kink, one flips all spins to the left (or right) of a chosen site (Figure 3.18). A configuration with m kinks has the energy

$$E_m = -NJ + m2J$$

Represent schematically the configurations corresponding respectively to the states of minimum and maximum energy E_{min} and E_{max}. Give the number of kinks and the degeneracies associated with these two states.

B Fixed number of kinks

We fix the number of kinks to $m = n$ and, therefore, the energy is fixed at $E = E_n$.

1. Calculate Ω_n, the number of states with n kinks and find the expression for the corresponding entropy $S(n, N)$ assuming $n, N \gg 1$.

2. Calculate the equilibrium temperature $\beta = 1/kT$ as a function of n and N. Verify that it can also be written as[47]

$$\beta = -\frac{1}{J}\tanh^{-1}\left(\frac{E}{NJ}\right) \tag{3.151}$$

It seems natural to associate high temperature with high energy. Is this confirmed for this spin lattice? If no, why?

3. Calculate the probability p that two neighbouring spins are antiparallel.

C Arbitrary number of kinks

We now consider the problem in the canonical ensemble. The temperature is fixed but the energy is known only on average.

1. Compare the changes in the entropy and energy when the system goes from no kinks to one kink. Verify that, for $T = 0$, there is no spontaneous magnetization in this one-dimensional system (cf. Section 3.1.4).

2. Show that the partition function of the one-dimensional Ising model may be written as

$$Z = 2^{(N+1)}(\cosh \beta J)^N$$

Can one make an analogy with paramagnetism and explain why the solution of the one-dimensional Ising model is trivial?

3. Calculate the average energy E and the average number of kinks $\langle m\rangle$. Find an expression for the probability p defined above. Sketch and comment on the curve representing p as a function of T. Show that the maximal entropy is obtained in the high temperature limit.

4. How can we justify the statement that the points of view adopted in parts B and C are equivalent?

3.8.2 Negative temperatures

For some physical systems, during transient periods that can last several minutes, the nuclear spins and the crystalline lattice can separately reach independent states of thermodynamic equilibrium.[48] The necessary condition for this to happen is that

[47] We give the following relation: $\tanh^{-1} x = \frac{1}{2}\ln\frac{1+x}{1-x}$. This function appears systematically in all problems concerning magnetism in spin-1/2 systems. See also Problem 3.8.2.

[48] Contrary to what is often stated, this may happen without the need for each subsystem to pass many times through all accessible states. Such a condition would require a relaxation time several times the age of the universe since the number of states for the spin system is of the order of $2^{10^{23}}$.

the spin–lattice relaxation time τ be long compared to the separate spin and lattice relaxation times τ_s and τ_l respectively. If the system is in a state of global equilibrium, the sudden change of an external parameter (for example a magnetic field) is followed by a transient period during which the two subsystems are isolated one from the other and may thus attain independent equilibrium states characterized by different temperatures, even *negative* temperature for the spin lattice. In this problem we illustrate this *a priori* surprising property by using the Ising spin model of a paramagnetic crystal (Sections 3.1.3 and 3.1.4)

$$H = -\mu B \sum_{i=1}^{N} S_i$$

1. Calculate the partition function, $\zeta(\beta, B)$, of a single spin. Give the probability p_+ (p_-) that a spin is in an energy state $\varepsilon_+ = -\mu B$ ($\varepsilon_- = \mu B$) and the corresponding average number of atoms $\overline{N}_+$ and $\overline{N}_-$. Deduce the expression for the average energy E.

2. Calculate the partition function for the N atoms $Z_N(\beta, B)$ using arguments at fixed N_+. Find the previously established expression for E.

3. Define $m = p_+ - p_-$ and show that

$$\beta\mu B = \tanh^{-1} m = \frac{1}{2}\ln\frac{1+m}{1-m}$$

Verify that the entropy of a spin, S/N, may be written in the form

$$S/N = -k\left[\frac{1+m}{2}\ln\frac{1+m}{2} + \frac{1-m}{2}\ln\frac{1-m}{2}\right] \tag{3.152}$$

and show that this expression could have been obtained directly from (2.51).

4. We now assume that the energy of the spin system is fixed. Starting with (3.152), plot S as a function of $E/N\mu B$ and show that there is a zone of negative temperatures. Plot the magnetization $\mathcal{M} = Nm\mu$ as a function of $\beta\mu B$. Verify that the negative temperatures are related to negative magnetizations.

Calculate explicitly the equilibrium temperature $T(E)$ reached by the spin system and verify that it can be written as

$$\beta(E) = \frac{1}{kT(E)} = -\frac{1}{\mu B}\tanh^{-1}\frac{E}{N\mu B}$$

5. We have two such crystals, one with N_1 moments μ_1, the other with N_2 moments μ_2, and we take $N_1\mu_1 > N_2\mu_2$. The two crystals are initially isolated and independent at equilibrium with inverse temperatures $\beta_1(E_1)$ and $\beta_2(E_2)$. We put the crystals in thermal contact and *consider only the interactions between the two spin systems*. Plot on the same figure the curves $\beta_1(E_1)$ and $\beta_2(E_2)$. Use a graphical analysis like we did for Figure 1.5 to show that if we maintain the point of view that the heat flow takes place from the hotter to the colder system, we are led to the conclusion that negative temperatures are hotter than positive ones!

6. At equilibrium and in the limit $|\beta_1\mu_1 B| \ll 1$ and $|\beta_2\mu_2 B| \ll 1$, show that the energies E_1' and E_2' of the two crystals are related by

$$\frac{E_1'}{E_2'} = \frac{N_1}{N_2}\left(\frac{\mu_1}{\mu_2}\right)^2$$

Let Q_1 be the amount of energy absorbed by crystal 1 in the form of heat. Show that

$$Q_1 = \frac{N_2(\mu_2 B)^2}{1+\frac{N_2}{N_1}\left(\frac{\mu_2}{\mu_1}\right)^2}(\beta_1 - \beta_2) \qquad (3.153)$$

Verify that this expression gives the temperature order established previously.

3.8.3 Diatomic molecules

We reconsider the problem of diatomic molecules (Section 3.3.2) by taking into account nuclear degrees of freedom due to nuclear spins. Equation (3.64) for the partition function should be modified to include the contributions of the spins embodied in ζ_{nucl}

$$Z_N(\beta) = \frac{1}{N!}\left(\zeta(\beta)\zeta_{\text{rot}}\zeta_{\text{vib}}\zeta_{\text{nucl}}\right)^N \qquad (3.154)$$

The spins of the two nuclei forming the molecule are written $\vec{F}$ and $\vec{F}'$. These spins interact with each other and with the electrons giving energy levels that depend on the total spin $\vec{F}_{\text{tot}} = \vec{F} + \vec{F}'$. However, the energies involved in these levels are of the order of 10^{-6} eV.

1. Show that for temperatures of the order of a kelvin or higher, the entropy of the nuclear spins is $kN\ln\left[(2F+1)(2F'+1)\right]$, where F and F' are integers or half integers. Deduce from this the expression for ζ_{nucl} when $T \geq 1$ K.

2. We now consider exclusively the hydrogen molecule where the nuclei are either protons H (spin-1/2) or deuterons D (spin-1). We recall that the D mass is twice the H mass with a precision of 10^{-3}. *We will consider only temperatures ranging from about 20 K to the ambient temperature (300 K).* For the H_2 molecule, the rotation temperature T_{rot} $(kT_{\text{rot}} = \hbar^2/2I)$ is about 85.3 K whereas the vibration temperature T_{vib} ($kT_{\text{vib}} = \hbar\omega_{\text{vib}}$) is about 6125 K. Show that in the temperature range of interest, we may take $\zeta_{\text{vib}} = \text{const}$. Calculate T_{rot} and T_{vib} for the HD and D_2 molecules.

3. The fact that the H_2 molecule has two identical nuclei leads to restrictions on the allowed values of J, the rotational angular momentum, due to the symmetry properties of the total wave function under exchange of the nuclei. The total nuclear spin can take the values $F_{\text{tot}} = 1$ and $F_{\text{tot}} = 0$. Show that $F_{\text{tot}} = 1$ leads to odd values for J (ortho-hydrogen) while $F_{\text{tot}} = 0$ leads to even values for J (para-hydrogen). Show that, in the temperature range of interest, the partition function may now be written as

$$Z_N(\beta) = \frac{1}{N!}\left(\zeta(\beta)\zeta_{\text{rot,nucl}}\right)^N \tag{3.155}$$

and give the expression for $\zeta_{\text{rot,nucl}}$ with explicit calculation.

4. Show that for $T \ll T_{\text{rot}}$ we can write

$$\zeta_{\text{rot,nucl}} \simeq 1 + A\exp(-B/T) \tag{3.156}$$

and determine A and B. Calculate $\zeta_{\text{rot,nucl}}$ for $T \gg T_{\text{rot}}$. Show that the resulting expression is half of what we would have obtained had we ignored the indistinguishability of the two nuclei. Can you interpret this result? Give the expressions for the specific heat for $T \ll T_{\text{rot}}$ and $T \gg T_{\text{rot}}$.

5. Let $r(T)$ be the ratio of the para-hydrogen population to the total number of molecules

$$r(T) = \frac{N_{\text{para}}}{N_{\text{para}} + N_{\text{ortho}}} \tag{3.157}$$

What are the limiting values of $r(T)$ for $T \ll T_{\text{rot}}$ and $T \gg T_{\text{rot}}$? Calculate numerically $r(T)$ for $T = T_{\text{rot}}$. Sketch a representative curve for $r(T)$.

3.8.4 Models of a boundary surface

We wish to model an interface, for example the interface between the liquid and gaseous phases. For simplicity, we consider a two-dimensional model where *the interface is a line* and not a surface (Figure 3.19).

A Continuum model

1. The interface is a continuous line given by the differentiable single valued function $y(x)$ (Figure 3.19), and we assume that $(\mathrm{d}y/\mathrm{d}x)^2 \ll 1$. The energy of the interface is proportional to its length $\mathcal{L}$: $H = \alpha(\mathcal{L} - L)$, where α is a positive

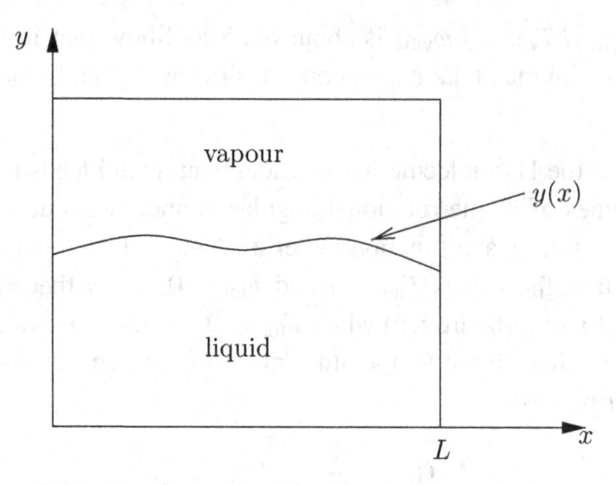

Figure 3.19 Continuum model of a one-dimensional interface.

constant, L the length of the container and αL is the energy of the shortest interface which is taken as the reference energy. Express $\mathcal{L}$ in terms of L and $(\mathrm{d}y/\mathrm{d}x)^2$.

2. We take the boundary conditions: $y(0) = y(L) = 0$. We can then write $y(x)$ in terms of its Fourier components

$$y(x) = \sum_{n=1}^{\infty} A_n \sin\left(\frac{\pi n x}{L}\right) \tag{3.158}$$

Show that H can be put in the form

$$H = \frac{\alpha\pi^2}{4L} \sum_{n=1}^{\infty} n^2 A_n^2 \tag{3.159}$$

3. Find, up to a multiplicative constant, the probability $P(A_n)$ to observe the mode n and the joint probability $P(A_n, A_m)$ to observe two modes. Use the result to obtain $\langle A_n \rangle$ and $\langle A_n A_m \rangle$.

4. Let $\Delta y = y(x) - y(x')$ be the interface height difference at the two points x and x'. When the two points x and x' are well separated, the average $\langle(\Delta y)^2\rangle^{1/2}$ may be considered as the width of the interface. Calculate $\langle y(x)y(x')\rangle$, say for $x' \geq x$, and use that to obtain $\langle(\Delta y)^2\rangle$. The following identity may be useful

$$\sum_{n=1}^{\infty} \frac{1}{n^2} \cos nx = \frac{\pi^2}{6} - \frac{\pi|x|}{2} + \frac{x^2}{4} \qquad 0 \leq x \leq 2\pi \tag{3.160}$$

B Discrete Ising model

We now use a two-dimensional Ising model where the interface is the line separating a region of positive magnetization (Ising spins +) from a region of negative

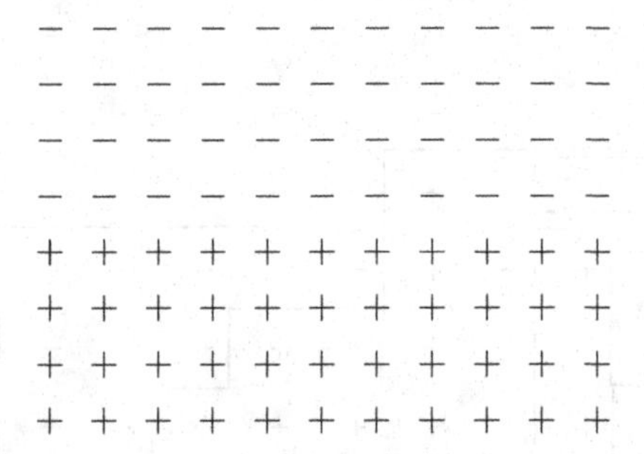

Figure 3.20 The ground state of the model.

magnetization (Ising spins −) (Figure 3.20). The spins are arranged on an $N \times N$ square lattice with unit spacing and $N \gg 1$. The system Hamiltonian is

$$H = -J \sum_{\langle i,j \rangle} S_i S_j \qquad S_i, S_j = \pm 1$$

where J is a positive constant and the sum $\sum_{\langle i,j \rangle}$ runs over near neighbours.

1. What does the quantity $\sum_{p=1}^{N} (1 + |y_p|)$ represent, where y_p is the height of the step at point x_p (for example, in Figure 3.21, $y_1 = 2$, $y_2 = 1$, $y_3 = 0$, $y_4 = -2$, etc.)? Show that the energy due to the presence of an interface (i.e. the energy difference between the ground state and a generic situation) is given by

$$H = 2JN + 2J \sum_{p=1}^{N} |y_p|$$

2. We now assume that the left extremity of the interface is fixed while the right one is free to move. The heights y_p are assumed to be integer random variables ranging from $-\infty$ to $+\infty$. Show that the partition function for the interface is given by

$$Z = \zeta^N$$

where ζ is the partition function for a single step and is given by

$$\zeta = \mathrm{e}^{-2\beta J} \coth(\beta J)$$

Calculate the surface tension, i.e. the free energy per unit length $f = F/N$, and the internal energy per unit length $\varepsilon = E/N$. Calculate the limits of f and ε for $\beta J \ll 1$ and $\beta J \gg 1$ and comment on the results.

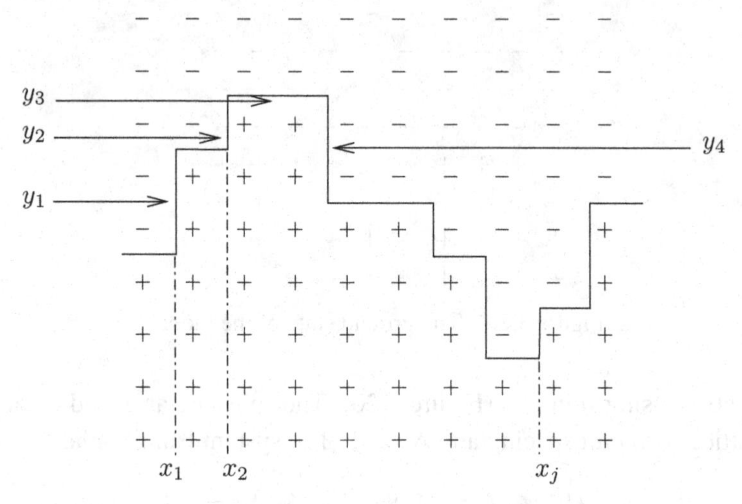

Figure 3.21 An arbitrary configuration of the system.

3. What is the probability that the step y_p be equal to $|p|$? Deduce the average values $\langle|y_p|\rangle$ and $\langle y_p^2\rangle$. Verify that

$$\varepsilon = 2J(1 + \langle|y|\rangle)$$

and that

$$\langle y_p^2\rangle = \frac{c}{\sinh^2 \beta J}$$

where c is a constant to be determined.

4. Since the y_p are independent variables, we can use the central limit theorem to evaluate the height difference Δy between two points on the interface at x_p and x_{p+q}

$$\Delta y = \sum_{i=p+1}^{p+q} y_i$$

When q is large, the average $\langle(\Delta y^2)\rangle^{1/2}$ may be considered as the thickness of the interface. Calculate $\langle(\Delta y^2)\rangle^{1/2}$. Useful relations

$$\coth x = \frac{1 + \mathrm{e}^{-2x}}{1 - \mathrm{e}^{-2x}} \qquad (\coth x)' = -\frac{1}{\sinh^2 x} \qquad 2 \sinh x \cosh x = \sinh 2x$$

3.8.5 Debye–Hückel approximation

An electrolyte is made of an equal number of positive and negative ions (charge $\pm q$) thus assuring global electric neutrality of the solution. On average over the whole volume of the solution, the densities of the two types of ions are equal (n), which means that the total ion density is $2n$. However, there are local fluctuations in the densities. To take them into account, we choose one particular ion as the origin of a reference frame. This ion will attract ions of opposite charge in its neighbourhood while repelling those with the same charge. Thus, the ion at the origin creates around it an 'ion cloud', which is spherically symmetric on average but inside of which the charge distribution is not uniform. We introduce the local charge densities $n^+(\vec{r})$ of ions of the same charge as the ion at the origin, and $n^-(\vec{r})$ of ions of opposite charge. The vector $\vec{r}$ gives the position with respect to the ion at the origin. We assume that deviations from the uniform situation are small, i.e. $|n^\pm(\vec{r}) - n| \ll n$, and to fix ideas we consider the ion at the origin to have the charge $+q$.

1. Let $\Phi(\vec{r})$ be the average electric potential inside the ionic cloud. The assumption of small deviations translates into the following condition: $|q\Phi(\vec{r})|/kT \ll 1$. Justify qualitatively this approximation and show that it allows us to write

$$n^\pm(\vec{r}) \simeq n\left(1 \mp \frac{q}{kT}\Phi(\vec{r})\right)$$

2. The charge densities are also related by the Poisson equation, which we can write in the following two forms

$$-\nabla^2\left(\Phi(\vec{r}) - \frac{q}{4\pi\varepsilon_0 r}\right) = \frac{q}{\varepsilon_0}(n^+(\vec{r}) - n^-(\vec{r}))$$

$$-\nabla^2\Phi(\vec{r}) = \frac{q}{\varepsilon_0}(n^+(\vec{r}) - n^-(\vec{r})) + \frac{q}{\varepsilon_0}\delta(\vec{r})$$

Comment on this equation and use the results of the previous question to rewrite it as a closed equation for Φ. Let $\widetilde{\Phi}(\vec{p})$ be the Fourier transform of $\Phi(\vec{r})$

$$\widetilde{\Phi}(\vec{p}) = \int \mathrm{d}^3 r\, \mathrm{e}^{\mathrm{i}\vec{p}\cdot\vec{r}}\, \Phi(\vec{r})$$

$$\Phi(\vec{r}) = \frac{1}{(2\pi)^3}\int \mathrm{d}^3 p\, \mathrm{e}^{-\mathrm{i}\vec{p}\cdot\vec{r}}\, \widetilde{\Phi}(\vec{p})$$

Calculate $\widetilde{\Phi}(\vec{p})$ and use the result to obtain $\Phi(\vec{r})$

$$\Phi(\vec{r}) = \frac{q}{4\pi\varepsilon_0 r}\mathrm{e}^{-r/b} \qquad b = \sqrt{\frac{\varepsilon_0 kT}{2nq^2}}$$

Note: It is easy to calculate the inverse transform of $\widetilde{\Phi}(\vec{p})$ using the method of residues. However, we give the following integral

$$\frac{1}{(2\pi)^3}\int \mathrm{d}^3 p\, \frac{\mathrm{e}^{-\mathrm{i}\vec{p}\cdot\vec{r}}}{\vec{p}^{\,2}+1} = \frac{1}{4\pi r}\mathrm{e}^{-r} \tag{3.161}$$

3. What is the total charge inside a sphere of radius R centred on the ion at the origin? It may be useful to note that $r\,\mathrm{e}^{-r/b} = b^2(\partial/\partial b)\mathrm{e}^{-r/b}$. What happens when $R \gg b$? Give a physical interpretation of your results, in particular the parameter b, called the *Debye length*.

4. Let $n_2^+(\vec{r},\vec{r}\,')$ be the pair density for ions of the same charge and $n_2^-(\vec{r},\vec{r}\,')$ be that for opposite charges (cf. Section 3.4.1). Justify the following expression for the electrostatic potential energy

$$E_{\text{pot}} = \frac{q^2}{4\pi\varepsilon_0}\int \mathrm{d}^3 r\, \mathrm{d}^3 r' \frac{1}{|\vec{r}-\vec{r}\,'|}\left(n_2^+(|\vec{r}-\vec{r}\,'|) - n_2^-(|\vec{r}-\vec{r}\,'|)\right) \tag{3.162}$$

Show that

$$n_2^\pm(\vec{r},\vec{r}\,') = n n^\pm(\vec{r}\,'-\vec{r})$$

and use this result to calculate explicitly the potential energy

$$E_{\text{pot}} = -nV\frac{q^2}{4\pi\varepsilon_0 b}$$

5. Let $\bar{r} \sim n^{-1/3}$ be the average distance between the ions. Verify that the condition $|q\Phi(\bar{r})|/kT \ll 1$ is equivalent to $\bar{r}/b \ll 1$. What is the physical interpretation of this relation?

3.8.6 Thin metallic film

We use a simple model to study the formation of thin metallic films deposited on a flat substrate. We can, for example, make thin gold films deposited on silicon by using an oven whose walls are made of silicon and into which we introduce a gas of gold whose temperature and pressure may be varied.

We use the following model to describe the thin film (Figure 3.22):

(*i*) the substrate is made of A sites ($A \gg 1$) on which the metallic atoms can be deposited,
(*ii*) the deposited atoms are spinless and their position completely determines their state,
(*iii*) on a given site, the atoms can pile up on top of each other without limit; each deposited atom has an energy $-\varepsilon_0$ ($\varepsilon_0 > 0$),
(*iv*) we ignore interactions between atoms deposited on neighbouring sites,
(*v*) each deposited atom contributes a height a to the thickness h of the film below it.

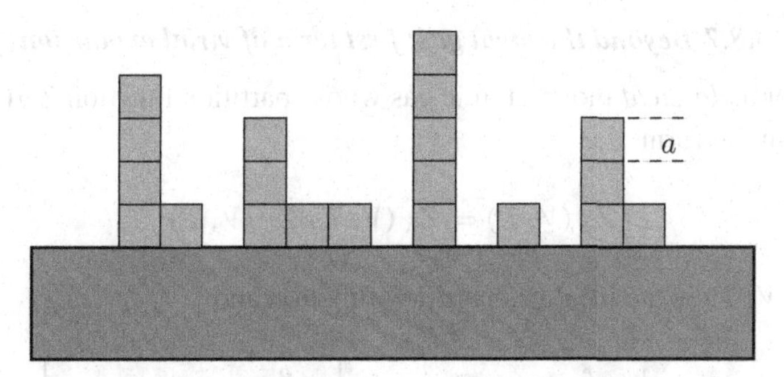

Figure 3.22 Deposition of a thin metallic film.

The gas of metal atoms (mass m and zero spin) is assumed to be ideal and classical and provides the system with a reservoir of atoms at chemical potential μ and temperature T.

1. Express the chemical potential μ of the gas in terms of its temperature T and pressure P.

2. Let $\mathcal{Q}$ be the grand canonical partition function for all of the sites and ξ that of a single site. Justify the following relation

$$\mathcal{Q} = \xi^A \tag{3.163}$$

and calculate ξ. Show that the chemical potential cannot exceed a limiting value to be determined and give the expression for $P_0(T)$, the pressure at this limit value.

3. Calculate $\overline{N}_0$, the average number of empty sites and $\langle h \rangle$, the average thickness of the film. We define the relative average deviation of the height of the film by

$$\frac{\Delta h}{\langle h \rangle} = \frac{\left(\langle h^2 \rangle - \langle h \rangle^2\right)^{1/2}}{\langle h \rangle}$$

Calculate this quantity.

4. Express the average height $\langle h \rangle$ in terms of P and P_0 and plot its behaviour as a function of P for two temperatures T_1 and T_2 ($T_1 < T_2$).

5. Express $\overline{N}_0$ as a function of P and P_0 and plot the behaviour as a function of P at a given temperature. Do the same for the quantity $\Delta h / \langle h \rangle$.

6. It can be shown that the metal film deposited on an insulating substrate conducts electricity when the number of empty sites is less than a known value x_0 called the percolation threshold. What is the minimum average thickness of the film for it to be conducting?

3.8.7 Beyond the ideal gas: first term of virial expansion

We consider a *classical* mono-atomic gas whose partition function $Z_N(V, T)$ may be written in the form

$$Z_N(V,T) = Z_K(V,T)\, Z_U(V,T) \tag{3.164}$$

where $Z_K(V, T)$ is the ideal gas partition function and

$$Z_U(V,T) = \frac{1}{V^N}\int \mathrm{d}^3r_1 \ldots \mathrm{d}^3r_N \exp\left[-\frac{\beta}{2}\sum_{i\neq j} U(|\vec{r}_i - \vec{r}_j|)\right] \tag{3.165}$$

The potential energy $U(|\vec{r}_i - \vec{r}_j|)$ of two molecules at $\vec{r}_i$ and $\vec{r}_j$ is a function only of the distance between them $|\vec{r}_i - \vec{r}_j|$. We shall study below the corrections to the ideal gas equation of state by obtaining a power series in the density $n = 1/v = N/V$. In this problem we stop at first order in $1/v$ but it is possible to derive a systematic expansion, called the virial expansion.

1. Show that the equation of state giving the pressure P as a function of v is

$$\frac{Pv}{kT} = 1 + \frac{V}{N}\frac{\partial \ln Z_U}{\partial V} = 1 + \frac{v}{N}\frac{\partial \ln Z_U}{\partial v} \tag{3.166}$$

2. We consider Z_U under the assumption that the gas is very dilute. It is convenient to write Z_U in the form

$$Z_U(V,T) = 1 + \frac{1}{V^N}\int \mathrm{d}^3r_1 \ldots \mathrm{d}^3r_N \left(\exp\left[-\frac{\beta}{2}\sum_{i\neq j} U(|\vec{r}_i - \vec{r}_j|\right] - 1\right) \tag{3.167}$$

From the form of the intermolecular potential (see Figure 3.23), it is clear that $\exp[-\beta U(r)]$ is different from 1 only for distances less than the range of the potential r_0. Sketch qualitatively $\exp[-\beta U(r)]$ for $T_1 \ll T$ and $T_2 \gg T$ where T is the temperature we used to obtain the curves in Figure 3.23. Let us consider a gas molecule and ignore interactions. What is the average number of molecules in the sphere of radius r_0 centered on this molecule? If the gas is very dilute, this number is $\ll 1$. What is the probability of finding another molecule at a distance $< r_0$ of the molecule at the centre? Under these conditions, what is the average number of pairs of molecules with a separation between the constituent molecules $< r_0$? What is the condition for this average number to be $\ll 1$? Show that if this

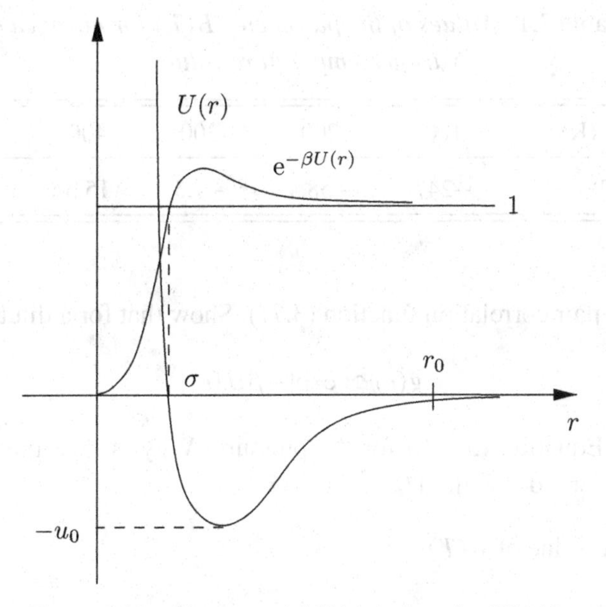

Figure 3.23 The potential $U(r)$ and $\exp[-\beta U(r)]$.

condition is satisfied we have

$$Z_U(V,T) \simeq 1 + \frac{N^2}{2V}\int d^3r \left(e^{-\beta U(r)} - 1\right) \tag{3.168}$$

$$= 1 - \frac{N^2}{V}B(T) \tag{3.169}$$

where $B(T)$, the second virial coefficient, is defined by the second equation. Show that the free energy is extensive and verify that, with the above approximations for Z_U, the equation of state becomes

$$\frac{Pv}{kT} = 1 + \frac{B(T)}{v} \tag{3.170}$$

3. Show that in the low density limit, the van der Waals equation (1.66)

$$\left(P + \frac{a}{v^2}\right)(v - b) = kT \tag{3.171}$$

reduces to an equation of the above type. Calculate $B(T)$ as a function of a, b and T. Can you give a physical interpretation of the result?

4. Another demonstration of (3.170). We start with Equation (3.87) for the exact expression for the pressure

$$P = nkT - \frac{1}{6}n^2 \int d^3r\, rU'(r)g(r) \tag{3.172}$$

Table 3.1 *Values of the parameter $B(T)$ for nitrogen at various temperature values.*

T (K)	100	200	300	400	500
$B(T)$ (Å^3)	−247	−58.4	−7.5	15.3	28.1

where g is the pair correlation function (3.77). Show that for a dilute gas we have

$$g(r) \simeq \exp[-\beta U(r)]$$

and re-obtain Equation (3.170) for the pressure. Why is the above equation for $g(r)$ not valid for a dense fluid?

5. What is the value of $B(T)$

(*i*) for a gas of hard spheres

$$U(r) = \begin{cases} +\infty & \text{if } r \le \sigma \\ 0 & \text{if } r > \sigma \end{cases}$$

(*ii*) for an inter-molecular potential given by

$$U(r) = \begin{cases} +\infty & \text{if } r \le \sigma \\ -u_0 & \text{if } \sigma \le r \le r_0 \\ 0 & \text{if } r > r_0 \end{cases}$$

How does the sign of $B(T)$ depend on the temperature? Can you give an interpretation for the change in the sign of $B(T)$? Hint: Examine the cases $T \to 0$ and $T \to \infty$.

(*iii*) for a Lennard-Jones potential

$$U(r) = 4u_0\left[\left(\frac{\sigma}{r}\right)^{12} - \left(\frac{\sigma}{r}\right)^{6}\right]$$

Show that $B' = B/\sigma^3$ is a universal function of the dimensionless parameter $\theta' = kT/u_0$, independent of the gas. Experiments show that hydrogen and helium do not fall on this universal curve at low temperature. Why?

6. The coefficient $B(T)$ has been measured for nitrogen as a function of T (although nitrogen is diatomic, it can be shown that the above arguments still hold). Calculate numerically $B(T)$ for the Lennard-Jones potential with the values $u_0/k = 95$ K and $\sigma = 3.74 \times 10^{-10}$ m and compare with the values in Table 3.1.

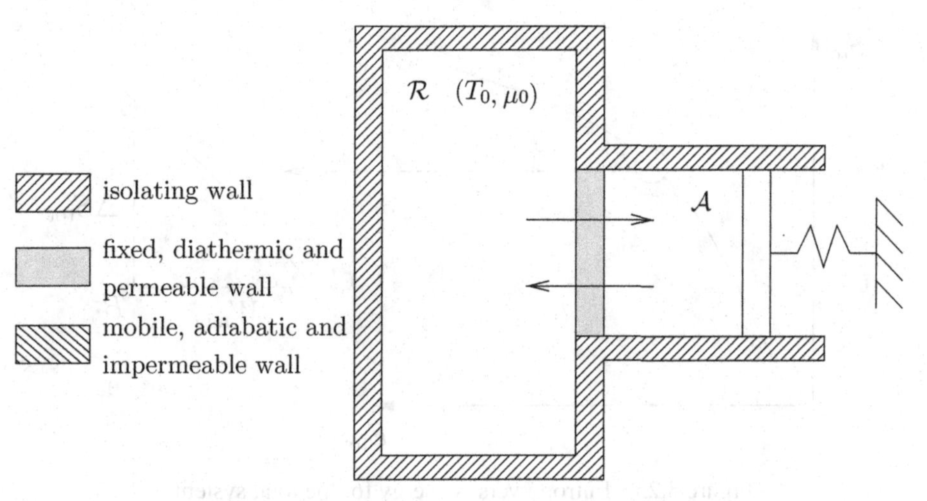

Figure 3.24 Schematic representation of the system under consideration.

3.8.8 Theory of nucleation

We consider a system $\mathcal{A}$ that can exchange particles and heat with a reservoir $\mathcal{R}$ at temperature T_0 and chemical potential μ_0 (Figure 3.24). The transformations of the reservoir will always be considered quasi-static. The system can, in addition, exchange energy in the form of work with the external world (represented by the spring in Figure 3.24).

1. Consider the case where the total system '$\mathcal{A} + \mathcal{R}$' is isolated from the external world, i.e. it does not exchange work with it. Show that the change in the quantity $(E - T_0S - \mu_0N)$ is negative when the system $\mathcal{A}$ is taken from an initial equilibrium state (i) to a final equilibrium state (f):

$$\Delta(E - T_0S - \mu_0N) \leq 0$$

In this equation, E, S and N are respectively the internal energy, the entropy and the number of particles of the system $\mathcal{A}$ (we use the same notations as in Sections 3.5.3 and 3.5.4).

2. Suppose that during a transformation $(i) \rightarrow (f)$ the system $\mathcal{A}$ receives an amount of work W from the external world. Show that

$$W \geq W_{\text{min}} = \Delta(E - T_0S - \mu_0N) \qquad (3.173)$$

where W_{min} is the work received during a reversible transformation $(i) \rightarrow (f)$.

3. Figure 3.25 shows, as a function of its total internal energy E_{tot}, the entropy S_{tot} of the total system at equilibrium. Justify qualitatively the form of the curve $S_{\text{tot}}(E_{\text{tot}})$.

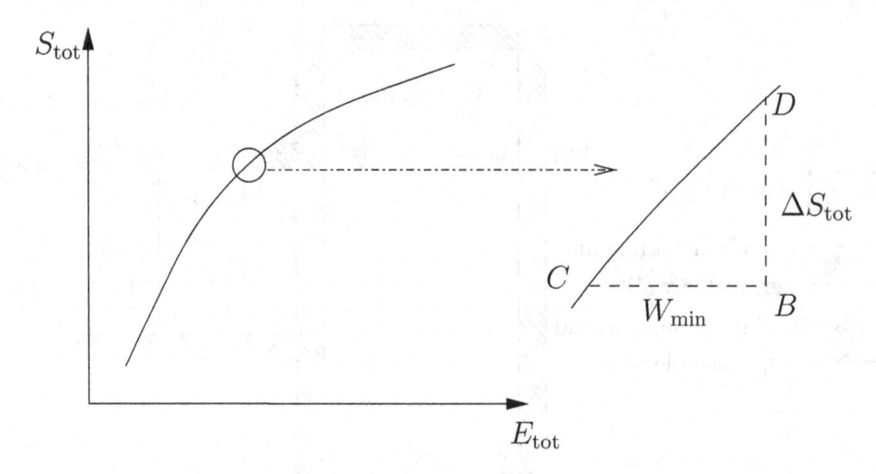

Figure 3.25 Entropy versus energy for the total system.

4. Starting in a situation of global equilibrium (D), a fluctuation in the total entropy can take the system to a non-equilibrium state (B). The point B may also be reached via an isentropic transformation by providing the system $\mathcal{A}$ (now isolated from $\mathcal{R}$) an amount of work W_{min} along the path $C \to B$. Assume that W_{min} is small compared to $E_{tot}(D)$. By using the entropy–energy curve, establish a relation between W_{min} and ΔS_{tot}. Show that the probability $\mathcal{P}$ of such a fluctuation is

$$\mathcal{P} \propto \exp[-W_{min}/kT_0]$$

It will be useful to recall that for an isolated system, the entropy is given by k times the logarithm of the number of accessible configurations.

5. We now consider the formation of a droplet in a supersaturated vapour in metastable equilibrium at temperature T. At this temperature, the pressure at the liquid–vapour transition is $\tilde{P}$, the vapour pressure is P_2 and the pressure at the interior of the liquid droplet is P_1 (Figure 3.26). Show that the formation of a droplet of volume V_1 is accompanied by a change in the grand potential given by

$$\Delta \mathcal{J} = V_1(P_2 - P_1)$$

The above expression corresponds to a situation where the gas–liquid interface is plane. To include the effect of the shape of the droplet, one should include a surface energy term, σA, where σ is the surface tension

$$\Delta \mathcal{J} = V_1(P_2 - P_1) + \sigma A$$

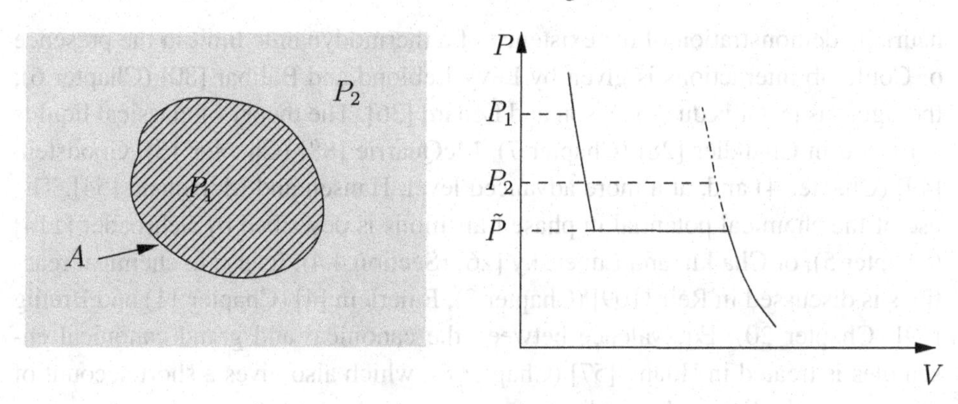

Figure 3.26 Phase diagram.

Argue that $\Delta\mathcal{J}$ corresponds to the minimum work to be furnished to the vapour to form the droplet

$$W_{\min} = \Delta\mathcal{J}$$

6. Calculate $W_{\min}$ for a spherical droplet of radius R. Plot $W_{\min}$ as a function of R and verify that it passes through a maximum for $R = R^*$ to be determined. What happens if the radius of the droplet is larger than the critical radius? Smaller?

7. Calculate the probability for the formation of a droplet of critical radius.

8. We take $\delta P_1 = P_1 - \tilde{P}$ and $\delta P_2 = P_2 - \tilde{P}$ ($|\delta P_1|, |\delta P_2| \ll \tilde{P}$). By using the Gibbs–Duhem relation, show that

$$v_1\,\delta P_1 = v_2\,\delta P_2$$

and that

$$R^* = \frac{2\sigma v_1}{\mu_2(P_2, T) - \mu_1(P_2, T)} \tag{3.174}$$

where v_1 and v_2 are the specific volumes of the liquid and gas phases respectively.

3.9 Further reading

The results presented in this chapter are classic and appear in all books on statistical mechanics. For additional reading we recommend Reif [109] (Chapter 7). The Ising model in one dimension is treated in detail by Le Bellac [72] (Chapter 1); the solution in two dimensions (difficult!) is described in Onsager's article [97]. See also Huang [57] (Chapter 17) and Landau and Lifshitz [70] (Section 151); the historical aspects of the model are discussed by Brush [22]. For a demonstration of the Lee–Yang theorem see Huang [57] (Appendix C) or Ma [85] (Chapter 9). A heuristic demonstration of the existence of a thermodynamic limit in the presence of Coulomb interactions is given by Lévy-Leblond and Balibar [80] (Chapter 6); the rigorous proof is due to Dyson and Lenard [36]. The theory of classical liquids is treated in Chandler [28] (Chapter 7), McQuarrie [88] (Chapter 13), Goodstein [49] (Chapter 4) and, at a more advanced level, Hansen and McDonald [54]. The use of the chemical potential in phase transitions is described by Schroeder [114] (Chapter 5), or Chaikin and Lubensky [26] (Section 4.4). Its use in chemical reactions is discussed in Reif [109] (Chapter 8), Baierlein [4] (Chapter 11) and Brenig [19] (Chapter 20). Equivalence between the canonical and grand canonical ensembles is treated in Huang [57] (Chapter 8), which also gives a short account of second quantization in Appendix A.

Additional information on the problems in this chapter can be found in the following references. The original experiment on negative temperatures was done by Purcell and Pound [105]. The Debye–Hückel approximation is discussed in Landau and Lifshitz [70] (Section 78). Further information on the virial expansion is found in Landau and Lifshitz [70] (Section 74) and in McQuarrie [88] (Chapter 12). For the theory of nucleation, consult Landau and Lifshitz [70] (Section 162).

4
Critical phenomena

In Chapter 3, we examined two examples of phase transitions, the paramagnetic–ferromagnetic transition (Section 3.1.4) and the liquid–vapour transition (Section 3.5.4). In the latter case, thermodynamic functions such as entropy or specific volume are discontinuous at the transition. When, for example, one varies the temperature T at constant pressure, the transition takes place at a temperature T_c where the Gibbs potentials of the two phases are equal, but the two phases coexist at T_c and are either stable or metastable in the vicinity of T_c. Each of the phases carries with it its own entropy, specific volume, etc., which are in general different at $T = T_c$, hence the discontinuities. Such a phase transition is called a *first order phase transition*. The picture is quite different in the paramagnetic–ferromagnetic phase transition: the thermodynamic quantities are continuous at the transition, the transition is not linked to the crossing of two thermodynamic potentials and one never observes metastability. Such a transition is called a *second order, or continuous, phase transition*.[1] The transition temperature is called the *critical temperature* and is denoted by T_c.

The new and remarkable feature which we shall encounter in continuous phase transitions is the existence of cooperative phenomena. To be specific, think of a spin system, for example an Ising or a Heisenberg model, where the interactions between spins are limited to nearest neighbours. More generally we shall consider *short range interactions*, which decrease rapidly as a function of the distance between spins, and we exclude from our study all long range interactions. We shall see that close to the critical point the correlations between spins extend to very large distances, even though the basic microscopic interaction between spins is

[1] 'Continuous phase transition' seems to be favoured in modern terminology, which we shall follow. A classification of phase transitions was proposed some eighty years ago by Ehrenfest, based on a hypothetical behaviour of the derivatives of the Gibbs potential: first order, second order, third order . . . , and is still to be found in some books. This classification has been now abandoned, because it does not correspond to physical reality, and nowadays one refers only to first order and continuous phase transitions.

short range. In other words, the correlation length, to be defined precisely in (4.30), tends to infinity at the critical temperature. Because of this cooperative character, some of the features of the transition become independent of the details of the basic interaction. Close to the critical temperature, *a priori* completely different systems such as a magnet and a fluid may behave in a very similar way. This allows one to classify continuous phase transitions in very general universality classes. This remarkable property of universality explains why a whole (and long!) chapter is devoted to the description of continuous phase transitions.

We have already encountered two important concepts in the theory of phase transitions, namely symmetry breaking and order parameter. It often happens that the symmetry group of the low temperature phase is a subgroup of that of the high temperature phase. An example is the paramagnetic–ferromagnetic phase transition of Section 3.1.4 in zero magnetic field. The paramagnetic phase is invariant under any space rotation, while the ferromagnetic phase is invariant only under rotations around the direction of the magnetization. This is the phenomenon called symmetry breaking, already briefly evoked in Section 3.1.4, and which will be examined in detail later in this chapter. The order parameter is a quantity which vanishes in the higher symmetry phase and is non-zero in the lower symmetry phase. In the paramagnetic–ferromagnetic phase transition, one can choose the magnetization, or any quantity proportional to it, as the order parameter. This order parameter vanishes continuously when $T \to T_c$. Although symmetry breaking and order parameters are more generally associated with continuous phase transitions, they may also occur in first order transitions, but in such a case the order parameter is discontinuous at $T = T_c$.[2] There also exist continuous phase transitions without symmetry breaking and order parameter: an example is the two-dimensional XY model.

The present chapter will address the theory of continuous phase transitions, also called *critical phenomena*, but we shall limit ourselves to the case of transitions with symmetry breaking. We shall first give a rather detailed account of symmetry breaking, taking the Ising model as a pedagogical example. Section 4.2 will be devoted to a very important approximation method, mean field theory, and Section 4.3 to a variant of this approximation, Landau's theory of continuous phase transitions.

Until the early 1970s, the only theoretical results on critical phenomena came from exact calculations (essentially Onsager's solution of the two-dimensional Ising model) and from the mean field or Landau's theory. Both Onsager's solution and experimental data pointed clearly toward the quantitative failure of Landau's

[2] A standard example is the ferro-electric transition in barium titanate. An exception where the order parameter is continuous is given by the Bose–Einstein transition in the non-interacting case, see Section 5.5, but, as pointed out in this section, the non-interacting Bose–Einstein transition is on the verge of being pathological.

theory, and a proper understanding of critical phenomena was considered at that time as one of the most important challenges of theoretical physics. The difficulty of the problem is that, close to a critical point, a huge number of degrees of freedom interact in a complicated way. How to effectively reduce this number of degrees of freedom was understood by Wilson, who introduced the renormalization group (RG) idea into the game. However, one should be aware that the renormalization group is not a recipe able to give almost automatically a solution to a physical problem, but rather a way of thinking about this problem, and in many instances the use of the RG framework involves a large amount of guesswork that is not always transparent to the non-expert (and even sometimes to the expert...). Wilson understood that the RG flow was able to give a qualitative picture of critical phenomena, which could be translated into a quantitative picture by using methods of quantum field theory (although quantum mechanics is completely irrelevant to critical phenomena, except at zero temperature). In Sections 4.4 and 4.5, we shall illustrate with the example of critical phenomena the notion of renormalization group flow, which is at the basis of RG methods.[3] We shall also give a self-contained and detailed calculation of critical exponents, which includes a flavour of field-theoretical methods, but without going into details (Feynman graphs) of the perturbative expansion.

4.1 Ising model revisited

4.1.1 Some exact results for the Ising model

In Chapter 3 we introduced the Ising model as the simplest non-trivial model of ferromagnetism; the Hamiltonian was written in (3.30) and the partition function in (3.31). From an easy computation in space dimension[4] $D = 1$ we came to the conclusion that no phase transition occurs, a somewhat disappointing result at first sight.[5] We are now going to prove that this result is specific to dimension one, and that a ferromagnetic phase transition does exist for any dimension $D \geq 2$. In an ideal scenario, one would hope to show the existence of a phase transition from a direct calculation of the partition function. Unfortunately, as was mentioned in Chapter 3, the analytic calculation of the partition function is possible only in dimension $D = 2$ in zero external magnetic field, $B = 0$. This calculation is extremely long and involved, although all steps are elementary, and it is clearly

[3] One problem in explaining the renormalization group is that simple examples are too simple to be really convincing, and convincing examples require a large amount of technicalities.

[4] We hope that, despite the identity of notations, no confusion will arise between the space dimension and the density operator.

[5] This negative result was first obtained by Ising, as a Ph.D. student of Lenz's. Ising became so discouraged by this result that he quit physics.

beyond the scope of this book; we refer the interested reader to the literature. A simpler argument, first put forward by Peierls, will allow us to prove the existence of a phase transition for any $D \geq 2$. We have shown in Section 3.1.5 the existence of a thermodynamic limit for the free energy per spin[6]

$$f = \lim_{N\to\infty} \frac{1}{N} F_N = \lim_{N\to\infty} -\frac{kT}{N} \ln Z_N$$

which is independent of the boundary conditions. N is the number of spins, Z_N the partition function and $F_N = -kT \ln Z_N$ the free energy. If we assume a zero external magnetic field, the symmetry of the Ising Hamiltonian under the operation $S_i \to -S_i$ entails that the magnetization per spin[7]

$$M = \frac{1}{N} \sum_{i=1}^{N} \langle S_i \rangle \tag{4.1}$$

must vanish whatever the number N of spins. It is thus impossible to obtain a non-zero magnetization without imposing some peculiar boundary conditions. For the sake of simplicity, we work from now on in dimension $D = 2$, but all the following arguments generalize to higher dimensions in a straightforward way. We shall impose the following boundary conditions: *all the spins along the boundary must be up*. A more physical condition with identical results will be used later on, by putting the spin system in a vanishing magnetic field. Notice that the ratio of the spins whose orientation is fixed *a priori* over the total number of spins tends to zero as $1/\sqrt{N}$. At $T = 0$, the spin configuration is in its ground state, with all spins up, and the energy is taken to be zero, by convention.

We shall give below a rigorous proof that a non-zero magnetization follows from these boundary conditions in the $N \to \infty$ limit. Before going into the details of the proof, let us begin with a handwaving argument to show that, in contrast to the one-dimensional case, the spontaneous magnetization does not vanish at low temperatures for $D \geq 2$. In $D = 1$, the preceding boundary conditions correspond to the first and last spin on the line being fixed in the up position. At $T = 0$, all spins are up, in the ground state configuration. The first excited state is obtained by creating two kinks (Problem 3.8.1), the energy cost being $4J$, *which does not depend on the number of down spins*. Since the kinks may be located anywhere on the line, the corresponding entropy is $\sim 2k \ln N$, and the free energy of the configurations with two kinks is

$$F \simeq 4J - 2kT \ln N$$

[6] The number of spins is now denoted by N, and not by N^2 as in Section 3.1.5.

[7] In the present chapter, in order to avoid cumbersome multiplicative factors, it will be convenient to define the site magnetization M_i as a dimensionless number, the average value of the spin S_i: $M_i = \langle S_i \rangle$, rather than $M_i = \mu \langle S_i \rangle$.

If N is large enough, then no matter how small T is, the free energy of configurations with two kinks will be lower than that of the ground state. These configurations will be more probable, and since their average magnetization is zero, the average magnetization of the one-dimensional Ising model will vanish. Let us contrast this argument with the corresponding one in the two-dimensional case, where we shall try to create a 'bubble' of down spins in the ground state up spins. These down spins will be enclosed by a line of length b, the perimeter of the bubble, and the energy cost will be $2bJ$. Let us find the entropy of the configuration with a bubble of perimeter b. Call $\nu(b)$ the number of configurations; we have the bound

$$\nu(b) \leq N\, 3^{b-1} \tag{4.2}$$

The factor N takes into account the fact that the polygon may be located anywhere on the lattice. To explain the factor 3^{b-1}, one remarks that to draw the polygon starting from some point on the lattice, one can take steps of length a, where a is the lattice spacing, in three directions at each lattice site: the drawing of a polygon is analogous to a random walk in which the walker never makes a backward step.[8] The free energy follows

$$F \simeq 2bJ - kT[\ln N + b \ln 3]$$

The crucial difference with the one-dimensional case is that the energy cost is proportional to b, and if we want the free energy to be lower than that of the ground state, then

$$b \lesssim \frac{kT \ln N}{2J - kT \ln 3}$$

Hence, for $T < 2J/(k \ln 3)$, configurations with a bubble of down spins will have a negligible probability for $b \geq b_{\max}$, where $b_{\max} \sim \ln N$, and the percentage of down spins will be $\sim \ln^2 N/N \to 0$ in the limit of large N. This argument makes likely the existence of a spontaneous magnetization for $T \lesssim 2J/(k \ln 3)$.

Let us now give the rigorous argument due to Peierls. This argument relies on the observation that there exists a one-to-one correspondence between a spin configuration and a set of polygons.[9] This is illustrated in Figure 4.1, where one draws lines of length one – from now on in this section we take the lattice spacing a to be unity – between up spins and down spins, starting from the boundary. The spin orientation changes every time one crosses one of the lines. These lines form

[8] A more precise bound would be $\nu(b) \leq N3^{b-1}(4/2b)$, as the first step can be taken in four different directions and one may start from an arbitrary point on the polygon, going either clockwise or counterclockwise. This improvement does not make any difference for the argument which follows.

[9] This observation is also the starting point for the simplest analytic calculation of the partition function. The original Onsager solution uses a transfer matrix approach.

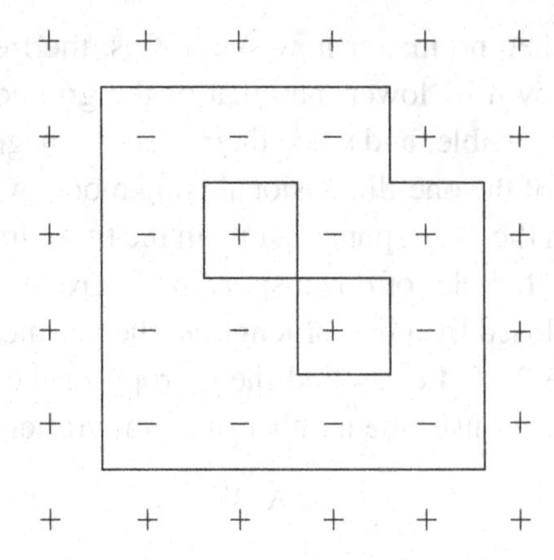

Figure 4.1 Correspondence between a spin configuration and a set of polygons.

closed polygons, which are characterized by their perimeter b and an index j: $j = 1, \ldots, \nu(b)$. The index j specifies the type of polygon as well as its position on the lattice. A polygon configuration $\mathcal{C}$ (or equivalently a spin configuration) will be denoted by

$$\mathcal{C} = \{P_{b_1}^{(j_1)}, \ldots, P_{b_n}^{(j_n)}, \ldots\}$$

and a sum over configurations is equivalent to a sum over all values of the spins

$$\sum_{\mathcal{C}} = \sum_{S_1=\pm 1} \cdots \sum_{S_N=\pm 1}$$

Let $N_b^{(j)}$ denote the number of sites contained inside polygon $P_b^{(j)}$ and let $N_+(\mathcal{C})$ and $N_-(\mathcal{C})$ be the number of up spins and of down spins of the configuration $\mathcal{C}$; the value of M for this configuration is

$$M(\mathcal{C}) = \frac{1}{N}\Big[N_+(\mathcal{C}) - N_-(\mathcal{C})\Big] = 1 - \frac{2N_-(\mathcal{C})}{N}$$

Our strategy will be to prove that $\langle N_-(\mathcal{C})\rangle/N < 1/2$ if the temperature is low enough. The following inequality holds for an arbitrary configuration $\mathcal{C}$

$$N_-(\mathcal{C}) \le \sum_b \sum_{1\le j\le \nu(b)} \chi_b^{(j)} N_b^{(j)} \tag{4.3}$$

where $\chi_b^{(j)}$ is the characteristic function of $P_b^{(j)}$

$$\chi_b^{(j)} = \begin{cases} 1 & \text{if } P_b^{(j)} \in \mathcal{C} \\ 0 & \text{if } P_b^{(j)} \notin \mathcal{C} \end{cases} \tag{4.4}$$

and $N_b^{(j)}$ the number of sites within $P_b^{(j)}$. A strict inequality holds as soon as there are boxed in polygons as in Figure 4.1, since in that case some of the polygons contain up spins. The equality sign in (4.3) holds only when there are no boxed in polygons. We also notice the following inequality

$$N_b^{(j)} \le \left(\frac{b}{4}\right)^2 \tag{4.5}$$

since for a given perimeter the square is the polygon with the largest area. Let us now bound the average $\langle \chi_b^{(j)} \rangle$

$$\langle \chi_b^{(j)} \rangle = \frac{1}{Z} \sum_{P_b^{(j)} \in \mathcal{C}} \mathrm{e}^{-\beta E(\mathcal{C})} = \frac{1}{\sum_{\mathcal{C}} \mathrm{e}^{-\beta E(\mathcal{C})}} \sum_{P_b^{(j)} \in \mathcal{C}} \mathrm{e}^{-\beta E(\mathcal{C})} \tag{4.6}$$

In the numerator, one sums over all configurations which contain the polygon $P_b^{(j)}$, for which $\chi_b^{(i)} = 1$. Let us define the configuration $\mathcal{C}^*(P_b^{(j)})$ which is obtained from $\mathcal{C}(P_b^{(j)})$ by flipping all spins inside $P_b^{(j)}$. As in the preceding discussion, the energy difference between the two configurations comes from the boundary

$$E(\mathcal{C}^*) = E(\mathcal{C}) - 2bJ \tag{4.7}$$

A lower bound on the denominator is obtained by summing only over the configurations belonging to $\mathcal{C}^*$, so that

$$\langle \chi_b^{(j)} \rangle \le \frac{1}{\sum_{\mathcal{C}^*(P_b^{(j)})} \mathrm{e}^{-\beta E(\mathcal{C}^*)}} \sum_{P_b^{(j)} \in \mathcal{C}} \mathrm{e}^{-\beta E(\mathcal{C})} = \mathrm{e}^{-2\beta bJ} \tag{4.8}$$

From (4.3) and (4.5) we get

$$N_-(\mathcal{C}) \le \sum_b \left(\frac{b}{4}\right)^2 \sum_j \chi_b^{(j)}$$

and from (4.2) and (4.8)

$$\begin{aligned}\langle N_- \rangle &\le \sum_b \left(\frac{b}{4}\right)^2 \sum_j \langle \chi_b^{(j)} \rangle \le \sum_b \left(\frac{b}{4}\right)^2 \mathrm{e}^{-2\beta bJ} \sum_{j=1}^{\nu(b)} 1 \\ &\le \sum_b \left(\frac{b}{4}\right)^2 \nu(b)\, \mathrm{e}^{-2\beta bJ} \le N \sum_b \left(\frac{b}{4}\right)^2 3^{b-1}\, \mathrm{e}^{-2\beta bJ}\end{aligned}$$

We finally get the following bound for $\langle N_- \rangle / N$

$$\frac{\langle N_- \rangle}{N} \le \sum_{b=4,6,\ldots} \left(\frac{b}{4}\right)^2 3^{b-1}\, \mathrm{e}^{-2\beta b J} \tag{4.9}$$

The series is convergent provided $3\exp(-2\beta J) < 1$. Defining the temperature $T_0 = 1/k\beta_0$ by

$$\sum_{b=4,6,\ldots} \left(\frac{b}{4}\right)^2 3^{b-1}\, \mathrm{e}^{-2\beta_0 b J} = \frac{1}{2} \tag{4.10}$$

we see that for $T < T_0$ we have $\langle N_- \rangle / N < 1/2$ whatever N, and thus $M > 0$.

Another important exact result is the Lee–Yang theorem. Let us modify in a trivial way the Hamiltonian (3.30)

$$H \to H' = -J \sum_{\langle i,j \rangle} S_i S_j - \mu B \sum_i (S_i - 1) = H_0 - \mu B \sum_i (S_i - 1)$$

The modification is trivial because this Hamiltonian differs from (3.30) by a constant $\mu B N$. Let us define

$$z = \mathrm{e}^{-2\beta\mu B}$$

The partition function Z_N is a polynomial of order N in z. Indeed

$$Z_N = \sum_{\mathcal{C}} \mathrm{e}^{-\beta H_0[S_i]} \prod_i \mathrm{e}^{\beta\mu B(S_i - 1)} \tag{4.11}$$

If n is the number of down spins in the configuration $\mathcal{C}$, one has

$$\sum_i (S_i - 1) = -2n$$

and

$$Z_N = \sum_{n=0}^{N} z^n Q_n \tag{4.12}$$

where Q_n is a positive number. For finite N, Z_N is an analytic function of z which does not vanish, so that $\ln Z$ and all thermodynamic functions are analytic functions of z. Since a phase transition is characterized by non-analytic behaviour of the thermodynamic functions, *it can only occur in the thermodynamic limit* $N \to \infty$ (or $V \to \infty$).[10] In this limit one can prove the following theorem

[10] The reader, who has undoubtely observed an ice cube floating in a glass of water, may find this statement a bit surprising. What is meant by this statement is that the *mathematical* signature of a phase transition can only be seen in the infinite volume limit.

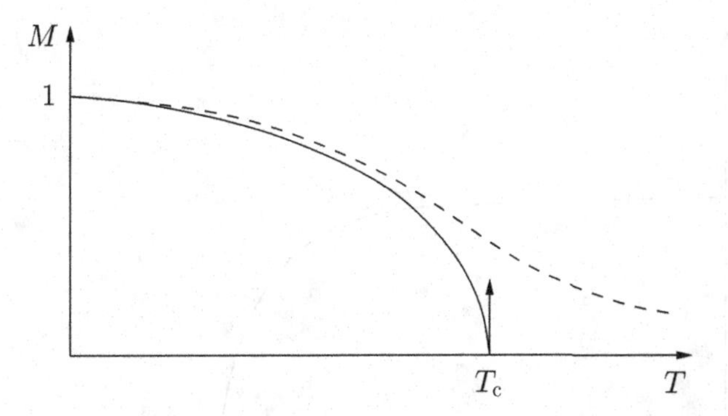

Figure 4.2 The magnetization as a function of T for $B = 0$ (full line) and $B \neq 0$ (dashed line). Note the vertical tangent at $T = T_c$.

Theorem (Lee and Yang) Let $Z = \lim_{N\to\infty}(Z_N)^{1/N}$. Z may vanish only if $|z| = 1$ and the thermodynamic functions are analytic functions of z if z does not lie on the circle $|z| = 1$, namely if $B \neq 0$.

As a consequence, *a phase transition can only occur if* $B = 0$. This result can be understood intuitively by looking at Figure 4.2. The zero field magnetization clearly has non-analytic behaviour as a function of the temperature, with a vertical derivative at $T = T_c$, but it varies continuously in a non-zero field. The occurrence of a phase transition can be checked in the exact $D = 2$ Onsager solution, which gives the free energy per spin f in zero magnetic field as

$$f = kT \ln \frac{1-x^2}{2} - \frac{kT}{8\pi^2}\int_0^{2\pi} \mathrm{d}p_x\,\mathrm{d}p_y \ln\left[(1+x^2)^2 - 2x(1-x^2)(\cos p_x + \cos p_y)\right] \quad (4.13)$$

with $x = \tanh(J/kT)$. The singularity can only arise from the vanishing of the argument of the logarithm which occurs for $\cos p_x = \cos p_y = 1$ and $x = \sqrt{2} - 1$ since

$$(1+x^2)^2 - 2x(1-x^2) = (x^2 - 2x - 1)^2 = 0 \text{ if } x = \sqrt{2} - 1$$

The critical value of x is thus $x_c = \sqrt{2} - 1$. Useful formulae for T_c are

$$\boxed{\sinh \frac{2J}{kT_c} = 1 \text{ or } kT_c = \frac{2J}{\ln(1+\sqrt{2})} \simeq 2.27J} \quad (4.14)$$

The value of T_c is deduced from a duality argument in Exercise 4.6.1. From the

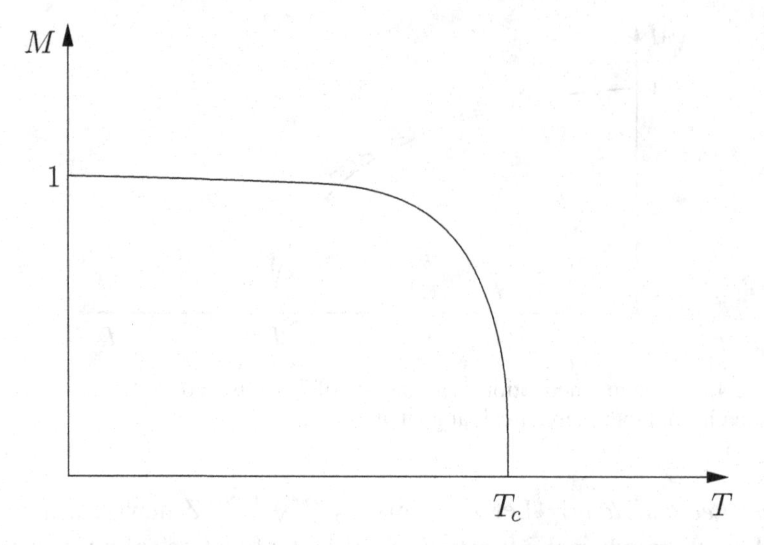

Figure 4.3 The zero-field magnetization of the two-dimensional Ising model.

expression of f one readily derives the behaviour of the specific heat close to $T = T_c$

$$C \propto \ln |T - T_c| \tag{4.15}$$

Since no analytic calculation of the partition function is available for $B \neq 0$, one has to rely on the expression for the correlation functions[11] to derive the expression for the zero field magnetization for $T \leq T_c$. The result, first derived by Yang, reads

$$M = \left(1 - \left[\sinh\left(\frac{2J}{kT}\right)\right]^{-4}\right)^{1/8} \tag{4.16}$$

and is sketched in Figure 4.3. Close to the critical temperature the magnetization behaves as

$$M \propto (T_c - T)^{1/8} \tag{4.17}$$

4.1.2 Correlation functions

We have mentioned in the introduction the cooperative character of continuous phase transitions. A quantitative measure of this cooperative character is given by the correlation length, which is deduced from the *correlation function*. In the case of Ising spins, the correlation function of two Ising spins S_i and S_j is defined as

[11] More precisely, in the limit $|\vec{r}_i - \vec{r}_j| \to \infty$, $\langle S_i S_j \rangle \to \langle S \rangle^2 = M^2$, as in this limit $\langle S_i S_j \rangle_c \to 0$ (see (4.25)).

the average value of the product $S_i S_j$ of the spins located at sites i and j, taken with the normalized Boltzmann weight $Z^{-1}\exp(-\beta H)$

$$\langle S_i S_j \rangle = \frac{1}{Z(B)} \sum_{\mathcal{C}} S_i S_j \mathrm{e}^{-\beta H} \tag{4.18}$$

where H is given by (3.30). To develop some familiarity with the correlation function, let us compute it explicitly in space dimension $D = 1$ and in zero magnetic field. Let Z_N denote the partition function for N Ising spins with periodic boundary conditions. We shall use the transfer matrix method described in Section 3.1.4. The correlation function is obtained by summing over all spin configurations (as in Section 3.1.4, $K = \beta J$)

$$\langle S_i S_j \rangle_N = \frac{1}{Z_N} \sum_{S_1=\pm 1} \cdots \sum_{S_N=\pm 1} \cdots \exp(K S_{i-1} S_j)[S_i] \exp(K S_i S_{i+1}) \cdots$$
$$\times \exp(K S_{j-1} S_j)[S_j] \exp(K S_j S_{j+1}) \cdots$$

Let us introduce the Pauli matrix σ_z

$$\sigma_z = \begin{pmatrix} 1 & 0 \\ 0 & -1 \end{pmatrix}$$

The product $\sigma_z \mathcal{T}$ of this matrix with the transfer matrix $\mathcal{T}$, which is given for $B = 0$ by

$$\mathcal{T} = \begin{pmatrix} \mathrm{e}^{K} & \mathrm{e}^{-K} \\ \mathrm{e}^{-K} & \mathrm{e}^{K} \end{pmatrix} \tag{4.19}$$

has matrix elements

$$(\sigma_z \mathcal{T})_{S_k S_l} = S_k \mathcal{T}_{S_k S_l}$$

Let R be the matrix which diagonalizes $\mathcal{T}$

$$R\mathcal{T}R^{-1} = \begin{pmatrix} \lambda_+ & 0 \\ 0 & \lambda_- \end{pmatrix} = \mathcal{D} \qquad R = R^{-1} = \frac{1}{\sqrt{2}} \begin{pmatrix} 1 & 1 \\ 1 & -1 \end{pmatrix}$$

The calculation of the partition function in Section 3.1.4 may be cast into the form

$$\mathrm{Tr}\, \mathcal{T}^N = \mathrm{Tr}\, (R^{-1}\mathcal{D}RR^{-1}\mathcal{D} \cdots \mathcal{D}R) = \mathrm{Tr}\, \mathcal{D}^N = \lambda_+^N + \lambda_-^N$$

We must now compute, assuming for definiteness $j > i$

$$\mathrm{Tr}\left[\mathcal{T}^i \sigma_z \mathcal{T}^{(j-i)} \sigma_z \mathcal{T}^{(N-j)}\right] = \mathrm{Tr}\left[\cdots \mathcal{D}R\sigma_z R^{-1}\mathcal{D} \cdots \mathcal{D}R\sigma_z R^{-1}\mathcal{D} \cdots\right]$$

As

$$R\sigma_z R^{-1} = \sigma_x = \begin{pmatrix} 0 & 1 \\ 1 & 0 \end{pmatrix}$$

we see that, in contrast to the calculation of the partition function, the eigenvalues λ_+ and λ_- of $\mathcal{T}$ will be interchanged for all spins lying between sites i and j

$$\sigma_x \mathcal{D}^{(j-i)} \sigma_x = \sigma_x \begin{pmatrix} \lambda_+^{(j-i)} & 0 \\ 0 & \lambda_-^{(j-i)} \end{pmatrix} \sigma_x = \begin{pmatrix} \lambda_-^{(j-i)} & 0 \\ 0 & \lambda_+^{(j-i)} \end{pmatrix}$$

so that in the limit $N \to \infty$ we have

$$\begin{aligned} \langle S_i S_j \rangle &= \lim_{N\to\infty} \lambda_+^{-N} \left[\lambda_+^{N(j-i)} \lambda_-^{(j-i)} + \lambda_-^{N(j-i)} \lambda_+^{(j-i)} \right] \\ &= \lambda_+^{-N} \lambda_+^{N-(j-i)} \lambda_-^{(j-i)} = \left(\frac{\lambda_-}{\lambda_+} \right)^{(j-i)} \end{aligned}$$

This result illustrates a general property: in the large N limit, the largest eigenvalue of the transfer matrix controls the partition function while its ratio to the second largest controls the correlation function. From the expression of the eigenvalues derived in Section 3.1.4 we derive (the absolute value $|i-j|$ takes care of the case $i > j$)

$$\langle S_i S_j \rangle = (\tanh K)^{|i-j|} = \left(\tanh \frac{J}{kT} \right)^{|i-j|} = \exp\left(-|i-j| \ln \tanh \frac{J}{kT} \right) \tag{4.20}$$

Equation (4.20) allows us to introduce the important concept of *correlation length*, for which the standard notation is ξ. Since the lattice spacing is taken to be unity, $|i-j|$ is nothing other than the distance r_{ij} between the sites i and j, and we have

$$\langle S_i S_j \rangle = \exp\left(-\frac{r_{ij}}{\xi} \right) \tag{4.21}$$

In the one-dimensional Ising model ξ is obtained from (4.20)

$$\xi = \frac{1}{|\ln \tanh J/kT|}$$

We observe that ξ is a decreasing function of the temperature, and that $\xi \to 0$ if $T \to \infty$, $\xi \to \infty$ if $T \to 0$. There is a competition between energy and entropy: energy tends to favour spin alignment, entropy spin disorder.

We have shown in Section 2.4.3 that the partition function plays the rôle of a generating function for average values and correlation functions. Let us generalize the Hamiltonian (3.30), taken for an arbitrary space dimension D, by using a site-dependent magnetic field B_i

$$H = -J\sum_{\langle i,j\rangle} S_i S_j - \mu \sum_i B_i S_i \tag{4.22}$$

A straightforward application of (2.67) gives for the site magnetization $M_i = \langle S_i \rangle$

$$M_i = \frac{1}{\beta\mu Z}\frac{\partial Z}{\partial B_i} = \frac{1}{\beta\mu}\frac{\partial \ln Z}{\partial B_i} \tag{4.23}$$

while the correlation function is obtained from (2.69)

$$\langle S_i S_j \rangle = \frac{1}{(\beta\mu)^2 Z}\frac{\partial^2 Z}{\partial B_i \partial B_j} \tag{4.24}$$

Equation (2.70) allows us to obtain the *connected correlation function*[12] G_{ij}, also noted $\langle S_i S_j\rangle_c$, where c stands for 'connected'

$$\boxed{G_{ij} = \langle S_i S_j\rangle_c = \langle S_i S_j\rangle - \langle S_i\rangle\langle S_j\rangle = \frac{1}{(\beta\mu)^2}\frac{\partial^2 \ln Z}{\partial B_i \partial B_j}} \tag{4.25}$$

It is obvious that the procedure can be generalized to the p-point correlation functions, namely the average of the product of p Ising spins. Differentiating the partition function to order p gives the p-point correlation functions, while differentiating the free energy $\ln Z$ gives the connected correlation functions.[13] The partition function Z is the generating function of correlations, while the free energy $\ln Z$ is the generating function of connected correlation functions. Even when the external field is zero, it may prove convenient to compute for example G_{ij} from (4.25) by setting $B_i = 0$ after the differentiations have been performed

$$G_{ij}\Big|_{B=0} = \langle S_i S_j\rangle - \langle S_i\rangle\langle S_j\rangle = \frac{1}{(\beta\mu)^2}\frac{\partial^2 \ln Z}{\partial B_i \partial B_j}\Big|_{B=0} \tag{4.26}$$

The fluctuation-response theorem (2.70) reads in the case of the Ising model

$$\boxed{\frac{\partial M_i}{\partial B_j} = \frac{1}{\beta\mu}\frac{\partial^2 \ln Z}{\partial B_i \partial B_j} = \beta\mu\, G_{ij}} \tag{4.27}$$

We conclude this subsection with two remarks.

[12] We have adopted the standard notation G for the correlation function, and in this chapter, the Gibbs potential will be denoted by Γ, instead of G.

[13] Connected correlation functions are also called cumulants, as they are the analogues of the cumulants of probability theory. See also Footnote 3, Chapter 3.

- The spin susceptibility χ is positive: indeed in a uniform magnetic field $B_i = B$

$$\frac{\partial M_i}{\partial B} = \sum_j \frac{\partial M_i}{\partial B_j}\frac{\partial B_j}{\partial B} = \beta\mu \sum_j G_{ij}$$

whence

$$\chi = \frac{\partial M}{\partial B} = \beta\mu \sum_j G_{ij} = \frac{\beta\mu}{N}\sum_{i,j} G_{ij} \tag{4.28}$$

where $M = \langle S_i \rangle$ is now independent of i from translation invariance, which also entails that $\sum_j G_{ij}$ is independent of i. Now

$$\sum_{i,j} G_{ij} = \sum_{ij} \langle (S_i - M)(S_j - M)\rangle = \langle (S_{\rm tot} - NM)^2 \rangle \geq 0 \tag{4.29}$$

where the total spin $S_{\rm tot} = \sum_i S_i$.[14] It follows from (4.28) and (4.29) that $\chi \geq 0$.
- Equation (4.21) suggests a general definition of the correlation length. For a translation invariant system, the connected correlation function is expected to behave as an exponential of the distance r between the two sites, and this exponential is controlled by the correlation length ξ

$$\boxed{r \to \infty : \quad G(\vec{r}\,) \sim C(r) \exp\left(-\frac{r}{\xi}\right)} \tag{4.30}$$

where C is some slowly varying function of r.

4.1.3 Broken symmetry

Let us examine in some detail our spin system in a vanishing magnetic field when $T < T_{\rm c}$. The problem which we encounter is that the symmetry of the state may be lower than the symmetry of the Hamiltonian. The Hamiltonian is invariant under the operation $S_i \to -S_i$, while the state with a positive magnetization, for example, is not invariant under the same operation. The symmetry group of the Hamiltonian is the group with two elements Z_2, which is *not* a symmetry group of the state. This phenomenon is known by the name of *(spontaneous) broken symmetry*, which underlies many developments in modern physics. Note that in order to obtain a non-zero magnetization, we had to break the symmetry by imposing a well-defined orientation of the spins on the boundary, although only a fraction $1/\sqrt{N} \to 0$ of the spins was *a priori* affected by this boundary condition. The spontaneous magnetization, which breaks the Z_2-symmetry, is called the *order*

[14] One can also derive (4.29) by computing the average value of $S_{\rm tot}$ and its dispersion in a uniform magnetic field.

parameter of the transition, a concept already introduced in Section 3.1.4. The order parameter is zero when the state has the same symmetry as the Hamiltonian, and it is non-zero when the state has a lower symmetry. Order parameters may also occur in first order phase transitions, but some transitions do not have order parameters, for example the gas–liquid phase transition. This absence of an order parameter may also happen in continuous phase transitions, an example being the two-dimensional XY model.[15]

From the fluctuation-reponse theorem (4.27), we see that the connected correlation function (up to a factor $\beta\mu$) represents the response of the average $M_i = \langle S_i \rangle$ of the spin at site i to a variation of the magnetic field at site j. However symmetry breaking must be taken into account for $T < T_c$, and this leads us to make an important observation on the state of the spin system – or simply the state – at $T < T_c$; the state is nothing other than the probability distribution which defines the averages, see (4.36). On physical grounds, we expect that the response of a spin at site i to a variation of the magnetic field at site j should vanish when the distance $|\vec{r}_i - \vec{r}_j| \to \infty$. This will be the case if the connected correlation function tends to zero in this limit. One then says that the state obeys the *clustering property*.[16] Now we must be very careful in zero external field for $T < T_c$, since, as we have seen in the previous section, the state depends on the boundary conditions. With all spins up on the boundary we obtained a positive zero-field magnetization per spin M_0, but we would have obtained a negative one, $-M_0$, with all spins down on the boundary. It can be shown that the preceding boundary conditions lead to states that obey the clustering property, namely to states whose connected correlation functions go to zero when the distance between sites goes to infinity.

In order to understand broken symmetry better, let us examine the behaviour of the free energy per spin $f(B)$ as a function of a uniform external magnetic field. We start from the following information:

(*i*) From (4.23) the magnetization per spin M is given by the derivative of the free energy per spin, within a factor $-1/\mu$

$$M = -\frac{1}{\mu}\frac{\partial f}{\partial B} \tag{4.31}$$

(*ii*) From (4.29) the magnetic susceptibility is positive

$$\chi = \frac{\partial M}{\partial B} \geq 0 \Longrightarrow -\frac{\partial^2 f}{\partial B^2} \leq 0 \tag{4.32}$$

[15] In the XY model, the spin is a vector in a two-dimensional space, which is in general different from the coordinate space. The model has a phase transition in $D = 2$, but no spontaneous magnetization.

[16] The clustering property also ensures that the free energy is an extensive quantity: see any book on quantum field theory.

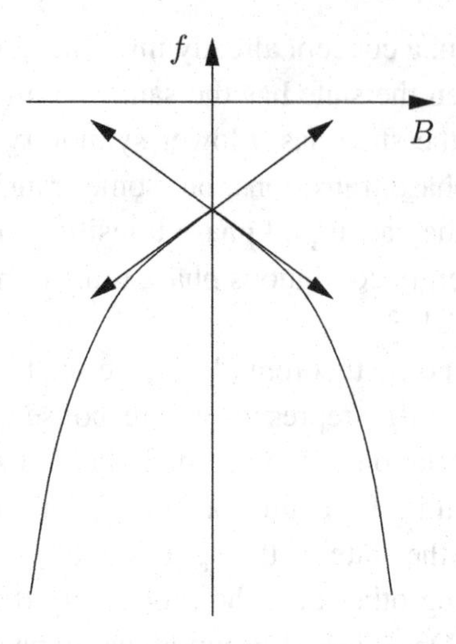

Figure 4.4 The free energy per spin as a function of B for $T < T_c$.

Thus f is a concave function of B. Moreover f is an even function of B from the symmetry $S_i \to -S_i$, $B \to -B$ of the Hamiltonian.

(*iii*) The Lee–Yang theorem states that f can only be singular at $B = 0$.

One may then sketch the behaviour of f as a function of B at a fixed temperature (Figure 4.4). There are two possible cases:

(*i*) f is differentiable at $B = 0$: then the spontaneous magnetization M_0 vanishes, $M_0 = 0$.

(*ii*) f is not differentiable at $B = 0$: then there are two possible values of the spontaneous magnetization with opposite signs

$$M = \lim_{B\to 0^+} -\frac{1}{\mu}\frac{\partial f}{\partial B} = M_0 \quad \text{or} \quad M = \lim_{B\to 0^-} -\frac{1}{\mu}\frac{\partial f}{\partial B} = -M_0 \tag{4.33}$$

We now use Peierls' argument to rule out possibility (*i*) for $T \le T_c$. Let $\hat{f}_N(B)$ be the free energy per spin computed for $B > 0$ and $\hat{M}_N(B)$ the corresponding magnetization, where the hat refers to the Peierls boundary conditions of Section 4.1.1. For all values of N there exists a strictly positive constant α such that $\hat{M}_N(0) \ge \alpha > 0$. On the other hand, as $\hat{f}_N$ is a concave function of B

$$\hat{f}_N(B) \le \hat{f}_N(0) - \alpha\mu B$$

Going to the thermodynamic limit $N \to \infty$ and using the fact that in this limit the free energy per spin is independent of the boundary conditions, we may drop the hat on the various quantities and get

$$f(B) \le f(0) - \alpha\mu B \tag{4.34}$$

which entails

$$\lim_{B\to 0^+} \frac{\partial f}{\partial B} \le -\alpha\mu$$

One thus gets a spontaneous magnetization *provided the limits $B \to 0$ and $N \to \infty$ are taken in the right order*

$$\boxed{M_0 = \lim_{B\to 0^+} \lim_{N\to\infty} M_N(B) = \lim_{B\to 0^+} \lim_{N\to\infty} -\frac{1}{\mu}\frac{\partial f_N}{\partial B}} \tag{4.35}$$

where $M_N(B)$ is computed for N spins in the presence of a uniform external field B without imposing any boundary condition. If the order of limits is interchanged, a zero magnetization is obtained from the symmetry of the Hamiltonian. In zero field, one can only obtain a vanishing magnetization for finite N if no boundary conditions or no external fields are imposed.

One can show that the zero-field magnetization M_0 obtained with Peierls' boundary condition and the limit (4.35) are identical. The existence of two possible values $\pm M_0$ of the spontaneous magnetization is less surprising when one realizes that there is an infinite energy difference $\simeq 2\mu N M_0 B$ between the two spin configurations in the infinite volume limit when B is non-zero, even if B is very small. One of the two configurations dominates over the other one depending on the sign of B. The Boltzmann distribution should be modified in order to obtain the average value of a quantity A. The expectation value $\langle\bullet\rangle_\pm$ of the probability distribution of the so-called *pure states* $\pm$ is defined by[17]

$$\langle A\rangle_\pm = \lim_{B\to 0^\pm} \lim_{V\to\infty} \frac{1}{Z}\sum_{\mathcal{C}} A\,\mathrm{e}^{-\beta H} \tag{4.36}$$

If the system is translation invariant, a mixture of pure states does not lead to the clustering property. Indeed, assume that one has a mixture of pure states with probabilities p and $(1-p)$

$$p\langle\bullet\rangle_+ + (1-p)\langle\bullet\rangle_-$$

and compute the connected correlation function $\langle S_i S_j\rangle_{c,p}$ when $|\vec{r}_i - \vec{r}_j| \to \infty$

$$\begin{aligned}\langle S_i S_j\rangle_p &= pM^2 + (1-p)M^2 = M^2\\ \langle S_i\rangle_p &= pM - (1-p)M = (2p-1)M\\ \langle S_i S_j\rangle_{c,p} &= 4p(1-p)M^2\end{aligned}$$

If $p \neq 0$, $\langle S_i S_j\rangle_c$ does not vanish when $|\vec{r}_i - \vec{r}_j| \to \infty$, which means that the mixed state does not have the clustering property, which only holds for the pure

[17] One says that the measure (4.36) *breaks ergodicity*: the Gibbs measure does not explore the whole phase space.

states $p = 0$ or $p = 1$. Indeed one can show that clustering states are in one-to-one correspondence with pure states.

4.1.4 Critical exponents

For the rest of this chapter, we shall be mostly interested in the behaviour of physical systems close to the critical temperature T_c of a continuous phase transition. We shall see that physical quantities such as the susceptibility or the specific heat diverge at the critical point, and that the divergence is characterized by a few numbers called critical exponents. The importance of critical exponents is their universality, namely systems in the same universality class have the same critical exponents. Let us begin with the correlation length. Experiment shows that the correlation length is large in the vicinity of the phase transition, and that the correlation function varies smoothly for distances on the order of the lattice spacing. One then uses a continuum approximation, where the vector $\vec{r}_i - \vec{r}_j$ between two lattice sites may be taken to vary continuously. It is convenient to introduce the Fourier transform $\tilde{G}(\vec{q})$ of the connected correlation function for a translation invariant system

$$\tilde{G}(\vec{q}) = \int \mathrm{d}^D r \, \mathrm{e}^{\mathrm{i}\vec{q}\cdot\vec{r}} \, G(\vec{r}) \tag{4.37}$$

In the vicinity of a continuous phase transition, experiment shows that the correlation function depends only on the distance $r = |\vec{r}_i - \vec{r}_j|$ for $r \gg a$ (remember that a is the lattice spacing) and not on the orientation of the vector $\vec{r}$.[18] Experimental results on the Fourier transform $\tilde{G}$,[19] which depends only on $q = |\vec{q}\,|$ can be parametrized as follows for $qa \ll 1$

$$\boxed{\tilde{G}(\vec{q}) = \frac{1}{q^{2-\eta}} f(q\xi)} \tag{4.38}$$

$\tilde{G}$ is a function of the product $q\xi$ times a power law in q. The correlation length itself behaves as

$$\boxed{\xi \simeq K|T - T_c|^{-\nu}} \tag{4.39}$$

where K is a constant factor. The above two equations define the *critical exponents*

[18] However, the correlation function depends in general on the orientation of $\vec{r}$ far from the transition or for $r \sim a$, as the lattice is not isotropic.

[19] It is worth recalling that experiment directly measures the Fourier transform of the correlation function, and not the correlation function in ordinary space. This can be understood from a derivation quite analogous to that given in Section 3.4.2.

η and ν.[20] The function $f(x)$ tends to a finite limit K' if $x \to \infty$, so that at exactly the critical temperature

$$\boxed{\tilde{G}(\vec{q}) = \frac{K'}{q^{2-\eta}} \qquad T = T_c} \tag{4.40}$$

At this point we return to ordinary space by taking the inverse Fourier transform

$$G(\vec{r}) = \int \frac{d^D q}{(2\pi)^D} e^{-i\vec{q}\cdot\vec{r}} \frac{1}{q^{2-\eta}} f(q\xi) \tag{4.41}$$

From a dimensional argument, one derives

$$\boxed{G(\vec{r}) = \frac{1}{r^{D+\eta-2}}\, g\left(\frac{r}{\xi}\right)} \tag{4.42}$$

As the form (4.38) has been assumed to be valid for $q \ll 1/a$, the form (4.41) will hold if $r \gg a$, namely when the distance between sites is much larger than the lattice spacing. As in (4.21), the asymptotic behaviour for $r \gg a$ of g is exponential

$$g\left(\frac{r}{\xi}\right) \sim \exp\left(-\frac{r}{\xi}\right) \tag{4.43}$$

When $T \neq T_c$, we can use the preceding results to establish a relation between critical exponents; such a relation is called a *scaling law*. When $T \neq T_c$, $\tilde{G}(0)$ is finite. In order to compensate for the divergent factor $q^{-2+\eta}$, it must therefore be the case that $f(q\xi) \sim (q\xi)^{2-\eta}$ as $q \to 0$. This entails

$$\tilde{G}(0) \sim \xi^{2-\eta} \sim |T - T_c|^{-\nu(2-\eta)} \tag{4.44}$$

The critical exponent γ is defined from the behaviour of the susceptibility

$$\boxed{\chi \sim |T - T_c|^{-\gamma}} \tag{4.45}$$

Since $\tilde{G}(0) \propto \chi$, the exponent γ is related to ν and η by the *scaling law*

$$\boxed{\gamma = \nu(2 - \eta)} \tag{4.46}$$

If $\eta < 2$, which is the usual case, the susceptibility diverges at $T = T_c$.

[20] In the older literature, one may find 'primed critical exponents' $\eta', \nu' \ldots$ defined below the critical temperature. It is known today that the values of the critical exponents are the same on both sides of T_c, and these primed exponents have been abandoned.

4.2 Mean field theory

4.2.1 A convexity inequality

Since the analytical solution of the Ising model is either complicated or out of reach, one must develop approximation methods. The most useful approximation, introduced by Weiss in 1907, is the so-called *mean field approximation*,[21] which can be used not only in the Ising model, but also in many other cases. We shall derive the approximation from a general convexity property of the partition function. Let us consider two different density operators D and D_λ, the first one corresponding to the usual Boltzmann distribution

$$D = \frac{1}{Z}\,\mathrm{e}^{-\beta H} \qquad Z = \mathrm{Tr}\,\mathrm{e}^{-\beta H} \tag{4.47}$$

and the second corresponding to the density operator of an approximate Hamiltonian H_λ, which later on will play the rôle of a trial Hamiltonian in a variational method

$$D_\lambda = \frac{1}{Z_\lambda}\,\mathrm{e}^{-\beta H_\lambda} \qquad Z_\lambda = \mathrm{Tr}\,\mathrm{e}^{-\beta H_\lambda} \tag{4.48}$$

Let us start from the inequality (2.56)[22]

$$-\mathrm{Tr}\, D_\lambda \ \ln D_\lambda \le -\mathrm{Tr}\, D_\lambda \ \ln D \tag{4.49}$$

which translates into

$$\mathrm{Tr}\left[D_\lambda(\beta H_\lambda + \ln Z_\lambda)\right] \le \mathrm{Tr}\left[D_\lambda(\beta H + \ln Z)\right] \tag{4.50}$$

Taking the logarithm leads to an inequality on the free energy

$$\boxed{F \le F_\lambda + \langle H - H_\lambda\rangle_\lambda = \Phi(\lambda)} \tag{4.51}$$

where the notation $\langle \bullet \rangle_\lambda$ stands for an average taken with the probability distribution D_λ (4.48): $\langle A\rangle_\lambda = \mathrm{Tr}(AD_\lambda)$. This inequality is the starting point of a variational estimate of the partition function. It ensures that the exact free energy is smaller than any variational estimate based on a trial Hamiltonian H_λ, and we may assess the quality of two successive approximations by checking that the second gives a smaller free energy than the first. It is shown in Exercise 4.6.4 that if H and H_λ differ by terms of order ε, then $\Phi(\lambda)$ and F differ by terms of order ε^2: this is a well-known property of the variational method.

[21] Sometimes called the molecular field approximation.

[22] In the case of the Ising model

$$\mathrm{Tr} \to \sum_{C} = \sum_{S_1=\pm 1} \cdots \sum_{S_N=\pm 1}$$

We have kept the notation 'trace' to emphasize that (4.51) is also valid for quantum systems.

4.2.2 Fundamental equation of mean field theory

It will be convenient to rewrite the Ising Hamiltonian (3.30) as

$$H = -\frac{1}{2}\sum_{i,j} J_{ij} S_i S_j - \mu \sum_i B_i S_i \tag{4.52}$$

where $J_{ij} = J$ if sites i and j are nearest neighbours and zero otherwise. We choose a trial Hamiltonian H_λ of the paramagnetic kind (3.20)

$$H_\lambda = -\mu \sum_i \lambda_i S_i \tag{4.53}$$

The parameter λ_i will be interpreted as an effective site-dependent magnetic field. As all sites are independent, the partition function Z_λ is given from (3.24) by

$$Z_\lambda = \prod_i [2\cosh(\beta\mu\lambda_i)]$$

and the free energy by

$$F_\lambda = -\frac{1}{\beta}\sum_i \ln[2\cosh(\beta\mu\lambda_i)]$$

The site magnetization M_i is obtained from (3.27)

$$M_i = -\frac{1}{\mu}\frac{\partial F_\lambda}{\partial \lambda_i} = \tanh(\beta\mu\lambda_i) \tag{4.54}$$

Since H_λ has a paramagnetic form, $\langle S_i S_j\rangle_\lambda = \langle S_i\rangle_\lambda \langle S_j\rangle_\lambda = M_i M_j$. Then

$$\langle H - H_\lambda\rangle_\lambda = -\frac{1}{2}\sum_{i,j} J_{ij} M_i M_j - \mu\sum_i (B_i - \lambda_i) M_i$$

and the function $\Phi(\lambda)$ in (4.51) is

$$\Phi(\lambda) = -\frac{1}{\beta}\sum_i \ln[2\cosh(\beta\mu\lambda_i)] - \frac{1}{2}\sum_{i,j} J_{ij} M_i M_j - \mu\sum_i (B_i - \lambda_i) M_i \tag{4.55}$$

The variational method instructs us to look for the minimum of $\Phi(\lambda)$. However, instead of differentiating with respect to λ_i, it is more convenient to differentiate with respect to M_i, which is allowed since M_i is an increasing function of λ_i. Using

$$\frac{\partial F}{\partial M_i} = \frac{\partial F}{\partial \lambda_i}\frac{\partial \lambda_i}{\partial M_i} = -\mu M_i \frac{\partial \lambda_i}{\partial M_i}$$

and setting $\Psi(M) = \Phi(\lambda(M))$we get

$$\frac{\partial \Psi}{\partial M_i} = -\mu M_i \frac{\partial \lambda_i}{\partial M_i} - \sum_j J_{ij} M_j - \mu B_i + \mu \lambda_i + \mu M_i \frac{\partial \lambda_i}{\partial M_i}$$

and the extremum condition $\partial \Psi / \partial M_i = 0$ gives the fundamental equation of mean field theory

$$\mu \lambda_i = \mu B_i + \sum_j J_{ij} M_j \tag{4.56}$$

or equivalently

$$\boxed{\frac{1}{\beta} \tanh^{-1} M_i = \mu B_i + \sum_j J_{ij} M_j} \tag{4.57}$$

This equation allows us to derive the expression for the correlation function in the mean field approximation. Differentiating (4.57) with respect to B_k and making use of (4.27) leads to

$$\frac{1}{1 - M_i^2} \langle S_i S_k \rangle_c = \delta_{ik} + \beta \sum_j J_{ij} \langle S_j S_k \rangle_c$$

In order to solve this equation analytically for $\langle S_i S_k \rangle_c$ we must restrict ourselves to the case of a uniform magnetic field B, so that the magnetization is site independent, $M_i = M$

$$\sum_j \left(\frac{\delta_{ij}}{1 - M^2} - \beta J_{ij} \right) \langle S_j S_k \rangle_c = \delta_{ik}$$

This equation has the form of a (discrete) convolution and is solved by Fourier transformation. We define the discrete Fourier transform $\tilde{f}(\vec{q})$ of a function $f_i = f(\vec{r}_i)$ by

$$\tilde{f}(\vec{q}) = \sum_i \mathrm{e}^{\mathrm{i}\vec{q} \cdot \vec{r}_i} f(\vec{r}_i) \tag{4.58}$$

In a D-dimensional space, let us draw from a site i D unit vectors $\hat{e}_\mu$ along the axes; site i is linked to its nearest neighbours by vectors $\pm a\hat{e}_\mu$ and the Fourier transform of the coupling $J_{ij} = J(\vec{r}_i, \vec{r}_j)$ is

$$\tilde{J}(\vec{q}) = \sum_i \mathrm{e}^{\mathrm{i}\vec{q} \cdot \vec{r}_i} J(\vec{r}_i, \vec{r}_j) = J \sum_{\pm\mu} \mathrm{e}^{\mathrm{i}aq_\mu} = 2J \sum_\mu \cos q_\mu$$

where $q_\mu = \vec{q} \cdot \hat{e}_\mu$ is the component of $\vec{q}$ along the axis $\hat{e}_\mu$. Inserting the Fourier transform $\tilde{G}(\vec{q})$ of the correlation function

$$\tilde{G}(\vec{q}) = \sum_i e^{i\vec{q}\cdot\vec{r}_i} G(\vec{r}_i, \vec{r}_j)$$

in the mean field equation[23] we get its mean field approximation

$$\boxed{\tilde{G}(\vec{q}) = \frac{1 - M^2}{1 - 2\beta J(1 - M^2)\sum_\mu \cos aq_\mu}} \tag{4.59}$$

Let us now take the long wavelength limit $aq \ll 1$

$$\begin{aligned}\sum_\mu \cos aq_\mu &= 1 - \frac{1}{2}\sum_\mu (aq_\mu)^2 + \frac{1}{24}\sum_\mu (aq_\mu)^4 + \cdots \\ &= 1 - \frac{1}{2}a^2\vec{q}^{\,2} + \frac{1}{24}\sum_\mu a^4 q_\mu^4 + \cdots \end{aligned} \tag{4.60}$$

The first two terms in the expansion (4.60) are isotropic, but the third and beyond are not. In the long wavelength limit we obtain an *isotropic* correlation function

$$\tilde{G}(\vec{q}) \simeq \frac{1 - M^2}{1 - 2D\beta J(1 - M^2) + \beta J(1 - M^2)q^2} \tag{4.61}$$

In zero magnetic field for $T > T_c$ the magnetization vanishes and (4.61) simplifies to

$$\tilde{G}(\vec{q}) \simeq \frac{1}{(1 - 2D\beta J) + \beta J q^2} \tag{4.62}$$

It is useful to make the following remarks:

- The result of the preceding calculation may look paradoxical: the mean field Hamiltonian (4.53) is of a paramagnetic type and the spins are independent of each other. One would thus expect a vanishing connected correlation function. Actually, had we computed the correlation function directly from the probability distribution D_λ, we would have obtained $\langle S_i S_j \rangle_c = 0$, but this is not how we have proceeded: we have computed the free energy $\Psi(M)$ and used the fluctuation-response theorem to obtain $\langle S_i S_j \rangle_c$ from $\partial M_i / \partial B_j$. Our result is then different from $\mathrm{Tr}\,[D_\lambda \langle S_i S_j \rangle_c]$: the direct calculation of the correlation function and the fluctuation-response theorem give identical answers only if the same Boltzmann weight is used in both calculations. It is explicitly shown in Exercise 4.6.4 that the fluctuation-response theorem gives better results than the direct calculation. Indeed, if H and H_λ differ by terms of order ε, we know that the error on

[23] $\tilde{G}(\vec{q})$, as well as $\tilde{J}(\vec{q})$, are independent of $\vec{r}_j$ from translation invariance.

$\Psi(M)$ is of order ε^2 and $\Psi(M)$ correctly gives the correlation function to this order, while $\mathrm{Tr}\,[D_\lambda(S_i S_j)_c]$ and the true correlation function differ by terms of order ε.

- The function $\Psi(M)$ has an interesting physical interpretation. The equation giving $\partial\Psi/\partial M_i$ is easily integrated to give

$$\Psi = -\sum_{i,j} J_{ij} M_i M_j - \mu \sum_i B_i M_i + \frac{1}{\beta} \sum_i \left(\frac{1+M_i}{2} \ln \frac{1+M_i}{2} + \frac{1-M_i}{2} \ln \frac{1-M_i}{2} \right) \tag{4.63}$$

The first term represents the energy, and the second one the statistical entropy S (2.46) (not to be confused with an Ising spin!) of the probability distribution

$$P = \prod_i \left(\frac{1+M_i}{2} \delta_{S_i\, 1} + \frac{1-M_i}{2} \delta_{S_i\, -1} \right)$$

Φ may be written as an average over the probability distribution P

$$\Phi = \langle H \rangle_P - \frac{1}{\beta} S[P] = F[P] \tag{4.64}$$

The mean field approximation is equivalent to a minimization of the 'free energy' $F[P]$.

4.2.3 Broken symmetry and critical exponents

We now apply a uniform external magnetic field B. The fundamental equation (4.57) becomes

$$\boxed{\frac{1}{\beta} \tanh^{-1} M = \mu B + 2DJM} \tag{4.65}$$

Given the magnetic field, this equation determines the magnetization; however it must be solved numerically. For a general lattice, $2D$ in (4.65) should be replaced by the number κ of nearest neighbours. It is convenient to define the function $g(M)$

$$g(M) = \tanh^{-1} M - \beta\kappa J M \tag{4.66}$$

and the mean field equation reads $g(M) = \beta\mu B$. The sign of the derivative $g'(M)$ at $M = 0$ will be crucial to the behaviour of the magnetization. For $M \to 0$, $g(M) \simeq M - \beta\kappa J M$ so that

$$g'(M=0) > 0 \quad \text{if} \quad \beta\kappa J < 1 \quad \text{or} \quad T > \kappa J/k$$
$$g'(M=0) < 0 \quad \text{if} \quad \beta\kappa J > 1 \quad \text{or} \quad T < \kappa J/k$$

Let us define $T_c = \kappa J/k$: T_c will be identified with the critical temperature in the mean field approximation. The function $g(M)$ is shown in Figure 4.5(a) for

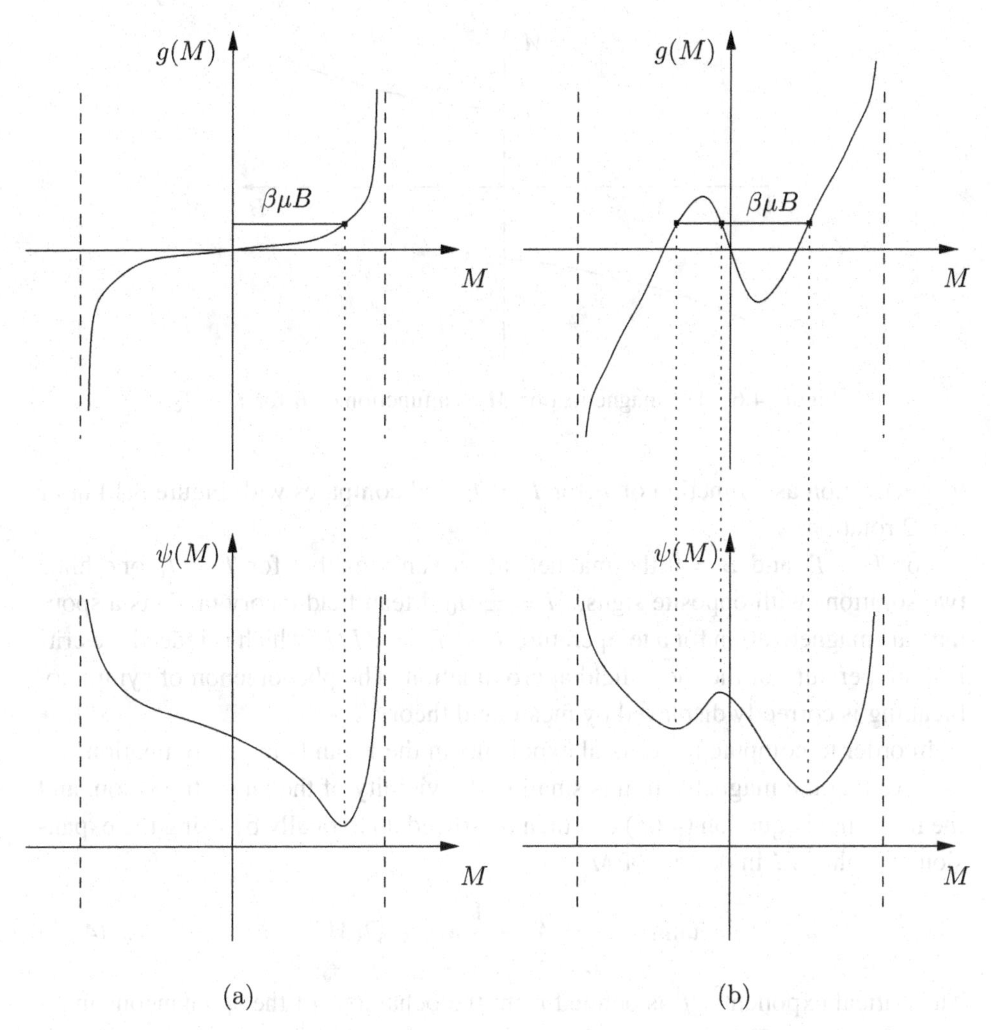

Figure 4.5 The functions $g(M)$ and $\psi(M)$ of mean field theory. (a) $T > T_c$, (b) $T < T_c$.

$T > T_c$ and 4.5(b) for $T < T_c$; one can see that there is only one solution to the mean field equation for the magnetization when $T > T_c$, while there are one or three solutions for $T < T_c$. When three solutions are found, the variational method tells us to choose the solution which corresponds to the minimum of the function $\psi(M) = \Psi/N$; this function is shown below $g(M)$ in Figure 4.5. The correct solution corresponds to a magnetization aligned with the field, in agreement with physical intuition. One of the other two solutions leads to a metastable state, while the last one corresponds to a maximum of Φ and thus to an unstable state. There is a remarkable parallel between mean field theory and the theory of phase transitions of Section 3.5.4. This parallel becomes obvious when one draws in Figure 4.6 the

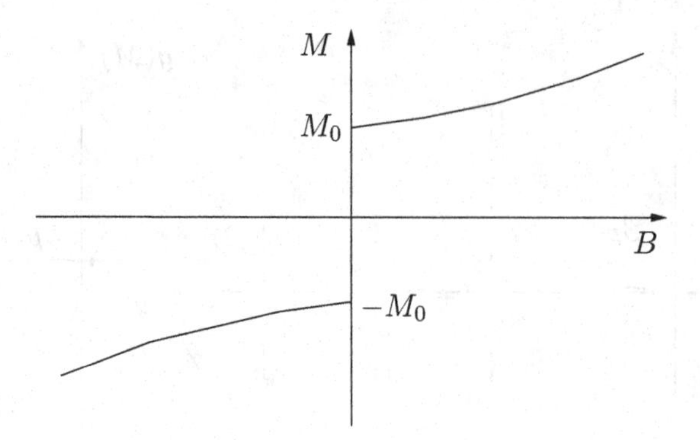

Figure 4.6 The magnetization M as a function of B for $T < T_c$.

magnetization as a function of B for $T < T_c$ and compares with Figure 3.14 after a $\pi/2$ rotation.

For $T > T_c$ and $B = 0$ the magnetization vanishes, but for $T < T_c$ one finds two solutions with opposite signs, $M = \pm M_0$. Mean field theory predicts a spontaneous magnetization for a temperature $T < T_c = \kappa J/k$, which is indeed the critical temperature of the mean field approximation. The phenomenon of symmetry breaking is correctly displayed by mean field theory.

In order to compute the critical exponents in the mean field approximation, we observe that the magnetization is small in the vicinity of the phase transition, and the mean field equation (4.65) can then be solved analytically by using the expansion of $\tanh^{-1} M$ in powers of M

$$\tanh^{-1} M = M + \frac{1}{3} M^3 + \mathcal{O}(M^5) \tag{4.67}$$

The critical exponent[24] $\tilde{\beta}$ is defined from the behaviour of the spontaneous magnetization M_0 as $T \to T_c$

$$M_0 \propto |T_c - T|^{\tilde{\beta}} \qquad T \to T_c \tag{4.68}$$

The mean field equation gives for small M_0

$$M_0 + \frac{1}{3} M_0^3 = \beta \kappa J M_0 = \frac{T_c}{T} M_0$$

whence

$$M_0(T) \simeq \sqrt{\frac{3}{T_c}} (T_c - T)^{1/2} \tag{4.69}$$

[24] The standard notation for this exponent is β; we have used $\tilde{\beta}$ to avoid a possible confusion with $\beta = 1/kT$.

The mean field approximation predicts $\tilde{\beta} = 1/2$. The critical exponent γ has been defined from the behaviour of the susceptibility in (4.45). Let us compute the susceptibility for $T > T_c$ and $B \to 0$

$$M \simeq \frac{T_c}{T} M - \frac{\mu B}{kT}$$

whence

$$M \simeq \frac{\mu B}{k(T - T_c)} \simeq \chi B \tag{4.70}$$

This entails the following behaviour of the zero field susceptibility

$$\chi \propto (T - T_c)^{-1} \tag{4.71}$$

and thus $\gamma = 1$. The same exponent is found if $T < T_c$ (Exercise 4.6.3). Another critical exponent, δ, describes the behaviour of the critical isotherm

$$M \propto B^{1/\delta} \qquad T = T_c \tag{4.72}$$

Mean field theory leads at once to

$$B = \frac{kT_c}{3\mu} M^3$$

and thus $\delta = 3$. The behaviour of the specific heat C in zero magnetic field defines the critical exponent α

$$C \propto (T - T_c)^{-\alpha} \qquad B = 0 \tag{4.73}$$

In the mean field approximation the internal energy E is

$$E = -\frac{1}{2}\kappa J N M_0^2 = \frac{3}{2} kN(T - T_c) \qquad T < T_c$$
$$E = 0 \qquad T > T_c$$

where we have used the expression (4.69) of M_0. The specific heat is $3kN/2$ for $T < T_c$ and zero for $T > T_c$: it is discontinuous but not infinite at $T = T_c$. One conventionally assigns the exponent $\alpha = 0$ to such a behaviour.

Finally the exponents η and ν are derived from the behaviour (4.61) or (4.62) of the correlation function. The case $T > T_c$ is the easier one; one remarks in (4.62) that

$$1 - 2D\beta J = 1 - \kappa\beta J = 1 - \frac{\beta}{\beta_c} \simeq \frac{T - T_c}{T_c}$$

Table 4.1 *Mean field (MFT) critical exponents compared to analytical ($D = 2$) and numerical ($D = 3$) results.*

	MFT	$D = 2$	$D = 3$
$T_c/\kappa J$	1	0.57	0.75
α	discont.	$\ln\|T - T_c\|$	0.110 ± 0.005
β	0.5	0.125	0.312 ± 0.003
γ	1	1.75	1.238 ± 0.002
δ	3	15	5.0 ± 0.05
η	0	0.25	0.0375 ± 0.0025
ν	0.5	1	0.6305 ± 0.0015

so that

$$\tilde{G}(q) \simeq \frac{1}{J\beta_c q^2 \left[1 + \left(\frac{T - T_c}{Jq^2}\right)\right]} \tag{4.74}$$

By identification with (4.38) one finds

$$\eta = 0 \qquad \nu = \frac{1}{2} \tag{4.75}$$

It is left as Exercise 4.6.3 to show that the same values are obtained for $T < T_c$.

Let us now summarize the results of mean field theory and assess the validity of the approximation by comparing its results with analytical or numerical results. Our first observation is that the existence of a phase transition and the values of the critical exponents are *independent of the space dimension D* in mean field theory. In dimension $D = 1$, mean field theory is *qualitatively wrong*, since it predicts a phase transition where there is none! Table 4.1 compares the values of the critical temperature and of the critical exponents in the mean field approximation with the analytical results in $D = 2$ and numerical results in $D = 3$. It is worth noting that the scaling law (4.46) is satisfied by the mean field exponents. It is clear from this table that the quality of the mean field approximation improves when D increases. We shall see that the mean field critical exponents become correct in $D \geq 4$,[25] but not the critical temperature;[26] the critical temperature is given correctly by mean field theory only when $D \to \infty$.

We conclude with a general remark: in order to apply mean field theory, one *must* have some knowledge of the phases one is looking for. It may well happen

[25] However the power law behaviour is supplemented with powers of $\ln|T - T_c|$ for $D = 4$.
[26] As we shall see later on, critical exponents are universal quantities, while the critical temperature is not.

that there are many phases in competition; then for all of these phases, it is necessary to write a mean field approximation relying on a trial Hamiltonian which is adapted to each particular phase, and a minimization procedure gives the phase diagram in some parameter space. For each point in this parameter space, one can find the lowest free energy, and thus the stable phase, within the approximation. The qualitative breakdown of mean field theory in the $D = 1$ Ising model and in the $D = 2$ XY model is due to an inadequate assumption on the phase diagram. The possibility of symmetry breaking is in some sense built into the mean field approximation as discussed in this section.[27] Therefore, it is not surprising that it fails qualitatively when there can be no symmetry breaking at all!

4.3 Landau's theory

4.3.1 Landau functional

Landau's theory is equivalent to mean field theory, but represents a more convenient formulation because it perfoms a coarse graining that washes out irrelevant details. On the other hand it is only valid in the vicinity of the phase transition, while there is no limitation of this kind for mean field theory.[28] Close to the phase transition the magnetization is small, and one can expand the statistical entropy in (4.63)[29]

$$\frac{1+M_i}{2}\ln\frac{1+M_i}{2}+\frac{1-M_i}{2}\ln\frac{1-M_i}{2}=\frac{1}{2}M_i^2+\frac{1}{12}M_i^4+\mathcal{O}(M_i^6)$$

An approximate expression for the function Ψ is then

$$\Psi \simeq -\frac{1}{2}\sum_{i,j} J_{ij}M_iM_j - \mu\sum_i B_iM_i + \frac{1}{\beta}\sum_i\left(\frac{1}{2}M_i^2+\frac{1}{12}M_i^4\right) \tag{4.76}$$

By defining

$$M_i = M(\vec{r}_i) \qquad M_{i+\mu} = M(\vec{r}_i + a\hat{e}_\mu)$$

the first term in (4.76) is rewritten as

$$\begin{aligned}-\frac{1}{2}\sum_{i,j} J_{ij}M_iM_j &= -J\sum_{i,\mu} M_iM_{i+\mu}\\ &= \frac{1}{2}J\sum_{i,\mu}(M_i - M_{i+\mu})^2 - DJ\sum_i M_i^2\end{aligned}$$

[27] It is possible to apply mean field theory even without assuming symmetry breaking: see for example [10, 11, 35].

[28] However, it would be possible to keep the exact expression using $\tanh^{-1} M$ instead of its power expansion, but the simplicity of Landau's theory would then be lost.

[29] This expansion is most easily obtained from the expression of $\partial\Psi/\partial M_i$.

Then (4.76) becomes

$$\Psi = \frac{1}{2}J\sum_{i,\mu}(M_i - M_{i+\mu})^2 + \left(\frac{1}{2\beta} - DJ\right)\sum_i M_i^2 + \frac{1}{12}\sum_i M_i^4 - \mu\sum_i B_i M_i \tag{4.77}$$

One remarks that

- the coefficient of $\sum_i M_i^2$ vanishes at the (inverse) critical temperature β_0 of the mean field approximation $\beta_0 = 1/(2JD)$, where, from now on, we distinguish between the true critical temperature T_c and its mean field approximation T_0; we rewrite this coefficient as

$$\frac{1}{2}\overline{r}_0(T - T_0) = \frac{1}{2}r_0(T)$$

- the coefficient of $\sum_i M_i^4$ is positive; we rewrite it as $(1/4!)u_0$, where $u_0 > 0$ is independent of the temperature.

We now perform a coarse graining in order to get a continuum theory. We multiply the expression of Ψ by a^D to display the Riemann approximation of an integral. The terms $\sum_i M_i^2$, $\sum_i M_i^4$ and $\sum_i B_i M_i$ have a simple continuum limit, if site i is located at point $\vec{r}$

$$a^D \sum_i M_i^2 \Longrightarrow \int \mathrm{d}^D r\, M^2(\vec{r})$$

$$a^D \sum_i M_i^4 \Longrightarrow \int \mathrm{d}^D r\, M^4(\vec{r})$$

$$a^D \mu \sum_i B_i M_i \Longrightarrow \mu \int \mathrm{d}^D r\, B(\vec{r}) M(\vec{r})$$

while the last term involves a derivative

$$a^D \sum_{i,\mu}(M_i - M_{i+\mu})^2 = a^{D-2}\sum_{i,\mu}\left(\frac{M_i - M_{i+\mu}}{a}\right)^2 \to a^2 \int \mathrm{d}^D r\, (\vec{\nabla} M)^2$$

One finally makes a rescaling of the magnetization $\sqrt{Ja^2}M \to M$ and redefines r_0 and u_0 to get the final form of the *Landau functional* $\mathcal{L}[M(\vec{r})]$[30]

$$\boxed{\mathcal{L}[M(\vec{r})] = \int \mathrm{d}^D r \left(\frac{1}{2}(\vec{\nabla} M(\vec{r}))^2 + \frac{1}{2}r_0 M^2(\vec{r}) + \frac{1}{4!}u_0 M^4(\vec{r}) - \mu B(\vec{r}) M(\vec{r})\right)} \tag{4.78}$$

[30] In order to distinguish carefully functionals from functions, we put the argument of functionals between square brackets.

Since $\mathcal{L}[M]$ was obtained from a lattice version of mean field theory, the variational equation $\partial\Psi/\partial M_i = 0$ reads in the case of Landau's theory

$$\frac{\delta \mathcal{L}[M]}{\delta M(\vec{r})} = -\nabla^2 M(\vec{r}) + r_0 M(\vec{r}) + \frac{u_0}{3!} M^3(\vec{r}) - \mu B(\vec{r}) = 0 \tag{4.79}$$

The ordinary partial derivative with respect to M_i has been replaced by a functional derivative (see Section A.6) with respect to $M(\vec{r})$; the function $M(\vec{r})$ thus appears as a variational parameter. The original Landau's derivation of (4.78) did not rely on mean field theory, but rather on the following observations:

- If one assumes the magnetization to be small, one may expand $\mathcal{L}[M]$ in powers of M and keep only the lowest powers.
- The Z_2 symmetry of the problem requires that $\mathcal{L}[M, B] = \mathcal{L}[-M, -B]$.
- The term $\frac{1}{2}(\nabla M)^2$ ensures that the magnetization is uniform in a uniform magnetic field since this term is minimum for an $\vec{r}$-independent magnetization. It is a *stiffness* term, which tends to prevent space variations of the magnetization.

In a uniform magnetic field, $\mathcal{L}$ becomes an ordinary function $\mathcal{L}(M)$ and Equation (4.79) simplifies to

$$\bar{r}_0(T - T_0)M + \frac{1}{3!}M^3 = \mu B \tag{4.80}$$

which is nothing other than the small-M approximation of the mean field equation (4.65). The shape of the function $\mathcal{L}(M)$ is qualitatively similar to that of the function $\psi(M)$ in Figure 4.5. In particular there is only one solution for M if $T > T_0$ and one or three solutions for $T < T_0$.

The connected correlation function is easily derived from (4.79), by taking the functional derivative of this equation with respect to $B(\vec{r}\,')$

$$\left[-\nabla_r^2 + r_0(T) + \frac{1}{2}\, u_0\, M^2(\vec{r})\right] G(\vec{r}, \vec{r}\,') = \delta(\vec{r} - \vec{r}\,') \tag{4.81}$$

because from (4.27)

$$G(\vec{r}, \vec{r}\,') = \frac{\delta M(\vec{r})}{\partial B(\vec{r}\,')} = \frac{\delta^2 \mathcal{L}[M]}{\delta M(\vec{r}) \delta M(\vec{r}\,')} \tag{4.82}$$

If the magnetic field is uniform, translation invariance entails that $G(\vec{r}, \vec{r}\,') = G(\vec{r} - \vec{r}\,')$, and since M is $\vec{r}$-independent, the Fourier transform of (4.81) takes the simple form

$$\left(q^2 + r_0(T) + \frac{1}{2}\, u_0 M^2\right) \tilde{G}(\vec{q}) = 1$$

In zero external field and for $T > T_0$ the magnetization vanishes and the correlation function is given by

$$\boxed{\tilde{G}(\vec{q}) = \frac{1}{q^2 + r_0(T)} = \frac{1}{q^2\left(1 + \dfrac{\overline{r}_0(T - T_0)}{q^2}\right)}} \tag{4.83}$$

From the definitions of the critical exponents and of the correlation length ξ we recover of course the mean field results

$$\xi = [\overline{r}_0(T - T_0)]^{-1/2} \qquad \eta = 0 \qquad \nu = \frac{1}{2}$$

If $T < T_0$, the magnetization M is equal to the spontaneous magnetization $\pm M_0$; as $M_0^2 = -6r_0/u_0$, the correlation function is

$$\tilde{G}(\vec{q}) = \frac{1}{q^2\left(1 + \dfrac{2\overline{r}_0(T_0 - T)}{q^2}\right)} \tag{4.84}$$

The critical exponents are unchanged, but the multiplicative coefficient in ξ is different

$$\xi = [2\overline{r}_0(T_0 - T)]^{-1/2}$$

The inverse Fourier transform is computed in Exercise 4.6.7 for $r/\xi \gg 1$

$$G(\vec{r}) \simeq \frac{1}{2\, r^{D-2}} \left(\frac{1}{2\pi}\right)^{\frac{D-1}{2}} \left(\frac{\xi}{r}\right)^{\frac{3-D}{2}} \mathrm{e}^{-r/\xi} \tag{4.85}$$

This result is exact in dimension $D = 3$ for all values of r.[31]

One of the main virtues of Landau's theory is that it allows one to take into account in a convenient way spatial variations of the order parameter. For example one can study a domain wall between two regions with opposite magnetizations. Let us assume that for $x \to -\infty$ the spontaneous magnetization M is negative while for $x \to \infty$ it is positive. There will be an Ising wall between the two regions. Landau's theory allows one to compute the shape of this wall; one finds

[31] One should not attempt to extract the critical exponent η from the exponent of r as (4.85) is valid only for $r \gg \xi$, except for $D = 3$, where $D - 2 = (D - 1)/2$! To find η, one must compute the Fourier transform of $1/q^2$, which is $\propto r^{-(D-2)}$, showing that $\eta = 0$ for any D.

(Exercise 4.6.5)

$$M(x) = M_0 \tanh\left(\frac{x}{2\xi}\right) \tag{4.86}$$

where we have assumed that the wall is centred at $x = 0$. In general, Landau's theory is a very useful tool for the study of interfaces. The interface energy σ, namely the energy per unit area due to the presence of the interface, is easily calculated (Exercise 4.6.5)

$$\sigma = \frac{2}{3}\frac{M_0^2}{\xi} \propto (T_0 - T)^{3/2} \tag{4.87}$$

In general, the behaviour of σ close to T_c, $\sigma \propto (T_c - T)^\mu$ defines a new critical exponent μ, whose mean field approximation is $\mu = 3/2$ from (4.87), with $T_c = T_0$.

4.3.2 Broken continuous symmetry

The Ising model and its mean field (or Landau) approximation are well suited for the description of magnets with an easy axis of magnetization: the magnetization lies essentially along a fixed axis with respect to the crystal lattice and components perpendicular to this axis may be neglected. This is not the general case: the magnetization is a vector $\vec{M}$ in three-dimensional space, which is the mean value of a classical spin $\vec{S}$ or of a quantum spin $\vec{\sigma}$. One says that the *dimension of the order parameter* is $n = 3$. The order parameter can also be a vector in some symmetry space which is not necessarily ordinary space, or even a tensor as in the case of superfluid helium-3. We shall limit ourselves to vector order parameters and denote by n their dimension, namely the dimension of the symmetry space. The dimension of the magnetization in the case of the Ising model is $n = 1$; it is $n = 3$ for the Heisenberg model (3.29).

A novel phenomenon arises when $n \geq 2$; we shall study it in the case $n = 2$, since the generalization to other values of n is straightforward. The order parameter is then a vector $\vec{M}$ of some two-dimensional space with components M_1 and M_2 and the generalization of the Landau functional (4.78) reads, defining $w_0 = u_0/3!$

$$\begin{aligned}
\mathcal{L}[\vec{M}] &= \int \mathrm{d}^D r \left(\frac{1}{2}(\vec{\nabla}\vec{M})^2 + \frac{w_0}{4}\left[\vec{M}^2 + \frac{r_0}{w_0}\right]^2\right) \\
&= \int \mathrm{d}^D r \left(\frac{1}{2}(\vec{\nabla}M_1)^2 + \frac{1}{2}(\vec{\nabla}M_2)^2 + \frac{w_0}{4}\left[(M_1^2 + M_2^2) + \frac{r_0}{w_0}\right]^2\right)
\end{aligned} \tag{4.88}$$

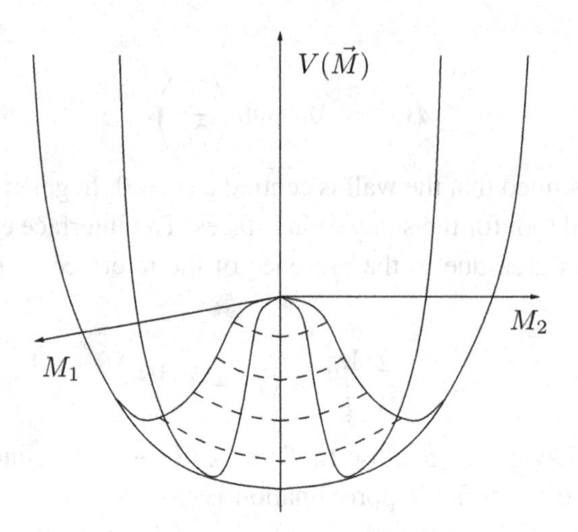

Figure 4.7 A potential with a 'Mexican hat' shape.

It is convenient to define a 'potential' $V(\vec{M})$

$$\begin{aligned} V(\vec{M}) &= \frac{w_0}{4}\left[(M_1^2 + M_2^2)^2 + \frac{r_0}{w_0}\right]^2 \\ &= \frac{1}{2}r_0(M_1^2 + M_2^2) + \frac{w_0}{4}(M_1^2 + M_2^2)^2 + \text{const} \end{aligned} \tag{4.89}$$

The Landau functional (and the potential V) have a $O(2)$ symmetry: they are invariant under rotations in the symmetry space

$$\begin{aligned} M_1' &= M_1 \cos\theta + M_2 \sin\theta \\ M_2' &= -M_1 \sin\theta + M_2 \cos\theta \end{aligned} \tag{4.90}$$

If $r_0 > 0$, the potential is minimum for $\vec{M} = 0$, but if $r_0 < 0$, or equivalently $T < T_0$, the potential has a minimum for[32]

$$\vec{M}^2 = -\frac{r_0}{w_0} = \frac{|r_0|}{w_0} = v^2$$

The potential $V(M)$ for $T < T_0$ has a 'Mexican hat' shape (see Figure 4.7): all points on the circle $\vec{M}^2 = v^2$ correspond to a minimum of V. In the case of a scalar order parameter, only two values of M were possible; now we have an infinity of possible values of the magnetization! The physical interpretation is that the magnetization may lie along any direction in the symmetry space. The direction of the magnetization is chosen by putting the system in an external two-dimensional

[32] In quantum field theory, v is called the 'vacuum expectation value' of the field.

field $\vec{B}$ with a fixed orientation, coupled to the magnetization through

$$-\mu\vec{B}\cdot\int \mathrm{d}^D r\, \vec{M}(\vec{r})$$

and by letting $\vec{B}\to 0$ in the infinite volume limit. For the sake of definiteness, let us take the magnetic field along the 1 axis, $\vec{B}=(B,0)$. Then the magnetization takes the value

$$M_1=\sqrt{\frac{|r_0|}{w_0}}=v \qquad M_2=0$$

This value of $\vec{M}$ minimizes the Landau functional. Let us define $\tilde{M}=M_1-v$; then

$$M_1^2+M_2^2=\tilde{M}^2+2v\tilde{M}+v^2+M_2^2$$

and

$$V(\vec{M})=|r_0|\tilde{M}^2+vw_0\tilde{M}(\tilde{M}^2+M_2^2)+\frac{w_0}{4}(\tilde{M}^2+M_2^2)^2 \tag{4.91}$$

The connected correlation function is computed from (4.82)

$$G_{ij}(\vec{r},\vec{r}\,')=\left.\frac{\delta^2\mathcal{L}[M]}{\delta M_i(\vec{r})\delta M_j(\vec{r}\,')}\right|_{M_1=v,M_2=0} \tag{4.92}$$

Since

$$\left.\frac{\delta}{\delta M_1}\right|_{M_1=v}=\left.\frac{\delta}{\delta\tilde{M}}\right|_{\tilde{M}=0}$$

we get

$$G_{11}(\vec{q})=\frac{1}{q^2+2|r_0|} \qquad G_{12}=G_{21}=0 \qquad G_{22}=\frac{1}{q^2} \tag{4.93}$$

In the case of G_{11} we recover the correlation length of the scalar model, $\xi=|2r_0(T)|^{-1/2}$, but the expression of G_{22} shows that the corresponding correlation length is infinite. The first correlation length is denoted by $\xi_\parallel$ and the second one by $\xi_\perp$

$$\xi_\parallel=|2r_0(T)|^{-1/2}=[2\overline{r}_0(T_0-T)]^{-1/2} \qquad \xi_\perp=\infty \tag{4.94}$$

In quantum field theory, an infinite correlation length is the signature for a massless particle, which is called a *Goldstone boson*. The reason for a zero mass is easy to understand: moving along the circle $\vec{M}^2=v^2$ does not cost any energy, while moving in the perpendicular direction leads to the same energy cost as the $n=1$ case.

Instead of writing the order parameter $\vec{M}$ as a two-dimensional vector, it is often convenient to build from it a complex number, $M = M_1 + \mathrm{i}M_2$, and this allows one to couple M to a vector potential, see Exercise 4.6.6. This coupling to a vector potential, and more generally to a non-Abelian vector potential, is at the basis of two very important models of physics: the Ginzburg–Landau model of superconductivity (Exercise 4.6.6) and the Glashow–Weinberg–Salam model of electroweak interactions.

4.3.3 Ginzburg–Landau Hamiltonian

We now change our point of view on Landau's theory: instead of deriving it from mean field theory, we try to obtain it as a first approximation to the exact calculation of a partition function with a well-defined Boltzmann weight given in (4.95). We return, for the sake of simplicity, to the case of a scalar order parameter, but there would be no difficulty in extending the following arguments to an order parameter with dimension n. We assume that the partition function $Z[B(\vec{r})]$, considered as a functional of $\vec{B}(\vec{r})$, may be computed as

$$\begin{aligned} Z[B(\vec{r})] &= \int \mathcal{D}\varphi(\vec{r}) \exp\Big(-H[\varphi(\vec{r})] - \int \mathrm{d}^D r\, B(\vec{r})\varphi(\vec{r})\Big) \\ &= \int \mathcal{D}\varphi(\vec{r}) \exp\Big(-H_1[\varphi(\vec{r}), B(\vec{r})]\Big) \end{aligned} \tag{4.95}$$

where H is a functional of $\varphi(\vec{r})$, called the *Ginzburg–Landau Hamiltonian*

$$\boxed{H[\varphi(\vec{r})] = \int \mathrm{d}^D r \left(\frac{1}{2}[\vec{\nabla}\varphi(\vec{r})]^2 + \frac{1}{2} r_0 \varphi^2(\vec{r}) + \frac{1}{4!} u_0 \varphi^4(\vec{r})\right)} \tag{4.96}$$

Some explanations are needed to clarify these equations. First $\varphi(\vec{r})$ is, for a fixed value of $\vec{r}$, a random variable, taking real values in $[-\infty, \infty]$; $\varphi(\vec{r})$ is called a *random field*, or simply a *field*. Its probability distribution is given by the Boltzmann weight $\exp(-H_1[\varphi, B])$. We have set to one some parameters: (*i*) $\beta = 1$ as β varies slowly close to the critical temperature and can be replaced by a constant β_0, which we put equal to one, and (*ii*) $\mu = 1$ to avoid trivial multiplicative factors. The only temperature dependent parameter is $r_0(T) = \bar{r}_0(T - T_0)$; $\int \mathcal{D}\varphi(\vec{r})$ is a shorthand notation

$$\int \mathcal{D}\varphi(\vec{r}) = \int \prod_i \mathrm{d}\varphi(\vec{r}_i)$$

The reader should not be misled by the continuum notation: there is always an underlying lattice in the Ginzburg–Landau Hamiltonian. One should always remember that only fluctuations with wavelengths greater than some lattice spacing

a must be kept. In other words, if one takes Fourier transforms, the wave vector q is limited by a cut-off $\Lambda \sim 1/a : q \le \Lambda$. Since this cut-off corresponds to short wavelengths, it is often called an *ultraviolet cut-off*. In quantum field theory, the ultraviolet cut-off is sent to infinity, as space is really continuous, but in statistical mechanics there is always a minimum length, the lattice spacing. The continuum notation is convenient, but it should never mask the underlying lattice.[33]

It is not easy to give a satisfactory derivation of the Hamiltonian (4.96). One can try to obtain it from a generalization of the Ising model together with a coarse graining procedure; one then argues that critical phenomena should be insensitive to microscopic details (Exercise 4.6.14). Maybe the best argument in favour of (4.96) is that all three terms in it are absolutely necessary: omitting any one of them would change the physics in an essential way.

We are going to show that Landau's theory is recovered when one computes the integral in (4.95) with a saddle point approximation. The equation that gives the maximum of the integrand is

$$B(\vec{r}) = \left.\frac{\delta H}{\delta\varphi(\vec{r})}\right|_{\varphi(\vec{r})=\varphi_0(\vec{r})} \tag{4.97}$$

This equation defines a configuration $\varphi_0(\vec{r})$ of the field, which is a functional of $B(\vec{r})$. In the saddle point approximation, the partition function is simply given by the maximum of the integrand

$$Z[B] = \exp\Big(-H[\varphi_0] + \int \mathrm{d}^D r\, B(\vec{r})\varphi_0(\vec{r})\Big)$$

so that the free energy $\ln Z$ is given (up to a multiplicative factor) by

$$\ln Z = -H[\varphi_0] + \int \mathrm{d}^D r\, B(\vec{r})\varphi_0(\vec{r}) \tag{4.98}$$

From the free energy we compute the magnetization

$$\begin{aligned} M(\vec{r}) = \frac{\delta \ln Z}{\delta B(\vec{r})} &= -\int \mathrm{d}^D r' \left.\frac{\delta H}{\delta\varphi(\vec{r}\,')}\right|_{\varphi_0} \frac{\delta\varphi_0(\vec{r}\,')}{\delta B(\vec{r})} + \varphi_0(\vec{r}) \\ &+ \int \mathrm{d}^D r' \,\frac{\delta\varphi_0(\vec{r}\,')}{\delta B(\vec{r})} B(\vec{r}\,') = \varphi_0(\vec{r}) \end{aligned}$$

where we have used (A.65) and (4.97). In the saddle point approximation, the magnetization is equal to the solution $\varphi_0(\vec{r})$ of (4.97), but this identification is not

[33] The effect of the lattice can be taken into account within a continuum formalism if one introduces for example a term

$$\frac{1}{2}\int \mathrm{d}^D r\, (\vec{\nabla}^2\varphi(\vec{r}))^2$$

in the Ginzburg–Landau Hamiltonian: this term acts as an ultraviolet cut-off.

valid beyond this approximation. Let us now compute the Legendre transfom of $\ln Z$, namely the Gibbs potential Γ,[34] which is a functional of the magnetization

$$\Gamma[M(\vec{r})] = \int \mathrm{d}^D r \, [M(\vec{r}) B(\vec{r}) - \ln Z] \tag{4.99}$$

Using (4.96) one finds

$$\Gamma[M(\vec{r})] = H[M(\vec{r})] = \int \mathrm{d}^D r \left(\frac{1}{2} [\vec{\nabla} M(\vec{r})]^2 + \frac{1}{2} r_0 M^2(\vec{r}) + \frac{1}{4!} u_0 M^4(\vec{r}) \right) \tag{4.100}$$

We have recovered the Landau functional (4.88) as the approximate Gibbs potential corresponding to the Ginzburg–Landau partition function (4.95). In the saddle point approximation, the Gibbs potential Γ has the same functional form as the Hamiltonian, although its meaning is of course completely different. This identity of form does not persist beyond the saddle point approximation, as we are now going to show.

4.3.4 Beyond Landau's theory

In our calculation of the partition function, we have obtained Landau's theory by replacing the integral of the Boltzmann weight by the maximum value of the integrand. The next step is to take into account the width of the peak around this maximum. The validity of the saddle point approximation relies on the fact that the peak around the maximum of the integrand is narrow. We may make the peak around the maximum $\varphi_0(\vec{r})$ as narrow as we wish by introducing artificially a small parameter $\hbar \to 0$.[35] Our goal is to obtain an expansion in powers of $\hbar$, although at the end of the calculations we shall set $\hbar = 1$. We thus write instead of (4.95)

$$\begin{aligned} Z[B(\vec{r})] &= \int \mathcal{D}\varphi(\vec{r}) \exp\left[-\frac{1}{\hbar} \left(H[\varphi(\vec{r})] + \int \mathrm{d}^D r \, B(\vec{r}) \varphi(\vec{r}) \right) \right] \\ &= \int \mathcal{D}\varphi(\vec{r}) \exp\left(-\frac{1}{\hbar} H_1[\varphi(\vec{r}), B(\vec{r})] \right) \end{aligned} \tag{4.101}$$

We expand H_1 around $\varphi_0(\vec{r})$, setting $\psi(\vec{r}) = \varphi(\vec{r}) - \varphi_0(\vec{r})$ and keeping only the lowest non-trivial term in the expansion

$$H_1 \simeq H[\varphi_0(\vec{r})] - \int \mathrm{d}^D r \, B(\vec{r}) \, \varphi_0(\vec{r}) + \frac{1}{2} \int \mathrm{d}^D r \, \mathrm{d}^D r' \psi(\vec{r}) \mathcal{H}(\vec{r}, \vec{r}\,') \psi(\vec{r}\,') \tag{4.102}$$

[34] As already mentioned, the Gibbs potential is denoted by Γ in the present chapter.

[35] The notation looks a bit odd, as our problem is a classical one, and Planck's constant is *a priori* not involved in it; however, the corresponding small parameter in quantum field theory *is* $\hbar$, and it is instructive to use the same notation.

where

$$\mathcal{H}(\vec{r},\vec{r}\,') = \frac{\delta^2 H[\varphi]}{\delta\varphi(\vec{r})\delta\varphi(\vec{r}\,')}\Big|_{\varphi=\varphi_0} \tag{4.103}$$

The integral over $\psi(\vec{r})$ is Gaussian and, from (A.47), ignoring unimportant multiplicative factors,

$$\int \mathcal{D}\psi \exp\left[-\frac{1}{2}\int \mathrm{d}^D\vec{r}\,\mathrm{d}^D r'\,\psi(\vec{r})\mathcal{H}(\vec{r},\vec{r}\,')\psi(\vec{r}\,')\right] \propto [\det\mathcal{H}]^{-1/2}$$
$$= \exp\left[-\frac{1}{2}\mathrm{Tr}\ln\mathcal{H}\right]$$

In order to compute the trace, we remark that $\mathcal{H}$ is given from (4.103) by

$$\mathcal{H}(\vec{r},\vec{r}\,') = \left(-\nabla_r^2 + r_0(T) + \frac{1}{2}u_0\varphi_0^2\right)\delta^{(D)}(\vec{r}-\vec{r}\,') \tag{4.104}$$

where $M = \varphi_0$ in the present approximation. The trace of a function $f(\vec{r},\vec{r}\,') = f(\vec{r}-\vec{r}\,')$ is computed by using a Fourier transformation which diagonalizes f

$$\begin{aligned}\mathrm{Tr}\, f = \int \mathrm{d}^D r\, f(\vec{r},\vec{r}) &= \int \mathrm{d}^D r\,\mathrm{d}^D r'\,\delta^{(D)}(\vec{r}-\vec{r}\,')f(\vec{r}-\vec{r}\,')\\ &= \int \mathrm{d}^D r\,\mathrm{d}^D r' \int \frac{\mathrm{d}^D q}{(2\pi)^D}\,\mathrm{e}^{\mathrm{i}\vec{q}\cdot(\vec{r}-\vec{r}\,')} f(\vec{r}-\vec{r}\,')\\ &= V\int \frac{\mathrm{d}^D q}{(2\pi)^D}\,\tilde{f}(\vec{q})\end{aligned}$$

where the last line has been obtained by integrating over $\vec{r}$ and $\vec{r}\,'' = \vec{r}-\vec{r}\,'$. The Fourier transform of (4.104) allows one to diagonalize $\mathcal{H}$ if the magnetic field is uniform, and the trace of $\ln\mathcal{H}$ is

$$\mathrm{Tr}\ln\mathcal{H}(\vec{r},\vec{r}\,') = V\int \frac{\mathrm{d}^D q}{(2\pi)^D}\ln\left(q^2 + r_0(T) + \frac{1}{2}u_0 M^2\right) \tag{4.105}$$

Our approximate free energy is now

$$\hbar \ln Z = H_1[\varphi_0] - \frac{\hbar}{2}\mathrm{Tr}\ln\mathcal{H}$$

and the magnetization M differs from φ_0 by terms of order $\hbar$

$$M = \frac{\delta(\hbar\ln Z)}{\delta B(\vec{r})} = \varphi_0(\vec{r}) + \mathcal{O}(\hbar)$$

Now we can write

$$H_1[\varphi_0] = H_1[M] + (H_1[\varphi_0] - H_1[M])$$

and the term between brackets is of order $\hbar^2$, because $\delta H_1/\delta\varphi|_{\varphi=\varphi_0} = 0$. We finally get the Gibbs potential Γ to order $\hbar$

$$\boxed{\Gamma[M] = H[M] + \frac{\hbar}{2}\, V \int \frac{\mathrm{d}^D q}{(2\pi)^D} \ln\left(q^2 + r_0(T) + \frac{1}{2}\, u_0 M^2\right) + \mathcal{O}(\hbar^2)} \tag{4.106}$$

The preceding equation has been proved only in the case of a uniform field, and Γ in (4.106) is in fact a function of M, and not a functional, but this equation can be generalized to non-uniform magnetization. Our approximation corresponds to taking into account the width of the peak around the maximum of the integrand in the saddle point approximation. The analogue of the Landau approximation in quantum field theory is the classical solution to the field equations, and the analogue of our approximation is the integration over the quantum fluctuations around the classical solution in a path integral approach. The expansion in powers of $\hbar$, whose first two terms are written in (4.106), is called the *loop expansion*.

4.3.5 Ginzburg criterion

We obtain from (4.106) the magnetic field B by differentiation

$$\begin{aligned}\frac{1}{V}\frac{\partial\Gamma}{\partial M} = B = r_0(T)M + \frac{u_0}{6}M^3 \\ + \frac{u_0\hbar}{2}\, M \int \frac{\mathrm{d}^D q}{(2\pi)^D} \frac{1}{q^2 + r_0(T) + \frac{1}{2}\, u_0 M^2} + \mathcal{O}(\hbar^2)\end{aligned} \tag{4.107}$$

From this equation we compute the susceptibility χ. For simplicity we limit ourselves to the case $T > T_\mathrm{c}$, $B = M = 0$

$$\left.\frac{\partial B}{\partial M}\right|_{M=0} = \frac{1}{\chi} = \rho = r_0(T) + \frac{u_0\hbar}{2} \int \frac{\mathrm{d}^D q}{(2\pi)^D} \frac{1}{q^2 + r_0(T)} + \mathcal{O}(\hbar^2) \tag{4.108}$$

where we have defined $\rho = 1/\chi$. A dimensional analysis of (4.108) reveals interesting features. Let us assign an arbitrary dimension to $\hbar$. Since $H/\hbar$ must be dimensionless, and since the dimensions of $\mathrm{d}^D r$ and $\vec{\nabla}$ are ℓ^D and ℓ^{-1}, where ℓ is a length, the dimension of φ is $\hbar^{1/2}\ell^{1-D/2}$. It follows at once that r_0 has dimension ℓ^{-2}, while u_0 has dimension $\hbar^{-1}\ell^{D-4}$. Equation (4.108) gives the first two terms of ρ as an expansion in powers of $\hbar$

$$\rho = c_0 + c_1'\hbar + c_2'\hbar^2 + \cdots$$

The parameters of the problem are a, r_0 and u_0, but only the dimension of u_0 depends on $\hbar$, so that dimensional analysis shows that the coefficient of $\hbar^L$ must also be proportional to u_0^L. At this point we may set our artificial parameter $\hbar = 1$ and write the expansion of ρ as an expansion in powers of u_0[36]

$$\rho = c_0 + c_1 u_0 + c_2 u_0^2 + \cdots$$

and interpret (4.108) with $\hbar = 1$ as the beginning of *perturbative series*, or a *perturbative expansion* in u_0.

Since ρ vanishes at the critical temperature, Equation (4.108) gives T_c, or equivalently $r_{0c} = r_0(T_c)$

$$r_{0c} + \frac{u_0}{2} \int \frac{d^D q}{(2\pi)^D} \frac{1}{q^2 + r_{0c}} = 0 \tag{4.109}$$

This equation shows that $(T_c - T_0)$ is of order u_0. We have computed explicitly the first two terms of the perturbative expansion, but it should in principle be possible to compute higher order terms by integrating more precisely the fluctuations around the saddle point, although the calculation becomes technically more and more demanding as the power of u_0 increases. It would seem that we can push the corrections to Landau's theory to an arbitrary accurracy, at least in principle. However, we are going to argue that the perturbative expansion is bound to fail when $T \to T_c$. The argument is essentially dimensional. It is conventional to take as the fundamental dimension the inverse of a length: a gradient then has dimension $+1$. Denoting by $[A]$ the dimension of a quantity A we get from the previous analysis with $\hbar = 1$

$$[\varphi] = d_\varphi^0 = \frac{D}{2} - 1 \qquad [r_0] = 2 \qquad [u_0] = 4 - D \tag{4.110}$$

d_φ^0 is called the *canonical dimension* of the field φ. Subtracting (4.109) from (4.108) yields

$$\rho = \left(r_0 - r_{0c}\right)\left(1 + \frac{u_0}{2} \int \frac{d^D q}{(2\pi)^D} \frac{1}{(q^2 + r_0)(q^2 + r_{0c})}\right)$$

Now, in a perturbative expansion in powers of u_0, we are allowed to replace r_0 and r_{0c} in the integral by their actual values ρ and 0, as this substitution would introduce corrections of order u_0^2 or higher. Thus we may write consistently with the expansion

$$\rho = \left(r_0 - r_{0c}\right)\left(1 + \frac{u_0}{2} \int \frac{d^D q}{(2\pi)^D} \frac{1}{q^2(q^2 + \rho)}\right) \tag{4.111}$$

[36] This is a familiar feature of the so-called $\hbar \to 0$ classical limit of quantum mechanics. Of course one never sends $\hbar$ to zero, see Footnote 28 of Chapter 3: it is always possible to work in a system of units where $\hbar = 1$. The classical limit is reached when a typical dimensionless physical parameter of the problem, for example a coupling constant, goes to zero. Here the small parameter is u_0.

If the space dimension D is bigger than 4, the integral is convergent at $q = 0$ even for $\rho = 0$ and

$$\rho = (r_0 - r_{0c})(1 + u_0 A) = B(T - T_c)$$

where A and B are unimportant constant factors. The critical exponent $\gamma = 1$ of Landau's theory is not modified! This result generalizes to all terms of the perturbative expansion and to all critical exponents. As we shall confirm in Section 4.5.1, the critical exponents are given correctly by Landau's (or mean field) theory if $D > 4$. The situation is quite different if $D < 4$: then we may rewrite, with the change of variables $q = \sqrt{\rho}\, q'$

$$\rho = \left(r_0 - r_{0c}\right)\left(1 + \frac{u_0}{2}\rho^{\frac{D-4}{2}} \int \frac{\mathrm{d}^D q'}{(2\pi)^D} \frac{1}{q'^2(q'^2 + 1)}\right) \tag{4.112}$$

It thus appears that the true expansion parameter is not u_0, but the *dimensionless* parameter $u_0 \rho^{\frac{D-4}{2}}$. Even if u_0 is chosen very small, but fixed, $\rho^{\frac{D-4}{2}} \to \infty$ for $T \to T_c$, and the effective expansion parameter diverges in this limit, which makes a straightforward perturbative expansion impossible: in contrast to the case $D > 4$, it is not consistent to assume that $\rho \propto (T - T_c)$. That the behaviour of the cases $D > 4$ and $D < 4$ is fundamentally different was first understood by Ginzburg, and this observation is known as the Ginzburg criterion on the space dimension.

An expression for the critical temperature can be obtained from the same line of argument. Equation (4.109) becomes

$$r_{0c} = \overline{r}_0(T_c - T_0) = -\frac{u_0}{2} \int \frac{\mathrm{d}^D q}{(2\pi)^D} \frac{1}{q^2} = -\frac{u_0}{2} \frac{S_D \Lambda^{D-2}}{(2\pi)^D (D-2)} \tag{4.113}$$

where

$$S_D = \frac{2\pi^{D/2}}{\Gamma(D/2)}$$

is the area of the unit sphere in dimension D, see (2.123), and Γ is Euler's gamma function. The equation is convergent for $q \to 0$ (or *infrared convergent*) only if $D > 2$. The validity of the Ginzburg–Landau Hamiltonian is limited to $D > 2$, and $D = 2$ is called the *lower critical dimension* of the model. There is no convergence problem for $q \to \infty$ because of the ultraviolet cut-off $\Lambda \sim 1/a$. Equation (4.112) shows that the true critical temperature T_c – or at least the temperature computed in the present approximation – is lower than that of the mean field theory. Taking fluctuations into account tends to make the transition more difficult and thus to lower the critical temperature.

Let us rephrase the previous results in a slightly different, but equivalent, way. In Section 4.5.2 we shall need to work at u_0 small, but fixed.[37] Then r_{0c} in (4.113) is proportional to a^{-2} and may be large if a is small. In quantum field theory we even have to take the limit $a \to 0$ and $r_{0c} \to \infty$. The small parameter near a critical point is $(r_0 - r_{0c})$, and at fixed u_0, we use r_0 as a control parameter in order to reach criticality. From (4.112), $\rho \simeq (r_0 - r_{0c})$ to lowest order, and an alternative dimensionless expansion parameter is also $u_0(r_0 - r_{0c})^{(D-4)/2}$, which becomes infinite for $D < 4$ when $r_0 \to r_{0c}$.

4.4 Renormalization group: general theory

4.4.1 Spin blocks

We have seen in the preceding section that naïve perturbation theory is divergent in the vicinity of a critical point and cannot be used directly to study critical phenomena for space dimension $D \le 4$. Nevertheless, in the following section, we shall show that perturbation theory can be reorganized in such a way that a calculation of critical exponents becomes possible: this is called 'renormalization group improved perturbation theory'. In order to understand the reorganization of the perturbative expansion, we have first to explain what the *renormalization group* (RG) is. At this point, we must emphasize that the renormalization group is a quite general strategy, which goes far beyond its use in perturbation theory. Indeed it is basically a 'non-perturbative technique', which can be used for example in the framework of numerical simulations. However, one should be aware that the renormalization group is an extremely general framework that must be adapted to the nature of the problem at hand. The RG does not provide one with an automatic recipe to compute for example critical exponents! A good deal of guesswork, intuitive reasoning, mathematical and numerical skill must be put into the game before one gets physically relevant results. We shall illustrate this general framework on the example of continuous phase transitions, a case where the use of the renormalization group is well understood. Before going into technical details, it is useful to give an intuitive picture of the behaviour of a physical (or model) system close to a critical point.

A priori, there are two important length scales in the system. First there is the *microscopic scale a*, which is characteristic of the range of the basic interaction, for example the interaction between Ising spins, and of the lattice spacing. In the Ising model (3.30), the range of the interaction coincides with the lattice spacing, but in general we can envisage systems where the interaction extends over a few

[37] More precisely we shall require that the dimensionless quantity $u_0 a^{4-D} \ll 1$.

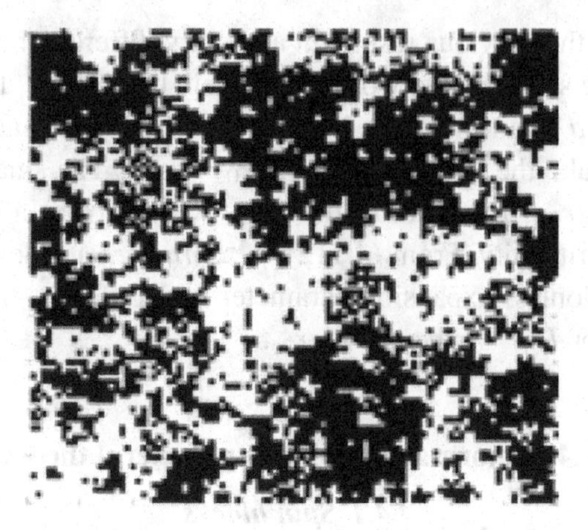

Figure 4.8 Ising spins near the critical temperature on a 100×100 lattice. Black shows the up spins.

lattice spacings. Long range interactions[38] would lead to a qualitatively different behaviour and are not included in our analysis. The second length scale is the correlation length ξ: in an Ising model, it gives a measure of the distance over which spins are correlated. Close to the critical temperature T_c, spins tend to be ordered in clusters of up spins and down spins, and ξ is very roughly the average size of these clusters. A more precise, but by no means unique, definition of ξ would be

$$\xi = \frac{\left\langle \sum_j |\vec{r}_i - \vec{r}_j| S_i S_j \right\rangle}{\langle \sum_j S_i S_j \rangle} \tag{4.114}$$

which differs by an unimportant numerical factor from the definition (4.30) based on the exponential decrease of the correlation function, if the slowly varying $C(r)$ is given for example by a power law. We recall that when $T \to T_c$, the correlation length diverges from (4.39) as a power of the *reduced temperature* t

$$\frac{\xi}{a} = \xi_{\pm} |t|^{-\nu} \qquad t = \frac{T - T_c}{T_c} \tag{4.115}$$

where the numerical coefficients ξ_+ (ξ_-) refer to the $t > 0$ ($t < 0$) behaviour.

Exactly at the critical point, the correlation length is infinite, and one finds clusters of all possible sizes: a snapshot of the system, with up (down) spins represented by black (white) squares would display a fractal structure (Figure 4.8). In

[38] What is a long range interaction can be given a precise mathematical meaning. We shall assume here that the interaction has a finite range, or decreases exponentially, but a rapid power law decrease could be acceptable.

oceans of up spins, one finds continents of down spins, which themselves contain lakes of up spins etc., all length scales being represented. In other words, one finds fluctuations of all possible scales, and the system is called *scale invariant*. Away from the critical point, one finds fluctuations of all possible scales between a and ξ. Actually, except in the thermodynamic limit, there is a third scale, the size L of the system. If the system has finite size, the preceding picture will only be valid if $L \gg \xi$. Finite size scaling (Section 7.6) teaches us how to handle the situation where $L \sim \xi$.

If we look at the system using a microscope with resolution $\lambda \leq a$, all details of the spin fluctuations will be revealed. Suppose instead that we use a microscope whose resolution is only $\lambda = ba$, where $1 \ll b \ll \xi/a$. All fluctuations with wavelengths $\leq ba$ will be averaged out, but the picture will still be that of a system close to the critical point, since $\xi/b \gg a$: we shall still observe large clusters of up spins and down spins. If we now reduce the magnification of the second microscope by a factor b and compare the images given by the two microscopes, the average size of the clusters of spins with the same orientation appears to be reduced by a factor b, and the same property holds for the correlation length, with a new correlation length $\xi' = \xi/b$. In other words, the unit of length has been rescaled by a factor b. If the microscope has a really bad resolution, $b \sim \xi/a$, then we do not observe large clusters of up and down spins, and the picture shown by the microscope is no longer that of a system close to the critical point.

In order to implement mathematically the action of microscopes with various resolutions, we average out the fluctuations of size $\leq b$ by building *spin blocks*. If I is a domain of size b, the spin block $S_I(b)$ of this domain will be defined by taking the average of the b^D spins belonging to the domain (Figure 4.9)[39]

$$S_I(b) = \frac{1}{b^D} \sum_{i \in I} S_i \tag{4.116}$$

One calls b the *scaling factor*. If we start with Ising spins $S_i = \pm 1$, the spin block $S_I(b)$ will take $(b^D + 1)$ values in the interval $[-1, +1]$, and will become in practice a continuous variable in this interval. Building the spin blocks is the first operation of a *renormalization group transformation* (RGT). An important observation is that a first RGT with scaling factor b followed by a second one with scaling factor b' is equivalent to a single RGT with scaling factor $b'' = bb'$. Although b

[39] This way of defining spin blocks is not unique. It has the advantage of being a linear operation. Another way of building spin blocks would be for instance to decide that the spin block is equal to $+1$ if the majority of the spins in the domain are up, and -1 in the opposite case, which corresponds to a non-linear operation. Although the non-linear construction may prove useful in some circumstances, we shall be concerned only with the linear one.

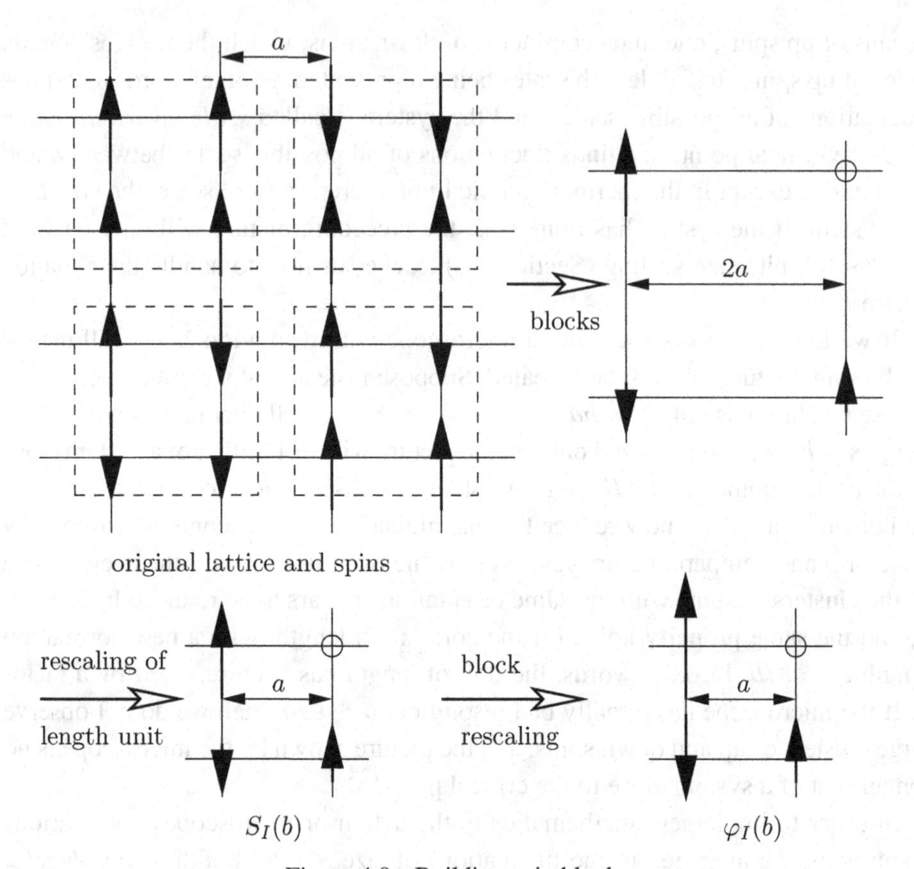

Figure 4.9 Building spin blocks.

is in principle an integer, $b = n$, its values are usually much larger than one and it can be considered as a continuous variable.[40]

The second operation, which is suggested by the change of magnification of the second microscope, is a *rescaling of the unit of length*: $L \to L/b$, so that the spin blocks live on a lattice with the same spacing as the original lattice. This rescaling allows us to compare the spin block system and the original system on the same lattice. The third and last operation is a *rescaling (or renormalization) of the spin blocks*. That such an operation is useful can be understood from the following argument. If we compare two different spin block systems with scaling factors b_1 and b_2, the range of the interval $[-1, +1]$ where the spin blocks $S_I(b_1)$ and $S_J(b_2)$

[40] It would not be difficult to give b continuous values. One could for example define a spin block by

$$S_I(b) = \left(\frac{1}{2\pi b^2}\right)^{D/2} \sum_j \left[S_j \exp\left(\frac{(\vec{r}_i - \vec{r}_j)^2}{2b^2} \right) \right]$$

where S_i is the spin located at the centre $\vec{r}_i$ of the block.

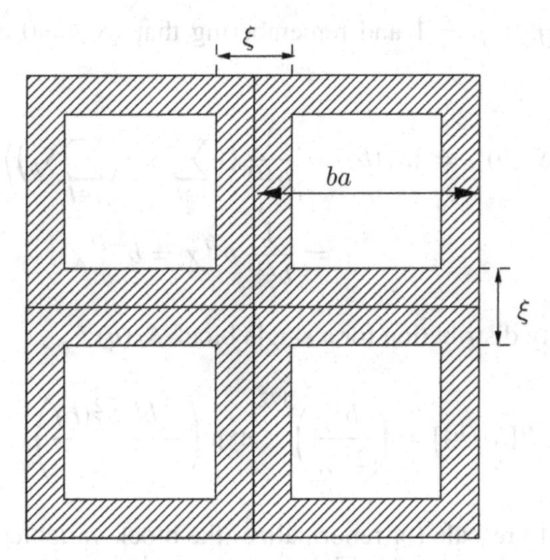

Figure 4.10 Interactions between blocks when $\xi/b \ll a$.

take non-zero values with a significant probability has no reason to be the same in both cases. In order to compare the two systems, it will be useful to rescale the blocks, in such a way that both ranges of variation are similar[41]

$$S_I(b) \to \varphi_I(b) = Z_1(b) S_I(b) \tag{4.117}$$

In order to illustrate our construction in a simple case, let us examine RGT in the high temperature limit, $T \gg T_c$. In this limit, the correlation length tends to zero. The correlation length of the spin block system will tend to zero if we choose a scaling factor $b \gg \xi/a$, so that the block correlation length $\xi' = \xi/b \ll a$. Such a situation corresponds to a high temperature limit for the blocks, and then the block variables become almost independent. Indeed, there are b^D spins in a block, and only the spins located in a neighbourhood of its boundary interact with the spins of neighbouring blocks. Within a block, the number of the spins that interact with neighbouring blocks is on the order of $(\xi/a)\, b^{D-1} = b^D(\xi/ba)$ and the fraction of interacting spins is only $\xi/ba \ll 1$ (Figure 4.10). Each block is the sum of a large number of random variables and has negligible interactions with its neighbours. The joint probability distribution of the blocks is then a product of independent distributions and, from the central limit theorem, each of these distributions is a Gaussian whose width is controlled by the magnetic susceptibility χ. Indeed,

[41] Assume that up spins are represented by black squares, down spins by white ones. In order to represent the blocks, we need a range of grey: values of $S(b)$ close to -1 will be light grey, and the grey will become darker and darker as the block spin value increases to $+1$. The picture of the blocks will be less contrasted than the original picture of the spins, and we must rescale (renormalize) the blocks in order to recover the original contrast.

from (4.28), with $\beta = \mu = 1$ and remembering that $\langle S_i \rangle = 0$ above the critical temperature, we get

$$\langle S_I^2(b)\rangle_c = \langle S_I^2(b)\rangle = \frac{1}{b^{2D}} \Big\langle \Big(\sum_{i\in I} S_i\Big)\Big(\sum_{j\in I} S_j\Big)\Big\rangle$$
$$= \frac{1}{b^{2D}} b^D \chi = b^{-D}\chi \tag{4.118}$$

Then the probability distribution for a single block $S_I(b)$ is

$$P[S_I(b)] = \left(\frac{b^D}{2\pi\chi}\right)^{1/2} \exp\left(-\frac{b^D S_I^2(b)}{2\chi}\right) \tag{4.119}$$

If we now choose to rescale (or renormalize) the block variables in the following way (we use the notations $\overline{\varphi}_I$ and $\overline{Z}_1$ instead of φ_I and Z_1 as the block rescaling is not yet in its final form)

$$S_I(b) \to \overline{\varphi}_I(b) = b^{D/2} S_I(b) = \overline{Z}_1(b) S_I(b) \qquad \overline{Z}_1(b) = b^{D/2} \tag{4.120}$$

the probability distribution P_+ of the rescaled blocks $\overline{\varphi}_I(b)$ is

$$P_+[\overline{\varphi}_I(b)] = \frac{1}{\sqrt{2\pi\chi}} \exp\left(-\frac{\overline{\varphi}_I^2(b)}{2\chi}\right) \tag{4.121}$$

and is independent of b; the notation P_+ indicates that $T > T_c$ or $t > 0$. Thus in the limit of large b, $b \gg \xi/a$, the probability distribution of $\overline{\varphi}_I(b)$ tends to a b-independent, finite limit. This is the first example of what we shall call a *fixed point* of the renormalization group. In our example, the fixed point is a limiting distribution that may be obtained in the limit $b \to \infty$ thanks to a suitable choice of the spin block rescaling.

We now turn to the low temperature limit $T \ll T_c$ in a clustering configuration, by imposing for example up spins on the boundary. The average value $\langle S_i \rangle$ of spin S_i is the magnetization per spin $M = \langle S_i \rangle$, independent of i from translation invariance. From the definition (4.116) we also have $\langle S_I(b)\rangle = M$ and moreover

$$\langle S_I^2(b)\rangle = M^2 + \langle S_I^2(b)\rangle_c = M^2 + b^{-D}\chi \tag{4.122}$$

If we choose the trivial rescaling

$$S_I(b) \to \overline{\varphi}_I(b) = b^0 S_I(b) = \overline{Z}_1(b) S_I(b) \qquad \overline{Z}_1(b) = 1$$

we get a Gaussian distribution centred at $\varphi_I = M$ whose limit is a delta-function if $b \to \infty$

$$P_{-}[\overline{\varphi}_I(b)] = \frac{1}{\sqrt{2\pi \chi b^{-D}}} \exp\left(-\frac{\left[\overline{\varphi}_I(b) - M\right]^2}{2\chi\, b^{-D}} \right) \to \delta\left[\overline{\varphi}_I(b) - M\right] \quad (4.123)$$

We conclude that the following choices of the block scaling factors

$$T > T_{\mathrm{c}}:\ \overline{Z}_1(b) = b^{D/2} \qquad T < T_{\mathrm{c}}:\ \overline{Z}_1(b) = b^0 = 1$$

lead to well-defined limiting probability distributions for the suitably rescaled blocks if $\xi/b \ll a$.

The last possibility is that our starting point is exactly the critical temperature $T = T_{\mathrm{c}}$. Then the formation of spin blocks does not change the correlation length, since it is infinite: we start from the critical point and we remain there. We shall see that it will be possible once again to obtain a limiting probability distribution by choosing a block scaling factor b^{ω_1} with an exponent ω_1 intermediate between 0 and $D/2$

$$\varphi_I(b) = Z_1(b) S_I(b) = b^{\omega_1} S_I(b) \quad (4.124)$$

4.4.2 Critical exponents and scaling transformations

The probability distributions (4.121) and (4.123) still depend on the initial temperature t through the susceptibility χ or the magnetization M, which are functions of t. In other words the fixed point still depends on the initial temperature of the system. Let us take one step further by eliminating this temperature dependence, first in the high temperature case. Consider the initial spin system at two different reduced temperatures t_1 and t_2, corresponding to two different correlation lengths ξ_1 and ξ_2 with $\xi_1, \xi_2 \gg a$ and choose the scaling factors b_1 and b_2 in such a way that the ratio ξ_i/b_i, $i = 1, 2$ is a constant length c'

$$\frac{\xi_1}{b_1} = \frac{\xi_2}{b_2} = c' \ll a \quad (4.125)$$

We can eliminate the dependence on the susceptibility χ in (4.121) by redefining the scaling factors. Since we know from (4.44) and (4.45) that

$$\chi = c''\, \xi^{2-\eta}$$

where c'' is a constant, we may define a new scaling factor $Z_1(b_i)$ taking care of χ as

$$Z_1(b_i) = \frac{1}{\sqrt{c''}}\, b_i^{D/2}\, \xi_i^{-(2-\eta)/2} \quad (4.126)$$

If the renormalized block is defined by $\varphi(b_i) = Z_1(b_i)S(b_i)$, we have for the probability distribution of $\varphi(b_i)$

$$P_+^*[\varphi(b_i)] = \frac{1}{\sqrt{2\pi}} \exp\left(-\frac{1}{2}\varphi^2(b_i)\right) \tag{4.127}$$

where we have suppressed the subscript I, $\varphi_I(b) \to \varphi(b)$. Thus, starting from two *different* reduced temperatures t_1 and t_2, $t_1, t_2 > 0$, we get the *same* probability distribution for the rescaled blocks provided $\xi_1/b_1 = \xi_2/b_2$: we get a t-independent fixed point. From (4.125) and (4.126) we can write $Z_1(b_i)$ as a power of b_i

$$Z_1(b_i) = c\, b_i^{(D-2+\eta)/2} = c\, b_i^{\omega_1} \tag{4.128}$$

with

$$c = [c'c''^{(2-\eta)}]^{-1/2} \text{ and } \omega_1 = \frac{1}{2}(D-2+\eta)$$

The next step is to *assume* that the property we have observed at high temperature is a general one. We start from two different initial reduced temperatures t_1 and t_2, and thus from two different correlation lengths ξ_1 and ξ_2, and apply RGTs such that $\xi_1/b_1 = \xi_2/b_2$, with $b_1, b_2 \gg 1$. Our fundamental assumption will be that the probability distribution of the block spins $S(b_i)$ are identical, provided they are properly rescaled with a factor $Z_1(b_i)$

$$P[S(b)] = Z_1(b)P_\pm^*\left[\frac{\xi}{b},\, Z_1(b)S(b)\right] = Z_1(b)P_\pm^*\left[\frac{\xi}{b},\, \varphi(b)\right] \tag{4.129}$$

where $P_\pm^*$ is a universal probability distribution of the rescaled block variable $\varphi(b) = Z_1(b)S(b)$; the subscripts $+$ and $-$ refer to $t > 0$ and $t < 0$ respectively. The assumption contained in (4.129) has been checked on numerical simulations of the Ising model (Figure 4.11) and will be shown to be a consequence of the renormalization group analysis. Our assumptions are not restricted to a single model: we shall see that all models (or physical systems) within a given *universality class* lead to the same probability distribution, and the rescalings $Z_1(b)$ of two different models differ only by constant factors. For example the two-dimensional Ising model on a square lattice and the Ising model on a hexagonal lattice belong to the same universality class. On the contrary, the two-dimensional and three-dimensional Ising models belong to different universality classes. The notion of universality class will be explained soon in the framework of the renormalization group.

The factor $Z_1(b)$ in front of $P_\pm^*$ has a trivial explanation: it ensures the correct normalization of P if $P_\pm^*$ is normalized to unity

$$\int P(S)\, \mathrm{d}S = \int P_\pm^*(\varphi)\, \mathrm{d}\varphi = 1$$

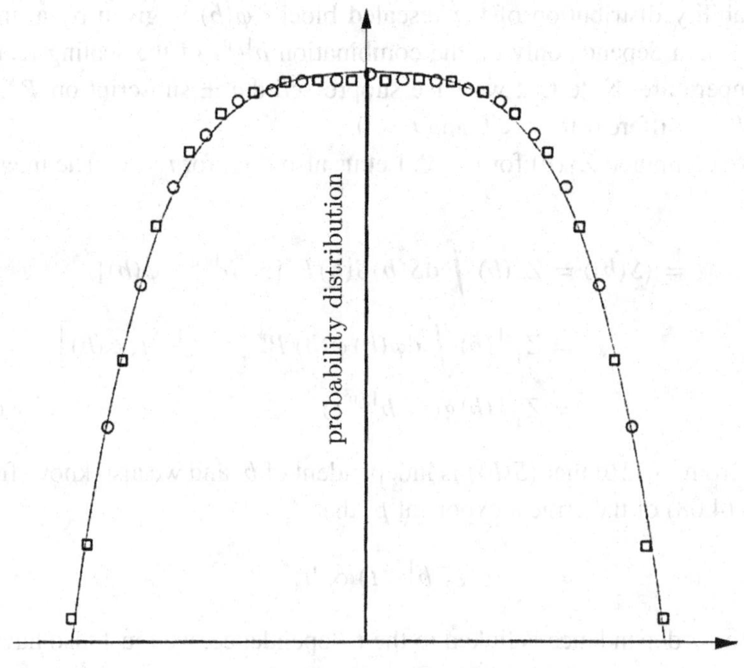

Figure 4.11 Probability distribution of spin blocks in the case of the two-dimensional Ising model. Squares (circles) correspond to $b_2 = 16\,(b_1 = 8)$ and reduced temperatures $t_2 = 0.0471\ (t_1 = 2t_2)$. Since $\nu = 1$ (see Table 4.1), $\xi_2/\xi_1 = t_1/t_2 = 2$. The exponent ω_1 is equal to 1/8 because $\eta = 1/4$. From Bruce and Wallace [21].

This remark being made, let us draw conclusions from (4.129). We rewrite this equation in a slightly more convenient, but completely equivalent, form. From (4.115) we have

$$\frac{\xi}{ba} = \xi_\pm |t|^{-\nu} b^{-1} = \left(\xi_\pm^{-1/\nu} |t| \, b^{1/\nu}\right)^{-\nu}$$

and we define $Z_2(b)$ through[42]

$$Z_2(b) = \xi_\pm^{-1/\nu} \, b^{1/\nu} = z_2^\pm \, b^{1/\nu} \tag{4.130}$$

Then (4.129) becomes

$$\boxed{P[S(b)] = Z_1(b) P^*[Z_2(b)t;\, Z_1(b)S(b)]} \tag{4.131}$$

[42] In quantum field theory, $Z_1(b)$ is related to field renormalization (formerly called wave function renormalization) and $Z_2(b)$ to mass renormalization.

The probability distribution of the rescaled blocks $\varphi(b)$ is given by a universal function, which depends only on the combination $b^{1/\nu}t$ of the scaling factor and of the temperature. Note that we have suppressed the $\pm$ subscript on P^*, as the function P^* is different for $t < 0$ and $t > 0$.

We have computed $Z_1(b)$ for $t > 0$. Let us also do it for $t < 0$. The magnetization M is

$$\begin{aligned} M = \langle S(b) \rangle &= Z_1(b) \int \mathrm{d}S(b) S(b) P^*[z_2^- b^{1/\nu} t; \varphi(b)] \\ &= Z_1^{-1}(b) \int \mathrm{d}\varphi(b)\, \varphi(b)\, P^* \left[z_2^- b^{1/\nu} t; \varphi(b) \right] \\ &= Z_1^{-1}(b) g(z_2^- b^{1/\nu} t) \end{aligned} \tag{4.132}$$

We know from (4.116) that $\langle S(b) \rangle$ is independent of b, and we also know, from the definition (4.68) of the critical exponent $\tilde{\beta}$, that

$$g(z_2^- b^{1/\nu} t) \propto |t|^{\tilde{\beta}}$$

Then, as the b-dependence is linked to the t-dependence, we must also have

$$g(z_2^- b^{1/\nu} t) \propto |t|^{\tilde{\beta}} b^{\tilde{\beta}/\nu}$$

In order to compensate for the $b^{\tilde{\beta}/\nu}$ factor so that $\langle S(b) \rangle = M$ is independent of b, the scaling factor $Z_1(b)$ must be given by a power law

$$Z_1(b) = z_1 b^{\tilde{\beta}/\nu} = z_1 b^{\omega_1}$$

and we identify the exponent ω_1 with $\tilde{\beta}/\nu$. Let us summarize our results for the block spin scaling factor $Z_1(b)$ and the temperature scaling factor $Z_2(b)$ for $t < 0$

$$Z_1(b) = z_1 b^{\omega_1} \qquad \omega_1 = \frac{\tilde{\beta}}{\nu} \tag{4.133}$$

$$Z_2(b) = z_2^- b^{\omega_2} \qquad \omega_2 = \frac{1}{\nu} \tag{4.134}$$

and we have *defined* $\omega_2 = 1/\nu$. It is instructive to check the consistency of these results. Starting from (4.122) with $\xi/b = \text{const}$

$$\begin{aligned} \langle S^2(b) \rangle = M^2 + b^{-D}\chi &= (m_0')^2 |t|^{2\tilde{\beta}} + b^{-D} \chi_0' |t|^{-\nu(2-\eta)} \\ &= m_0^2 b^{-2\tilde{\beta}/\nu} + \chi_0 b^{-(D-2+\eta)} \end{aligned}$$

Now, in order to obtain a temperature-independent limit distribution, the two contributions to $\langle S^2 \rangle$ must have the same b-dependence, which implies, as in (4.128), that $\omega_1 = (D - 2 + \eta)/2$ and that

$$P^*_-(\varphi(b)) = \frac{1}{\sqrt{2\pi\chi_0}} \exp\left(-\frac{(\varphi - m_0)^2}{2\chi_0}\right)$$

The requirement that we get a temperature-independent limit distribution has allowed us to show that the exponent ω_1 is the same in the high and low temperature limits, and that the critical exponents $\tilde{\beta}$, ν and η are not independent.

Let us summarize and comment on the preceding results.

(*i*) The probability distribution $P[S(b); t]$ can be expressed in terms of a unique function P^* within a universality class

$$P[S(b); t] = z_1 b^{\omega_1} P^*[z_2^{\pm} b^{\omega_2} t; z_1 b^{\omega_1} S(b)] \tag{4.135}$$

where b is the size of the spin blocks and $b \gg 1$: one must build spin blocks which are large enough in order to average out details which are specific to a given temperature or a given model within a universality class. The short range details of the models are contained in the multiplicative factors z_1 and $z_2^{\pm}$ which are *non-universal features*, and depend specifically on the model.

(*ii*) At the critical point two different values b_1 and b_2 of the scaling factor give

$$P[S(b_1); t = 0] = z_1 b_1^{\omega_1} P^*[0; z_1 b_1^{\omega_1} S(b_1)]$$
$$P[S(b_2); t = 0] = z_1 b_2^{\omega_1} P^*[0; z_1 b_2^{\omega_1} S(b_2)] \tag{4.136}$$

The range of variation of $S(b_2)$ is rescaled with respect to that of $S(b_1)$ by a factor $(b_1/b_2)^{\omega_1}$.

(*iii*) If $b_1^{\omega_2} t_1 = b_2^{\omega_2} t_2$, the probability distributions at t_1 and t_2 are identical provided one rescales the block variables by a factor (see Figure 4.11)

$$\left(\frac{b_1}{b_2}\right)^{\omega_1} = \left(\frac{t_2}{t_1}\right)^{\omega_1/\omega_2} = \left(\frac{t_2}{t_1}\right)^{\omega_1 \nu}$$

(*iv*) The joint probability distributions of two or more spin blocks are also given by universal distributions; for example in the case of the joint distribution of two blocks S_I and S_J

$$P_2[S_I(b), S_J(b); t] = (z_1)^2 b^{2\omega_1} P_2^* \left[z_2^{\pm} b^{\omega_2} t; z_1 b^{\omega_1} S_I(b), z_1 b^{\omega_1} S_J(b); \vec{r}_I - \vec{r}_J\right] \tag{4.137}$$

4.4.3 Critical manifold and fixed points

We have postulated the existence of limit distributions for the rescaled spin blocks in (4.131). This assumption can be reformulated by looking at the transformation of the Boltzmann weight in a RGT. This will lead to a geometrical picture, the renormalization group flow, and to a better understanding of our assumptions. If the Boltzmann weight of the original spin system is given by some Hamiltonian $H[S_i]$[43]

$$P_B[S_i] \propto \exp(-H[S_i])$$

then the Boltzmann weight of the blocks is given in terms of a block Hamiltonian $H'[S'_I]$, which can be formally written in terms of the original spin Hamiltonian $H[S_i]$. It will be convenient to denote renormalization group transformed quantities with a prime: $\varphi_I(b) \to S'_I$, $H \to H'$. From now on we use only rescaled blocks, so that no ambiguity should arise. From the definitions (4.116) and (4.117) of the rescaled blocks $\varphi_I(b) = S'_I$, we can write

$$\exp(-\mathcal{G})\exp\big(-H'[S'_I]\big) = \sum_{[S_i]}\prod_I \delta\left(S'_I - \frac{Z_1(b)}{b^D}\sum_{i\in I} S_i\right)\exp\big(-H[S_i]\big) \tag{4.138}$$

The factor $\exp(-\mathcal{G})$ stems from the fact that the RGT leads to a term in H' that is independent of the blocks, and it is convenient to isolate this term from those which depend explicitly on the block variables. We also note that the partition function of the spins is the same as that of the blocks

$$\sum_{[S'_I]}\exp\big(-\mathcal{G} - H'[S'_I]\big) = \sum_{[S_i]}\exp\big(-H[S_i]\big)$$

because it is equivalent to first building the blocks and then summing over the block configurations, or to summing directly over the spins.

If we start for example from an Ising model, the block Hamiltonian will not have an Ising form: first the blocks are no longer Ising spins, they look more like continuous variables, and second there is no reason for their interactions to be limited to nearest neighbours. In fact a block interacts with a large number of other blocks, and we have to write a completely general form for H', the only constraint being the original symmetry of the Ising Hamiltonian in zero magnetic field $H[-S_i] = H[S_i]$, which is translated into $H'[-S'_I] = H'[S'_I]$ (Figure 4.12)

$$-H' = K_1\sum_{nn} S'_I S'_J + K_2\sum_{nnn} S'_I S'_J + K_3\sum_{p} S'_I S'_J S'_K S'_L + \cdots \tag{4.139}$$

The first summation runs over nearest neighbours (nn), the second over next-nearest neighbours (nnn), the third one over 'plaquettes' (p) ... One important restriction is that the couplings remain short range, so that the number of important couplings is effectively limited. The coefficients K_α in (4.139) are called *coupling*

[43] The temperature dependence is hidden in the Hamiltonian.

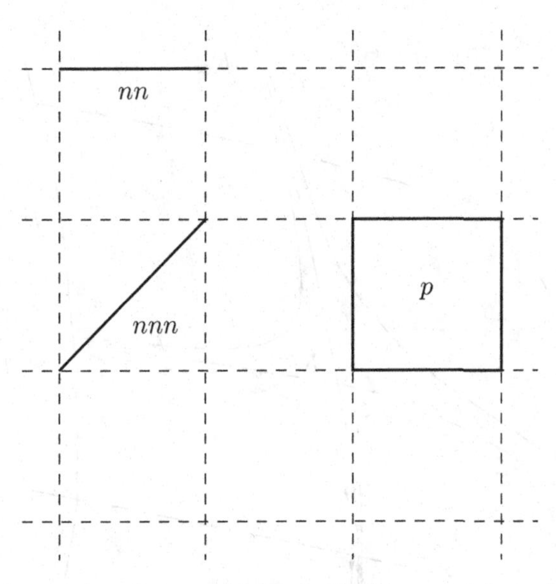

Figure 4.12 Interactions between blocks in the Ising model. *nn*: Nearest neighbours. *nnn*: Next-nearest neighbours. *p*: Plaquettes.

constants and the RGT can be written as a transformation law on the coupling constants

$$K'_\alpha = f_\alpha(K_\beta) \text{ or } K_\alpha(b) = f_\alpha(K_\beta) \tag{4.140}$$

The coupling constants span a *parameter space* of huge dimension, and a RGT is a set of functions in this parameter space. It is important to understand that the temperature is hidden in the coupling constants: for example in the case of the Ising model, where only $K_1 \neq 0$, $K_1 < 0.44$ corresponds to $t > 0$ and $K_1 > 0.44$ to $t < 0$ (see (4.14)). Thus, when the temperature changes, the coupling constants of a given model describe a curve in parameter space, which is called the *physical line of the model* (Figure 4.13). If the scaling factor b varies continuously, we shall be able to follow the RGT as *trajectories in parameter space*, not to be confused with the physical line of a model. During the preceding discussion, we have proved two properties of the RGT in the high and low temperature limits, by appealing to the central limit theorem: if we start from a positive (reduced) temperature $t > 0$, after a sufficient number of transformations ($\xi/b \ll a$), we reach a limiting probability distribution for the blocks, and the same is true, but with a *different* limiting distribution, if we start from a negative (reduced) temperature $t < 0$. These two distributions are *fixed points* of the RGT: the RGT trajectories converge toward the high temperature fixed point K_∞ if we start from $t > 0$, and toward the low temperature fixed point K_0 if we start from $t < 0$. One says that points in parameter space with $t > 0$ belong to the *attraction basin of the high temperature fixed*

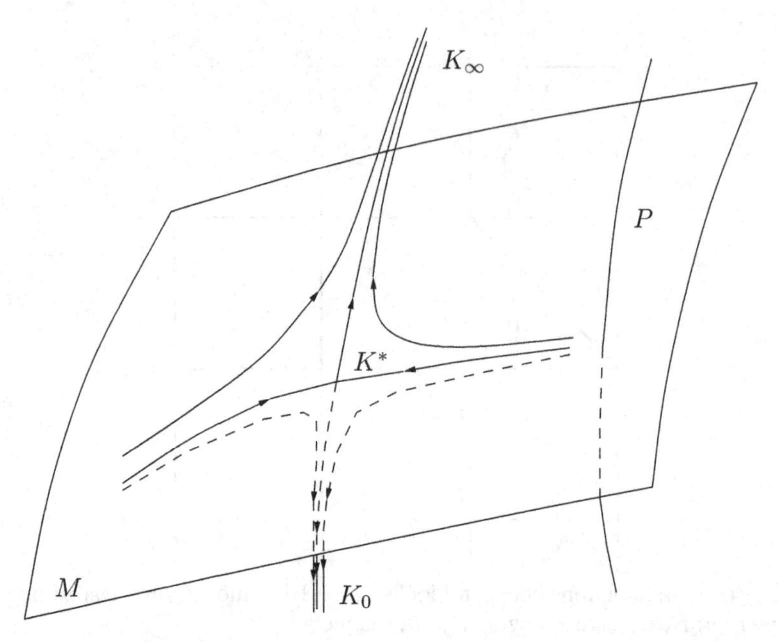

Figure 4.13 Renormalization group flow, critical manifold and physical line.

point, and that points with $t < 0$ belong to the attraction basin of the low temperature fixed point (Figure 4.13). The surface that divides the parameter space into the two attraction basins is called the *critical manifold*, and a point on this manifold corresponds to a critical system, whose temperature is $t = 0$. In the case of an ordinary critical point, such as that of the paramagnetic–ferromagnetic transition, the critical manifold has codimension one; multicritical points give rise to critical manifolds of higher codimension. In the preceding discussion, we have assumed that we were able to reach a limiting distribution starting from a point with $t = 0$, namely a point on the critical manifold. In our new language, this means that a RGT trajectory that starts from a point located on the critical manifold converges toward a fixed point K^* on the critical manifold, and the parameters K^*_α define a probability distribution which is the limiting probability distribution P^* of the previous subsection when $t = 0$. This assumption is a strong one, for there could be many possibilities other than a fixed point, for example a line of fixed points, as in the two-dimensional XY model, a limit cycle or a strange attractor. The *renormalization group flow*, namely the set of trajectories of the RGT, is drawn qualitatively on Figure 4.13. Note that K^*, K_∞ and K_0 are the only possible fixed points of the RGT, because $\xi = 0$ and $\xi = \infty$ are the only possible solutions of $\xi' = \xi/b$. If one starts from a point located close to, but not on, the critical manifold, the trajectory first stays close to the critical manifold, but begins to diverge and finally converges toward the high or low temperature fixed points.

Let $\{K^*_\alpha\}$ be the set of coupling constants at the fixed point on the critical manifold, and H^* the corresponding Hamiltonian. This set of coupling constants obeys

$$K^*_\alpha = f_\alpha(K^*_\beta) \tag{4.141}$$

In general RGT are complicated non-linear transformations, but it is possible to linearize in the neighbourhood of the fixed point if we define

$$\delta K_\alpha = K_\alpha - K^*_\alpha$$

and assume that this quantity is small. Then we can write

$$\delta K'_\alpha = \sum_\beta T_{\alpha\beta}\delta K_\beta \qquad T_{\alpha\beta} = \left.\frac{\partial f_\alpha}{\partial K_\beta}\right|_{K^*} \tag{4.142}$$

Let φ^i be the right eigenvectors of $T_{\alpha\beta}$ and λ_i the corresponding right eigenvalues. Although there is no reason for $T_{\alpha\beta}$ to be a symmetric matrix, we assume that its eigenvalues are real and that its eigenvectors form a basis in parameter space

$$\sum_\beta T_{\alpha\beta}\varphi^i_\beta = \lambda_i\varphi^i_\alpha = b^{\omega_i}\varphi^i_\alpha \tag{4.143}$$

The eigenvalue λ_i may be written as b^{ω_i} because the product of two RGT with scaling factors b and b' is a RGT with scaling factor $b'' = bb'$: $\lambda(b)\lambda(b') = \lambda(bb')$,[44] which implies that $\lambda(b)$ is a power of b. Thanks to our completeness assumption, we may decompose δK_α on the eigenvectors φ_i

$$\delta K_\alpha = \sum_i t_i\,\varphi^i_\alpha$$

and

$$\delta K'_\alpha = \sum_{i,\beta} T_{\alpha\beta}\varphi^i_\beta t_i = \sum_i b^{\omega_i} t_i\varphi^i_\alpha = \sum_i t'_i\varphi^i_\alpha$$

The variable t_i which transforms under a RGT as

$$t'_i = b^{\omega_i}\,t_i \tag{4.144}$$

is called a *scaling field*. In the vicinity of the fixed point, one may write the Hamiltonian as

$$H = H^* + \delta H = H^* + \sum_\alpha O_\alpha\,\delta K_\alpha$$

The quantities O_α are called *operators*, a terminology borrowed from quantum

[44] Provided $Z_1(bb') = Z_1(b)Z_1(b')$, which is the case if $Z_1(b)$ is a power of b.

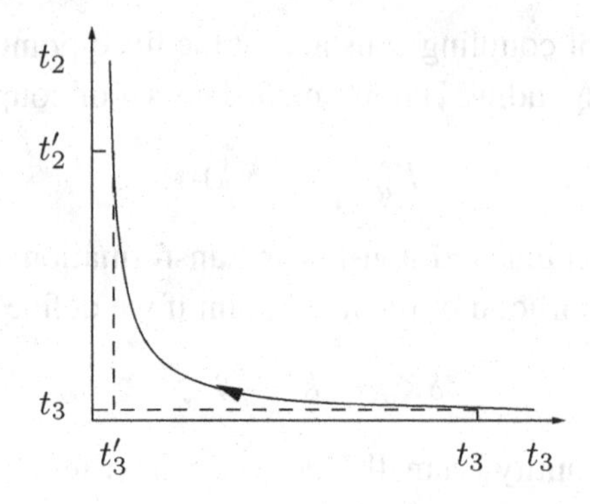

Figure 4.14 Flow diagram in the (t_2, t_3) plane.

field theory. The decomposition of H is rewritten in terms of the eigenvectors φ^i_α

$$H = H^* + \sum_{i,\alpha} \varphi^i_\alpha t_i O_\alpha = H^* + \sum_i t_i O_i$$

and the RGT reads

$$H' = H^* + \sum_{i,\alpha} \varphi^i_\alpha t_i' O_\alpha = H^* + \sum_i (b^{\omega_i} t_i) O_i \tag{4.145}$$

The operators O_i

$$O_i = \sum_\alpha \varphi^i_\alpha O_\alpha \tag{4.146}$$

are called *scaling operators* conjugate to the scaling variables t_i.

If the exponent ω_i in (4.144) is positive, the scaling variable and the conjugate operator will be called *relevant*, while they will be called *irrelevant* if $\omega_i < 0$, and *marginal* if $\omega_i = 0$. In Figure 4.13, there is one relevant variable which is labelled t_2,[45] and two irrelevant ones, t_3 and t_4. Let us examine the vicinity of the fixed point in the (t_2, t_3) plane. If $\delta H = H - H^*$ is written in terms of the scaling operators O_2 and O_3

$$\delta H = t_2 O_2 + t_3 O_3$$

its evolution in a RGT is given by

$$\delta H' = t_2\, b^{\omega_2} O_2 + t_3\, b^{\omega_3} O_3$$

and is represented schematically on Figure 4.14. If t_2 is small, the evolution first follows the critical manifold from which it diverges when the scaling variable t_2' is large enough.

[45] The reason for labelling this variable t_2 and not t_1 will be clear soon.

The plane tangent to the critical manifold at K^* is characterized by $t_2 = 0$. Thus, one must adjust one variable in order to be on the critical manifold. This variable is nothing other than the reduced temperature t, within a multiplicative factor. If n variables must be fixed to a particular value, then one has a multicritical point of order n.[46] Starting from a point in the vicinity of the critical surface, the RGT trajectory will first stay close to this manifold and the system will stay close to critical. When the relevant variable t_2 becomes large enough, the trajectory leaves the critical surface and converges to the low or high temperature fixed point depending on the sign of t_2. The transition region corresponds to a value of b such that $t_2' \sim 1$, namely $t_2 b^{\omega_2} \sim 1$. This transition region corresponds also to a situation where the correlation length is of the order of ba, $\xi/b \sim a$. Then we have

$$b \sim \frac{\xi}{a} \sim |t_2|^{-1/\omega_2} = |t|^{-1/\omega_2} \tag{4.147}$$

so that the critical exponent ν is to be identified with $1/\omega_2$

$$\boxed{\nu = \frac{1}{\omega_2}} \tag{4.148}$$

Thus the exponent ω_2 introduced in (4.134) is the same as that found from the RG behaviour of the relevant field t_2: in both cases ω_2 is determined by requiring that $\xi/b = a$, and this was anticipated by using the same notation.

4.4.4 Limit distributions and correlation functions

For a given value of the scaling factor b, the RGT define probability distributions $P[\varphi_I(b)]$ for the blocks. Let us write for example the one-block probability distribution in terms of the scaling fields t_i

$$P[\varphi_I(b)] = P\left[\varphi_I(b); z_2^{\pm} b^{\omega_2} t, b^{\omega_3} t_3, \ldots\right] \tag{4.149}$$

where we have used the property $t_2 \propto t$ and written explicitly the proportionality factor $z_2^{\pm}$. Taking $1 \ll b \ll \xi/a$ we have

$$z_2^{\pm} b^{\omega_2} t \simeq \left(\frac{\xi}{b}\right)^{-1/\nu}$$

and for the irrelevant fields

$$b^{\omega_i} t_i = b^{-|\omega_i|} t_i \to 0 \text{ if } i \geq 3$$

[46] For reasons to be explained later, a critical point of order $n = 2$ is called a *tricritical point*.

so that the probability distribution of $\varphi_I(b)$ tends to a limiting distribution in agreement with (4.131)

$$P[\varphi_I(b)] \to P^*\left[\varphi_I(b); z_2^{\pm} b^{\omega_2} t\right] \tag{4.150}$$

Thus the renormalization group allows us to justify the assumptions (4.129) or (4.131) made in the preceding subsection, *provided we can neglect the irrelevant fields*, a point to which we shall soon return. The same reasoning holds for the joint probability distribution of several blocks, and for example in the case of two blocks

$$P_2[\varphi_I(b), \varphi_J(b)] \to P_2^*\left[\varphi_I(b), \varphi_J(b); z_2^{\pm} b^{\omega_2} t; |\vec{r}_I - \vec{r}_J|\right] \tag{4.151}$$

From this equation we derive the two-point correlation function of the blocks

$$\langle \varphi_I(b) \varphi_J(b) \rangle \simeq G(|\vec{r}_I - \vec{r}_J|; z_2^{\pm} b^{\omega_2} t)$$

In order to relate this correlation function to that of the original spins $\langle S_i S_j \rangle$, we observe that we can neglect the variation of the spin correlation function when selecting different spins in the blocks, provided the centres of the blocks are sufficiently far apart: if i and j are located at the centres of blocks I and J, then we have

$$\langle S_{i'} S_{j'} \rangle \simeq \langle S_i S_j \rangle \ \text{ if } i; i' \in I; \ j, j' \in J$$

Taking into account the relation between the (renormalized) blocks and the spins

$$\varphi_I = b^{\omega_1 - D} \sum_{S_i \in I} S_i \tag{4.152}$$

leads to

$$\langle \varphi_I(b) \varphi_J(b) \rangle \simeq b^{2\omega_1} \langle S_i S_j \rangle$$

The rescaling of the unit of length allows us to derive, with $r = |\vec{r}_i - \vec{r}_j|$

$$\langle S_i S_j \rangle_c = G_{ij}(r) \simeq b^{-2\omega_1} G_{IJ}\left(\frac{r}{b}, z_2^{\pm} b^{\omega_2} t\right) \tag{4.153}$$

We now choose $b = \xi/a$, $z_2^{\pm} b^{\omega_2} t = \pm 1$, and write the final result in the form valid for $r \gg a$

$$\boxed{r \gg a: \ \ G(r) = \frac{1}{r^{2\omega_1}} f_{\pm}\left(\frac{r}{\xi}\right)} \tag{4.154}$$

From this equation we derive the following scaling law for the correlation function $G(r)$ if r and $\lambda r \ll \xi$

$$a \ll r,\ \lambda r \ll \xi: \quad G(\lambda r) = \lambda^{-2\omega_1} G(r) \tag{4.155}$$

The correlation function displays a power law behaviour identical to that found on the critical manifold provided $r \ll \xi$. The exponent ω_1 is directly related to the critical exponent η. By comparison with (4.42) we express ω_1 as a function of η

$$\boxed{\omega_1 = \frac{1}{2}(D - 2 + \eta)} \tag{4.156}$$

in agreement with our result of Section 4.4.2. This scaling property is easily generalized to N-point correlation functions. Let us write explicitly the case $N = 4$

$$G^{(4)}(\lambda\vec{r}_1, \lambda\vec{r}_2, \lambda\vec{r}_3, \lambda\vec{r}_4) = \lambda^{-4\omega_1} G(\vec{r}_1, \vec{r}_2, \vec{r}_3, \vec{r}_4) \tag{4.157}$$

provided all distances $a \ll |\vec{r}_i - \vec{r}_j| \ll \xi$ where, of course, from translation invariance, $G^{(4)}$ depends only on the differences $(\vec{r}_i - \vec{r}_j)$. The behaviour (4.155) and (4.157) of the correlation functions is called *scale invariance*. *Naïve scale invariance* follows from the assumption that correlation functions can only depend on the differences $|\vec{r}_i - \vec{r}_j|$ and on no other length scale. This implies that G must be given by a power law

$$G(r) = \frac{C}{r^{\alpha}}$$

where C is a dimensionless constant and $\alpha = D - 2 = 2d_\varphi^0$ from dimensional analysis. This power law is dictated by the dimension (4.110) of the field, since $G(\vec{r} - \vec{r}\,') = \langle \varphi(\vec{r})\varphi(\vec{r}\,')\rangle$. However, dimensional analysis cannot be the last word, because ω_1 is not equal to the canonical dimension $D/2 - 1$ of the field, but rather to (4.156). Thus ω_1 is called the *anomalous dimension* of the field and is often noted $\omega_1 = d_\varphi \neq d_\varphi^0$. The fact that the exponent is not that obtained from dimensional analysis means that the length scale a has not been eliminated from the game. It remains absolutely necessary to build the correct correlation function.

Let us summarize what the renormalization group flow of Figure 4.13 has taught us. Starting from a point close to the critical manifold, RGT lead us first toward the fixed point K^*, and then either toward the fixed point K_∞ or the fixed point K_0, depending on whether $t > 0$ or $t < 0$. Starting on the critical manifold leads us to the fixed point K^*, if the initial point is in the attraction basin of K^*. Two different points in parameter space correspond to two different models, excluding the exceptional case where the two points are on the physical line of a single model. The exponents ω_1 and ω_2, or equivalently the critical exponents η and ν, are given by the linearized RGT, so that all models in the attraction basin of K^*

have the same critical exponents. We have thus given a precise meaning to the concept of universality: models in the same universality class are all models that belong to the attraction basin of a given fixed point. Of course, the linearization is only valid close to K^*, but this does not spoil the general picture: starting from a point which is close enough to the critical manifold, RGT lead us in the vicinity of K^*, where linearization is possible, even though the initial point may belong to a region where linearization is not valid. However, in order to compare different models, we first have to perform RGT with large enough b, $1 \ll b \ll \xi/a$, so that all details which are not universal, namely all details which are specific to a given model, are averaged out. This is the reason why Equation (4.154), for example, is only valid for $r \gg a$: if one looks at short distances $r \sim a$, one is sensitive to details of the model. Such short distance details are included for example in the coefficients z_1 and $z_2^{\pm}$ of the renormalization factors $Z_1(b)$ and $Z_2(b)$. Similarly one can define the *width of the critical region*, which is the region where one observes the behaviour (4.154) of the correlation function. For simplicity, we write only one irrelevant field, t_3. From the requirement that, in order to reach the limit distribution P^*, the irrelevant fields must be negligible, a first condition is $|t_3|(r/a)^{\omega_3} \ll 1$, which ensures that the correlation function is given by a power law for $a \ll r \ll \xi$. A second condition is that $\xi^{\omega_3}|t_3| \ll 1$. These two conditions define the width of the critical region. If $|\omega_3|$ is small, or if $|t_3|$ is large, it may be difficult in practice to observe the critical behaviour.

4.4.5 Magnetization and free energy

So far in this section, we have worked in zero external field. The coupling of the spins to a uniform external field B reads $B\sum_i S_i$, from which we derive the coupling to the blocks

$$B\sum_i S_i = B\, b^{D-\omega_1}\sum_I \varphi_I(b) = B'\sum_I \varphi_I(b)$$

Under the renormalization group transformation, the transformed magnetic field B' is given by

$$B \to B' = b^{D-\omega_1} B = b^{\omega_B} B \tag{4.158}$$

The exponent

$$\omega_B = D - \omega_1 = \frac{1}{2}(D + 2 - \eta) \tag{4.159}$$

is positive in all cases of physical interest, and B is a relevant field. Criticality is ensured by the vanishing of the *two* relevant fields t and B

$$\text{Criticality:} \quad t = 0 \ \text{and} \ B = 0$$

In order to reach criticality, one must adjust the values of *two relevant fields*, the temperature and the magnetic field. A *tricritical point* is reached by adjusting the values of *three relevant fields*. The behaviour of the magnetization is derived from that of the magnetic field, where in a RGT we have

$$M(t', B'; t_2', t_3', \ldots) = b^{\omega_1} M(t, B; t_2, t_3, \ldots)$$

The factor b^{ω_1} arises because M is the average value of the spin. If the irrelevant fields $t_2', t_3', \ldots \ll 1$ we can write

$$\begin{aligned} M(t, B) &\simeq b^{-\omega_1} M(b^{\omega_1} t, b^{\omega_B} B) \\ &= b^{-\omega_1} M\left(\pm\left(\frac{\xi}{b}\right)^{-1/\nu}, \ b^{\omega_B} B\right) \end{aligned}$$

For $t = 0$ one chooses $b = B^{-1/\omega_B}$ and gets

$$M(0, B) = B^{\omega_1/\omega_B} M(0, 1) \propto B^{\omega_1/\omega_B}$$

This behaviour allows us to identify the critical exponent δ defined in (4.72)

$$\boxed{\delta = \frac{\omega_B}{\omega_1} = \frac{D + 2 - \eta}{D - 2 + \eta}} \tag{4.160}$$

For $t < 0$ and $B = 0$ one sets $b = \xi$

$$M(t, 0) = \xi^{-\omega_1} M(-1, 0) \propto |t|^{\nu\omega_1}$$

and the result for exponent $\tilde{\beta}$ defined in (4.68) confirms that already found in (4.138)

$$\boxed{\tilde{\beta} = \frac{\omega_1}{\omega_2} = \nu\omega_1 = \frac{\nu}{2}(D - 2 + \eta)} \tag{4.161}$$

It remains to determine the exponent α (4.73), which controls the behaviour of the specific heat. To this end we study the free energy in zero external field. In contrast to the correlation functions, which do not depend on the factor $\mathcal{G}$ in (4.138),[47] the free energy does involve the spin independent part $\mathcal{G}$ of the renormalized Hamiltonian. It is worth noticing that the argument which follows does not depend on

[47] In an equation like (4.138), $\exp(-\mathcal{G})$ cancels between the numerator and the denominator.

the linearity of the RGT, and is also valid for non-linear RGTs. If Z' denotes the partition function of H'

$$Z' = Z(K') = \sum_{[S'_I]} \exp(-H'[S'_I])$$

and F' the corresponding free energy

$$Z' = \sum_{[S']} \exp(-H'[S']) \qquad F' = F(K') = -\ln Z'$$

we have from (4.138) and the identity of the block and spin free energies

$$F(K) = F(K') + \mathcal{G}(K)$$

Define the free energy density and $g(K)$ as

$$f(K) = \frac{1}{L^D} F(K) \qquad g(K) = \frac{1}{L^D} \mathcal{G}(K)$$

where L is the size of the sample, and take for definiteness RGTs with scaling factor $b = 2$. We then have

$$f(K') = \left(\frac{2}{L}\right)^D F(K')$$

and thus

$$f(K) = g(K) + 2^{-D} f(K') \tag{4.162}$$

Let us iterate (4.162), starting from an initial set of parameters K_0

$$\begin{aligned}
f(K_0) &= g(K_0) + \ 2^{-D} f(K_1) \\
f(K_1) &= g(K_1) + \ 2^{-D} f(K_2) \\
&\vdots \\
f(K_n) &= g(K_n) + 2^{-D} f(K_{n+1}) \\
&\vdots
\end{aligned}$$

Multiply the second equation by 2^{-D}, the third by 2^{-2D}, ..., the nth by 2^{-nD}, ... and add. One finds

$$F(K_0) = \sum_{n=0}^{\infty} 2^{-nD} g(K_n)$$

or reverting to a continuous variable b

$$f(K_0) = \frac{1}{\ln 2} \int_1^{\infty} \frac{\mathrm{d}b}{b} b^{-D} g(K(b)) \tag{4.163}$$

The integral is controlled by the transition region between the critical regime and the high (or low) temperature regime, because the trajectory spends a long 'time' in the vicinity of the fixed point. In this region we have $b \sim |t|^{-1/\omega_2}$, $g(K(b))$ is slowly varying since we are in the vicinity of the fixed point, and we obtain

$$f(K_0) \sim b^{-D} \sim |t|^{D/\omega_2} = |t|^{\nu D} \tag{4.164}$$

The specific heat is $\sim \mathrm{d}^2 f/\mathrm{d}t^2$, which gives the value of the critical exponent α

$$\boxed{\alpha = 2 - \nu D} \tag{4.165}$$

The renormalization group analysis has allowed us to derive the four critical exponents $\alpha, \tilde{\beta}, \gamma$ and δ in terms of two basic exponents η and ν, or equivalently ω_1 and ω_2

$$\boxed{\alpha = 2 - \nu D \qquad \tilde{\beta} = \frac{\nu}{2}(D - 2 + \eta) \qquad \delta = \frac{D + 2 - \eta}{D - 2 + \eta} \qquad \gamma = \nu(2 - \eta)} \tag{4.166}$$

In other words, we have derived four scaling laws.

4.5 Renormalization group: examples

4.5.1 Gaussian fixed point

We are now going to apply renormalization group methods to the Ginzburg–Landau Hamiltonian (4.96). There exists no exact calculation of the partition function, and correspondingly there is no way of writing exact RG equations in closed analytic form. The difficulty stems from the different character of the terms $(\vec{\nabla}\varphi)^2/2$ and $u_0\varphi^4/4!$ in (4.96). As we shall see, the first term, which is complicated in $\vec{r}$-space, is diagonalized in Fourier space, while the opposite is true of the latter term, which is called the *interaction* and is denoted by V. If we set $u_0 = 0$, $H \to H_0$, we get the *free Hamiltonian*, or the *Gaussian Hamiltonian* H_0. The free Hamiltonian can be straightforwardly expressed in terms of the Fourier components $\tilde{\varphi}(\vec{q})$ of the field $\varphi(\vec{r})$

$$\tilde{\varphi}(\vec{q}) = \int \frac{\mathrm{d}^D r}{L^{D/2}} \, \mathrm{e}^{\mathrm{i}\vec{q}\cdot\vec{r}} \, \varphi(\vec{r}) \tag{4.167}$$

or conversely

$$\varphi(\vec{r}) = \frac{1}{L^{D/2}} \sum_{\vec{q}} \mathrm{e}^{-\mathrm{i}\vec{q}\cdot\vec{r}}\, \varphi(\vec{q}) \to \int \frac{\mathrm{d}^D q}{L^{D/2}(2\pi)^{D/2}}\, \mathrm{e}^{-\mathrm{i}\vec{q}\cdot\vec{r}}\, \tilde{\varphi}(\vec{q}) \tag{4.168}$$

where L is, as previously, the size of the sample. The Fourier transform $\tilde{G}(\vec{q})$ of the correlation function is easily seen to be

$$\tilde{G}(\vec{q}) = \int \mathrm{d}^D r\, \mathrm{e}^{\mathrm{i}\vec{q}\cdot\vec{r}}\, G(\vec{r}) = \langle \tilde{\varphi}(\vec{q})\, \tilde{\varphi}(-\vec{q}) \rangle \tag{4.169}$$

Let us write H_0 in terms of the Fourier components $\tilde{\varphi}(\vec{q})$

$$H_0 = \sum_{0 \le q \le \Lambda} (r_0 + q^2) |\tilde{\varphi}(\vec{q})|^2 \tag{4.170}$$

where $\Lambda \sim 2\pi/a$ is the ultraviolet cut-off. As promised, the Fourier transformation diagonalizes H_0. The Boltzmann weight $\exp(-H_0)$ is a product of independent Gaussian distributions, and the correlation function is immediately obtained from (A.45)

$$\tilde{G}_0(\vec{q}) = \frac{1}{r_0 + q^2} \tag{4.171}$$

In Fourier space, a RGT with scaling factor b will correspond to integrating over all Fourier components of the field such that $\Lambda/b \le q \le \Lambda$. In terms of wavelengths, this means that all fluctuations with wavelengths $a \le \lambda \le ba$ will be averaged out. Let us write

$$\varphi(\vec{r}) = \varphi_1(\vec{r}) + \overline{\varphi}(\vec{r}) \tag{4.172}$$

where $\varphi_1(\vec{r})$ has Fourier components in the interval $0 \le q \le \Lambda/b$ and $\overline{\varphi}(\vec{r})$ in the interval $\Lambda/b \le q \le \Lambda$. Writing $H = H_0 + V$ with

$$V = \frac{u_0}{4!} \int \mathrm{d}^D r\, \varphi^4(\vec{r})$$

we have

$$H(\varphi_1, \overline{\varphi}) = H_0(\varphi_1) + H_0(\overline{\varphi}) + V(\varphi_1, \overline{\varphi})$$

because the Fourier components of H_0 are decoupled, while this is not true for V. If we ignore, for the time being, the rescaling of the field φ_1, which will be taken care of later on, the RGT leads to

$$\exp(-H_1'[\varphi_1]) = \exp(-H_0[\varphi_1]) \frac{\int D\overline{\varphi}\, \exp(-H_0[\overline{\varphi}])\, \exp(-V[\varphi_1, \overline{\varphi}])}{\int D\overline{\varphi}\, \exp(-H_0[\overline{\varphi}])}$$

If we restrict ourselves to first order in u_0, the transformed Hamiltonian is[48]

$$H_1'[\varphi_1] = H_0[\varphi_1] + \frac{\int \mathcal{D}\overline{\varphi}\, \exp(-H_0[\overline{\varphi}])\, V[\varphi_1, \overline{\varphi}]}{\int \mathcal{D}\overline{\varphi}\, \exp(-H_0[\overline{\varphi}])} + \mathcal{O}(u_0^2) \tag{4.173}$$

The Hamiltonian H_0 is Gaussian and $V(\varphi_1, \overline{\varphi})$ is a polynomial in $\overline{\varphi}$. In order to evaluate the mean values, we need the correlation

$$\overline{G}(\vec{r} - \vec{r}\,') = \langle \overline{\varphi}(\vec{r})\, \overline{\varphi}(\vec{r}\,') \rangle = \frac{\int \mathcal{D}\overline{\varphi}\, \exp(-H_0[\overline{\varphi}])\, \overline{\varphi}(\vec{r})\, \overline{\varphi}(\vec{r}\,')}{\int \mathcal{D}\overline{\varphi}\, \exp(-H_0[\overline{\varphi}])}$$

The calculation is exactly the same as that of the full correlation function $G(\vec{r})$, except that the q-integration is restricted to the interval $\Lambda/b \le q \le \Lambda$, instead of $0 \le q \le \Lambda$. Then, from (4.172) we have

$$\overline{G}(\vec{r} - \vec{r}\,') = \int_{\Lambda/b \le q \le \Lambda} \frac{\mathrm{d}^D q}{(2\pi)^D} \frac{\mathrm{e}^{-\mathrm{i}\vec{q}\cdot(\vec{r} - \vec{r}\,')}}{q^2 + r_0} \tag{4.174}$$

Reverting to the calculation of $\langle V(\varphi_1, \overline{\varphi}) \rangle$ leads to

$$\langle (\varphi_1(\vec{r}) + \overline{\varphi}(\vec{r}))^4 \rangle = \varphi_1^4(\vec{r}) + 6\varphi_1^2(\vec{r}) \langle \overline{\varphi}(\vec{r})\, \overline{\varphi}(\vec{r}) \rangle + \langle \overline{\varphi}^4(\vec{r}) \rangle$$

The last term in the preceding equation is a constant that can be dropped, since it contributes only to the '$\mathcal{G}$-term' in (4.138). The second term is $6\,\varphi_1^2(\vec{r})\overline{G}(0)$, and we therefore obtain for H_1'

$$H_1' = \int \mathrm{d}^D r \left[\frac{1}{2} (\vec{\nabla}\varphi_1)^2 + \frac{1}{2} r_0 \varphi_1^2 + \frac{u_0}{4!} \varphi_1^4 + \frac{u_0}{4} \varphi_1^2\, \overline{G}(0) \right]$$

It remains to implement the dilatation of the unit of length and the renormalization of the field

$$\mathrm{d}^D r \to b^D \mathrm{d}^D r' \qquad \vec{\nabla} \to b^{-2} \vec{\nabla}' \qquad \varphi_1 \to b^{-\omega_1} \varphi'$$

in order to find the renormalized Hamiltonian H'

$$H' = \int \mathrm{d}^D r' b^{D-2\omega_1-2} \left[\frac{1}{2} (\vec{\nabla}'\varphi')^2 + \frac{1}{2} b^2 \left(r_0 + \frac{u_0}{2} \overline{G}(0) \right) \varphi'^2 + b^{2-2\omega_1} \frac{u_0}{4!} \varphi'^4 \right] \tag{4.175}$$

Let us examine field renormalization. The correlation function of the free Hamiltonian H_0 behaves as $1/q^2$ at the critical point, which means that ω_1 is the canonical dimension, $D/2 - 1$, of φ. With $\omega_1 = D/2 - 1$ in (4.175), we see that, *to this order in* u_0, the RGT has not affected the coefficient of $(\vec{\nabla}\varphi)^2$. Setting $\varepsilon = 4 - D$,

[48] In order to get higher order terms, one must use an expansion in terms of cumulants.

the transformation laws for r_0 and u_0 are

$$r_0' = b^2 \left(r_0 + \frac{u_0}{2} \overline{G}(0) \right) \tag{4.176}$$

$$u_0' = b^{4-D} u_0 = b^{\varepsilon} u_0 \tag{4.177}$$

These equations are linear and exhibit a fixed point at $r_0 = u_0 = 0$. It is easy to draw the renormalization group flow in the plane $\{r_0, u_0\}$ (Exercise 4.6.10) and to see that the two scaling fields have exponents 2 and ε. Thus, for $D > 4$, the fixed point has the characteristics of the fixed point K^* described in the preceding discussion, with a relevant field $r_0 \propto t$ and a corresponding exponent $\omega_2 = 2$, and an irrelevant field with exponent $\varepsilon < 0$. Moreover the field renormalization exponent is $\omega_1 = D/2 - 1$, so that $\eta = 0$. The conclusion of the renormalization group analysis is that for $D > 4$, *the critical exponents η and ν are given by their mean-field values*: $\eta = 0$, $\nu = 1/\omega_2 = 1/2$. The fixed point at $r_0 = u_0 = 0$ is called the *Gaussian fixed point*. It corresponds to a non-interacting Hamiltonian at $T = T_c$. However, for $D < 4$, to the Gaussian fixed point correspond two relevant fields, because $\varepsilon > 0$, and the Gaussian fixed point becomes unstable. At $D = 4$, it crosses another fixed point, the non-Gaussian fixed point, which becomes the relevant one for the determination of the critical exponents when $D < 4$. We could pursue the analysis leading to (4.174) to order u_0^2 in order to find this new fixed point, but the calculations are cumbersome and we prefer to follow a powerful method inspired by quantum field theory, which is simpler and gives a flavour of more general computations.

4.5.2 Non-Gaussian fixed point

We have seen in Section 4.3.5 that the effective dimensionless expansion parameter in the Ginzburg–Landau Hamiltonian is not u_0, but rather $u_0(r_0 - r_{0c})^{-\varepsilon/2}$, and that this parameter tends to infinity at the critical point for $D < 4$. The Ginzburg–Landau Hamiltonian depends in fact upon two parameters that, in field-theoretic language, are the 'bare mass' squared[49] $(r_0 - r_{0c})$ (more precisely the difference between the bare mass squared and the critical mass squared) and the 'bare coupling constant' u_0. In a perturbative approach, r_{0c} is a function of u_0 and of the cut-off $\Lambda \sim 1/a$. Two important points need be emphasized:

- These bare parameters are really *short distance parameters*: they are defined at the microscopic scale a and depend on the short distance $\sim a$ behaviour of the model.

[49] In a system of units where $\hbar = c = 1$, energy and mass have dimension of an inverse length, and the field φ has dimension $(\text{mass})^{D/2-1}$. Then, if one interprets the Ginzburg–Landau Hamiltonian as a Hamiltonian of relativistic quantum field theory, the coefficient r_0 in front of φ^2 has the dimension of a mass squared.

- As explained in Section 4.3.5, we shall work at fixed u_0 and reach the critical point by using $(r_0 - r_{0c})$ as a control parameter.

What we want to do is to reshuffle the perturbative expansion in powers of u_0 sketched in Section 4.3.5 by using instead of the bare parameters the renormalized ones: the renormalized mass ξ^{-1} (see Footnote 42) and a renormalized coupling constant g, to be defined later. The new basic length scale ξ is no longer a short distance one, but a rather long distance, or macroscopic, scale since $\xi \gg a$. It turns out that in the vicinity of $D = 4$, g is small, on the order of ε, and a perturbative expansion in powers of g becomes possible. This perturbative expansion allows one to express the critical exponents as power series of ε. This is the famous 'epsilon-expansion' of Wilson and Fisher. When g is not so small, for example when $D = 3$ (or $\varepsilon = 1$), more sophisticated techniques may be used to get reliable results.

In order to define g, we start from the four-point function

$$\Gamma^{(4)}(\vec{r}_1, \vec{r}_2, \vec{r}_3, \vec{r}_4) = \frac{V^{-1}\,\delta^4\Gamma[M(\vec{r})]}{\delta M(\vec{r}_1)\delta M(\vec{r}_2)\delta M(\vec{r}_3)\delta M(\vec{r}_4)}$$

where $\delta/\delta(M(\vec{r}))$ is a functional derivative with respect to $M(\vec{r})$ and $V^{-1}\Gamma$ is density of the Gibbs potential (4.99). This density can be written in terms of the four-point correlation function (4.157) and of the two-point correlation function G, but we shall not need this relation.[50] All we shall need is the Fourier transform $\tilde{\Gamma}^{(4)}$ of $\Gamma^{(4)}$ taken at $\vec{q}_i = 0$. This quantity is obtained by differentiating four times the uniform magnetization Gibbs potential $\Gamma(M)$ whose approximate form is given in (4.106). In fact a uniform magnetization has only a single Fourier component $\vec{q} = 0$. The differentiations are easily performed by expanding the logarithm in (4.106) into powers of u_0

$$\ln\Big(q^2 + r_0 + \frac{1}{2}u_0 M^2\Big) = \ln(q^2 + r_0) + \frac{1}{2}u_0 M^2 (q^2 + r_0)^{-1} - \frac{1}{8}u_0^2 M^4 (q^2 + r_0)^{-2} + \cdots$$

Assuming $t > 0$ and zero external field, we must set $M = 0$ at the end of the calculation and get

$$\tilde{\Gamma}^{(4)}(\vec{q}_i = 0) = u_0 - \frac{3}{2}u_0^2 \int \frac{\mathrm{d}^D k}{(2\pi)^D} \frac{1}{(q^2 + r_0)^2} + \mathcal{O}(u_0^3) \tag{4.178}$$

[50] In Fourier space this relation is

$$\tilde{\Gamma}^{(4)}(\vec{q}_i) = \tilde{G}^{(4)}(\vec{q}_i)\Big(\prod_{i=1}^{4} \tilde{G}(\vec{q}_i)\Big)^{-1}$$

We shall need the transformation law of $\tilde{\Gamma}^{(4)}$ under a RGT. We note that, from (A.7), $\tilde{\Gamma}^{(2)}(\vec{q}=0)$ is the inverse of the two-point correlation function $\tilde{G}(\vec{q}=0)$

$$\tilde{\Gamma}^{(2)}(\vec{q}=0) = \frac{\mathrm{d}^2\Gamma(M)}{\mathrm{d}M^2} = \frac{1}{\tilde{G}(\vec{q}=0)}$$

The transformation law of $\tilde{G}$ is obtained from (4.153), which may be written

$$G\left(\frac{r}{b}, K(b)\right) = b^{2\omega_1} G(r, K)$$

and is easily translated to Fourier space[51]

$$\tilde{G}(bq, K(b)) = b^{2\omega_1 - D} G(q, K)$$

The transformation law of $\tilde{\Gamma}^{(2)}$, which is the inverse of $\tilde{G}$, is then

$$\tilde{\Gamma}^{(2)}(\vec{q}=0; K(b)) = b^{D-2\omega_1}\tilde{\Gamma}^{(2)}(\vec{q}=0; K)$$

To go from $\tilde{\Gamma}^{(2)}$ to $\tilde{\Gamma}^{(4)}$, we must perform two more differentiations with respect to M, giving the transformation law of $\tilde{\Gamma}^{(4)}$

$$\tilde{\Gamma}^{(4)}(\vec{q}_i=0; K(b)) = b^{D-4\omega_1}\,\tilde{\Gamma}^{(4)}(\vec{q}_i=0; K) \tag{4.179}$$

It is instructive to perform a dimensional check of this equation. According to (4.177) the dimension of $\tilde{\Gamma}^{(4)}(\vec{q}_i=0)$ is the same as that of u_0, namely $4-D$. The dimension of M is the same as that of φ, $(D/2-1)$, so that the dimension of $\tilde{\Gamma}^{(4)}(\vec{q}_i=0)$ is

$$-D+4\left(\frac{D}{2}-1\right) = D-4$$

In a RGT, the canonical dimension of the field, $D/2-1$, must be replaced by its anomalous dimension ω_1, and instead of $D-4$ in the previous equation we find $(-D+4\omega_1)$. The transformation law of $\tilde{\Gamma}^{(4)}(\vec{q}_i=0)$ is equivalent to

$$\tilde{\Gamma}^{(4)}(\vec{q}_i=0) \sim \xi^{4\omega_1-D} = \xi^{D-4+2\eta} \tag{4.180}$$

At this point we are in a position to define accurately the field renormalization. Since the two-point correlation behaves as $\exp(-r/\xi)$ for $r\to\infty$, its Fourier transform has poles at $q=\pm\mathrm{i}\xi^{-1}$ (or more generally branch points issuing from these points), and can thus be written for $q^2 \lesssim \mathcal{O}(\xi^{-2})$

$$q^2 \lesssim \xi^{-2}:\ \ \tilde{G}(\vec{q}) = \frac{Z}{q^2+\xi^{-2}+\mathcal{O}(q^4)} \tag{4.181}$$

[51] One writes

$$\int \mathrm{d}^D r\, \mathrm{e}^{\mathrm{i}\vec{q}\cdot\vec{r}}\, G(r,K) = \int \mathrm{d}^D r\, \mathrm{e}^{\mathrm{i}(b\vec{q})\cdot(\vec{r}/b)}\, b^{-2\omega_1} G\left(\frac{r}{b}, K(b)\right) = b^{D-2\omega_1}\int \mathrm{d}^D r'\, \mathrm{e}^{\mathrm{i}(b\vec{q})\cdot\vec{r}'}\, G(r', K(b))$$

The renormalization constant Z is closely linked to our previous scaling factor $Z_1(b)$. It could also be defined more formally by

$$Z^{-1} = \frac{\mathrm{d}^2}{\mathrm{d}q^2}\,\tilde{G}(\vec{q})\bigg|_{q^2=0}$$

We observe that Z is the multiplicative factor which relates the susceptibility χ to the correlation length squared ξ^2, because from (4.44)

$$\chi = \tilde{G}(\vec{q}=0) = Z\xi^2 \sim \xi^{2-\eta}$$

We have thus discovered that the renormalization constant Z behaves as $\xi^{-\eta}$, or more precisely $(\xi/a)^{-\eta}$ if we want to write explicit dimensionless relations.[52] When $u_0 = 0$, in the case of the Gaussian Hamiltonian, we have $Z = 1$. We have also seen in the previous subsection that there is no field renormalization at order u_0: $Z = 1 + \mathcal{O}(u_0^2)$.

Let us now build the dimensionless quantity g from

$$g = \xi^{4-D} Z^2\, \tilde{\Gamma}^{(4)}(\vec{q}_i = 0) \tag{4.182}$$

To lowest order in u_0, we have $g = r_0^{-\varepsilon/2} u_0$ because from (4.83) the correlation length is $r_0^{-1/2}$ with the Gaussian Hamiltonian: this shows that g is indeed related to a coupling constant. The crucial property of g is that, from (4.178), it stays *finite* at the critical point, even though $\xi \to \infty$. We shall call g the *renormalized coupling constant*.

Our original Ginzburg–Landau Hamiltonian depended on two bare parameters r_0 and u_0; now, in the vicinity of the critical point, the theory still depends on two parameters, ξ and g, but these are *renormalized parameters*. If we are able to reshuffle the new perturbative expansion written in terms of the diverging bare dimensionless parameter $r_0^{-\varepsilon/2} u_0$ as an expansion in powers of g, we shall have made a big step forward in the solution of our difficulties, since g remains finite. To this end, let us examine the relation betwen g and u_0. For $u_0 \to 0$ at fixed ξ we have $g \simeq u_0 \xi^{\varepsilon}$ while close to the critical point, at fixed u_0 and $\xi \to \infty$, $g \to g^*$. The evolution of g between the two limits $g \to 0$ and $g \to g^*$ is described by a very important function, the so-called *beta-function*, defined by

$$\boxed{\beta(g, \varepsilon) = -\xi\, \frac{\mathrm{d}g}{\mathrm{d}\xi}\bigg|_{u_0}} \tag{4.183}$$

It is also convenient to define the dimensionless bare coupling constant g_0 by $g_0 = u_0 \xi^{\varepsilon}$. Being dimensionless, g can only be a function of g_0, so that an alternative

[52] The relation $Z \propto \xi^{-\eta}$ shows that, apart from a trivial dimensional factor $\xi^{D/2-1}$, $Z \propto Z_1^{-2}$.

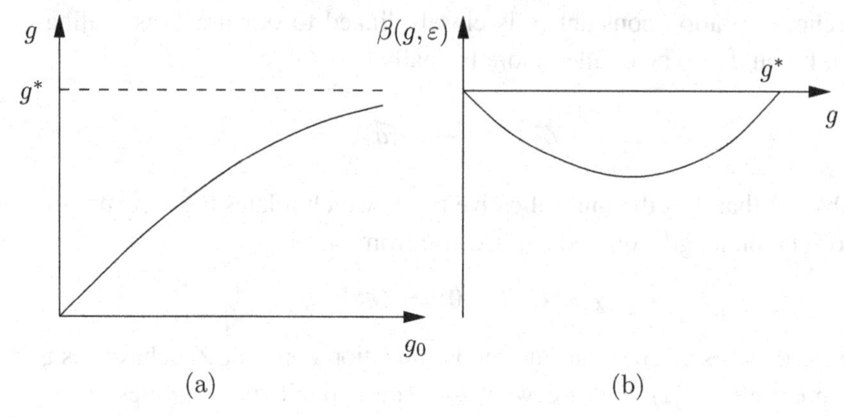

Figure 4.15 Behaviour of $g(g_0)$ and of $\beta(g, \varepsilon)$.

expression for $\beta(g, \varepsilon)$ is

$$\beta(g, \varepsilon) = -\varepsilon g_0 \frac{\mathrm{d}g}{\mathrm{d}g_0} \tag{4.184}$$

In fact g and $\beta(g, \varepsilon)$ are also functions of the other dimensionless ratio a/ξ of the problem, and the above statements only hold in the limit $a/\xi \to 0$. One can show that g contains terms of order $(a/\xi)^{\varepsilon}$ and $\beta(g, \varepsilon)$ terms of order $(a/\xi)^{2+\varepsilon}$, see Exercise 4.6.12. For small enough g_0, $g(g_0) \simeq g_0$, and $\beta(g, \varepsilon) \simeq -\varepsilon g$. For $g_0 \to \infty$, $g(g_0) \to g^*$, so that $g(g_0)$ and $\beta(g, \varepsilon)$ have the qualitative behaviour displayed on Figure 4.15. One notices that when the bare coupling constant g_0 varies in the interval $[0, +\infty]$, the renormalized one varies in $[0, g^*]$. The solution of the differential equation

$$\frac{\mathrm{d}g_0}{\mathrm{d}g} = -\frac{\varepsilon g_0}{\beta(g, \varepsilon)}$$

with the boundary condition $g(g_0) \to g$ when $g \to 0$ reads

$$\boxed{g_0 = g \exp\left(-\int_0^g \mathrm{d}g' \left(\frac{\varepsilon}{\beta(g', \varepsilon)} + \frac{1}{g'} \right) \right)} \tag{4.185}$$

As $\beta(g', \varepsilon) \to -\varepsilon g'$ when $g' \to 0$, the integral converges at its lower limit. The critical point corresponds to $g_0 \to \infty$, or to $g \to g^*$. Assuming that the zero of $\beta(g, \varepsilon)$ at $g = g^*$ is a simple one, we can parametrize $\beta(g, \varepsilon)$ by introducing its derivative $\beta'(g^*, \varepsilon) = \omega$ at $g = g^*$

$$\beta(g, \varepsilon) \simeq \omega(g - g^*), \quad g \to g^*$$

This leads to the behaviour

$$g_0 \sim |g^* - g|^{-\varepsilon/\omega} \tag{4.186}$$

or equivalently

$$\xi \sim |g^* - g|^{-1/\omega} \tag{4.187}$$

More accurate expressions can be found in Exercise 4.6.12 where it is also shown that $-\omega$ must be identified with the exponent ω_3 of the leading irrelevant eigenvalue: $\omega_3 = -\omega$. It should be clear that g^* corresponds really to a fixed point of the renormalization group, since the derivative $\xi(\mathrm{d}g/\mathrm{d}\xi)$ vanishes at $g = g^*$.

Having found the RG fixed point, our task is now to compute the critical exponents. In addition to Z, we shall need a second renormalization constant $\overline{Z}$

$$\overline{Z} = -\frac{1}{2}\xi^3 \frac{\mathrm{d}(r_0 - r_{0\mathrm{c}})}{\mathrm{d}\xi} \tag{4.188}$$

The reasons behind this definition are not very obvious at first sight. At the present stage of our discussion, we can only argue that $\overline{Z}$ is dimensionless and that it is equal to one in the (free) Gaussian model with Hamiltonian H_0, as $r_0 = \xi^{-2}$ and $r_{0\mathrm{c}} = 0$ in that case. Since they are dimensionless, the renormalization constants Z and $\overline{Z}$ are functions of g (or g_0) only. From these renormalization constants we define the dimensionless functions $\gamma(g)$ and $\overline{\gamma}(g)$

$$\gamma(g) = -\xi \frac{\mathrm{d}}{\mathrm{d}\xi} \ln Z(g)\Big|_{u_0} = \beta(g,\varepsilon)\frac{\mathrm{d}\ln Z}{\mathrm{d}g} \tag{4.189}$$

$$\overline{\gamma}(g) = -\xi \frac{\mathrm{d}}{\mathrm{d}\xi} \ln \overline{Z}(g)\Big|_{u_0} = \beta(g,\varepsilon)\frac{\mathrm{d}\ln \overline{Z}}{\mathrm{d}g} \tag{4.190}$$

The boundary conditions are $Z = \overline{Z} = 1$ for $g \to 0$, and the solutions of the above equations give Z and $\overline{Z}$ as functions of g

$$Z(g) = \exp\left(\int_0^g \frac{\gamma(g')}{\beta(g',\varepsilon)}\,\mathrm{d}g'\right) \tag{4.191}$$

$$\overline{Z}(g) = \exp\left(\int_0^g \frac{\overline{\gamma}(g')}{\beta(g',\varepsilon)}\,\mathrm{d}g'\right) \tag{4.192}$$

If we assume a generic situation where $\gamma(g)$ and $\overline{\gamma}(g)$ do not vanish at $g = g^*$ we get, close to the critical point

$$Z(g) \simeq |g - g^*|^{\gamma(g^*)/\omega} \sim \xi^{-\gamma(g^*)} \tag{4.193}$$

$$\overline{Z}(g) \simeq |g - g^*|^{\overline{\gamma}(g^*)/\omega} \sim \xi^{-\overline{\gamma}(g^*)} \tag{4.194}$$

The first equation allows us to identify the exponent η

$$\boxed{\eta = \gamma(g^*)} \tag{4.195}$$

In order to interpret $\overline{\gamma}(g^*)$, let us recall that $(r_0 - r_{0c}) \propto (T - T_c)$, and that from the definition of the exponent ν

$$r_0 - r_{0c} \sim \xi^{-1/\nu} \qquad \frac{\mathrm{d}(r_0 - r_{0c})}{\mathrm{d}\xi} \sim -\xi^{-\frac{1}{\nu}-1}$$

so that from (4.193) and (4.194)

$$\boxed{\nu = \frac{1}{2 + \overline{\gamma}(g^*)}} \tag{4.196}$$

4.5.3 Critical exponents to order ε

It remains to evaluate the quantities g^*, $\gamma(g^*)$ and $\overline{\gamma}(g^*)$. Note that all the relations derived in the preceding subsection and in particular (4.195) and (4.196) are quite general and *do not depend on a perturbative expansion*, to which we now turn. We start from (4.178) and notice that, at order u_0^2, we can set $Z = 1$ and replace r_0 by ξ^{-2} in the integral. Using

$$\int \frac{\mathrm{d}^D q}{(2\pi)^D} \frac{1}{q^2 + r_0} = S_D \int \frac{q^{D-1}\mathrm{d}q}{(2\pi)^D} \frac{1}{q^2 + r_0} = \frac{\Gamma(2 - D/2)}{(4\pi)^{D/2}} (r_0)^{D/2-2}$$

where Γ is Euler's factorial, S_D the surface of the unit sphere in dimension D (2.123) and the definition (4.182) of g, we get

$$g\xi^{-\varepsilon} = u_0 - \frac{3}{2} u_0^2 \frac{\Gamma(\varepsilon/2)}{(4\pi)^{D/2}} \xi^{\varepsilon}$$

Since the integral is convergent, we have taken the infinite cut-off limit $\Lambda \to \infty$, or $a \to 0$. Taking the derivative of both sides of the above equation with respect to ξ and using the definition (4.184) of $\beta(g, \varepsilon)$ leads to

$$\beta(g, \varepsilon) = -\varepsilon g + \frac{3}{2} \varepsilon \frac{\Gamma(\varepsilon/2)}{(4\pi)^{D/2}} g^2 + \mathcal{O}(g^3)$$

We have thus obtained the first two terms of the perturbative expansion of $\beta(g, \varepsilon)$ in powers of the *renormalized* coupling constant g. Note that, in the course of the derivation, we have replaced $u_0^2 \xi^{\varepsilon}$ by $g^2 \xi^{\varepsilon}$, which is correct to this order of perturbation theory: the next term in the expansion of $\tilde{\Gamma}^{(4)}(0)$ is of order $u_0^3 \xi^{2\varepsilon}$ and would lead to a g^3 term in $\beta(g, \varepsilon)$. For small values of ε we can use the leading behaviour of the Γ-function

$$x \to 0: \quad \Gamma(x) \simeq \frac{1}{x}$$

and get

$$\boxed{\beta(g,\varepsilon) = -\varepsilon g + \frac{3g^2}{16\pi^2}} \tag{4.197}$$

The fixed point g^* is located at

$$\boxed{g^* = \frac{16\pi^2\varepsilon}{3}} \tag{4.198}$$

For small values of ε, the renormalized coupling constant at the critical point is small, of order ε, and this observation lies at the basis of the epsilon-expansion.

We have seen that $Z = 1 + \mathcal{O}(g^2)$, and we cannot compute $\mathcal{O}(g^2)$ terms within the techniques developed so far. We have then to stay with the mean field value $\eta = 0$ for the critical exponent η, or in other words, we have to stay with the canonical dimension of the field. However, we are in a position to compute the correction of order ε to ν, starting from (4.111). To the order of perturbation theory we are working with, we can write this equation

$$r_0 - r_{0c} = \xi^{-2}\left(1 + \frac{u_0}{2}\int \frac{\mathrm{d}^D q}{(2\pi)^D}\,\frac{1}{q^2(q^2+\xi^{-2})}\right)$$

We have used the relation between $\rho = \chi^{-1}$ and ξ and the fact that $Z = 1 + \mathcal{O}(u_0^2)$. The integral is evaluated thanks to the identity

$$\int_0^\infty \frac{u^\alpha \mathrm{d}u}{(u+1)^\beta} = \frac{\Gamma(\alpha+1)\Gamma(\beta-\alpha-1)}{\Gamma(\beta)}$$

We take the $\varepsilon \to 0$ limit of the result

$$r_0 - r_{0c} = \xi^{-2}\left(1 + \frac{u_0\xi^\varepsilon}{16\pi^2\varepsilon}\right)$$

so that from the definitions of $\overline{Z}(g)$ and $\overline{\gamma}(g)$ we get

$$\overline{Z}(g) = 1 + \frac{g}{16\pi^2\varepsilon} + \mathcal{O}(g^2) \tag{4.199}$$

$$\overline{\gamma}(g) = -\frac{g}{16\pi^2\varepsilon} + \mathcal{O}(g^2) \tag{4.200}$$

Combining this last equation and (4.198) leads to

$$\overline{\gamma}(g^*) = -\frac{\varepsilon}{3} \tag{4.201}$$

and to

$$\boxed{\nu = \frac{1}{2} + \frac{\varepsilon}{12} + \mathcal{O}(\varepsilon^2)} \tag{4.202}$$

We have thus succeeded in computing our first correction to mean field theory! Instead of looking for an epsilon-expansion of the critical exponents, we could have worked at a fixed value of D. We would have then found $\nu = 0.6$ for $D = 3$, instead of $\nu = 0.57$ in the epsilon-expansion (Exercise 4.6.12).

The preceding calculation has been performed with a one-dimensional order parameter, $n = 1$. It would not be very difficult to generalize it to an order parameter with any value of n, see Exercise 4.6.11. One finds

$$\boxed{\nu = \frac{1}{2} + \frac{n+2}{4(n+8)}\,\varepsilon + \mathcal{O}(\varepsilon^2)} \tag{4.203}$$

a formula that displays explicitly the dependence of the critical exponent on the dimension of the order parameter. We see that models with the same D and n have the same critical exponents: they belong to the same universality class. To order ε, the critical exponent η keeps its mean field value $\eta = 0$ because there is no field renormalization to order u_0. The lowest non-trivial field renormalization appears at order u_0^2, which is translated into a term of order ε^2 in η

$$\boxed{\eta = \frac{n+2}{2(n+8)^2}\,\varepsilon^2 + \mathcal{O}(\varepsilon^3)} \tag{4.204}$$

The epsilon-expansion of the critical exponents is an asymptotic expansion. Indeed, the exponents cannot be analytic at $\varepsilon = 0$, because they are exactly equal to their mean field values for any $D > 4$, or $\varepsilon < 0$. Since one wants to extrapolate the results of the expansion valid for $\varepsilon \ll 1$ to the physical value $\varepsilon = 1$, one faces convergence problems, which may be treated thanks to sophisticated resummation methods, which are today well under control.

A final comment is that the general strategy of the renormalization group may have been lost in the technical details. We began with an infinite-dimensional parameter space, but we ended up by using a single parameter, the renormalized coupling constant g! What happened is that we followed in parameter space a very peculiar trajectory, the one linking the origin, corresponding to the trivial Gaussian fixed point $g = 0$, to the non-trivial fixed point at $g = g^*$ when the dimensionless bare coupling constant $g_0 = u_0\xi^\varepsilon$ varies from zero to infinity.

4.5.4 Scaling operators and anomalous dimensions

We have seen that all critical exponents can be expressed in terms of two exponents only, which we may choose to be η and ν, or equivalently ω_1 and ω_2. We are going to elaborate on this property, by writing a general form of the correlation functions. We write the Hamilonian in the vicinity of the fixed point as a function of $\vec{r}$-dependent scaling fields $t_a(\vec{r})$ and their conjugate scaling operators $O_a(\vec{r})$ (see (4.166))[53]

$$H = H^* + \sum_a \int \mathrm{d}^D r \; t_a(\vec{r}) O_a(\vec{r}) \tag{4.205}$$

Let us examine the correlation function of two scaling operators

$$G_{ab}(\vec{r}_i, \vec{r}_j) = \langle O_a(\vec{r}_i) O_b(\vec{r}_j) \rangle = \frac{1}{Z} \frac{\delta^2 Z}{\delta t_a(\vec{r}_i)\; \delta t_b(\vec{r}_j)} \Big|_{t=0} \tag{4.206}$$

where Z is the partition function calculated from the Hamiltonian (4.205). To compute the RG transformed correlation function, we remark first that $Z(t) = Z'(t')$ and furthermore that

$$\frac{\delta t'_a(\vec{r}_I)}{\delta t_a(\vec{r}_i)} = b^{\omega_a}$$

if $\vec{r}_i$ belongs to block I. We thus have for the correlation function of the two renormalization group transformed operators

$$\begin{aligned}
\langle O'_a(\vec{r}_I) O'_b(\vec{r}_J) \rangle &= \frac{\delta^2 \ln Z'}{\delta t'_a(\vec{r}_I) \delta t'_b(\vec{r}_J)} \Big|_{t'=0} \\
&= \sum_{\vec{r}_i \in I, \vec{r}_j \in J} \frac{\delta^2 \ln Z}{\delta t_a(\vec{r}_i) \delta t_b(\vec{r}_j)} \Big|_{t=0} \frac{\delta t_a(\vec{r}_i)}{\delta t'_a(\vec{r}_I)} \frac{\delta t_b(\vec{r}_j)}{\delta t'_b(\vec{r}_J)}
\end{aligned}$$

so that, taking into account translation invariance, we find in the critical region

$$G_{ab}\left(\frac{\vec{r}_i - \vec{r}_j}{b}; t b^{\omega_2}\right) = \left(b^{D-\omega_a}\right)\left(b^{D-\omega_b}\right) G_{ab}(\vec{r}_i - \vec{r}_j; t) \tag{4.207}$$

In the region $a \ll |\vec{r}_i - \vec{r}_j| \ll \xi$ this entails a scale invariant behaviour

$$G_{ab}\left(\frac{\vec{r}_i - \vec{r}_j}{b}\right) = \left(b^{D-\omega_a}\right)\left(b^{D-\omega_b}\right) G_{ab}(\vec{r}_i - \vec{r}_j) \tag{4.208}$$

which may be translated into

$$\langle O_a(\vec{r}_i) O_b(\vec{r}_j) \rangle = \frac{1}{r^{D-\omega_a}} \frac{1}{r^{D-\omega_b}} f_{ab}\left(\frac{r}{\xi}\right) \tag{4.209}$$

[53] We use the index a, rather than i, in order to avoid confusion with the site labelling.

where $r = |\vec{r}_i - \vec{r}_j|$. Let us recover the result (4.154) for the ordinary correlation function, which is the expectation value of a product of fields φ, conjugate to the magnetic field B, so that $t_a = t_b = B$

$$G_{BB}(\vec{r}_i, \vec{r}_j) = \langle \varphi(\vec{r}_i)\varphi(\vec{r}_j)\rangle$$

The exponents in (4.208) are $\omega_a = \omega_b = \omega_B = D - \omega_1$ and one recovers (4.154). From the expression of the Ginzburg–Landau Hamiltonian and the relation $r_0 \propto t \propto t_2$, the field conjugate to t_2 is φ^2 and we get for example

$$\langle \varphi^2(\vec{r}_i)\varphi^2(\vec{r}_j)\rangle = \frac{1}{r^{2(D-\omega_2)}}\, g\left(\frac{r}{\xi}\right) \tag{4.210}$$

As we have already seen, the anomalous dimension of the fields φ is $d_\varphi = D - \omega_B = (D - 2 + \eta)/2$, while (4.210) shows that the anomalous dimension of φ^2 is

$$d_{\varphi^2} = D - \omega_2 = D - \frac{1}{\nu} \tag{4.211}$$

In the absence of interactions, the critical exponents are given by their mean field values $\eta = 0$ and $\nu = 1/2$, so that one recovers the values (4.110) drawn from dimensional analysis $d_\varphi = d^0_\varphi = (D-2)/2$ and $d_{\varphi^2} = d^0_{\varphi^2} = 2d^0_\varphi = D - 2$. A straightforward generalization allows one to derive the behaviour of more complicated correlation functions, for example

$$\left\langle \varphi^2\left(\frac{\vec{r}_i}{b}\right)\varphi\left(\frac{\vec{r}_j}{b}\right)\varphi\left(\frac{\vec{r}_k}{b}\right)\right\rangle \simeq b^{d_{\varphi^2}+2d_\varphi}\langle \varphi^2(\vec{r}_i)\varphi(\vec{r}_j)\varphi(\vec{r}_k)\rangle$$

when $a \ll |\vec{r}_i - \vec{r}_j|, |\vec{r}_j - \vec{r}_k|, |\vec{r}_k - \vec{r}_i| \ll \xi$. One notes that *inside a correlation function* $\varphi^2(\vec{r})$ does not behave as $\varphi(\vec{r})\varphi(\vec{r})$. This is due to the fact that the scaling behaviour of $\langle \varphi(\vec{r})\varphi(\vec{r}\,')\rangle$ that is controlled by the anomalous dimension d_φ is valid only when $|\vec{r} - \vec{r}\,'| \gg a$ and cannot be extrapolated to $\vec{r}\,' \to \vec{r}$, and especially to $\vec{r}\,' = \vec{r}$. Of course, when there are no interactions, the naïve identification $\varphi^2(\vec{r}) = \varphi(\vec{r})\varphi(\vec{r})$ is valid, and as we have already checked, in that case $d^0_{\varphi^2} = 2d^0_\varphi$. An explicit illustration of these results is proposed in Exercise 4.6.13.

The generalization to higher powers of φ is far from trivial. In fact φ^4, for example, is *not* a scaling field: renormalization mixes it with φ^2, $\nabla^2\varphi^2$ and $(\vec{\nabla}\varphi)^2$.[54] The only simple result is that renormalization does not mix odd and even powers of φ. In the end, the reason why all exponents can be expressed in terms of η and ν is that these exponents are directly related to the anomalous dimensions of φ and φ^2.

[54] This happens because the relevant field theory is the four-dimensional field theory. It can be shown that an operator (called a composite operator in field theory) of dimension Δ is mixed by renormalization with all operators of dimension less than or equal to Δ. The operator φ^4, being of dimension four, mixes with φ^2, of dimension two, and with $\nabla^2\varphi^2$ and $(\vec{\nabla}\varphi)^2$ (or $\varphi\nabla^2\varphi$) which are of dimension four.

4.6 Exercises

4.6.1 High temperature expansion and Kramers–Wannier duality

1. Show that the partition function of the two-dimensional Ising model on a square lattice can be written as

$$Z(K) = (\cosh K)^L \sum_{\mathcal{C}} \prod_{\langle i,j\rangle} \left(1 + S_i S_j \tanh K\right)$$

where L is the number of pairs of nearest neighbours, or number of links, on the lattice ($L = 2N$ if N is the number of sites, provided boundary conditions are ignored) and $K = J/(kT)$. Consider a term in the expansion of $\prod_{\langle i,j\rangle}$ of the form

$$S_1^{n_1} \cdots S_i^{n_i} \cdots S_N^{n_N} (\tanh K)^b$$

Show that in the sum over all configurations $\mathcal{C}$ this term will give a vanishing contribution unless $n_i = 0,\ 2$ or 4. Show that there is a one-to-one correspondence between terms giving a non-zero contribution to $Z(T)$ and a set of closed polygons, with possible intersections, drawn on the lattice (Figure 4.16).

2. *High temperature expansion.* If $T \to \infty$, $K \to 0$ and one gets an approximate value of $Z(K)$ by keeping the lowest powers of $\tanh K$. The general form of the expansion is

$$Z(K) = 2^N (\cosh K)^L \left[1 + \sum_{b=4,6,8,\ldots} \nu(b)(\tanh K)^b\right]$$

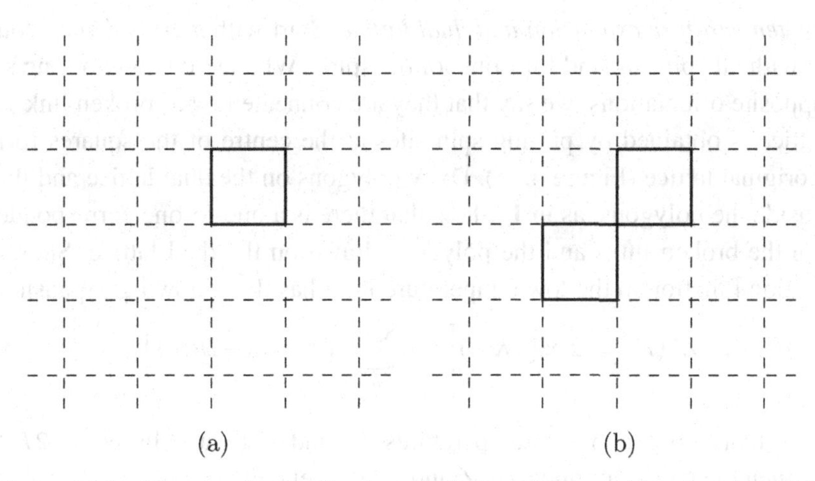

Figure 4.16 Closed polygons for the calculation of the partition function of the Ising model.

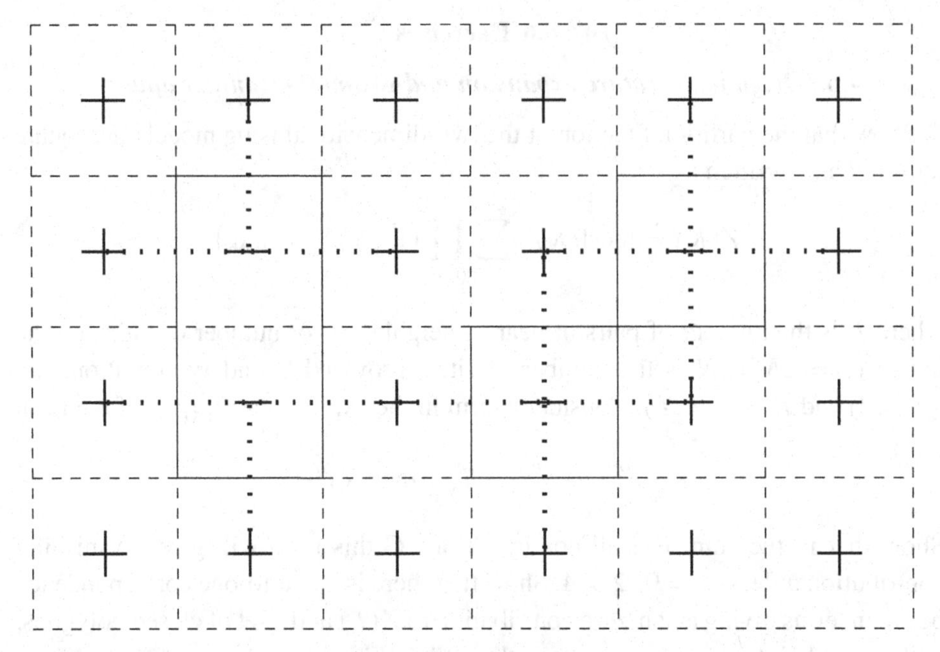

Figure 4.17 The original and dual lattices. Broken links are represented by dotted lines, the dual lattice by dashed lines and polygons drawn on the dual lattice by full lines.

where b is as Section 4.1.1 the perimeter of the polygon and $\nu(b)$ the number of polygons with perimeter b. Draw the polygons with perimeters 4, 6 and 8 and compute to leading order, namely without paying attention to the boundary conditions, the coefficient of the corresponding powers of $\tanh K$. Check that the free energy is extensive within this approximate calculation.

3. *Low temperature expansion and dual lattice.* Start with a ground state configuration with all spins up and flip some of the spins. When two neighbouring spins have opposite orientations, we say that they are connected by a 'broken link'. The dual lattice is obtained by putting spin sites at the centre of the squares formed by the original lattice (Figure 4.17). Draw polygons on the dual lattice and flip all spins inside the polygons, as in **1**. Show that there is a one-to-one correspondence between the broken links and the polygons drawn on the dual lattice. Show that the partition function in the low temperature limit has the following expansion

$$Z^*(K) = 2\exp(KL)\Big[1 + \sum_n \nu(n)\exp(-2nK)\Big]$$

4. Show that if one chooses temperatures T^* and T related by $\exp(-2K^*) = \tanh K$, then the partition functions Z and Z^* are related by

$$\frac{Z(K)}{2^N(\cosh K)^L} = \frac{Z^*(K^*)}{2\exp(-K^*L)}$$

Show that if one *assumes* that there is only one critical point, then the critical temperature $kT_c = J/K_c$ must be given by

$$\exp(-2K_c) = \tanh K_c \quad \text{or} \quad \sinh^2 K_c = 1$$

in agreement with (4.14).

4.6.2 Energy–energy correlations in the Ising model

1. In the Ising model, introduce for any space dimension the quantity

$$E_i = \frac{1}{2} \sum_{\langle i,j \rangle} S_i S_j$$

where the sum is taken at fixed i. Show that the specific heat per spin c can be related to the connected energy–energy correlation function $\langle E_i E_j \rangle_c$

$$c = k\beta^2 J \sum_j \langle E_i E_j \rangle_c = k\beta^2 J \sum_j \langle (E_i - \langle E_j \rangle)(E_j - \langle E_j \rangle) \rangle$$

2. Compute explicitly $\langle E_i E_j \rangle_c$ in the one-dimensional case. Hint: Calculate the four-point correlation functions $\langle S_i S_j S_k S_l \rangle$, $i \le j \le k \le l$ and examine separately the cases $j = i$, $j = i + 1$ and $j = i + p$, $p \ge 2$.

3. Check your result by computing the internal energy and the specific heat of the one-dimensional Ising model from (3.35).

4.6.3 Mean field critical exponents for $T < T_c$

1. By using the expression (4.69) of M_0, compute the critical exponent γ for $T < T_c$ in the mean field approximation.

2. Starting from (4.61) and still using the expression of M_0, compute $\tilde{G}(\vec{q})$ and the exponents η and ν for $T < T_c$.

4.6.4 Accuracy of the variational method

1. Assume that the trial Hamiltonian (4.53) differs from the true Hamiltonian by terms of order ε

$$H_\lambda = H + \varepsilon H_1$$

The free energy F is a function of ε, $F(\varepsilon)$. Show that

$$\frac{dF}{d\varepsilon} = \langle H_1 \rangle$$

and that the function Φ (4.59) obeys

$$\left.\frac{\mathrm{d}\Phi}{\mathrm{d}\varepsilon}\right|_{\varepsilon=0} = 0$$

Thus $\Phi = F + \mathcal{O}(\varepsilon^2)$, where $F = F(\varepsilon = 0)$.

2. Also show that

$$\left.\frac{\mathrm{d}^2\Phi}{\mathrm{d}\varepsilon^2}\right|_{\varepsilon=0} = \beta\langle H_1^2\rangle_c$$

4.6.5 Shape and energy of an Ising wall

Landau's theory is well suited to the study of the transition region between two regions of space with different magnetizations. Such a transition region is called a *domain wall.* We shall study an Ising wall in a one-dimensional geometry: the magnetization $M \to \pm M_0$ depending on whether $z \to \pm\infty$. The magnetization must then change sign (at least once, but we assume that it changes only once) at a point which we may locate at $z = 0$. What follows is peculiar to the Ising model where the spin has a fixed orientation. For vector spins, the magnetization can rotate from $-M_0$ to $+M_0$ and one has a Bloch wall, rather than an Ising wall.[55]

1. We write the Landau functional in the following form, integrating over the x and y directions

$$\begin{aligned}\mathcal{L} &= \int \mathrm{d}z \left(\frac{1}{2}\left(\frac{\mathrm{d}M}{\mathrm{d}z}\right)^2 + \frac{1}{2} r_0 M^2 + \frac{1}{4!} u_0 M^4\right) \\ &= \int \mathrm{d}z \left(\frac{1}{2}\left(\frac{\mathrm{d}M}{\mathrm{d}z}\right)^2 + V(M)\right)\end{aligned}$$

Show that $\delta\mathcal{L}/\delta M = 0$ leads to

$$\frac{\mathrm{d}^2 M}{\mathrm{d}z^2} = V'(M)$$

and that this equation can be interpreted as the equation of motion

$$\frac{\mathrm{d}^2 x}{\mathrm{d}t^2} = -U'(x)$$

[55] The rotation of vector spins is well known, for example in the theory of spin waves. There are no spin waves with Ising spins!

of a fictitious particle with unit mass on the x axis in a potential energy $U(x)$, with the substitutions $M \to x,\ z \to t,\ V(M) \to -U(x)$. From this analogy derive

$$\frac{1}{2}\left(\frac{\mathrm{d}M}{\mathrm{d}z}\right)^2 = V(M) - V(M_0)$$

Discuss qualitatively the motion of the fictitious particle.

2. Set $f(M) = M/M_0$ and note that the correlation length $\xi = (2|r_0|)^{-1/2}$. Show that

$$\xi^2\left(\frac{\mathrm{d}f}{\mathrm{d}z}\right)^2 = \frac{1}{4}\left(1 - f^2\right)$$

and derive

$$f(z) = \tanh\left(\frac{z}{2\xi}\right)$$

Sketch the shape of the transition region as a function of z.

3. The surface energy σ is the difference between the energy with a wall and without a wall. Show that

$$\sigma = \frac{M_0^2}{2\xi}\int_{-1}^{+1}\left(1 - f^2\right)\mathrm{d}f \propto (T_c - T)^{3/2}$$

4.6.6 The Ginzburg–Landau theory of superconductivity

The Ginzburg–Landau theory of superconductivity addresses the interaction between a *complex* superconducting wave function ψ and a static magnetic field $\vec{B} = \vec{\nabla} \times \vec{A}$. The wave function is not that of individual electrons, but rather that of a condensate (see Section 5.5) of Cooper pairs, which are pairs of loosely bound electrons. Its modulus square represents the density n_S of Cooper pairs: $|\psi|^2 = n_S$. The condensate wave function vanishes above the critical temperature T_c and is non-zero below T_c. One postulates a thermodynamic potential which generalizes (4.78)

$$\Gamma - \Gamma_N = \int \mathrm{d}^3r\left(a|\psi|^2 + b|\psi|^4 + \frac{1}{2m}\left|\left(-\mathrm{i}\hbar\vec{\nabla} - q\vec{A}\right)\psi\right|^2 + \frac{1}{2\mu_0}\vec{B}^2 - \frac{1}{\mu_0}\vec{H}\cdot\vec{B}\right)$$

where $\vec{H}$ is the induction, Γ_N is the thermodynamical potential of the normal phase, m and q are the mass and charge of the Cooper pairs. The coefficient b is always positive, but a changes sign at $T = T_c$ and becomes negative for $T < T_c$, so that $|\psi|^2$ is non-zero below the critical temperature.

1. Show that the preceding expression for Γ is equivalent to (4.88) if $\vec{A} = 0$.

2. Show that Γ is invariant under the local gauge transformation

$$\psi(\vec{r}) \to \psi'(\vec{r}) = \exp(-\mathrm{i}q\Lambda(\vec{r}))\psi(\vec{r})$$
$$\vec{A}(\vec{r}) \to \vec{A}'(\vec{r}) = \vec{A}(\vec{r}) - q\vec{\nabla}\Lambda(\vec{r})$$

3. Show that the minimization

$$\frac{\delta\Gamma}{\delta\psi(\vec{r})} = \frac{\delta\Gamma}{\delta\vec{A}(\vec{r})} = 0$$

leads to the equations of motion

$$0 = \frac{1}{2m}\left(-\mathrm{i}\hbar\vec{\nabla} - q\vec{A}\right)\psi + a\psi + 2b|\psi|^2\psi$$
$$\vec{\nabla}\times\vec{B} = -\frac{\mathrm{i}\hbar q\mu_0}{2m}\left[\psi^*\vec{\nabla}\psi - (\vec{\nabla}\psi)^*\psi\right] - \frac{\mu_0 q^2}{m}\vec{A}|\psi|^2$$

and to the boundary conditions

$$\hat{n}\times(\vec{B}-\vec{H}) = 0 \qquad \hat{n}\cdot(-\mathrm{i}\hbar\vec{\nabla} - q\vec{A})\psi = 0$$

where $\hat{n}$ is a unit vector perpendicular to the surface separating the normal and the superconducting phases. Hint:

$$\vec{\nabla}\cdot(\vec{V}\times\vec{W}) = \vec{W}\cdot(\vec{\nabla}\times\vec{V}) - \vec{V}\cdot(\vec{\nabla}\times\vec{W})$$

so that

$$\delta(\vec{\nabla}\times\vec{A})^2 = \delta\vec{A}\cdot(\vec{\nabla}\times\vec{B}) + \vec{\nabla}\cdot(\delta\vec{A}\times\vec{B})$$

4. In a uniform situation where $|\psi|^2 = -a/(2b)$, show that superconductivity is destroyed by a magnetic induction H if

$$H^2 \geq H_c^2(T) = \frac{\mu_0 a^2}{2b}$$

5. If $B = 0$ in a one-dimensional geometry, show that in a normal–superconducting junction the order parameter increases as

$$\psi(z) = \sqrt{\frac{|a|}{2b}}\tanh\left(\frac{z}{\xi\sqrt{2}}\right)$$

where $\xi = (4m|a|/\hbar^2)^{-1/2}$ is the *coherence length*, see Exercise 4.6.5.

6. Assuming now $\psi = \text{const}$, $\vec{A} \neq 0$, show that

$$\vec{j}(\vec{r}\,) = -\frac{q^2 n_S}{m}\vec{A}$$

Show that the magnetic field penetrates in the superconductor up to a distance called the *penetration length*

$$\kappa = \left(\frac{m}{\mu_0 q^2 n_S}\right)^{1/2}$$

The fact that a magnetic field does not penetrate in the bulk of a superconductor is called the *Meissner effect.*

4.6.7 Mean field correlation function in $\vec{r}$-space

We wish to compute the two-point correlation function in the mean field approximation from its Fourier transform (4.83) or (4.84)

$$G(\vec{r}\,) = \int \frac{\mathrm{d}^D q}{(2\pi)^D}\, \mathrm{e}^{\mathrm{i}\vec{q}\cdot\vec{r}} \frac{1}{q^2 + m^2}$$

where $m^2 = \overline{r}_0(T - T_0)$ for $T > T_0$ and $m^2 = 2\overline{r}_0(T_0 - T)$ for $T < T_0$. Write

$$\frac{1}{q^2 + m^2} = \int_0^\infty \mathrm{d}t\, \exp[-t(q^2 + m^2)]$$

and perform the angular integration to obtain

$$G(\vec{r}\,) = \frac{\pi^{D/2}}{(2\pi)^D} \int_0^\infty \frac{\mathrm{d}t}{t^{D/2}} \exp\left[-\left(tm^2 + \frac{\vec{r}^{\,2}}{4t}\right)\right]$$

Evaluate $G(\vec{r}\,)$ by using a saddle point approximation and show that

$$G(\vec{r}\,) = \frac{1}{2m}\left(\frac{m}{2\pi r}\right)^{(D-1)/2} \mathrm{e}^{-mr}\left(1 + \mathcal{O}\left(\frac{1}{mr}\right)\right)$$

4.6.8 Critical exponents for $n \gg 1$

When the dimension n of the order parameter is large, it is easy to give a simple approximate evaluation of the critical exponents. Let $\Phi(\vec{r})$ be an n-component random field $\Phi(\vec{r}) = \{\varphi_1(\vec{r}), \ldots, \varphi_n(\vec{r})\}$, $n \gg 1$. We define

$$\Phi^2(\vec{r}) = \sum_{i=1}^{n} \varphi_i^2(\vec{r}) \qquad (\vec{\nabla}\Phi(\vec{r}))^2 = \sum_{i=1}^{n} (\vec{\nabla}\varphi_i)^2$$

and we write the Ginzburg–Landau Hamiltonian as

$$H = \int \mathrm{d}^D r \left[\frac{1}{2} (\vec{\nabla}\Phi(\vec{r}))^2 + \frac{1}{2} r_0 \Phi^2(\vec{r}) + \frac{u_0}{4n} \Phi^4(\vec{r}) \right]$$

1. *Self-consistent equation for* $\Phi^2(\vec{r})$. We use a random phase approximation: since Φ^2 is a sum of random variables φ_i^2, if $\varphi_i^2 \sim 1$, we may expect that $\Phi^2 \sim n$, and that the fluctuation $\langle (\Phi^2 - \langle \Phi^2 \rangle)^2 \rangle$ is also $\sim n$. We write

$$(\Phi^2)^2 = (\Phi^2 - \langle \Phi^2 \rangle)^2 + 2\Phi^2 \langle \Phi^2 \rangle - \langle \Phi^2 \rangle^2$$

Show that if one neglects terms of order unity

$$H \simeq \int \mathrm{d}^D r \left[\frac{1}{2} (\vec{\nabla}\Phi(\vec{r}))^2 + \frac{1}{2} r_0 \Phi^2(\vec{r}) + \frac{u_0}{2n} \langle \Phi^2 \rangle \Phi^2(\vec{r}) \right]$$

Show then that the Fourier transform of the correlation function becomes

$$\tilde{G}_{ij}(q) = \frac{\delta_{ij}}{q^2 + r_0 + \dfrac{u_0}{n} \langle \Phi^2 \rangle}$$

and derive the self-consistent equation for $\langle \Phi^2 \rangle$

$$\langle \Phi^2 \rangle = n K_D \int_0^{\Lambda} \frac{q^{D-1} \mathrm{d}q}{q^2 + r_0 + \dfrac{u_0}{n} \langle \Phi^2 \rangle}$$

where $K_D = S_D/(2\pi)^D$. Why do we choose a u_0/n coupling rather than u_0/n^2 for instance?

2. *Equation for the susceptibility*. Let $\rho(T) = 1/\chi(T)$ be the inverse of the susceptibility. Show that the critical temperature T_c is determined by the equation

$$\rho(T_c) = r_0(T_c) + \frac{u_0}{n} \langle \Phi^2 \rangle = 0$$

Find the critical exponent η. Show that for $n \to \infty$, with an error of order $1/n$, we have

$$\rho(T) = \overline{r}_0 (T - T_c) - u_0 \rho(T) K_D \int_0^{\Lambda} \frac{q^{D-1}\, \mathrm{d}q}{q^2 + \rho(T)}$$

3. *Critical exponents*. Deduce from the preceding equation the critical exponent γ in the two cases

(i) $D > 4$ (ii) $2 < D < 4$

Also find the critical exponent ν.

4.6.9 Renormalization of the Gaussian model

1. In this exercise, we give a direct proof of the RG equations when $u_0 = 0$. For notational simplicity we work in one dimension, but the results are trivially generalized to any D. Write a Hamiltonian of the following form on a one-dimensional lattice, where φ_i is the field at site i

$$H = \frac{1}{2}\sum_{i=0}^{N-1} \varphi_i^2 - \frac{1}{2}\beta J \sum_i (\varphi_{i-1}\varphi_i + \varphi_i\varphi_{i+1})$$

The second term couples neighbouring sites. Show that in Fourier space

$$\begin{aligned} H &= \frac{1}{2}\sum_q [1 - 2\beta J \cos q]|\tilde{\varphi}(q)|^2 \\ &= \frac{1}{2}\sum_q [(1 - 2\beta J) + \beta J q^2 + \mathcal{O}(q^4)]|\tilde{\varphi}(q)|^2 \\ &= \frac{1}{2}\sum_q [r_0 + cq^2 + dq^4 + \cdots]|\tilde{\varphi}(q)|^2 \end{aligned}$$

where $q = 2\pi p/N$ and

$$\tilde{\varphi}(q) = \frac{1}{\sqrt{N}}\left(\varphi_0 + \mathrm{e}^{\mathrm{i}q}\varphi_1 + \mathrm{e}^{2\mathrm{i}q}\varphi_2 + \cdots\right)$$

2. Build a block φ'_I of two neighbouring sites by writing

$$\varphi'_I = 2^{-\omega_1}[\varphi_i + \varphi_{i+1}]$$

Show that for $q \ll 1$

$$\varphi'(q') = 2^{-\omega_1 + 1/2}\varphi(q)$$

and derive the RG equations

$$r'_0 = 2^{1-2\omega_1} r_0 \qquad c' = 2^{-2\omega_1 - 1} c$$

What are the fixed points? Give their physical interpretation and show that $\nu = 1/2$. Show that d is an irrelevant field.

3. Generalize to dimension D and show that one always finds $\nu = 1/2$. Can you explain the isotropy of the correlation function close to $T = T_c$ for $r \gg 1$?

4.6.10 Scaling fields at the Gaussian fixed point

1. We start from Equations (4.176)–(4.177). Show that

$$\overline{G}(0) = \frac{K_D \Lambda^{D-2}}{D-2}(1 - b^{2-D}) + \mathcal{O}(r_0) = 2B(1 - b^{2-D}) + \mathcal{O}(r_0)$$

where $K_D = S_D/(2\pi)^D$ and that the matrix $T(b)$ (4.142) is

$$T(b) = \begin{pmatrix} b^2 & B(b^2 - b^\varepsilon) \\ 0 & b^\varepsilon \end{pmatrix}$$

2. Find the eigenvalues and eigenvectors of T for $D > 4$ ($\varepsilon < 0$). Show that the scaling fields are $t_2 = r_0 + u_0 B$ and $t_3 = u_0$. What are the scaling operators O_2 and O_3? Draw the critical surface in the (u_0, r_0) plane and find the critical temperature.

4.6.11 Critical exponents to order ε for $n \neq 1$

We assume that the order parameter has dimension n. Using the notation of Exercise 4.6.8, the Hamiltonian is written

$$H = \int \mathrm{d}^D r \left[\frac{1}{2}(\vec{\nabla}\Phi(\vec{r}))^2 + \frac{1}{2} r_0 \Phi^2(\vec{r}) + \frac{u_0}{4!}\Phi^4(\vec{r})\right]$$

The magnetization has n components and is denoted by $\mathcal{M} = \{M_1, \ldots, M_n\}$.

1. Compute the Gibbs potential $\Gamma(\mathcal{M})$ using the methods of Section 4.3.4. Note that the matrix $\mathcal{H}$ (4.103) now carries indices, $\mathcal{H}_{ij}(\vec{r}, \vec{r}\,')$, and that it obeys

$$\mathcal{H}_{ij}(\vec{r}, \vec{r}\,') = \left[\left(-\vec{\nabla}_r^2 + r_0\right)\delta_{ij} + \frac{u_0}{6}\left(\delta_{ij}\Phi^2 + 2\varphi_i\varphi_j\right)\right]\delta^{(D)}(\vec{r} - \vec{r}\,')$$

Calculate $\ln\det\mathcal{H}_{ij}(\vec{q})$ for a uniform Φ. One may notice that the calculation simplifies by choosing $\mathcal{M} = \{M, 0, \ldots, 0\}$, $\mathcal{M}^2 = M^2$. Show that, to order $\hbar$, $\Gamma(\mathcal{M})$ becomes for $T > T_c$

$$\frac{1}{V}\Gamma(\mathcal{M}) = \frac{1}{2}r_0\mathcal{M}^2 + \frac{u_0}{4!}\mathcal{M}^4 + \frac{\hbar}{2}\int \frac{\mathrm{d}^D q}{(2\pi)^D}\ln\left(q^2 + r_0(T) + \frac{1}{2}u_0\mathcal{M}^2\right)$$
$$+ \frac{\hbar(n-1)}{2}\int \frac{\mathrm{d}^D q}{(2\pi)^D}\ln\left(q^2 + r_0(T) + \frac{1}{6}u_0\mathcal{M}^2\right)$$

2. Use the same techniques to show ($\hbar = 1$)

$$r_0 - r_{0c} = \rho\left(1 + \frac{u_0(n+2)}{6}\int \frac{\mathrm{d}^D q}{(2\pi)^D}\frac{1}{q^2(q^2 + \rho)}\right)$$

and

$$\tilde{\Gamma}^{(4)}(0) = u_0 - \frac{u_0^2(n+8)}{6} \int \frac{\mathrm{d}^D q}{(2\pi)^D} \frac{1}{(q^2+\rho)^2}$$

Repeat the reasoning of Sections 4.5.2 and 4.5.3 to find the value (4.203) of ν.

4.6.12 Irrelevant exponents

1. Show that the bare coupling constant u_0 is given as a function of g by

$$u_0 \simeq \xi^{-\varepsilon} g^* A(g^*) \left(\frac{g^*}{|g^* - g|} \right)^{\varepsilon/\omega}$$

with

$$A(g) = \exp\left[- \int_0^g \mathrm{d}g' \left(\frac{\varepsilon}{\beta(g', \varepsilon)} + \frac{1}{g'} - \frac{\varepsilon}{\omega(g' - g^*)} \right) \right]$$

Deduce from this equation that

$$g \simeq g^* \left(1 + \mathcal{O}(\xi^{-\omega})\right)$$

and identify ω with the leading irrelevant exponent $-\omega_3$. Compute ω by using the approximate form (4.197) for $\beta(g, \varepsilon)$.

2. Repeat the calculations of Sections 4.5.2 and 4.5.3 without taking the $\varepsilon \to 0$ limit. Show that $\overline{\gamma}(g^*) = (D-4)/3$. Find ν for $D = 3$.

4.6.13 Energy–energy correlations

1. In the Ginzburg–Landau model one often calls the scaling operator $\varphi^2(\vec{r})$ the *energy operator* since it is conjugate to the temperature. Show that the specific heat C is related to the volume integral of the (connected) energy–energy correlation $\langle \varphi^2(\vec{r})\varphi^2(\vec{r}\,')\rangle_c$; this is a fluctuation-response theorem.

2. Assume that $\langle \varphi^2(\vec{r})\varphi^2(\vec{r}\,')\rangle_c$ obeys a scaling law

$$\langle \varphi^2(\vec{r})\varphi^2(\vec{r}\,')\rangle_c = |\vec{r} - \vec{r}\,'|^{-2\sigma} f\left(\frac{|\vec{r} - \vec{r}\,'|}{\xi} \right)$$

Use the results of **1** to determine the value of σ as a function of the critical exponent α and show that

$$\sigma = D - \frac{1}{\nu}$$

$\sigma = d_{\varphi^2}$ is the anomalous dimension of the scaling operator φ^2.

3. Consider the correlation function $\langle\varphi^2(\vec{r}_1)\varphi(\vec{r}_2)\varphi(\vec{r}_3)\rangle$, taking $T > T_c$ for simplicity. Show that the volume integral of this correlation function may be related to the derivative of a thermodynamic function. Show that one gets the correct critical exponents for this thermodynamic function by writing for $T = T_c$

$$\left\langle\varphi^2\left(\frac{\vec{r}_1}{b}\right)\varphi\left(\frac{\vec{r}_2}{b}\right)\varphi\left(\frac{\vec{r}_3}{b}\right)\right\rangle = b^{d_{\varphi^2}+2d_\varphi}\langle\varphi^2(\vec{r}_1)\varphi(\vec{r}_2)\varphi(\vec{r}_3)\rangle$$

Generalize to any product of φ and φ^2s. The preceding results also apply to the Ising model where the energy operator is defined in Exercise 4.6.2.

4.6.14 'Derivation' of the Ginzburg–Landau Hamiltonian from the Ising model

Let us write the partition function of the Ising model as

$$Z[B] = \int \prod_i \mathrm{d}P_0\,[S_i]\,\mathrm{e}^{-\beta H[S_i]}$$

where

$$\mathrm{d}P_0[S_i] = 2\delta(S_i^2 - 1)\mathrm{d}S_i$$

The Ginzburg–Landau model is obtained by the substitution

$$\mathrm{d}P_0\,[S_i] \to \mathrm{d}P[S_i] = \exp\left[-\frac{1}{2}\,S_i^2 - g(S_i^2 - 1)^2\right]\mathrm{d}S_i$$

where g is a positive number. Note that, for the computation of the correlation functions, it is not necessary to fix the normalization of the probability distribution.

1. Show that one recovers the Ising model in the limit $g \to \infty$.

2. Show that $g = 0$ corresponds to the Gaussian model. Hint: transform the sum over nearest neighbours $\sum_{\langle i,j\rangle} = \sum_{i,\mu}$ as in Section 4.3.1 and note that $S_{i+\mu} - S_i \simeq \partial_\mu S_i$. Finally rescale the spin by introducing a field $\varphi(\vec{r}) = \sqrt{Ja^2}\,S(\vec{r})$.

3. Obtain the Ginzburg–Landau Hamiltonian (4.96) for finite, non-zero values of g, by an appropriate rescaling of S and a redefinition of the various parameters. Thus the Ginzburg–Landau Hamiltonian represents a kind of interpolation between the Ising and Gaussian models.

4.7 Further reading

Excellent general introductions to critical phenomena can be found in Bruce and Wallace [21] and in Cardy [25]. There are many detailed books on the subject, among them: Ma [84], Pfeuty and Toulouse [101], Le Bellac [72]. At a more advanced level, the use of field theoretical methods is described by Amit [1], Parisi [99], Zinn-Justin [125] or Itzykson and Drouffe [58]. Applications to physical systems or models (percolation, polymers, etc.) may be found for example in Chaikin and Lubensky [26] or in Schwabl [115].

5

Quantum statistics

In Chapter 3 we exhibited the limitations of a purely classical approach. For example, if the temperature is below a threshold value, some degrees of freedom become 'frozen' and the equipartition theorem is no longer valid for them. The translational degrees of freedom of an ideal gas appear to escape this limitation of the classical (or more precisely, semi-classical) approximation. We shall see in this chapter that, in fact, this is not so: if the temperature continues to decrease below some reference temperature, the classical approximation will deteriorate progressively. However, in this case, the failure of the classical approximation is not related to freezing degrees of freedom but rather to the symmetry properties of the wave function for identical particles imposed by quantum mechanics. A rather spectacular consequence is that *the kinetic energy is no longer a measure of the temperature*. In a classical gas, even in the presence of interactions, the average kinetic energy is equal to $3kT/2$, but this result does not hold when the temperature is low enough, even for an ideal gas. For example, if we consider the conduction electrons in a metal as an ideal gas, we shall show that the average kinetic energy of an electron is not zero even at zero temperature. In addition, this kinetic energy is about 100 times kT at normal temperature. Let us consider another example. In a gaseous mixture of helium-3 and helium-4 at low temperature, the average kinetic energies of the two isotopes are different: the average kinetic energy of helium-3 is larger than $3/2kT$ while that of helium-4 is smaller. Helium-3 is a fermion and obeys the Fermi–Dirac statistics whereas helium-4, a boson, satisfies the Bose–Einstein statistics. In this chapter we study the Fermi–Dirac and Bose–Einstein statistics and as important applications we consider electrons in metals, photon gas, vibrations in solids and the Bose–Einstein condensation. These examples illustrate the new, and often spectacular, effects due to quantum statistics.

5.1 Bose–Einstein and Fermi–Dirac distributions

A fundamental postulate of quantum mechanics is that there exist only two classes of particles: the *fermions* and the *bosons*. Equivalently, we can say that there are only two kinds of *quantum statistics*: Fermi–Dirac statistics for fermions and Bose–Einstein statistics for bosons. The state vector of a system of identical fermions must be antisymmetric under the exchange of all the coordinates (spin and space) of two particles, whereas for a system of bosons, the state vector must be symmetric under the same exchange. For fermions, the antisymmetry of the state vector leads to the Pauli exclusion principle: it is not possible to put two identical fermions in the same quantum state. A profound result in relativistic quantum mechanics, but whose demonstration requires advanced concepts that are beyond our scope here, is the 'spin-statistics theorem' which says that particles with integer spins $(0, 1, 2, \dots)$ are bosons and particles with half-integer spins $(1/2, 3/2, \dots)$ are fermions. Consequently, electrons, protons and neutrons (spin $1/2$) are fermions, while π-mesons (spin 0) and photons (spin 1) are bosons. These properties generalize to composite particles as long as we can neglect their internal structure: protons and neutrons are made of three quarks of spin $1/2$ and, therefore, necessarily have half-integer spin (in fact the spin is $1/2$) and are thus fermions. Generally, a particle formed from an odd (even) number of fermions is a fermion (boson). In this way, the helium-4 atom which contains two protons, two neutrons and two electrons is a boson but helium-3 with only one neutron is a fermion. In spite of the fact that the electronic wave functions are identical for these two atoms, this 'detail' leads to fundamentally different behaviour at low temperature: helium-4 is superfluid at a temperature of about 2 K, but helium-3 becomes superfluid only below 3 mK!

We can understand the differences between these situations with the help of the 'quantum ideal gas' model. Although this model neglects the inter-particle interactions, it turns out to be very useful in diverse physical situations. Let us examine the case of identical and independent particles placed in a potential well. The energy levels of these particles are designated by ε_ℓ where ℓ labels the value of the energy and all other quantum numbers necessary to specify the quantum state.[1] For identical particles, giving the *occupation numbers* n_ℓ for each level, i.e. the number of particles in each level, ε_ℓ, determines the energy level E_r of the system. For example, consider a total of four particles in the following configuration: one particle in the ground state ε_0, two in the first level ε_1, none in the second and one in the third ε_3, in other words the configuration $r = r_1 \equiv \{n_0 = 1, n_1 = 2,$

[1] Energy levels that are degenerate but specified by different quantum numbers will be labelled by different values for ℓ (see Footnote 15 in Chapter 2). Furthermore, for convenience, we choose the ground state level as the reference energy $\varepsilon_0 = 0$.

$n_2 = 0, n_3 = 1, n_{\ell>3} = 0\}$, then the energy of the system is given by

$$E_{r_1} = \varepsilon_0 + 2\varepsilon_1 + \varepsilon_3$$

Generally, a configuration of the system will be specified equivalently by the index r or by the values of the occupation numbers $\{n_\ell\}$

$$E_r = E(\{n_\ell\}) = n_0\varepsilon_0 + \cdots + n_\ell\varepsilon_\ell + \cdots = \sum_{\ell=0}^{\infty} n_\ell\varepsilon_\ell \tag{5.1}$$

Recall (cf. (2.62)) that these energy levels can be highly degenerate.

By using the relation (5.1) between a microstate and the single particle occupation states, we can write in different ways the trace that appears in the definition (3.1b) of the partition function

$$\mathrm{Tr} = \sum_r = \sum_{\{n_\ell\}} \tag{5.2}$$

The partition function for N identical independent particles subject to the constraint $N = \sum_\ell n_\ell$ is then written as

$$\begin{aligned} Z_N &= \sum_r \exp(-\beta E_r) \\ &= \sum_{\substack{\{n_\ell\} \\ \sum_\ell n_\ell = N}} \exp(-\beta E(\{n_\ell\}) = \sum_{\substack{\{n_\ell\} \\ \sum_\ell n_\ell = N}} \prod_{\ell=0}^{\infty} \exp(-\beta n_\ell\varepsilon_\ell) \end{aligned} \tag{5.3}$$

5.1.1 Grand partition function

In practice, the constraint $N = \sum_\ell n_\ell$ is not easily taken into account. Therefore, instead of using the canonical partition function (5.3) it is more convenient to use the grand partition function where the number of particles is not fixed. From Equation (3.135) we have

$$\mathcal{Q}(z, V, T) = \sum_{N=0}^{\infty} z^N Z_N(V, T) = \sum_{N=0}^{\infty} \sum_{\substack{\{n_\ell\} \\ \sum_\ell n_\ell = N}} \prod_{\ell=0}^{\infty} \left[z^{n_l} \exp(-\beta n_\ell\varepsilon_\ell) \right] \tag{5.4}$$

This expression for $\mathcal{Q}$ can be simplified, thanks to the summation over N, by using

$$\sum_{N=0}^{\infty} \sum_{\substack{\{n_\ell\} \\ \sum_\ell n_\ell = N}} \equiv \sum_{\{n_\ell\}} \sum_N \delta_{N, \sum n_\ell} \equiv \sum_{\{n_\ell\}}$$

For a fixed $\{n_\ell\}$, the Kronecker delta contributes for only one value of N. We are now able to perform the sum in (5.4) over all the $\{n_\ell\}$ independently

$$\mathcal{Q}(z, V, T) = \sum_{\{n_\ell\}} \prod_\ell \left(z\mathrm{e}^{-\beta\varepsilon_\ell}\right)^{n_\ell} = \prod_\ell \sum_{n_\ell} \left(z\mathrm{e}^{-\beta\varepsilon_\ell}\right)^{n_\ell} \tag{5.5}$$

The final expression for the grand partition function depends on the statistics. For Fermi–Dirac (FD) statistics, $n_\ell = 0$ or 1 because we cannot put more than one particle in quantum level ε_ℓ

$$\sum_{n_\ell=0}^{1} z^{n_\ell}\mathrm{e}^{-\beta n_\ell \varepsilon_\ell} = 1 + z\mathrm{e}^{-\beta\varepsilon_\ell}$$

On the other hand, for Bose–Einstein (BE) statistics, all integer values of n_ℓ are allowed. We thus need to sum a geometric series which converges[2] for $\mu \le \varepsilon_0$

$$\sum_{n_\ell=0}^{\infty} z^{n_\ell}\mathrm{e}^{-\beta n_\ell \varepsilon_\ell} = \frac{1}{1 - z\mathrm{e}^{-\beta\varepsilon_\ell}}$$

The final results are

$$\mathcal{Q}_{\mathrm{FD}}(z, V, T) = \prod_\ell \left(1 + z\,\mathrm{e}^{-\beta\varepsilon_\ell}\right) \tag{5.6}$$

$$\mathcal{Q}_{\mathrm{BE}}(z, V, T) = \prod_\ell \frac{1}{1 - z\,\mathrm{e}^{-\beta\varepsilon_\ell}} \qquad \mu \le \varepsilon_0 \tag{5.7}$$

These equations allow us to determine the average number of particles (or the average occupation number) in level ε_ℓ

$$\langle n_\ell \rangle = -\frac{1}{\beta}\frac{\partial \ln \mathcal{Q}}{\partial \varepsilon_\ell}$$

which gives

$$\langle n_\ell \rangle_{\mathrm{FD}} = \frac{1}{z^{-1}\mathrm{e}^{\beta\varepsilon_\ell} + 1} = \frac{1}{\mathrm{e}^{\beta(\varepsilon_\ell - \mu)} + 1} \tag{5.8a}$$

$$\langle n_\ell \rangle_{\mathrm{BE}} = \frac{1}{z^{-1}\mathrm{e}^{\beta\varepsilon_\ell} - 1} = \frac{1}{\mathrm{e}^{\beta(\varepsilon_\ell - \mu)} - 1} \tag{5.8b}$$

The convergence condition for the series (5.7), $\mu \le \varepsilon_0$ or equivalently $z \le e^{\beta\varepsilon_0}$ is equivalent to the physically expected condition $\langle n_\ell \rangle_{\mathrm{BE}} \ge 0$. There is no restriction on μ for Fermi–Dirac statistics.

[2] Strictly speaking, the series converges only for $\mu < \varepsilon_0$. The case $\mu = \varepsilon_0$ is possible in the thermodynamic limit (see Section 5.5). For finite volume, we must stay with the strict inequality $\mu < \varepsilon_0$.

5.1.2 Classical limit: Maxwell–Boltzmann statistics

It is instructive to recover the Maxwell-Boltzmann (MB) statistics as the limit of quantum statistics. The classical limit is reached when the constraints imposed by quantum mechanics (which led from (5.5) to (5.8)) no longer distinguish between the two kinds of statistics, in other words when $\langle n_\ell \rangle_{\rm FD} \simeq \langle n_\ell \rangle_{\rm BE}$. Consequently, we need to be able to neglect the ± 1 in Equations (5.8) and therefore

$$\langle n_\ell \rangle_{\rm FD} = \langle n_\ell \rangle_{\rm BE} \ll 1$$

The classical limit, then, corresponds to

$$z \ll 1$$

To make the connection with the classical ideal gas, we impose the following *ad hoc* (or Maxwell–Boltzmann) prescription: the statistical weight of a state (ℓ) must be weighted by $1/n_\ell!$. With this prescription, the partition function for N MB particles becomes

$$\begin{aligned} Z_N^{\rm MB} &= \sum_{\sum n_\ell = N} \frac{{\rm e}^{-n_0 \beta \varepsilon_0}}{n_0!} \cdots \frac{{\rm e}^{-n_\ell \beta \varepsilon_\ell}}{n_\ell!} \cdots \\ &= \frac{1}{N!} \left(\sum_\ell {\rm e}^{-\beta \varepsilon_\ell} \right)^N \end{aligned} \tag{5.9}$$

We easily go from the first line to the second in Equation (5.9) by using the multinomial theorem (Exercise 5.6.1). It is, however, more instructive to calculate the grand partition function

$$\mathcal{Q}_{\rm MB}(z, V, T) = \sum_N z^N Z_N^{\rm MB} = \exp\Big(z \sum_\ell {\rm e}^{-\beta \varepsilon_\ell} \Big) \tag{5.10}$$

We obtain the expression for the average occupation number

$$\langle n_\ell \rangle_{\rm MB} = -\frac{1}{\beta} \frac{\partial \ln \mathcal{Q}_{\rm MB}}{\partial \varepsilon_\ell} = z {\rm e}^{-\beta \varepsilon_\ell} \le z \tag{5.11}$$

and we confirm that indeed the three statistics coincide

$$\langle n_\ell \rangle_{\rm MB} = \langle n_\ell \rangle_{\rm FD} = \langle n_\ell \rangle_{\rm BE} \ll 1$$

for $z \ll 1$. It is important to understand that the Maxwell–Boltzmann statistics does not correspond to physical particles. Rather, it is defined through a prescription chosen such that the classically calculated results agree with the correct quantum results in the limit of very small average occupation number. This necessitates the normalization $1/N!$ in Equation (5.9).

For a classical ideal gas, $z \simeq (\lambda/d)^3$ (cf. (3.139)), where λ is the thermal wavelength and d the average distance between particles. Thus, we obtain again the condition $\lambda \ll d$, which was established qualitatively in Section 3.1.2 for the classical limit. We also re-obtain from Equation (5.10) the grand partition function (3.137) for the classical ideal gas since

$$\sum_{\ell} \mathrm{e}^{-\beta\varepsilon_\ell} \to \frac{V}{h^3}\int \mathrm{d}^3 p\, \mathrm{e}^{-\beta p^2/(2m)} = \frac{V}{\lambda^3}$$

This justifies *a posteriori* the $1/N!$ factor introduced in (3.13).

5.1.3 Chemical potential and relativity

When the number of particles can change, which is allowed in relativistic quantum mechanics where particles can be destroyed and created, the chemical potential of the different particle species is in general constrained by conservation laws, e.g. conservation of electric charge. However, the simplest case occurs when particle production/destruction is not constrained by any conservation law, which means, as we shall demonstrate, that the chemical potential is zero. This is the case, for example, for a gas of photons in thermal equilibrium: the photon chemical potential is zero. Since the number of photons is not fixed *a priori*, the number of particles N should be viewed as an internal variable to be determined by minimizing the free energy F, which is also a function of the controlled parameters V and T,

$$\left.\frac{\partial}{\partial N} F(T, V; N)\right|_{T,V} = 0$$

In other words we have $\mu = 0$.

We now examine a case with a conservation law, for example a system of electrons and positrons in thermal equilibrium. If these are the only charged particles whose number can change, then charge conservation means that the number of electrons, N_-, minus the number of positrons, N_+, is constant: $N_- - N_+ = N_0$. The free energy F is a function of the controlled parameters V, T and N_0 as well as the internal variable N_- (or N_+). It may also be considered as a function of the controlled parameters T and V and the two internal variables N_+ and N_- which are related by a conservation law

$$F(T, V, N_0; N_-) \equiv F(T, V; N_-, N_+ = N_- - N_0)$$

At equilibrium, the value of N_- (and therefore N_+) is given by the minimization of the free energy

$$\left.\frac{\partial F}{\partial N_-}\right|_{T,V,N_0} = \left.\frac{\partial F}{\partial N_-}\right|_{T,V,N_+} + \left.\frac{\partial F}{\partial N_+}\right|_{T,V,N_-} \frac{\mathrm{d}N_+}{\mathrm{d}N_-} = \mu_- + \mu_+ = 0$$

The chemical potentials of the electrons and positrons therefore obey

$$\mu_+ = -\mu_- \tag{5.12}$$

The above result, (5.12), can be interpreted using the results of Chapter 3 on chemical reactions. In fact, a thermal bath of electrons (e^-) and positrons (e^+) also contains photons (γ). We may thus interpret the thermal and chemical equilibrium as the result of the following reaction

$$\gamma \leftrightharpoons e^- + e^+$$

Since the photonic chemical potential is zero, we conclude from (3.114) that $\mu_- = -\mu_+$.

As a final remark, we note that taking into account the mass energy of the particles amounts to shifting the zero of energy by mc^2, consequently adding mc^2 to the chemical potential. The relativistic energy of a free particle is given by

$$\varepsilon_p = \sqrt{m^2c^4 + p^2c^2} \tag{5.13}$$

Using this relativistic expression of the energy for a gas of free massive spin zero bosons, the average occupation number is, according to (5.8b),

$$\langle n_{\vec{p}} \rangle = \frac{1}{e^{\beta(\varepsilon_p - \mu)} - 1}$$

Thus, the condition $\mu \leq \varepsilon_0$ becomes here $\mu \leq mc^2$.

5.2 Ideal Fermi gas

5.2.1 Ideal Fermi gas at zero temperature

At zero temperature, one must minimize the energy. This is done by filling the lowest possible energy levels ε_ℓ, starting with the ground state, see Figure 5.1. The

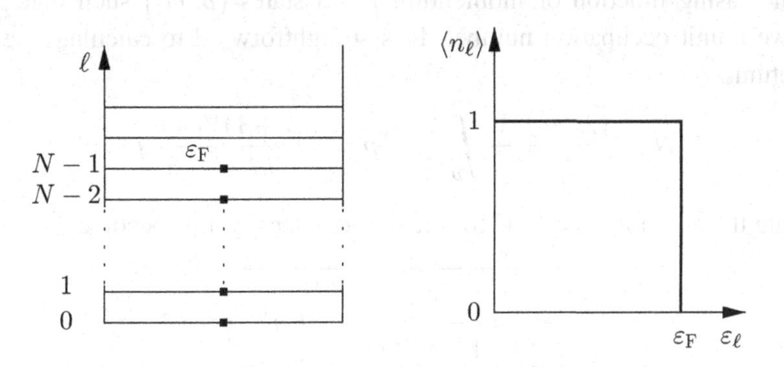

Figure 5.1 Level filling and occupation numbers at $T = 0$.

last level to be filled is called the *Fermi level* or the *Fermi energy* $\varepsilon_{\rm F}$. The average occupation numbers are therefore given by

$$\langle n_\ell \rangle = \begin{cases} 1 & \text{for } \ell \text{ such that } \varepsilon_\ell \le \varepsilon_{\rm F} \\ 0 & \text{for } \ell \text{ such that } \varepsilon_\ell > \varepsilon_{\rm F} \end{cases} = \theta(\varepsilon_{\rm F} - \varepsilon_\ell) \tag{5.14}$$

We may obtain this same result by taking the limit $T \to 0$ of the Fermi distribution

$$\lim_{\beta\to\infty} \langle n_\ell \rangle = \lim_{\beta\to\infty} \frac{1}{{\rm e}^{\beta(\varepsilon_\ell - \mu)} + 1} = \theta(\mu - \varepsilon_\ell) \tag{5.15}$$

Comparing (5.14) and (5.15) shows that, at zero temperature, the chemical potential is the same as the Fermi energy: $\mu(T=0) = \varepsilon_{\rm F}$. The physical interpretation of this result is clear. At zero temperature, $\mu(T=0) = \partial E/\partial N$, which for closely spaced levels and $N \gg 1$ is equivalent to

$$\mu(T=0) = E(N) - E(N-1) = \varepsilon_{\rm F}$$

In fact, adding a particle corresponds to increasing the total energy by $\varepsilon_{\rm F}$. It is very important to keep in mind that the equality $\mu = \varepsilon_{\rm F}$ is valid only for $T = 0$. We shall see below that $\mu(T) \le \mu(T=0)$.

We now focus attention on a free Fermi gas of particles with mass m and spin s, in a box of volume V. A quantum state, ℓ, of a single particle is then specified by its momentum, $\vec{p}$, and the z component of its spin: $m_z = -s, -s+1, \ldots, s$, i.e. $\ell \equiv \{\vec{p}, m_z\}$. Clearly, in the absence of an external field, the energy is purely kinetic and depends only on $p = |\vec{p}|$. For each value of $\vec{p}$ there are $(2s+1)$ degenerate states (same energy) which we denote by ε_p. The sum over ℓ then becomes

$$\sum_\ell = \sum_{\vec{p}, m_z} = (2s+1)\sum_{\vec{p}} \to \frac{(2s+1)V}{h^3}\int {\rm d}^3p$$

The Fermi energy $\varepsilon_{\rm F}$ defines what is called the *Fermi momentum* $p_{\rm F}$: Since energy is an increasing function of momentum p, all states $\{\vec{p}, m_z\}$ such that $p \le p_{\rm F}$ will have a unit occupation number. It is straightforward to calculate the Fermi momentum

$$N = \frac{(2s+1)V}{h^3}\int_{p \le p_{\rm F}} {\rm d}^3p = \frac{(2s+1)V}{h^3}\frac{4\pi}{3}p_{\rm F}^3 \tag{5.16}$$

and using the notation $n = N/V$ for the particle density, this becomes

$$\boxed{p_{\rm F} = \left[\frac{6\pi^2}{(2s+1)}\right]^{1/3} \hbar\, n^{1/3}} \tag{5.17}$$

This equation is valid for the non-relativistic as well as the relativistic regimes. The sphere of radius p_F is called the *Fermi sphere* and its surface is the *Fermi surface*. Equation (5.17) gives the Fermi energy in the non-relativistic limit where $\varepsilon_p = p^2/2m$

$$\boxed{\varepsilon_F = \frac{p_F^2}{2m} = \left[\frac{6\pi^2}{(2s+1)}\right]^{2/3} \frac{\hbar^2}{2m} n^{2/3}} \tag{5.18}$$

The case one usually encounters is that of spin $1/2$ $(2s+1=2)$. The Fermi energy ε_F is the characteristic energy scale for the quantum problem and, as in Chapter 3, it allows us to define a characteristic temperature, called the *Fermi temperature* T_F by $\varepsilon_F = kT_F$. The Fermi temperature gives a scale for comparison: for a temperature $T \gg T_F$ we expect the quantum effects not to play a rôle, whereas for $T \lesssim T_F$ quantum effects are important.

It is useful to give an order of magnitude for a very important example of a Fermi gas, namely that of conduction electrons in metals. Let us consider copper with a mass density of 8.9 g cm^{-3} and atomic mass 63.5. This yields a density of 8.4×10^{22} atoms/cm^3, which is also the density of electrons since copper has one conduction electron per atom. Putting this and $s = 1/2$ in Equation (5.18), we find the Fermi energy $\varepsilon_F = 7.0$ eV which corresponds to the Fermi temperature $T_F \simeq 8 \times 10^4$ K. This is a rather typical scale for metals, ε_F is of the order of a few eV and, consequently, for ordinary temperatures we have $T_F \gg T$. In such a case we say that the Fermi gas is *degenerate*.

Let us now calculate the internal energy and the pressure. From Equation (5.16) with $s = 1/2$, we find for the number of particles

$$N = \frac{V}{\pi^2\hbar^3} \int_0^{p_F} \mathrm{d}p\, p^2 = \frac{V}{3\pi^2\hbar^3}\, p_F^3 \tag{5.19}$$

and for the energy

$$E = \frac{V}{\pi^2\hbar^3} \int_0^{p_F} \mathrm{d}p \left(\frac{p^2}{2m}\right) p^2 = \frac{3}{5} N\varepsilon_F \tag{5.20}$$

Another useful expression is obtained by using Equation (5.18) for ε_F

$$E = \left(3\pi^2\right)^{2/3} \frac{3N\hbar^2}{10m} n^{2/3} \tag{5.21}$$

This equation along with (2.28) gives the expression for the pressure

$$P = \frac{2}{3}\frac{E}{V} = (3\pi^2)^{2/3}\frac{\hbar^2}{5m}n^{5/3} \tag{5.22}$$

The pressure of the Fermi gas is non-zero even at $T = 0$! This result is at variance with what happens with a classical ideal gas. Another important remark is that the average kinetic energy of a particle increases as a power of the density $E/N \propto n^{2/3}$. For an electron gas, the potential energy per particle is of the order of $e^2/d \propto n^{1/3}$, where e is the charge and $d \sim n^{-1/3}$ the average distance between electrons. The denser the Fermi gas, the more the kinetic energy dominates the potential energy or, in other words, the higher the density of a degenerate Fermi gas, the closer it is to an ideal gas! Again, this behaviour is the opposite of the classical gas where the ideal gas approximation is better the smaller the density.

We end with an intuitive picture of the 'Heisenberg–Pauli principle':[3] the order of magnitude of the uncertainty in p is $\Delta p \sim p_{\rm F}$, while the uncertainty in the position x is of the order of $\Delta x \sim V^{1/3}$. Using Equation (5.17) for $p_{\rm F}$ we get

$$\Delta p\, \Delta x \sim \hbar N^{1/3} \tag{5.23}$$

Qualitatively, due to the Pauli exclusion principle, the $\hbar$ in the Heisenberg uncertainty principle is replaced by $\hbar N^{1/3}$.

5.2.2 Ideal Fermi gas at low temperature

By low temperature we mean $T \ll T_{\rm F}$. In this case we shall see that $\mu \neq \varepsilon_{\rm F}$, but that the correction is small so that the condition $T \ll T_{\rm F}$ also gives $kT \ll \mu$. From the discussion in the previous section, the average fermion occupation number (5.8a) in a momentum state $\vec{p}$ is given by

$$\langle n_{\vec{p}}\rangle = \frac{2s+1}{\mathrm{e}^{\beta(\varepsilon_p - \mu)} + 1} = (2s+1) f(\varepsilon_p)$$

This equation defines the *Fermi distribution*, or the *Fermi function*, $f(\varepsilon)$. By setting $x = \beta(\varepsilon - \mu)$, this function is written

$$f(\varepsilon) = \frac{1}{\mathrm{e}^{\beta(\varepsilon-\mu)} + 1} \to f(x) = \frac{1}{\mathrm{e}^x + 1} = \frac{1}{2}\left(1 - \tanh\frac{x}{2}\right) \tag{5.24}$$

The property $\tanh v = -\tanh(-v)$ immediately gives a symmetry property of the function $f(\varepsilon)$ (Figure 5.2). Its slope is given by

$$\frac{\partial f}{\partial \varepsilon} = \frac{\partial x}{\partial \varepsilon}\frac{\partial f}{\partial x} = \frac{1}{kT}\left(-\frac{1}{4\cosh^2(x/2)}\right)$$

[3] See reference [80].

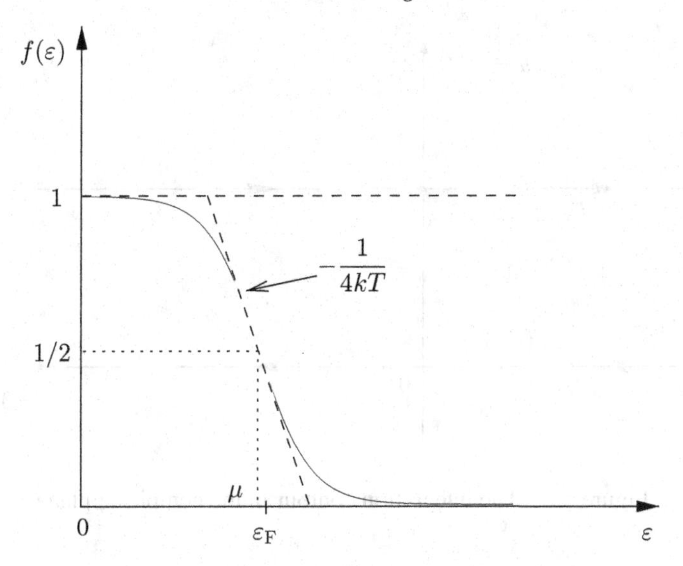

Figure 5.2 The Fermi distribution at $T \neq 0$.

In particular, at the symmetry point $x = 0$, or equivalently $\varepsilon = \mu$, we have

$$\left.\frac{\partial f}{\partial \varepsilon}\right|_{\varepsilon=\mu} = -\frac{1}{4kT} \tag{5.25}$$

The function $f(\varepsilon)$ decreases from a value of 1 to 0 within an interval of $\sim 4kT$. We have, therefore, a 'smoothed θ function' with a width of about $kT \ll \mu \simeq \varepsilon_F$ (Figure 5.2). Equation (5.29), which we now demonstrate, approximates the Fermi function with a Heaviside distribution plus corrections in the form of derivatives of the Dirac distribution. To show this, we need the following integral

$$I_m = \int_{-\infty}^{\infty} \mathrm{d}x \, \frac{x^m \, \mathrm{e}^x}{(\mathrm{e}^x + 1)^2} = \int_{-\infty}^{\infty} \mathrm{d}x \, \frac{x^m}{(\mathrm{e}^x + 1)\left(\mathrm{e}^{-x} + 1\right)} \tag{5.26}$$

To evaluate this integral, we use the following expression

$$J(p) = \int_{-\infty}^{\infty} \mathrm{d}x \, \frac{\mathrm{e}^{\mathrm{i}px}}{(\mathrm{e}^x + 1)\left(\mathrm{e}^{-x} + 1\right)} = \sum_{m=0}^{\infty} \frac{(\mathrm{i}p)^m}{m!} I_m$$

This integral may be calculated with the residue method.[4] By using the contour C in Figure 5.3, and calling the integrand $f(z)$, we can easily show that the integral

$$\int_C \mathrm{d}z \, f(z) = \left(1 - \mathrm{e}^{-2\pi p}\right) J(p)$$

[4] One can also use the integrals I_m given in Section A.5.2. It is also possible to establish (5.29) starting with (5.86).

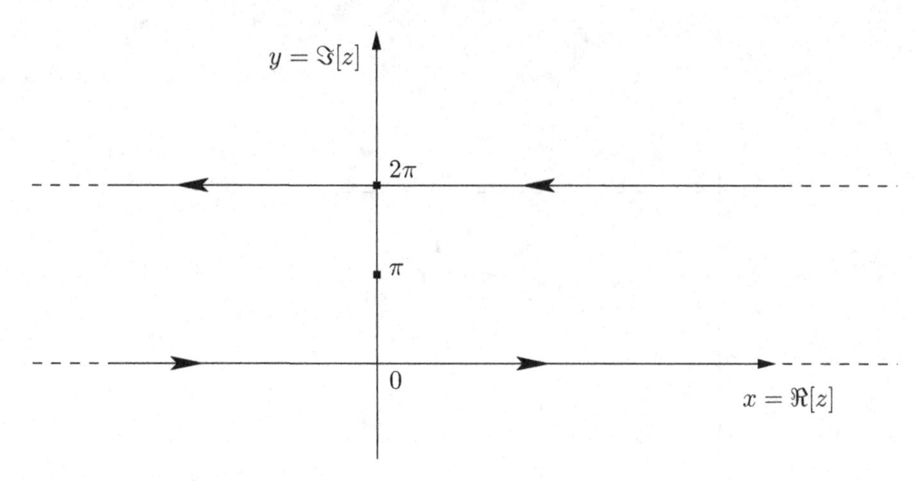

Figure 5.3 The integration contour in the complex z plane.

equals $2i\pi$ times the residue of the pole at $z = i\pi$ which is equal to $-ipe^{-\pi p}$. This gives

$$J(p) = \frac{\pi p}{\sinh \pi p} = 1 - \frac{1}{6}(\pi p)^2 + \mathcal{O}\left((\pi p)^4\right) \tag{5.27}$$

from which we get $I_0 = 1$, $I_2 = \pi^2/3$ etc.
We will now use Equation (5.27) to evaluate integrals of the form

$$I(\beta) = \int_0^\infty d\varepsilon \, \frac{\varphi'(\varepsilon)}{e^{\beta(\varepsilon-\mu)} + 1} = \beta \int_0^\infty d\varepsilon \, \frac{\varphi(\varepsilon) e^{\beta(\varepsilon-\mu)}}{\left(e^{\beta(\varepsilon-\mu)} + 1\right)^2}$$

where $\varphi(\varepsilon)$ satisfies $\varphi(0) = 0$, and where the second integral was obtained by integrating by parts. Since we assume $\beta\mu \gg 1$, we may extend the limits of the integral from $-\infty$ to $+\infty$ with an error of the order of $\exp(-\beta\mu)$. Finally, expand $\varphi(\varepsilon)$ in a Taylor series near $\varepsilon = \mu$

$$\begin{aligned}
I(\beta) &= \beta \sum_{m=0}^{\infty} \frac{\varphi^{(m)}(\mu)}{m!} \int_{-\infty}^{\infty} d\varepsilon \, \frac{(\varepsilon - \mu)^m e^{\beta(\varepsilon-\mu)}}{\left(e^{\beta(\varepsilon-\mu)} + 1\right)^2} \\
&= \sum_{m=0}^{\infty} \frac{\varphi^{(m)}(\mu)}{m!} \beta^{-m} \int_{-\infty}^{\infty} dx \, \frac{x^m e^x}{(e^x + 1)^2} \\
&= \sum_{m=0}^{\infty} \frac{\varphi^{(m)}(\mu)}{m!} \beta^{-m} I_m \\
&= \varphi(\mu) + \frac{\pi^2}{6}(kT)^2 \varphi''(\mu) + \mathcal{O}\left((kT/\mu)^4\right)
\end{aligned} \tag{5.28}$$

This expansion leads to the following expression for the Fermi distribution

$$\boxed{f(\varepsilon) = \frac{1}{e^{\beta(\varepsilon-\mu)}+1} = \theta(\mu-\varepsilon) - \frac{\pi^2}{6}(kT)^2\delta'(\varepsilon-\mu) + \mathcal{O}\Big((kT)^4\delta^{(3)}(\varepsilon-\mu)\Big)} \tag{5.29}$$

This is the *Sommerfeld formula*, which is a good approximation of the Fermi function $f(\varepsilon)$ in the form of a 'smoothed θ function'. As we saw, this was obtained by using an expansion in terms of the small parameter $(kT/\mu)^2$ or, equivalently, (kT/ε_F) since $\mu \simeq \varepsilon_F$.

The derivation which follows emphasizes that the thermodynamic properties of the Fermi gas at low temperatures $T \ll T_F$ *are governed exclusively by the neighbourhood of the Fermi surface*. Moreover this derivation may be easily adapted to the case of an interacting Fermi gas (see Section 5.2.3 and Problem 8.6.7). Let us use the Sommerfeld formula to evaluate, when $T \ll T_F$ and at constant V and N, the derivative with respect to T of a physical quantity, $\Phi(T)$, defined as the average value of a function $\varphi(\varepsilon)$

$$\Phi(T) = \int_0^\infty d\varepsilon\, f(\varepsilon)\rho(\varepsilon)\varphi(\varepsilon)$$

where $\rho(\varepsilon)$ is the level density. Substituting Equation (5.29) in the above expression for $\Phi(T)$ yields

$$\Phi(T) \simeq \int_0^\mu d\varepsilon\, \rho(\varepsilon)\varphi(\varepsilon) + \frac{\pi^2}{6}k^2T^2[\rho'(\mu)\varphi(\mu) + \rho(\mu)\varphi'(\mu)]$$

which allows us to evaluate $(\partial\Phi/\partial T)_{V,N}$

$$\left.\frac{\partial\Phi}{\partial T}\right|_{V,N} \simeq \frac{\partial\mu}{\partial T}\rho(\varepsilon_F)\varphi(\varepsilon_F) + \frac{\pi^2}{3}k^2T\rho'(\varepsilon_F)\varphi(\varepsilon_F) + \frac{\pi^2}{3}k^2T\rho(\varepsilon_F)\varphi'(\varepsilon_F) \tag{5.30}$$

The replacement of μ by ε_F is allowed because the error is of higher order in the small parameter $(kT/\mu)^2 \simeq (kT/\varepsilon_F)^2$. Let us first examine the case $\varphi(\varepsilon) = 1$, $\varphi'(\varepsilon) = 0$, $\Phi(T) = N$, $(\partial N/\partial T)_{V,N} = 0$, for which Equation (5.30) yields

$$\left.\frac{\partial\mu}{\partial T}\right|_{V,N} \simeq -\frac{\pi^2}{3}k^2T\frac{\rho'(\varepsilon_F)}{\rho(\varepsilon_F)} = -\frac{\pi^2}{6}\frac{k^2T}{\varepsilon_F}$$

The first expression for $\partial\mu/\partial T$ may be adapted to an interacting Fermi gas (Problem 8.6.7) while in the second we have used the explicit expression for the density of states for the *ideal* spin 1/2 Fermi gas in three dimensions[5]

$$\rho(\varepsilon) = \frac{8\pi V}{h^3}\, p^2 \frac{\mathrm{d}p}{\mathrm{d}\varepsilon} = 4\pi V \left(\frac{2m}{h^2}\right)^{3/2} \sqrt{\varepsilon} = AV\sqrt{\varepsilon}$$

We now obtain μ by integration subject to the boundary condition $\mu(T=0) = \varepsilon_{\rm F}$

$$\mu(T) \simeq \varepsilon_{\rm F}\left(1 - \frac{\pi^2}{12}\left(\frac{kT}{\varepsilon_{\rm F}}\right)^2\right) \tag{5.31}$$

This result shows that for $T \neq 0$ the chemical potential is no longer equal to the Fermi energy, and that the relative difference is of the order of $(kT/\varepsilon_{\rm F})^2$. Since the first two terms in (5.30) cancel out, we obtain a formula easily generalized to the interacting case

$$\left.\frac{\partial \Phi}{\partial T}\right|_{V,N} \simeq \frac{\pi^2}{3}\, k^2 T \varphi'(\varepsilon_{\rm F})\rho(\varepsilon_{\rm F}) \tag{5.32}$$

The derivative $(\partial\Phi/\partial T)_{V,N}$ is proportional to the density of states at the Fermi level $\rho(\varepsilon_{\rm F})$: *it depends only on the neighbourhood of the Fermi surface* through $\varphi'(\varepsilon_{\rm F})$ and $\rho(\varepsilon_{\rm F})$. On the contrary, $\Phi(T)$ depends on all the energies $\varepsilon \lesssim \varepsilon_{\rm F}$, and consequently requires a much more detailed knowledge of the Fermi gas. Of course, we can calculate easily $\Phi(T)$ for a free Fermi gas, but our demonstration allows us to control $(\partial\Phi/\partial T)_{V,N}$, even in the presence of interactions, as long as we know the neighbourhood of the Fermi surface. As an important application of Equation (5.32), we calculate the specific heat C_V by taking $\varphi(\varepsilon) = \varepsilon$, $\Phi(T) = E$, $(\partial\Phi/\partial T)_{V,N} = (\partial E/\partial T)_{V,N} = C_V$. This gives

$$\boxed{C_V = \frac{\pi^2 k^2 T}{3}\rho(\varepsilon_{\rm F}) = \frac{Vk^2T}{3\hbar^3}\, m p_{\rm F} = N\frac{\pi^2 k^2 T}{2\varepsilon_{\rm F}}} \tag{5.33}$$

In order to go from the first to the second expression for C_V, we have used

$$\rho(\varepsilon_{\rm F}) = \frac{8\pi V}{h^3}\, p_{\rm F}^2 \left.\frac{\mathrm{d}p}{\mathrm{d}\varepsilon}\right|_{p=p_{\rm F}} = \frac{Vmp_{\rm F}}{\pi^2\hbar^3}$$

and to go from the second to the third, we have used Equation (5.17) which gives $p_{\rm F}$ as a function of the density. Equation (5.33) shows that the specific heat is proportional to T and therefore vanishes at $T = 0$ as required by the third law. We can interpret physically the first of Equations (5.33) if we consider conduction electrons in a metal, as an example to fix ideas. Then, only the electrons lying within kT of the Fermi surface can be thermally excited since the other electrons,

[5] In two dimensions, $\rho(\varepsilon) \propto$ const, $\rho'(\varepsilon) = 0$. It is then easy to show that there are no corrections to μ in powers of $(kT/\varepsilon_{\rm F})^2$. See Exercise 5.6.3.

deeper in the *Fermi sea*, are blocked from moving up by the Pauli exclusion principle. An energy of the order of kT would only allow these deeper electrons to reach already occupied levels. The number of electrons that can be thermally excited is of the order of $\Delta N \simeq kT\rho(\varepsilon_F)$, and for these electrons the equipartition theorem is approximately valid

$$C_V \sim k\Delta N \simeq k(kT)\rho(\varepsilon_F)$$

The factor $\pi^2/3$ (see (5.33)) can only be obtained from the exact calculation. A similar argument leads to the following approximate form for the third expression for C_V: $\Delta N/N \sim kT/\varepsilon_F$, whereas the second expression shows the mass dependence since the Fermi momentum p_F depends only on the density n according to (5.17).

5.2.3 Corrections to the ideal Fermi gas

We have just calculated the contribution of conduction electrons ($\propto T$) to the specific heat in a metal at low temperature. There are several other contributions to C_V. For example, we should add to this the contribution of lattice vibrations, which we shall find in Section 5.3 to be proportional to T^3. In a ferromagnet (iron, cobalt, etc.) the contribution of spin waves is proportional to $T^{3/2}$ (Exercise 5.6.6), and there could be even more contributions at very low temperature, e.g. nuclear. For a non-ferromagnetic metal, it is convenient to examine C_V/T as a function of T^2. This gives the straight line

$$\frac{C_V}{T} = \gamma + AT^2$$

which permits the easy determination of the parameters γ and A (Figure 5.4). The theoretical value of γ (per conduction electron) is given by (5.33): $\gamma_{\text{th}} = \pi^2k^2/(2\varepsilon_F)$.

Table 5.1 shows that the ratio $\gamma_{\text{exp}}/\gamma_{\text{th}}$ can be quite different from 1, which leads us to question whether the ideal Fermi gas can indeed model correctly the behaviour of the conduction electrons in a metal. The implicit assumption we have

Table 5.1 *Comparison of the experimental and theoretical values of the contribution to the specific heat of conduction electrons in a metal.*

	Li	Na	Cu	Ag	Au
$\gamma_{\text{exp}}/\gamma_{\text{th}}$	2.3	1.3	1.3	1.1	1.1

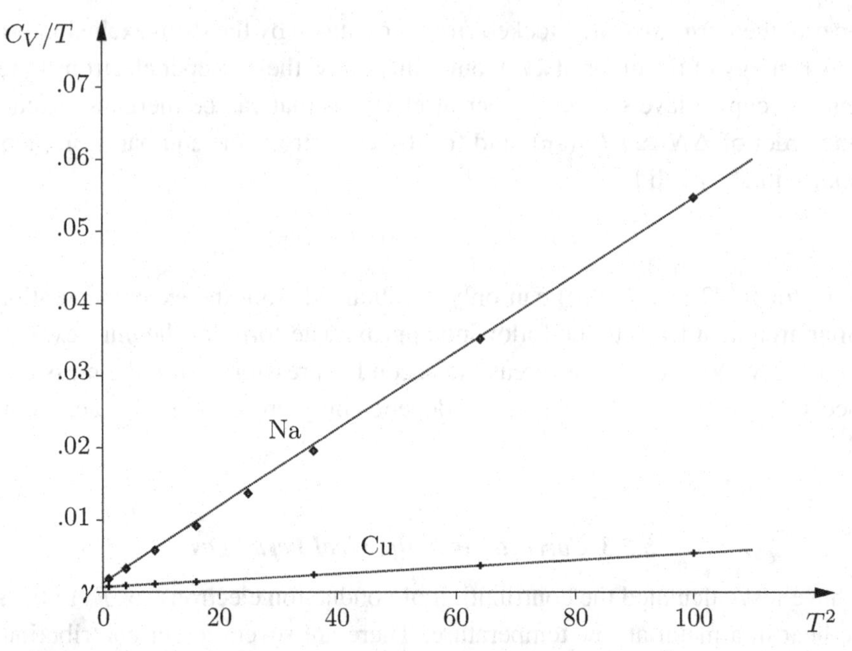

Figure 5.4 The straight lines giving C_V/T as a function of T^2 for sodium and copper. Note the very small contribution of the conduction electrons.

made is that the interactions of the electrons with the periodic potential of the ions and the phonons, as well as the interactions among the electrons themselves, are small and give corrections not exceeding the order of 10 to 20%. Table 5.1 shows that these 'corrections' can be as big as a factor of 2 or more, and thus cannot be considered small.

In order to clarify the physics of the Fermi gas with interactions, let us start with liquid helium-3 (Problem 5.7.6) which is *a priori* simpler than the case of electrons in a metal. The helium-3 atoms are fermions, as are electrons, but they are not in a periodic potential and the only interactions to take into account are interactions between the atoms themselves. For temperatures between that of the superfluid transition (around 3 mK) and 100 mK, helium-3 is well described by the 'Fermi liquid[6] theory of Landau' which postulates that the spectrum of energy levels in the presence of interactions is qualitatively the same as for the free theory (see Problem 8.6.7). In particular, the Fermi surface is assumed to remain well defined at $T = 0$ in spite of the interactions. The interactions are included by replacing the particles with quasi-particles whose properties and effective interactions are different from those of the initial particles.[7] Let us consider the calculation of the

[6] We will use interchangeably the terms 'Fermi liquid' and 'Fermi gas'.

[7] In the most favourable cases, the interactions between quasi-particles are weak but this is not always so. For helium-3 the interactions between the quasi-particles are very important.

specific heat. The reasoning that led to Equation (5.32) remains valid[8] if we replace the energy $\varepsilon_{\rm F} = p_{\rm F}^2/(2m)$ by a Fermi energy $\varepsilon_0 = \mu(T = 0)$ and at the same time introduce an effective mass m^* defined *only* on the Fermi surface by

$$\left.\frac{\mathrm{d}\varepsilon}{\mathrm{d}p}\right|_{p=p_{\rm F}} = \frac{p_{\rm F}}{m^*}$$

The density of states $\rho(\varepsilon_0)$ then becomes

$$\rho(\varepsilon_0) = \frac{8\pi V}{h^3}\, p_{\rm F}^2 \left.\frac{\mathrm{d}p}{\mathrm{d}\varepsilon}\right|_{p=p_{\rm F}} = \frac{V m^* p_{\rm F}}{\pi^2 \hbar^3}$$

It is enough to replace m with m^* in the expression for the density of states of an ideal gas. It follows from this that the second expression for C_V in (5.33) becomes

$$C_V = \frac{V k^2 T}{3\hbar^3}\, m^* p_{\rm F}$$

We see that C_V is still proportional to T. However, the coefficient of T in Equation (5.33), which is specific to an ideal gas, is multiplied by m^*/m. For helium-3, m^*/m takes the values $3 \lesssim m^*/m \lesssim 5$ as the pressure changes from zero to about 30 atm (the solidification pressure), and consequently the correction factor is not small! However, the linear dependence of C_V on T is well satisfied for $T \lesssim 100$ mK.[9] The success of the Landau theory is due to three properties: (i) the number of quasi-particles is small when $T \to 0$, (ii) because of the Pauli principle, it can be shown that the quasi-particles have an average lifetime that diverges when their energy is close to ε_0 and $T \to 0$, which makes them very well-defined objects, (iii) the physics of interest is controlled by the immediate neighbourhood of the Fermi surface.

By way of application, let us calculate the correction to Equation (5.31) for the chemical potential. The specific heat, C_V, gives the entropy, which in turn gives the free energy using $\partial F/\partial T = -S$. The chemical potential is then obtained by differentiating the free energy density, $f = F/V$, with respect to the density n

$$\mu = \left.\frac{\partial f}{\partial n}\right|_T = \varepsilon_0 - \frac{\pi^2 k^2 T^2}{12\varepsilon_{\rm F}}\,\frac{m^*}{m}\left(1 + \frac{3n}{m^*}\frac{\partial m^*}{\partial n}\right)$$

The effective mass is not the only parameter in the Landau theory. For example, the compressibility and the magnetic susceptibility depend on additional parameters

[8] It can be shown that interactions between the quasi-particles can be neglected in the calculation of C_V, see Problem 8.6.7.

[9] It might appear that this temperature is very low, but, since $T_{\rm F} \sim 5$ K for helium-3, 100 mK corresponds to a ratio $T/T_{\rm F} \simeq 2 \times 10^{-2}$. This is comparable to the ratio $T/T_{\rm F}$ for an electron gas in a metal at room temperature.

that describe the interactions between quasi-particles and are defined on the Fermi surface.

The case of electrons in metals is in principle more complicated due to (i) the interactions between the electrons and the periodic potential and the phonons, and (ii) the long range nature of the Coulomb interactions between the electrons. Despite these difficulties, the model of quasi-free electrons gives good results if we use, for the free electrons, an effective mass that approximates the combined effects of electron–lattice and electron–electron interactions. However, the Landau theory does not hold everywhere. In the case of a superconducting metal or superfluid helium-3, the energy spectrum has nothing to do with that of a free Fermi gas. An arbitrarily weak attractive interaction leads to a superfluid state for $T \to 0$. Another example comes from systems of strongly correlated electrons. The fractional Hall effect (see Problem 8.6.5 for an introduction to the quantum Hall effect) exhibits effective particles with fractional electric charge $1/3$, $1/5$, etc. of the electronic charge. High temperature superconductors also seem to defy the Landau theory. Therefore, the success of the quasi-free electron model in describing their properties in metals is not obvious *a priori*.

5.3 Black body radiation

The study of the bosonic ideal gas follows in a natural way that of the fermionic one. Because bosons are not subject to such a strong constraint as the Pauli principle, one might suspect their behaviour to be not as rich as that of fermions. The diversity of subjects we shall discuss below will show this not to be the case at all. We need to distinguish between two cases: (i) bosonic gas with fluctuating number of particles and (ii) bosonic gas with fixed number of bosons. The former is encountered, for example, in the study of the 'cosmic microwave background radiation' as well as the specific heat in solids, while the latter is found in the phenomenon of Bose–Einstein condensation, treated in Section 5.5.

5.3.1 Electromagnetic radiation in thermal equilibrium

Consider an empty cavity of volume V at temperature T. The walls of this cavity continually emit and absorb electromagnetic radiation or, in quantum mechanical terms, photons which are in thermal equilibrium with the walls. We thus have a photon gas in the cavity. Photons are bosonic particles whose number is not constrained by a conservation law and, consequently, they have zero chemical potential (Section 5.1.3). According to Equation (5.8b) with $z = 1$, the average number of photons in the momentum state $\vec{p}$ is given by

$$\langle n_{\vec{p}} \rangle = \frac{2}{e^{\beta \varepsilon_p} - 1} \qquad \varepsilon_p = c|\vec{p}| = cp \tag{5.34}$$

The factor 2 comes from the two possible independent polarizations of the photon.[10] This yields the energy, $E(V, T)$, which depends only on the temperature and the volume

$$E(V,T) = \sum_{\vec{p}} \langle n_{\vec{p}} \rangle \varepsilon_p = \frac{2V}{h^3} \int \mathrm{d}^3 p \, \frac{cp}{\mathrm{e}^{\beta cp} - 1}$$

It is convenient to introduce the dimensionless variable $v = \beta cp$ to obtain the energy density[11] $\epsilon(T) = E(V, T)/V$

$$\epsilon(T) = \frac{(kT)^4}{\pi^2 \hbar^3 c^3} \int_0^\infty \mathrm{d}v \, \frac{v^3}{\mathrm{e}^v - 1} = \frac{\pi^2 (kT)^4}{15 \hbar^3 c^3} \tag{5.35}$$

where we have used the integral (A.53). From relation (2.29) between pressure and energy, we have for an ultra-relativistic gas $P = \epsilon/3$, which we use to obtain

$$\boxed{\epsilon(T) = \sigma' T^4 \qquad \sigma' = \frac{\pi^2 k^4}{15 \hbar^3 c^3} \qquad P = \frac{1}{3} \sigma' T^4} \tag{5.36}$$

The pressure and energy of a photon gas are proportional to T^4. One often uses for photons the frequency rather than the momentum or energy. So, if we write the energy density $\epsilon(T)$ as an integral over ω of a density, $\epsilon(\omega, T)$, per unit frequency and volume,

$$\epsilon(T) = \int_0^\infty \mathrm{d}\omega \, \epsilon(\omega, T)$$

we obtain from (5.35)

$$\boxed{\epsilon(\omega, T) = \frac{\hbar}{\pi^2 c^3} \frac{\omega^3}{\mathrm{e}^{\beta \hbar \omega} - 1}} \tag{5.37}$$

This equation was first written by Planck in 1900 and is known as 'Planck's black body radiation law' (Figure 5.5). This terminology will be justified in Section 5.3.2.

It is now instructive to appeal to dimensional analysis. The only quantities available to construct $\epsilon(T)$ (dimension $ML^{-1}T^{-2}$) are $\hbar$ (dimension ML^2T^{-1}),

[10] The photon is a spin-1 particle and one therefore expects three polarization states. However, the photon has zero mass and for such a particle with spin s, m_z can take only two values, $m_z = \pm s$ with the quantization axis taken in the direction of the momentum. When the interactions of a massless particle (fermion or boson) do not conserve parity, only one of these two projections is allowed. Thus, at least in the old picture (before 1999), the neutrino exists only in the state $m_z = -1/2$ while the antineutrino has $m_z = 1/2$. However, there is today mounting evidence that the neutrino is massive and this simple picture is spoiled. The spin properties of massless particles were analyzed by Wigner.

[11] One should not confuse the energy density ϵ with ε which represents an individual energy.

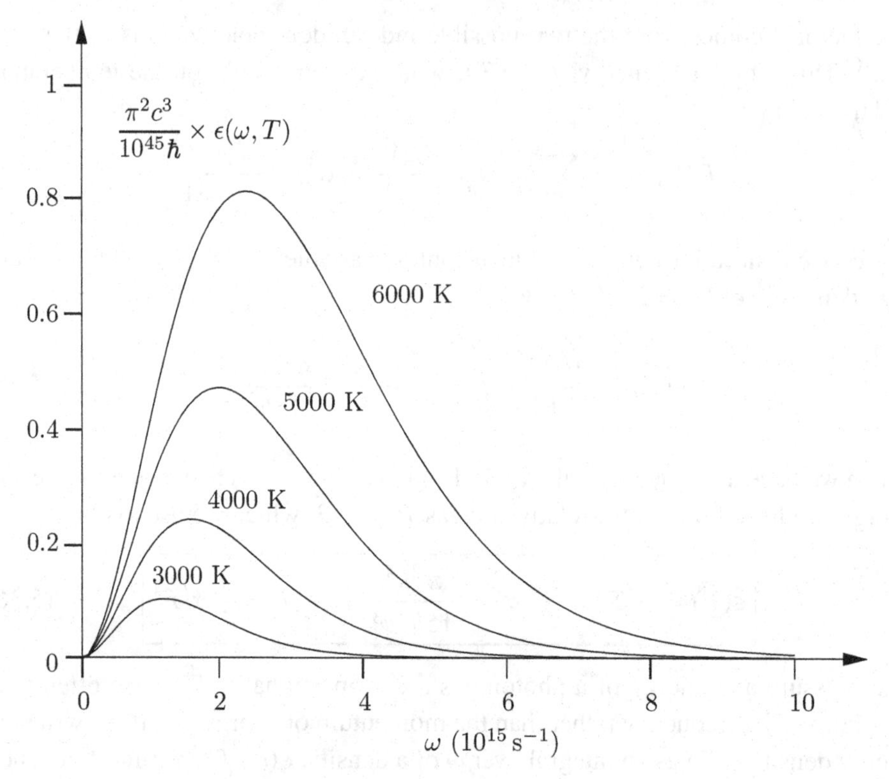

Figure 5.5 The spectral density $\epsilon(\omega, T)$.

$c\,(LT^{-1})$ and $kT(ML^2T^{-2})$. Only one solution is possible; it has the form $\epsilon(T) = A\hbar^{-3}c^{-3}(kT)^4$, where A is a numerical coefficient given explicitly in (5.36). We remark that, without Planck's constant, it is simply impossible to construct $\epsilon(T)$! Now let us consider $\epsilon(\omega, T)$ which has the dimension $ML^{-1}T^{-1}$. Before the discovery of quantum mechanics, c, ω and kT were available to construct $\epsilon(\omega, T)$. The only possible solution was then $\epsilon(\omega, T) = Bc^{-3}(kT)\omega^2$, where B is a dimensionless constant and this is indeed what we obtain from Equation (5.37) in the classical limit $\hbar\omega \ll kT$. But with such an expression for $\epsilon(\omega, T)$, the integral which gives $\epsilon(T)$ does not converge for $\omega \to \infty$: once again we find that we cannot construct $\epsilon(T)$ without $\hbar$. This simple argument shows that the theory of electromagnetic radiation at thermal equilibrium cannot exist in classical physics. It was this difficulty that led Planck in 1900 to introduce the constant that carries his name. We see in (5.37) that quantum mechanics introduces a convergence factor $\exp(-\beta\hbar\omega)$ in the calculation of $\epsilon(T)$ when $\omega \to \infty$.

The most famous example of black body radiation is the 'microwave background radiation' that fills the universe. In fact, the frequency distribution of this radiation follows Planck's law (5.37) with an effective temperature of 3 K, but it

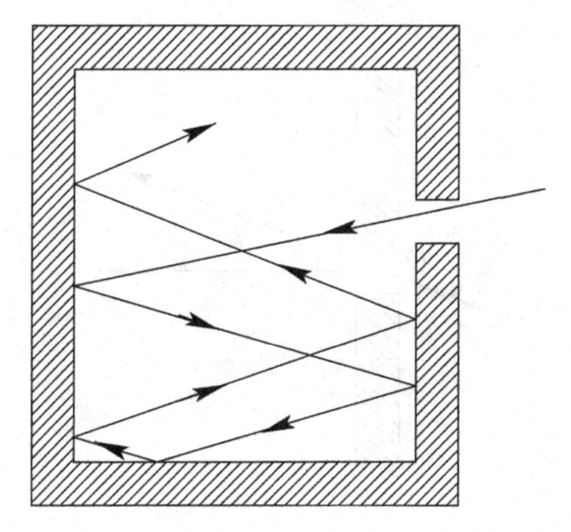

Figure 5.6 Experimental realization of a black body.

is no longer at thermal equilibrium. It is thought to have decoupled from matter about 500 000 years after the big bang, i.e. the birth of the universe. When this decoupling took place, the temperature was about 10^4 K, the continuing expansion of the universe has since reduced it to the current value of 3 K.

5.3.2 Black body radiation

A *black body* is a body which absorbs 100% of incident electromagnetic radiation. A good realization of a black body for electromagnetic radiation in the visible range is a small hole in a box whose interior is covered with carbon black (Figure 5.6). At every collision with a wall, the incident photon has a good chance of being absorbed and the probability of it exiting the box is almost zero. The interior of the box contains electromagnetic radiation in equilibrium with the walls at temperature T. Let us calculate the energy (per unit time and unit area) emitted by the hole, which, by definition, is the energy flux $\Phi(T)$ of black body radiation. We need $f(\vec{p})$, the number of photons per unit volume in $\mathrm{d}^3 p$, which is given by (5.34)

$$f(\vec{p}) = \frac{2}{h^3} \frac{1}{\mathrm{e}^{\beta c p} - 1} \tag{5.38}$$

Since $f(\vec{p}) = f(p)$, the energy density $\epsilon(T)$ becomes

$$\epsilon(T) = \int \mathrm{d}^3 p \, f(\vec{p}) pc = 4\pi c \int_0^\infty \mathrm{d}p \, f(p) p^3 \tag{5.39}$$

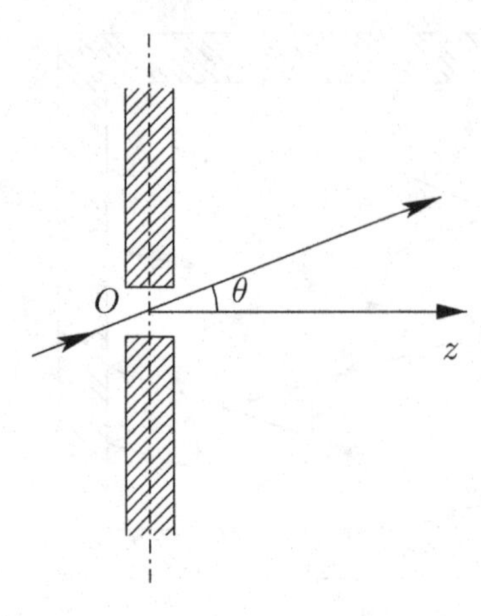

Figure 5.7 Convention for the axis in the calculation for black body radiation.

To fix ideas, we take the z axis, Oz, perpendicular to the opening. The energy flux for a photon of energy $\varepsilon_p = cp$ is (Figure 5.7)

$$\varepsilon_p v_z = c^2 p \cos\theta$$

where θ is the angle between the photon momentum $\vec{p}$ and Oz. Keeping in mind that escaping photons have $v_z > 0$, we get for the total flux $\Phi(T)$

$$\Phi(T) = \int_{p_z>0} \mathrm{d}^3 p\, f(\vec{p}) c^2 p \cos\theta$$

$$= \left[\int_0^\infty \mathrm{d}p\, c^2 p f(p) p^2\right]\left[2\pi \int_0^{\pi/2} \mathrm{d}\theta\, \cos\theta \sin\theta\right]$$

Comparing this with Equation (5.39), and noting that the value of the integral in the second bracket is $1/2$, gives

$$\boxed{\Phi(T) = \frac{c}{4}\epsilon(T) = \sigma T^4 \qquad \sigma = \frac{\pi^2 k^4}{60\hbar^3 c^2}} \tag{5.40}$$

The constant σ is called the Stefan–Boltzmann constant and, in the MKS system, has the value

$$\sigma = 5.67 \times 10^{-8}\ \text{watt m}^{-2}\,\text{K}^{-4}$$

The energy flux per unit frequency is proportional to

$$\frac{\omega^3}{\mathrm{e}^{\hbar\omega/kT} - 1} \propto \frac{v^3}{\mathrm{e}^v - 1}$$

with $v = \hbar\omega/(kT)$. The curve for the flux as a function of ω has a maximum at $v_m = 2.82$, which can also be written as $\omega_m = 2.82(kT)/\hbar$. For two temperatures T_1 and T_2, the two maxima are related by

$$\frac{\hbar\omega_{1m}}{kT_1} = \frac{\hbar\omega_{2m}}{kT_2}$$

This is called the 'Wien displacement law', which shows that the position of the maximum of the curve is proportional to the temperature, while the value of the maximum is proportional to T^3 (Figure 5.5). If we observe a hot object which can be approximated by a black body, then by measuring the flux as a function of ω and determining the maximum, we can determine its temperature. In this way, we can measure the surface temperature of the Sun or of a star. If we also know the *luminosity* L of the star, i.e. the luminous power it emits, then using Planck's law (5.40) we can determine its radius R

$$L = 4\pi R^2 \sigma T_s^4 \tag{5.41}$$

For the Sun, we have an independent determination of R which allows us to check the validity of this method. In practice, however, we are not dealing here with a true black body and some corrections must be included.

5.4 Debye model

5.4.1 Simple model of vibrations in solids

Another important application of the case where the number of bosons is not fixed (recall that in this case the chemical potential $\mu = 0$) concerns the thermodynamics of vibrations in solids. To deal with this problem it is instructive to consider a simple one-dimensional model of vibrations. Let us consider a chain of identical springs whose ends are attached to particles of mass m, which can move without friction along the x-axis (Figure 5.8). Let x_n be the position of the nth mass, $\overline{x}_n$ its equilibrium position and $q_n = x_n - \overline{x}_n$. For convenience, we use periodic boundary conditions, $x_n = x_{n+N}$, where N is the total number of masses (or springs), n ranges from 0 to $N-1$ and $\overline{x}_n = na$, where a is the distance between two consecutive masses at equilibrium, i.e. the lattice spacing.
The Hamiltonian of our system is then

$$H = \sum_{n=0}^{N-1} \frac{p_n^2}{2m} + \frac{1}{2} K \sum_{n=0}^{N-1} (q_{n+1} - q_n)^2 \tag{5.42}$$

where the momentum p_n of a mass m is $p_n = m\dot{q}_n$ and K is the spring constant. The first term in Equation (5.42) is the kinetic energy term while the second is the potential energy. To diagonalize this Hamiltonian, we find the normal modes q_k

$$q_k = \frac{1}{\sqrt{N}} \sum_{n=0}^{N-1} e^{ik\overline{x}_n} q_n = \sum_n U_{kn} q_n \qquad k = j \times \frac{2\pi}{Na} \qquad j = 0, \ldots, N-1 \tag{5.43}$$

The subscripts k and n allow us to distinguish between normal modes (k) and position (n). The U_{kn} defined by Equation (5.43) are the matrix elements of the discrete (or lattice) Fourier transform. The matrix U is unitary

$$\begin{aligned}\sum_n U_{kn} U^\dagger_{nk'} &= \frac{1}{N} \sum_n e^{ik\overline{x}_n} e^{-ik'\overline{x}_n} = \frac{1}{N} \sum_n \exp\left[\frac{2i\pi}{Na}(k-k')\overline{x}_n\right] \\ &= \frac{1}{N} \frac{1-\exp(2i\pi(k-k'))}{1-\exp(2i\pi(k-k')/N)} = \delta_{kk'}\end{aligned}$$

By noting that $U^\dagger_{nk} = U^*_{kn} = U_{-kn}$, the above equation becomes

$$\sum_n U_{kn} U^\dagger_{nk'} = \sum_n U_{kn} U_{-k'n} = \delta_{kk'} \tag{5.44}$$

The unitarity of U_{kn} allows us to write the inverse Fourier transform[12]

$$q_n = \frac{1}{\sqrt{N}} \sum_{k=-\pi/a}^{\pi/a} e^{-ik\overline{x}_n} q_k = \sum_k U^\dagger_{nk} q_k = \sum_k U_{-kn} q_k \tag{5.45}$$

The Fourier transform (5.43) and its inverse (5.45) apply equally well to the momentum; it suffices to make the substitution, for example in (5.43), $q_n \to p_n$ and $q_k \to p_k$. We can, therefore, find the diagonal form of the Hamiltonian by using (5.44), (5.45) and the analogue of (5.45) for the momentum. Let us first examine the kinetic term in (5.42)

$$\sum_n p_n^2 = \sum_n \sum_{k,k'} U_{-kn} U_{-k'n} p_k p_{k'} = \sum_{k,k'} \delta_{k\,-k'} p_k p_{k'} = \sum_k p_k p_{-k}$$

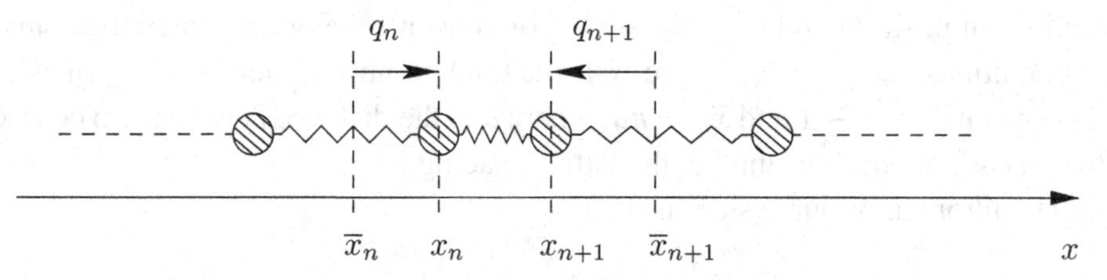

Figure 5.8 A model for vibrations in a solid.

[12] We are considering situations with $N \gg 1$. It is convenient to take k in the first Brillouin zone: $k = -\frac{\pi}{a}, \ldots, \frac{\pi}{a}$, and to ignore boundary effects.

This is nothing more than Parseval's theorem. Now we transform the potential energy term,

$$\sum_n (q_{n+1} - q_n)^2 = \sum_n \sum_{k,k'} \left(U_{-k,\,n+1} - U_{-k,\,n}\right)\left(U_{-k',\,n+1} - U_{-k',\,n}\right) q_k q_{k'}$$
$$= \sum_n \sum_{k,k'} \left(\mathrm{e}^{-\mathrm{i}ka} - 1\right)\left(\mathrm{e}^{-\mathrm{i}k'a} - 1\right) U_{-k,n} U_{-k',n} q_k q_{k'}$$
$$= \sum_k \left(\mathrm{e}^{-\mathrm{i}ka} - 1\right)\left(\mathrm{e}^{\mathrm{i}ka} - 1\right) q_k q_{-k} = 4 \sum_k \sin^2\left(\frac{ka}{2}\right) q_k q_{-k}$$

Combining these two equations we express the Hamiltonian in the following form

$$H = \sum_k \frac{p_k p_{-k}}{2m} + \frac{1}{2} K \sum_k 4 \sin^2\left(\frac{ka}{2}\right) q_k q_{-k} = \sum_k \frac{p_k p_{-k}}{2m} + \frac{1}{2} m \sum_k \omega_k^2 q_k q_{-k} \tag{5.46}$$

where we have defined the normal mode frequency of mode k by

$$\omega_k = 2\sqrt{\frac{K}{m}} \sin\frac{|ka|}{2} \tag{5.47}$$

Equation (5.47), which gives ω_k in terms of k, is called a *dispersion law* for the normal modes (Figure 5.9). Equation (5.47) was obtained in the framework of classical physics, but it can be generalized without modification to the quantum case where q_n and p_n are operators. We now consider the quantum case $q_n \to \mathsf{q}_n$ and $p_n \to \mathsf{p}_n$.

To diagonalize H, we still need to decouple the modes k and $-k$. To achieve this, we define respectively the *annihilation and creation operators* a_k *and* $\mathsf{a}_k^\dagger$ for the

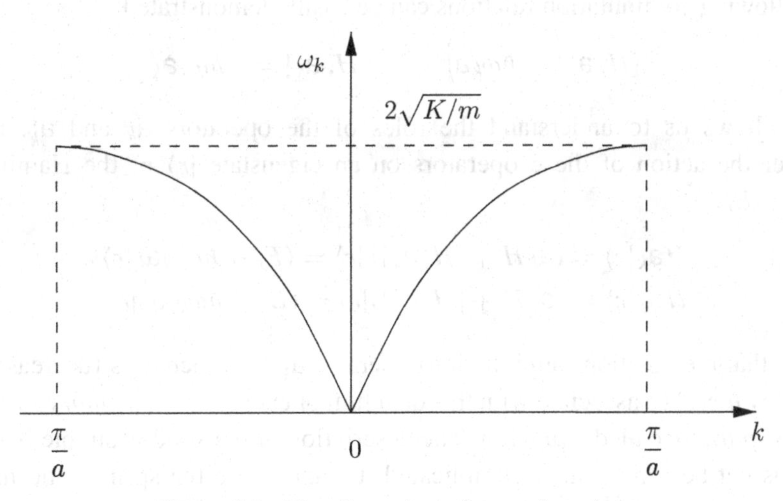

Figure 5.9 The dispersion of normal modes.

harmonic oscillator. Let us first derive the canonical commutation relations for q_k and $\mathsf{p}_{k'}$ starting with $[\mathsf{q}_n, \mathsf{p}_{n'}] = \mathrm{i}\hbar\delta_{nn'}$

$$[\mathsf{q}_k, \mathsf{p}_{k'}] = \sum_{nn'} U_{kn} U_{k'n'} [\mathsf{q}_n, \mathsf{p}_{n'}] = \mathrm{i}\hbar \sum_n U_{kn} U_{k'n} = \mathrm{i}\hbar\delta_{k\,-k'} \tag{5.48}$$

where we have used (5.44). We now define the operators a_k and $\mathsf{a}_k^\dagger$ by

$$\mathsf{q}_k = \sqrt{\frac{\hbar}{2m\omega_k}}\left(\mathsf{a}_k + \mathsf{a}^\dagger_{-k}\right) \qquad \mathsf{p}_k = \frac{1}{\mathrm{i}}\sqrt{\frac{\hbar m\omega_k}{2}}\left(\mathsf{a}_k - \mathsf{a}^\dagger_{-k}\right) \tag{5.49}$$

It is easy to show that the commutation relations (5.48) are satisfied if we have

$$[\mathsf{a}_k, \mathsf{a}_{k'}^\dagger] = \delta_{kk'} \tag{5.50}$$

Note that the RHS of (5.48) has $\delta_{k\,-k'}$ while (5.50) has $\delta_{kk'}$. By substituting the Equations (5.49) in Equation (5.46) for H and using the commutation relations (5.50), we get

$$H = \sum_{k=-\pi/a}^{\pi/a} (\mathsf{a}_k^\dagger \mathsf{a}_k + \mathsf{a}^\dagger_{-k}\mathsf{a}_{-k} + 1)\frac{\hbar\omega_k}{2}$$

By noting that $\sum_k \mathsf{a}_k^\dagger \mathsf{a}_k = \sum_k \mathsf{a}^\dagger_{-k}\mathsf{a}_{-k}$, we can put the Hamiltonian in the form of a sum over independent harmonic oscillators

$$\boxed{H = \sum_{k=-\pi/a}^{\pi/a} \left(\mathsf{a}_k^\dagger \mathsf{a}_k + \frac{1}{2}\right)\hbar\omega_k} \tag{5.51}$$

The following commutation relations can be easily demonstrated

$$[H, \mathsf{a}_k^\dagger] = \hbar\omega_k \mathsf{a}_k^\dagger \qquad [H, \mathsf{a}_k] = -\hbar\omega_k \mathsf{a}_k$$

which allows us to understand the rôles of the operators a_k and $\mathsf{a}_k^\dagger$. Let us consider the action of these operators on an eigenstate $|r\rangle$ of the Hamiltonian $(H|r\rangle = E_r|r\rangle)$

$$H\mathsf{a}_k|r\rangle = (\mathsf{a}_k H + [H, \mathsf{a}_k])\,|r\rangle = (E_r - \hbar\omega_k)\mathsf{a}_k|r\rangle$$
$$H\mathsf{a}_k^\dagger|r\rangle = (\mathsf{a}_k^\dagger H + [H, \mathsf{a}_k^\dagger])|r\rangle = (E_r + \hbar\omega_k)\mathsf{a}_k^\dagger|r\rangle$$

We see that the creation (annihilation) operator $\mathsf{a}_k^\dagger$ (a_k) increases (decreases) the energy by $\hbar\omega_k$. We associate with this quantum of energy *an elementary excitation* or *quasi-particle* called a *phonon*. The description of the solid is simple if we use phonons but becomes rather complicated if want to use the springs and masses. The operator $\mathsf{n}_k = \mathsf{a}_k^\dagger \mathsf{a}_k$, which commutes with H, counts the number of particles in mode k. Let $|0_k\rangle$ be the ground state of the harmonic oscillator in mode k. Using the familiar properties of the quantum harmonic oscillator, we construct the excited state $|n_k\rangle$ of mode k as follows

$$|n_k\rangle = \frac{1}{\sqrt{n_k!}}\,(\mathsf{a}_k^\dagger)^{n_k}|0_k\rangle$$

This state satisfies

$$\mathsf{a}_k^\dagger \mathsf{a}_k|n_k\rangle = \mathsf{n}_k|n_k\rangle = n_k|n_k\rangle \tag{5.52}$$

An eigenstate, $|r\rangle$, of H is constructed by taking the tensor product of the states $|n_k\rangle$ of the different modes k

$$|r\rangle = \bigotimes_{k=-\pi/a}^{k=\pi/a} |n_k\rangle$$

$$H|r\rangle = \sum_{k=-\pi/a}^{\pi/a}\left(n_k + \frac{1}{2}\right)\hbar\omega_k|r\rangle$$

The Hilbert space thus constructed is called a *Fock space*. The state r is completely specified by giving the occupation numbers: $r \equiv \{n_k\}$. The average $\langle \mathsf{n}_k\rangle \equiv \langle n_k\rangle$ is the average occupation number of mode k or, equivalently, the average number of phonons in mode k. We obtain the average energy by using (5.51)

$$E = \sum_k\left\langle\left(\mathsf{a}_k^\dagger \mathsf{a}_k + \frac{1}{2}\right)\right\rangle\hbar\omega_k = \sum_k\left(\langle n_k\rangle + \frac{1}{2}\right)\hbar\omega_k \tag{5.53}$$

The formalism we have just developed permits the description of situations where the number of particles (in this case the phonons) is variable. It is (improperly) called *second quantization*.

Real situations are more complicated than this simple example with only harmonic interactions. When anharmonic interactions are included, the phonon description is simple only at low temperature where the number of phonons is small. In fact, anharmonic interactions introduce interactions among the phonons, the normal modes are no longer decoupled and the number of phonons in a mode is no longer constant.[13]

When $|ka| \ll 1$, the dispersion law (5.47) becomes linear

$$|k|a \ll 1 \qquad \omega_k \simeq \sqrt{\frac{K}{m}}\,|k|a = c_s k \tag{5.54}$$

[13] Without anharmonic coupling between the modes or with the environment (similar to the coupling of photons to the walls in the case of black body radiation) no thermal equilibrium is possible.

where $c_s = a\sqrt{K/m}$ is the speed of sound for low frequencies. In fact, recalling the relations between K and Young's modulus Y, $K = aY$,[14] and between the mass m and the mass density ρ, $m = \rho a^3$, we obtain the classic result $c_s = \sqrt{Y/\rho}$.

5.4.2 Debye approximation

The preceding calculation can be generalized easily to three dimensions. To fix ideas, let us consider a simple cubic lattice; the normal modes are thus labelled by a wave vector $\vec{k}$, which we can choose in the first Brillouin zone

$$-\frac{\pi}{a} \leq k_x, k_y, k_z \leq \frac{\pi}{a}$$

We also assume that we can ignore the dependence of the frequency $\omega_{\vec{k}}$ on the direction of $\vec{k}$, in other words, $\omega_{\vec{k}} = \omega_k$.[15] In the Debye approximation we assume that the relation $\omega = c_s k$ is valid for all frequencies. It is then convenient to express the density of modes in terms of the frequencies,

$$\frac{V}{(2\pi)^3}\,\mathrm{d}^3k \to \frac{V}{2\pi^2}\,k^2\,\mathrm{d}k = \frac{V}{2\pi^2 c_s^3}\,\omega^2\,\mathrm{d}\omega$$

Furthermore, here we need to take into account the fact that we have three independent oscillation directions for sound waves, instead of the two for light waves. In addition to the two transverse polarizations, i.e. perpendicular to the wave vector (shear waves), there is a longitudinal polarization parallel to the wave vector and which corresponds to a compression mode. Since each atom can oscillate in three independent directions, the total number of vibration modes, which is equal to the total number of degrees of freedom, is three times the number of atoms. This will allow us to fix the maximum frequency, ω_D, called the *Debye frequency* (provisionally lost in going from (5.47) to (5.54)) by the condition

$$\frac{3V}{2\pi^2 c_s^3}\int_0^{\omega_D} \mathrm{d}\omega\,\omega^2 = 3N$$

This result is intuitively obvious if we interpret it in the framework of the Einstein model where atoms vibrate independently around their equilibrium positions. Since there are three independent vibration directions, there are in total $3N$

[14] The relation $K = aY$ can be easily derived by approximating an elastic rod with a collection of parallel spring chains.

[15] Note that here $k \geq 0$, whereas in Section 5.4.1 k was an algebraic number. In fact we could have used the notation k_x instead of k in this section.

harmonic oscillators and therefore $3N$ normal modes. We then obtain the Debye frequency as a function of the speed of sound c_s and density n

$$\omega_D = c_s\left(6\pi^2 n\right)^{1/3} \tag{5.55}$$

The density of modes $\rho_D(\omega)$ can finally be written as

$$\boxed{\rho_D(\omega) = \frac{3V}{2\pi^2 c_s^3}\,\omega^2\theta(\omega_D - \omega)} \tag{5.56}$$

We associate with the Debye frequency a characteristic energy, $\hbar\omega_D$, and therefore a characteristic temperature T_D, the *Debye temperature*, given by[16]

$$\boxed{kT_D = \hbar\omega_D} \tag{5.57}$$

Like T_F for the Fermi gas, T_D fixes a temperature scale; low temperature is $T \ll T_D$ whereas high temperature is $T \gg T_D$.

To complete this discussion we first consider some orders of magnitude. In the case of steel, the Young modulus is $Y \simeq 2\times 10^{11}\,\mathrm{N\,m^{-2}}$ and the mass density $\rho \simeq 8\times 10^3\ \mathrm{kg\,m^{-3}}$. This gives for the speed of sound $c_s \simeq 5\times 10^3\,\mathrm{m\,s^{-1}}$ and for the density $n \simeq 8.6\times 10^{28}\,\mathrm{m^{-3}}$. We therefore have $\omega_D \simeq 8.6\times 10^{13}\,\mathrm{s^{-1}}$ from (5.55) and $T_D \simeq 650\,\mathrm{K}$ from (5.57). Generally, Debye temperatures are of the order of a few hundred degrees Kelvin. We now return to the polarizations of sound waves. In real physical systems, the speed of longitudinal waves is different from that of transverse ones. In addition to Young's modulus, the physics of vibrations in isotropic solids introduces the Poisson coefficient $\sigma, 0 \le \sigma \le 1/2$, the propagation speeds of longitudinal (c_L) and transverse (c_T) waves are given by

$$c_L^2 = \frac{Y}{\rho}\,\frac{1-\sigma}{(1-2\sigma)(1+\sigma)} \qquad c_T^2 = \frac{Y}{\rho}\,\frac{1}{2(1+\sigma)} \le c_L^2 \tag{5.58}$$

Since $\rho_D(\omega)$ is proportional to c_s^{-3} (Equation (5.56)), we may define an average speed c_s by

$$\frac{3}{c_s^3} = \frac{2}{c_T^3} + \frac{1}{c_L^3} \tag{5.59}$$

This description is valid for an isotropic solid. It becomes more complicated if we take into account crystal symmetries.[17] Similarly, the real mode density $\rho(\omega)$ is quite different from the Debye mode density (5.56) (see Figure 5.10). However,

[16] Here k is the Boltzmann constant, not the wave vector.

[17] When c_s depends on the direction of $\vec{k}$, we use an equation similar to (5.59) to calculate the average speed. This is done by averaging c_s^{-3} over the directions of $\vec{k}$.

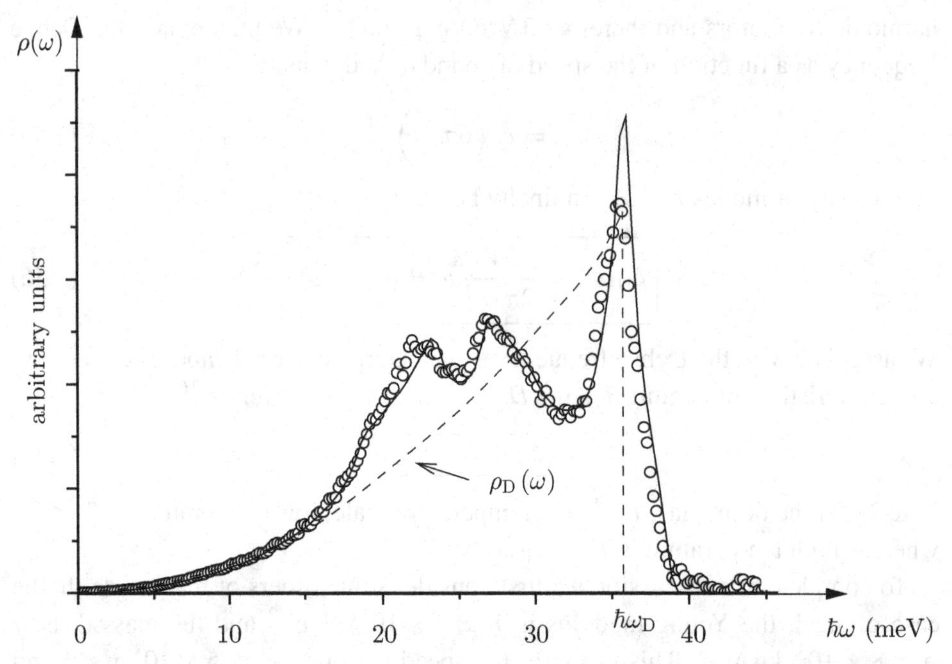

Figure 5.10 Density of modes $\rho(\omega)$ for α-iron measured by inelastic nuclear absorption of X-rays (circles, data taken at the Advanced Photon Source, Argonne) or inelastic neutron scattering (solid line) [29]. This is compared with Debye density $\rho_{\rm D}(\omega)$ ($\hbar\omega_{\rm D} = 36\,\text{meV}$). The ω^3 behaviour is observed only at very low frequencies. Note that the areas under the solid and dashed lines are equal: $\int \mathrm{d}\omega\, \rho(\omega) = \int \mathrm{d}\omega\, \rho_{\rm D}(\omega) = 3N$.

the Debye model does become reliable at low temperature where long wavelength phonons dominate the physics and $\omega(k) = c_{\rm s}k$ is valid.

5.4.3 Calculation of thermodynamic functions

The Hamiltonian (5.51) in one dimension has allowed us to interpret each normal mode as a harmonic oscillator. This can be easily generalized to the case of an isotropic solid in three dimensions. Each mode is labelled by a wave vector $\vec{k}$ and the summation over these wave vectors is done in the first Brillouin zone. The average number of phonons in mode $\vec{k}$ is obtained from Equation (3.67) with $\omega \to \omega_k$ because each mode is an independent harmonic oscillator

$$\langle n_{\vec{k}} \rangle = \frac{3}{\mathrm{e}^{\beta\hbar\omega_k} - 1} \tag{5.60}$$

The factor 3 takes into account the three polarizations. This equation emphasizes the fact that phonons are bosons with zero chemical potential since their number is not conserved.

According to Equation (5.53), we calculate the average energy by summing over all modes

$$E = \sum_{\vec{k}} \left(\langle n_{\vec{k}} \rangle + \frac{3}{2} \right) \hbar\omega_k = \frac{3V}{(2\pi)^3} \int \mathrm{d}^3k \left[\frac{1}{\mathrm{e}^{\beta\hbar\omega_k} - 1} + \frac{1}{2} \right] \hbar\omega_k$$

With the Debye approximation and the change of notation $\omega_k \to \omega$ this becomes

$$E = \int_0^\infty \mathrm{d}\omega\, \rho_\mathrm{D}(\omega) \left[\frac{1}{\mathrm{e}^{\beta\hbar\omega} - 1} + \frac{1}{2} \right] \hbar\omega = \frac{3V}{2\pi^2 c_\mathrm{s}^3} \int_0^{\omega_D} \mathrm{d}\omega\, \omega^2 \left[\frac{1}{\mathrm{e}^{\beta\hbar\omega} - 1} + \frac{1}{2} \right] \hbar\omega$$

The second term in the bracket in the above equation is the 'zero point energy' E_0, which is obtained when all normal modes are in the ground state

$$E_0 = \frac{9}{8} N\hbar\omega_\mathrm{D} \tag{5.61}$$

This is an additive constant which we may take as the origin (the zero of energy). With this convention, the internal energy is given only by the first term in the brackets. It is convenient to introduce the dimensionless integration variable $v = \beta\hbar\omega$. The energy integral is then

$$E = \frac{3V}{6\pi^2} \frac{(kT)^4}{(\hbar c_\mathrm{s})^3} \left(3 \int_0^{T_D/T} \mathrm{d}v \, \frac{v^3}{\mathrm{e}^v - 1} \right)$$

Defining the function $D(y)$ by

$$D(y) = 3 \int_0^y \mathrm{d}v \, \frac{v^3}{\mathrm{e}^v - 1} \tag{5.62}$$

the energy becomes

$$E = 3NkT \left(\frac{T}{T_D} \right)^3 D\left(\frac{T_D}{T} \right) \tag{5.63}$$

In general the function $D(y)$ should be evaluated numerically. However, we can easily obtain the high and low temperature limits:

(*i*) $y = T_\mathrm{D}/T \ll 1$, $D(y) \simeq y^3$

$$E \simeq 3NkT \qquad C_V \simeq 3Nk$$

Not surprisingly, we find in the high temperature limit ($T \gg T_\mathrm{D}$) the Dulong–Petit law and the equipartition theorem.

(*ii*) $y = T_D/T \gg 1$. In this case, we can replace y by ∞ in Equation (5.62) for $D(y)$ since the integral is convergent. Using (A.53)

$$\boxed{E = \frac{3}{5}\pi^4 NkT\left(\frac{T}{T_D}\right)^3 \qquad C_V = \frac{12}{5}\pi^4 kN\left(\frac{T}{T_D}\right)^3} \tag{5.64}$$

The proportionality of C_V to T^3 is called the *Debye law*. We remark that the behaviour of C_V when $T_D/T \gg 1$ is very similar to what we found for the photons: the cut-off frequency, ω_D, does not play any rôle. The above results (5.64) can be obtained from those for photons simply by replacing the speed of light in (5.35) by the speed of sound and multiplying by $3/2$ to take into account the fact that here we have three polarizations instead of the two for photons.

The discussion in Section 5.2 showed that the specific heat of a non-ferromagnetic metal at low temperature is of the form (Figure 5.4)

$$\frac{C_V}{T} = \gamma + AT^2$$

if we take into account conduction electrons and phonons. The coefficient A is given by (5.64) and can be measured experimentally, thus giving the Debye temperature T_D. But T_D can also be calculated using elastic constants Y and σ (see Equations (5.55), (5.58) and (5.59)). The good agreement between these two independent methods (Table 5.2) shows the relevance of the Debye theory and underlines the close relation between elastic and thermodynamic properties of a solid.

The specific heat is shown schematically in Figure (5.11) for arbitrary temperatures. Note the cubic behavior, T^3, for low temperatures and the asymptotic behaviour for $T \to \infty$ given by the Dulong–Petit law or, equivalently, the equipartition theorem.

Table 5.2 *Comparison of the Debye temperatures for different solids obtained from specific heat measurements, $T_D^{C_V}$, and calculated from elastic properties, T_D^{elas}*

	$T_D^{C_V}$	T_D^{elas}
NaC*l*	308 K	320 K
KC*l*	230 K	246 K
Ag	225 K	216 K
Zn	308 K	305 K

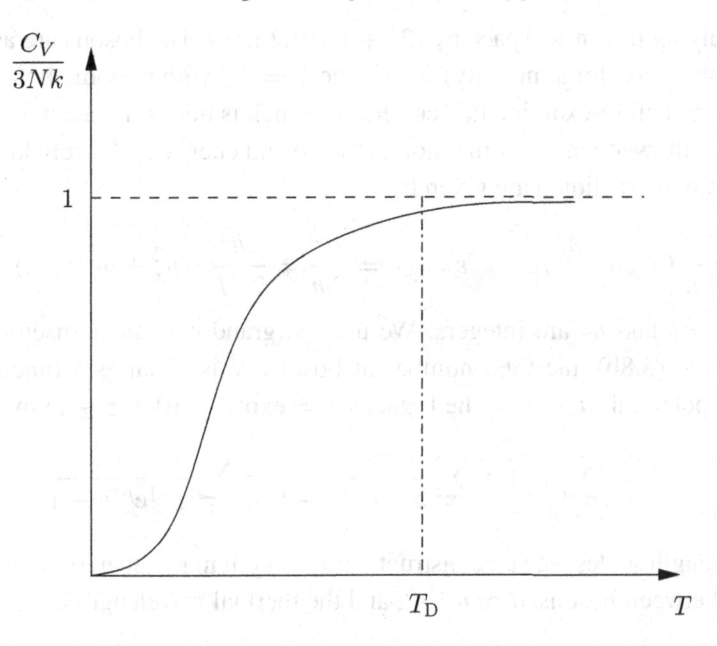

Figure 5.11 The specific heat as a function of temperature in the Debye model. Note that the classical limit is essentially reached at $T = T_D$.

The reader will notice the great similarity, and few differences, between phonons and photons. The formalism developed in Section 5.4.1 can be applied to photons by quantizing the electromagnetic field in a cavity. Then, the normal modes correspond to the eigenmodes of the cavity. The basic difference from the phonons is that there is no analogue of the Debye temperature for the photons; the frequencies of the eigenmodes can be arbitrarily high.[18]

5.5 Ideal Bose gas with a fixed number of particles

5.5.1 Bose–Einstein condensation

In the two preceding sections we dealt with bosons, photons and phonons, whose numbers were not fixed and therefore with vanishing chemical potential, $\mu = 0$. In this section, we shall consider an ensemble of a fixed number of bosons. This is the case, for example, of helium-4 in a closed container: the atoms can be neither created nor destroyed, unlike phonons and photons! Although this is a bad approximation for helium-4, we assume that the bosons do not interact. We also assume that the spin is zero, but it would be easy to generalize to an arbitrary integer spin s

[18] The zero point energy (5.61) is infinite for the photons. It should, therefore, be subtracted in such a way that the vacuum energy (no photons) is zero. This is an example of 'renormalization'. The zero point energy does have observable effects, notably the 'Casimir effect': two metallic plates placed in vacuum will attract each other due to the modification of the photon dispersion law in the space between them.

by multiplying the phase space by $(2s+1)$, cf. (5.16). The bosons are assumed to be in a cubic box (for simplicity) of volume $V = L^3$ with periodic boundary conditions. We shall re-examine this condition, which is not as innocent as one might think. The allowed values of the momentum $\vec{p}$ and energy ε_p (purely kinetic since there are no interactions) are given by

$$\vec{p} = \frac{h}{L}(n_x, n_y, n_z) \qquad \varepsilon_p = \varepsilon = \frac{\vec{p}^{\,2}}{2m} = \frac{h^2}{2mL^2}(n_x^2 + n_y^2 + n_z^2) \tag{5.65}$$

where n_x, n_y and n_z are integers. We use the grand canonical ensemble where, according to (5.8b), the total number of bosons N is given as a function of the chemical potential, $\mu \le 0$, or the fugacity $z = \exp(\beta\mu)$ $(0 < z \le 1)$ by

$$N = \sum_{\vec{p}} \langle n_{\vec{p}} \rangle = \sum_{\vec{p}} \frac{1}{\mathrm{e}^{\beta(\varepsilon_p - \mu)} - 1} = \sum_{\vec{p}} \frac{1}{z^{-1}\mathrm{e}^{\beta\varepsilon_p} - 1} \tag{5.66}$$

The two length scales we can construct with the system parameters are the average distance between bosons, $d \sim n^{-1/3}$, and the thermal wavelength λ

$$\lambda = \frac{h}{(2\pi mkT)^{1/2}} \sim T^{-1/2} \tag{5.67}$$

Let us start with a classical ideal gas where $d \gg \lambda$ or $n\lambda^3 \ll 1$. By reducing the temperature or increasing the density, we expect to observe quantum effects when d and λ are of the same order of magnitude, i.e. when $n\lambda^3 \sim 1$. As bosons like to occupy the same quantum state, eventually a finite fraction will occupy the ground state $\vec{p} = 0$ or $\varepsilon = 0$. This is the phenomenon of *Bose–Einstein condensation*,[19] which is a phase transition and is characterized by a critical temperature T_c. More precisely, for a fixed density n, this condensation happens for $T \le T_c$, or for a fixed temperature, it happens for $n \ge n_c$. We use the usual transformation to change from the variable $\vec{p}$ to the energy ε in phase space

$$\frac{V}{h^3}\,\mathrm{d}^3p \to \rho(\varepsilon)\,\mathrm{d}\varepsilon \qquad \rho(\varepsilon) = \frac{2\pi V(2m)^{3/2}}{h^3}\,\varepsilon^{1/2}$$

The zero of the energy is fixed by the ground state $\varepsilon_{\vec{p}=0}$. In the sum over $\vec{p}$ in Equation (5.66) we separate and write explicitly the contribution of the ground state

$$N = \frac{z}{1-z} + \int_0^\infty \mathrm{d}\varepsilon \frac{\rho(\varepsilon)}{z^{-1}\mathrm{e}^{\beta\varepsilon} - 1} = N_0 + N_1 \tag{5.68}$$

[19] There is no such Bose–Einstein condensation for photons because the number of photons goes to zero as the temperature is reduced.

$N_0 = \langle n_0 \rangle$ is the number of bosons in the ground state $\vec{p} = 0$, and N_1 is the number of bosons in all the higher excited states $\vec{p} \neq 0$. Since $z \leq 1$, we have $z^{-1}e^{\beta\varepsilon} \geq e^{\beta\varepsilon}$ and N_1 is bounded by

$$N_1 \leq \int_0^\infty d\varepsilon \frac{\rho(\varepsilon)}{e^{\beta\varepsilon} - 1} \tag{5.69}$$

The integral in (5.68) converges for $\varepsilon \to 0$ and of course also for $\varepsilon \to \infty$.[20] The maximum number of bosons in the excited states is obtained for $\mu = 0$. If the upper bound on N_1 becomes less than N for a temperature below a critical temperature $kT_c = 1/\beta_c$, then a finite fraction of bosons must necessarily occupy the ground state. Assuming for the moment that $\mu = 0$ (this will be justified soon) this critical temperature is obtained by taking $N = N_1$, then from (5.69) we have

$$n = \frac{N}{V} = \frac{1}{V} \int_0^\infty d\varepsilon \frac{\rho(\varepsilon)}{e^{\beta_c\varepsilon} - 1} = \frac{2\pi(2m)^{3/2}}{h^3}(kT_c)^{3/2} \int_0^\infty dx \frac{x^{1/2}}{e^x - 1}$$

Using integral (A.48) this becomes

$$\boxed{n = \left(\frac{2\pi mkT_c}{h^2}\right)^{3/2} \zeta(3/2) = \frac{\zeta(3/2)}{\lambda_c^3}} \tag{5.70}$$

where λ_c is the thermal wavelength (5.67) at the critical temperature, $T = T_c$. The criticality condition is then given by $n\lambda^3 = \zeta(3/2)$ which justifies our previous discussion and makes precise the heuristic condition $n\lambda^3 \sim 1$. For a temperature $T < T_c$, we shall always have from Equation (5.68) and $z = 1$

$$N = N_0 + V\frac{\zeta(3/2)}{\lambda^3}$$

By dividing this equation by N, using Equation (5.70) and the fact that $\lambda \sim T^{-1/2}$, we obtain

$$\boxed{\frac{N_0}{N} = 1 - \left(\frac{T}{T_c}\right)^{3/2}} \tag{5.71}$$

Figure 5.12 shows N_0/N as a function of T. It is reminiscent of the magnetization versus T curve for a ferromagnet but with an important difference: the tangent at $T = T_c$ has a finite slope here whereas it is vertical for a ferromagnet. As we shall soon see, the Bose–Einstein transition is first order (discontinuous) while the

[20] In two dimensions, the integral diverges for $\varepsilon = 0$ and Bose–Einstein condensation does not take place at finite temperature (Exercise 5.6.5).

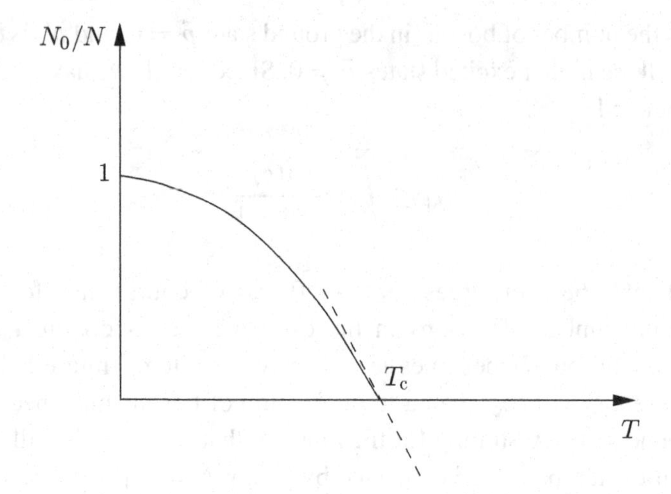

Figure 5.12 The condensate fraction N_0/N as a function of T. Note that the slope of the tangent at $T = T_c$ is finite.

ferromagnetic transition is continuous. The condensate fraction N_0/N is the *order parameter* for this transition: it is zero above the critical temperature and takes on a finite value below T_c, exactly like the magnetization which is the order parameter in the ferromagnetic case.

In the above discussion we took $\mu = 0$. It is now necessary to justify this assumption by examining what happens in a finite volume. We rewrite Equation (5.68) as

$$N = \frac{z}{1-z} + \frac{V}{\lambda^3} G(z) \qquad G(z) = \frac{2}{\sqrt{\pi}} \int_0^\infty \mathrm{d}x \frac{x^{1/2}}{z^{-1}\mathrm{e}^x - 1} \tag{5.72}$$

The multiplicative factor $2/\sqrt{\pi}$ was chosen so that $G(1) = \zeta(3/2)$. Equation (5.72) gives N_0/N in the form

$$\frac{N_0}{N} = \frac{z}{N(1-z)} = 1 - \frac{1}{n\lambda^3} G(z) \tag{5.73}$$

We can solve this graphically for z (Figure 5.13) by finding the intersection of the hyperbola $z/(N(1-z))$ with the curve $1 - G(z)/(n\lambda^3)$. In the thermodynamic limit $N \to \infty$ (or $V \to \infty$) at fixed n, the hyperbola tends to the horizontal axis for $0 < z < 1$, and vertically to $+\infty$ for $z \to 1$. Figure 5.13(a) shows that for $\zeta(3/2)/(n\lambda^3) < 1$, the limiting value of z as $N \to \infty$ is indeed $z = 1$, which corresponds to $\mu = 0$. On the other hand, we see in Figure 5.13(b) that for $\zeta(3/2)/(n\lambda^3) > 1$, z is strictly less than 1. Therefore, the condition $\mu = 0$ (or $z = 1$) that we have used in the preceding discussion is justified in the

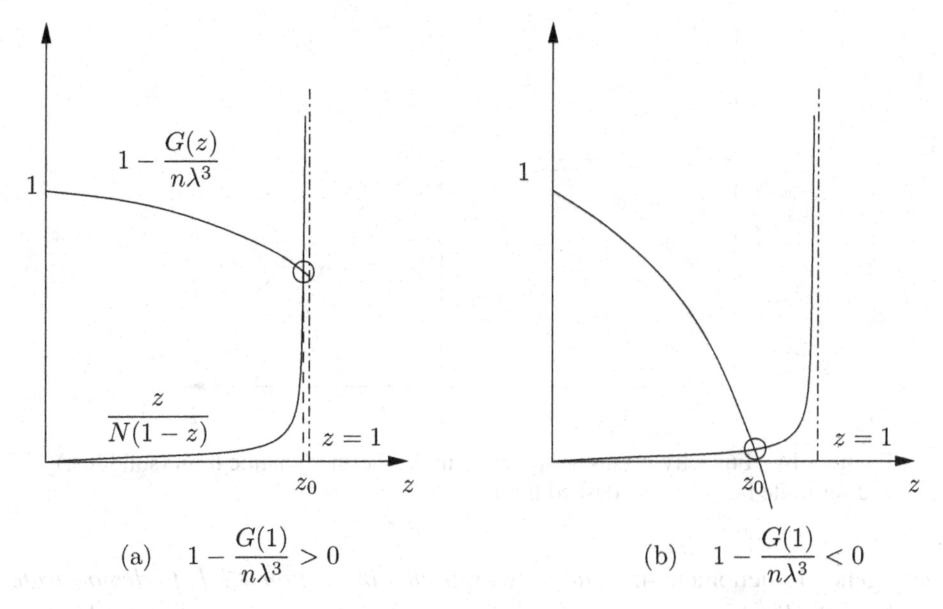

Figure 5.13 Graphical solution of Equation (5.73). The two cases: (a) $z \to 1$ in the thermodynamic limit and (b) $z \to z_0 < 1$.

thermodynamic limit $V \to \infty$ for $T < T_c$, but for $T > T_c$ z is given by $G(z) = n\lambda^3$. Let us now estimate μ for V large but finite for $T < T_c$. Since N_0/V must remain finite for $V \to \infty$, we must have $(1 - z) \sim 1/V$, in other words $|\mu|$ tends to zero like $1/V = L^{-3}$. Examining now the average occupation number of a state $\vec{p} \neq 0$, given by (5.65), we see that $\varepsilon_p \propto L^{-2} \gg \mu$. With this and Equation (5.66) we get the average occupation number

$$\langle n_{\vec{p}} \rangle = \frac{1}{e^{\beta(\varepsilon_p - \mu)} - 1} \sim \frac{1}{\beta(\varepsilon_p - \mu)} \propto L^2$$

and therefore

$$\langle n_{\vec{p} \neq 0} \rangle \propto L^2 \ll \langle n_{\vec{p}=0} \rangle = N_0 \propto L^3$$

Therefore, in Equation (5.66), once we have separated the $\vec{p} = 0$ term in the sum over $\vec{p}$, the sum for $\vec{p} \neq 0$ may be replaced by an integral

$$\sum_{\vec{p} \neq 0} \to \frac{V}{h^3} \int d^3 p$$

This result justifies Equation (5.68). Figure 5.14 gives z as a function of temperature for finite and infinite V. For finite V, z is an analytic function of T and the phase transition is not visible. But in the thermodynamic limit, z is no longer an analytic function of T at $T = T_c$, which signals a phase transition. This is a

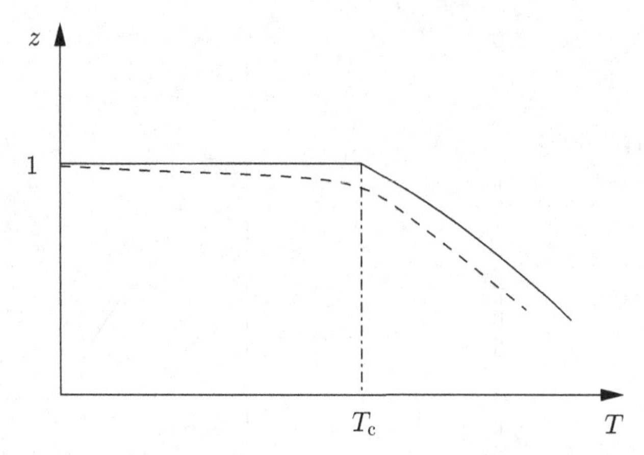

Figure 5.14 Fugacity versus temperature in the thermodynamic limit (solid line) and for finite but large N (dashed line).

very general phenomenon: as *already explained in Section 4.1.1, to demonstrate mathematically the existence of a phase transition, it is necessary to take the thermodynamic limit.*

In summary, in the thermodynamic limit, Bose–Einstein condensation takes place when the following condition relating the density and the thermal wavelength

$$n\lambda^3 = \zeta(3/2) \tag{5.74}$$

is satisfied. With relation (5.67) between λ and T in mind, Equation (5.74) can be interpreted in two ways: at fixed density, and starting at high temperature, we can reach the phase transition by lowering T to T_c, or at fixed T, and starting at low density, the transition is reached by increasing n to n_c.

5.5.2 Thermodynamics of the condensed phase

Using Equation (5.7) for the grand partition function $\mathcal{Q}(z, V, T)$, we obtain the pressure from Equation (3.130)

$$\ln \mathcal{Q}(z, V, T) = \beta PV = -\sum_{\vec{p}} \ln\left(1 - z e^{-\beta\varepsilon_{\vec{p}}}\right) \tag{5.75}$$

As before, we separate the $\vec{p} = 0$ term which is equal to

$$-\ln(1 - z) \sim -\ln\left(\frac{1}{V}\right)$$

by using the estimate for z that we have just obtained. Since

$$\lim_{V\to\infty} \frac{1}{V} \ln \frac{1}{V} = 0$$

we conclude that this term does not contribute to $\ln \mathcal{Q}$, and thus we can replace the sum over $\vec{p}$ directly by an integral. Changing to polar coordinates, we obtain the following expressions for the pressure

$$\beta P = -\frac{4\pi}{h^3} \int_0^\infty \mathrm{d}p\, p^2 \ln\left(1 - z\mathrm{e}^{-\beta\varepsilon_p}\right) \qquad T > T_c \tag{5.76}$$

$$\beta P = -\frac{4\pi}{h^3} \int_0^\infty \mathrm{d}p\, p^2 \ln\left(1 - \mathrm{e}^{-\beta\varepsilon_p}\right) \qquad T \leq T_c \tag{5.77}$$

where we have used $z = 1$ for $T \leq T_c$.

We shall concentrate on the only case where simple calculations are possible, namely $T \leq T_c$. Performing the variable change $p \to \varepsilon$ followed by $\beta\varepsilon \to x$ in Equation (5.77), and integrating by parts, we get

$$\beta P = \frac{4\pi}{3h^3} (2mkT)^{3/2} \int_0^\infty \mathrm{d}x\, \frac{x^{3/2}}{\mathrm{e}^x - 1} = \frac{\zeta(5/2)}{\lambda^3} \tag{5.78}$$

For $T \leq T_c$, the pressure is independent of the density! The isotherms for a boson gas are sketched in Figure 5.15. If we compare this with the isotherms for a liquid–gas transition, we see that here things happen as though the specific volume of the condensed phase were zero. This can be understood intuitively if we recall that for an ideal Bose gas (no interactions), as we have here, the bosons are point

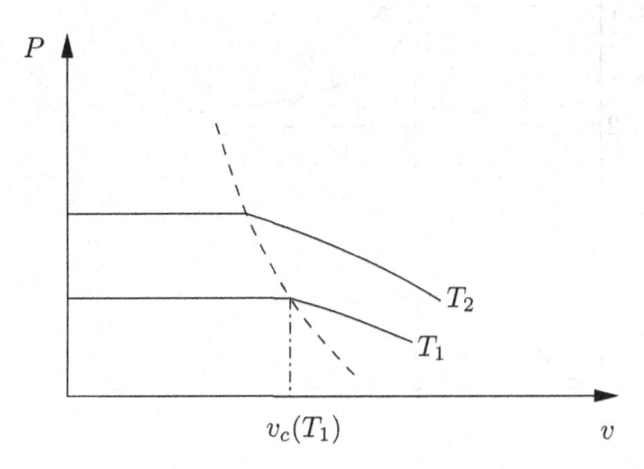

Figure 5.15 The isotherms of the free boson gas.

particles with no size. Therefore, we can put an arbitrary number of particles in a given volume. Clearly this demonstrates the limits on the validity of the ideal gas approximation. The phase transition line in the (T, P) plane is given by Equation (5.78)

$$P_0(T) = kT\,\frac{\zeta(5/2)}{\lambda^3} = \alpha T^{5/2} \tag{5.79}$$

The internal energy is obtained from $E = (3/2)PV \propto T^{5/2}$, which gives the specific heat C_V

$$\frac{C_V}{Nk} = \frac{15}{4}\frac{\zeta(5/2)}{\zeta(3/2)}\left(\frac{T}{T_c}\right)^{3/2} \tag{5.80}$$

and the entropy per particle

$$\check{s} = \frac{5}{2}\frac{\zeta(5/2)}{\zeta(3/2)}\left(\frac{T}{T_c}\right)^{3/2} k$$

The specific heat is shown in Figure 5.16. Note the $T^{3/2}$ behaviour for $T \to 0$, which is directly related to the dispersion law $\omega(l) \propto k^2$ (Exercise 5.6.6).

The physical picture of Bose–Einstein condensation is that of a mixture of two phases: a normal boson gas and a 'condensate' which contains bosons with zero momentum and specific volume. The entropy, energy and pressure are accounted for entirely by the normal component. The latent heat of transition per particle, ℓ,

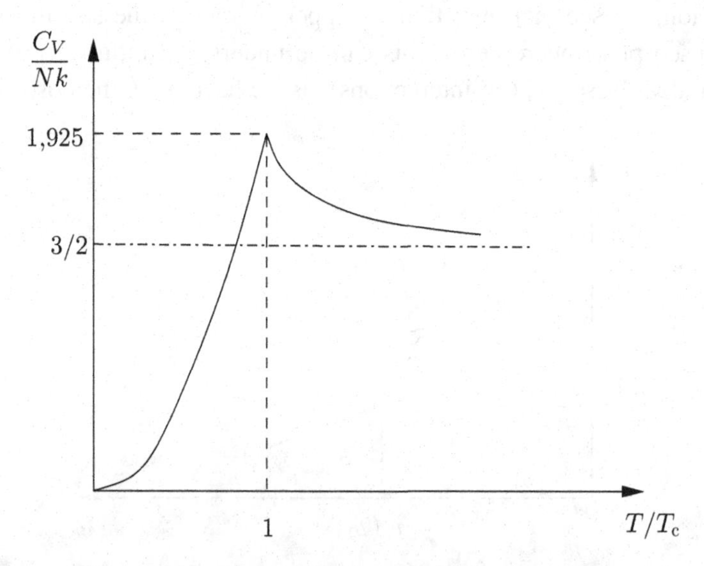

Figure 5.16 The specific heat of the free boson gas.

may be obtained with the help of the Clapeyron relation and (5.79)

$$\ell = T\Delta v \frac{\mathrm{d}P_0}{\mathrm{d}T} = \frac{5}{2}\frac{1}{n_c(T)}\alpha T^{5/2} = \frac{5}{2}\frac{\zeta(5/2)}{\zeta(3/2)}kT \tag{5.81}$$

because the volume change is simply the critical volume $v_\mathrm{c} = 1/n_\mathrm{c}$. We also remark that $\ell = \check{s}_\mathrm{c}T$ where $\check{s}_\mathrm{c}$ is the entropy per particle at $T = T_\mathrm{c}$. Since the latent heat of the transition is non-zero, the phase transition in the absence of interactions is first order.

As we have already mentioned, the free theory (no interactions) is problematic. For example, instead of periodic boundary conditions we could have imposed infinite walls at the boundaries forcing the wave function to vanish there. In that case, the ground state wave functions would become a product of sines and, consequently, the particle density in the condensed state would no longer be constant and the thermodynamic limit would fail to exist: the density would not be uniform in a large container. Furthermore, it does not seem realistic to take a vanishing condensate volume. In fact, only the inclusion of repulsive interactions allows us to stabilize the bosonic system and gives it a thermodynamic limit. For example, if the bosons were hard spheres of volume v_0, the volume of the condensate would be at least $N_0 v_0$. To describe interactions between the bosons, it is convenient to represent the interaction potential with a delta-function (or pseudo-potential) acting on the two-particle wave function $\Psi(\vec{r}_{12})$ as follows[21]

$$\begin{aligned} V(\vec{r}_1 - \vec{r}_2)\Psi(\vec{r}_{12}) &= g\delta(\vec{r}_1 - \vec{r}_2)\frac{\partial}{\partial r_{12}}\left(r_{12}\Psi(\vec{r}_{12})\right) \\ g = \frac{4\pi a\hbar^2}{m} &\qquad r_{12} = |\vec{r}_1 - \vec{r}_2| \end{aligned} \tag{5.82}$$

In fact, bosons have low momenta and their wave functions are, therefore, widely spread out in space. Consequently, they are not sensitive to the short distance details of the interaction potential which may then be approximated by a delta-function. The parameter a in Equation (5.82) is called the *scattering length*, which is another length scale in addition to d and λ for the problem with interactions. The total scattering cross section at low energy is given by $\sigma = 4\pi a^2$, and the differential cross section is isotropic because scattering takes place only in the zero angular momentum channel, $l = 0$. The cross section does not give information on the sign of a which is very important. If $a < 0$ the interaction is attractive, if $a > 0$ the interaction is repulsive, while for hard spheres, a is simply the diameter of the sphere. When $|a| \ll d$, the interactions may be treated perturbatively, which was done by the Soviet School in the 1950s. Interactions change the form of the

[21] The results discussed at the end of this subsection are shown without proof. The interested reader should consult the references.

dispersion law for the elementary excitations giving

$$\omega(k) = \sqrt{\left(\frac{\hbar k^2}{2m}\right)^2 + \frac{4\pi a n}{m^2}\hbar^2 k^2} \tag{5.83}$$

Of course this equation has no meaning unless $a > 0$; in other words for repulsive interactions. For small values of k, Equation (5.83) gives the speed of sound

$$\omega(k) \simeq c_{\rm s} k \qquad c_{\rm s} = \frac{\hbar}{m}\sqrt{\frac{4\pi a n}{m}} \tag{5.84}$$

The elementary excitations in the condensed state are phonons, and so the behaviour of the specific heat for $T \to 0$ is no longer $T^{3/2}$, as in the free theory, but rather T^3 as in the Debye model. Furthermore, at $T = 0$, the number of bosons in the condensate is no longer 100%, instead it is given by

$$\frac{N_0}{N} = 1 - \frac{8}{3}\sqrt{\frac{a^3 n}{\pi}} \tag{5.85}$$

We see that the introduction of repulsive interactions changes qualitatively the properties of a condensed boson gas. These interactions stabilize the system, allow it to have a thermodynamic limit, and the phase transition to the condensed state becomes continuous. For attractive interactions, condensates are possible but only for a finite number of bosons. The free theory, at the border between $a > 0$ and $a < 0$, is thus marginally stable and it is not difficult to uncover its limitations.

5.5.3 Applications: atomic condensates and helium-4

Until recently, only superfluid helium-4 could be taken as a possible example of Bose–Einstein condensation (although superconductivity can also be regarded as the condensation of Cooper pairs and superfluidity of helium-3 as the condensation of pairs of atoms). However, the interactions between helium atoms are strong, and so the relationship between superfluidity and Bose–Einstein condensation is not clear. In 1995, two groups in the USA, one at Boulder, the other at MIT, succeeded in producing atomic condensates using rubidium for the first group, and sodium for the second. These condensates are called *atomic gas condensates* because, unlike helium, the atoms stay in the gaseous phase.[22]

The atoms are first cooled by lasers and placed in a magneto-optic trap. They are then transferred to a harmonic magnetic trap where evaporative cooling is used

[22] These condensates are, in fact, metastable because three-body interactions lead to the formation of molecules. However, the system is dilute enough that the corresponding time scale is on the order of minutes. This leaves enough time to manipulate the condensates in experiments.

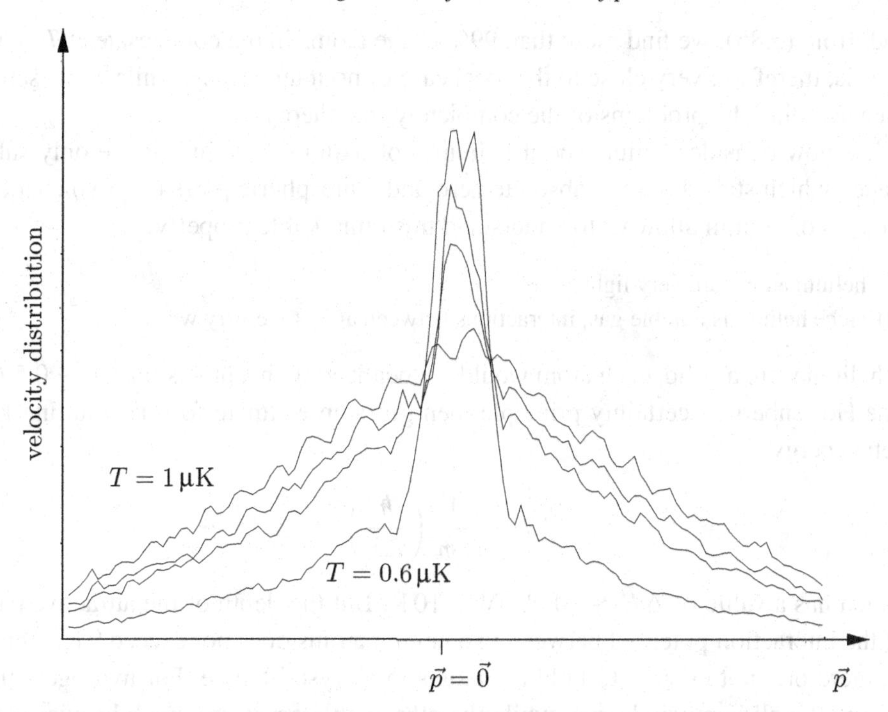

Figure 5.17 Bose–Einstein condensation: the spatial extent of a condensate after a time of flight is proportional to the initial velocity distribution. Courtesy of J. Dalibard and F. Chevy.

to lower their temperature even more. This last procedure is analogous to cooling a cup of coffee by blowing on it, which removes the most energetic molecules. For condensates, blowing is replaced with a 'magnetic knife', which eliminates the faster atoms leaving behind atoms that rethermalize at a lower temperature.[23] In the MIT experiment, the atomic density in the centre of the condensate was of the order of 10^{20} atoms/m^3, and the temperature was in the μK range. Problem (5.7.5) shows that these conditions lead to the formation of a condensate where the atoms occupy the ground state of the harmonic oscillator. The condensate can be visualized by switching off the trap and doing time-of-flight measurements, which allows the reconstruction of the velocity distribution in the condensate. Figure 5.17 shows several vertical cuts through each of the images of a rubidium condensate where the transition temperature is 0.9 μK. Above the transition ($T = 1\ \mu$K), the velocity distribution has Maxwellian form whereas below the transition there is a distinct peak at $\vec{p} = 0$. Its form is determined by the distribution of velocities in the ground state of the harmonic oscillator. The scattering length for rubidium is $a \simeq 5.4$ nm

[23] This procedure is rather wasteful: typically starting with about 10^9 atoms in the first trap, the number of atoms which is eventually found in the condensate after evaporative cooling is around 10^6.

and, from (5.85), we find more than 99% of the atoms in the condensate at $T = 0$. This is, therefore, very close to the ideal case of no interactions, while at the same time avoiding the problems of the completely free theory.

We now consider helium: be it helium-3 or helium-4, helium is the only substance which stays liquid at absolute zero and atmospheric pressure. Two characteristics of helium allow us to understand this remarkable property:

(*i*) helium atoms are very light,
(*ii*) since helium is a noble gas, interactions between atoms are very weak.

If helium were a solid, each atom would be localized with a precision $\Delta x \sim 0.5$ Å. The Heisenberg uncertainty principle then gives an estimate for the resulting kinetic energy

$$\Delta E \sim \frac{1}{2m}\left(\frac{\hbar}{\Delta x}\right)^2$$

which has a value of $\Delta E \sim 10^{-3}$ eV ~ 10 K. But the depth of the attractive part of the interaction potential between two helium atoms does not exceed 9 K, which is, therefore, not enough to hold the atoms in a crystal lattice. For hydrogen, the potential well is much deeper, while the other noble gases are much heavier than helium and so they cannot be liquids at zero temperature. This argument is independent of statistics and is valid both for helium-3 and helium-4. At sufficiently high pressure, 25 atmospheres for helium-4 and 30 atmospheres for helium-3, helium is solidified.

Nonetheless, one isotope is fermionic, the other bosonic, and they will display radically different behaviour at low temperatures. Helium-3 is the only true Fermi liquid available for experiments since electrons in a metal are subject to the periodic potential of the ions (some stars also give examples of Fermi liquids, cf. Exercise 5.6.2 and Problem 5.7.3). Helium-3 remains a Fermi liquid down to a temperature of 3 mK below which it becomes superfluid. Helium-4, to which we confine our attention now, is a normal liquid down to $T = 2.18$ K below which it becomes superfluid. This transition is called the 'lambda point'. The phase diagram is shown in Figure 5.18. At the transition, the specific volume of helium-4 is $v_\lambda \simeq 46.2$ Å^3, which corresponds to a density $n \simeq 2.16$ atom/Å^3. If we now boldly assume that we can neglect the interactions and use Equation (5.74) we find the Bose–Einstein condensation temperature of 3.14 K. This is in reasonable agreement with the temperature of the lambda point. However, this agreement must be considered accidental: the interactions are so strong that at absolute zero the condensate fraction does not exceed 10%. This value of 10% comes from Monte Carlo simulations and has never been confirmed experimentally. It seems plausible that

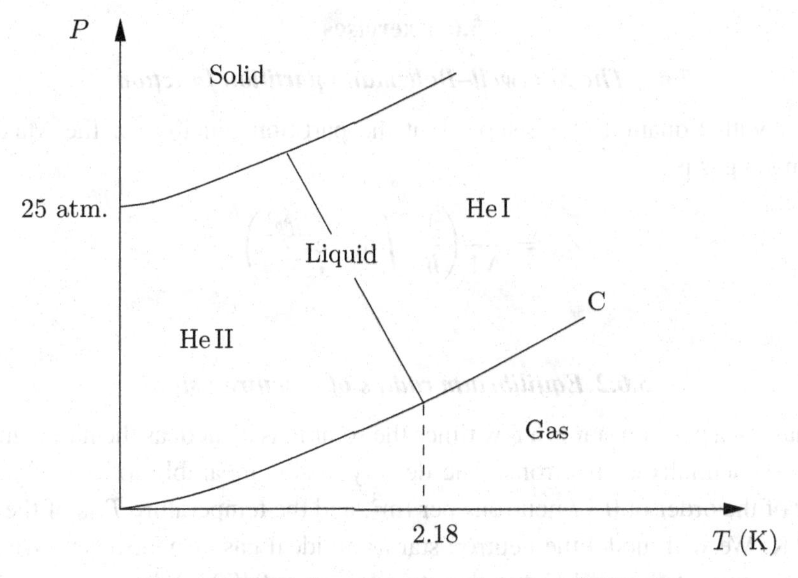

Figure 5.18 Schematic phase diagram of helium.

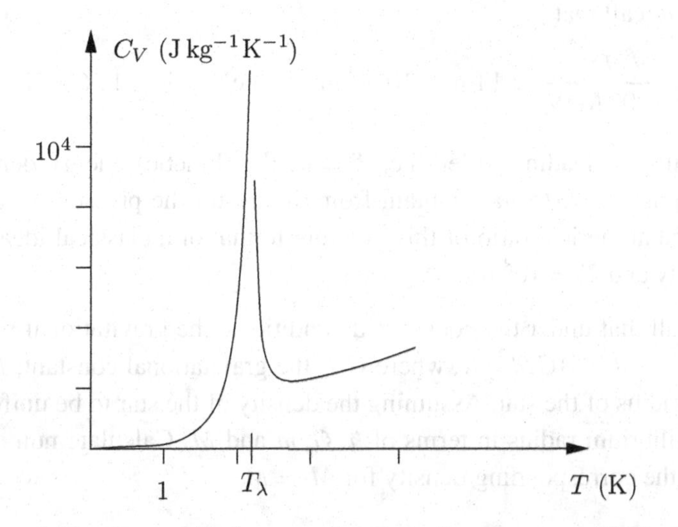

Figure 5.19 The specific heat as a function of temperature of helium-4.

the superfluidity of helium-4 is related to Bose–Einstein condensation, but there is no formal proof of this. Finally, and this is also the case for the atomic condensates, the transition is continuous; the specific heat diverges logarithmically at the critical temperature (Figure 5.19). Other remarkable properties of superfluid helium-4 will be described in Problem 8.6.6.

5.6 Exercises

5.6.1 The Maxwell–Boltzmann partition function

Starting with Equation (5.1), show that the partition function of the Maxwell–Boltzmann gas is

$$Z_N = \frac{1}{N!}\left(\frac{V}{h^3}\int \mathrm{d}^3 p\, \mathrm{e}^{-\frac{\beta p^2}{2m}}\right)^N$$

5.6.2 Equilibrium radius of a neutron star

The mass of a neutron star is a few times the solar mass[24] and, as the name implies, is made essentially of neutrons. The density n is comparable to that of nuclear matter of the order of 0.17 neutrons per fm^3, and the temperature T is of the order of 10^8 K. We will model the neutron star as an ideal gas of neutrons.[25] We recall that the neutron has spin $1/2$ and that its mass, m, is $940\,\mathrm{MeV/c^2}$.

1. Show that we may consider the neutron gas as a non-relativistic degenerate Fermi gas. Recall that

$$\frac{\hbar c}{200\,\mathrm{MeV}} \simeq 1\,\mathrm{fm} = 10^{-15}\,\mathrm{m} \qquad 1\,\mathrm{eV} \simeq k \times 11\,600\,\mathrm{K}$$

2. Show that to leading order (i.e. $T = 0$) the (kinetic) energy density of the neutron gas is $\epsilon = (3/5)n\varepsilon_{\mathrm{F}}$. Obtain from this result the pressure of the gas and calculate the numerical ratio of this pressure to that of a classical ideal gas at the same density and $T = 10^8$ K.

3. We recall that under the considered conditions, the gravitational potential energy is $E_{\mathrm{G}} = -(3/5)GM^2/R$, where G is the gravitational constant, M the mass and R the radius of the star. Assuming the density of the star to be uniform, calculate its equilibrium radius in terms of $\hbar, G, m$ and M. Calculate numerically this radius and the corresponding density for $M = M_\odot$.

5.6.3 Two-dimensional Fermi gas

Consider N electrons of mass m, momentum $\vec{p}$ and energy $\varepsilon = p^2/2m$, constrained to move on a surface of area S. These electrons form a two-dimensional gas, which we will assume to be ideal.

[24] The theoretical limit is 3.5 $M_\odot$ with some uncertainty due to imprecise knowledge of the equation of state of nuclear matter. The dozen or so neutron stars of well-measured masses have values of 1.3–1.8 $M_\odot$.
[25] This model is clearly very approximate, but to go further requires general relativity.

1. Verify that the level density $\rho(\varepsilon)$ is a constant, D, which you should determine.

2. Calculate the Fermi energy ε_F of the gas. Calculate its energy E in terms of N and ε_F. Show that the condition for the classical approximation (non-degenerate gas) to hold may be written as $N \ll DkT$, where k is the Boltzmann constant and T the temperature of the gas. Give a physical interpretation of this condition.

3. We now consider the case of a highly degenerate gas, i.e. $\beta\mu \gg 1$ ($\beta = 1/kT$). Verify that in this limit, integrals of the form

$$\int_{\mu}^{+\infty} \mathrm{d}x \, \frac{1}{\mathrm{e}^{\beta x} + 1} \quad \text{and} \quad \int_{\mu}^{+\infty} \mathrm{d}x \, \frac{x}{\mathrm{e}^{\beta x} + 1}$$

are of the order of $\mathrm{e}^{-\beta\mu}$. Show the following equality

$$\int_{-\mu}^{\mu} \mathrm{d}x \, \frac{\varphi(x)}{\mathrm{e}^{\beta x} + 1} = \int_{-\mu}^{0} \mathrm{d}x \, \varphi(x) + \int_{0}^{\mu} \mathrm{d}x \, \frac{\varphi(x) - \varphi(-x)}{\mathrm{e}^{\beta x} + 1} \tag{5.86}$$

where $\varphi(x)$ is an arbitrary function.

4. Calculate the chemical potential μ and average energy E at temperature T ignoring terms of the order $\mathrm{e}^{-\beta\mu}$.

5. Verify that by using the Sommerfeld expansion (5.29) of the Fermi distribution, we regain, up to terms of order $\mathrm{e}^{-\beta\mu}$, the expression for μ and E just obtained. What particularity of the system studied here explains this result?

5.6.4 Non-degenerate Fermi gas

We consider a very weakly degenerate ideal Fermi gas of spin s.

1. We recall that for a classical ideal gas, the grand partition function is given in terms of the thermal wavelength $\lambda = h/(2\pi mkT)^{1/2}$ by

$$\ln \mathcal{Q} = (2s + 1)\frac{zV}{\lambda^3} \tag{5.87}$$

We define the specific volume $v = V/N$ and introduce for convenience the quantity $v' = (2s + 1)v$. Express the condition for the classical approximation $z \ll 1$ in terms of v and λ. Interpret this condition.

2. For a Fermi gas, the grand partition function is given by

$$\ln \mathcal{Q} = AV \int_{0}^{\infty} \mathrm{d}\varepsilon \, \sqrt{\varepsilon} \ln\left(1 + z\mathrm{e}^{-\beta\varepsilon}\right) \tag{5.88}$$

Give the expression for the constant A. By expanding this expression for $\ln \mathcal{Q}$ in powers of z for $z \ll 1$, show that to order z^2 we have

$$\ln \mathcal{Q} \simeq N\frac{v'}{\lambda^3}z\left(1-\frac{z}{2^{5/2}}\right) \tag{5.89}$$

From $\ln \mathcal{Q} = \beta PV$ and (5.89), obtain the expansion in powers of z of the equation of state of a weakly degenerate Fermi gas. By following the general treatment in Section 3.6.2, we eliminate z in favour of the density. This allows us to obtain the equation of state in terms of powers of n, i.e. a kind of virial expansion.

3. Show that

$$z = \frac{\lambda^3}{v'}\left[1+\frac{1}{2^{3/2}}\frac{\lambda^3}{v'}+\mathcal{O}\left(\left(\frac{\lambda^3}{v'}\right)^2\right)\right]$$

Verify that the equation of state takes the form

$$\beta P = \frac{1}{v}\left[1+\frac{1}{2^{5/2}}\frac{\lambda^3}{v'}+\mathcal{O}\left(\left(\frac{\lambda^3}{v'}\right)^2\right)\right] \tag{5.90}$$

By comparing with the virial expansion for a gas of hard spheres (Problem 3.8.7), give a physical interpretation of this result in terms of effective repulsion.

4. How should the previous results be modified for a boson gas with N fixed?

5. Consider an equal mixture (in number of atoms) of ^{3}He, ^{4}He. Calculate the difference in the kinetic energies between the two isotopes at $T = 10\,\text{K}$ and $P = 2$ atm.

5.6.5 Two-dimensional Bose gas

We consider a system of N spinless bosons constrained to move on a surface of area S. Express, in terms of the density N/S and thermal wavelength $\lambda = h/(2\pi mkT)^{1/2}$, the conditions that fix the chemical potential. Solve these equations graphically and show that a two-dimensional boson gas does not exhibit Bose–Einstein condensation.

5.6.6 Phonons and magnons

As important applications of Bose–Einstein statistics with variable numbers of particles we mention phonons and magnons:

- phonon: the quantum of vibration (sound wave) in a solid,
- magnon: the quantum of a spin wave in a ferromagnetic crystal.

These waves have different dispersion relations. Consequently, each of these waves will have a different signature on the thermodynamic properties of the material, notably on its specific heat. We assume the following generic form for the dispersion relation, valid for small q

$$\omega(q) = aq^{\alpha}$$

where a is a constant and $\vec{q}$ the wave vector. For phonons we have $\alpha = 1$ and for magnons $\alpha = 2$.

1. Show that the mode density may be written as

$$\rho(\omega)\,\mathrm{d}\omega = g\frac{V}{2\pi^2}\frac{1}{\alpha}\frac{1}{a^{\frac{3}{\alpha}}}\omega^{(\frac{3}{\alpha}-1)}\,\mathrm{d}\omega$$

The factor g is the number of possible polarizations of the wave or, equivalently, the spin degeneracy of the bosons.

2. Verify that at low temperature the dominant behaviour of the specific heat is given by

$$C_V \propto T^{3/\alpha} \tag{5.91}$$

Can one obtain this result from a dimensional argument?

5.6.7 Photon–electron–positron equilibrium in a star

In a star, the electron density fluctuates because of creation and annihilation of pairs $(\mathrm{e}^+, \mathrm{e}^-)$. Therefore, the densities of the electrons (n_-) and positrons (n_+) are internal variables whose difference $(n_- - n_+)$ remains constant and equal to some value n_0, which is determined by the previous evolution of the star.

1. Assuming that there is a chemical equilibrium in the star,

$$\gamma \leftrightarrows \mathrm{e}^+ + \mathrm{e}^- \tag{5.92}$$

write the simple relation which fixes the difference in the chemical potentials for the electrons and positrons.

2. We first consider the situation where the rate of pair creation is low, i.e. the low temperature regime $kT \ll mc^2$. However, we do assume that the temperature is high enough to prevent atoms from forming. Under these conditions, the electrons and positrons form two classical (non-degenerate) ideal gases. Express the densities, n_- and n_+, at equilibrium in terms of n_0 and T.

3. Now consider the regime $kT \gg mc^2$: the electron and positron gases now are degenerate, ultra-relativistic ideal Fermi gases. In addition, the rate of pair production is so high that we may ignore the initial electron density n_0. This means that the electron and positron densities are equal, $n_- = n_+ = n$. Verify that $\mu_+ = \mu_- = 0$ and calculate the densities n_+ and n_-.

Show that the energy density of the electron (or positron) gas has the same temperature dependence as a photon gas

$$\epsilon(T) = \alpha T^4$$

Compare this expression with (5.36) and show that

$$\alpha = \frac{7}{8}\sigma'$$

5.7 Problems

5.7.1 Pauli paramagnetism

We have seen in Section 3.1.3 that a paramagnetic crystal in a magnetic field $\vec{B} = B\hat{z}$ acquires a magnetization given by (3.27). The magnetic susceptibility in zero field, χ_0, is from Equation (3.27)

$$\chi_0 = \lim_{B \to 0} \frac{1}{V} \frac{\partial \mathcal{M}}{\partial B} = \frac{n\tilde{\mu}^2}{kT} \tag{5.93}$$

where $n = N/V$ is the density and $\tilde{\mu}$ the magnetic moment. This result is valid for localized ions. The situation is different for electrons in a metal. On the one hand, the electronic magnetic moments tend to align with the magnetic field, like in the normal paramagnetic case, but on the other hand, Fermi statistics will profoundly change the law (5.93). This kind of paramagnetism, whose properties we shall examine, is called *Pauli paramagnetism*. Consider the electrons of a metal in a constant magnetic field $\vec{B} = B\hat{z}$. Their energy

$$\varepsilon' = \frac{p^2}{2m} \mp \tilde{\mu} B$$

has two contributions:

- a kinetic energy $\varepsilon = p^2/2m$,
- a magnetic energy $-\tilde{\mu}B$ ($+\tilde{\mu}B$) for electrons with spins parallel (antiparallel) to the magnetic field.

1. Show that the density of electrons n_+ (n_-) with spin parallel (antiparallel) to the magnetic field is given by

$$n_\pm = \frac{A}{2} \int_0^\infty \frac{\sqrt{\varepsilon}\, \mathrm{d}\varepsilon}{\exp[\beta(\varepsilon \mp \tilde{\mu}B - \mu)] + 1} \tag{5.94}$$

where μ is a chemical potential to be determined and $A = 4\pi(2m/h^2)^{3/2}$.

2. First consider $T = 0$. Show that for $B \to 0$, the difference $(n_+ - n_-)$ is given by

$$n_+ - n_- \simeq \frac{3}{2}\frac{n\tilde{\mu}B}{\varepsilon_F}$$

where ε_F is the Fermi energy of the electron gas in the absence of the magnetic field. Deduce from this the zero field susceptibility

$$\chi_0 = \frac{3}{2}\frac{n\tilde{\mu}^2}{\varepsilon_F} = \frac{3}{2}\frac{n\tilde{\mu}^2}{kT_F} \tag{5.95}$$

If we compare this result to (5.93), we see that the Fermi temperature T_F plays the rôle of T.

3. Still at $T = 0$, show that μ is not equal to ε_F and that we can write the leading correction in the form

$$\mu \simeq \varepsilon_F\left(1 - \lambda\left(\frac{\tilde{\mu}B}{\varepsilon_F}\right)^2\right) \tag{5.96}$$

Determine λ explicitly. Is the expansion in powers of $\tilde{\mu}B/\varepsilon_F$ justified for a magnetic field of 1 T? Write $n_\pm$ in the form

$$n_\pm = \frac{n}{2}(1 \pm r)$$

and show that r is given by the equation

$$(1+r)^{2/3} - (1-r)^{2/3} = \frac{2\tilde{\mu}B}{\varepsilon_F} \tag{5.97}$$

which can be solved numerically for all B. Obtain the results of Question 2 for $B \to 0$.

4. We now consider the case $T \neq 0$ and $B \to 0$. Using a Taylor expansion to first order in B,

$$\frac{1}{\exp[\beta(\varepsilon \mp \tilde{\mu}B - \mu)] + 1} \simeq \frac{1}{\exp[\beta(\varepsilon - \mu)] + 1} \mp \tilde{\mu}B\frac{\partial}{\partial\varepsilon}\left(\frac{1}{\exp[\beta(\varepsilon - \mu)] + 1}\right) \tag{5.98}$$

give the expression for $(n_+ - n_-)$ to this order in B. Use this result to obtain the zero-field susceptibility

$$\chi_0 = \frac{3}{2}\frac{n\tilde{\mu}^2}{\varepsilon_F}\left(1 - \frac{\pi^2}{12}\left(\frac{kT}{\varepsilon_F}\right)^2\right) \tag{5.99}$$

5.7.2 Landau diamagnetism

We reconsider the situation described in the previous problem and focus on the motion of conduction electrons. We shall thus demonstrate a diamagnetic effect (negative susceptibility) called *Landau diamagnetism*. We consider the following model:

- we consider the conduction electrons to form an ideal gas of fermions, mass m, spin $1/2$ and enclosed in a box of volume $V = L_x L_y L_z$,
- we ignore the effect of the magnetic field on the spin magnetic moment of the electron; we only include the effect of the field on the orbital magnetic moment.

With this model, the energy levels of the conduction electrons are of the form

$$\varepsilon_{n,n_z} \equiv \varepsilon_n(k_z) = \frac{\hbar^2 k_z^2}{2m} + \left(n + \frac{1}{2}\right) 2\mu_B B$$

where n is a positive integer (or zero) and $k_z = 2\pi n_z / L_z$ with n_z an integer (positive, negative or zero) and $\mu_B = e\hbar/2m$ is the Bohr magneton.

1. By comparing the energy spectrum at fixed k_z in zero field with that in the presence of a field, show that the degeneracy g of these energy levels is

$$g = \frac{2L_x L_y}{\pi \hbar^2} m \mu_B B \tag{5.100}$$

2. The magnetization is obtained, as usual, from the free energy

$$\mathcal{M} = -\left.\frac{\partial F}{\partial B}\right|_{T,V,N}$$

Obtain an explicit expression starting with the grand potential $\mathcal{J}(T, V, \mu; B)$ and verify that the zero field magnetic susceptibility is given by

$$\chi_0 = -\frac{1}{V} \lim_{B \to 0} \left.\frac{\partial^2 \mathcal{J}}{\partial^2 B}\right|_{T,V,\mu} \tag{5.101}$$

3. Starting with the expression for the grand partition function ξ_λ for an individual state, $(\lambda) \equiv \{n, n_z\}$, and by considering k_z as a continuous variable, show that the grand potential is given by

$$\mathcal{J}(T, V, \mu; B) = -2\frac{mVkT}{\pi^2 \hbar^2} \mu_B B \times \int_0^{+\infty} \mathrm{d}k_z \sum_{n=0}^{+\infty} \ln\left(1 + z \exp\left[-\frac{\hbar^2 k_z^2}{2mkT}\right] \exp\left[-\frac{(2n+1)\mu_B B}{kT}\right]\right) \tag{5.102}$$

4. Use the Euler–Maclaurin formula

$$\int_0^\infty \mathrm{d}x\, f(x) = \sum_{n=0}^{\infty} \alpha f\left(\left(n+\tfrac{1}{2}\right)\alpha\right) - \frac{\alpha^2}{24} f'(x=0) + \mathcal{O}(\alpha^3)$$

to write down the expansion of $\mathcal{J}$ to second order in B. Use this to obtain the diamagnetic (or Landau) susceptibility in the form of a definite integral featuring the temperature and chemical potential.

5. By using Equation (5.98), it can be easily shown that the magnetic susceptibility due to the coupling between the magnetic field and the electron spins (i.e. Pauli susceptibility) may be written as

$$\chi_0^{\mathrm{P}}(T) = \frac{1}{V}\mu_{\mathrm{B}}^2 \int_0^\infty \mathrm{d}\varepsilon\, \rho'(\varepsilon) f(\varepsilon)$$

where $\rho(\varepsilon)$ is the individual density of states in zero field and $f(\varepsilon)$ the Fermi factor. Show for a free electron gas at any temperature that, up to a sign, the diamagnetic susceptibility is minus one third of the paramagnetic one

$$\chi_0^{\mathrm{L}}(T) = -\frac{1}{3}\chi_0^{\mathrm{P}}(T)$$

5.7.3 White dwarf stars

A white dwarf is the final phase in the life of a star that has exhausted its usable nuclear fuel if its initial mass is not too large. Our purpose here is to calculate the maximum mass M_{C} of a white dwarf, called the *Chandrasekhar mass*. A star tends to collapse under the influence of its own gravitational force and, unless there is a physical effect opposing this collapse, it will collapse into a black hole. In an active star like the Sun, gravitational collapse is opposed and balanced by the pressure of the thermonuclear reactions. We shall show that in a white dwarf, where thermonuclear reactions have been extinguished, the collapse is balanced by the pressure of the electrons, if the mass M is less than M_{C}. The orders of magnitude characteristic of a white dwarf are:

- mass density at the centre: $\rho \sim 10^7$ g cm^{-3}
- mass: $M \sim 2\times 10^{30}$ kg
- radius: $R \sim 5\times 10^6$ m
- temperature at the centre: $T \sim 10^7$ K

A white dwarf has a mass close to that of the Sun and a radius close to that of the Earth.

1. By assuming that a white dwarf is made of (helium-4)$^{++}$ ions and electrons, and that its density is uniform, use the above data to calculate the electron density, the Fermi momentum p_F and the Fermi energy. Take into account the fact that the electron gas is relativistic. By comparing the Fermi temperature to that of the star, verify that we are dealing here with a very good approximation of a degenerate Fermi gas.

2. We shall now calculate the kinetic energy E_C of the electrons by assuming a uniform density and a degenerate Fermi gas. It will be convenient to include in E_C the mass energy of the electrons which is constant (Nmc^2) and which will have no influence on our arguments. Using the dimensionless variables $x = p/(mc)$ and $x_F = p_F/(mc)$, where m is the electron mass and c the speed of light, show that E_C (including the rest mass energy) is given by

$$E_C = \frac{8\pi V m^4 c^5}{h^3} \int_0^{x_F} \mathrm{d}x\, x^2 \sqrt{1+x^2} = \frac{8\pi V m^4 c^5}{h^3} f(x_F)$$

The function f is given by (see a table of integrals)

$$f(x_F) = \frac{1}{8}\left[x_F\sqrt{1+x_F^2}\,(1+2x_F^2) - \ln\left(x_F + \sqrt{1+x_F^2}\right)\right]$$

It is instructive to consider the non-relativistic ($x_F \ll 1$) and ultra-relativistic ($x_F \gg 1$) limits by expanding the integrand in powers of x and $1/x$ respectively (what are the limitations of the latter expansion?). Calculate in the two cases the first two terms in the expansion and verify the result by expanding $f(x_F)$ in powers of x_F and $1/x_F$. Hint: Expand the derivative of the logarithm for $x_F \ll 1$. Show that in the non-relativistic case, the first term gives the mass energy. *For the rest of the problem, we focus exclusively on the ultra-relativistic case.* Show that E_C is given by

$$E_C \simeq \frac{\hbar c V}{4\pi^2}(3\pi^2 n)^{4/3}\left(1 + \frac{m^2c^2}{\hbar^2}\left(\frac{1}{3\pi^2 n}\right)^{2/3}\right) \tag{5.103}$$

Estimate the pressure due to the electrons and show that the pressure of the ions is negligible.

3. When the mass M of the star is given, its radius R can be determined by minimizing the total energy E (which is equivalent to minimizing the free energy since $T \simeq 0$). E is the sum of the kinetic energy E_C and the gravitational potential energy E_G

$$E = E_C + E_G$$

Let Z and A be the average charge and atomic mass of the ions and $\mu = (A/Z)m_\mathrm{p}$, where m_p is the proton mass. For (helium-4)$^{++}$ ions we have $\mu = 2m_\mathrm{p}$. Using (5.103), show that E_C is of the form

$$E_\mathrm{C} = \frac{C_1}{R}(1 + C_2R^2) = \frac{C_1' M^{4/3}}{R}(1 + C_2R^2)$$

and E_G is of the form

$$E_\mathrm{G} = -C_3 \frac{GM^2}{R}$$

where G is the gravitational constant. Give explicitly the constants C_1', C_2, C_3. Examine the curve giving E as a function of R at fixed M and show that it admits a minimum only if $C_3GM^{2/3} < C_1'$. Show that in this case

$$R^2 = \frac{1}{C_2}\left(1 - \left(\frac{M}{M_0}\right)^{2/3}\right) \qquad M_0 = \left(\frac{C_1'}{C_3G}\right)^{3/2}$$

Plot the curve giving R in terms of M.

4. Show that the white dwarf can exist only if

$$M \le M_0 = \left[\frac{\hbar c}{3\pi}\left(\frac{9\pi}{4\mu}\right)^{4/3}\frac{5}{3G}\right]^{3/2}$$

Give the numerical value of M_0 for $A/Z = 2$. Calculate the radius and mass density for $M = M_\odot$.

The calculation of the maximum mass was done assuming a uniform density. Chandrasekhar did not make this assumption and gave the more precise value for the maximum mass: $M_\mathrm{C} = 1.4\, M_\odot$.

5.7.4 Quark–gluon plasma

In this problem we use a system of units where $\hbar = c = k = 1$.

A Gas of π mesons

The π meson is a spinless boson which can exist in three states of charge: π^+, π^0, and π^-. We assume its mass to be zero, which is a good approximation at high enough temperatures. We therefore are in the ultra-relativistic case where the dispersion relation is given by $\varepsilon_p = |\vec{p}| = p$. We shall study the properties of a π meson gas at equilibrium at temperature T ($T = 1/\beta$) with a zero chemical potential (it is easy to see that the chemical potential for a massless boson is necessarily zero).

1. Express the logarithm of the grand partition function $\ln \mathcal{Q} = \beta PV$ in terms of T and V. (Hint: Integrate by parts.) Express then the pressure $P(T)$ and the energy density $\epsilon(T)$ as functions of T.

2. Give the expression for the constant volume specific heat per unit volume. Obtain from this result the entropy density $s(T)$. Calculate the particle density $n(T)$ and show that the densities $n(T)$ and $s(T)$ are proportional

$$s(T) = \frac{2\pi^4}{45\,\zeta(3)} n(T) \tag{5.104}$$

3. Calculate the average energy per particle and show that for this gas the classical approximation is invalid for all temperatures.

B Ultra-relativistic fermion gas

Consider now a gas of massless spin-1/2 fermions. This gas contains, in general, particles and antiparticles (for example electrons and positrons) whose number is not fixed. However, charge conservation fixes the difference in the numbers of particles (N_+) and antiparticles (N_-):

$$N_+ - N_- = \text{const} = N_0$$

1. Let $F(N_0, V, T; N_+)$ be the total free energy of the gas. By exploiting the fact that at equilibrium F must be a minimum with respect to variations of the internal variable N_+, show that the chemical potential for particles, μ_+, and that for antiparticles, μ_-, are equal and opposite in sign. We take $\mu_+ = -\mu_- = \mu$.

2. Show that $\ln \mathcal{Q}$ is given by

$$\ln \mathcal{Q} = \frac{VT^3}{3\pi^2} \int_0^{+\infty} \mathrm{d}x\, x^3 \left[\frac{1}{\mathrm{e}^{(x-\mu')} + 1} + \frac{1}{\mathrm{e}^{(x+\mu')} + 1} \right] \tag{5.105}$$

with $\mu' = \beta\mu$. Calculate $\ln \mathcal{Q}$ in the case $\mu = 0$ and deduce the expression for the pressure $P(T)$.[26] Calculate the entropy density s and particle density n.

3. Calculate $T \ln \mathcal{Q}$ when $T = 0$ and $\mu \neq 0$. Obtain P, ϵ and n. What is the entropy in this case?

4. Show that when both T and μ are not equal to zero, we have

$$T \ln \mathcal{Q} = \frac{V}{6\pi^2} T^4 \left(\frac{7\pi^4}{30} + \pi^2 \left(\frac{\mu}{T}\right)^2 + \frac{1}{2} \left(\frac{\mu}{T}\right)^4 \right) \tag{5.106}$$

[26] This regime corresponds to the physical situation described in Question 3 of Exercise 5.6.7.

C Quark–gluon gas

The π mesons are made of quarks and antiquarks which are spin-1/2 fermions that can be considered massless. The quarks are of two types, 'up' and 'down' of respective charges 2/3 and −1/3 in units of the proton charge.[27] Furthermore, there is another quantum number (called 'colour') which takes three values for the quarks and eight values for the gluons. At low temperature, the quarks and gluons do not exist as free particles, they are confined inside the mesons (and more generally confined in the hadrons, the particles that participate in the strong interaction). As the temperature rises, a transition takes place that frees the quarks and gluons. We thus obtain a gas, assumed ideal, of quarks and gluons: The *quark–gluon plasma.*

1. Show that for a zero quark chemical potential, the pressure of this ideal gas is

$$P_{\text{plasma}}(T) = \frac{37\pi^2}{90} T^4 \tag{5.107}$$

2. It can be shown that the above expression should be modified: one has to add to the free energy of the quark–gluon plasma a volume contribution BV where B is a positive constant. What is the new expression for the pressure P?

3. Compare the expressions for the pressure of the π meson gas and the quark–gluon plasma. Show that at low temperature the meson phase is stable while at high temperature it is the quark–gluon plasma that is stable (it may be helpful to plot P as a function of T in the two cases). What is the value of the transition temperature T_c as a function of B? Calculate T_c numerically for $B^{1/4} = 200$ MeV.

4. Show that the transition is of first order and calculate the latent heat.

5.7.5 Bose–Einstein condensates of atomic gases

We consider an ideal gas of non-relativistic spinless bosons of mass m in an external potential $U(\vec{r})$ slowly varying in space.

1. Considering the level counting in a box of volume ℓ^3 such that $U(\vec{r})$ changes slowly over the distance ℓ, show that the density of states is

$$\rho(\varepsilon) = \frac{2\pi(2m)^{3/2}}{h^3} \int \mathrm{d}^3r \sqrt{\varepsilon - U(\vec{r})}\, \theta(\varepsilon - U(\vec{r}))$$

27 In fact, there are four other types (flavours) of quarks: strange (s), charm (c), bottom (b) and top (t); but these quarks are heavy and do not appear in the quark–gluon plasma.

where $\theta(x)$ is the Heaviside function. Another method is to use a semi-classical approximation for the density of states

$$\rho(\varepsilon) = \int \frac{\mathrm{d}^3 p\, \mathrm{d}^3 r}{h^3}\, \delta(\varepsilon - H(\vec{p}, \vec{r}))$$

Write the Hamiltonian $H(\vec{p}, \vec{r})$ for a single particle and obtain again the above result for $\rho(\varepsilon)$. In the case of free bosons confined in a box of volume V, what is $U(\vec{r})$? Find in this case the usual expression for $\rho(\varepsilon)$ (used in Section 5.5.1).

2. Take for $U(\vec{r})$ the potential energy of the harmonic oscillator

$$U(\vec{r}) = \frac{1}{2} m\, \omega^2 r^2$$

Calculate $\rho(\varepsilon)$. We give the useful integral

$$\int_0^1 \mathrm{d}u\, u^2 (1 - u^2)^{1/2} = \frac{\pi}{16} \tag{5.108}$$

3. Verify that with a judicious choice of the zero of energy, the chemical potential can be chosen to vanish at the Bose–Einstein transition. Show that the presence of the confining potential modifies the critical temperature of the condensation and that expression (5.70) should be replaced by

$$kT_{\mathrm{c}} = \hbar\omega \left(\frac{N}{\zeta(3)} \right)^{1/3}$$

Also show that for $T \leq T_{\mathrm{c}}$, the fraction N_0/N of atoms in the ground state satisfies a different law from (5.71)

$$\frac{N_0}{N} = 1 - \left(\frac{T}{T_{\mathrm{c}}} \right)^3 \tag{5.109}$$

4. In the experiment of Mewes *et al.* [92], $N = 15 \times 10^6$ sodium atoms are confined in a magnetic trap, their potential energy being that of a three-dimensional harmonic oscillator of frequency $\omega = 2\pi \times 122\,\mathrm{Hz}$. What is the transition temperature? (In the experiments, the trap is asymmetric and ω is an average value.)

5. Now considering an arbitrary potential energy, show that for $T = T_{\mathrm{c}}$ the density of particles as a function of $\vec{r}$ is given by

$$n(\vec{r}) = \frac{2\pi (2m)^{3/2}}{h^3} \int_{U(\vec{r})}^{\infty} \mathrm{d}\varepsilon\, \frac{\left(\varepsilon - U(\vec{r})\right)^{1/2}}{\mathrm{e}^{\beta\varepsilon} - 1} \tag{5.110}$$

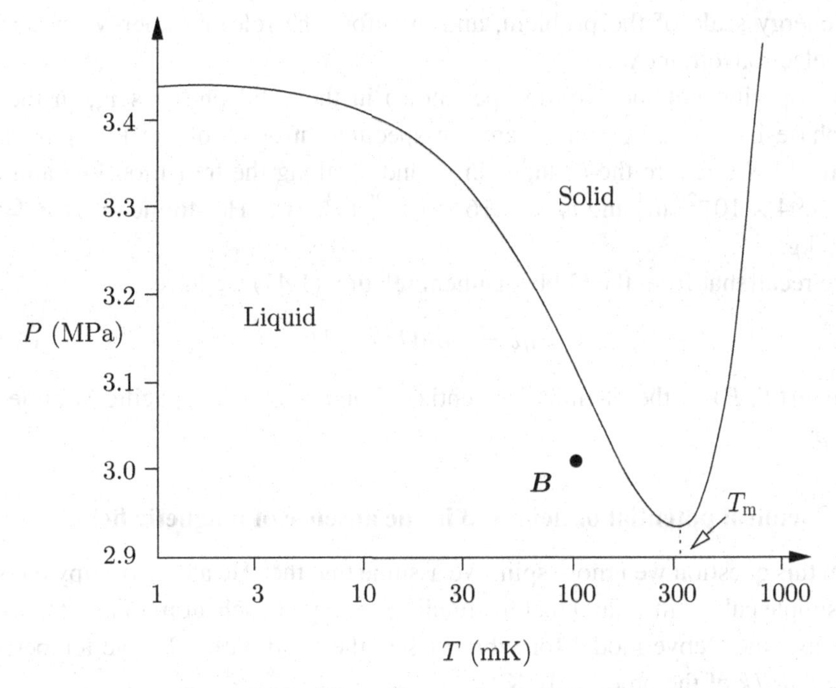

Figure 5.20 Phase diagram in the P–T plane.

What choice have you made for the zero of the energy? Show that the density $n(\vec{r} = 0)$ at the transition is identical to that given by (5.70) if $U(\vec{r})$ has a global minimum at $r = 0$. Calculate the numerical value of this density in the experiment of Mewes *et al.* The mass of the sodium atom used in the experiment is $m = 3.8 \times 10^{-26}$ kg.

5.7.6 *Solid–liquid equilibrium for helium-3*

Helium-3 is an isotope whose nucleus contains two protons and a neutron and has spin $1/2$ and, consequently, the nucleus is a fermion. As for helium-4, the total angular momentum of the two electrons is zero (the nuclear spin of ^{4}He also vanishes), and therefore the total angular momentum of the ^{3}He atom is $1/2$ and is of nuclear origin. At very low temperature ($T \lesssim 1$ K), helium-3 is liquid for pressures less than about 30 atm (1 atm $\simeq 10^5$ Pa), and solid for higher pressures. The phase diagram is shown in Figure 5.20. The reference pressure will be taken to be $P_0 \simeq 3.4 \times 10^6$ Pa, which is the transition pressure at $T = 0$. In all this problem, except Question C.5, the temperature range considered is $10^{-2}\,\mathrm{K} \lesssim T \lesssim 1\,\mathrm{K}$.[28]

[28] For even lower temperatures, of the order of 3 mK, ^{3}He becomes superfluid. This superfluid phase is beyond the scope of the current problem which focuses on temperatures higher than a few mK.

The energy scale of the problem, and therefore the relevant energy unit, is the milli-electronvolt, meV.

The specific volume (volume per atom) in the solid phase is v_s, in the liquid phase is v_ℓ and σ_s and σ_ℓ are the specific entropies of the two phases respectively. We ignore the changes in v_s and v_ℓ along the transition line and take $v_s = 3.94 \times 10^{-29}\,\mathrm{m}^3$ and $v_\ell = 4.16 \times 10^{-29}\,\mathrm{m}^3$. The ^{3}He atomic mass $m = 5 \times 10^{-27}$ kg.

We recall that from the Gibbs–Duhem relation (1.41) we have

$$d\mu = -\sigma\, dT + v\, dP \tag{5.111}$$

where $\mu(T, P)$ is the chemical potential, v and σ are the specific volume and entropy.

A Chemical potential of helium-3 in the absence of magnetic fields

1. In this question we ignore spin. We assume that the ^{3}He atoms occupy the sites of a simple cubic lattice and that the binding energy of each atom to its site is $-u_0$. We adopt the Debye model for vibrations in the solid with a Debye temperature $T_D = \hbar\omega_D/k$ of the order of 16 K.

Using (5.64) and (5.61), show that in the temperature range of interest, the free energy per atom is given by

$$f = -u_0 + \frac{9}{8}\,\hbar\omega_D - \frac{1}{5}\,\pi^4 (kT)\left(\frac{T}{T_D}\right)^3 \tag{5.112}$$

Argue that dropping the last term in (5.112) is an excellent approximation.

2. Now we include the spin 1/2 of the atom. In the temperature range considered here, the spins are totally disordered in the absence of a magnetic field. What should we then add to f to take the spins into account?

3. Show that the chemical potential of the solid phase is

$$\mu_s(T, P) = -u_0 + \frac{9}{8}\,\hbar\omega_D - kT \ln 2 + v_s P \tag{5.113}$$

Does this expression for μ_s obey the third law of thermodynamics? Consider Figure 5.21. What do you expect to happen for $T \lesssim 10^{-2}$ K?

In what follows we assume the solid to be incompressible.

B Helium-3 as an ideal Fermi gas

Assume that we may ignore interactions between the helium-3 atoms, in other words we consider helium-3 to be an ideal Fermi gas. Calculate the logarithm of

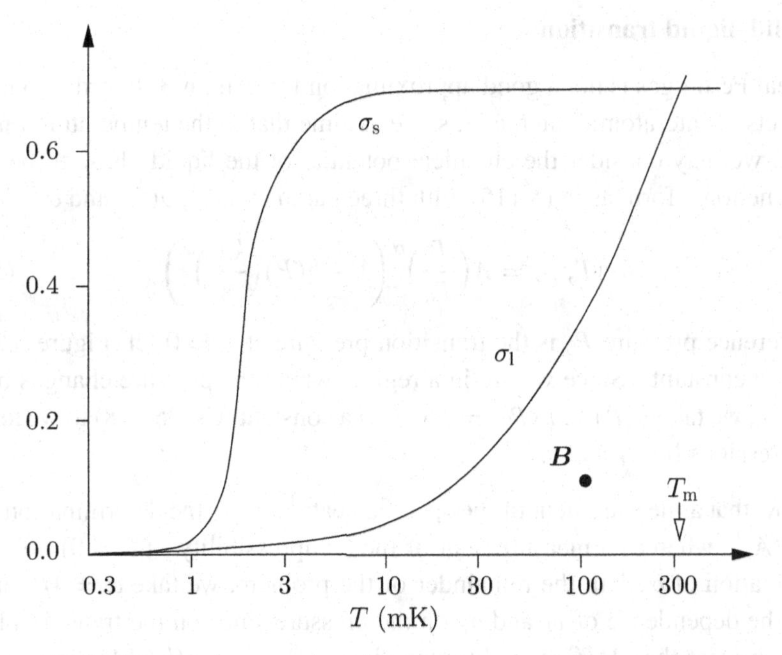

Figure 5.21 The entropy as a function of temperature in the solid and liquid phases.

the grand canonical partition function, $\ln \mathcal{Q}$, and show that to order $(kT/\mu)^2$, the pressure can be written in terms of the chemical potential as

$$P = \frac{1}{\beta V} \ln \mathcal{Q} = \frac{2}{15} \frac{(2m)^{3/2}}{\pi^2 \hbar^3} \mu^{5/2} \left(1 + \frac{5\pi^2}{8} \left(\frac{kT}{\mu}\right)^2\right) \tag{5.114}$$

Consider a reference pressure P_0 to which corresponds the Fermi energy $\varepsilon_F(P_0)$. Show that to order T^2, the chemical potential of the Fermi ideal gas may be written in the form

$$\mu(T, P) = \varepsilon_F(P_0) \left(\frac{P}{P_0}\right)^{2/5} \left(1 - \left(\frac{kT}{\varepsilon_F(P_0)}\right)^2 \left(\frac{P_0}{P}\right)^{4/5}\right) \tag{5.115}$$

Also show that the compressibility coefficient per unit volume κ_T at $T = 0$ is given in terms of the density of states at the Fermi level $\rho(\varepsilon_F)$ by

$$\rho(\varepsilon_F) = \frac{1}{\pi^2 \hbar^3} p_F^2 \frac{dp}{d\varepsilon}\bigg|_{p=p_F} = \frac{m p_F}{\pi^2 \hbar^3} \tag{5.116}$$

and the density n by

$$\kappa_T(T = 0) = \frac{1}{n^2} \rho(\varepsilon_F) \tag{5.117}$$

C Solid–liquid transition

The ideal Fermi gas is not a good approximation for helium-3. In order to include the effects of interatomic interactions, we assume that in the temperature range of interest, we may consider the chemical potential of the liquid phase to be of the same functional form as in (5.115) with three parameters A, $b(P)$ and α

$$\mu_\ell(T,p) = A\left(\frac{P}{P_0}\right)^\alpha \left(1 - b(P)\left(\frac{kT}{A}\right)^2\right) \tag{5.118}$$

The reference pressure P_0 is the transition pressure at $T = 0$ (cf. Figure 5.20), A and α are constants. Since we are in a regime where the pressure changes by less than 15%, we take $b(P) \simeq b(P_0) = b$ with b a constant. Use (5.118) to write down the expressions for σ_ℓ and v_ℓ.

1. Show that a measurement of the specific heat[29] allows the determination of the ratio b/A,[30] whereas a measurement of the compressibility $\kappa(T=0)$ allows the determination of α.[31] In the remainder of the problem, we take $\alpha = 4/5$ and we ignore the dependence of v_ℓ and σ_ℓ on the pressure since on the transition line P changes by less than 15%. In addition we ignore the term $b(kT/A)^2$ in v_ℓ since it is much smaller than 1 (this will be justified in Question 4).

2. Write the condition that gives the pressure $P(T)$ at the transition between the liquid and solid phases in terms of T and the other parameters in (5.118). Use this to obtain $\mathrm{d}P/\mathrm{d}T$, or equivalently, use the Clapeyron formula. The derivative $\mathrm{d}P/\mathrm{d}T$ vanishes at $T = T_\mathrm{m} \simeq 0.32\,\mathrm{K}$. Determine b/A.

3. Using the value of this derivative at $T = 0.01\,\mathrm{K}$

$$\left.\frac{\mathrm{d}P}{\mathrm{d}T}\right|_{T=0.01\mathrm{K}} \simeq -3.7\times 10^6\,\mathrm{Pa\,K^{-1}}$$

determine $(v_\ell - v_\mathrm{s})$. Verify that your results are in good agreement with the numerical values given in the introduction.

4. Determine A and b in terms of v_ℓ, P_0 and T_m. Justify the approximation $b(kT/A)^2 \ll 1$. Give the numerical values of A in meV, A/k in K, and b (use $v_\ell P_0 = 0.884\,\mathrm{meV}$). Compare these values with those for the ideal Fermi gas at

[29] For $T \to 0$, the constant volume and constant pressure specific heats become equal, cf. (1.59).

[30] Experimentally, the results have the form $c/T \simeq a + bT^2 \ln T$, and c/T decreases with T instead of staying constant.

[31] Our parametrization is based on a simplified version of the Landau theory of Fermi liquids (Section 5.2.3 and Problem 8.6.7). To include the effect of interactions between atoms, this theory introduces two parameters: (i) an effective mass m^* so that the density of states and the specific heat are multiplied by m^*/m compared with the free theory, (ii) a parameter Λ_0 which modifies the compressibility (5.117), which becomes $\kappa(T=0) = \rho(\varepsilon_\mathrm{F})/(n^2(1+\Lambda_0))$. The density of states is calculated with the effective mass.

the same density. With $u_0 \simeq 0.58$ meV, calculate the Debye frequency $\hbar\omega_D$ in K. Estimate the speed of sound in solid helium-3.

5. Starting at point B in the liquid phase with $T < T_m$ (cf. Figure 5.20), we compress isothermally until we reach the transition line. We then remove the heat reservoir and compress adiabatically and quasi-statically. Sketch in Figure 5.20 the path followed by the representative point of the transformation. Show that the liquid–solid transition for $T < T_m$ is accompanied by cooling. This is the 'Pomeranchuk effect', which is used to cool down to temperatures of the order of mK. Explain qualitatively why the liquid is more ordered than the solid.

6. We now consider the regime $T \lesssim 10^{-2}$ K (Figure 5.21). Why does the derivative dP/dT vanish at $T = 0$? Sketch on Figure 5.21 the path of the representative point for the transformation discussed in the previous question. Determine graphically the minimum temperature we can reach.

D Polarized helium-3

1. The nuclear magnetic moment of ^{3}He is $\mu_{He} = -2.13 \times \mu_N$, where $\mu_N = e\hbar/(2m_p)$ is the nuclear magneton. If solid helium-3 at $T = 0.01$ K is subjected to a magnetic field of 10 T parallel to the z axis, what will be the average value of the z component of the spin $S_z/\hbar$?
2. By using optical laser pumping methods, we can obtain average values $\langle S_z \rangle/\hbar \simeq 1/2$. In such a case, what does $P(T)$ look like qualitatively? Show in particular that the curve no longer has a minimum (the new transition pressure at $T = 0$ is around 2.8×10^6 Pa).

5.7.7 *Superfluidity for hardcore bosons*

We have seen that the ideal Bose gas (no interactions) undergoes Bose–Einstein condensation at finite temperature in three dimensions. However, this does not mean that the ideal Bose gas in the condensed phase is superfluid. As soon as the fluid is set in motion the condensate and the superfluid state are destroyed.

In order to have a stable superfluid, interactions are necessary. We shall not study the conditions for stability in this problem but we will consider a system of strongly interacting bosons. It turns out that it is straightforward to analyze the properties of the *hardcore* boson gas in the mean field approximation at *zero* temperature $T = 0$.[32]

In this problem we consider the simple case of a two-dimensional lattice with periodic boundary conditions: the bosons occupy lattice sites and are allowed to

[32] It is also straightforward to consider *spin-wave corrections* to the mean field result. These corrections take into account low lying quasi-particle excitations and change the results quantitatively but not qualitatively.

jump from site to site. Due to the hardcore interaction between the bosons, multiple occupation of a site is not allowed.

Note: There is no true condensate in two dimensions at *finite* temperature, but at $T = 0$ (the present case) the condensate can exist.

The system is described by the following Hamiltonian

$$\mathsf{H} = -t \sum_{\langle ij \rangle} (\mathsf{a}_j^\dagger \mathsf{a}_i + \mathsf{a}_i^\dagger \mathsf{a}_j) - \mu \sum_i \mathsf{a}_i^\dagger \mathsf{a}_i \tag{5.119}$$

where i and j label the two-dimensional lattice sites and $\langle ij \rangle$ denotes nearest neighbours. The operator $\mathsf{a}_i^\dagger$ creates a boson on site i and a_i destroys one at site i. The parameter t $(= \hbar^2/2m)$ is called the hopping parameter and is a measure of the kinetic energy per jump. μ is the chemical potential and $\mathsf{a}_i^\dagger \mathsf{a}_i = \mathsf{n}_i$ is the number operator on site i and the sum over all sites gives the total number operator: $\sum_i \mathsf{n}_i = \mathsf{N}_\mathrm{b}$. The expectation value of this operator is the average number of bosons, N_b, on the lattice.

The creation and destruction operators, $\mathsf{a}_i^\dagger$ and a_i, obey the commutation relation $[\mathsf{a}_i^\dagger, \mathsf{a}_j] = 0$ for $i \neq j$ and the anticommutation relation $\{\mathsf{a}_i^\dagger, \mathsf{a}_i\} = \mathsf{a}_i^\dagger \mathsf{a}_i + \mathsf{a}_i \mathsf{a}_i^\dagger = 0$ that forbids multiple occupancy.

It is easier to deal with this system by mapping it onto a system of interacting spins 1/2:[33]

$$\begin{aligned} \mathsf{a}_i^\dagger \mathsf{a}_i &= \mathsf{S}_i^z + \frac{1}{2} \\ \mathsf{a}_i &= \mathsf{S}_i^- \\ \mathsf{a}_i^\dagger &= \mathsf{S}_i^+ \end{aligned} \tag{5.120}$$

where S_i^z is the z component of the spin operator and where we have taken $\hbar = 1$, $\mathsf{S}_i^+, \mathsf{S}_i^-$ are the spin raising and lowering operators on site i:

$$\mathsf{S}_i^+ = \mathsf{S}_i^x + i\mathsf{S}_i^y, \qquad \mathsf{S}_i^- = \mathsf{S}_i^x - i\mathsf{S}_i^y \tag{5.121}$$

This is a spin-1/2 system, $[\mathsf{S}_i^-, \mathsf{S}_j^+] = -2\delta_{ij}\mathsf{S}_i^z$, $\mathsf{S}_i^+|-\rangle_i = |+\rangle_i$, $\mathsf{S}_i^-|+\rangle_i = |-\rangle_i$, $\mathsf{S}_i^-|-\rangle_i = 0$, where $|-\rangle_i$ and $|+\rangle_i$ are respectively the spin down and spin up states at site i. Also recall that S_i^z is diagonal, $\langle + |\mathsf{S}_i^z| - \rangle = 0$.

1. Write down the Hamiltonian in the spin variables. With this mapping, it should be evident that the occupation of a given site is either 0 or 1 and that, therefore, the hardcore constraint is automatically satisfied.

We now analyze this Hamiltonian in the mean field approximation. To this end we need to choose the mean field state. To fix ideas we consider the system in two dimensions, i.e. the site label i (or j) denotes the x and y coordinates of the site.

[33] This is a special case of the Holstein–Primakoff transformation for $s = 1/2$. See Reference [63] for more detail.

2. The empty state (no bosons) is clearly

$$|0\rangle = \prod_i |-\rangle_i \qquad (5.122)$$

where $|-\rangle_i$ is a down spin at site i and $\prod_i$ denotes a product over all the lattice sites. Notice that the sites are independent. Show that this state contains no bosons.

3. Now we take a more general mean field state. We assume that the spins are independent and all make the same angle θ with the positive z axis (i.e. the system is magnetized). The state at site i is:[34]

$$|\psi_i\rangle = [u e^{-i\phi_i/2} + v \mathsf{S}_i^+ e^{i\phi_i/2}]|-\rangle_i \qquad (5.123)$$

where $u = \sin\theta/2$ ($v = -\cos\theta/2$) is the probability amplitude that the spin is down (up). ϕ_i is the azimuthal angle which may depend on the site i. In this part we will take $\phi_i = 0$ everywhere.

The mean field state for the whole system is given by a product of the independent $|\psi_i\rangle$

$$|\Psi\rangle = \prod_i |\psi_i\rangle \qquad (5.124)$$

The angle θ can be easily related to the density as follows. The total z component of the spin is $\mathsf{S}^z_{\text{tot}} = \sum_i \mathsf{S}^z_i$. Using the first of Equations (5.120), relate $\mathsf{S}^z_{\text{tot}}$ to $\mathsf{N}_b = \sum_i \mathsf{n}_i$ and N. Now use the mean field state to calculate the expectation values directly

$$\langle \mathsf{S}^z_{\text{tot}}\rangle = \langle\Psi|\sum_i \mathsf{S}^z_i|\Psi\rangle \qquad (5.125)$$

Use these two results to show that the particle density $\rho = N_b/N$ is related to θ via

$$\cos\theta = 2\left(\rho - \frac{1}{2}\right) \qquad (5.126)$$

4. The number of bosons in the condensate is given by the number of bosons in the ground state, i.e. the zero momentum state,

$$N_0 = \langle\Psi|\tilde{\mathsf{a}}^\dagger(\vec{k}=0)\tilde{\mathsf{a}}(\vec{k}=0)|\Psi\rangle \qquad (5.127)$$

[34] This can be obtained simply by rotating the down spin by an angle $(\theta - \pi)$ about the y axis, then by an angle ϕ about the z axis:

$$|\psi_i\rangle = e^{i\phi_i \mathsf{S}^z_i} e^{i(\theta-\pi)\mathsf{S}^y_i}|-\rangle_i$$

where

$$\tilde{\mathsf{a}}^\dagger(\vec{k}) = \frac{1}{\sqrt{N}} \sum_{\vec{j}} \mathrm{e}^{-\mathrm{i}\vec{k}.\vec{j}2\pi/L} \mathsf{a}_j^\dagger \tag{5.128}$$

$$\tilde{\mathsf{a}}(\vec{k}) = \frac{1}{\sqrt{N}} \sum_{\vec{j}} \mathrm{e}^{\mathrm{i}\vec{k}.\vec{j}2\pi/L} \mathsf{a}_j \tag{5.129}$$

The Fourier transforms $\tilde{\mathsf{a}}^\dagger(\vec{k})$ and $\tilde{\mathsf{a}}(\vec{k})$ create or destroy a particle with momentum $\vec{k}2\pi/L$ where $\vec{k} = (k_x, k_y)$, $0 \le k_x < (L-1)$ and $0 \le k_y < (L-1)$ are integers and where we have taken a square lattice with $N = L^2$ sites.

Express N_0 in terms of S^+ and S^- and show that in the mean field approximation the condensate fraction is given by

$$\frac{\langle \mathsf{N}_0 \rangle}{N} = \rho_0 = \rho(1-\rho) \tag{5.130}$$

Comment on this result and compare with the ideal gas case. What is the limit of this expression for $\rho \ll 1$? Can you explain this?

5. The condensate fraction, ρ_0, is related to the order parameter of the system: $\langle \mathsf{a}_i \rangle = \langle \mathsf{a}_i^\dagger \rangle^\star$. We demonstrate this with the mean field model. Using Equations (5.123) and (5.124), show that

$$|\langle \Psi | \frac{1}{N} \sum_i \mathsf{a}_i | \Psi \rangle| = uv \frac{1}{N} |\sum_{i=1}^{N} \mathrm{e}^{\mathrm{i}\phi_i}| \tag{5.131}$$

This shows clearly that if the phase is random, i.e. if there is no long range order in the phase, called 'Off-Diagonal Long Range Order' (ODLRO), the order parameter vanishes. Assume $\phi_i = 0$ everywhere, i.e. perfect ODLRO, and calculate the order parameter and express it in terms of ρ. Compare the result with (5.130).

6. The grand potential is given by[35]

$$\mathcal{J} = \langle \Psi | \mathsf{H} | \Psi \rangle \tag{5.132}$$

Calculate this quantity in the mean field approximation and show that it is given by

$$\mathcal{J} = -\frac{\mu N}{2} - \frac{\mu N}{2} \cos\theta - tN \sin^2\theta \tag{5.133}$$

[35] Recall from Equation (3.131) that

$$TS = \frac{1}{\beta} \ln \mathcal{Q} + E - \mu N_\mathrm{b}$$

where, of course, $\mathcal{J} = -(\ln \mathcal{Q})/\beta$. Then at $T = 0$, we have $\mathcal{J} = E - \mu N_\mathrm{b}$. This gives Equation (5.132).

7. The angle θ is a variational parameter, its optimum value is obtained by minimizing the grand potential. Obtain θ as a function of μ and express the condensate fraction, the grand potential and the boson density in terms of μ.

8. To calculate the *superfluid density*, i.e. the density of the part of the fluid that flows coherently without dissipation, we recall that the average momentum of a quantum state is given by

$$\vec{p} = \langle \Psi | - \mathrm{i}\vec{\nabla} | \Psi \rangle \tag{5.134}$$

Calculating this quantity in the mean field state (5.124) with (5.123) we see that, since θ is constant, $\vec{p} = 0$ if $\phi =$ constant or $\phi =$ random with zero average.

So the only way to obtain superfluid flow in this system is to have a *gradient in the phase* of the quantum state. But since we have periodic boundary conditions, and since the wave function must be single valued, we must have the following condition for the phase gradient $\nabla\phi$

$$\sum_i \nabla\phi_i = 2\pi\ell \tag{5.135}$$

where ℓ is an integer.[36] In other words, if we go around the system, the wave function should return to its original value. We can satisfy this condition by taking the mean field state to be

$$|\Psi\rangle_T = \prod_i \mathrm{e}^{\mathrm{i}\phi_i}[u + v\mathrm{S}_i^+]|-\rangle_i \tag{5.136}$$

where ϕ_i increases by $\delta\phi = 2\pi/L$ per lattice spacing as we move in the positive x direction. We will take ϕ_i to be constant in the y direction. Clearly the total phase change as we go around the system once in the x direction is 2π and the velocity quantization condition is satisfied. Show that the grand potential $\mathcal{J}_\mathrm{T}$ for this 'twisted' mean field state is given by

$$\mathcal{J}_\mathrm{T} = -\frac{\mu N}{2} - \frac{\mu N}{2}\cos\theta - \frac{tN}{2}(1 + \cos\delta\phi)\sin^2\theta \tag{5.137}$$

Keeping $\delta\phi$ constant, minimize $\mathcal{J}_\mathrm{T}$ with respect to θ and find the chemical potential in terms of θ.

9. The difference in the two energies $\mathcal{J}_\mathrm{T}$ and $\mathcal{J}$ is due to the fact that in the first case the fluid is in motion with a velocity $v = \delta\phi(\hbar/m) = \delta\phi/m$ (since we take $\hbar = 1$) while in the second case the fluid is at rest. Therefore, the difference is due only to the kinetic energy of the superfluid, $(\Delta\mathcal{J})/N = \rho_\mathrm{s} m v^2/2$ where ρ_s is the density of 'superfluid particles'. Calculate ρ_s as a function of ρ in the limit of very large size L, i.e. $\delta\phi = 2\pi/L \to 0$. Don't forget: $t = \hbar^2/(2m)$.

[36] This gives the velocity quantization condition for superfluid flow.

10. Compare ρ_s with ρ_0. This result is valid only in the mean field limit. If we include quasi-particle excitations (spin waves) this relation changes. Can you give a physical argument of how collisions will change this? For more details see [16].

5.8 Further reading

The symmetrization postulate for the state vector is remarkably well explained and commented in Lévy-Leblond and Balibar [80] (Chapter 7); see also Cohen-Tannoudji *et al.* [30] (Chapter XIV) or Messiah [89] (Chapter XIV). The spin-statistics theorem is demonstrated at an advanced level by Streater and Wightman [117] (Chapter 4). Further discussion of quantum statistics may be found, for example, in Reif [109] (Chapter 9), Baierlein [4] (Chapters 8 and 9), Huang [57] (Chapters 9, 11 and 12) or Goodstein [49] (Chapter 2). The Landau theory of Fermi liquids is discussed in Pines and Nozières [102] (Chapter 1) or Baym and Pethick [15]. The direct measurement of fractional charges in the quantum Hall effect was performed by Saminadayar *et al.* [112]. An excellent book on the big bang and the origin of the universe is Weinberg's [122]. The criterion for superfluidity was first elucidated by Onsager and Penrose [98], see also Leggett [77, 78], Goodstein [49] (Chapter 5), Nozières and Pines [96] and Ma [85] (Chapter 30). The first gaseous atomic condensates were observed by Anderson *et al.* [2] and Mewes *et al.* [92]; Bose–Einstein condensation has also been observed for hydrogen [45]. See also the review articles by Ketterle [62] and Burnett *et al.* [23] and, at an advanced level, Dalfovo *et al.* [32]. Interactions in a Bose–Einstein condensate are treated in Landau and Lifshitz [70] (Section 78) and in Huang [57] (Chapter 19). Applications of quantum statistics to condensed matter can be found in Ashcroft and Mermin [3] and in Kittel [64].

The following references may be useful for the problems. Quantum mechanical treatment of the motion of a charged particle in a magnetic field can be found in Cohen-Tannoudji *et al.* [30] (Appendix E_{VI}). A review of applications of statistical mechanics to the physics of stars is in Balian and Blaizot [8]. For further reading on the physics of stars we recommend Rose [111]. An introduction to the quark–gluon plasma is in Le Bellac [73] (Chapter 1). For the properties of liquid helium-3, consult Wilks [123] (Chapter 17) and Tilley and Tilley [118] (Chapter 9).

6

Irreversible processes: macroscopic theory

In the preceding chapters, we have limited our analysis to equilibrium situations. This is rather restrictive since non-equilibrium phenomena, such as heat conduction or diffusion, are of great interest and cannot be ignored. To remedy this, we focus in this chapter on an introduction to non-equilibrium phenomena. Further developments of the subject will be found in Chapters 8 and 9.

We have seen that equilibrium statistical mechanics is built on a general and systematic approach, namely the Boltzmann–Gibbs distribution. No such general approach is available for non-equilibrium situations; instead, we find a large variety of methods suited to particular cases and situations. What we are able to control well are cases close to equilibrium where we can rely on rather general methods like linear response theory, which will be described in Sections 9.1 and 9.2. In the present chapter, we shall consider a macroscopic approach, that of transport coefficients, which is the non-equilibrium analogue of equilibrium thermodynamics. At this stage, we shall not attempt a calculation of these transport coefficients from a microscopic theory. We shall only show that these coefficients satisfy a number of general properties, their actual values being taken from experiments. This parallels equilibrium thermodynamics where we uncovered a number of general relations between thermodynamic quantities while we did not attempt to calculate, for example, the specific heat from a microscopic theory but took its value from experiments. In Chapters 8 and 9 we shall demonstrate, for some simple cases, how to calculate transport coefficients starting with a microscopic theory (kinetic theory in Chapter 8 and linear response in Chapter 9) just as we calculated the specific heats in some simple cases by using equilibrium statistical mechanics.

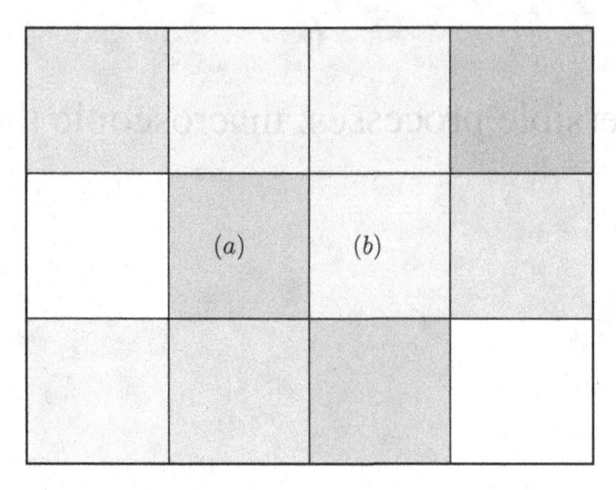

Figure 6.1 Dividing a system into cells.

6.1 Flux, affinities, transport coefficients

6.1.1 Conservation laws

A physical system can be so far out of equilibrium that quantities like the temperature or pressure cannot even be defined. We shall not consider such extreme cases and concentrate, instead, on cases where we can define locally thermodynamic variables. We start with an idealized case where we assume the system to be composed of homogeneous cells, small on the macroscopic scale but large on the microscopic one, and labelled $(a, b, \dots)$ (Figure 6.1).[1] We also assume that cells interact weakly with their neighbours so that each cell independently attains a local equilibrium with a microscopic relaxation time, τ_{micro}, which is very small compared to the macroscopic relaxation time, τ_{macro}, needed to achieve global equilibrium:[2] $\tau_{\text{micro}} \ll \tau_{\text{macro}}$. We say we have a situation of *local equilibrium* when

(*i*) each subsystem is at equilibrium independently of the other subsystems,
(*ii*) interactions between neighbouring subsystems are weak.

Let $A_i(a, t)$ be an extensive quantity (e.g. energy, number of particles, momentum etc.) labelled by the index i and contained in cell a at time t. We call *flux* $\Phi_i(a \to b)$ the amount of A_i transferred from cell a to cell b per unit time. Note that $\Phi_i(a \to b)$ is the net flux between a and b and therefore

$$\Phi_i(a \to b) = -\Phi_i(b \to a) \tag{6.1}$$

[1] In some cases (see Problem 6.5.4) these cells are not spatial.

[2] Key to the existence of these two time scales are conservation laws that forbid certain physical quantities (called slow variables) to return to global equilibrium on short time (and space) scales. As will be explained in Section 9.3, non-equilibrium statistical mechanics distinguishes fast variables, characterized by microscopic time and space scales, from slow variables characterized by macroscopic time and space scales. In Section 9.3 and in Section 2.6.3, τ_{micro} is denoted by τ^* and τ_{macro} by τ.

Cell a may have a source of quantity A_i. For example, if our system is a nuclear reactor and $A_i(a,t)$ the number of neutrons in cell a, neutrons are produced when uranium or plutonium atoms fission and therefore act as sources. On the other hand, the moderator absorbs neutrons, and therefore acts as a sink. The change per unit time of $A_i(a,t)$, $\mathrm{d}A_i(a,t)/\mathrm{d}t$, is then the sum of the contributions of the other cells and those of the sources. This leads to the equation

$$\boxed{\frac{\mathrm{d}A_i(a,t)}{\mathrm{d}t} = -\sum_{b\neq a} \Phi_i(a \to b) + \Phi_i(\text{sources} \to a)} \tag{6.2}$$

This is a *conservation equation* for the quantity A_i.

Clearly, this decomposition of the system into independent cells is an idealization. We can obtain a more realistic formulation if we express Equation (6.2) in local form. We assume that in the neighbourhood of each point in space reigns a local equilibrium, so that we can locally define thermodynamic quantities, for example a local temperature $T(\vec{r},t)$ or a local chemical potential $\mu(\vec{r},t)$. For such a local equilibrium to be possible, we need, in addition to $\tau_{\text{micro}} \ll \tau_{\text{macro}}$, a condition on length scales. If l_{micro} is a characteristic length scale over which is established a local equilibrium, and l_{macro} a length characteristic of the variations of thermodynamic quantities like the temperature, then we must have $l_{\text{micro}} \ll l_{\text{macro}}$. A regime of local equilibrium is also called a *hydrodynamic regime*: the hydrodynamic description of a fluid relies crucially on the condition of local equilibrium. So, from this construction, we see that the hydrodynamic regime, i.e. local equilibrium, describes the dynamics of a system subject to perturbations of *long wavelength* λ (long compared to microscopic lengths, $\lambda \gg l_{\text{micro}}$) and *low frequencies* ω (low compared to microscopic frequencies $\omega \ll 1/\tau_{\text{micro}}$). A local equilibrium regime will often be obtained by imposing external constraints on the system, for example a temperature gradient. If these constraints are time-independent, the system will reach a *stationary* (i.e. *time independent) non-equilibrium regime.*

Let $\rho_i(\vec{r},t)$ be the density of quantity A_i, e.g. density of energy, of particles of momentum etc. $A_i(a,t)$ is then the integral of the density over the volume $V(a)$ of the cell[3]

$$A_i(a,t) = \int_{V(a)} \mathrm{d}^3r\, \rho_i(\vec{r},t) \tag{6.3}$$

We can therefore define a corresponding current density, or more briefly a current, $\vec{j}_i(\vec{r},t)$. If $\overrightarrow{\Delta S}$ is the small oriented surface separating cells a and b (Figure 6.2(a)),

[3] The definition of densities supposes a process of 'coarse graining': the cell size needs to be large enough for the concepts of local temperature etc. to have a meaning. We also note that we are using the so-called Eulerian description: the cells are fixed in space and do not follow the movement of a given mass of fluid, which would correspond to the Lagrangian description.

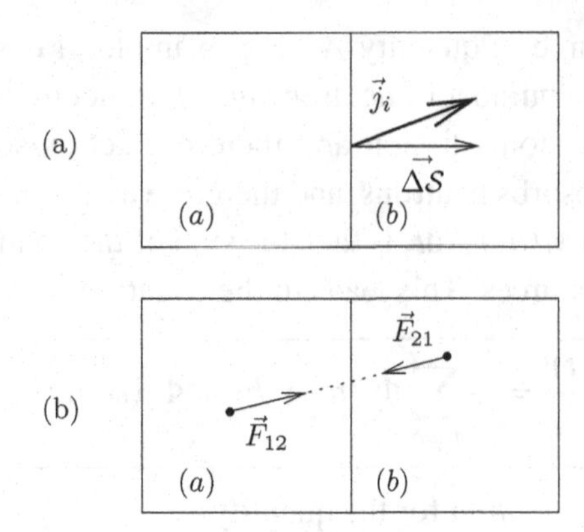

Figure 6.2 (a) Current and flux, (b) momentum exchange between two cells.

the flux $\Phi_i(a \to b)$ may be written as

$$\Phi_i(a \to b) \simeq \overrightarrow{\Delta \mathcal{S}} \cdot \vec{j}_i$$

Now let $\mathcal{S}(a)$ be the surface surrounding cell (a): the total flux crossing $\mathcal{S}(a)$ is given by the surface integral of the current

$$\Phi_i^{\mathcal{S}}(t) = \sum_b \Phi_i(a \to b) = \int_{\mathcal{S}(a)} \overrightarrow{\mathrm{d}\mathcal{S}} \cdot \vec{j}_i(\vec{r}, t) \tag{6.4}$$

The current $\vec{j}_i$ characterizes the flow of quantity i across a surface. Let us consider a simple example. If $\rho_N(\vec{r}, t) \equiv n(\vec{r}, t)$ is the particle density and $\vec{u}(\vec{r}, t)$ the average particle velocity, the current $\vec{j}_N$ is of course given by (a microscopic demonstration is proposed in Exercise 6.4.1)

$$\vec{j}_N(\vec{r}, t) = n(\vec{r}, t)\vec{u}(\vec{r}, t) \tag{6.5}$$

One should be careful, however, since currents are not always the product of a density and an average velocity. In fact, (6.5) implies that, in the absence of sources, the number of particles in cell a can change only because particles enter and leave this cell by crossing the surface $\mathcal{S}(a)$ and that the only contributions to the current are of the form (6.5). In the case of energy or momentum, there can also be exchanges between two cells even if no particles cross the surface separating them. Figure 6.2(b) illustrates that two cells can exchange momentum due to the principle of action and reaction. If we focus on the forces between the two particles, the momenta $\vec{p}_1$ and $\vec{p}_2$ of the two particles belonging to two different cells satisfy

$$\frac{\mathrm{d}\vec{p}_1(t)}{\mathrm{d}t} + \frac{\mathrm{d}\vec{p}_2(t)}{\mathrm{d}t} = 0$$

In general, the momentum current cannot be put in the form (6.5) and the right hand side of (6.2) will have three contributions from

(*i*) particles entering and leaving the volume $V(a)$ of the cell surrounded by the surface $\mathcal{S}(a)$,
(*ii*) forces applied by the other particles of the fluid outside the cell in question,[4]
(*iii*) external forces, such as gravity, which play in this case the rôle of sources.

The local version of the conservation equation (6.2) is obtained by using the Green theorem. The flux $\Phi_i^{\mathcal{S}}$ is written as a volume integral

$$\Phi_i^{\mathcal{S}} = \int_{\mathcal{S}} \mathrm{d}\vec{\mathcal{S}} \cdot \vec{j}_i = \int_V \mathrm{d}^3 r\, \vec{\nabla} \cdot \vec{j}_i$$

whereas the source term becomes the integral of a density σ_i

$$\Phi_i(\text{sources} \to \text{a}) = \int_V \mathrm{d}^3 r\, \sigma_i$$

The conservation equation (6.2) becomes

$$\int_V \mathrm{d}^3 r\, \frac{\partial \rho_i}{\partial t} = -\int_V \mathrm{d}^3 r\, \vec{\nabla} \cdot \vec{j}_i + \int_V \mathrm{d}^3 r\, \sigma_i$$

Since this equation is valid for all volumes V, we obtain from it the *local conservation equation* (or the continuity equation)

$$\boxed{\frac{\partial \rho_i}{\partial t} + \vec{\nabla} \cdot \vec{j}_i = \sigma_i} \tag{6.6}$$

Physical quantities obeying Equation (6.6) are called *conserved quantities* and will play a fundamental rôle in what follows.

6.1.2 Local equation of state

Returning to the cell picture, we can attribute an entropy $S(a)$ to each of these cells since each is at local equilibrium. Since the interactions among the cells are assumed weak, the total entropy S_{tot} is obtained as a simple sum of the individual entropies[5]

$$S_{\text{tot}} = \sum_a S(a) \tag{6.7}$$

[4] We assume that all forces have short range. Ambiguities may appear for long range forces. A remark along those lines has already been made in Chapter 1.

[5] The reader should have noted the special rôle played by the entropy compared to the other extensive variables. This special rôle will be confirmed throughout this chapter.

Let $\gamma_i(a)$ be the intensive conjugate variable of $A_i(a)$ defined by

$$\gamma_i(a) = \frac{\partial S_{\text{tot}}}{\partial A_i(a)} = \frac{\partial S(a)}{\partial A_i(a)} \tag{6.8}$$

For the five extensive quantities we consider in this chapter, namely the particle number N, the energy E, the three components of the momentum[6] $\vec{P}$ (with components $P_\alpha, \alpha = (x, y, z)$) the intensive conjugate variables are

$$A_i = N \qquad \gamma_N = -\frac{\mu}{T} \tag{6.9}$$

$$A_i = E \qquad \gamma_E = \frac{1}{T} \tag{6.10}$$

$$A_i = P_\alpha \qquad \gamma_{P_\alpha} = -\frac{u_\alpha}{T} \tag{6.11}$$

The first two equations are the classic thermodynamic relations (1.9) and (1.11); the third involves the average local velocity $\vec{u}$, or the *flow velocity* of the fluid in the considered cell. To demonstrate the third equation, we note that collective motion does not change the entropy of a fluid:[7] a glass of water on a plane in uniform motion has the same entropy as the same glass of water back on the ground. Let E' be the energy of the fluid at rest and S_0 its entropy,

$$S_0(E') = S(E, \vec{P} = 0)$$

To go from E' to E, we need to add the kinetic energy of the mass M of the fluid which is in motion

$$E = E' + \frac{\vec{P}^{\,2}}{2M}$$

and

$$S(E, \vec{P}) = S_0\left(E - \frac{\vec{P}^{\,2}}{2M}\right)$$

This yields

$$\gamma_{P\alpha} = \frac{\partial S}{\partial P_\alpha} = -\frac{P_\alpha}{M}\frac{\partial S_0}{\partial E'} = -\frac{u_\alpha}{T}$$

Let us consider all the extensive variables and use a reasoning based on the extensivity of the entropy which we already encountered in Section 1.3.3:

[6] To avoid all confusion between the momentum and the pressure, the pressure will be noted by $\mathcal{P}$ in Chapters 6 and 8. The lower case letter $\vec{p}$ will denote the momentum of a single particle while $\vec{P}$ that of a collection of particles.

[7] We can show this rigorously in statistical mechanics (Exercise 2.7.7).

$S(\lambda A_i) = \lambda S(A_i)$, where λ is a scale factor. Differentiating this equation with respect to λ and putting $\lambda = 1$, we find

$$\sum_i A_i(a) \frac{\partial S(a)}{\partial A_i(a)} = \sum_i \gamma_i(a) A_i(a) = S(a)$$

which gives

$$\boxed{\sum_{i,a} \gamma_i(a) A_i(a) = S_{\text{tot}}} \tag{6.12}$$

The local version of (6.12) is

$$\boxed{\sum_i \int \mathrm{d}^3 r\, \gamma_i(\vec{r}, t) \rho_i(\vec{r}, t) = S_{\text{tot}}(t)} \tag{6.13}$$

which we can also write in the form of functional derivatives (Section A.6)

$$\gamma_i(\vec{r}, t) = \frac{\delta S_{\text{tot}}(t)}{\delta \rho_i(\vec{r}, t)} \tag{6.14}$$

6.1.3 Affinities and transport coefficients

When two neighbouring cells a and b have different intensive variables γ_i, an exchange of A_i will take place between them. For example, a difference in temperature engenders an exchange of heat. The difference $\gamma_i(b) - \gamma_i(a)$, which measures the deviation from equilibrium, is called the *affinity* $\Gamma_i(a, b)$

$$\boxed{\Gamma_i(a, b) = \gamma_i(b) - \gamma_i(a)} \tag{6.15}$$

Let us consider the example of the temperature and suppose $\gamma_E(a) < \gamma_E(b)$ or $T(a) > T(b)$. In order to re-establish equality of temperatures, i.e. thermal equilibrium, a heat flux is established from a to b. In general, a system responds to differences in affinities between cells by attempting to establish an equilibrium state via the exchange of A_i between them. For sufficiently small affinities, we may use a linear approximation that expresses the flux in terms of the affinities

$$\boxed{\Phi_i(a \to b) = \sum_j L_{ij}(a, b) \Gamma_j(a, b)} \tag{6.16}$$

Note (*i*) the coupled character of these equations: an affinity Γ_j can produce a flux of the quantity A_i, and (*ii*) the equality $L_{ij}(a, b) = L_{ij}(b, a)$. The proportionality coefficients which relate the flux and the affinities are called response coefficients.

Let us now examine the local version of (6.16). If the centre of cell a is at point $\vec{r}$ and that of cell b at point $\vec{r} + \mathrm{d}\vec{r}$, the differences $\gamma_i(b) - \gamma_i(a)$ can be written as gradients

$$\gamma_i(b) - \gamma_i(a) \simeq \mathrm{d}\vec{r} \cdot \vec{\nabla}\gamma_i$$

However, care must be taken because, in general, currents also contain contributions from equilibrium currents which do not lead to any net exchange of A_i: only the difference $\vec{j}_i - \vec{j}_i^{\,\mathrm{eq}}$ is affected by the gradients of γ_i. For the current components j_α^i,[8] the local version of (6.16) then becomes,

$$\boxed{j_\alpha^i(\vec{r}, t) - j_\alpha^{i,\mathrm{eq}} = \sum_{j,\beta} L_{ij}^{\alpha\beta}\, \partial_\beta \gamma_j(\vec{r}, t)} \tag{6.17}$$

Equation (6.17) is called a *transport equation* and the coefficients $L_{ij}^{\alpha\beta}$ are called *transport coefficients*.

6.1.4 Examples

We illustrate the preceding formal definition with a few simple examples of transport equations.

Heat diffusion in an insulating solid (or a simple fluid)

Heat transport in an insulating solid is accomplished entirely by lattice vibrations, there is no net transport of particles. We thus write a transport equation for the heat (or energy) current $\vec{j}_E$[9]

$$\boxed{\vec{j}_E = -\kappa \vec{\nabla} T} \tag{6.18}$$

where κ is the *coefficient of thermal conductivity*. Since κ is positive, the heat current flows in a direction opposite to that of the temperature gradient. To make the connection between (6.18) and the general transport equation (6.17), we remark that, with $\gamma_E = 1/T$, the latter becomes

$$j_\alpha^E = \sum_\beta L_{EE}^{\alpha\beta}\, \partial_\beta \left(\frac{1}{T}\right) = \sum_\beta L_{EE}\delta_{\alpha\beta}\, \partial_\beta \left(\frac{1}{T}\right) = -L_{EE}\frac{1}{T^2}\, \partial_\alpha T$$

[8] There is no difference between upper and lower indexes. Our convention is to write $\vec{j}_i$ for the current and j_α^i for a component. Also, as in A.4, we write ∂_x for the partial derivative $\partial/\partial x$.

[9] For a conductor, however, conduction electrons play the dominant rôle in heat transfer. For reasons to be explained in Section 6.3.3, Equation (6.18) is also valid for a simple fluid at rest.

Since the only available vector is the temperature gradient $\vec{\nabla} T$, $\vec{j}_E$ is necessarily parallel to it which results in the Kronecker $\delta_{\alpha\beta}$.[10] Comparing with (6.18) yields

$$L_{EE}^{\alpha\beta} = \kappa T^2 \delta_{\alpha\beta} \tag{6.19}$$

We now use the exact continuity equation (6.6). In the absence of heat sources, and using $\epsilon = \rho_E$ for the energy density, we have

$$\frac{\partial \epsilon}{\partial t} = -\vec{\nabla} \cdot \vec{j}_E = \kappa \vec{\nabla} \cdot (\vec{\nabla} T) = \kappa \nabla^2 T$$

By assuming that the specific heat per unit volume,[11] C, is independent of T, i.e. $\epsilon = CT$, we obtain the diffusion equation for T

$$\boxed{\frac{\partial T}{\partial t} = \frac{\kappa}{C} \nabla^2 T} \tag{6.20}$$

Equation (6.20) is also called a heat equation. The general form of a diffusion equation for a quantity $A(\vec{r}, t)$ is

$$\boxed{\frac{\partial A}{\partial t} = D \nabla^2 A} \tag{6.21}$$

where D is the *diffusion coefficient*. From the heat equation (6.20) we therefore have $D = \kappa / C$. This equation plays a very important rôle in physics and it is worthwhile to discuss briefly its solution. By taking the spatial Fourier transform of $A(\vec{r}, t)$

$$\tilde{A}(\vec{k}, t) = \int \mathrm{d}^3 r \, \mathrm{e}^{-\mathrm{i}\vec{k}\cdot\vec{r}} A(\vec{r}, t) \tag{6.22}$$

we obtain

$$\frac{\partial}{\partial t} \tilde{A}(\vec{k}, t) = -Dk^2 \tilde{A}(\vec{k}, t)$$

whose solution is

$$\tilde{A}(\vec{k}, t) = \mathrm{e}^{-Dk^2 t} \tilde{A}(\vec{k}, 0)$$

We obtain $A(\vec{r}, t)$ by performing the inverse Fourier transform

$$A(\vec{r}, t) = \int \frac{\mathrm{d}^3 k}{(2\pi)^3} \mathrm{e}^{\mathrm{i}\vec{k}\cdot\vec{r}} \, \mathrm{e}^{-Dk^2 t} \, \tilde{A}(\vec{k}, 0) \tag{6.23}$$

[10] In general terms, the proportionality to the gradient is due to rotation invariance, see Section A.

[11] For a solid, the specific heats at constant volume and pressure are almost the same. It is therefore not necessary to specify which we are using.

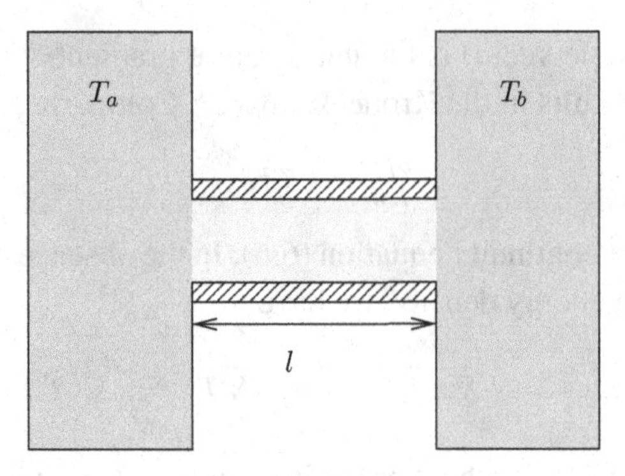

Figure 6.3 A heat conductor of length l connecting two heat reservoirs.

If the initial condition is $A(\vec{r}, 0) = \delta(\vec{r})$, that is we introduce a certain amount of A at time $t = 0$ at point $\vec{r} = 0$, then $\tilde{A}(\vec{k}, 0) = 1$ and the Fourier transform in (6.23) is that of a Gaussian

$$A(\vec{r}, t) = \frac{1}{(4\pi Dt)^{3/2}} \exp\left(-\frac{\vec{r}^{\,2}}{4Dt}\right) \tag{6.24}$$

It is easy to obtain from (6.24) the solution of the diffusion equation for an arbitrary initial condition. This is done in Exercise 6.4.2, which also shows the remarkable connection between a random walk and diffusion: Equation (6.24) shows that $\langle \vec{r}^{\,2} \rangle = 6Dt$. The property $\langle \vec{r}^{\,2} \rangle \propto t$ is a common characteristic of random walks and diffusion.[12]

As another example illustrating a simple solution of the diffusion equation, we take the case of a heat conducting bar of length l connecting two heat reservoirs at temperatures T_a and T_b with $T_a > T_b$. The entire system is thermally insulated from the outside world (Figure 6.3). We assume the heat flux to be small enough for the reservoirs to remain at constant temperatures and we consider the stationary (but not equilibrium!) situation $\partial T/\partial t = 0$. The heat equation (6.20) is thus reduced to $\partial^2 T/\partial x^2 = 0$. Applying the correct boundary conditions, we obtain the temperature of the bar as a function of position

$$T(x) = T_a + \frac{x}{l}(T_b - T_a) \tag{6.25}$$

Particle diffusion

We now examine the non-equilibrium case of a solute with inhomogeneous concentration $n(\vec{r}, t)$ in a solvent, where the temperature of the whole system is uniform. To return to equilibrium, the system will homogenize the concentration by establishing a particle current $\vec{j}_N$, thus transporting particles from higher to lower

[12] In the case of so-called anomalous diffusion, the exponent of t is different from unity.

concentration regions. To leading approximation, the current is proportional to the concentration gradient (Fick's law)

$$\boxed{\vec{j}_N(\vec{r}, t) = -D\vec{\nabla} n(\vec{r}, t)} \tag{6.26}$$

The positive proportionality coefficient D is the diffusion constant. Equation (6.26) shows that the current is in the opposite direction to the concentration gradient, as is expected. Let us relate Fick's law to the general formulation in (6.17). By using (1.42) that relates $(\partial\mu/\partial n)_T$ to the coefficient of isothermal compressibility, κ_T,

$$\left(\frac{\partial\mu}{\partial n}\right)_T = \frac{1}{\kappa_T n^2}$$

we obtain

$$\vec{\nabla}\gamma_N = \vec{\nabla}\left(-\frac{\mu}{T}\right) = -\frac{1}{T}\left(\frac{\partial\mu}{\partial n}\right)_T \vec{\nabla} n = -\frac{1}{T}\frac{1}{\kappa_T n^2}\vec{\nabla} n$$

Comparing the two expressions for the current

$$\vec{j}_N = -D\vec{\nabla} n = L_{NN}\vec{\nabla}\gamma_N$$

allows us to identify the transport coefficients $L_{NN}^{\alpha\beta}$

$$L_{NN}^{\alpha\beta} = \delta_{\alpha\beta} D T \kappa_T n^2 \tag{6.27}$$

6.1.5 Dissipation and entropy production

Let us examine the rate of change of the entropy $S(a)$ in cell (a). In the absence of sources, (6.2) gives together with (6.8)

$$\frac{\mathrm{d}S(a)}{\mathrm{d}t} = \sum_i \frac{\partial S(a)}{\partial A_i(a)}\frac{\mathrm{d}A_i(a)}{\mathrm{d}t} = \sum_i \gamma_i(a)\frac{\mathrm{d}A_i(a)}{\mathrm{d}t} = -\sum_{i,b\neq a}\gamma_i(a)\Phi_i(a\to b) \tag{6.28}$$

We then have, using the definition (6.15) of the affinity $\Gamma_i(a, b)$,

$$\begin{aligned}\gamma_i(a) &= \frac{1}{2}(\gamma_i(a) + \gamma_i(b)) + \frac{1}{2}(\gamma_i(a) - \gamma_i(b)) \\ &= \frac{1}{2}(\gamma_i(a) + \gamma_i(b)) - \frac{1}{2}\Gamma_i(a, b)\end{aligned}$$

The evolution equation of $S(a)$ becomes

$$\frac{\mathrm{d}S(a)}{\mathrm{d}t} + \sum_{b\neq a}\Phi_S(a\to b) = \frac{1}{2}\sum_{i,b\neq a}\Gamma_i(a, b)\Phi_i(a\to b) \tag{6.29}$$

with

$$\Phi_S(a \to b) = \frac{1}{2}\sum_i (\gamma_i(a) + \gamma_i(b))\Phi_i(a \to b) = -\Phi_S(b \to a) \quad (6.30)$$

If we calculate the evolution of the total entropy, $(\mathrm{d}S_{\mathrm{tot}}/\mathrm{d}t)$, the sum over a will cause the Φ_S to cancel in pairs due to their antisymmetry (6.30) under the exchange of a and b. The Φ_S part in (6.29) does not contribute to $(\mathrm{d}S_{\mathrm{tot}}/\mathrm{d}t)$ and, therefore, corresponds to reversible exchanges. Only the right hand side of (6.29), which vanishes for reversible processes, contributes to total entropy production. This production takes place at the interfaces between cells and corresponds to dissipation. *In general, entropy production is called dissipation: all physical phenomena which are accompanied by entropy production will be called* dissipative.

At this point, it is appropriate to mention two fundamental properties of the response coefficients L_{ij} defined in (6.16). If the system is isolated, we know that the total entropy can only increase, and therefore

$$\sum_{i,a,b\neq a} \Gamma_i(a,b)\Phi_i(a \to b) \geq 0 \quad (6.31)$$

Using techniques of advanced non-equilibrium statistical mechanics, one can prove the more detailed property[13]

$$\sum_i \Gamma_i(a,b)\Phi_i(a \to b) \geq 0 \quad (6.32)$$

and from (6.16), this becomes

$$\boxed{\sum_{i,j} \Gamma_i(a,b)L_{ij}(a,b)\Gamma_j(a,b) \geq 0} \quad (6.33)$$

This equation shows that the symmetric matrix $[L_{ij}(a,b) + L_{ji}(a,b)]$ is positive.[14] The rate of change of total entropy is

$$\frac{\mathrm{d}S_{\mathrm{tot}}}{\mathrm{d}t} = \frac{1}{2}\sum_{i,j,a,b} \Gamma_i(a,b)L_{ij}(a,b)\Gamma_j(a,b) \geq 0 \quad (6.34)$$

The second essential property of L_{ij} is the Onsager symmetry relation

$$L_{ij}(a,b) = L_{ji}(b,a)\ (= L_{ji}(a,b)) \quad (6.35)$$

This symmetry property is the result of time reversal invariance, or micro-reversibility. The proof of (6.35) is similar to that given in (9.67) for the dynamical

[13] However, the principle of maximum entropy is, in a sense, stronger than (6.32) because it applies even if the intermediate states, between the initial and final states, are not at local equilibrium.
[14] See Section 1.4.2 for the definition and properties of positive matrices.

susceptibility. The most precise form of (6.35) is as follows. Suppose that A_i has definite parity, $\varepsilon_i = \pm 1$, under time reversal as discussed in Sections 2.6 and 9.2.4

$$A_i^\theta(t) = \varepsilon_i A_i(-t) \tag{6.36}$$

where $A_i^\theta(t)$ is the time reversed $A_i(t)$. For example, position, particle number and energy all have parity $+1$ while velocity and momentum have parity -1. If, in addition, the system is in a magnetic field $\vec{B}$, the precise version of (6.35) is

$$L_{ij}(a, b; \{\gamma_k\}; \vec{B}) = \varepsilon_i \varepsilon_j L_{ji}(b, a; \{\varepsilon_k \gamma_k\}; -\vec{B}) \tag{6.37}$$

because time reversal implies inversion of the electrical currents that create the magnetic field. In the local formulation, the transport coefficients $L_{ij}^{\alpha\beta}$ (6.17) will therefore satisfy

$$\boxed{L_{ij}^{\alpha\beta}(\{\gamma_k\}; \vec{B}) = \varepsilon_i \varepsilon_j L_{ji}^{\beta\alpha}(\{\varepsilon_k \gamma_k\}; -\vec{B})} \tag{6.38}$$

In this local formulation, we define the entropy density s as the entropy per unit volume and, from (6.7) and (6.8) in the absence of external sources for A_i, we can write $\mathrm{d}S_{\mathrm{tot}} = \sum_{i,a} \gamma_i \, \mathrm{d}A_i(a)$ and thus $\mathrm{d}s = \sum_i \gamma_i \, \mathrm{d}\rho_i$. The local version of (6.28) is then

$$\frac{\partial s}{\partial t} = \sum_i \gamma_i \frac{\partial \rho_i}{\partial t} = -\sum_i \gamma_i (\vec{\nabla} \cdot \vec{j}_i) \tag{6.39}$$

where the sum runs over all conserved quantities of the system. To obtain the analog of (6.29) we rewrite (6.39) in the form

$$\frac{\partial s}{\partial t} + \vec{\nabla} \cdot \left(\sum_i \gamma_i \vec{j}_i \right) = \sum_i \vec{j}_i \cdot \vec{\nabla} \gamma_i \tag{6.40}$$

If we integrate over the volume of the system to calculate $\mathrm{d}S_{\mathrm{tot}}/\mathrm{d}t$, the second term on the left hand side of (6.40) does not contribute. In fact, using the Green theorem

$$\int \mathrm{d}^3 r \, \vec{\nabla} \cdot \left(\sum_i \gamma_i \vec{j}_i \right) = \sum_i \int \left(\overrightarrow{\mathrm{d}S} \cdot \vec{j}_i \right) \gamma_i = 0$$

because the currents vanish at the surface of the system. Only the term $\sum_i \vec{j}_i \cdot \vec{\nabla} \gamma_i$ contributes to the creation of total entropy: in other words, it acts as an entropy source.

We have seen in (6.17) that we should distinguish between total currents and equilibrium currents: equilibrium currents do not contribute to entropy production because at equilibrium the entropy remains constant. Then the quantity

$$\sum_i \vec{j}_i^{\,\mathrm{eq}} \cdot \vec{\nabla}\gamma_i$$

must be written as a divergence. We can show (Problem 6.5.1) for a simple fluid, i.e. a fluid with one type of molecule, that

$$\sum_{i=1}^{5} \vec{j}_i^{\,\mathrm{eq}} \cdot \vec{\nabla}\gamma_i = -\vec{\nabla} \cdot \left(\frac{\mathcal{P}}{T}\vec{u}\right) \tag{6.41}$$

where $\mathcal{P}$ is the pressure. The index i goes from 1 to 5: particle number, energy and the three components of the momentum. Define the entropy current as

$$\boxed{\vec{j}_S = \sum_i \gamma_i \vec{j}_i + \frac{\mathcal{P}}{T}\vec{u}} \tag{6.42}$$

Then, (6.40) gives

$$\frac{\partial s}{\partial t} + \vec{\nabla} \cdot \vec{j}_S = \sum_i \left[(\vec{j}_i - \vec{j}_i^{\,\mathrm{eq}}) \cdot \vec{\nabla}\gamma_i \right]$$

which, upon using (6.17), becomes

$$\boxed{\frac{\partial s}{\partial t} + \vec{\nabla} \cdot \vec{j}_S = \sum_{i,j,\alpha,\beta} (\partial_\alpha \gamma_i)\, L_{ij}^{\alpha\beta}\, (\partial_\beta \gamma_j)} \tag{6.43}$$

The right hand side of this equation describes entropy production at point $\vec{r}$.

We illustrate entropy production with the simple example of the heat conducting bar (Figure 6.3) in a stationary regime. Let Q be the heat transferred per unit time from a to b and $\mathcal{S}$ the cross-sectional area of the bar. The only current is the energy current flowing along the x axis and which is equal to $j_E = Q/\mathcal{S}$. Therefore, (6.42) gives the entropy current as

$$j_S(x) = \frac{1}{T(x)}\, j_E = \frac{1}{T(x)} \frac{Q}{\mathcal{S}} \tag{6.44}$$

where the temperature, $T(x)$, is given by (6.25). Now consider a section of the bar $[x, x + \mathrm{d}x]$, and the entropy current entering and leaving it. The position dependence of the entropy current leads to a negative entropy balance

$$j_S(x) - j_S(x + \mathrm{d}x) = \frac{Q}{\mathcal{S}} \left(\frac{1}{T(x)} - \frac{1}{T(x + \mathrm{d}x)} \right) = \frac{1}{T^2} \frac{Q}{\mathcal{S}} \frac{\mathrm{d}T}{\mathrm{d}x}\, \mathrm{d}x < 0 \tag{6.45}$$

In this stationary situation, the entropy of the slice must remain constant, which cannot happen unless entropy is produced at every point of the bar. The entropy production is given by

$$-\frac{1}{T^2}\frac{Q}{\mathcal{S}}\frac{\mathrm{d}T}{\mathrm{d}x}\,\mathrm{d}x > 0 \tag{6.46}$$

This term corresponds to the entropy source in the continuity equation (6.43). It is instructive to obtain the above result using this equation. For the present case, the right hand side of (6.43) is given by

$$L_{EE}\left(\frac{\partial \gamma_E}{\partial x}\right)^2 = \kappa T^2 \frac{1}{T^4}\left(\frac{\mathrm{d}T}{\mathrm{d}x}\right)^2 = -\frac{\kappa}{T^2}\frac{\mathrm{d}T}{\mathrm{d}x}\frac{Q}{\kappa \mathcal{S}} = -\frac{1}{T^2}\frac{Q}{\mathcal{S}}\frac{\mathrm{d}T}{\mathrm{d}x}$$

where we have used (6.19) for L_{EE} and (6.18) in the form $Q/\mathcal{S} = -\kappa\,\mathrm{d}T/\mathrm{d}x$, which indeed gives (6.46). We also verify that the rate of total entropy production per unit time corresponds to that obtained from the entropy change of the reservoirs

$$\frac{\mathrm{d}S_{\text{tot}}}{\mathrm{d}t} = -Q\int_0^l \mathrm{d}x\,\frac{1}{T^2}\frac{\mathrm{d}T}{\mathrm{d}x} = Q\left(\frac{1}{T_b} - \frac{1}{T_a}\right) \tag{6.47}$$

This example illustrates clearly that entropy is produced at every point of the bar.

6.2 Examples

6.2.1 Coupling between thermal and particle diffusion

A simple but instructive model for transport is that of a gas of light particles in motion, scattering elastically off randomly located scattering centres. This model, which we will identify as the Boltzmann–Lorentz model in Chapter 8, applies, for example, in the following situations:

- neutrons in a nuclear reactor,
- electrons in a semiconductor or conductor,
- light solute molecules in a solvent with very heavy molecules,
- impurities in a solid at high temperature.

In general, we also use the ideal gas approximation, classical or quantum, for the light particles, which often have a very small density. In this model, energy is conserved since the collisions are elastic, but momentum is not conserved because it is absorbed in collisions with scattering centres. We therefore only consider the conserved particle and energy densities, n and ϵ, as well as the associated currents $\vec{j}_N$ and $\vec{j}_E$. This gives equations which couple the diffusion of heat and particles: a temperature gradient can produce a particle flux and, conversely, a density gradient can cause heat flow. Assuming the medium is isotropic, which means

$L_{ij}^{\alpha\beta} = \delta_{\alpha\beta} L_{ij}$ (see Section A.4.2), we can write Equation (6.17) as

$$\vec{j}_E = L_{EE} \vec{\nabla} \frac{1}{T} + L_{EN} \vec{\nabla} \left(\frac{-\mu}{T} \right) \tag{6.48}$$

$$\vec{j}_N = L_{NE} \vec{\nabla} \frac{1}{T} + L_{NN} \vec{\nabla} \left(\frac{-\mu}{T} \right) \tag{6.49}$$

The Onsager relations (6.35) imply $L_{EN} = L_{NE}$. The thermal conductivity is defined in the absence of particle current: $\vec{j}_N = 0$,

$$L_{NE} \vec{\nabla} \frac{1}{T} + L_{NN} \vec{\nabla} \left(\frac{-\mu}{T} \right) = 0$$

Expressing $\vec{\nabla}(-\mu/T)$ in terms of $\vec{\nabla}(1/T)$, we find

$$\vec{j}_E = L_{EE} \vec{\nabla} \frac{1}{T} - \frac{L_{NE}}{L_{NN}} L_{EN} \vec{\nabla} \left(\frac{1}{T} \right)$$

Comparing with Equation (6.18) we obtain the coefficient of thermal conductivity κ

$$\kappa = \frac{1}{T^2 L_{NN}} \left(L_{EE} L_{NN} - L_{EN}^2 \right) \tag{6.50}$$

We note that the positivity of the 2×2 matrix of transport coefficients implies that κ is positive. It is instructive to emphasize the differences between (6.50) and (6.19). In an insulating solid, heat transport is effected by the vibrations of the lattice and it is tempting to interpret it as the result of particle transport where the particles are phonons. However, unlike for molecules, the phonon density does not obey a continuity equation since phonons can be destroyed and created with no constraints. Therefore, heat transport in an insulating solid cannot be interpreted as being due to particle transport. The diffusion coefficient is defined at constant temperature and results of Section 6.1.4 remain unchanged.

6.2.2 Electrodynamics

As another example, we examine the case where the gas of light particles is a gas of charge carriers to which correspond an electric charge density, $\rho_{\rm el}$, and an electric current density, $\vec{j}_{\rm el}$,

$$\rho_{\rm el} = qn \qquad \vec{j}_{\rm el} = q \vec{j}_N \tag{6.51}$$

where q is the charge of the carriers. We assume that these charges are placed in an average electric potential $\Phi(\vec{r})$, and thus an average electric field $\vec{E} = -\vec{\nabla}\Phi$. We ignore magnetic and polarization effects of the medium. The electric current is given by the local Ohm's law

$$\vec{j}_{\text{el}} = \sigma_{\text{el}} \vec{E} = -\sigma_{\text{el}} \vec{\nabla} \Phi \tag{6.52}$$

The *electrical conductivity* σ_{el} is a transport coefficient, in fact one of the most familiar transport coefficients, which gives rise to dissipation via the Joule effect. We shall show that electrical conductivity and diffusion are intimately connected. To this end, we study the effect of the potential Φ on the entropy. Placing the system in a macroscopic force field that changes very slowly at the microscopic scale does not change its entropy because each energy level of a particle is simply shifted by $q\Phi$. The densities of charge carriers and energy in the absence of the potential, n' and ϵ', are related to those in the presence of the potential, n and ϵ, by

$$n' = n \qquad \epsilon' = \epsilon - nq\Phi$$

and the entropy density satisfies

$$s(\epsilon, n; \Phi) = s(\epsilon - nq\Phi, n; \Phi = 0) = s'(\epsilon', n') \tag{6.53}$$

where s' is the entropy density in the absence of the potential. This yields the chemical potential[15]

$$\mu = -T \frac{\partial}{\partial n} s(\epsilon, n; \Phi) = -T \frac{\partial}{\partial n} s'(\epsilon - nq\Phi, n) = \mu' + q\Phi \tag{6.54}$$

where μ' is the chemical potential in the absence of the electric potential

$$\mu' = \mu(\epsilon - nq\Phi, n; \Phi = 0) \tag{6.55}$$

For uniform temperature, we deduce from (6.54) the particle current density

$$\vec{j}_N = L_{NN} \vec{\nabla} \left(-\frac{\mu}{T} \right) = -\frac{1}{T} L_{NN} \left. \frac{\partial \mu'}{\partial n} \right|_T \vec{\nabla} n + \frac{q}{T} L_{NN} \vec{E} \tag{6.56}$$

The current density $\vec{j}_N$ has a component due to diffusion which tends to make the system uniform, and an electric component produced by the externally applied electric field. Both components are governed by the same microscopic mechanism and controlled by the transport coefficient L_{NN}. For a uniform density we have

$$\vec{j}_{\text{el}} = \sigma_{\text{el}} \vec{E} \qquad \sigma_{\text{el}} = \frac{q^2}{T} L_{NN} \tag{6.57}$$

Anticipating results in Chapter 8, we now follow a simple argument from kinetic theory. If τ^* is the time between two successive collisions of a charge carrier, which we call collision time, and if collisions are not correlated (no memory), a

[15] The transformation rule for μ is obvious in the grand canonical ensemble where nothing changes if we shift all energy levels and μ by $q\Phi$.

particle of mass m will be accelerated while between collisions and its velocity will increase by[16]

$$\vec{u} = \frac{q\vec{E}}{m}\tau^* = \mu_{\rm el}(q\vec{E}) = \mu_{\rm el}\vec{F} \tag{6.58}$$

The quantity $\mu_{\rm el}$ is called electric mobility. We see in Equation (6.58) that it is the velocity, not the acceleration, that is proportional to the force! Of course this is due to the fact that the particle has to restart its motion from zero after each collision since it has no memory of its previous condition. The concept of mobility also appears in problems involving viscous forces. For example, an object falling in a viscous fluid will attain a limiting speed $v_L = g/\gamma$, where g is the gravitational acceleration, and the viscous force is given by $-\gamma m v$ (Exercise 6.4.3). The velocity v_L is proportional to the force mg.

In general, the electric current is given as a function of the mobility by

$$\vec{j}_{\rm el} = qn\vec{u} = q^2 n\mu_{\rm el}\vec{E}$$

Comparing this with (6.57) we find

$$\sigma_{\rm el} = q^2 n\mu_{\rm el} \qquad L_{NN} = Tn\mu_{\rm el} \tag{6.59}$$

For a classical ideal gas, we have $\kappa_T = 1/\mathcal{P}$ and $\mathcal{P} = nkT$ and, therefore, Equation (6.27) becomes

$$\boxed{D = \mu_{\rm el}kT} \tag{6.60}$$

This is an Einstein relation. Another Einstein relation, relating the diffusion coefficient D of a spherical particle of radius R and the viscosity (defined in Section 6.3.3), is the object of Exercise 6.4.3 where we show

$$\boxed{D = \frac{kT}{6\pi\eta R}} \tag{6.61}$$

It is instructive to obtain (6.60) using an argument from equilibrium statistical mechanics. We suppose that the potential Φ depends only on the position x. Clearly, this position dependence of the potential creates an electric current

$$\vec{j}_{\rm el} = -\sigma_{\rm el}\vec{\nabla}\Phi$$

corresponding to a particle current

$$\frac{1}{q}\vec{j}_{\rm el} = -qn\mu_{\rm el}\frac{\partial\Phi}{\partial x}\hat{x}$$

[16] The velocity is written $\vec{u}$ because it is a collective velocity on top of the velocities due to thermal fluctuations.

When the particles are at equilibrium in an external force field, $q\Phi(x)$, the density $n(x)$ follows a Boltzmann law

$$n(x) \propto \exp(-\beta q \Phi(x))$$

The non-uniform density entails a diffusion current

$$\vec{j}_D = -D\frac{\partial n}{\partial x}\hat{x} = D\beta q \frac{\partial \Phi}{\partial x} n\hat{x}$$

However, at equilibrium the total particle current must vanish

$$\vec{j}_N = \frac{1}{q}\vec{j}_{\rm el} + \vec{j}_D = \vec{0}$$

This condition of vanishing total current gives Equation (6.60). A similar argument is applied in Exercise 6.4.3 to obtain (6.61).

6.3 Hydrodynamics of simple fluids

6.3.1 Conservation laws in a simple fluid

Our aim in this section is not to study hydrodynamic flow, for which the interested reader is referred to the appropriate references. Instead, our goal is to illustrate with an important example the concepts defined in Section 6.1. A *simple fluid* is a fluid made of one type of structureless molecule, the archetypal example being argon in liquid form (dense fluid) or gaseous form (dilute fluid). In such a case there are only five conserved densities: particle density, energy density and the three components of the momentum density. Instead of using the particle density, we shall use the mass density, $\rho = mn$, where m is the mass of the molecule. The momentum density is $\vec{g}$ and the energy density is ϵ as before. To these densities correspond currents and conservation laws in the form of continuity equations like (6.6). We shall see that the mass current, $\vec{j}_M$, is nothing more than the momentum density, $\vec{g}$. The momentum current is a tensor of components $T_{\alpha\beta}$ since the momentum density itself is a vector. The energy current is a vector $\vec{j}_E$. We shall use in this section the convention that repeated indexes are summed (see Section 10.4). With this convention, the divergence of a vector $\vec{A}$ is written

$$\vec{\nabla} \cdot \vec{A} = \partial_\alpha A_\alpha \tag{6.62}$$

We now examine the conservation of mass and momentum.[17]

[17] We recommend that the reader who has no previous experience with hydrodynamics solves Problem 6.5.2 before continuing.

Mass conservation

Let the flow velocity $\vec{u}(\vec{r},t)$ be the velocity of a fluid element (also called 'fluid particle' in hydrodynamics) at point $\vec{r}$ and time t. Recall that the hydrodynamic approximation assumes that at each point reigns a local equilibrium corresponding to a local temperature $T(\vec{r},t)$, a local entropy density $s(\vec{r},t)$ etc. According to (6.6), the mass current is given by (Exercise 6.4.1 provides a microscopic demonstration)

$$\frac{\partial \rho}{\partial t} + \vec{\nabla}\cdot(\rho\vec{u}) = 0 \tag{6.63}$$

With our convention on repeated indexes, we write this as

$$\boxed{\frac{\partial \rho}{\partial t} + \partial_\beta(\rho u_\beta) = 0} \tag{6.64}$$

We can rewrite this mass conservation equation by expanding the second term

$$\frac{\partial \rho}{\partial t} + u_\beta\,\partial_\beta\rho + \rho\,\partial_\beta u_\beta = \frac{D\rho}{Dt} + \rho(\vec{\nabla}\cdot\vec{u}) = 0$$

where we have introduced the *material* (*or convective*) *derivative* D/Dt

$$\boxed{\frac{D}{Dt} = \frac{\partial}{\partial t} + u_\beta\,\partial_\beta = \frac{\partial}{\partial t} + \vec{u}\cdot\vec{\nabla}} \tag{6.65}$$

The physical interpretation of this material derivative is very important. Suppose we follow the motion of a fluid element between t and $t+\mathrm{d}t$ and where $\vec{r}$ is a fixed point in space. During the time interval $\mathrm{d}t$, the fluid element moves a distance $\mathrm{d}\vec{r} = \vec{u}\,\mathrm{d}t$, and the change in the α component of its velocity is (see Figure 6.4)

$$\begin{aligned} u_\alpha(\vec{r}+\mathrm{d}\vec{r}, t+\mathrm{d}t) - u_\alpha(\vec{r},t) &= \frac{\partial u_\alpha}{\partial t}\mathrm{d}t + \mathrm{d}\vec{r}\cdot\vec{\nabla}u_\alpha \\ &= \left(\frac{\partial u_\alpha}{\partial t} + (\vec{u}\cdot\vec{\nabla})u_\alpha\right)\mathrm{d}t = \frac{Du_\alpha}{Dt}\,\mathrm{d}t \end{aligned}$$

$D\vec{u}/Dt$ is then the acceleration of the element of fluid. The term $(\vec{u}\cdot\vec{\nabla})$ is called the *advection* or *convection* term.

Conservation of momentum

We assume that the fluid element is in the form a parallelepiped whose faces are parallel to the coordinate axes. Consider the face with the normal vector (pointing toward the outside of the parallelepiped) parallel to the β axis, and let $\mathcal{S}_\beta$ be the oriented surface labelled by its normal vector. Let us introduce the stress tensor $\mathcal{P}_{\alpha\beta}$ where $-\mathcal{P}_{\alpha\beta}$ is the α component of the force per unit area applied by the fluid outside the parallelepiped on the surface $\mathcal{S}_\beta$ (see Figure 6.5). Clearly we may have $\alpha = \beta$. To obtain the total force applied by the rest of the fluid on the fluid enclosed

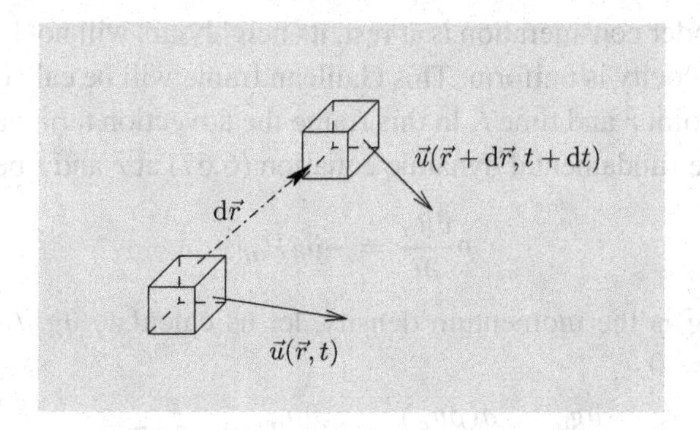

Figure 6.4 Motion of an element of fluid.

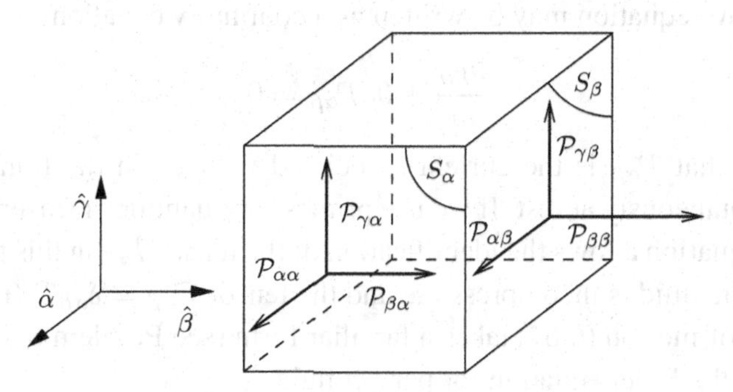

Figure 6.5 Definition of $\mathcal{P}_{\alpha\beta}$.

in the parallelepiped, we need to integrate over all the faces. Since the mass of the enclosed fluid is ρV, where V is the volume of the parallelepiped, Newton's law is given by

$$\rho V \frac{Du_\alpha}{Dt} = -\int_S \mathrm{d}^2 \mathcal{S}_\beta \mathcal{P}_{\alpha\beta} = -\int_V \mathrm{d}^3 r\, \partial_\beta \mathcal{P}_{\alpha\beta} \tag{6.66}$$

Green's theorem allows us to express the surface integral giving the flux of $\mathcal{P}_{\alpha\beta}$ across the parallelepiped as a volume integral. The above equation becomes

$$\boxed{\rho \frac{Du_\alpha}{Dt} = -\partial_\beta \mathcal{P}_{\alpha\beta}} \tag{6.67}$$

The α component of the force exerted by the rest of the fluid on a unit volume element is given by $-\partial_\beta \mathcal{P}_{\alpha\beta}$. This result leads to the momentum conservation law. If at time t we use a Galilean reference frame moving at a constant uniform velocity $\vec{u}$ with respect to the laboratory frame, the fluid element is instantaneously at rest in this frame. We say 'instantaneously' because, whereas the velocity of the reference frame does not change, the velocity of the fluid element itself will, in general, change. Furthermore, while this frame was chosen such that at time t the fluid element under consideration is at rest, its neighbours will not be at rest, unless the fluid flow velocity is uniform. This Galilean frame will be called the *rest frame of the fluid*, at point $\vec{r}$ and time t. In this frame the advection term vanishes, and by using (6.65), the fundamental dynamic equation (6.67) at $\vec{r}$ and t becomes

$$\rho \frac{\partial u_\alpha}{\partial t} = -\partial_\beta \mathcal{P}_{\alpha\beta} \tag{6.68}$$

Recalling that $\vec{g}$ is the momentum density, let us calculate $\partial g_\alpha/\partial t$ in this same frame where $\vec{u} = 0$

$$\frac{\partial g_\alpha}{\partial t} = \frac{\partial(\rho u_\alpha)}{\partial t} = \rho \frac{\partial u_\alpha}{\partial t} = -\partial_\beta \mathcal{P}_{\alpha\beta}$$

Since the above equation may be written as a continuity equation

$$\frac{\partial g_\alpha}{\partial t} + \partial_\beta \mathcal{P}_{\alpha\beta} = 0 \tag{6.69}$$

we conclude that $\mathcal{P}_{\alpha\beta}$ is the current associated with g_α in the frame where the fluid is instantaneously at rest. If we use primes for quantities measured in the rest frame, this equation allows the identification of the tensor $T_{\alpha\beta}$ in this frame: $T'_{\alpha\beta} = \mathcal{P}_{\alpha\beta}$. When the fluid is incompressible and the tensor $\mathcal{P}_{\alpha\beta} = \delta_{\alpha\beta}\mathcal{P}$ (i.e. diagonal), the equation of motion (6.67) takes a familiar form (see Problem 6.5.2). In fact, it is reduced to the Euler equation for perfect fluids

$$\boxed{\rho \frac{D\vec{u}}{Dt} = -\vec{\nabla}\mathcal{P}} \tag{6.70}$$

which confirms that $\mathcal{P}$ should be interpreted as the pressure.

Viscous effects, which are ignored in the Euler equations, make the tensor $\mathcal{P}_{\alpha\beta}$ non-diagonal. This can be illustrated by the following simple example. A horizontal plate moves at a height $z = L$ with a velocity $\vec{u} = u_0\hat{x}$ in the x direction (see Figure 6.6). The fluid between $z = 0$ and $z = L$ acquires a horizontal velocity $u_x(z)$, which depends on the height z. In the laminar regime, the relative velocity of the fluid with respect to the walls vanishes at the walls, $u_x(0) = 0$, $u_x(L) = u_0$, and varies linearly in between. The fluid above a plane of constant z applies on

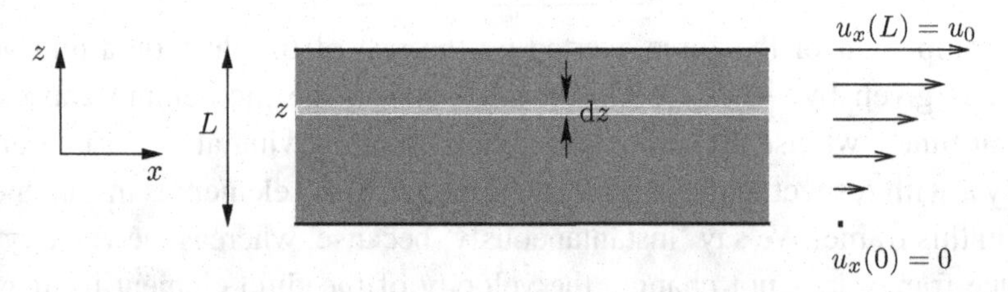

Figure 6.6 A fluid in simple shear motion.

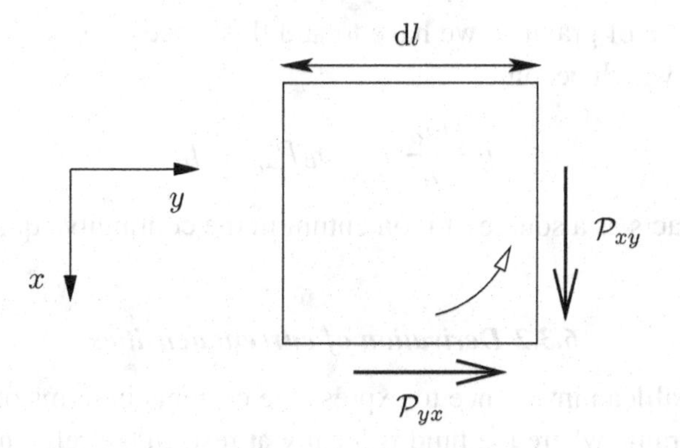

Figure 6.7 Symmetry of $\mathcal{P}_{\alpha\beta}$: if $\mathcal{P}_{yx} \neq \mathcal{P}_{xy}$, the cube will start rotating. The curved arrow shows the rotation direction when $\mathcal{P}_{yx} > \mathcal{P}_{xy}$.

the fluid below it a force per unit area equal to $-\mathcal{P}_{xz}$ where, from (6.66), x is the direction of the force and z that of the normal to the surface of separation. The *shear viscosity* η is defined by the transport equation

$$-\mathcal{P}_{xz} = \eta \frac{\mathrm{d}u_x(z)}{\mathrm{d}z} \tag{6.71}$$

We easily show an important symmetry property of $\mathcal{P}_{\alpha\beta}$. Consider the z component of the torque $\vec{\Gamma}$ applied on an infinitesimal cube of fluid of side dl (Figure 6.7)

$$\Gamma_z \propto (\mathrm{d}l)^3(\mathcal{P}_{yx} - \mathcal{P}_{xy})$$

The z component of the angular velocity $\vec{\omega}$ satisfies

$$I \frac{\mathrm{d}\omega_z}{\mathrm{d}t} = \Gamma_z \tag{6.72}$$

where I is the moment of inertia. Writing I as a function of dl and the mass M of the cube, we obtain

$$I \propto M(\mathrm{d}l)^2 = \rho(\mathrm{d}l)^3(\mathrm{d}l)^2$$

We see from (6.72) that if d$l \to 0$, we have

$$\mathrm{d}\omega_z/\mathrm{d}t = \Gamma_z/I \propto (\mathrm{d}l)^{-2} \to \infty$$

unless $\mathcal{P}_{yx} = \mathcal{P}_{xy}$, in other words $\mathcal{P}_{\alpha\beta}$ is a symmetric tensor

$$\boxed{\mathcal{P}_{\alpha\beta} = \mathcal{P}_{\beta\alpha}} \tag{6.73}$$

Finally, we may introduce external sources to the continuity equation. If the fluid is subjected to external forces, like gravity, with the external force density $\vec{f}$ (=

$-\rho g\hat{z}$ in the case of gravity), we have to add this force density to the equation of motion (6.67), which becomes

$$\rho \frac{Du_\alpha}{Dt} = -\partial_\beta \mathcal{P}_{\alpha\beta} + f_\alpha \tag{6.74}$$

The new term acts as a source of momentum in the continuity equation (6.69).

6.3.2 Derivation of current densities

We now use Galilean invariance to express the currents in terms of their values in the reference frame where the fluid is locally at rest. In the reference frame of the laboratory, a fluid element moves with a velocity $\vec{u}$. Equivalently, in the rest frame of the fluid element, the laboratory moves with a velocity $-\vec{u}$. More generally, we use a Galilean frame $R(-\vec{v})$ moving at some velocity $-\vec{v}$ relative to the rest frame of the fluid. In the frame $R(-\vec{v})$, the densities are given in terms of the (primed) densities in the rest frame by

$$\text{mass density} \qquad \rho = \rho' \tag{6.75}$$

$$\text{momentum density} \qquad \vec{g} = \rho\vec{v} \tag{6.76}$$

$$\text{energy density} \qquad \epsilon = \epsilon' + \frac{1}{2}\rho\vec{v}^{\,2} \tag{6.77}$$

To derive the current conservation law, consider a density χ and its associated current $\vec{j}_\chi$. Going from a frame at velocity $-\vec{v}$ to one at $-(\vec{v} + \mathrm{d}\vec{v})$ has two effects on the current.

(*i*) A first effect comes from the change $\mathrm{d}\chi$ in the density χ when we go from $-\vec{v}$ to $-(\vec{v} + \mathrm{d}\vec{v})$

$$\mathrm{d}\chi = \chi(\vec{v} + \mathrm{d}\vec{v}) - \chi(\vec{v}) = \zeta_\alpha(\vec{v})\, \mathrm{d}v_\alpha \tag{6.78}$$

which defines $\zeta_\alpha(\vec{v})$. If $\vec{j}_{\zeta_\alpha}$ is the current associated with ζ_α, the change in the density (6.78) causes a change in the β component of the current

$$\mathrm{d}^{(1)} j^\chi_\beta = j^{\zeta_\alpha}_\beta \mathrm{d}v_\alpha \tag{6.79}$$

(*ii*) The second effect comes from the fact that a flux is calculated relative to a fixed surface $\mathcal{S}$ in a given reference frame. But a fixed surface in the frame $R(-(\vec{v} + \mathrm{d}\vec{v}))$ moves with a velocity $-\mathrm{d}\vec{v}$ in the frame $R(-\vec{v})$, which induces an additional flux

$$\mathrm{d}\Phi_\chi = \chi(\vec{v})\vec{\mathcal{S}} \cdot \mathrm{d}\vec{v} = \chi(\vec{v})\mathcal{S}_\beta\, \mathrm{d}v_\beta$$

and a contribution to the current

$$\mathrm{d}^{(2)} j^\chi_\beta = \chi(\vec{v})\, \mathrm{d}v_\beta \tag{6.80}$$

In the case of mass current, only the second effect is present since the mass is a Galilean invariant: $\mathrm{d}\vec{j}_M = \rho\,\mathrm{d}\vec{v}$, or $\vec{j}_M = \rho\vec{v} = \vec{g}$ as expected. We now choose χ as the α component of the momentum density, $\chi = g_\alpha = \rho v_\alpha$. For an infinitesimal Galilean transformation we have $\mathrm{d}g_\alpha = \rho\,\mathrm{d}v_\alpha$. The β component of the mass density current is ρv_β, and if $T_{\alpha\beta}$ is the β component of the current associated with g_α, we have

$$\mathrm{d}^{(1)}T_{\alpha\beta} = \rho v_\beta\,\mathrm{d}v_\alpha$$

whereas the second contribution is

$$\mathrm{d}^{(2)}T_{\alpha\beta} = \rho v_\alpha\,\mathrm{d}v_\beta$$

Adding the two effects we obtain

$$\mathrm{d}T_{\alpha\beta} = \rho v_\beta\,\mathrm{d}v_\alpha + \rho v_\alpha\,\mathrm{d}v_\beta$$

This can be integrated from $\vec{v} = 0$ to $\vec{v} = \vec{u}$

$$\boxed{T_{\alpha\beta} = \mathcal{P}_{\alpha\beta} + \rho u_\alpha u_\beta} \tag{6.81}$$

subject to the boundary conditions

$$T_{\alpha\beta}(\vec{v} = 0) = T'_{\alpha\beta} = \mathcal{P}_{\alpha\beta}$$

The term $\rho u_\alpha u_\beta = g_\alpha u_\beta$ is a convection term: all currents $\vec{j}_\chi$ include a convection term $\chi\vec{u}$ due to the transport of the density χ with a velocity $\vec{u}$.

We now consider the energy density $\chi = \epsilon$. The infinitesimal transformation law for ϵ is

$$\mathrm{d}\epsilon = \rho v_\alpha\,\mathrm{d}v_\alpha = g_\alpha\,\mathrm{d}v_\alpha$$

The β component of the current associated with g_α is $T_{\alpha\beta}$. This gives

$$\mathrm{d}^{(1)}j^E_\beta = T_{\alpha\beta}\,\mathrm{d}v_\alpha$$

whereas

$$\mathrm{d}^{(2)}j^E_\beta = \epsilon\,\mathrm{d}v_\beta = \left(\epsilon' + \frac{1}{2}\rho\vec{v}^{\,2}\right)\mathrm{d}v_\beta$$

Adding the two effects and using (6.81) we obtain

$$\mathrm{d}j^E_\beta = (\rho v_\alpha v_\beta + \mathcal{P}_{\alpha\beta})\mathrm{d}v_\alpha + \left(\epsilon' + \frac{1}{2}\rho\vec{v}^{\,2}\right)\mathrm{d}v_\beta$$

which we can write in the form of a differential equation

$$\frac{\partial j^E_\beta}{\partial v_\alpha} = \left[\mathcal{P}_{\alpha\beta} + \epsilon'\delta_{\alpha\beta}\right] + \left[\rho v_\alpha v_\beta + \frac{1}{2}\rho\vec{v}^{\,2}\delta_{\alpha\beta}\right] \tag{6.82}$$

The term in the first bracket in (6.82) is measured in the rest frame and is independent of $\vec{v}$. The integration of (6.82) in the general case is left for Exercise 6.4.4. Here we consider only the one-dimensional case

$$\frac{\partial j^E}{\partial v} = (\mathcal{P} + \epsilon') + \frac{3}{2}\rho v^2$$

which, when integrated from $v = 0$ to $v = u$, gives

$$j^E = j^{E'} + (\mathcal{P} + \epsilon')u + \frac{1}{2}\rho u^3 = j^{E'} + \mathcal{P}u + \epsilon u$$

In the general case (Exercise 6.4.4) we obtain

$$\boxed{j^E_\beta = j^{E'}_\beta + \mathcal{P}_{\alpha\beta}u_\alpha + \epsilon u_\beta} \tag{6.83}$$

The term ϵu_β is a convection term whereas $\mathcal{P}_{\alpha\beta}u_\alpha$ is the work done by the pressure. The current $\vec{j}^{\,\prime}_E$ in the rest frame is the heat current. In addition to the mass conservation equation (6.64), we have, in the absence of external sources, the momentum conservation equations

$$\boxed{\frac{\partial g_\alpha}{\partial t} + \partial_\beta T_{\alpha\beta} = 0} \tag{6.84}$$

as well as energy conservation

$$\boxed{\frac{\partial \epsilon}{\partial t} + \partial_\beta j^E_\beta = 0} \tag{6.85}$$

The expressions for the momentum current $T_{\alpha\beta}$ and energy current $\vec{j}_E$ are given in (6.81) and (6.83) respectively. Adding to these equations the transport equations, (6.87) and (6.88), and the local equations of state one obtains a closed set of equations as shown in Table 6.1.

6.3.3 Transport coefficients and the Navier–Stokes equation

We conclude this brief discussion of hydrodynamics by writing down the corresponding transport equations (6.17). In principle, we expect a 5×5 matrix of transport coefficients, in other words a total of 15 coefficients after taking into

Table 6.1 *Densities and currents for a simple fluid.*

Density	Current	Equilibrium current	Continuity equation
ρ	$\vec{j}_M = m\vec{j}_N = \vec{g} = \rho\vec{u}$	$\vec{j}_M^{\,\mathrm{eq}} = \vec{j}_M$	$\dfrac{\partial \rho}{\partial t} + \vec{\nabla}\cdot\vec{g} = 0$
g_α	$T_{\alpha\beta} = \mathcal{P}_{\alpha\beta} + \rho u_\alpha u_\beta$	$T_{\alpha\beta}^{\mathrm{eq}} = \mathcal{P}\delta_{\alpha\beta} + \rho u_\alpha u_\beta$	$\dfrac{\partial g_\alpha}{\partial t} + \partial_\beta T_{\alpha\beta} = 0$
ϵ	$j_\beta^E = j_\beta^{E'} + \mathcal{P}_{\alpha\beta}u_\alpha + \epsilon u_\beta$	$\vec{j}_E^{\,\mathrm{eq}} = (\epsilon + \mathcal{P})\vec{u}$	$\dfrac{\partial \epsilon}{\partial t} + \partial_\beta j_\beta^E = 0$

account the Onsager symmetry relations (6.38). Fortunately, there are further simplifications that reduce this number to only three transport coefficients. A first crucial simplification comes from the fact that the mass current $\vec{j}_M$ is always equal to the equilibrium current $\rho\vec{u}$ regardless of whether or not the system is at equilibrium. Consequently, $L_{Mi} = L_{iM} = 0$ for all i; a spatial variation of the chemical potential does not cause a particle flux (nor an energy one), at least not directly. There is no diffusion coefficient in a simple fluid; return to equilibrium is accomplished by a complex mechanism. The number of possible affinities remains twelve: $\partial_\beta(1/T)$ and $\partial_\beta(u_\alpha/T)$. However, we can always choose to write the transport equations in a frame where the fluid is locally at rest and use the Galilean transformations (6.81) and (6.83) to obtain the currents in an arbitrary frame. Since in the rest frame we have $\vec{u} = 0$, $\partial_\beta(u_\alpha/T)$ reduces to $(1/T)\partial_\beta u_\alpha$.[18] It remains to exploit rotation and parity invariance. The energy current $\vec{j}_E'$ must be proportional to a vector and the only candidate at hand is the affinity $\vec{\nabla}(1/T)$. In fact another possible candidate would be the other affinity $\vec{\nabla} \times \vec{u}$, but this is a pseudo-vector and parity invariance does not allow a vector to be proportional to a pseudo-vector. The difference $\mathcal{P}_{\alpha\beta} - \delta_{\alpha\beta}\mathcal{P}$ is a two-dimensional symmetric tensor. The two possible constructions, using the affinities, that give such a quantity are

$$\partial_\alpha u_\beta + \partial_\beta u_\alpha \quad \text{and} \quad \delta_{\alpha\beta}(\partial_\gamma u_\gamma) = \delta_{\alpha\beta}(\vec{\nabla}\cdot\vec{u})$$

Instead of these combinations, it is convenient to introduce the traceless symmetric tensor $\Delta_{\alpha\beta}$ ($\Delta_{\alpha\alpha} = 0$)

$$\Delta_{\alpha\beta} = \frac{1}{2}\left(\partial_\alpha u_\beta + \partial_\beta u_\alpha\right) - \frac{1}{3}\,\delta_{\alpha\beta}(\partial_\gamma u_\gamma) \tag{6.86}$$

[18] It is important to keep in mind the local nature of the rest frame: $\vec{u} = 0$ only at one point and $\partial_\beta u_\alpha \neq 0$ at this point.

Then, the transport equations can be written in terms of the three independent coefficients L_{EE}, η and ζ

$$\boxed{\begin{aligned} \vec{j}'_E &= L_{EE}\vec{\nabla}\frac{1}{T} \\ \mathcal{P}_{\alpha\beta} - \delta_{\alpha\beta}\mathcal{P} &= -\zeta\, \delta_{\alpha\beta}(\partial_\gamma u_\gamma) - 2\eta\, \Delta_{\alpha\beta} \end{aligned}} \qquad \begin{aligned} (6.87) \\ (6.88) \end{aligned}$$

The coefficient of thermal conductivity κ is defined for a fluid uniformly at rest ($\vec{u} = 0\ \forall\vec{r}$) where there is a temperature gradient. In a fluid (uniformly) at rest, there are no shear forces and $\mathcal{P}_{\alpha\beta} = \delta_{\alpha\beta}\mathcal{P}$. The equation of motion (6.68) then gives $\partial_\beta\mathcal{P}_{\alpha\beta} = \partial_\alpha\mathcal{P} = 0$; thus the pressure is uniform. This implies that the density variations are such that the pressure gradient remains zero. For example, for a dilute gas we have $\mathcal{P} \simeq nkT$ and so the product nT must remain constant. As we have seen above, a density gradient does not cause a particle current: in a simple fluid, it is the pressure gradient that produces fluid motion. According to (6.19), the coefficient L_{EE} is related to the coefficient of thermal conductivity by $L_{EE} = \kappa T^2$.

In order to identify the coefficient η, consider the definition of shear viscosity and Figure 6.6. Since only the velocity component u_x is non-zero and since it depends only on z, we have $\vec{\nabla}\cdot\vec{u} = 0$. Consequently, the tensor $\Delta_{\alpha\beta}$ simplifies to

$$\Delta_{xz} = \frac{1}{2}\left(\frac{\partial u_z}{\partial x} + \frac{\partial u_x}{\partial z}\right) = \frac{1}{2}\frac{\mathrm{d}u_x}{\mathrm{d}z} \qquad (6.89)$$

and Equation (6.88) becomes

$$\mathcal{P}_{xz} = -\eta\frac{\mathrm{d}u_x}{\mathrm{d}z} \qquad (6.90)$$

The transport coefficient ζ is called bulk viscosity and in general plays a rather minor rôle. In fact it is absent for an incompressible fluid ($\vec{\nabla}\cdot\vec{u} = 0$) and we will see in Section 8.3.2 that it vanishes for a dilute mono-atomic gas.

The combination of the equations of motion with the transport equations allows us to write the fundamental equation for the dynamics of a simple fluid, the Navier–Stokes equation. The pressure tensor is written in terms of the viscosities η and ζ as

$$\mathcal{P}_{\alpha\beta} = \delta_{\alpha\beta}\mathcal{P} - \zeta\delta_{\alpha\beta}(\vec{\nabla}\cdot\vec{u}) - \eta(\partial_\alpha u_\beta + \partial_\beta u_\alpha) + \frac{2}{3}\eta\,\delta_{\alpha\beta}(\vec{\nabla}\cdot\vec{u})$$

and its divergence is

$$\partial_\beta\mathcal{P}_{\alpha\beta} = \partial_\alpha\mathcal{P} - \zeta\partial_\alpha(\vec{\nabla}\cdot\vec{u}) - \eta(\partial_\beta^2 u_\alpha) - \eta\partial_\alpha(\vec{\nabla}\cdot\vec{u}) + \frac{2}{3}\eta\partial_\alpha(\vec{\nabla}\cdot\vec{u})$$

The right hand side is the α component of the vector

$$\vec{\nabla}\mathcal{P} - \zeta\vec{\nabla}(\vec{\nabla}\cdot\vec{u}) - \eta\nabla^2\vec{u} - \frac{1}{3}\eta\vec{\nabla}(\vec{\nabla}\cdot\vec{u})$$

which then leads to the Navier–Stokes equation[19]

$$\boxed{\frac{\partial\vec{u}}{\partial t} + (\vec{u}\cdot\vec{\nabla})\vec{u} + \frac{1}{\rho}\vec{\nabla}\mathcal{P} = \frac{\eta}{\rho}\nabla^2\vec{u} + \frac{1}{\rho}\left(\frac{\eta}{3}+\zeta\right)\vec{\nabla}(\vec{\nabla}\cdot\vec{u})} \tag{6.91}$$

In the absence of dissipation, when the transport coefficients vanish, we regain the Euler equation

$$\frac{\partial\vec{u}}{\partial t} + (\vec{u}\cdot\vec{\nabla})\vec{u} + \frac{1}{\rho}\vec{\nabla}\mathcal{P} = 0 \tag{6.92}$$

To achieve the construction of Section 6.1 in this particular case, we need to establish the form of the entropy current. We start in the rest frame where, according to the general construction (6.42), we have

$$\vec{j}_S^{\,\prime} = \sum_i \gamma_i'\vec{j}_i^{\,\prime} = \frac{1}{T}\vec{j}_E^{\,\prime} \tag{6.93}$$

Since the entropy density is a Galilean invariant, the entropy current is given in general by (Problem 6.5.1)

$$\vec{j}_S = \vec{j}_S^{\,\prime} + s\vec{u} = \frac{1}{T}\vec{j}_E^{\,\prime} + s\vec{u} \tag{6.94}$$

According to (6.43), we have

$$\begin{aligned}\frac{\partial s}{\partial t} + \vec{\nabla}\cdot\vec{j}_S &= \vec{\nabla}\left(\frac{1}{T}\right)\cdot\vec{j}_E^{\,\prime} + \sum_{\alpha,\beta}\left(-\frac{1}{T}\partial_\alpha u_\beta\right)(\mathcal{P}_{\alpha\beta} - \delta_{\alpha\beta}\mathcal{P}) \\ &= \kappa T^2\left(\vec{\nabla}\frac{1}{T}\right)^2 + \frac{\zeta}{T}\left(\vec{\nabla}\cdot\vec{u}\right)^2 + \frac{2\eta}{T}\sum_{\alpha,\beta}(\Delta_{\alpha\beta})^2 \geq 0\end{aligned} \tag{6.95}$$

which implies the positivity of the coefficients κ, η and ζ.

[19] The reader will remark that $(\eta/\rho)\,\nabla^2\vec{u}$ in Equation (6.91) is a diffusion term. In the absence of advection, and if $\vec{\nabla}\mathcal{P} = 0$ and $\vec{\nabla}\cdot\vec{u} = 0$, the Navier–Stokes equation becomes a diffusion equation for the velocity. This equation allows us to justify the linear dependence of $u_x(z)$ on z in Figure 6.6.

6.4 Exercises

6.4.1 Continuity equation for the density of particles

The distribution function $f(\vec{r}, \vec{p}, t)$ of a collection of particles is the ensemble average

$$f(\vec{r}, \vec{p}, t) = \Big\langle \sum_{j=1}^{N} \delta(\vec{r} - \vec{r}_j(t))\delta(\vec{p} - \vec{p}_j(t)) \Big\rangle \tag{6.96}$$

$f(\vec{r}, \vec{p}, t)\mathrm{d}^3r\,\mathrm{d}^3p$ is the number of particles in the phase space volume element $\mathrm{d}^3r\,\mathrm{d}^3p$. Show that the density n and the current $\vec{j}_N$ are given by

$$n(\vec{r}, t) = \int \mathrm{d}^3p\, f(\vec{r}, \vec{p}, t) \qquad \vec{j}_N(\vec{r}, t) = \int \mathrm{d}^3p\, \vec{v} f(\vec{r}, \vec{p}, t) \tag{6.97}$$

Use this result to obtain Equation (6.6). Also demonstrate the continuity equation.

6.4.2 Diffusion equation and random walk

1. Let $A(\vec{r}, t)$ be a quantity obeying the diffusion equation (6.21) with the initial condition

$$A(\vec{r}, t = 0) = A_0(\vec{r}) \tag{6.98}$$

Give the expression for $A(\vec{r}, t)$ $\forall t > 0$.

2. During a random walk (in one dimension for simplicity), a walker takes a step at regular time intervals of ε. With a 50% probability, the walker jumps a distance a left or right on the x axis. Each step is independent of preceding steps and one says the random walk is a Markovian process. The walker leaves point $x = 0$ at time $t = 0$. Calculate the average distance $\langle x \rangle$ travelled in N steps as well as $\langle x^2 \rangle$. Show that $\langle x^2 \rangle$ is proportional to N, or in other words to $t = N\varepsilon$. Show that for $N \to \infty$, the probability $P(x, t)$ of finding the walker at point x at time t is a Gaussian.[20] Identify the diffusion coefficient.

6.4.3 Relation between viscosity and diffusion

Consider very small particles (but macroscopic compared to atomic scales) suspended in a fluid in thermal equilibrium at temperature T. The particle density as a function of height z is $n(z)$, m is the particle mass, k the Boltzmann constant and g the gravitational acceleration.

[20] $P(x, t)$ is in fact the conditional probability of finding the walker at x at time t knowing that at $t = 0$ it was at $x = 0$.

1. Show that $n(z)$ has the form

$$n(z) = n_0 \mathrm{e}^{-\lambda z}$$

Determine λ as a function of m, g, k and T. If we want to have observable effects over distances of the order of a centimetre, what should be the order of magnitude of the mass m at $T = 300\,\mathrm{K}$?

2. The particles are now under the influence both of gravity and a viscous force proportional to velocity

$$\vec{F} = -\alpha\vec{v}$$

where $\alpha = 6\pi\eta R$, η is the fluid viscosity and R is the radius of the particles, which are assumed spherical.[21] What is the limiting velocity v_L of the particles in the gravitational field?

3. The particles are subjected to two mutually opposing influences: on the one hand, they move down with a velocity v_L and on the other hand, diffusion tries to re-establish a uniform density. Let D be the coefficient of diffusion of the particles in the fluid. What is the diffusion current j_z^N? Why is it directed toward $z > 0$?

4. By considering that at equilibrium the gravitational and diffusion effects balance each other, establish the Einstein relation (6.61) between the viscosity and the diffusion coefficient.

6.4.4 Derivation of the energy current

Derive Equation (6.83) for the energy current. Hint: Treat the cases $\alpha = \beta$ and $\alpha \neq \beta$ separately.

6.4.5 Lord Kelvin's model of Earth cooling

Lord Kelvin's assumptions to compute the age of the Earth were the following.

(*i*) Earth was formed initially with a uniform temperature equal to the fusion temperature θ_0. In this exercise, temperatures will be measured in degrees Celsius and denoted by θ.

(*ii*) Cooling is due to diffusion, heat being transported by thermal diffusion to the Earth surface at a temperature $\theta = 0\,°\mathrm{C}$, and then dissipated in the atmosphere and in outer space. Being unaware of radioactivity, Lord Kelvin could not include in his assumptions the heat generated by radioactivity in the centre of the Earth.

[21] This law was demonstrated by Stokes. Its proof can be found in many volumes on fluid mechanics. See for example [37] or [53].

(*iii*) He also approximated Earth by a semi-infinite medium limited by the plane $x = 0$, the interior of the Earth corresponding to $x \leq 0$. Thus one is led to a one-dimensional model with a single coordinate, x.

1. Show that the following form of the temperature for $x \leq 0$

$$\theta(x,t) = \frac{\theta_0}{\sqrt{\pi D t}} \int_{-\infty}^{0} \mathrm{d}x' \exp\left(-\frac{(x-x')^2}{4Dt}\right)$$

obeys the heat equation (6.20) with the correct boundary conditions. $D = \kappa / C$ is the thermal diffusivity.

2. Compute the temperature gradient at point x inside the Earth, and at the Earth surface $x = 0$

$$\left.\frac{\partial \theta}{\partial x}\right|_{x=0} = -\frac{\theta_0}{\sqrt{\pi D t}}$$

Hint: Use the change of variables $u = x - x'$.

3. From the following value of the temperature gradient at the Earth surface as measured today

$$\left.\frac{\partial \theta}{\partial x}\right|_{x=0} = -3 \times 10^{-2}\,{}^\circ\mathrm{C\,m^{-1}}$$

and the values $4D = 1.2 \times 10^6\,\mathrm{m^2\,s^{-1}}$, $\theta_0 = 3800\,{}^\circ\mathrm{C}$, estimate the age of the Earth. Given your numerical result and the estimate of 4.5 billion years for the age of the Earth, what do you think is wrong with Lord Kelvin's assumptions? Hint: The answer is *not* in radioactivity!

4. Write an expression for $\theta(r,t)$ assuming spherical symmetry in terms of the Earth radius R. The integral that you will obtain cannot be computed analytically, but you should be able to show that the previous one-dimensional approximation is valid provided $R \gg \sqrt{Dt}$.

6.5 Problems

6.5.1 Entropy current in hydrodynamics

Consider a simple fluid characterized by the following five densities and conjugate intensive variables

$$\rho_1 = \epsilon \qquad \rho_2 = n \qquad \rho_3 = g_x \qquad \rho_4 = g_y \qquad \rho_5 = g_z$$

$$\gamma_1 = \frac{1}{T} \quad \gamma_2 = -\frac{\mu}{T} \quad \gamma_3 = -\frac{u_x}{T} \quad \gamma_4 = -\frac{u_y}{T} \quad \gamma_5 = -\frac{u_z}{T}$$

where $\vec{g} = mn\vec{u} = \rho\vec{u}$ is the momentum density. Each fluid element has a flow velocity $\vec{u}(\vec{r})$ measured in the laboratory reference frame R. We define a local Galilean frame, $R'(\vec{u}(\vec{r}))$, in which the element d^3r is at rest. All quantities in this frame are primed.

1. Starting with the Galilean invariance of the entropy density,

$$s(\epsilon, n, \vec{g}) = s'(\epsilon - \vec{g}^{\,2}/2mn, n, 0)$$

which you should justify briefly, show that the transformation law for the chemical potential is

$$\mu = \mu' - \frac{1}{2}m\vec{u}^{\,2}$$

2. Derive the following relation

$$\sum_i \gamma_i \rho_i = \sum_i \gamma_i' \rho_i' = s - \frac{\mathcal{P}}{T} \tag{6.99}$$

3. We recall that the rate of entropy dissipation is given by (6.40) and that Table 6.1 gives the definitions of the currents and equilibrium currents. At equilibrium, the total entropy of the system remains constant. This implies that the contribution of the equilibrium currents to the local entropy balance

$$\sum_i \vec{j}_i^{\,\mathrm{eq}} \cdot \vec{\nabla}\gamma_i$$

may be written in the form of a divergence of a vector. Show that

$$\sum_i \vec{j}_i^{\,\mathrm{eq}} \cdot \vec{\nabla}\gamma_i = (\vec{u} \cdot \vec{\nabla}) \sum_i \gamma_i \rho_i - \sum_i \gamma_i (\vec{u} \cdot \vec{\nabla}) \rho_i - \frac{\mathcal{P}}{T}\vec{\nabla} \cdot \vec{u} \tag{6.100}$$

Justify the following expression for the gradient of the entropy density

$$\vec{\nabla}s = \sum_i \gamma_i \vec{\nabla}\rho_i$$

and use this result as well as (6.99) to show that (6.100) can be written as

$$\sum_i \vec{j}_i^{\,\mathrm{eq}} \cdot \vec{\nabla}\gamma_i = -\vec{\nabla} \cdot \left(\frac{\mathcal{P}}{T}\vec{u}\right) \tag{6.101}$$

4. The contributions of equilibrium currents are taken into account in the definition of the entropy current

$$\vec{j}_S = \sum_i \gamma_i \vec{j}_i + \frac{\mathcal{P}}{T}\vec{u}$$

Show that

$$\sum_i \gamma_i(\vec{j}_i - \rho_i\vec{u}) = \sum_i \gamma_i'\vec{j}_i' = \frac{\vec{j}_E'}{T} \tag{6.102}$$

and establish the transformation law of the entropy density current

$$\vec{j}_S = \vec{j}_S' + s\vec{u} = \frac{\vec{j}_E'}{T} + s\vec{u}$$

Interpret these two expressions.

6.5.2 Hydrodynamics of the perfect fluid

Consider a fluid where the only internal force is the pressure; the tensor $\mathcal{P}_{\alpha\beta}$ has the form $\mathcal{P}\delta_{\alpha\beta}$. We also assume that there is no thermal conduction. In such a fluid, dissipation is absent and the fluid is called a 'perfect fluid': the right hand side of the transport equations (6.87) and (6.88) vanishes.

1. Show that the force acting on a fluid volume element $\mathrm{d}V$ is $-\vec{\nabla}\mathcal{P}\,\mathrm{d}V$. Hint: Study first the case of pressure changing only along the z axis, and take the volume element to be a parallelepiped with faces parallel to the axes. Obtain the Euler equation (6.92).

2. For what is to follow, we need a thermodynamic identity. We will derive this by considering a volume V that follows the current locally and is thus at rest in the fluid's frame. Let $s = S/V$ and $h' = \overline{H}/V$ be the entropy and enthalpy per unit volume in this reference frame (since the entropy is a Galilean invariant, there is no need to distinguish between s and s'). The entropy and enthalpy per particle are given by $\check{s} = S/N$ and $\check{h}' = \overline{H}/N$ (see Section 3.5.1). Using (3.95), show that

$$\mathrm{d}\epsilon' = T\,\mathrm{d}s + \left(\frac{h' - Ts}{\rho}\right)\mathrm{d}\rho \tag{6.103}$$

3. Define the kinetic energy density $k = \frac{1}{2}\rho\vec{u}^{\,2}$ and the corresponding current $\vec{j}_K = \frac{1}{2}\rho\vec{u}^{\,2}\vec{u}$. By using (6.92) and the equation for mass conservation, (6.64), show that

$$\frac{\partial k}{\partial t} + \partial_\beta\left[j_\beta^K + \mathcal{P}u_\beta\right] = \mathcal{P}(\partial_\beta u_\beta) \tag{6.104}$$

4. The absence of dissipation in an ideal fluid is expressed by the entropy conservation law

$$\frac{\partial s}{\partial t} + \partial_\beta(su_\beta) = 0 \tag{6.105}$$

Use this equation and (6.103) to put the time derivative of the total energy density

$$\epsilon = k + \epsilon' = \frac{1}{2}\rho\vec{u}^{\,2} + \epsilon'$$

in the form

$$\frac{\partial\epsilon}{\partial t} + \partial_\beta\left[j_\beta^K + \mathcal{P}u_\beta\right] = \mathcal{P}(\partial_\beta u_\beta) - T\partial_\beta(su_\beta) - \left(\frac{h' - Ts}{\rho}\right)\partial_\beta(\rho u_\beta) \quad (6.106)$$

Again use (6.103) to obtain the final result

$$\frac{\partial\epsilon}{\partial t} + \partial_\beta\left(\left[\frac{1}{2}\rho\vec{u}^{\,2} + h'\right]u_\beta\right) = 0 \quad (6.107)$$

Give the physical interpretation of this result.

6.5.3 *Thermoelectric effects*

A metal is a binary system made of a lattice of fixed positive ions and mobile electrons. We can observe a charge current in response to a temperature gradient and, conversely, heat is produced when current is flowing. The metal is put at a uniform potential Φ (cf. Section 6.2.2). Primed quantities correspond to the case $\Phi = 0$ and the unprimed quantities to the case $\Phi \neq 0$. We recall that in this situation $T' = T$, $n' = n$ and (6.54) $\mu = \mu' + q\Phi$.

1. *Transformation laws of L_{ij}.* We consider a non-equilibrium situation where the linear response model is relevant. We know, for the isolated system, the coefficients of linear response

$$\vec{j}_E^{\,\prime} = L'_{EE}\vec{\nabla}\left(\frac{1}{T}\right) + L'_{EN}\vec{\nabla}\left(-\frac{\mu'}{T}\right) \quad (6.108a)$$

$$\vec{j}_N^{\,\prime} = L'_{NE}\vec{\nabla}\left(\frac{1}{T}\right) + L'_{NN}\vec{\nabla}\left(-\frac{\mu'}{T}\right) \quad (6.108b)$$

The metal is now put at the potential $\Phi(\vec{r})$, which depends on space but *varies very slowly at the microscopic scale.* The transformation laws established for the uniform case are still valid locally. Establish the following transformation laws for the particle and energy current densities

$$\vec{j}_N = \vec{j}_N^{\,\prime} \qquad \vec{j}_E = \vec{j}_E^{\,\prime} + q\Phi\vec{j}_N$$

Use this result to show that in the presence of the potential $\Phi(r)$, the linear response coefficients transform as

$$L_{NN} = L'_{NN} \tag{6.109a}$$

$$L_{EE} = L'_{EE} + 2q\Phi\, L'_{NE} + q^2\Phi^2\, L'_{NN} \tag{6.109b}$$

$$L_{NE} = L_{EN} = L'_{NE} + q\Phi\, L'_{NN} \tag{6.109c}$$

Verify, with the help of these results, that the thermal conductivity coefficient of the metal κ (6.50) is the same in the presence or absence of the external potential.

2. *Seebeck effect.* The first thermoelectric effect to be studied was the Seebeck effect: in an open circuit, a temperature gradient produces an electromotive force (i.e. a gradient of the electrochemical potential μ/q). We define the 'Seebeck coefficient' (or coefficient of thermoelectric power) $\bar{\varepsilon}$ of a material as the electromotive force created in an open circuit by a unit temperature gradient

$$\vec{\nabla}\left(\frac{\mu}{q}\right) = -\bar{\varepsilon}\vec{\nabla}T \qquad \vec{j}_N = \vec{0}$$

Express the Seebeck coefficient in terms of the linear response coefficients.

We construct a thermocouple with two metallic wires A and B of different Seebeck coefficients (Figure 6.8). The junctions are maintained at two different temperatures T_1 and T_2. A capacitor is introduced in the metal B and maintained at temperature T. Calculate the potential difference across the capacitor in terms of T_1, T_2 and the Seebeck coefficients $\bar{\varepsilon}_A$ and $\bar{\varepsilon}_B$ of metals A and B respectively.

3. *Joule effect.* In an electrically neutral metal where the temperature, electric field, density of charge carriers and current densities are uniform, an electric current produces the well-known thermoelectric phenomenon, the Joule effect. Verify

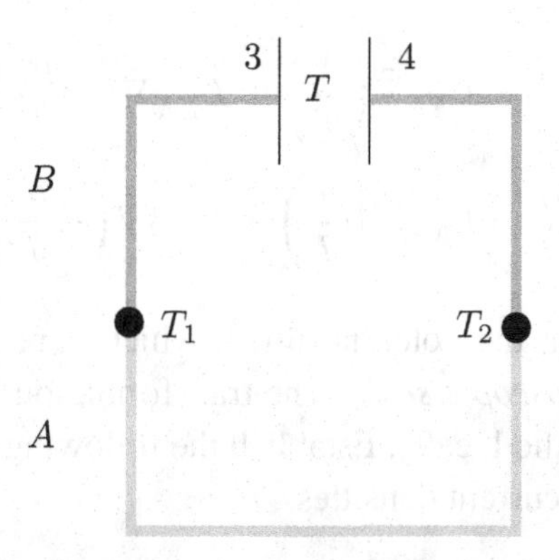

Figure 6.8 Schematic representation of a thermocouple.

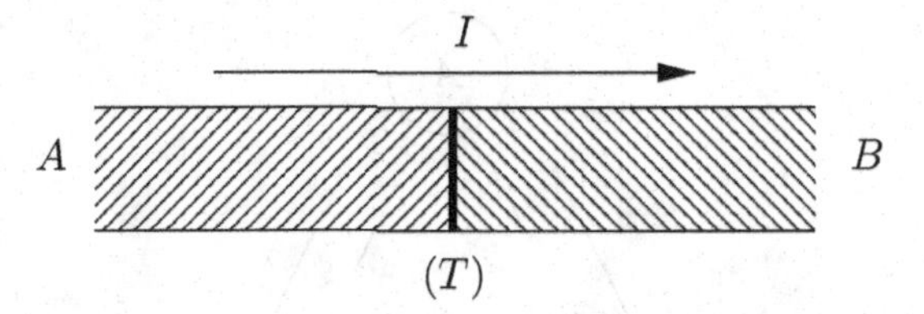

Figure 6.9 A junction between two different metals.

that the particle (charge carriers) flux is accompanied by an energy flux

$$\vec{j}_E = -\frac{q}{T} L_{EN} \vec{\nabla} \Phi = \frac{L_{EN}}{q L_{NN}} \vec{j}_{\text{el}} \tag{6.110}$$

where $\vec{j}_{\text{el}} = -\sigma_{\text{el}} \vec{\nabla} \Phi$. Find expression (6.57) for the coefficient of electrical conductivity σ_{el}.

Calculate the electric power density $\partial \epsilon / \partial t$ and the entropy current density $\vec{j}_S$. What is the value of $\vec{\nabla} \cdot \vec{j}_S$? Verify that the rate of entropy dissipation is given by

$$\frac{\partial s}{\partial t} = \frac{1}{T} \frac{\vec{j}_{\text{el}}^{\,2}}{\sigma_{\text{el}}} \tag{6.111}$$

The above equation shows that the 'Joule power', $\vec{j}_{\text{el}}^{\,2}/\sigma_{\text{el}}$, is directly related to the rate of entropy dissipation. In the model considered here, the charge carriers do not exchange energy with the medium, which, therefore, in turn cannot exchange heat with the outside world. In a realistic situation, what is the mechanism that permits energy to be transferred to the outside thus allowing the Joule effect to be observed?

4. *Peltier effect.* Consider a junction between two different metals A and B through which flows a uniform current of intensity I (Figure 6.9).

The system is maintained at temperature T. Show that at the junction a thermoelectric effect, heating or cooling depending on the direction of the current, will be observed. What power W must we supply to the junction to keep its temperature at T?

If the current I flows from A to B, we define the 'Peltier coefficient', Π_{AB}, as the power per unit current absorbed by the circuit, $\Pi_{AB} = W/I$. Relate this coefficient to the Seebeck coefficients of the two materials.

6.5.4 Isomerization reactions

A chemical exists in three isomeric forms A, B and C related by the triangle of isomerization reactions schematically shown in Figure 6.10.

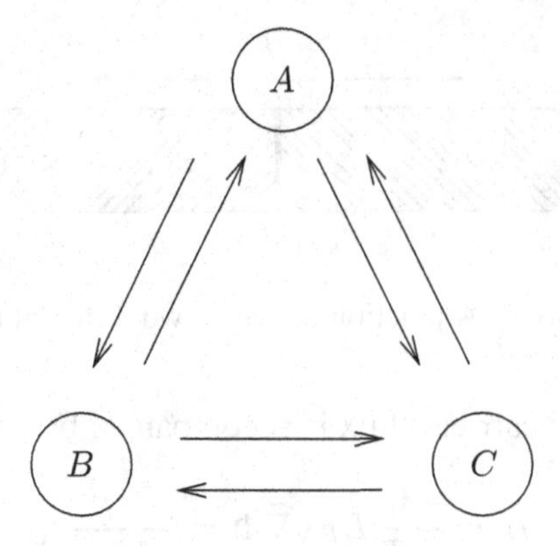

Figure 6.10 Possible exchange directions between three isomeric forms.

Designate by $N_A(t)$, $N_B(t)$ and $N_C(t)$ the numbers of A, B, and C molecules present at time t.

1. If k_{ij} is the spontaneous transformation rate between isomers i and j, write the three equations which govern the kinetics of the reactions. We assume that a stationary situation can be established, i.e. that equilibrium currents ensure that the numbers of molecules N_A^0, N_B^0 and N_C^0 remain constant. Verify that for any isomer i we have

$$\sum_{j \neq i} (k_{ij} N_j^0 - k_{ji} N_i^0) = 0$$

Applying detailed balance (see Section 7.1), which is based on time reversal invariance of microscopic processes, we can demonstrate the sufficient but not necessary relation

$$k_{ij} N_j^0 = k_{ji} N_i^0$$

2. We now move a little away from the stationary situation and designate the deviations by $x_i = N_i - N_i^0$ ($|x_i| \ll N_i^0$). The isomers are assumed to be ideal solutions (or ideal gases). Show that the mixing entropy may be written as

$$S - S_0 = -k \sum_i \left(N_i \ln N_i - N_i^0 \ln N_i^0 \right) \tag{6.112}$$

where S_0 is the entropy at equilibrium. It is sufficient to calculate the mixing entropy of two ideal gases at fixed pressure.

3. This problem cannot be analyzed using weakly coupled spatial cells (Section 6.1) and we therefore have to define the affinities (6.15) differently. Since the affinities are a measure of departure from equilibrium, it is natural to write

$$\Gamma_i = \frac{\partial(S - S_0)}{\partial x_i}$$

We want to re-express the kinetic equations of the first question in the form of linear relations between the fluxes $j_i = \dot{x}_i$ and the affinities Γ_i. Express the entropy change in terms of the deviations from equilibrium by expanding $S - S_0$ to leading order in x_i. Verify that the entropy is indeed maximum at equilibrium. Calculate the affinities and show that the kinetic equations can be written in the form

$$j_i = L_{ii}\Gamma_i + \sum_{j \neq i} L_{ij}\Gamma_j$$

Identify the phenomenological coefficients L_{ij} and verify that detailed balance implies the Onsager symmetry relations $L_{ij} = L_{ji}$.

6.6 Further reading

Sections 6.1 and 6.2 follow closely (including notation) Chapter 14 in Balian [5], which the reader should consult for further reading. Other useful references are Reif [109] (Chapter 15), Landau and Lifshitz [70] (Chapter XII), Kubo [68] (Chapter 6) and Kreuzer [67] (Chapters 1 to 3). At a more advanced level, the book by Foerster [43] is indispensable: in particular, one finds there an extensive discussion of the relation between time scales and conservation laws. A discussion of hydrodynamics geared for physicists can be found in Guyon *et al.* [53] and Faber [37].

7

Numerical simulations

As we have seen on several occasions, a great many of the physical phenomena of interest concern collective behaviour or strongly correlated particles. While it is sometimes possible to obtain useful insight using approximation methods, such as perturbation theory, reliable controlled approximations do not exist. In such situations, numerical simulations have become an indispensable tool.

We shall not attempt in this chapter to discuss computer programming nor shall we discuss many of the powerful methods, both algorithmic and in data analysis, that have been developed since numerical simulations have become an important research tool. Instead, we shall assume that the reader already has enough familiarity with computer programming to apply the notions which we discuss here. Our goal is to explain enough of the basics of the Monte Carlo method (classical and quantum) to allow the reader to apply it readily to interesting equilibrium statistical physics problems of the type discussed in this book. The problems in this chapter have been tailored with this in mind. The programs needed are relatively simple and the physics illustrates many of the phenomena already seen: phase transitions with and without spontaneous symmetry breaking, critical exponents, diverging susceptibilities, scaling, vortices, superfluids etc. Consequently, some of the problems are somewhat long and may be better considered as mini-projects.

7.1 Markov chains, convergence and detailed balance

Even relatively innocent looking models are impossible to solve exactly. The exact solution of the two-dimensional Ising model is known but not in the presence of an external magnetic field. In three dimensions, even in the absence of a magnetic field, the exact solution is not known. In fact, relatively few models can be solved exactly; see Reference [14] for an excellent review of the subject. In the absence of exact solutions we may resort to approximations such as mean field, low temperature expansion, high temperature expansion, perturbation etc. To go beyond such approximations, numerical simulations have proven to be a very powerful tool.

The history of numerical simulations is long and interesting, but its use has really come of age with the advent of the electronic computer in the late forties. It has since become a mature and fast developing subject used in most scientific fields.

In this chapter, we discuss the basics of the Monte Carlo numerical simulation method for both classical and quantum systems. In addition, we restrict our attention to equilibrium situations.[1] Our goal is to present enough detail to permit a reader who is new to numerical simulations to get started on some of the problems discussed in this book. To do more would require much more than one chapter; we therefore refer the reader to appropriate references as the need arises.

The partition function was introduced in Chapter 2 and further developed in Chapter 3. In particular we have seen that the average of a physical quantity, A, is given by

$$\langle A \rangle = \frac{1}{Z} \mathrm{Tr}\, A \mathrm{e}^{-\beta H} \tag{7.1}$$

where $Z = \mathrm{Tr} \mathrm{e}^{-\beta H}$ is the partition function. Taking, for the moment, Equation (7.1) to describe a classical system[2] such as the Ising model, the above equation may be interpreted as the average of the quantity A over all configurations of the system weighted by the probability of each configuration. This probability is given by the Boltzmann weight, $\mathrm{e}^{-\beta E}$, normalized by the partition function,

$$P_\nu = \frac{1}{Z} \mathrm{e}^{-\beta E_\nu} \tag{7.2}$$

where the index ν labels the configuration so that E_ν is the energy of the ν configuration.

It is then clear that if we have the means to generate a very large number of configurations distributed according to Equation (7.2), the measurement of averages may be effected by visiting every such configuration, calculating the physical quantity of interest and performing an arithmetic average over the contributions of all the configurations we have. We may, for example, generate configurations at random and then decide to keep them or not based on their weight given by (7.2). However, this method will not work in practice since there are so many configurations, the likelihood of generating at random a configuration that will be accepted is very small. It is much more efficient to define a dynamic process[3] that allows us

[1] The non-equilibrium situation in the form of the Langevin equation will be discussed in Chapter 9.
[2] The discussion presented here also applies to the quantum case which will be addressed later in this chapter.
[3] Since our goal here is to study equilibrium properties, we choose *any* convenient dynamic process that takes the system to the appropriate equilibrium.

to sample mostly the 'important' configurations with occasional visits to 'unimportant' ones.[4] Therefore, our goal is to construct algorithms that will generate configurations with the correct probability distribution (7.2).

To this end we introduce the master equation that describes the time evolution of the probability density

$$\frac{\mathrm{d}P_\nu(t)}{\mathrm{d}t} = \sum_\sigma \big(P_\sigma(t)W(\sigma \to \nu) - P_\nu(t)W(\nu \to \sigma)\big) \tag{7.3}$$

In this equation, $P_\nu(t)$ is the probability of having configuration ν at time t and $W(\sigma \to \nu)$ is the probability per unit time for the system to go from configuration σ to ν. The first term on the right hand side is the rate at which the state ν is being populated while the second term is the rate of its depopulation. This innocuous looking equation is very important: all the physics is hidden in the W terms. Since if the system is in some configuration ν, it will necessarily make a transition to *some* other configuration, we must have

$$\sum_\sigma W(\nu \to \sigma) = 1 \tag{7.4}$$

where the sum over σ includes the configuration ν. In addition, since the system at any time t must be in one of its available configurations, we have the normalization condition

$$\sum_\nu P_\nu(t) = 1 \tag{7.5}$$

The solution of Equation (7.3) subject to the condition (7.5) gives the time evolution of the probability distribution $P_\nu(t)$. We require that in the long time limit, the stationary solution be $P_\nu(t \to \infty) = P_\nu$, the Boltzmann probability density. Of course the dynamics are determined by the choice of $W(\nu \to \sigma)$.

To use the master equation to generate configurations, it is convenient to use a Markov process. For our purposes, a Markov process satisfies the following two important properties. (*i*) The transition probability $W(\sigma \to \nu)$ does not depend on time and (*ii*) it only depends on the configurations ν and σ, in particular it does not depend on what happened before σ. So, given a configuration σ, the Markov process will randomly generate a new configuration ν: it will not always generate the same new configuration given the same σ.

In addition, we have to impose another condition on our Markov process: it must be ergodic, which means that any possible configuration of the system may be reached from any other possible configuration (see Footnote 17 Chapter 4). In other words, the dynamics we choose (determined by $W(\sigma \to \nu)$) must not split

[4] Important and unimportant refer to the relative weight as given by Equation (7.2) which determines the probability of visiting the configurations.

configuration space into non-communicating regions. This is necessary if we are to generate configurations distributed according to the correct Boltzmann distribution.

In the stationary limit $\mathrm{d}P_\nu(t)/\mathrm{d}t = 0$ and therefore

$$\sum_\sigma P_\sigma W(\sigma \to \nu) = \sum_\sigma P_\nu W(\nu \to \sigma) \tag{7.6}$$

We see from Equation (7.4) that the right hand side of (7.6) is just P_ν. By choosing dynamics that satisfy (7.6) we assure the convergence, in the long time limit, to the required probability distribution. However, this condition is not simple to implement due to the sum over all configurations on the left hand side. This problem may be overcome by imposing the sufficient but not necessary condition

$$P_\sigma W(\sigma \to \nu) = P_\nu W(\nu \to \sigma) \tag{7.7}$$

which is called the condition of *detailed balance*: satisfying this condition automatically leads to (7.6). The detailed balance condition has already been encountered in Exercise 6.5.4 and is a consequence of time reversal invariance (see Chapter 9). Our goal is to generate configurations according to the Boltzmann distribution, which we know to satisfy

$$\frac{P_\sigma}{P_\nu} = \mathrm{e}^{-\beta(E_\sigma - E_\nu)} \tag{7.8}$$

Equations (7.7) and (7.8) allow us to define the transition probabilities, which must now satisfy

$$\boxed{\frac{W(\nu \to \sigma)}{W(\sigma \to \nu)} = \frac{P_\sigma}{P_\nu} = \mathrm{e}^{-\beta(E_\sigma - E_\nu)}} \tag{7.9}$$

The choice of $W(\nu \to \sigma)$ is not at all unique. One of the most widely used choices is that of Metropolis,

$$W(\nu \to \sigma) = 1 \quad \text{for } P_\sigma \geq P_\nu \ \ (E_\sigma \leq E_\nu) \tag{7.10a}$$

$$W(\nu \to \sigma) = \frac{P_\sigma}{P_\nu} \quad \text{for } P_\sigma < P_\nu \ \ (E_\sigma > E_\nu) \tag{7.10b}$$

which can be easily shown to satisfy detailed balance. This is the choice we shall use in this chapter without forgetting that others are possible too. In principle, the implementation of this choice in Monte Carlo simulations is quite straightforward. Starting from some configuration, ν, we attempt a random change that is accepted with probability one if the Boltzmann weight of the resulting configuration σ is larger than that of the original. If the weight of the new configuration is less than that of the original, we accept the change with probability P_σ/P_ν. This will be discussed in more detail in the next section.

Notice that if we only accept changes that increase the Boltzmann weight, we shall be seeking the configuration with the lowest energy. Accepting moves that increase the energy, albeit with reduced probability, allows the system to sample fluctuations away from the lowest energy configuration thus giving the correct distribution.

7.2 Classical Monte Carlo

We now discuss the basic application of the Metropolis Monte Carlo algorithm to classical statistical mechanics problems. To fix ideas, we consider the example of the Ising model on a square lattice. However, we emphasize that the method is very general: it is restricted neither to discrete variables nor to discrete space (lattices).

7.2.1 Implementation

The classical Ising model was introduced in Chapter 3, Equation (3.30),

$$H = -J \sum_{\langle i,j \rangle} S_i S_j - \mu B \sum_{i=1}^{N} S_i \tag{7.11}$$

where J is the interaction strength between a spin at site i and its nearest neighbour j, $\sum_{\langle i,j \rangle}$ indicates the sum over such nearest neighbours, B is an external magnetic field, μ the magnetic moment of the spins and N the total number of spins. The partition function is (3.31)

$$Z = \sum_{\{S_i\}} \exp\left(\beta J \sum_{\langle i,j \rangle} S_i S_j + \beta \mu B \sum_i S_i \right) \tag{7.12}$$

where $\beta = 1/kT$ and $\sum_{\{S_i\}}$ is shorthand for $\sum_{S_1=\pm 1} \cdots \sum_{S_N=\pm 1}$. We are usually interested in the thermodynamic properties of our system, but in numerical simulations we can only consider finite system sizes. In order to minimize finite size effects and extract results that are relevant to the thermodynamic limit, we typically use periodic boundary conditions.[5] Consider a D-dimensional lattice with L sites on a side, i.e. total number of sites $N = L^D$. Then applying periodic boundary conditions means that $S_i = S_{i+\hat{x}L}$ where $\hat{x}$ is a unit vector in the x direction,

[5] We also use *finite size scaling*, which will be discussed in Section 7.6 and illustrated in the problems in this chapter. This very important method rests on the understanding of critical phenomena and the scaling of physical quantities as a transition point is approached. This was discussed in Chapter 4.

with similar conditions in the other directions. In other words, a translation of L in any direction returns us to the starting point:[6] The lattice forms a torus. This also means that, for example, the nearest neighbour in the forward x direction of a spin with $x = L$ is a spin with $x = 1$.[7]

The implementation is now straightforward: we take $P_\nu(t)$ to be the time independent Boltzmann weight in Equation (7.12) and $W(\nu \to \sigma)$ is given by Equations (7.10a) and (7.10b). We first choose an arbitrary initial configuration of the system. Since our algorithm, that is based on a Markov chain, is ergodic and satisfies detailed balance, it will reach an equilibrium that is independent of the initial configuration. Any configuration is acceptable as an initial choice. We then choose a site and change it, $S_i \to -S_i$.[8] The Boltzmann weights of the two configurations, before (P_{old}) and after (P_{new}) the change, are compared and the change accepted or rejected according to the criterion in Equations (7.10a) and (7.10b). To implement this criterion, we generate a random number r (see Section 7.7) uniformly distributed between 0 and 1, then

$$\frac{P_{\text{new}}}{P_{\text{old}}} = \frac{\mathrm{e}^{-\beta E_{\text{new}}}}{\mathrm{e}^{-\beta E_{\text{old}}}} \geq r \qquad \text{Accept change} \tag{7.13a}$$

$$\frac{P_{\text{new}}}{P_{\text{old}}} = \frac{\mathrm{e}^{-\beta E_{\text{new}}}}{\mathrm{e}^{-\beta E_{\text{old}}}} < r \qquad \text{Reject change} \tag{7.13b}$$

where E_{new} (E_{old}) is the total energy after (before) the change is suggested. Clearly, to calculate the difference between these two energies, we do not need to calculate the total energy. Since we are only changing one spin, the only changes in the energy are due to the spin involved and its nearest neighbours. So, we only have $2D$ spins contributing to $E_{\text{new}} - E_{\text{old}}$. Accepting the change means that the new value of the spin is kept, otherwise the spin is restored to its original value and we proceed to the next site where these steps are repeated. When all the spins have been visited and changes attempted, we shall have performed a 'sweep' or a Monte Carlo step (MCS). After enough sweeps have been performed, the resulting configurations will be correctly distributed according to the Boltzmann weight as demanded by the master equation (7.3).

7.2.2 Measurements

Once the configurations are generated with the correct Boltzmann distribution we may measure many physical quantities and correlations by visiting these

[6] See Figure 3.2 for an illustration of the one-dimensional case.
[7] Lattice sites are labelled with a single index $i = (x, y, \ldots)$, with $1 \leq x \leq L$, $1 \leq y \leq L$ etc.
[8] Sites may be visited in any order, in particular we may visit them sequentially.

configurations, performing the measurements and then the arithmetic averages. Clearly, a simple arithmetic average takes into account the Boltzmann weight since the configurations are already weighted correctly. Of course, in practice we do not save the configurations in order to perform the measurements at the end. Once the system has thermalized (see Section 7.2.3) we perform the measurements as the configurations are generated (see below).

We have already encountered various physical quantities in the Ising model such as the average of the total energy, $E = \langle H \rangle$, the specific heat,

$$C = \frac{1}{NkT^2}(\langle H^2 \rangle - \langle H \rangle^2) \tag{7.14}$$

the magnetization per spin, $M = \langle S_i \rangle$, and the magnetic susceptibility,

$$\chi = \frac{N}{kT}(\langle S_i^2 \rangle - \langle S_i \rangle^2) \tag{7.15}$$

where $N = L^D$ is the total number of spins. The measurement of the average energy (or its square) is very simple, we simply evaluate Equation (7.11) (or its square) for every configuration generated and find the average. The magnetization per spin is obtained by measuring for each configuration the quantity $|\sum_i S_i|/N$ and then averaging over all the available configurations. Usually, during a simulation, the magnetization can change direction. For example for the Ising model, positive magnetization can flip over and become negative. The absolute value in the expression for M ensures that such events do not give the 'false' result that the magnetization vanishes.

Another very important quantity to measure is the correlation function between variables at different locations. For example, the two-point connected correlation function, G_{ij}, is given by Equation (4.25),

$$\begin{aligned} G_{ij} &= \langle S_i S_j \rangle - \langle S_i \rangle \langle S_j \rangle \\ &= \langle S_i S_j \rangle - M^2 \end{aligned} \tag{7.16}$$

The properties and uses of this function are discussed in Chapter 4.[9] In particular, it allows us to calculate numerically the correlation length, ξ, Equation (4.30), which diverges in the vicinity of a critical point. Monte Carlo simulations allow us to study the divergence of the correlation length and to determine the correlation length critical exponent, ν, and other critical exponents. Some of these issues will be illustrated in the problems.

[9] Care must be taken when measuring G_{ij} on a finite lattice with periodic boundary conditions, since the largest separation between two sites, i and j, is $L/2$ not L. The correlation function G_{ij} is symmetric with respect to $L/2$.

7.2.3 Autocorrelation, thermalization and error bars

As we have seen above, the Monte Carlo simulation starts by choosing an initial configuration, typically, at random. However, it is extremely likely that the configuration thus chosen is not an 'equilibrium' configuration in the sense that its Boltzmann weight is very small. Therefore we should not start measuring physical quantities from the very start: we should first do an appropriate number of 'thermalization' sweeps thus allowing the configurations to be generated with the correct weight. The number of thermalization sweeps depends on the system, its size and its proximity to a critical point. To determine this, we need to calculate the autocorrelation function of a physical quantity. The autocorrelation function is quite similar to the two-point function defined above except that the separation is in 'time' not space. In the present context, time is the Monte Carlo sweep: two measurements of a physical quantity, A, separated by time t are performed on two configurations separated by t Monte Carlo sweeps,

$$C(t) = \frac{\langle A(t_0)A(t_0 + t)\rangle - \langle A\rangle^2}{\langle A^2\rangle - \langle A\rangle^2} \tag{7.17}$$

The normalization is such that $C(0) = 1$. In addition, as the time separation between the two configurations increases, all correlations will be lost, $\langle A(t_0)A(t_0 + t)\rangle \to \langle A\rangle^2$, and thus

$$C(t \to \infty) = 0 \tag{7.18}$$

In general, the correlation function decreases exponentially as a function of time

$$C(t) = \mathrm{e}^{-t/\tau} \tag{7.19}$$

where τ is taken to be the decorrelation (relaxation) time. Figure 7.1 shows the autocorrelation function of the magnetization for 32×32 (dashed) and 100×100 (solid) Ising systems at $T = 2.4J$. Figure 7.2 is the same but on semilog scale to illustrate clearly the exponential decay. The dashed lines in Figure 7.2 are exponential fits that give the values $\tau \approx 20$ for 32×32 and $\tau \approx 40$ for 100×100. Clearly, the magnetization is correlated over a number of sweeps which increases with the size of the system at fixed temperature.

We can now make more precise the notion of thermalization sweeps we mentioned previously. The number of thermalization sweeps should be several times τ to allow the system to forget its starting point and start sampling its equilibrium configurations.

In order to estimate the error bars, we rely on the central limit theorem, which says that if we make a large number of independent measurements, the results will follow a Gaussian distribution centred at the (unknown) exact value. The standard

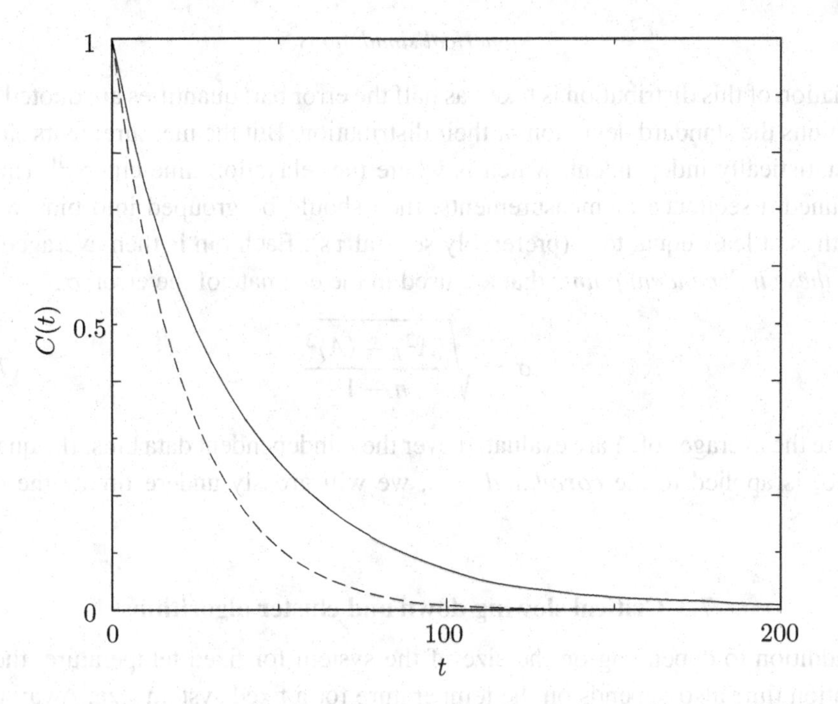

Figure 7.1 The magnetization autocorrelation function as a function of Monte Carlo time for 32×32 (dashed) and 100×100 (solid) Ising lattices at $T = 2.4J$.

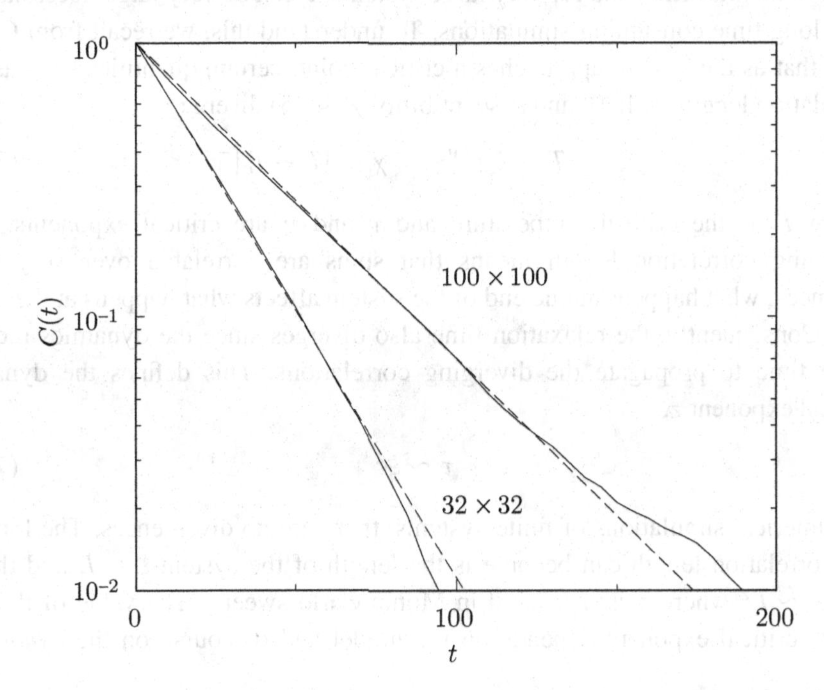

Figure 7.2 Same as Figure 7.1 on semilog scale. The early part of the curves is linear indicating exponential decay. The dashed lines are exponential fits $C(t) = \exp(-t/\tau)$ giving $\tau \approx 20$ (32×32) and $\tau \approx 40$ (100×100).

deviation of this distribution is taken as half the error bar; quantities are quoted plus or minus the standard deviation of their distribution. But the measurements should be statistically independent, which is where the relaxation time enters.[10] Having obtained a sequence of measurements, they should be grouped into bins whose width is at least equal to τ (preferably several τs). Each bin is then averaged and it is *these independent points* that are used in the estimate of the error, σ,

$$\sigma = \sqrt{\frac{\langle A^2 \rangle - \langle A \rangle^2}{n-1}} \tag{7.20}$$

where the averages of A are evaluated over the n independent data bins. If Equation (7.20) is applied to the *correlated* data, we will grossly underestimate the error bars.

7.3 Critical slowing down and cluster algorithms

In addition to depending on the size of the system for fixed temperature, the relaxation time also depends on the temperature for a fixed system size. Away from the transition, τ is small and the simulation does not pose a challenge. However, close to the transition and for very large systems, τ can be very large necessitating very long time consuming simulations. To understand this, we recall from Chapter 4 that as the system approaches a critical point, certain quantities such as the correlation length ξ (4.30) and susceptibility χ (4.45) diverge

$$\xi \sim |T - T_c|^{-\nu} \qquad \chi \sim |T - T_c|^{-\gamma} \tag{7.21}$$

where T_c is the critical temperature and ν and γ are critical exponents. The diverging correlation length means that spins are correlated over very long distances; what happens at one end of the system affects what happens at the other end. Consequently, the relaxation time also diverges since the dynamics require more time to propagate the diverging correlations. This defines the dynamic critical exponent z,

$$\tau \sim \xi^z \tag{7.22}$$

In numerical simulations of finite systems, there are no divergences. The longest the correlation length can become is the length of the system $\xi \sim L$ and therefore $\tau \sim L^z$ where τ is measured in Monte Carlo sweeps. The value of the dynamic critical exponent depends on the model and of course on the *dynamics*,

[10] There are many relaxation times for a system, in fact there are as many relaxation times as there are states. The technique we are outlining here is the most basic and is sufficient to get a reasonable idea of the relaxation time.

i.e. the simulation algorithm, used.[11] The best known value of z for the two-dimensional Ising model using the Metropolis Monte Carlo algorithm is [95] $z = 2.1665 \pm 0.0012$. Since the number of operations per sweep scales with the volume of the system, L^D, we see that the computer time necessary to maintain the same statistical accuracy grows as

$$\tau_{\text{cpu}} \sim L^{z+D} \sim L^{2.17+2} \tag{7.23}$$

Therefore, near a critical point the relaxation time can become enormous, making it difficult to generate the large number of decorrelated measurements necessary to reduce statistical errors. Physically, what happens is that the increasing correlation length manifests itself in the formation of domains of parallel (i.e. strongly correlated) spins. It is very difficult to flip these domains because the probability of flipping a spin in the middle of the domain is very small since it is surrounded by parallel spins that strongly favour keeping it parallel. The mechanism for flipping them in the Metropolis algorithm is to eat away at the domain from the boundary one spin at a time. This is a very slow process.

In order to reduce the dynamic critical exponent we need to change the simulation dynamics in such a way that the dynamic universality class is changed. Choosing another local dynamics where sites are updated locally and without taking correlations into account will not help. From the above discussion, it is clear that it is desirable to flip entire clusters of correlated spins. The construction and implementation of such algorithms, known as *cluster algorithms*, has become an important activity in the field of Monte Carlo simulations, for both classical and quantum systems.

The simplest idea is to choose a spin at random as the seed from which the cluster will be grown. It is, perhaps, tempting to grow the cluster in the following way: having chosen the seed spin (to fix ideas say it is up), find which of its neighbours are also up and add them to the cluster, then find the neighbours of the neighbours which are also up. These up spins are then added to the cluster and so on iteratively until the cluster stops growing, and then flip all the spins thus added to the cluster. However, the problem with this is that just because two neighbouring spins are parallel does not mean they are correlated. For example, at very high temperature spins are not correlated, yet, since on average half the spins are up and half are down, there will be many instances of parallel neighbours. Such parallel but uncorrelated spins should not belong to the same cluster. On the other hand, at very low temperature where the system is magnetized ($D \geq 2$ for the Ising model) parallel spins are indeed strongly correlated and the clusters extend over the size of the system.

[11] As discussed in Chapter 4, critical exponents define universality classes. The dynamic critical exponent then depends on the universality class of the model *and* the dynamics used.

Therefore, a parallel neighbour should be added to the cluster with some probability, P_b, for forming a bond and rejected with the probability $(1 - P_b)$. This P_b depends on the temperature and may be obtained via an interesting and insightful formulation of the Ising model called the Fortuin–Kasteleyn transformation [44]. However, we determine it here by using the condition of detailed balance (7.9).

Consider starting with a configuration ν whose Boltzmann weight is P_ν, choose a site at random and grow a cluster $\mathcal{C}$ by activating bonds with the probability P_b as described above. Flipping this cluster yields a new configuration σ with Boltzmann weight P_σ. Detailed balance, Equation (7.9), concerns choosing the exact same cluster in σ and going back to configuration ν. We need the probabilities of choosing the same cluster for each of the configurations ν and σ. The probability for having $\mathcal{C}$ in ν is the product of the Boltzmann weight[12] P_ν with the probability for constructing $\mathcal{C}$, $P_b^\ell (1 - P_b)^{n_\nu}$ where ℓ is the number of activated bonds in the cluster and n_ν is the number of candidate bonds on the *boundary* of $\mathcal{C}$ that were rejected with probability $(1 - P_b)$ each. Similarly, the probability of $\mathcal{C}$ in σ is the product of the Boltzmann weight P_σ with the probability for constructing $\mathcal{C}$, $P_b^\ell (1 - P_b)^{n_\sigma}$. Notice that since we have the same cluster, the number of activated bonds in both cases is the same. However, the number of rejected bonds on the boundary is different. Therefore, Equation (7.9) becomes

$$\frac{W(\nu \to \sigma)}{W(\sigma \to \nu)} = \frac{P_\sigma (1 - P_b)^{n_\sigma}}{P_\nu (1 - P_b)^{n_\nu}} \tag{7.24}$$

This is illustrated in Figure 7.3 where in (a) is shown a cluster of down spins (open circles) between the inner and outer contours. This cluster is then flipped to give the final configuration shown in (b). The rejected bonds in the initial configuration (a) correspond to two nearest neighbour open circles separated by a contour line segment. It is easy to see that $n_\nu = 10$. Similarly for (b), rejected bonds correspond to two nearest neighbour full circles separated by a contour line segment and we see that now $n_\sigma = 26$.

We want the algorithm to be as efficient as possible, therefore, once the cluster is grown, we flip it with unit probability. This means that $W(\nu \to \sigma)/W(\sigma \to \nu) = 1$. Therefore

$$\frac{P_\sigma (1 - P_b)^{n_\sigma}}{P_\nu (1 - P_b)^{n_\nu}} = 1 \tag{7.25}$$

The only difference between the configurations ν and σ is the flipped cluster. It is then easy to verify that (see Section 4.1.1)

$$\frac{P_\sigma}{P_\nu} = \mathrm{e}^{-\beta(E_\sigma - E_\nu)} = \mathrm{e}^{+2J\beta(n_\sigma - n_\nu)} \tag{7.26}$$

[12] Of course the Boltzmann weight must be normalized by the partition function Z to obtain a probability. But since Z is constant, it will drop out from the equations. To avoid cluttering the discussion we do not show it.

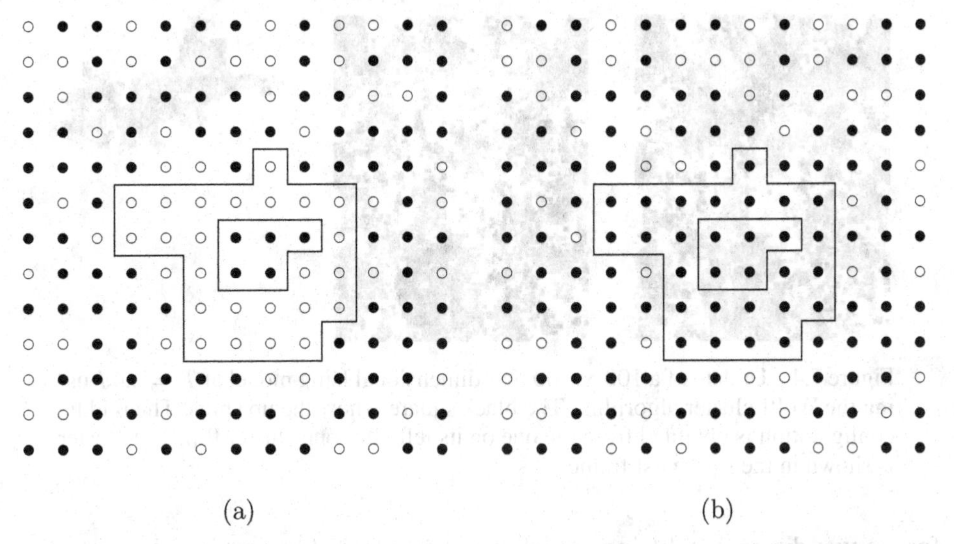

Figure 7.3 (a) A cluster of down spins (open circles) is shown between the inner and outer closed contours. (b) The cluster of down spins is flipped, its members becoming up spins (full circles).

since the only energy difference will arise from the boundary of $\mathcal{C}$. Equation (7.25) therefore becomes

$$\left(e^{2J\beta}(1 - P_b)\right)^{n_\sigma - n_\nu} = 1 \tag{7.27}$$

whose solution in zero magnetic field is

$$\boxed{P_b = 1 - e^{-2J\beta}} \tag{7.28}$$

which is the probability of adding a spin to $\mathcal{C}$ if it has the same value as the cluster spins. This single cluster update is called the Wolff algorithm.

At high temperature, $\beta \ll 1$, $P_b \ll 1$ and the typical cluster size is one spin. So, this cluster algorithm reduces to the Metropolis algorithm in that limit. At very low temperature, $\beta \gg 1$, $P_b \approx 1$ and all parallel spins belong to the cluster. In this case too, the algorithm is not efficient because flipping all the spins back and forth does not get the system closer to equilibrium. In these two limit cases one might just as well use a local update algorithm like Metropolis. The power of this cluster algorithm is evident near a critical point where the correlation length is of the order of the size of the system. Consider, for example, the two-dimensional Ising model at the (thermodynamic) critical point $\beta_c = \ln(1 + \sqrt{2})/2J$ (Equation 4.14) which gives $P_b = \sqrt{2}/(1 + \sqrt{2}) \approx 0.586$. Therefore, we see that at this temperature P_b is large enough to form large clusters but still reject about half the candidate sites. This results in large clusters and leads to a dynamic critical exponent $z \approx 0.25$

Figure 7.4 Update of a 100×100 two-dimensional Ising model at $T = 2.4J$ using the Wolff cluster algorithm. The black squares show the up spins. The middle configuration is obtained from the one on its left after one cluster flip. The cluster is shown in the rightmost frame.

for the two-dimensional Ising model, which results in huge savings in simulation times for large systems near the critical temperature.

In Figure 7.4 we illustrate for the two-dimensional 100×100 Ising model at $T = 2.4J$ (i.e. near the critical temperature) how the cluster update can change the configuration substantially. The black (white) shows the up (down) spins. In the first configuration on the left, a site is chosen as the seed to grow the cluster, which is then flipped resulting in the configuration in the middle figure. The cluster itself is shown in the figure on the right. We see that the cluster can be very large.

This cluster algorithm is not restricted to the Ising model. It easily applicable to other discrete models like the Potts model and the Z_N clock models[13] and even to continuous spin models like the Heisenberg model. Such cluster ideas have also been extended to simulations of quantum systems. It should be noted, however, that it is not always possible to define appropriate clusters to update as described above: frustrated systems are an important example.

7.4 Quantum Monte Carlo: bosons

The term quantum Monte Carlo (QMC) should be understood to mean Monte Carlo of a quantum system. The simulation itself, as we shall see, proceeds via classical Monte Carlo. However, as we have seen in the previous section, in a Monte Carlo simulation of a classical system, the Boltzmann weight is considered as a probability density and used to decide whether or not to accept changes in

[13] The Ising model is in fact the Z_2 clock model. In the Z_N clock model, the spin variables take the values $S = \exp(\mathrm{i}2n\pi/N)$ where $n = 0, 1, \ldots, N-1$. It is clear that the Ising model is the $N = 2$ case. These models will be discussed in Problem 7.9.3.

the configurations. The same cannot be immediately done for a quantum system since the weight $\exp(-\beta H)$ is an operator. The partition function needs first to be expressed in terms of c-numbers before a Monte Carlo algorithm can be applied to it.

In this section we shall outline how this may be done for a bosonic system. However, there are many possible algorithms and to discuss them all here is beyond our scope. Instead, we sketch a general approach for lattice bosons[14] which will culminate in what is known as the world line algorithm. While not necessarily the fastest algorithm, it is straightforward to implement and offers an elegant intuitive picture of the quantum system.

7.4.1 Formulation and implementation

To explain this algorithm, it is best to consider a specific Hamiltonian. We shall see other examples in the problems. To fix ideas, we consider the one-dimensional lattice boson model described by the Hamiltonian (5.119)

$$H = -t \sum_{\langle ij \rangle} (\mathsf{a}_j^\dagger \mathsf{a}_i + \mathsf{a}_i^\dagger \mathsf{a}_j) - \mu \sum_i \mathsf{n}_i + \mathsf{U} \tag{7.29}$$

where i and j label the one-dimensional lattice sites[15] and $\langle ij \rangle$ denotes nearest neighbours that are separated by the lattice spacing, a, taken to be unity. The operator $\mathsf{a}_i^\dagger$ (a_i) creates (destroys) a boson on site i and $[\mathsf{a}_i, \mathsf{a}_j^\dagger] = \delta_{ij}$. The parameter t $(= \hbar^2/2ma^2)$ is called the hopping parameter and is a measure of the kinetic energy per jump. μ is the chemical potential, $\mathsf{a}_i^\dagger \mathsf{a}_i = \mathsf{n}_i$ is the number operator on site i and the sum over all sites gives the total number operator: $\sum_i \mathsf{n}_i = \mathsf{N}_\mathrm{b}$. The expectation value of this operator is the average number of bosons, N_b, on the lattice. The potential energy operator, U, is a function only of number operators, for example

$$\mathsf{U} = V_0 \sum_i \mathsf{n}_i(\mathsf{n}_i - 1) + V_1 \sum_{\langle ij \rangle} \mathsf{n}_i \mathsf{n}_j \tag{7.30}$$

where V_0 is the contact interaction[16] and V_1 the nearest neighbour interaction. As is clear, V_0 does not contribute when a site has one or no bosons. Of course it is possible to add more terms, such as next nearest neighbours, depending on the physics of interest.

[14] Bosons, such as helium atoms, can also be simulated in the continuum, but that is beyond the scope of this chapter.

[15] This algorithm can be generalized without difficulty to higher dimensions. The only difference is that there are new world line moves that are possible in higher dimensions and should be included.

[16] In the limit $V_0 \to \infty$ we obtain the hardcore boson model that was studied in the mean field approximation in Problem 5.7.7.

Most QMC algorithms start with the identity

$$Z = \mathrm{Tr} e^{-\beta H} \tag{7.31}$$

$$= \mathrm{Tr}\left(e^{-\Delta\tau H} e^{-\Delta\tau H} \dots e^{-\Delta\tau H}\right) \tag{7.32}$$

where $\Delta\tau \equiv \beta/L_\tau$ and L_τ is the number of exponentials in the trace chosen so that $\Delta\tau \ll 1$.[17] The reason for using Equation (7.32), which is clearly an identity and involves no approximations, will become clear shortly. It will be useful to write $H = H_1 + H_2$ where H_1 connects site 1 to site 2, 3 to 4, 5 to 6 etc. while H_2 connects site 2 to site 3, 4 to 5, 6 to 7 etc.

$$H_1 = -t \sum_{i\ \mathrm{odd}} (\mathsf{a}_i^\dagger \mathsf{a}_{i+1} + \mathsf{a}_{i+1}^\dagger \mathsf{a}_i) + \frac{1}{2}\mathsf{U} \tag{7.33}$$

$$H_2 = -t \sum_{i\ \mathrm{even}} (\mathsf{a}_i^\dagger \mathsf{a}_{i+1} + \mathsf{a}_{i+1}^\dagger \mathsf{a}_i) + \frac{1}{2}\mathsf{U} \tag{7.34}$$

This allows us to write the first approximation, called the Trotter–Suzuki approximation,

$$e^{\Delta\tau H} \approx e^{\Delta\tau H_1} e^{\Delta\tau H_2} + \mathcal{O}(\Delta\tau)^2 \tag{7.35}$$

In writing Equation (7.35) the Baker–Hausdorf identity, $e^{\mathsf{A}} e^{\mathsf{B}} = e^{\mathsf{A}+\mathsf{B}+[\mathsf{A},\mathsf{B}]+\cdots}$ was used with $\mathsf{A} = \Delta\tau H_1$ and $\mathsf{B} = \Delta\tau H_2$. In addition, since this algorithm is most conveniently formulated in the canonical ensemble, the chemical potential is dropped.

Inserting Equation (7.35) in (7.32) expresses the partition function as a trace over the product of $2L_\tau$ exponentials. The trace may be evaluated in any representation since it is invariant. It is convenient to do this in the occupation number representation, $|\mathbf{n}\rangle \equiv |n_1, n_2, \dots, n_i, \dots, n_S\rangle$ where $i = 1, 2, \dots, S$ is the spatial label of the site. Then

$$\mathsf{a}_i^\dagger |n_1, n_2, \dots, n_i, \dots, n_S\rangle = \sqrt{n_i + 1}\, |n_1, n_2, \dots, n_i + 1, \dots, n_S\rangle \tag{7.36}$$

$$\mathsf{a}_i |n_1, n_2, \dots, n_i, \dots, n_S\rangle = \sqrt{n_i}\, |n_1, n_2, \dots, n_i - 1, \dots, n_S\rangle \tag{7.37}$$

$$\mathsf{n}_i |n_1, n_2, \dots, n_i, \dots, n_S\rangle = n_i\, |n_1, n_2, \dots, n_i, \dots, n_S\rangle \tag{7.38}$$

[17] The alert reader will notice that $\Delta\tau$ has the dimensions of inverse energy. Therefore it is meaningless to say $\Delta\tau \ll 1$. This, in fact, should be taken as shorthand for the following: since, in general, we want to ignore terms of order $(\Delta\tau)^2[\mathsf{A}, \mathsf{B}]$, where A and B are non-commuting parts of the Hamiltonian, we want the product of $(\Delta\tau)$ with the largest energy scale (coupling parameter) in A or B to be very small.

To evaluate the trace, introduce between each pair of exponentials the identity $I = \sum_{\mathbf{n}} |\mathbf{n}\rangle\langle\mathbf{n}|$ which leads to

$$Z = \sum_{\{\mathbf{n}\}} \langle \mathbf{n}^1 | \mathrm{e}^{-\Delta\tau H_2} | \mathbf{n}^{2L_\tau} \rangle \langle \mathbf{n}^{2L_\tau} | \mathrm{e}^{-\Delta\tau H_1} | \mathbf{n}^{2L_\tau - 1} \rangle \cdots \langle \mathbf{n}^3 | \mathrm{e}^{-\Delta\tau H_2} | \mathbf{n}^2 \rangle \langle \mathbf{n}^2 | \mathrm{e}^{-\Delta\tau H_1} | \mathbf{n}^1 \rangle \tag{7.39}$$

Several comments are now in order.

(*i*) The operator $\exp(-\Delta\tau H_\ell)$, $\ell = 1, 2$, has the same form as the *time evolution* operator in quantum mechanics, $\exp(-\mathrm{i}\Delta t H)$, which evolves a quantum state at time Δt if we make the substitution $\Delta\tau \to \mathrm{i}\Delta t$. We therefore interpret $\exp(-\Delta\tau H_\ell)$ as an evolution operator in *imaginary time*. Then Equation (7.39) may be interpreted as follows: $\exp(-\Delta\tau H_1)$ acts on $|\mathbf{n}^1\rangle$, which is the state at the first imaginary time slice, and evolves it by $\Delta\tau$ at which point its overlap with $|\mathbf{n}^2\rangle$ is taken. Then $\exp(-\Delta\tau H_2)$ takes over and evolves $|\mathbf{n}^2\rangle$ by $\Delta\tau$. When both H_1 and H_2 act in this way, the system will have evolved by one time slice $\Delta\tau$ since it takes the action of both Hamiltonians to evolve all sites. Therefore, Z is now written as a 'path integral' over all allowed configurations of the bosons as they are evolved from the first time slice to the last.

(*ii*) The meaning of the superscripts in the state vectors should now be clear: they label the time slices. The first bra on the left is labelled by 1, the same as the first ket on the right, because this is a trace.

(*iii*) The partition function, originally written as a trace of an operator, is now formulated as a sum of a product of matrix elements of Hermitian operators. These matrix elements are c-numbers, and the sum over $\{\mathbf{n}\}$ sums all possible boson configurations connected by the imaginary time evolution operator. For example, all possible configurations[18] of $|\mathbf{n}^\ell\rangle$ are summed over, but only those which have overlap with the preceding and succeeding time evolved states will contribute. Notice that the periodic boundary conditions in the imaginary time direction are imposed by the trace: this is not just a convenient choice!

(*iv*) Since $\Delta\tau \ll 1$, the matrix elements can be evaluated to a reasonably high order in $\Delta\tau$. As a result, the product of matrix elements in Equation (7.39) is known and may be used as a 'Boltzmann weight' for the configurations. *It is crucial to remark the following: the summand in Equation (7.39) may be used as a probability density in the Monte Carlo simulation because it is positive semi-definite. It never becomes negative for this bosonic problem, as is clear for example in Equation (7.40), since the matrix elements of* $\exp(\Delta\tau t(a_j^\dagger a_{j+1} + a_{j+1}^\dagger a_j))$ *are never negative.* We shall see below two very important examples where this is not the case!

(*v*) It should now be clear why $\Delta\tau$ was introduced by the seemingly trivial Equation (7.32): it gives a small parameter that allows *the breakup of the complicated many body problem into a product of independent two-site problems* and the evaluation of

[18] An example of a configuration of $|\mathbf{n}^\ell\rangle$ is given by Equation (7.38).

the matrix elements in (7.39). It is crucial to note, though, that this calculation is *non-perturbative*. At no point was it assumed that any of the interaction terms in the Hamiltonian is small.

(*vi*) As a consequence of item (v), the algorithm thus constructed is exact except for the systematic errors committed when terms of order $(\Delta\tau)^2$ were ignored, for example in Equation (7.35). These are called Trotter errors. It has been proven rigorously that the limit of Equation (7.39) as $\Delta\tau \to 0$ is the correct partition function. In other words, if one wants to eliminate the Trotter errors, one can do the same simulation at different values of $\Delta\tau$ and then extrapolate the results to $\Delta\tau = 0$.[19] As it stands, this algorithm gives averages of physical quantities with a Trotter error of $\mathcal{O}(\Delta\tau)^2$.

As mentioned in item (i), the partition function is now expressed as a path integral (7.39). This equation has an appealing intuitive geometrical interpretation in terms of *world lines*, i.e. the space–imaginary time path traced by the bosons as they evolve. To see this, consider, for example, the matrix element $\langle \mathbf{n}^2 | \mathrm{e}^{-\Delta\tau H_1} | \mathbf{n}^1 \rangle$. As is clear from its definition (7.33), the hopping term (kinetic energy) in H_1 couples only the pairs (1,2), (3,4), (5,6) etc. but not the pairs (2,3), (4,5), (6,7) etc., which are connected by H_2. Consequently, the matrix elements of $\mathrm{e}^{-\Delta\tau H_1}$ are products of *independent* two-site problems: (1,2), (3,4), (5,6) ... Similarly, a matrix element of $\exp(-\Delta\tau H_2)$ is a product of independent two-site problems (2,3), (4,5), (6,7) ... This is shown in Figure 7.5 where the shaded squares represent the imaginary time evolution of connected pairs. The thick lines represent a possible configuration of boson world lines. Because of the checkerboard division of $H = H_1 + H_2$, a boson can jump only across a shaded square. The world line algorithm consists of suggesting deformations of the world lines (the arrow and dashed line in Figure 7.5), and accepting or rejecting them such that detailed balance is satisfied. For the case of onsite interaction only, i.e. $V_1 = 0$ in (7.30), a move affects four matrix elements, the four shaded plaquettes surrounding the arrow in Figure 7.5 and labelled A, B, C and D. The ratio of the product of these four matrix elements after and before the move is calculated. The move is accepted if this ratio is larger than a random number drawn from a uniform distribution between zero and one, just like in the case of classical Monte Carlo. For longer range interaction (but still nearest neighbour hopping) more plaquettes are involved but the method remains the same.

The problem has thus been reduced to a product of independent two-site problems where one needs to evaluate the matrix element

$$\langle n'_j, n'_{j+1} | \mathrm{e}^{-\Delta\tau H_\ell} | n_j, n_{j+1} \rangle \approx \mathrm{e}^{-\Delta\tau U'/4} \mathrm{e}^{-\Delta\tau U/4} \langle n'_j, n'_{j+1} | \mathrm{e}^{-\Delta\tau \mathrm{K}_\ell} | n_j, n_{j+1} \rangle \tag{7.40}$$

[19] There are more elaborate algorithms which work directly in continuous imaginary time.

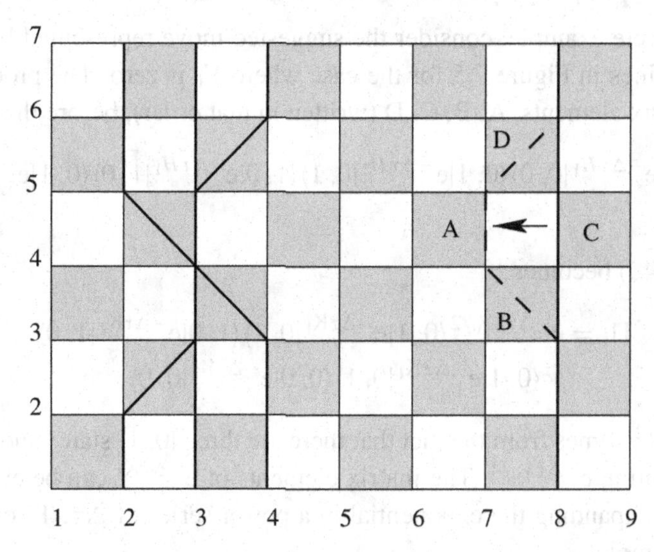

Figure 7.5 The decomposition of the full Hamiltonian H into H_1 and H_2 results in a 'checkerboard' pattern for the space–imaginary time lattice. Periodic boundary conditions are used in space, so that site 1 is identified with site 9. Furthermore, the trace in the partition function imposes periodic boundary conditions in the imaginary time direction so that world lines have to close on themselves. Only for even time slices are bosons allowed to hop between sites $(2, 3)$, $(4, 5)$, $(6, 7), \ldots$, while only for odd time slices can they hop between sites $(1, 2)$, $(3, 4), \ldots$ The bold paths are boson world lines, the arrow shows a typical allowed move.

where $\mathsf{K}_\ell = -t(\mathsf{a}_j^\dagger \mathsf{a}_{j+1} + \mathsf{a}_{j+1}^\dagger \mathsf{a}_j)$ is the hopping term acting at imaginary time slice ℓ, (n_j, n_{j+1}) $((n'_j, n'_{j+1}))$ are the occupations of nearest neighbour sites (i, j) before (after) the action of the time evolution operator. We have suppressed the imaginary time superscript for the states. Since the potential energy operator is diagonal in this representation, the terms U and U' are evaluated by replacing the number operators with the actual particle occupations and then simply factored out of the matrix elements. Less trivial is the evaluation of the matrix elements of the exponential of the kinetic energy operator. Because in principle many bosons can occupy a single site, even the two-site problem involves an infinite dimensional Hilbert space. In practice however the Hilbert space may be truncated at an occupation that depends on the number density of the particles. For example if one is interested in a number density, ρ, of order unity, it suffices to truncate the Hilbert space at a maximum site occupation of about five or so.

The matrix elements in this truncated space may be calculated to all orders in $\Delta\tau$ by doing an exact numerical diagonalization of the kinetic energy operators. However, it is also useful and instructive, and in many cases sufficient, to do the calculations to some low order, for example $\mathcal{O}(\Delta\tau)^2$.

As a concrete example, consider the suggested move represented by the arrow and dashed lines in Figure 7.5 for the case where V_1 is zero. The product, Π_i, of the four matrix elements, A, B, C, D (written in that order), before the move is

$$\Pi_{\rm i} = \langle 0,0|e^{-\Delta\tau H_1}|0,0\rangle\langle 0,1|e^{-\Delta\tau H_2}|0,1\rangle\langle 1,0|e^{-\Delta\tau H_1}|1,0\rangle\langle 0,1|e^{-\Delta\tau H_2}|0,1\rangle \tag{7.41}$$

which by (7.40) becomes

$$\Pi_{\rm i} = e^{-\Delta\tau 3V_0/2}\langle 0,1|e^{-\Delta\tau K_1}|0,1\rangle\langle 1,0|e^{-\Delta\tau K_2}|1,0\rangle \times\langle 0,1|e^{-\Delta\tau K_1}|0,1\rangle\langle 0,0|e^{-\Delta\tau K_2}|0,0\rangle \tag{7.42}$$

The $e^{-\Delta\tau 3V_0/2}$ comes from the fact that there are three $|0,1\rangle$ states and three $\langle 0,1|$ each contributing $e^{-\Delta\tau V_0/4}$. The matrix elements of $e^{-\Delta\tau K_\ell}$ can be evaluated, for example, by expanding the exponential in a power series of $\Delta\tau$. To order $\Delta\tau$ we have, for example,

$$\langle 0,1|e^{-\Delta\tau K_i}|0,1\rangle \approx \langle 0,1|(1-\Delta\tau K_i)|0,1\rangle \tag{7.43}$$

But, since $\langle 0,1|a_2^\dagger a_1|0,1\rangle = 0$, we get that $\langle 0,1|K_i|0,1\rangle = 0$ and Equation (7.43) becomes

$$\langle 0,1|e^{-\Delta\tau K_i}|0,1\rangle = \langle 0,1|0,1\rangle = 1 \tag{7.44}$$

For these matrix elements, the kinetic energy term contributes at higher orders of $\Delta\tau$, for example at $(\Delta\tau)^2$, where the boson at a site can jump to its neighbour and back in one evolution step. Therefore, to order $\Delta\tau$, we have

$$\Pi_{\rm i} = e^{-\Delta\tau 3V_0/2} \tag{7.45}$$

The product of the matrix elements *after* the suggested move is

$$\Pi_{\rm f} = e^{-\Delta\tau 3V_0/2}\langle 1,0|e^{-\Delta\tau K_1}|0,1\rangle\langle 0,0|e^{-\Delta\tau K_2}|0,0\rangle \times\langle 0,1|e^{-\Delta\tau K_1}|1,0\rangle\langle 0,1|e^{-\Delta\tau K_2}|0,1\rangle \tag{7.46}$$

As before, expanding the exponentials to order $\Delta\tau$ and using Equations (7.36), (7.37) and (7.38), we obtain

$$\Pi_{\rm f} = e^{-\Delta\tau 3V_0/2}(t\Delta\tau)^2 \tag{7.47}$$

To satisfy detailed balance, the suggested move is accepted if $\Pi_{\rm f}/\Pi_{\rm i}$ is larger than a random number drawn from a uniform distribution between zero and one. Notice that, in this example, the contribution of the potential energy cancels out and does not contribute to the accept/reject decision. For the case where $V_1 = 0$, which applies to this example, the potential energy term enters only when a singly occupied site becomes multiply occupied or vice versa.

Note that in the example configuration in Figure 7.5, three bosons meet on one site, then one of them peels off while the other two continue straight up until they too separate. It is meaningless to ask which boson is doing what: bosons are indistinguishable particles, their wave function is symmetric, and their creation and annihilation operators satisfy an algebra that must always be respected. The symmetry of the wave function and the indistinguishability of the particles are taken into account automatically in the world line algorithm in Equation (7.40): while calculating the matrix elements of K_i one has to use the creation and annihilation operators and their commutator algebra, which ensures the proper symmetry. This is one of the big advantages of the second quantized form of the Hamiltonian. We shall consider a different approach in the Problem 7.9.8, the real space representation, where this point will be re-emphasized.

To summarize, the general procedure for the Monte Carlo calculation is as follows:

(*i*) Initialize the lattice to some beginning occupation state. This starting configuration must respect the various symmetries of the Hamiltonian as well as allowing the transfer of particle number only across shaded squares where the kinetic energy acts. For example, there can be no truncated world lines: world lines must wrap across the lattice in the imaginary time direction and close upon themselves.

(*ii*) Sweep through the lattice, suggesting moves to other allowed boson occupation states. In practice, a good way to visualize such moves is as pulling a world line segment across an unshaded plaquette as in Figure 7.5.

(*iii*) Monitor the correlation time and make measurements whenever statistically independent configurations are obtained. Only measurements that conserve particle number locally are easy to make.

(*iv*) Control the systematics of the $\Delta\tau \to 0$ and $\beta \to \infty$ (ground state) extrapolations.

Some comments about the general structure of this formulation are in order. We started with a one-dimensional quantum model and mapped it onto a *classical* model in two dimensions where the rôle of the new dimension is played by $\beta = (k_{\mathrm{B}}T)^{-1}$. This classical formulation allows the numerical simulation to be implemented. Because of the appearance of the 'imaginary time' evolution operator, the new dimension is interpreted as the imaginary time dimension. As a consequence, while in the quantum model a boson is represented by a point object, in the new formulation the boson traces a world line. This gives the appealing physical interpretation of the path integral representation of the partition function as a weighted sum over all allowed conformations of the boson world lines. This is a general feature of quantum models: a D-dimensional quantum model maps onto a $(D+1)$-dimensional classical model.

In addition, the presence of a time axis (albeit imaginary) allows the calculation of *dynamic quantities* with the proviso that a rotation to real time be performed at the end.[20] In general such rotations may be rather difficult and are beyond the scope of this chapter. However, see Problem 7.9.8 for an example and further discussion.

To obtain the ground state, $\beta \to \infty$, one needs the 'thermodynamic' limit in the imaginary time direction. However, for a very hot system, $\beta \ll 1$, the extent in the imaginary time direction is very short: the classical representation of the D-dimensional quantum model remains essentially D-dimensional. Real time is treated in a very similar manner in quantum field theories at finite temperature.

7.4.2 Measurements

In a classical Monte Carlo simulation, such as for the Ising model, once the configurations are obtained, correctly distributed according to the normalized Boltzmann weight, it is straightforward to calculate the averages of physical quantities such as energies and correlation functions. In quantum Monte Carlo one has the problem that the system is defined in terms of quantum operators that were mapped onto a classical system ('world lines' in our case) for the purpose of numerical simulations.

In order to calculate averages of physical quantities, one must start with their definitions in terms of quantum operators, and map these definitions to the classical system at hand. Consider calculating the expectation value of a physical quantity described by the operator O (constructed from the creation and annihilation operators). We then have

$$\langle \mathsf{O} \rangle \equiv \frac{1}{Z} \mathrm{Tr}\left(\mathsf{O} \mathrm{e}^{-\beta H} \right) \tag{7.48}$$

which, by following exactly the same steps that led from (7.32) to (7.39), is mapped onto

$$\begin{aligned} \langle \mathsf{O} \rangle = \frac{1}{Z} \sum_{\{\mathbf{n}\}} & \langle \mathbf{n}^1 | \mathrm{e}^{-\Delta\tau H_2} | \mathbf{n}^{2L} \rangle \langle \mathbf{n}^{2L} | \mathrm{e}^{-\Delta\tau H_1} | \mathbf{n}^{2L-1} \rangle \\ & \ldots \langle \mathbf{n}^{2i+1} | \mathrm{e}^{-\Delta\tau H_2} \mathsf{O} | \mathbf{n}^{2i} \rangle \langle \mathbf{n}^{2i} | \mathrm{e}^{-\Delta\tau H_1} | \mathbf{n}^{2i-1} \rangle \\ & \ldots \langle \mathbf{n}^3 | \mathrm{e}^{-\Delta\tau H_2} | \mathbf{n}^2 \rangle \langle \mathbf{n}^2 | \mathrm{e}^{-\Delta\tau H_1} | \mathbf{n}^1 \rangle \end{aligned} \tag{7.49}$$

The imaginary time step at which we choose to place the operator O is, of course, arbitrary. In practice, to reduce statistical fluctuations, the matrix element for O is

[20] This is known as a Wick rotation and is often needed in quantum field theories.

averaged at all the imaginary time steps. Define $\mathsf{O}[\{n_i\}]$ as the contribution of a *single* configuration to the expectation value of O

$$\begin{aligned}\mathsf{O}[\{n_i\}] = \frac{1}{Z}&\langle \mathbf{n}^1|\mathrm{e}^{-\Delta\tau H_2}|\mathbf{n}^{2L}\rangle\langle \mathbf{n}^{2L}|\mathrm{e}^{-\Delta\tau H_1}|\mathbf{n}^{2L-1}\rangle \\ &\ldots\langle \mathbf{n}^{2i+1}|\mathrm{e}^{-\Delta\tau H_2}\mathsf{O}|\mathbf{n}^{2i}\rangle\langle \mathbf{n}^{2i}|\mathrm{e}^{-\Delta\tau H_1}|\mathbf{n}^{2i-1}\rangle \\ &\ldots\langle \mathbf{n}^3|\mathrm{e}^{-\Delta\tau H_2}|\mathbf{n}^2\rangle\langle \mathbf{n}^2|\mathrm{e}^{-\Delta\tau H_1}|\mathbf{n}^1\rangle\end{aligned} \tag{7.50}$$

With the exception of the term $\langle \mathbf{n}^{2i+1}|\mathrm{e}^{-\Delta\tau H_2}\mathsf{O}|\mathbf{n}^{2i}\rangle$, the numerator in (7.50) is identical to the summand in (7.39), while (7.49) is a sum over all configurations of (7.50). In other words, if one simply calculates $\langle \mathbf{n}^{2i+1}|\mathrm{e}^{-\Delta\tau H_2}\mathsf{O}|\mathbf{n}^{2i}\rangle$ for the configuration $\{n_i\}$, one would be calculating the average of O but with the wrong Boltzmann weight because the term $\langle \mathbf{n}^{2i+1}|\mathrm{e}^{-\Delta\tau H_2}|\mathbf{n}^{2i}\rangle$ is missing from Equation (7.49). To remedy the situation, we multiply and divide by the missing term. Thus (7.49) becomes

$$\boxed{\langle \mathsf{O}\rangle = \sum_{\{\mathbf{n}\}} P_{\mathrm{B}}[\{n_i\}]\frac{\langle \mathbf{n}^{2i+1}|\mathrm{e}^{-\Delta\tau H_2}\mathsf{O}|\mathbf{n}^{2i}\rangle}{\langle \mathbf{n}^{2i+1}|\mathrm{e}^{-\Delta\tau H_2}|\mathbf{n}^{2i}\rangle}} \tag{7.51}$$

where $P_{\mathrm{B}}[\{n_i\}]$ is the normalized 'Boltzmann weight'

$$\begin{aligned}P_{\mathrm{B}}[\{n_i\}] = \frac{1}{Z}&\langle \mathbf{n}^1|\mathrm{e}^{-\Delta\tau H_2}|\mathbf{n}^{2L}\rangle \\ &\times\langle \mathbf{n}^{2L}|\mathrm{e}^{-\Delta\tau H_1}|\mathbf{n}^{2L-1}\rangle\ldots\langle \mathbf{n}^3|\mathrm{e}^{-\Delta\tau H_2}|\mathbf{n}^2\rangle\langle \mathbf{n}^2|\mathrm{e}^{-\Delta\tau H_1}|\mathbf{n}^1\rangle\end{aligned} \tag{7.52}$$

So, we finally see that to measure expectation values, one visits each configuration (which, we repeat, is weighted by the correct Boltzmann weight (7.52)) and calculates $\langle \mathbf{n}^{2i+1}|\mathrm{e}^{-\Delta\tau H_2}\mathsf{O}|\mathbf{n}^{2i}\rangle$ normalized by $\langle \mathbf{n}^{2i+1}|\mathrm{e}^{-\Delta\tau H_2}|\mathbf{n}^{2i}\rangle$. If O is diagonal in the occupation state representation, for example the potential energy, the measurement simplifies tremendously because the matrix element of O is immediately calculable and can be moved out of the matrix element. In other words, if $\mathsf{O} = F[\{\mathsf{n}\}]$, then $\langle \mathbf{n}^{2i+1}|\mathrm{e}^{-\Delta\tau H_2}F[\{\mathsf{n}\}]|\mathbf{n}^{2i}\rangle = F[\{n\}]\langle \mathbf{n}^{2i+1}|e^{-\Delta\tau H_2}|\mathbf{n}^{2i}\rangle$, where $F[\{n\}]$ is a c-number function of the occupations. Normalizing this with $\langle \mathbf{n}^{2i+1}|\mathrm{e}^{-\Delta\tau H_2}|\mathbf{n}^{2i}\rangle$, the matrix elements cancel and we are left only with $F[\{n\}]$. So, for such diagonal cases, all one needs to do is calculate $F[\{n\}]$. But for other quantities, such as the kinetic energy, one must follow the above prescription.

In addition to the kinetic and potential energies, a very useful quantity is the structure factor, i.e. the Fourier transform of the density–density correlation function, whose continuous version was introduced in Section 3.4.2,

$$S(\vec{k}) = \frac{1}{N} \sum_{\vec{j},\vec{l}} \mathrm{e}^{i\vec{k}\cdot\vec{l}} \langle \mathrm{n}(\vec{j})\mathrm{n}(\vec{j}+\vec{l})\rangle \tag{7.53}$$

$$= \frac{1}{N} \sum_{\vec{j},\vec{l}} \mathrm{e}^{i\vec{k}\cdot\vec{l}} \langle n(\vec{j})n(\vec{j}+\vec{l})\rangle \tag{7.54}$$

where the number operators were replaced by their eigenvalues since they are diagonal in this representation. $n(\vec{j})$ is the number of bosons at the site $\vec{j}$, $\vec{k}$ is the momentum, and N is the number of sites on the lattice. In the thermodynamic limit, long range order is signalled by linear growth of $S(\vec{k})$ as a function of N for a specific value, $\vec{k}^*$, of the momentum. This is the ordering vector and describes the geometry of the long range order.

7.4.3 Quantum spin-1/2 models

As we have seen in Problem 5.7.7, in the hardcore limit $V_0 \to \infty$, we can use the Holstein–Primakoff transformation to map the Hamiltonian (7.29) onto the quantum spin system

$$H = -\frac{J_x}{2} \sum_{\langle ij \rangle} (\mathrm{S}_j^+ \mathrm{S}_i^- + \mathrm{S}_j^- \mathrm{S}_i^+) + J_z \sum_{\langle ij \rangle} \mathrm{S}_j^z \mathrm{S}_i^z - h \sum_i \mathrm{S}_i^z \tag{7.55}$$

$$= -J_x \sum_{\langle ij \rangle} (\mathrm{S}_j^x \mathrm{S}_i^x + \mathrm{S}_j^y \mathrm{S}_i^y) + J_z \sum_{\langle ij \rangle} \mathrm{S}_j^z \mathrm{S}_i^z - h \sum_i \mathrm{S}_i^z \tag{7.56}$$

where $J_x = 2t$, $J_z = V_1$ and $h = \mu - V_1$, and where we have ignored a constant term. This model is also called the spin-1/2 XXZ model because the coupling of the x spin components is the same as that for the y components but different from that of the z components. Since $t > 0$, we see that the first sum in (7.56) represents *ferromagnetic* coupling. Now we may repeat exactly all the steps that led from (7.31) to (7.39) and (7.40) and construct a world line algorithm for this quantum spin model.

In the current language, the representation used is that which diagonalizes S_i^z. Then the Boltzmann weight used in the simulation (see (7.41) (7.46) and (7.52)) will contain matrix elements of the form

$$\langle + + |e^{-K}| + + \rangle = \langle - - |e^{-K}| - - \rangle = 1 \tag{7.57}$$

$$\langle + - |e^{-K}| + - \rangle = \langle - + |e^{-K}| - + \rangle = \cosh\left(\frac{J_x}{2}\Delta\tau\right) \tag{7.58}$$

$$\langle + - |e^{-K}| - + \rangle = \langle - + |e^{-K}| + - \rangle = \sinh\left(\frac{J_x}{2}\Delta\tau\right) \tag{7.59}$$

$$K = -\frac{J_x}{2}(S_j^+ S_i^- + S_j^- S_i^+) \tag{7.60}$$

where $|+\rangle$ ($|-\rangle$) is an up (a down) spin. As long as J_x is positive, i.e. for the *ferromagnetic* case, all matrix elements are positive and, consequently, so is the Boltzmann weight. The numerical simulation can proceed without difficulty because the quantity used as the probability of the accept/reject is never negative.

For the *antiferromagnetic*, $J_x < 0$, case this is no longer the case: the matrix element (7.59) becomes negative and this may lead to the 'Boltzmann weight' becoming negative too. In such a case, we cannot use it as a probability for accepting moves. This is what is called the *sign problem* in quantum Monte Carlo simulations.

For a bipartite lattice, such as the square lattice, this problem can be easily remedied when the interactions of the x and y components are between nearest neighbours only (as is the case here). Consider the antiferromagnetic model

$$H = +J_x \sum_{\langle ij \rangle} (S_j^x S_i^x + S_j^y S_i^y) + J_z \sum_{\langle ij \rangle} S_j^z S_i^z - h \sum_i S_i^z \tag{7.61}$$

on a square lattice with $J_x > 0$. This H gives a negative contribution for (7.59). Suppose the lattice is $L_x \times L_y$ and that the sites are labelled by the two integers[21] $1 \le r_x \le L_x$ and $1 \le r_y \le L_y$. It is easy to see that, if we define the parity of a site by $(-1)^{r_x + r_y}$, then neighbouring sites have opposite parities. We may therefore perform the following transformation

$$S_i^x \to S_i^x (-1)^{r_x + r_y} \tag{7.62}$$

$$S_i^y \to S_i^y (-1)^{r_x + r_y} \tag{7.63}$$

$$S_i^z \to S_i^z \tag{7.64}$$

This has the effect of preserving the correct commutator algebra of the spins and at the same time changing the sign of the J_x interaction! This reverts the Hamiltonian to that given by Equation (7.56), in other words to the ferromagnetic case, which is free of the sign problem.

[21] In Equation (7.61) we are using the one-index notation $i = r_x + (r_y - 1)L_x$ to represent the sites. The two notations are clearly equivalent.

Alas, it is not always possible to remedy the sign problem this easily. Many physically interesting cases, in fact, suffer from this problem. For example, it is easy to convince oneself that the antiferromagnetic model on a triangular lattice cannot be freed of the sign problem by such a transformation. For such frustrated[22] cases, the world line algorithm, and indeed all other algorithms known to date, fail. There are 'fixes' that help in some cases but no true solution has been found.

7.5 Quantum Monte Carlo: fermions

We developed in the previous section the world line algorithm, a simulation algorithm for bosonic and quantum spin systems. In fact, this algorithm is quite general and may, in principle be used for fermions.[23] We have also seen that in the case of competing interactions such as antiferromagnets on a triangular lattice, this method (and all quantum Monte Carlo methods for that matter) suffers from a sign problem in the Boltzmann weight.

In this section, we shall see that things are much worse for fermions. One aspect of this problem may be demonstrated on the simplest of models: 'scalar fermions'. Scalar fermions are spinless objects that satisfy Fermi–Dirac statistics. While these objects do not exist as fundamental particles, they are very useful constructs to study simplified fermionic systems. Consider the two-dimensional lattice system, with periodic boundary conditions, described by the Hamiltonian

$$H = -t \sum_{\langle ij \rangle} (\mathsf{c}_j^\dagger \mathsf{c}_i + \mathsf{c}_i^\dagger \mathsf{c}_j) - \mu \sum_i \mathsf{n}_i \tag{7.65}$$

where $\mathsf{c}_j^\dagger$ creates a fermion at site j, c_j destroys one at site j, μ is the chemical potential and $\mathsf{n}_i = \mathsf{c}_i^\dagger \mathsf{c}_i$ is the number operator on site i. This is the spinless fermion Hubbard model. The fermionic operators satisfy the *anticommutator* algebra

$$\{\mathsf{c}_i, \mathsf{c}_j^\dagger\} \equiv \mathsf{c}_i \mathsf{c}_j^\dagger + \mathsf{c}_j^\dagger \mathsf{c}_i = \delta_{ij} \tag{7.66}$$

$$\{\mathsf{c}_i, \mathsf{c}_j\} \equiv \mathsf{c}_i \mathsf{c}_j + \mathsf{c}_j \mathsf{c}_i = 0 \tag{7.67}$$

$$\{\mathsf{c}_i^\dagger, \mathsf{c}_j^\dagger\} \equiv \mathsf{c}_i^\dagger \mathsf{c}_j^\dagger + \mathsf{c}_j^\dagger \mathsf{c}_i^\dagger = 0 \tag{7.68}$$

[22] A system is said to be frustrated if one cannot satisfy all the interactions. For example, in an antiferromagnet, the spins on nearest neighbour sites want to be antiparallel. It is easy to see that going around an elementary square on a square lattice, we can make all the nearest neighbours antiparallel. However, going around an elementary triangle on a triangular lattice, we see that we can satisfy the antiferromagnetic interaction on two of the sides of the triangle but not the third. This is frustration [38].

[23] In fact this algorithm was developed with fermions in mind. As we shall see, it is not very well adapted for this application.

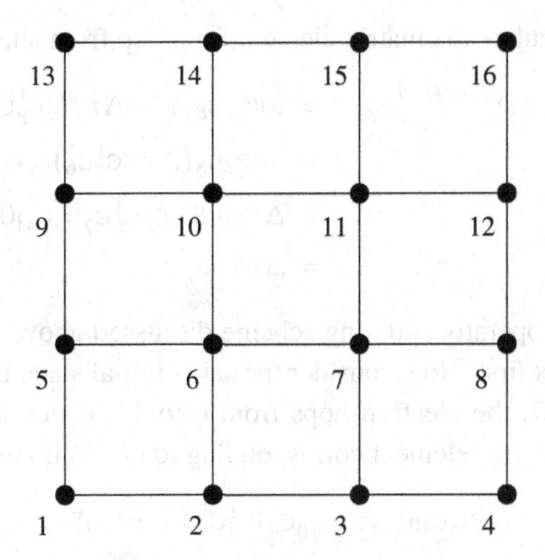

Figure 7.6 Convention for ordering the fermion creation operators on the lattice.

We immediately see from (7.68) that when $i = j$ we have $c_i^\dagger c_i^\dagger = 0$, in other words, two fermions cannot be created in the same state.

The Hamiltonian (7.65) in fact describes the ideal Fermi gas (Section 4.2) on a lattice: it is easy to diagonalize with a Fourier transformation and solve exactly. Nonetheless, this simple model exhibits an aspect of a problem that plagues numerical simulations of fermionic systems.

To illustrate this problem, consider, for example, populating sites 6 and 8 of the square lattice depicted in Figure 7.6. To do that, one needs to apply the creation operators on these sites, but the question is in what order this should be done. To see the problem, remark that we may accomplish that by applying on the empty state (vacuum) the creation operators first on site 8 and then site 6 or the other way around. But these two are not the same since

$$c_8^\dagger c_6^\dagger |0\rangle = -c_6^\dagger c_8^\dagger |0\rangle \tag{7.69}$$

We did not face this problem in the bosonic case because the operators commuted whereas here they anticommute. We therefore see the necessity of *choosing* an order for the application of the creation operators and following it in all subsequent calculations. One such choice is shown in Figure 7.6 where the numbers show the order in which the creation operator is applied to create a particle at a particular site.

Now consider a matrix element such as appears in Equation (7.39) with fermions on sites 6 and 8 in the initial state and sites 7 and 8 in the final state.

In other words, calculate the matrix element for a hop from site 6 to site 7

$$\langle 0|c_7 c_8 e^{-\Delta\tau H} c_8^\dagger c_6^\dagger|0\rangle = \langle 0|c_7 c_8 (1 - \Delta\tau H) c_8^\dagger c_6^\dagger|0\rangle \tag{7.70}$$

$$= \langle 0|c_7 c_8 (\Delta\tau\, t c_7^\dagger c_6) c_8^\dagger c_6^\dagger|0\rangle \tag{7.71}$$

$$= \Delta\tau t \langle 0|c_7 c_8 c_8^\dagger c_7^\dagger c_6 c_6^\dagger|0\rangle \tag{7.72}$$

$$= \Delta\tau\, t \tag{7.73}$$

where we used the operator ordering scheme discussed above: creation operators with lower index act first. Now consider the same initial state, but instead of a hop from site 6 to site 7, the electron hops from 6 to 10, which is permitted by the Hamiltonian. The matrix element corresponding to (7.73) is easily evaluated

$$\langle 0|c_8 c_{10} (\Delta\tau\, t c_{10}^\dagger c_6) c_8^\dagger c_6^\dagger|0\rangle = -\Delta\tau t \tag{7.74}$$

So, as in Section 7.4.3, we see that some matrix elements entering in the Boltzmann weight (7.52) can be negative. Consequently, the Boltzmann weight itself may become negative and once again we face the sign problem. Worse, this happens for *free* fermions, a trivial exactly solvable case.

The world line algorithm is, therefore, not at all well adapted to deal with fermions in more than one dimension.[24] There are several other simulation algorithms that do not suffer from the sign problem *for free fermions. Unfortunately, all fermion Monte Carlo algorithms suffer from the sign problem when interactions are included except for some very special cases.*[25] There is no known general solution for this problem: it remains one of the greatest challenges for numerical simulations.

Still, in some cases we may be able to perform Monte Carlo simulations for fermions if the sign problem is 'not very bad'. Consider evaluating numerically the expectation value of some physical quantity given by the operator A

$$\langle \mathsf{A}\rangle = \frac{\mathrm{Tr}\, \mathsf{A} e^{-\beta H}}{\mathrm{Tr}\, e^{-\beta H}} \tag{7.75}$$

Mapping the traces of operators onto integrals (or sums) over the configurations of an equivalent classical model (as we did above for the free case) gives

$$\langle \mathsf{A}\rangle = \frac{\int \mathrm{D}\phi\, A[\{\phi\}]\, P_\mathrm{B}[\{\phi\}]}{\int \mathrm{D}\phi\, P_\mathrm{B}[\{\phi\}]} \tag{7.76}$$

[24] In one dimension the sign problem does not appear in the world line algorithm as we saw when the fermion jumped from site 6 to site 8.

[25] We shall not discuss the interacting fermion case since it involves notions not discussed in this book such as calculating quantities of the form Tr $\exp(\sum_{i,j} c_i^\dagger M_{ij} c_j)$ where, unlike the ideal gas case, the matrix M_{ij} is not diagonalizable by a Fourier transformation.

where we represented the classical variables generically by ϕ. As mentioned above, the 'Boltzmann weight', $P_B[\{\phi\}]$, can become negative. This does not pose a problem from the formal point of view. The partition function itself, i.e. the denominotor of (7.76), is positive even if, for some configurations of ϕ, the integrand becomes negative. The problem however is numerical: how can the Monte Carlo method be applied when $P_B[\{\phi\}] < 0$?

The trick is to write $P_B[\{\phi\}] \equiv \text{sgn}(P_B)|P_B[\{\phi\}]|$ where $\text{sgn}(P_B)$ is the sign of the Boltzmann weight for that particular configuration. Then,

$$\langle A \rangle = \frac{Z'}{Z'} \frac{\int D\phi\, A[\{\phi\}]\, \text{sgn}(P_B)|P_B[\{\phi\}]|}{\int D\phi\, \text{sgn}(P_B)|P_B[\{\phi\}]|} \tag{7.77}$$

where we multiplied and divided by

$$Z' = \int D\phi\, |P_B[\{\phi\}]| \tag{7.78}$$

Defining $\langle \bullet \rangle'$ as the expectation value taken with the new partition function (7.78), i.e. using the *positive* semi-definite Boltzmann weight $|P_B[\{\phi\}]|$ we can write

$$\langle A \rangle = \frac{\langle A[\{\phi\}]\, \text{sgn}(P_B)\rangle'}{\langle \text{sgn}(P_B)\rangle'} \tag{7.79}$$

In other words, instead of calculating the expectation value directly, as one usually does, it is calculated as the ratio of two expectation values each calculated using a positive Boltzmann weight.

However, the sign problem has not been solved since $\langle \text{sgn}(P_B)\rangle'$ can be very small. In that case, $\langle A \rangle$, which need not be a small number, is being calculated as the ratio of two very small, and quite possibly statistically noisy, numbers. This yields very poor numerical results. It is possible to estimate how rapidly accuracy is lost. Recall that

$$\langle \text{sgn}(P_B)\rangle' = \frac{\int D\phi\, \text{sgn}(P_B)|P_B[\{\phi\}]|}{\int D\phi\, |P_B[\{\phi\}]|} = \frac{Z}{Z'} \tag{7.80}$$

which means we may write

$$\langle \text{sgn}(P_B)\rangle' = e^{-\beta(F-F')} \tag{7.81}$$

where F (F') is the free energy of the system described by Z (Z'). Even for small differences between F and F', $\langle \text{sgn}(P_B)\rangle'$ decreases *exponentially* with β thus preventing simulations at very low temperature. In addition, since the free energy is extensive, $\langle \text{sgn}(P_B)\rangle'$ also decreases exponentially with the system size, limiting simulations to rather small sizes.

Therefore, the sign problem makes numerical simulation of fermionic systems exponentially difficult!

7.6 Finite size scaling

In our discussion of critical phenomena in Chapter 4, we always assumed that the size L of the system is not a relevant parameter or, in other words, we have assumed the thermodynamic limit $L \to \infty$. In this limit, the properties of a subsystem should be identical to the bulk properties of the whole system. Let us then imagine that we have divided a large system into smaller systems, of size L/p, where p is some integer and try to assess whether they have the same thermodynamic properties. The answer will be positive if we can ignore the correlations between two neighbouring smaller systems, namely if the correlation length is much smaller than L/p, $\xi \ll L/p$, as in Figure 4.10. What happens for systems whose size is not much bigger than the correlation length ξ is thus an interesting problem that we are going to examine in this section. The results of our study will be particularly important for the interpretation of numerical simulations because the number of sites is then necessarily finite. One measures thermodynamic quantities in finite systems, and it is obviously essential to be able to extrapolate to the thermodynamic limit. How the scaling behaviour of an infinite system observed in the vicinity of a critical point translates to a finite system is called *finite size scaling*.

We assume the system to be restricted to a hypercube of volume L^D. Finite size scaling can also be used to study systems which are finite in some direction(s) and infinite in other direction(s). One could wish, for example, to obtain the properties of a two-dimensional Ising model in a strip infinite along the x-axis, while finite and of width L along the y-axis. We shall limit ourselves to truly finite systems, leaving the other case to Exercise 7.8.2.

Consider a thermodynamic quantity, for example the susceptibility, which is a function of L and of the reduced temperature t (4.115), $A_L(t)$. Taking the lattice spacing a to be unity, the number of sites in the hypercube is L^D and in a RGT this number is reduced to $L'^D = (L/b)^D$. Thus $1/L' = b/L$, and we can think of $1/L$ as a relevant variable with exponent $\omega = 1$ so that A_L transforms in a RGT as[26]

$$A_L(t) \equiv A\left(t, \frac{1}{L}\right) = b^{\omega_A}\, A\left(t', \frac{1}{L'}\right) = b^{\omega_A} A\left(t b^{1/\nu}, \frac{b}{L}\right)$$

with an anomalous dimension ω_A to be determined later on. We choose as usual $b = |t|^{-\nu}$ and get

$$A\left(t, \frac{1}{L}\right) = |t|^{-\nu\omega_A}\, A\left(\pm 1, \frac{|t|^{-\nu}}{L}\right)$$

[26] Clearly, this is a somewhat handwaving argument that can be made rigorous using field-theoretical methods. Indeed, the short distance singularities which govern the renormalization group equations in field theory are identical in a finite and in an infinite volume: short distances do not depend on finite volume effects. Thus the renormalization group equations remain unchanged in a finite volume [1, 125].

which can be rewritten

$$A_L(t) = L^{\omega_A} f(tL^{1/\nu}) \tag{7.82}$$

For finite values of L, all thermodynamic quantities and their derivatives are analytic functions of t, which implies that the function $f(x)$ is an analytic function of x. In the limit $L \to \infty$, $A_L(t) \simeq c_\pm |t|^{-\rho}$, where ρ is some critical exponent and c_+ (c_-) corresponds to $t > 0$ ($t < 0$). Then

$$\lim_{|x| \to \pm\infty} f(x) = c_\pm |x|^{-\rho}$$

and in this limit the factor L^{ω_A} is compensated if $\omega_A = \rho/\nu$

$$\lim_{L \to \infty} A_L(t) \simeq c_\pm L^{\omega_A} (|t| L^{1/\nu})^{-\rho} = c_\pm |t|^{-\rho}$$

so that the final formula for $A_L(t)$ is

$$\boxed{A_L(t) = L^{\rho/\nu} f(tL^{1/\nu})} \tag{7.83}$$

Since the correlation length is proportional to $|t|^{-\nu}$, another way of writing (7.83) is

$$A_L(t) = L^{\rho/\nu} g_\pm\left(\frac{L}{\xi(t)}\right) \tag{7.84}$$

In the thermodynamic limit, $A_\infty(t) = A(t)$ is singular at $t = 0$. For the sake of definiteness, let us assume that $\rho > 0$, so that $A(t)$ diverges at $t = 0$. This divergence is translated into a maximum of $A_L(t)$ at finite L and the function $f(x)$ must have a maximum at some value of x, $x = x_0$, so that $f'(x_0) = 0$. Then $A_L(t)$ will have a maximum, not at $t = 0$, but rather at

$$t_c(L) = x_0 L^{-1/\nu} \tag{7.85}$$

The position of the maximum approaches $t = 0$ (or $T = T_c$) when $L \to \infty$ and the peak becomes higher and narrower. An example is given in Figure 7.7, where the susceptibility of the two-dimensional Ising model has been computed in a numerical simulation (Problem 7.9.1).
Let us consider the quantity

$$B_L(t) = L^{-\rho/\nu} A_L(t) = f(tL^{1/\nu}) \tag{7.86}$$

If the reduced temperatures t_1 and t_2 and the sizes L_1 and L_2 are related by

$$t_1 L_1^{1/\nu} = t_2 L_2^{1/\nu}$$

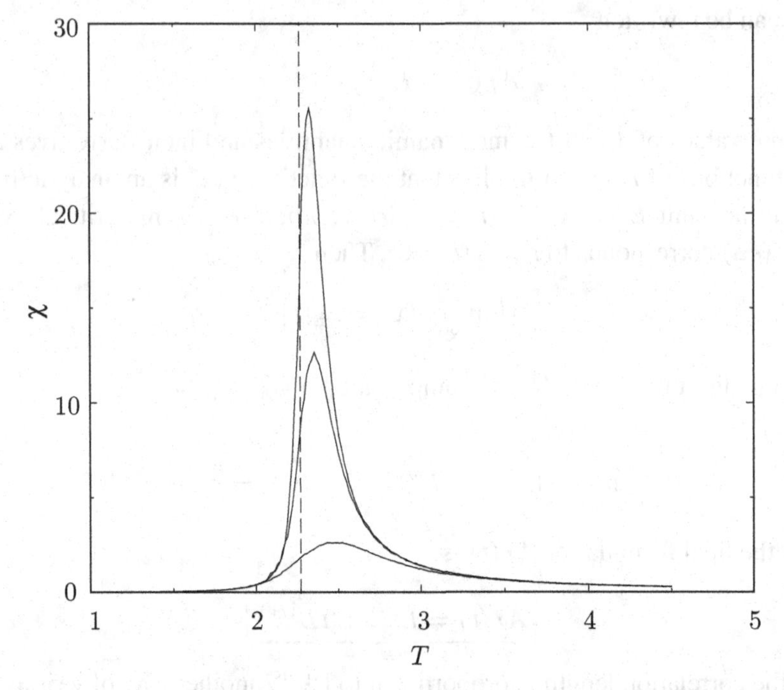

Figure 7.7 Behaviour of the magnetic susceptibility of the two-dimensional Ising model computed numerically for finite lattices with sizes (from lowest peak to the highest) 10×10, 24×24, 32×32. The dashed line gives the position of T_c in the thermodynamic limit.

then $B_{L_1}(t_1) = B_{L_2}(t_2)$, and if we plot $B_L(t)$ as a function of $tL^{1/\nu}$ for different sizes, the data from the numerical simulations should collapse on a single curve in the vicinity of $t = 0$. The drawback of this presentation is one needs to know the critical temperature, and it is more appropriate to use the reduced temperature t' related to $t_c(L)$ by

$$t' = t - t_c(L) = t - x_0 L^{-1/\nu}$$

Since we have

$$f(tL^{1/\nu}) = f(t'L^{1/\nu} + x_0) \tag{7.87}$$

where $f'(x_0) = 0$, the quantities B_L for different sizes will be related as previously by

$$B_{L_1}(t'_1) = B_{L_2}(t'_2) \quad \text{if} \quad t'_1 L_1^{1/\nu} = t'_2 L_2^{1/\nu}$$

Plotting $B_L(t)$ as a function of $t'L^{1/\nu}$ for different sizes L, all the numerical data should collapse on a single curve.

It will be convenient to rewrite the previous result by using T and T_c, introducing the 'critical temperature at size L', $T_c(L)$, which is given from (7.85) by

$$T_c(L) = T_c + (x_0 T_c) L^{-1/\nu} \tag{7.88}$$

Then we get from (7.87)

$$A_L(T) \sim L^{\rho/\nu} f(L^{1/\nu}(T - T_c(L))) \tag{7.89}$$

which allows us to make three related important remarks. The first is that if we plot $L^{-\rho/\nu} A_L(T)$ as a function of the scaling variable $L^{1/\nu}(T - T_c(L))$, the data from the simulation of systems of different sizes will collapse on the same curve in the vicinity of the critical temperature. The second remark is that at $T = T_c(L)$ we have

$$A_L(T = T_c(L)) \sim L^{\rho/\nu} f(0) \tag{7.90}$$

This important result tells us that the scaling of physical quantities measured at $T_c(L)$ is described by the bulk critical exponents. Therefore, the simulation of finite size systems does offer the means to determine bulk properties when analyzed correctly. From (7.86) this last result remains valid if instead of taking $T = T_c(L)$ we take $T = T_c$. This leads us to our third remark: if we plot $L^{-\rho/\nu} A_L(T)$ versus T, then at $T = T_c$ the curves from different lattice sizes will all intersect at one point. This offers another way to use finite size scaling to determine T_c. These three important remarks will be demonstrated in the problems, especially 7.9.1.

There are two important cases where the divergence of a physical quantity does not follow a power law. The first is that of a logarithmic, rather than power law, divergence. For example, the specific heat exponent of the two-dimensional Ising model is $\alpha = 0$. This means that the specific heat does not diverge as a power at $T = T_c$ but rather as a logarithm

$$A(t) \sim C \ln|t| \tag{7.91}$$

Using $\xi(T) \sim |t|^{-\nu}$ gives for T near T_c

$$A \sim \frac{C}{\nu} \ln \xi \tag{7.92}$$

which becomes for finite size systems

$$A_L(T_c(L)) \sim \frac{C}{\nu} \ln L \tag{7.93}$$

Another important case is that of the Kosterlitz–Thouless (KT) phase transition. For T above the KT transition, T_{KT}, we have for the correlation function

$$G(r) \sim e^{-r/\xi(T)} \tag{7.94}$$

while below the transition, $T < T_{\mathrm{KT}}$, we have

$$G(r) \sim \frac{1}{r^{\eta(T)}} \tag{7.95}$$

Therefore, for $T < T_{\mathrm{KT}}$ $\xi(T)$ is infinite and so this entire temperature range is critical. When T approaches T_{KT} from above, $T \to T_{\mathrm{KT}}^{+}$, the correlation length exhibits an essential singularity

$$\xi(T) \sim \exp\left(\frac{b}{t^{1/2}}\right) \tag{7.96}$$

where $t = (T - T_{\mathrm{KT}})/T_{\mathrm{KT}}$ and b is an unimportant constant. Other quantities, such as the magnetic susceptibility, may be expressed in terms of ξ

$$\chi \sim \xi^{2-\eta_{\mathrm{c}}} \tag{7.97}$$

$$\sim \exp\left(\frac{c}{t^{1/2}}\right) \tag{7.98}$$

where $\eta_{\mathrm{c}} = \eta(T = T_{\mathrm{KT}})$ and c is a constant.

For finite size scaling, the relevant quantity is still the scaling variable $x = L/\xi$. We can then repeat previous scaling arguments by taking $\xi(T) \sim L$ in Equation (7.96) and show for example that

$$T_{\mathrm{c}}(L) \sim T_{\mathrm{c}} + \frac{b^2}{(\ln L)^2} \tag{7.99}$$

where b is the same parameter appearing in (7.96). Equation (7.99) is the Kosterlitz–Thouless analogue of Equation (7.88). The KT transition will be illustrated in Problem 7.9.4.

7.7 Random number generators

From the discussions in this chapter it is clear that random numbers play a central rôle in Monte Carlo simulations. The random numbers discussed here are drawn from a uniform distribution between zero and one. Often one may need numbers drawn from different distributions, for example Gaussian as in the case of the Langevin equation (see Chapter 9). In such cases, there are methods to use uniformly distributed random numbers to produce the desired distribution. We therefore concentrate on this fundamental case.

It may, however, appear inconsistent to expect to be able to produce sequences of random numbers using a set of deterministic rules in the form of a computer program. This is indeed what we wish to do: during a Monte Carlo simulation, one calls a 'random number generator', which is basically a set of sufficiently complex rules so that the outcome is a sequence of apparently random numbers. The appearance of randomness is, of course, not enough. For a generator to be reliable, the sequences it produces must satisfy certain stringent conditions. We shall not enter here into the detailed mathematics of random number generators; for that we refer the reader to the literature. For an in-depth analysis of the issues involved see [65] and references therein, while for a utilitarian approach and the programs for several generators see [103]. Rather, our goal here is to present a general discussion and to caution the reader against some pitfalls.

It should be clear that one statistical test that must be satisfied by a generator is that the numbers be uniformly distributed, in other words, the interval [0, 1] be uniformly filled.[27] This condition is easily verified by generating a very long sequence, calculating the moments and comparing with the expected values.

Another rather intuitive condition is that, by virtue of being random, the numbers should be uncorrelated. It is surprising how little attention is sometimes paid to this very important condition. Computers usually come equipped with a random number generator, usually called rand(), of the linear congruential type

$$I_{i+1} = (aI_i + c)\mathrm{mod}(m) \tag{7.100}$$

where I_i is an integer between 0 and $m - 1$. When the parameters a (called the multiplier) and c (called the increment) are appropriately chosen, Equation (7.100) generates a non-recurring sequence of integers of length m. For this reason, m is chosen to be the largest integer that can be represented on a computer, $m = 2^{31} - 1$ for a 32 bit machine. Clearly, to obtain real numbers between 0 and 1, we simply take I_{i+1}/m. To start the sequence, i.e. to get I_1, one must supply the 'seed' for the sequence, I_0. Of course, in a numerical simulation the physical results should be independent of the choice of seed: any value for I_0 is as good as any other.

However, even when the parameters a, c are well chosen, the sequential values of I_i are correlated[28] and we strongly advise against the use of such generators. Two reliable and well tested generators that enjoy widespread use are ran2 and ran3 in Reference [103]. However, avoid using ran1: correlations appear for very long sequences.

In general, it is advisable to have more than one generator at one's disposal and to perform high precision test simulations using them. If the results are statistically the same, that adds to our confidence in the reliability of both. If the results differ, then one or both generators are defective.

[27] Often the interval is [0, 1] but this does not pose a problem.

[28] See Reference [103] Chapter 7 or [94] Chapter 16 for discussion and examples.

7.8 Exercises

7.8.1 Determination of the critical exponent ν

A possible determination of the critical exponent ν is given by the following method. Taking $A_L(t) = \xi_L(t)$, where $\xi_L(t)$ is the correlation length at size L, show that the ratio of correlation lengths at the same size but taken at two different reduced temperatures is given by

$$\frac{\xi_L(t')}{\xi_L(t)} = \frac{L'}{L}$$

provided $tL^{1/\nu} = t'L'^{1/\nu}$. Expanding $\xi_L(t)$ near $t = 0$ in powers of t

$$\xi_L(t) = AL + t\dot{\xi}_L(0) + \mathcal{O}(t^2)$$

show that

$$\left(1 + \frac{1}{\nu}\right) \ln \frac{L}{L'} = \ln \frac{\dot{\xi}_L(0)}{\dot{\xi}_{L'}(0)}$$

Observe that all quantities in this formula are measured at finite L.

7.8.2 Finite size scaling in infinite geometries

1. Consider a 2-D Ising model on an infinitely long two-dimensional strip of finite width L. Write down an expression for the susceptibility as a function of $T - T_{c2}$ and L, where T_{c2} is the critical temperature of the 2-D Ising model. Remember that the one-dimensional Ising model has no phase transition at finite temperature.

2. Consider now a 3-D Ising model, infinite in two dimensions but finite and of width L in the third dimension. Show that the susceptibility may be written as

$$\chi_L(T) = L^{\gamma/\nu} f\left(L^{1/\nu}(T - T_c)\right) g\left(L^{1/\nu}(T - T_c)\right)$$

where $f(x)$ is an infinitely differentiable function of x which obeys

$$\lim_{|x| \to \infty} f(x) = A_{\pm}|x|^{-(\gamma - \gamma_2)}$$

and

$$g(x) = \left[B_+\theta(x + b) + B_-\theta(-x - b)\right] |x + b|^{-\gamma_2}$$

In these equations, γ and ν are critical exponents of the (infinite) 3-D Ising model, γ_2 a critical exponent of the 2-D Ising model, b a positive number and θ the step function. What is the critical temperature at finite L? Explain why b must be positive.

3. The specific heat of the 2-D Ising model diverges logarithmically

$$C_2 \propto C_\pm \ln |T - T_{c2}|$$

Write a finite size scaling expression for the specific heat of the 3-D Ising model of Question 2.

7.8.3 Bosons on a single site

Consider the simple one site problem described by the quantum Hamiltonian,

$$H = \mathsf{a}^\dagger \mathsf{a} + b(\mathsf{a}^\dagger \mathsf{a})^2 \tag{7.101}$$

which describes the energy at a site due to the presence of n ($\mathsf{a}^\dagger \mathsf{a}|n\rangle = n|n\rangle$) particles. The operators $\mathsf{a}^\dagger$ and a create and destroy a boson at the site.

The two terms of the Hamiltonian commute: we can therefore write the partition function as

$$Z = \mathrm{Tr}\, \mathrm{e}^{-\beta H} \tag{7.102}$$

$$= \sum_{n=0}^{\infty} \mathrm{e}^{-\beta(n+bn^2)} \tag{7.103}$$

Write a Monte Carlo program to evaluate the average energy, E, the specific heat, C, Equation (7.14), and the average occupation, $\langle \mathsf{a}^\dagger \mathsf{a} \rangle$, as a function of temperature for various values of the parameter b.

While the partition function cannot be evaluated analytically, the series may be summed numerically to very high precision. By expressing E, C and $\langle \mathsf{a}^\dagger \mathsf{a} \rangle$ in terms of derivatives of Z, evaluate these quantities explicitly and compare with the results of the Monte Carlo simulation.

7.9 Problems

7.9.1 Two-dimensional Ising model: Metropolis

The goal of this problem is to simulate the two-dimensional Ising model and study its properties using the methods of finite size scaling discussed in Section 7.6.

Since this program will serve as the model for the programs needed in the following problems, it is worthwhile spending some time on making it as efficient as possible.

1. Show that the Metropolis algorithm, defined by Equations (7.10a) and (7.10b), satisfies detailed balance, Equation (7.7) or (7.9). To this end, consider the initial and final configurations i and f respectively. First assume $P_\mathrm{f} > P_\mathrm{i}$ and show that detailed balance is satisfied. In this case, of course, $W(\mathrm{i} \to \mathrm{f}) = 1$ and $W(\mathrm{f} \to \mathrm{i}) = P_\mathrm{i}/P_\mathrm{f}$. Do this also for the case $P_\mathrm{f} < P_\mathrm{i}$.

2. Write a program to simulate the two-dimensional Ising model with periodic boundary conditions. The system length L (total number of spins is L^2) should be an easily modifiable parameter as should the temperature.

Run your simulation as follows: 10^5 sweeps to thermalize the system, 3×10^5 measurement sweeps measuring every 10 sweeps. These choices give good statistical accuracy. Do the simulation from $T = 0.015$ to $T = 4.5$ in steps of 0.015 and for $L = 10, 16, 24, 36$.[29] For each of these temperatures and system sizes, measure the average total energy, $E = \langle H \rangle$, the specific heat, C,

$$C = \frac{1}{NkT^2}(\langle H^2 \rangle - \langle H \rangle^2) \tag{7.104}$$

the magnetization per site, $M = \langle S_i \rangle$, and the magnetic susceptibility, χ,

$$\chi = \frac{N}{kT}\left(\langle \frac{1}{N}|\sum S_i|^2 \rangle - \langle S_i \rangle^2\right) \tag{7.105}$$

Once the program is written and debugged, these simulations should take a few hours for the larger system sizes.[30]

3. Make a plot of the average energy, E, as a function of T for the four system sizes. Comment on what you observe for $2 \leq T \leq 3$.

4. Plot the susceptibility as a function of T for the four systems simulated. Note that both the position and the height of the peak in χ change with the system size. For each system size, determine the temperature $T_\mathrm{c}(L)$ at which the peak is observed as well as the height of the peak.

Verify Equation (7.88) by plotting $T_\mathrm{c}(L)$ versus L^{-1}. This assumes the exact value of $\nu = 1$. Compare the value you obtain for the thermodynamic critical temperature, T_c, with the exact value $T_\mathrm{c} = 2.269\,185\ldots$ To do this analysis well without assuming the exact value of ν, one would need results for larger system sizes: a two parameter fit through four points is not very reliable but it is worth trying.

5. Verify Equation (7.90) by plotting $\chi(T_\mathrm{c}(L))$ versus L on log-log scale. According to (7.90) we expect a power law (i.e. a straight line on the double logarithmic scale) which gives the exponent γ/ν. Compare the value you find for γ/ν with the exact value $\gamma/\nu = 1.75$.

Plot $L^{-\gamma/\nu}\chi$ as a function of the scaling variable $L^{1/\nu}(T - T_\mathrm{c}(L))$. According to Equation (7.89) the results from the different size systems should collapse on the same curve in the critical region if the exponents used are the right ones.

[29] T is taken in units of J.
[30] The simulations suggested in these problems have been verified to take a few hours on 2 GHz processors.

6. Plot the magnetization, M, versus T and observe the finite size effects near the transition. Verify Equation (7.90) again, this time by plotting $L^{\beta/\nu}M$ as a function of T for the different sizes simulated. To determine the best value for β/ν, we find the exponent that gives the best intersection at one point of all the magnetization curves. To see what this should look like, try the exact value $\beta/\nu = 0.25$. Compare the temperature at which the curves meet with the exact critical temperature T_c.

Now plot $L^{\beta/\nu}M$ versus the scaling variable $L^{1/\nu}(T - T_c')$ where T_c' is the value of the temperature where the curves meet. You should observe the magnetization curves collapse onto a single curve in the critical region (the region around T_c).

7. Plot C versus T for the systems simulated. Notice that both the position and the height of the peak in C change with the system size. Find $C_{\max}(L)$, the values of C at the peaks and the temperature, $T_c'(L)$, where these peaks are found.

Plot $T_c'(L)$ versus $L^{-1/\nu}$ and find the thermodynamic critical temperature T_c. Compare this with the exact value and the value obtained from the peaks of the magnetic susceptibility.

For the two-dimensional Ising model the exponent α describing the divergence of the specific heat vanishes, $\alpha = 0$. This means that the divergence of C is logarithmic. Verify this by plotting $C_{\max}(L)$ versus $\ln L$.

8. As we saw in Chapter 1, the entropy per spin can be obtained by integrating the specific heat,

$$S(T) = \int_0^T \mathrm{d}T' \frac{C(T')}{T'} \tag{7.106}$$

In particular, for the Ising model we have $S(T \to \infty) = \ln 2$. Verify this by integrating numerically your results from $T = 0.015$ to $T = 4.5$. Compare the $S(\infty)$ thus obtained with the exact result. The agreement should be quite good.

Calculate and plot $S(T)$ versus T for $0.015 \le T \le 4.5$. Using the relation

$$F = E - TS \tag{7.107}$$

calculate and plot the free energy as a function of temperature. Comment on the form of the free energy.

7.9.2 Two-dimensional Ising model: Glauber

The Metropolis algorithm is defined by Equations (7.10a) and (7.10b). However, as we mentioned in Section 7.1, there are other choices one may make for the transition probabilities, $W(\nu \to \sigma)$. One such alternative choice gives the Glauber dynamics,

$$W(\nu \to \sigma) = \frac{P_\sigma}{P_\nu + P_\sigma} \tag{7.108}$$

This is the expression for $W(\nu \to \sigma)$ regardless of whether the final state is more or less likely. To accept a move we simply compare the above expression to a random number drawn from a uniform distribution from 0 to 1. We accept if W is greater than this random number.

1. Show that Glauber dynamics satisfies detailed balance.

2. Modify your Metropolis Ising program to perform the Glauber dynamics. Repeat the simulations in the preceding problem and convince yourself that the physical results do not change with the change of Monte Carlo dynamics.

7.9.3 Two-dimensional clock model

The goal of this problem is to study a generalization of the Ising model examined in the two preceding problems.

The Ising Hamiltonian, in the absence of an external magnetic field, is invariant under the transformation $S_i \to -S_i$. This symmetry is called Z_2 and can also be seen clearly in an alternative formulation of the Ising Hamiltonian. Instead of writing the Hamiltonian as in Equation (7.11) (with $B = 0$), we can write

$$H = -J \sum_{\langle i,j \rangle} \cos(n_i - n_j)\pi \tag{7.109}$$

where the integer variables n_i take the values 0, 1. In this formulation we may consider the Ising variables to be angles that take on the values 0 or π. In this case the Z_2 invariance corresponds to the replacement $n_i \to n_i + 1$ everywhere on the lattice. It is clear that the Hamiltonian is invariant under this transformation. This immediately suggests a generalization of the Ising model to the Z_q clock model.[31] The variables now become angles that take on the q values $n_i 2\pi/q$ where $n_i = 0, 1, \ldots, q-1$. The Hamiltonian then becomes

$$H = -J \sum_{\langle i,j \rangle} \cos(n_i - n_j)\frac{2\pi}{q} \tag{7.110}$$

It is straightforward to adapt the program for the two-dimensional Ising model of Problem 7.9.1 to this Z_q model. However, care must be taken to keep the program

[31] This model is also known as the vector Potts model.

as efficient as possible and avoid unnecessary calculations. For example, the unit of angle $2\pi/q$ should be calculated only once in the program, rather than every time it is needed. More importantly, the $\cos(n_i - n_j)2\pi/q$ should not be calculated every time its value is needed as the simulation proceeds. Instead, one notes that there are only q possible values that could ever be invoked, so these values are calculated only once at the begining and stored in an array. Then we simply look up the value of the cosine or sine that corresponds to the current value of $n_i - n_j$. Failing to optimize in such a way will slow down the program significantly and make the simulations we shall discuss below too time consuming. With such optimizations the simulations will take of the order of some hours.

The simulation then proceeds along the same lines as the Ising case. Choose an arbitrary initial configuration of n_i, then visit the sites, suggest changes and decide to accept or reject according to the Metropolis criterion. The change suggested at a site i is to replace the current value n_i by any one of the other $q - 1$ values.

Make sure that in your program the size of the system, L, the temperature and the order of the symmetry, q, are all easily modifiable parameters. Of course, periodic boundary conditions should be used.

Also recall that in the Ising case, the magnetization *of a configuration* is given by $\mathcal{M} = |\sum_i S_i|$ because the spins are ± 1. However, for $q > 2$, the magnetization is given by

$$\begin{aligned}
\mathcal{M}_x &= \sum_i \cos\frac{2\pi}{q} n_i \\
\mathcal{M}_y &= \sum_i \sin\frac{2\pi}{q} n_i \\
\mathcal{M} &= \sqrt{\mathcal{M}_x^2 + \mathcal{M}_y^2} \qquad (7.111)
\end{aligned}$$

It is very important that the average of the total magnetization, $\mathcal{M}$, given by Equation (7.111) be measured, not $\mathcal{M}_x$ or $\mathcal{M}_y$. Here again one should avoid the repeated calculation of the trigonometric functions. They should be stored in an array and simply referred to as the need may arise.

1. Run your Metropolis program for a system of size 10×10, q ranging from 2 to 10 and for T ranging from 0.015 to 4.5 in steps of 0.015. Do 3×10^5 thermalization and 5×10^5 measurement sweeps. This should give small error bars and can be done in a matter of hours.

Measure the average total energy, E, the specific heat, C (7.14), the magnetization per spin, M, and the magnetic susceptibility, χ (7.15).

For each value of q, plot these quantities as functions of the temperature. Comment on the behaviour you observe.

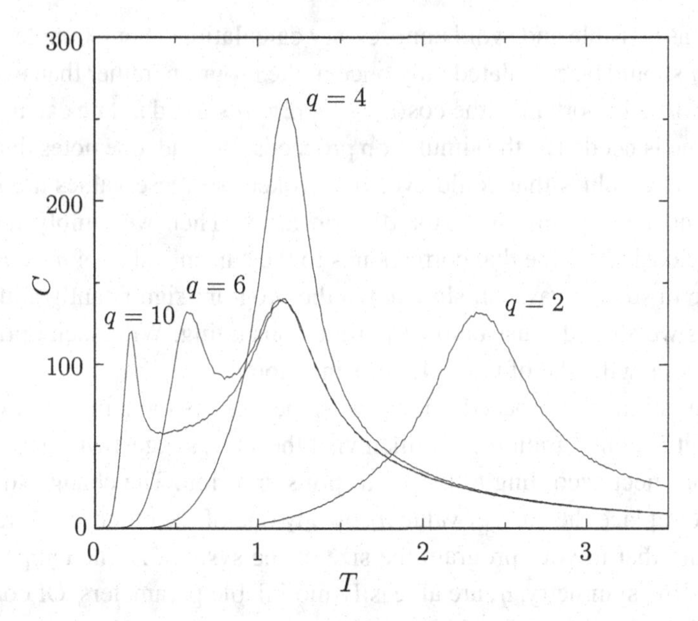

Figure 7.8 The specific heat, C, as a function of T for the Z_q models with $q = 2, 4, 6, 10$ on a 10×10 lattice.

2. The data will have shown (see Figure 7.8) that for $q \geq 5$ the specific heat shows two peaks indicating the presence of two phase transitions at temperatures T_1 and T_2 ($T_1 < T_2$). For $q \leq 4$ only one peak is present. For now, define $T_c^q(L)$ as the position of the specific heat peaks for the $L \times L$ Z_q system. Plot q (on the y-axis) versus T_c^q (on the x-axis) for the $L = 10$ system.

The plot will show that for $q \leq 4$ there is only one transition temperature while for $q \geq 5$ there are two transitions. This definition of T_c^q is of course approximate. To determine the critical temperatures more precisely one should use finite size scaling (see below). But this is certainly good enough to show the general features with some precision.

We have seen the duality transformation for the two-dimensional Ising model in Exercise 4.6.1. This duality transformation can be extended to the Z_q clock model [113], which is found to be self dual under certain approximations (except for $q = 2, 3, 4$ which are special, see below). This approximate self duality yields the result that for $q \geq 5$, $T_1 = 4\pi^2/(1.17q^2)$. Do you find this to be satisfied?

In fact, it can be shown that the Z_3 model is exactly self dual with the original and dual inverse temperatures related by

$$\beta = -\frac{2}{3}\ln\frac{e^{3\tilde{\beta}/2} - 1}{e^{3\tilde{\beta}/2} + 2} \tag{7.112}$$

Assuming a single phase transition, this gives the exact critical temperature by putting $\beta = \tilde{\beta}$

$$\beta_c = \frac{2}{3}\ln(1+\sqrt{3}) \tag{7.113}$$

for the Z_3 model. Compare this value with the numerical one you obtained.

It can also be shown that the partition function for the $q = 4$ model, $Z^{(q=4)}$, is in fact the product of two Ising model partition functions, $q = 2$, at half the β

$$Z^{(q=4)}(\beta) = \left(Z^{(q=2)}(\beta/2)\right)^2 \tag{7.114}$$

This very interesting relation tells us that $\beta_c(Z_4) = 2\beta_c(Z_2)$, in other words the T_c for the Z_4 model is half that for the Ising model. Compare this with your numerical results.

3. We will now do finite size scaling for Z_3 and Z_4. For $q = 3, 4$ do the simulations as above for $L = 8, 10, 12, 14, 16, 24$. In each case find the temperature, $T_c(L)$, corresponding to the maximum of the magnetic susceptibility. Plot $T_c(L)$ as a function of $L^{-1/\nu}$ assuming $\nu = 1$ as in the Ising case and obtain the thermodynamic critical temperature T_c for $q = 3, 4$. Compare with the above exact results.

Plot the peak value of the susceptibility, $\chi_{\max}$, versus L and use finite size scaling (see Problem 7.9.1) to obtain the critical exponent γ for $q = 3, 4$. Compare with the value found for the Ising model.

As in Problem 7.9.1, show that the susceptibility curves for the different system sizes collapse on the same curve when scaling variables are used.

Plot $L^{\beta/\nu}M$ as a function of T for all sizes and $q = 3, 4$. Take $\beta/\nu = 0.25$ as in the Ising case and show that for this choice all the Z_3 curves intersect at one point and so do the Z_4 curves (at a different point). Compare the critical temperatures thus obtained with those obtained by the above extrapolation $L \to \infty$ and with the exact values.

These results show that the critical exponents (ν, β, γ) for $q = 3, 4$ are the same as for the Ising model, $q = 2$. The models are in the same universality class. At least for the $q = 4$ case this should not be a surprise: according to Equation (7.114) the two partition functions are intimately related.

4. We shall now examine a case where there are two phase transitions, say Z_8. As is clear from Equation (7.110), the limit of the Z_q model as $q \to \infty$ is the XY (plane rotator) model which exhibits a single Kosterlitz–Thouless (KT) transition (see the discussion after Equation (7.94) and the following problem). It is also clear from the q versus T_c^q figure of Part 2 that as $q \to \infty$, $T_1 \to 0$ and T_2 stays finite, and in fact goes to T_{KT}.

It is therefore natural to ask for what value of q will the transitions become of the KT rather than the Ising type. For lattice sizes that can be done quickly, it is difficult to answer this question unambiguously. Large scale simulations show that for $q \geq 5$ all transitions in this model are of the KT variety. We assume this and do the finite size scaling accordingly.

Do the simulation as above for $q = 8$ and $L = 8, 10, 12, 14, 16, 24$. Determine $T_2(L)$ (the higher critical temperature) from the peaks of the magnetic susceptibility and plot this versus $(\ln L)^{-2}$ in accordance with Equation (7.99). Obtain the values of T_2 and b in the thermodynamic limit.

The finite size scaling assumption tells us that[32]

$$\chi = L^{\omega} f\left(\frac{\xi}{L}\right) \tag{7.115}$$

$$= L^{\omega} f\left(\frac{e^{b/\sqrt{t}}}{L}\right) \tag{7.116}$$

where $t = T - T_2$. Therefore, plotting $L^{-\omega}\chi$ versus the scaling variable ξ/L should collapse the data from the different system sizes onto one curve. First, to determine ω, plot $\chi_{\max}(L)$ (the susceptibility at the peak corresponding to $T_2(L)$) as a function of L. According to the above equation, this should diverge as a power of L whose exponent is ω, which you determine.

For $T > T_2$, and using the value of b found above, plot $L^{-\omega}\chi$ versus $L^{-1}\exp(b/\sqrt{t})$ for all systems simulated and verify the data collapse.

The data are therefore consistent with a KT phase transition at T_2.

The magnetization versus temperature curve clearly shows the two transitions. For T very low, M is close to unity and drops sharply to a smaller non-zero value as T_1 is approached. As T is further increased, M decreases rather steeply but then near T_2 it drops precipitously to a small value. Clearly, for $T < T_1$ we expect M to be finite. On the other hand, for $T > T_2$ we expect the system to be disordered and $M \to 0$. But what happens for the intermediate temperature range $T_1 < T < T_2$? To determine whether or not the system is magnetized, we resort to finite size scaling. Plot M versus L for representative temperatures in the three temperature ranges, say $T = 0.24, 0.84, 2.1$. Comment on what you find. This question will be addressed again for the XY model in the next problem.

5. As discussed in Problem 7.9.1, one can obtain the entropy from the specific heat, Equation (7.106). By performing this integral numerically, verify that for Z_q we have $S(T \to \infty) = \ln q$. Then calculate $S(T)$ and $F(T) = E(T) - TS(T)$.

[32] Since the scaling function, f, is a function of $x = L/\xi$, it may also be considered a function of $x^{-1} = \xi/L$, which is what we shall do here.

This problem shows the richness of behaviour found in a very simple looking model. It also shows the very important rôle played by the symmetry of the Hamiltonian: for $q = 2, 3, 4$ there is only one transition and the systems are in the same universality class as the Ising model. On the other hand changing the symmetry to $q \geq 5$, there are two phase transitions as the temperature is changed and they are in the Kosterlitz–Thouless universality class. See Reference [27] for more details.

7.9.4 Two-dimensional XY model: Kosterlitz–Thouless transition

The goal of this problem is to study in more detail the behaviour of the two-dimensional XY model which may be considered as the $q \to \infty$ limit of the Z_q model studied in Problem 7.9.3. In that problem we saw that for $q \geq 5$, the Z_q model exhibits two phase transitions at T_1 and T_2 ($T_1 < T_2$), both of the Kosterlitz–Thouless type. Approximate duality arguments show that $T_1 \propto q^{-2} \to 0$ as q increases leaving only one finite temperature phase transition at T_2.

In this problem we shall elaborate the nature of the transition at T_2 (which we refer to henceforth as T_{KT}) as well as the nature of the two phases above and below T_{KT}.

The Hamiltonian of the XY model is given by

$$H = -J \sum_{\langle i,j \rangle} \cos(\theta_i - \theta_j) \tag{7.117}$$

with its partition function

$$Z = \int_{-\pi}^{\pi} \prod_k \mathrm{d}\theta_k \, \exp[\beta J \sum_{\langle i,j \rangle} \cos(\theta_i - \theta_j)] \tag{7.118}$$

where $-\pi < \theta \leq +\pi$, and $\langle i, j \rangle$ denotes nearest neighbours. The product over $\mathrm{d}\theta_k$ runs over all sites of a two-dimensional lattice of size $L \times L$ with periodic boundary conditions.

We have two options for the simulations. We may simply use the program for the Z_q model with q taken sufficiently large, say $q = 1000$ or more, so that this discrete model mimics closely the true XY model. The other option is of course to modify the program so that the spin variables are indeed taken to be continuous angles. One advantage of the first option is that it uses an existing and debugged program. Another advantage is that, due to the discreteness of the spins, we can still use lookup tables for the trigonometric functions as we did in the preceding problems. This speeds up the program.

Either way, the simulation proceeds as before with sites visited and changes in the spins accepted or rejected with the Metropolis criterion (7.13a) and (7.13b). However, in this model we have the freedom of suggesting large or small changes. A good rule of thumb is to tune Δ, where the change $\delta\theta$ is given by $-\Delta \le \delta\theta \le \Delta$, such that the acceptance rate is about 50%. If the acceptance rate is too small or too large, the system evolves towards equilibrium very slowly.

Use your program to simulate the XY (or Z_{1000}) model for $L = 8, 10, 12, 14, 16, 20$ and $0.015 \le T \le 3$ in steps of $\Delta T = 0.015$. Do 2×10^5 Monte Carlo sweeps to thermalize the system and 3×10^5 measurement sweeps measuring every 50 sweeps, i.e. 6×10^3 measurements in total. The quantities to measure are the average total energy, E, the specific heat, C (Equation (7.14)), the magnetization, $\mathcal{M}$ (the XY analogue of Equations (7.111)[33]), the magnetic susceptibility, χ (Equation (7.15)), and the average number of vortex pairs per site (see Question 4 for the definition).

1. Make a plot of the energy per site as a function of the temperature for the systems simulated. Compare with the Ising model and comment on what is observed for $1 \le T \le 1.5$.

2. Make a plot of the specific heat, C, as a function of T for the various system sizes. Comment on what is observed.

Define $T_c^C(L)$ as the 'critical' temperature for system size $L \times L$ obtained from the peak in the specific heat. Plot $T_c^C(L)$ versus $(\ln L)^{-2}$ and find the thermodynamic critical temperature T_c^C.

3. Recall that this model has continuous symmetry and that such a symmetry does not break spontaneously in two dimensions at finite temperature. This means that the magnetization vanishes at all finite temperatures. Make a plot of the magnetization per site, M, versus T for all system sizes. Does the system magnetize for $T < T_c^C$? To answer this question, observe the dependence of M on the system size for $T < T_c^C$ and contrast this behaviour with that of the Ising model which we know to be magnetized for $T < T_c$.

Further insight on the magnetization, and other physical quantities, may be obtained at very low temperature with the help of the *spin wave* approximation. At very low temperature ($\beta \gg 1$) Equation (7.118) shows that only very small values of $|\theta_i - \theta_j|$ will contribute to the partition function. We may then approximate $\cos(\theta_i - \theta_j) \approx 1 - (\theta_i - \theta_j)^2/2$ in the exponential. The partition function then

[33] While calculating χ one should take $M = 0$ since, as we shall see, the magnetization vanishes in this model.

becomes[34]

$$Z = \int_{-\infty}^{+\infty} \prod_k \mathrm{d}\theta_k \, \exp\left[-\frac{\beta}{2} \sum_{\langle i,j \rangle} (\theta_i - \theta_j)^2 \right] \tag{7.119}$$

$$= \int_{-\infty}^{+\infty} \prod_k \mathrm{d}\theta_k \, \exp\left[\frac{\beta}{2} \sum_{ij} \theta_i G_{ij}^{-1} \theta_j \right] \tag{7.120}$$

where we have ignored an irrelevant constant and we extended the integration limits to $\pm\infty$. The lattice Green function, $G_{ij} = G(\vec{n})$, is given by

$$G(\vec{n}) = \frac{1}{L^2} \sum_{\vec{k}} \frac{e^{i\vec{k}.\vec{n}}}{4 - 2\cos k_x - 2\cos k_y} \tag{7.121}$$

where $(k_x, k_y) = (\ell_x 2\pi/L, \ell_y 2\pi/L)$ with $-L/2 \leq (\ell_x, \ell_y) < L/2 - 1$. The Gaussian form of the partition function (7.119) may now be used to calculate averages of quantities that depend on θ (see Section A.5.1.). Show that the magnetization per spin is given by

$$M = \langle \cos\theta \rangle \tag{7.122}$$

$$\approx 1 - \frac{1}{2}\langle \theta^2 \rangle \tag{7.123}$$

$$= 1 - \frac{T}{8\pi}(\ln L^2 + \ln 2) \tag{7.124}$$

Compare this with the magnetization from the numerical simulations and determine the domain of the validity of the spin wave approximation.

Note that both the spin wave approximation and the numerical simulations give M that depends on the system size for $T \ll T_c^C$. In particular, for fixed very low temperature, the magnetization decreases with system size contrary to what happens with the Ising model. This indicates that the magnetization vanishes in the thermodynamic limit.

Although $M = 0$ in the thermodynamic limit, it is easy to see that the measured magnetic susceptibility *diverges* as the transition is approached from above, and *remains infinite* for all lower temperatures. According to Kosterlitz–Thouless theory, the susceptibility diverges as in Equation (7.97). If your computer resources are sufficient, you may want to measure χ as a function of T for $1 < T < 1.4$ and $L = 100$ and fit the results to the form given in (7.97). This will give an estimate for T_{KT}.

[34] We have taken $J = 1$ and all energies are measured in units of J.

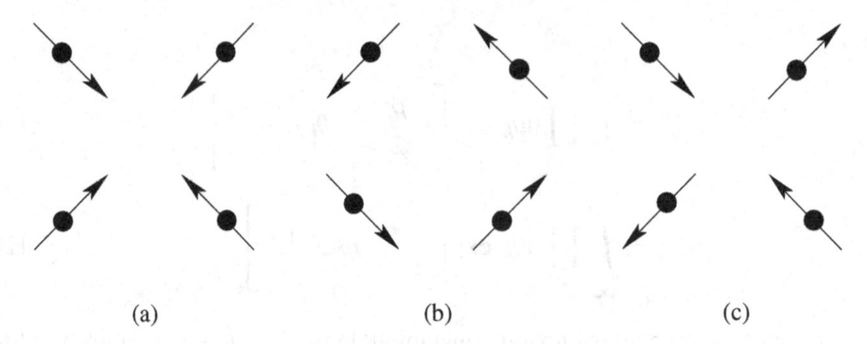

(a) (b) (c)

Figure 7.9 Typical vortex configurations. The solid circles represent sites around an elementary plaquette and the arrows show the directions of the spins. (a) and (b) show vortices while (c) shows an antivortex. Note that (b) can be obtained from (a) by displacing each arrow by one lattice spacing in the anticlockwise direction. Configuration (c) cannot be obtained from (a) or (b).

4. The magnetization is, therefore, not a good order parameter to characterize the phase transition in the two-dimensional XY model. It turns out (as shown by Kosterlitz and Thouless) that the phase transition is driven by the binding and unbinding of *vortices*. A vortex (or antivortex) describes what happens with the spin angles as we go around an elementary plaquette. Figure 7.9 shows typical vortex and antivortex configurations of unit vorticity. To calculate the vorticity, one goes around an elementary plaquette and measures the difference of the angles at the end and begining of each bond.[35] One should make sure that these differences are always between $\pm\pi$. Then the sum of these differences around the plaquettes is taken. For example, performing this operation in the anti-clockwise direction for vortex (a) in Figure 7.9 gives 2π. On the other hand, configuration (c) gives -2π.

Vortices are very easy to discern visually. To do this, modify your program to represent each spin by an arrow pointing in the direction prescribed by the angle. Then take a configuration corresponding to $T = \infty$ (i.e. a random configuration) and quench to $T = 0$. To quench to zero temperature, perform the Monte Carlo simulation and always accept the change in the spin if it reduces the energy and *always reject* the change if it does not. If the system size is relatively large, say 50×50, several vortex–antivortex pairs will be easily visible.

It will also be clear that there is an equal number of vortices as antivortices and that they are closely associated in pairs: for every vortex, there is a clearly associated antivortex nearby. In fact, at low temperature, the vortex–antivortex pairs are *bound* together: the energy of the pair diverges logarithmically as the separation grows.

[35] This should always be done in the same direction. We may choose to do this in the clockwise or anticlockwise directions, but once the choice is made it should be adhered to.

Modify your program to measure the number of vortex pairs and repeat the simulation for $L = 20$ and $0.5 \le T \le 1.5$ in temperature steps of 0.015. Do 3×10^5 sweeps to thermalize and 7×10^5 to measure (more if your computer resources permit). Plot the number of pairs/site versus temperature and comment on the results. Does this provide a physical interpretation of the Kosterlitz–Thouless transition?

If the data are accurate enough, they will show that the proliferation of vortices appears for $T < T_c^C$. In fact KT theory predicts that $T_{KT} < T_c^C$, i.e. the KT transition takes place at a temperature slightly lower than that of the peak of the specific heat. The transition takes place when the vortex–antivortex pairs *first start to dissociate* and are no longer bound.

This problem has shown that not all phase transitions are associated with the establishment of long range order as happens in a magnetized phase. There are other indicators of phase transitions besides long range order. For example, for this model, the correlation function decays exponentially with distance for $T > T_{KT}$ and as a power for $T < T_{KT}$. Another indicator is the behaviour of the vortices (topological excitations) as a function of the temperature.

7.9.5 Two-dimensional XY model: superfluidity and critical velocity

As we saw in Problem 5.7.7, the superfluid density, ρ_s, is calculated from the difference of the free energies of the system with and without a phase twist in the wave function. Since the wave function is single valued, this twist must be a multiple of 2π and the velocity quantization condition (5.135) must be satisfied. It was then argued that

$$\mathcal{J}_T \approx \mathcal{J} + \frac{N}{2}\rho_s m v^2 \tag{7.125}$$

where $\mathcal{J}$ and $\mathcal{J}_T$ are the grand potentials with and without a phase twist, N is the number of sites, m the particle mass, $v = \delta\phi(\hbar/m)$. The phase gradient is given by $\delta\phi = 2\pi n/L \ll 1$ where L is the length of the system and n an integer.

The calculations in Problem 5.7.7 were done at $T = 0$. It is not difficult to see that if we fix the number of particles (i.e. the angle θ in Problem 5.7.7), then at finite temperature, the phase ϕ will exhibit thermal fluctuations. Since the coupling between the phase angles at different sites is only through nearest neighbours, we see that this now looks similar to the XY model and thus motivates us to study the question of superfluidity in the two-dimensional XY model.[36]

[36] The relation between the XY model and bosonic systems can be made more rigorous. This is beyond the scope of the current presentation.

We examine then, in this problem, the Kosterlitz–Thouless transition from the viewpoint of superfluidity, and since we perform Monte Carlo simulations in the canonical ensemble, we use the free energy, F, rather than the grand potential. For small velocities, v, Equation (7.125) can be shown to reduce to

$$F_{\mathrm{T}} \approx F + \frac{N}{2}\rho_{\mathrm{s}} m v^2 = F + \frac{N}{2}\rho_{\mathrm{s}} m (v_x^2 + v_y^2) \tag{7.126}$$

from which immediately follow

$$\rho_{\mathrm{s}} v_x = \frac{1}{N}\frac{\partial F_{\mathrm{T}}}{\partial v_x} \tag{7.127}$$

and

$$\rho_{\mathrm{s}} = \frac{1}{N}\frac{\partial^2 F_{\mathrm{T}}}{\partial v_x^2} \tag{7.128}$$

with similar expressions using v_y. We see, therefore, that to calculate the superfluid momentum and density we need to calculate derivatives of the free energy with respect to the superfluid velocity $v = (\hbar/m)\delta\phi = \delta\phi$, where we take $m = \hbar = 1$. In this problem, the angle θ is the analog of the azimuthal angle ϕ in Problem 5.7.7.

1. The partition function for the XY model is given in Equation (7.118). To calculate the derivatives with respect to the phase gradient, we simply take $\delta_{ij} \equiv (\theta_i - \theta_j)$ as the local phase gradient and perform the differentiation with respect to all bonds, i.e. all nearest neighbour pairs. Show that

$$\frac{\partial F}{\partial v_x} = \frac{1}{Z}\int_{-\pi}^{\pi}\prod_k \mathrm{d}\theta_k \sum_{\langle i,j\rangle:x}\sin(\theta_i - \theta_j)\exp[\beta\sum_{\langle i,j\rangle}\cos(\theta_i - \theta_j)] \tag{7.129}$$

which is written as

$$\boxed{\frac{\partial F}{\partial v_x} = \Big\langle \sum_{\langle i,j\rangle:x}\sin(\theta_i - \theta_j)\Big\rangle} \tag{7.130}$$

Similarly, the second derivative is given by

$$\boxed{\frac{\partial^2 F}{\partial v_x^2} = \beta\Big\{\Big\langle \sum_{\langle i,j\rangle:x}\sin(\theta_i - \theta_j)\Big\rangle^2 - \Big\langle\Big(\sum_{\langle i,j\rangle:x}\sin(\theta_i - \theta_j)\Big)^2\Big\rangle\Big\} + \Big\langle \sum_{\langle i,j\rangle:x}\cos(\theta_i - \theta_j)\Big\rangle} \tag{7.131}$$

where the notation $\langle i, j\rangle : x$ means the sum is performed over nearest neighbours only in the x direction.

Therefore, although it is not easy to measure the free energy itself, it is straightforward to calculate its derivatives since they can be easily related to correlation functions as can be seen from Equations (7.129) and (7.131).

2. Modify your program to measure the second derivative of F. Note that in this case, with no applied external fields, $\langle \sum_{\langle i,j\rangle} \sin(\theta_i - \theta_j)\rangle = 0$ identically. Do the simulation for $L = 8, 10, 12, 14, 16, 20$ in the temperature range $0.015 \leq T \leq 2$ in temperature steps of 0.015. Do 3×10^5 sweeps to thermalize and 6×10^5 (more if computer resources allow) to measure the superfluid density[37]

$$\boxed{\rho_s = \frac{1}{L^2}\frac{\partial^2 F}{\partial v_x^2}} \tag{7.132}$$

Plot ρ_s as a function of T for all the sizes simulated. Discuss.

We can estimate ρ_s using the spin wave approximation (see Problem 7.9.4). Show that to leading order in this approximation we have,

$$\rho_s = \frac{1}{2L^2}\Big\langle \sum_{\langle i,j\rangle} \cos(\theta_i - \theta_j)\Big\rangle \tag{7.133}$$

$$\approx \frac{1}{2L^2}\sum_{\langle i,j\rangle}\Big(1 - \frac{1}{2}\langle(\theta_i - \theta_j)^2\rangle\Big) \tag{7.134}$$

$$= 1 - \frac{1}{4}T \tag{7.135}$$

Compare this result with the numerical simulations.

3. It turns out that in the thermodynamic limit, ρ_s jumps to zero exactly at the KT transition, T_{KT}. The value of T_{KT} is related to the value of ρ_s *at* T_{KT} by the so-called 'universal jump condition'

$$T_{KT} = \frac{\pi}{2}\rho_s(T_{KT}) \tag{7.136}$$

Solve this equation graphically by plotting the straight line $\rho_s(T) = 2T/\pi$ on the same figure as your results for $\rho_s(L)$ versus T. The intersection of the line with the ρ_s curves of the various sizes yields the value of T_{KT} as a function of system size.

Plot $T_{KT}(L)$ versus $(\ln L)^{-2}$ and find the thermodynamic critical temperature T_{KT}. Compare this value with that obtained from the specific heat in Problem 7.9.4.

4. In Question 2 we used (7.131) to calculate ρ_s. This was convenient since the superfluid momentum (7.127) and (7.130) was zero. In other words, the superfluid

[37] In your program it will be convenient and, in fact, better to perform the sums in Equation (7.131) over both x and y nearest neighbours. In that case, Equation (7.132) is replaced by

$$\rho_s = \frac{1}{2L^2}\Big(\frac{\partial^2 F}{\partial v_x^2} + \frac{\partial^2 F}{\partial v_y^2}\Big)$$

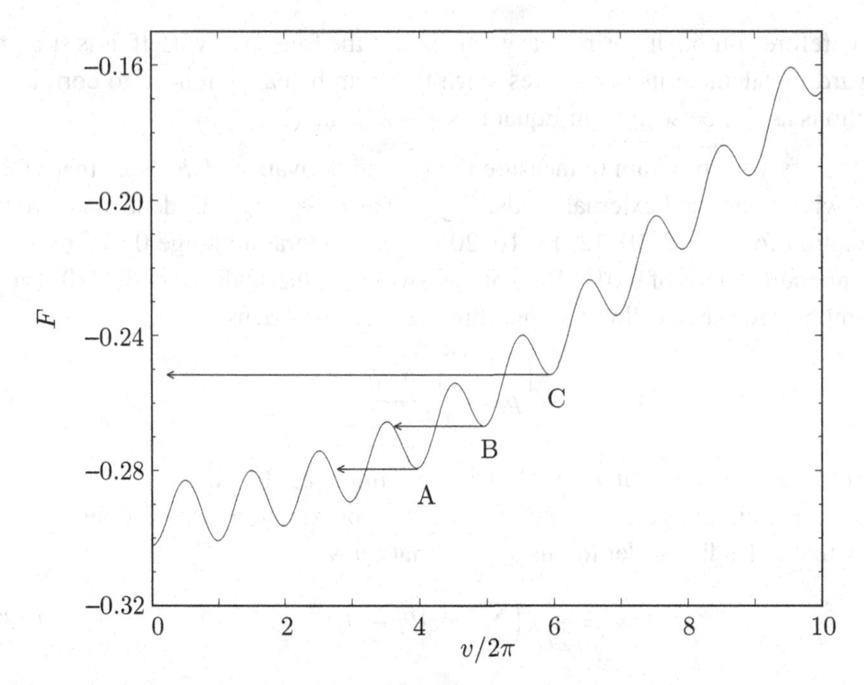

Figure 7.10 The free energy of 'stable' superfluid.

was at equilibrium: there was no 'persistent flow', i.e. no superflow. However, the fact that there is non-zero ρ_s when the fluid is stationary does not mean that the system is superfluid in the sense that one can produce superflow that will persist a long time. For example, below a critical temperature, the ideal Bose gas has non-zero superfluid density when at rest, which is destroyed when it is set in motion. If persistent flow can be established, we say the superfluid is 'stable' even though this is not true in the strict sense: recall that the superfluid density was defined (7.128) from the difference of the free energies of the system while in motion and at rest. Clearly, therefore, the free energy of the system when in motion is greater than that when at rest, and consequently the system cannot be stable: at best it is metastable.

The F versus v curve for a 'stable' superfluid is depicted schematically in Figure 7.10. The local minima in the free energy correspond to values of the superfluid velocity satisfying the quantization condition (5.135). It is clear that if the system is put in a deep local minimum, such as A in Figure 7.10, it may stay there for a very long time, thus exhibiting persistent flow. On the other hand, if the system is in B, it may tunnel out but get trapped in the next local minimum, thus remaining superfluid but at a smaller velocity. If the system is in C, persistent flow may last a little while and then get totally destroyed as the system tunnels out of the local minimum and all the way to $v = 0$. In addition, at even higher velocity, the flow

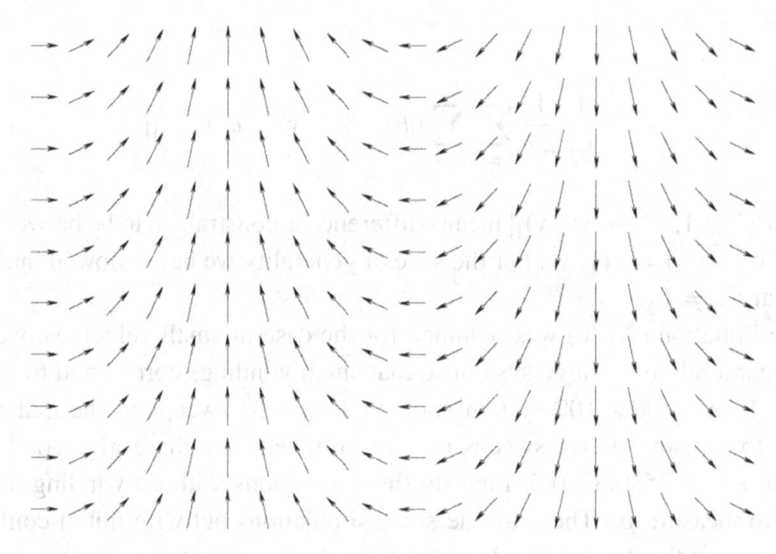

Figure 7.11 A spin configuration satisfying the velocity quantization condition (7.137) with $n = 1$ on a lattice with $L_x = 20$.

becomes *unstable* not just metastable. The heights of the barriers between local minima depend on the temperature, of course: the closer the system is to the critical point, the lower the barriers and the shorter the lifetimes of the metastable states.

We shall now study these issues numerically by placing the system in various local minima of its free energy. To place the system in the nth local minimum, the superfluid velocity in the x direction, v_x, must be such that

$$\sum_{x=1}^{L} v_x(x, y) = \sum_{x=1}^{L} \big(\theta(x+1, y) - \theta(x, y)\big) = 2\pi n \tag{7.137}$$

where, to avoid confusion, we have labelled the sites of the two-dimensional lattice with two indices (x, y). Figure 7.11 shows such a configuration for $n = 1$ on a system of length $L_x = 20$ in the x direction.

We therefore see that it is a simple matter to start the simulation with an initial configuration that puts the system in the local minimum of our choice. It is very important (and interesting) to keep in mind that by putting the system in any configuration other than $n = 0$, the system will not be at equilibrium: under such conditions we would be studying the flow properties of the superfluid in metastable states!

Modify your program to allow the choice of 'winding' initial spin configurations as discussed above. Also, since we are interested in transitions between local minima, you should be able to monitor the quantization number n, henceforth referred to as the winding number. Modify your program to measure the winding number using

$$n = \frac{1}{L_y} \frac{1}{2\pi} \sum_{y=1}^{L_y} \sum_{x=1}^{L_x} ||\theta(x+1, y) - \theta(x, y)|| \tag{7.138}$$

where $||\theta(x+1, y) - \theta(x, y)||$ means difference is constrained to be between $\pm\pi$, and $\theta(Lx+1, y) = \theta(1, y)$. For the sake of generality we have allowed the possibility that $L_x \neq L_y$.

Since Equation (7.126) was obtained for the case of small velocities, we need to consider a relatively large system so that small windings correspond to low velocities. Take a 100×100 system and run 1.5×10^6 sweeps to thermalize and 3×10^6 to measure (fewer sweeps may be sufficient, but these give good precision) for $T = 0.25, 0.33, 0.5$. First do the simulations with no winding and use (7.131) to measure ρ_s. Then run the same simulations but with initial configurations corresponding to $n = 1$ and use the superfluid momentum (7.130) to measure ρ_s. Compare the results.

For $T = 0.25$ you may also try $n = 2, 3$; the temperature is low enough for the system to stay in these metastable states for a very long time.

This makes very clear the existence of metastable states due to the presence of barriers in the free energy between configurations with different windings. This allows us to have persistent flow, a hallmark of superfluidity.

5. The dependence of the free energy on the velocity, Figure 7.11, tells us that the superfluid will flow at the nearest velocity which properly quantizes the flux. As a consequence, if we prepare the superfluid with an initial velocity less than π/L_x, the superfluid will come to rest.[38] To see this numerically, define the average velocity of a configuration as

$$\bar{v}_x = \frac{1}{L_y L_x} \sum_{y=1}^{L_y} \sum_{x=1}^{L_x - 1} ||\theta(x+1, y) - \theta(x, y)|| \tag{7.139}$$

This equation is very similar to (7.138) but note that the sum over x is stopped at $L_x - 1$. This is done to allow us to follow the evolution of velocities that do not quantize the flux.

Take a 100×100 system at $T = 0.25$, start with configurations with various incomplete (non-integer) windings and follow the evolution of $\bar{v}_x$. Do not thermalize

[38] This is the analogue of the Meissner effect. See Exercise 4.6.6 and References [77, 78, 85].

the system and measure after every sweep. You will find, for example, that starting with 'winding' less than 0.5, the system will unwind completely and $\bar{v}_x/2\pi \to 0$; on the other hand, for a winding greater than 0.5 but less than 1.5, the system will go to the nearest full winding $\bar{v}_x/2\pi \to 1$.

Examine this behaviour at various other temperatures and windings.

6. In the previous part we saw that there are local minima in the dependence of the free energy on the superfluid velocity. In this part we shall study very qualitatively the tunnelling between these minima and its relation to the 'critical velocity'.

The critical velocity is the flow velocity at which sufficient excitations are created in the superfluid to destroy the coherence in the phase of the wave function and thus superfluidity itself. The theoretical treatment of this question is beyond the scope of this book, but using our model, we are able to observe the destruction of superfluid flow as the initial velocity is increased.

Run the simulation for the 100×100 system at $T = 0.5$, no thermalization sweeps, 5×10^4 measurement sweeps and measure every 50 sweeps. Plot $\bar{v}_x$ (or the winding) as a function of 'time', i.e. Monte Carlo sweep. Do this for initial configurations with windings $n = 1, 4, 6, 8, 10, 12$. What do you estimate to be the critical velocity at this temperature?

Try the same at other temperatures, for example $T = 0.25, 0.8$. Does the critical velocity decrease or increase with temperature?

It is very instructive to visualize the simulation to see what kind of fluctuations destroy the superfluid flow in this model. Consider a 50×50 system at $\beta = 3$ and run your simulation, with no thermalization, visualizing the spins (for example as arrows) on the screen every 10 or so sweeps. Starting with a winding $n = 1$, one finds that the system stays in this configuration a very long time. On the other hand, starting with 5, we see that the system will rapidly unwind to lower n via the creation of vortices, some of which can be very large. Figure 7.12 shows the initial configuration of the system with $n = 5$ and Figure 7.13 shows the configuration 2580 sweeps later. Note the creation of large vortices which unwind the system and decrease its speed.

We therefore see, qualitatively, that higher velocity makes it easier for large vortices to form and disrupt phase coherence, thus disordering the system and destroying its superfluid motion.

7. So, this model allows a non-zero superfluid density below the KT transition. But is there a *condensate*? To answer this question we recall the result of Problem 5.7.7 Part 5 where we showed that

$$\rho_0 \propto \left| \frac{1}{N} \sum_{i=1}^{N} \mathrm{e}^{\mathrm{i}\phi_i} \right|^2 \tag{7.140}$$

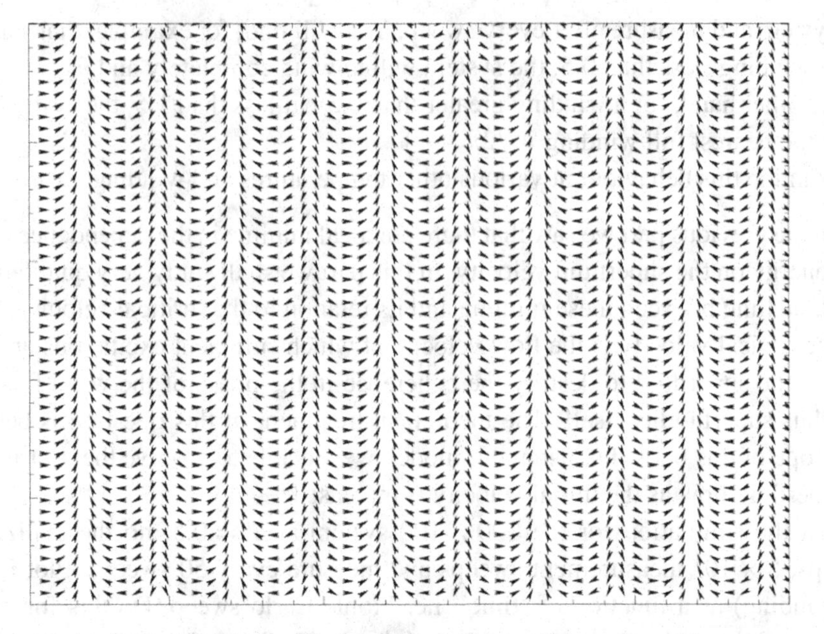

Figure 7.12 Initial spin configuration with $n = 5$ for a 50×50 system.

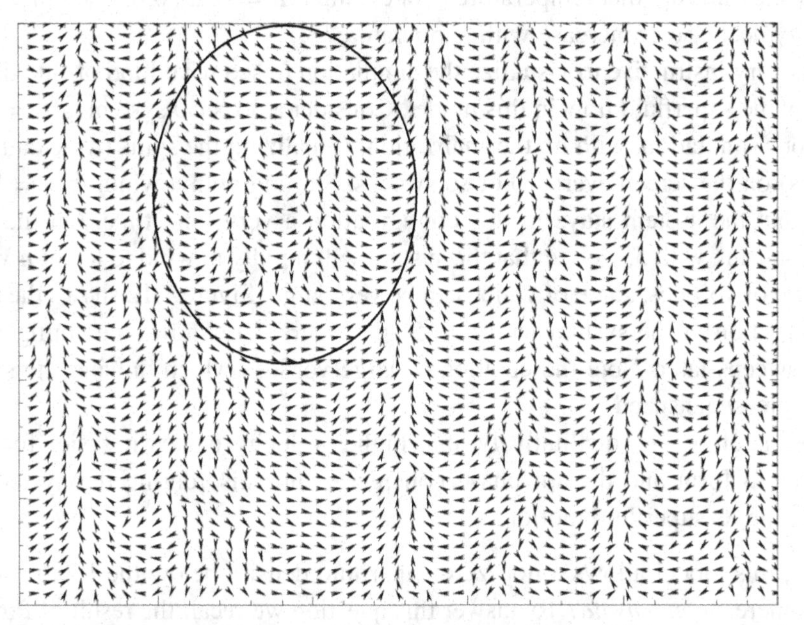

Figure 7.13 The same system as in Fig. 7.12 after evolving 2580 sweeps at $\beta = 3$. Note the large vortex enclosed in the ellipse. Such topological excitations may cause persistent flow to stop.

where N is the total number of sites. Furthermore, recalling that what we called ϕ_i in Problem 5.7.7 is called θ_i here, we see immediately that

$$\rho_0 \propto |M|^2 \tag{7.141}$$

where M is the magnetization per site (see Problem 7.9.4). In view of the results of Problem 7.9.4, does the two-dimensional XY model have a condensate below T_{KT}?

This is a beautiful example of a model that supports a stable superfluid, i.e. with finite critical velocity, but *no condensate*. In Problem 5.7.7 we found that the two-dimensional hardcore boson model at $T = 0$ has both $\rho_s \neq 0$ and $\rho_0 \neq 0$. The big difference is that, while in both cases we have two-dimensional models with continuous symmetry, in the current case T is finite, while in Problem 5.7.7, $T = 0$. In two dimensions, the continuous symmetry may be broken at $T = 0$ but not at finite T.

For the three-dimensional XY model, we expect both $\rho_s \neq 0$ and $\rho_0 \neq 0$.

7.9.6 Simple quantum model: single spin in transverse field

In this problem we shall consider a very simple exactly solvable quantum model and treat it with the Trotter–Suzuki approach discussed in Section 7.4.1. One goal is to show how a familiar classical model may arise from this approach even if the original Hamiltonian appears, at first sight, to be unrelated. The other goal is to analyze exactly the Trotter errors coming from the discretization of the inverse temperature β.

Consider the one spin system described by the Hamiltonian

$$H = -t\sigma_x - h\sigma_z \tag{7.142}$$

where σ_x and σ_z are Pauli matrices. This may be interpreted as a single quantum spin, σ_x, subjected to a magnetic field in the transverse direction, z.

1. This model is trivial to solve exactly. Evaluate the partition function, $Z = \mathrm{Tr}\,\mathrm{e}^{-\beta H}$, exactly by using the explicit expressions for the Pauli matrices.

2. Apply the breakup (7.32) and the Trotter–Suzuki approximation (7.35) where $-t\sigma_x$ is the analogue of the kinetic energy term and $-h\sigma_z$ that of the potential energy. To evaluate the trace, choose the representation that diagonalizes σ_z

$$\sigma_z|S_z\rangle = S_z|S_z\rangle \qquad 1 = \sum_{S_z=\pm 1} |S_z\rangle\langle S_z| \tag{7.143}$$

Insert the above resolution of unity between each pair of exponentials and, following the steps from (7.32) to (7.40), show that the partition function may be written in the form

$$Z = \sum_{\{S_z^\ell\}} \mathrm{e}^{\Delta \tau h \sum_\ell S_z^\ell} \prod_{\ell=1}^{L_\tau} \langle S_z^{\ell+1} | \mathrm{e}^{\Delta\tau\sigma_x} | S_z^\ell \rangle \tag{7.144}$$

where ℓ labels the imaginary time and $\Delta\tau = \beta/L_\tau$. Note that here only L_τ imaginary time slices are needed rather than $2L_\tau$ as in (7.39). This is because the origin of the factor 2 in (7.39) is the breaking of the Hamiltonian into H_1 and H_2, which connect different sites. That was done to reduce the many body problem to a product of independent two-site problems. This breakup does not intervene here.

By evaluating explicitly the matrix elements $\langle S_z^{\ell+1} | \mathrm{e}^{\Delta\tau\sigma_x} | S_z^\ell \rangle$, show that they may be written as

$$\langle S_z^{\ell+1} | \mathrm{e}^{\Delta\tau\sigma_x} | S_z^\ell \rangle = \mathrm{e}^{\gamma S_z^{\ell+1} S_z^\ell} \tag{7.145}$$

where

$$\gamma = -\frac{1}{2} \ln(\tanh \Delta\tau\, t) \tag{7.146}$$

and that, consequently, the partition function becomes

$$Z = \sum_{\{S_z^\ell\}} \exp\left[\gamma \sum_\ell S_z^\ell S_z^{\ell+1} + \Delta\tau h \sum_\ell S_z^\ell \right] \tag{7.147}$$

One recognizes the partition function of the one-dimensional Ising model in a magnetic field. We see again the pattern of a D-dimensional quantum model mapping onto a $(D+1)$-dimensional classical model: in this case a zero-dimensional quantum model is mapped onto a one-dimensional (Ising) model.

3. Write explicitly the 2×2 transfer matrix, $\mathcal{T}$, for the one-dimensional Ising model (Section 3.1.4)

$$\mathcal{T}_{S_z^{\ell+1} S_z^\ell} \equiv \mathcal{T}(S_z^{\ell+1}, S_z^\ell) = \exp\left(\gamma S_z^{\ell+1} S_z^\ell + \frac{\Delta\tau h}{2} (S_z^{\ell+1} + S_z^\ell) \right) \tag{7.148}$$

From Equation (3.34) we know that

$$Z = \mathrm{Tr}(\mathcal{T}^{L_\tau}) = \lambda_+^{L_\tau} + \lambda_-^{L_\tau} \tag{7.149}$$

where $\lambda_+ > \lambda_-$ are the eigenvalues of $\mathcal{T}$. Evaluate this expression exactly and, by comparing with the exact partition function from Part 1, study the Trotter errors. Show that for a fixed β and $L_\tau \to \infty$, the one-dimensional classical Ising partition function (7.149) gives the exact zero-dimensional quantum partition function.

This demonstrates that, at least for this model, the Trotter–Suzuki approximation is a controlled one. This result is general.

7.9.7 One-dimensional Ising model in transverse field: quantum phase transition

The goal of this problem is to have more practice with the world line formulation and the Trotter–Suzuki approximation and to see a quantum phase transition where a thermal transition is absent.

Consider the periodic N-site one-dimensional system described by the Hamiltonian

$$H = H_0 + H_1 \tag{7.150}$$

$$H_0 = J \sum_{i=1}^{N} \sigma_{i+1}^z \sigma_i^z \tag{7.151}$$

$$H_1 = -h \sum_{i=1}^{N} \sigma_i^x \tag{7.152}$$

Clearly, H_0 is the interaction of the z components of nearest neighbour spins on a one-dimensional lattice with N sites. It therefore describes a classical one-dimensional Ising model. H_1 is the contribution to H due to the application of a magnetic field h in the transverse x direction. Since $[H_0, H_1] \neq 0$, the model is not classical and must be treated by quantum methods. The partition function is, of course, given by

$$Z = \mathrm{Tr}\, \mathrm{e}^{-\beta H} \tag{7.153}$$

1. Derive the equivalent of Equation (7.39) for this model.[39] To this end use the representation $|\mathbf{S}_\ell\rangle = |S_{1\ell}^z, S_{2\ell}^z, ..., S_{i\ell}^z, \ldots, S_{N\ell}^z\rangle$, where the first subscript of $S_{i\ell}^z$ denotes the spatial site while the second is for the imaginary time slice

$$\sigma_i^z |\mathbf{S}_\ell\rangle = S_{i\ell}^z |\mathbf{S}_\ell\rangle \tag{7.154}$$

In this representation, the exponential of H_0 is diagonal and very easy to evaluate. By noting that

$$\mathrm{e}^{-\Delta\tau H_1} = \mathrm{e}^{\Delta\tau h \sum_i \sigma_i^x} = \prod_{i=1}^{N} \mathrm{e}^{\Delta\tau h \sigma_i^x} \tag{7.155}$$

show that

$$\langle \mathbf{S}_{l+1}|\mathrm{e}^{-\Delta\tau H_1}|\mathbf{S}_l\rangle = \prod_{i=1}^{N} \langle S_{il+1}|\mathrm{e}^{\Delta\tau h \sigma_i^x}|S_{il}\rangle \tag{7.156}$$

[39] In this problem, as in the previous one, we need only L_τ imaginary time slices rather than the $2L_\tau$ of (7.39). See the discussion in Part 2 of Problem 7.9.6.

Calculate these matrix elements and show that the partition function thus becomes

$$Z = \sum_{\{S^z_{i\ell}\}} \exp\Big(\sum_{i,\ell} (J\,\Delta\tau\, S^z_{i\ell} S^z_{i+1\ell} + \gamma\, S^z_{i\ell} S^z_{i\ell+1})\Big) \tag{7.157}$$

where

$$\gamma = -\frac{1}{2}\ln(\tanh \Delta\tau h) \tag{7.158}$$

is obtained when (7.156) is evaluated.[40]

We immediately remark that this is nothing but the classical two-dimensional Ising model with anisotropic coupling.[41] In the x direction $\beta_{\rm cl} J_x = J\Delta\tau$ and in the (imaginary time) y direction $\beta_{\rm cl} J_y = \gamma$, where $\beta_{\rm cl}$ is the inverse temperature of the *equivalent classical* two-dimensional model (see below).

2. We may now deduce the phase diagram of the one-dimensional quantum model from what is known about the classical model in two dimensions. Recall first that in Exercise 4.6.1 the Kramers–Wannier duality transformation was used to determine the exact critical temperature of the isotropic two-dimensional Ising model. You may show that in the anisotropic case this transformation yields

$$2\tanh(2J_x\beta^{\rm c}_{\rm cl})\,\tanh(2J_y\beta^{\rm c}_{\rm cl}) = 1 \tag{7.159}$$

where $\beta^{\rm c}_{\rm cl}$ is the (classical) critical inverse temperature, $\beta^{\rm c}_{\rm cl} = (k_{\rm B} T^{\rm c}_{\rm cl})^{-1}$. Equivalently, we may write

$$2\tanh(2J^{\rm c}\Delta\tau)\tanh(2\gamma^{\rm c}) = 1 \tag{7.160}$$

where $J^{\rm c}$ and $\gamma^{\rm c} = \gamma(h^{\rm c})$ are the critical values of the coupling parameters.

Apply this relation to the present model to deduce its phase diagram in the $J-h$ plane.

Discussion: It is important to distinguish the rôle of β in the original quantum model from its classical analogue $\beta_{\rm cl}$, which is deduced from the Boltzmann weight in (7.157) and which appears in Equation (7.159). The classical model exhibits a phase transition *in the thermodynamic limit*, that is, when the size of the lattice is infinite *in both directions* (space and imaginary time) and when $\beta_{\rm cl} J_x$ and $\beta_{\rm cl} J_y$ satisfy (7.159) or in other words when Equation (7.160) is satisfied. In the language of the one-dimensional Ising model in a transverse field whose partition function is given by (7.153), the condition that the classical lattice be infinite corresponds to taking β in (7.153) to infinity, since β gives one of the classical linear

[40] See Problem 7.9.6.

[41] In fact, had we started with the D-dimensional Ising model in a transverse field, we would have obtained the same result but in $(D+1)$ dimensions with the coupling in the D space dimensions $J\Delta\tau$ and the imaginary time coupling γ which is still given by (7.157). See [66] for more detail.

dimensions. $N \to \infty$ is also required of the original linear lattice size. In other words, the phase transition of the one-dimensional quantum model takes place at $T = 0$ by tuning J and h. This is a 'quantum phase transition' in the *ground state* of the one-dimensional Ising model in a transverse field. Since β is infinite, there are no thermal fluctuations that can drive a transition. The only remaining tunable parameters are J and h which control quantum fluctuations since they are the coefficients of the non-commuting variables.

In Chapter 4 the correlation length, ξ, was defined via the connected spatial Green function (4.30). In the present case, one must distinguish between correlations in the spatial and the imaginary time directions: the former give ξ and the latter give τ, the correlation time. The relation between τ and ξ *defines* the *dynamic critical exponent z*

$$\tau \sim \xi^z \tag{7.161}$$

However, near the critical temperature of the classical anisotropic Ising model, rotational symmetry is restored and the correlations in the x and y directions become the same. In other words, in this case

$$\tau \sim \xi \tag{7.162}$$

and consequently $z = 1$. We see that this world line approach to the quantum problem and its mapping onto an equivalent classical one yielded one of the critical exponents, z, as well as the phase diagram.

7.9.8 Quantum anharmonic oscillator: path integrals

The aim of this problem is to apply the Trotter–Suzuki breakup to write a world line formulation of a single anharmonic oscillator. In fact, this will turn out to be intimately related to the Feynman path integral formulation of quantum systems. For more detail see References [40] and [31]. We shall then simulate numerically various aspects of this system.

Consider a single one-dimensional anharmonic quantum oscillator [40, 31]. The partition function is,

$$Z = \mathrm{Tr}\ \mathrm{e}^{-\beta H} \tag{7.163}$$

$$H = \frac{\mathsf{P}^2}{2m} + \frac{1}{2}m\omega^2\mathsf{X}^2 + \frac{1}{4}\lambda\mathsf{X}^4 \tag{7.164}$$

where

$$[\mathsf{P}, \mathsf{X}] = \frac{\hbar}{\mathrm{i}} \tag{7.165}$$

1. The exponential of the total Hamiltonian cannot be evaluated exactly when $\lambda \neq 0$. However, it is possible to evaluate the exponential of the kinetic and potential energies separately. Unfortunately, since X and P do not commute, we cannot break up the exponential of H into a product of exponentials. We therefore proceed as we did in the other quantum cases where the partition function is first written in the form (7.32) and then the Trotter–Suzuki breakup is used. By using complete sets of states in the coordinate representation

$$1 = \int \mathrm{d}x |x\rangle\langle x|, \qquad \mathsf{X}|x\rangle = x|x\rangle \tag{7.166}$$

write the partition function in the form corresponding to Equation (7.32). Note that we have here only L_τ imaginary time slices, not $2L_\tau$ as in (7.32). In the present case there is only H and not H_1 and H_2.

By using the Trotter–Suzuki approximation

$$\mathrm{e}^{-\Delta\tau H} \approx \mathrm{e}^{-\Delta\tau \mathsf{P}^2/2m} \mathrm{e}^{-\Delta\tau m\omega^2 \mathsf{X}^2/2 - \Delta\tau\lambda \mathsf{X}^4/4} + \mathcal{O}(\Delta\tau)^2 \tag{7.167}$$

and the fact that X is diagonal in the coordinate representation (7.166), show that the partition function becomes[42]

$$\begin{aligned} Z \approx \int & \mathrm{d}x_1\, \mathrm{d}x_2 \ldots \mathrm{d}x_L \exp\Big(-\frac{1}{2} m\omega^2 \Delta\tau \sum_{\ell=1}^{L_\tau} x_\ell^2 - \frac{1}{4}\lambda\Delta\tau \sum_{\ell=1}^{L_\tau} x_l^4\Big) \\ & \langle x_1|\mathrm{e}^{-\Delta\tau \mathsf{P}^2/2m}|x_{L_\tau}\rangle\langle x_{L_\tau}|\mathrm{e}^{-\Delta\tau \mathsf{P}^2/2m}|x_{L_\tau - 1}\rangle \ldots \\ & \langle x_3|\mathrm{e}^{-\Delta\tau \mathsf{P}^2/2m}|x_2\rangle\langle x_2|\mathrm{e}^{-\Delta\tau \mathsf{P}^2/2m}|x_1\rangle \end{aligned} \tag{7.168}$$

By using

$$1 = \int \mathrm{d}p |p\rangle\langle p|, \qquad \mathsf{P}|p\rangle = p|p\rangle \tag{7.169}$$

and

$$\langle p|x_\ell\rangle = \mathrm{e}^{-\mathrm{i}px_\ell} \tag{7.170}$$

show that

$$\langle x_{\ell+1}|\mathrm{e}^{-\Delta\tau \mathsf{P}^2/2m}|x_\ell\rangle = \sqrt{\frac{2m\pi}{\Delta\tau}} \exp\big[-\frac{1}{2} m\Delta\tau[(x_{\ell+1} - x_\ell)/\Delta\tau]^2\big] \tag{7.171}$$

[42] We recall that the first bra on the extreme left must close the first ket on the extreme right since the partition function is a trace.

and that, consequently, the partition function can be written as

$$Z = \int \mathcal{D}x \, \mathrm{e}^{-\Delta\tau S_{\mathrm{cl}}} \tag{7.172}$$

$$S_{\mathrm{cl}} = \frac{1}{2}m\omega^2 \sum_\ell x_\ell^2 + \frac{1}{4}\lambda \sum_\ell x_\ell^4 + \frac{1}{2}m \sum_\ell \left(\frac{x_{\ell+1} - x_\ell}{\Delta\tau}\right)^2 \tag{7.173}$$

where $\int \mathcal{D}x \equiv \int \mathrm{d}x_1 \mathrm{d}x_2 \ldots \mathrm{d}x_{L_\tau}$. In this equation, S_{cl} is the effective classical 'action' of the quantum system. Again, remark that while the quantum system is one-dimensional, the classical action is two-dimensional.

This is the Feynman formulation of the quantum oscillator in terms of path integrals but in *imaginary time*. To obtain (7.172) from Feynman's formulation, one must substitute real time by imaginary time: $t \to i\Delta\tau$. All the variables are now classical and the Monte Carlo simulation can proceed.

2. As discussed in Section 7.4.2, before performing measurements in quantum Monte Carlo, one must derive the classical expression that corresponds to the operator of interest. In the present context, Equation (7.51) is written

$$\langle \mathsf{O} \rangle = \int \mathcal{D}x \, P_B[\{x_\ell\}] \frac{\langle x^{\ell+1} | \mathrm{e}^{-\Delta\tau H} \mathsf{O} | x^\ell \rangle}{\langle x^{\ell+1} | \mathrm{e}^{-\tau H} | x^\ell \rangle} \tag{7.174}$$

With the help of this equation and (7.167) show that

$$\left\langle \frac{1}{2}m\omega^2 \mathsf{X}^2 + \frac{1}{4}\lambda \mathsf{X}^4 \right\rangle = \frac{\int \mathcal{D}x \left(\frac{1}{2}m\omega^2 x_\ell^2 + \frac{1}{4}\lambda x_\ell^4\right) \mathrm{e}^{-\Delta\tau S_{\mathrm{cl}}}}{\int \mathcal{D}x \, \mathrm{e}^{-\Delta\tau S_{\mathrm{cl}}}} \tag{7.175}$$

$$\equiv \left\langle \left(\frac{1}{2}m\omega^2 x_\ell^2 + \frac{1}{4}\lambda x_\ell^4 \right) \right\rangle \tag{7.176}$$

where (7.176) defines our convention that when we write classical quantities between angular brackets, we mean the average is taken with respect to the classical partition function (7.172) and (7.173). In Equation (7.176), ℓ may have any value, in other words, the average may be evaluated at any imaginary time. In practice, to reduce statistical fluctuations one takes an average over all times.

We see that the expression for the potential energy is simple and in fact intuitive. We also see that this simple form appears in the classical action.

It might appear at first sight that the kinetic energy also appears in an intuitive form in the classical action. If one interprets $(x_{\ell+1} - x_\ell)/\Delta\tau)$ as the velocity at imaginary time ℓ, one sees immediately that the last term in S_{cl} has the form of a kinetic energy.

However, we shall now see that, because of the $\sqrt{2m\pi/\Delta\tau}$ pre-factor in (7.171), the kinetic energy term is in fact more subtle.

This may be seen in two ways. One may apply (7.174) where O is the kinetic energy operator. To this end, one needs to evaluate

$$\left\langle x_{\ell+1}\left|\frac{\mathsf{P}^2}{2m}\mathrm{e}^{-\Delta\tau\mathsf{P}^2/2m}\right|x_\ell\right\rangle = \int \mathrm{d}p\langle x_{\ell+1}|p\rangle\langle p|\mathsf{P}^2\mathrm{e}^{-\Delta\tau\mathsf{P}^2/2m}|x_\ell\rangle \tag{7.177}$$

Evaluate this expression and show that the kinetic energy is given by

$$\left\langle x_{\ell+1}\left|\frac{\mathsf{P}^2}{2m}\mathrm{e}^{-\Delta\tau\mathsf{P}^2/2m}\right|x_\ell\right\rangle = \frac{1}{2\Delta\tau} - \left\langle\frac{1}{2}m\left(\frac{x_{\ell+1}-x_\ell}{\Delta\tau}\right)^2\right\rangle \tag{7.178}$$

As for the potential energy, one usually does the average over all imaginary times. The other way to obtain the expression for kinetic energy is to recall that the average total energy may be obtained from

$$\langle H\rangle = -\frac{1}{Z}\frac{\partial Z}{\partial\beta} = -\frac{1}{L_\tau}\frac{1}{Z}\frac{\partial Z}{\partial\Delta\tau} \tag{7.179}$$

where to obtain the last term on the right we used $\beta \equiv L_\tau\Delta\tau$. Use this equation along with (7.172) and (7.173) to calculate the average total energy. Knowing the potential energy (7.176) yields the average kinetic energy. Show that (7.178) is obtained again.

Another interesting quantity is the average occupation number $\langle\mathsf{n}\rangle \equiv \langle\mathsf{a}^\dagger\mathsf{a}\rangle$ where the creation and destruction operators for the oscillator are defined in Equations (5.49) and (5.50)

$$\mathsf{X} = \sqrt{\frac{\hbar}{2m\omega_k}}\left(\mathsf{a}+\mathsf{a}^\dagger\right) \qquad \mathsf{P} = \frac{1}{\mathrm{i}}\sqrt{\frac{\hbar m\omega_k}{2}}\left(\mathsf{a}-\mathsf{a}^\dagger\right) \tag{7.180}$$

It is easy to show that the commutation relations (7.165) are satisfied if we have

$$[\mathsf{a},\mathsf{a}^\dagger] = 1 \tag{7.181}$$

Write the operator n in terms of X and P and then using Equations (7.176) and (7.178), express $\langle\mathsf{n}\rangle$ in terms of the classical variables.

3. Write a program to simulate this system using the Metropolis algorithm. The simulation proceeds as discussed before: choose an initial configuration $\{x_\ell\}$, visit each imaginary time slice and suggest a change $x_\ell \to x_\ell + \delta x$, accept/reject according to the Metropolis criterion which satisfies detailed balance. Choose δx such that the acceptance rate is about 50%.

Note that, as far as the simulation is concerned, one can think of this model as a one-dimensional lattice with L_τ sites, the site variables are the real numbers x_ℓ whose interactions are described by the classical action. So, in that sense, the program to simulate this model is very similar in structure to the programs you have written for the other lattice systems.

Test the program for $\lambda = 0$, which is exactly solvable. Verify that the potential and kinetic energies and $\langle \mathsf{n} \rangle$ from the program agree with the exact results. For the parameters we shall use here, typical runs have 2×10^5 thermalization iterations and 2×10^6 to 5×10^6 to measure. To simplify the numerical factors in the analytic expressions choose $m = 1$ and $\omega = 2$.

Note: Clearly, as the temperature is increased, the kinetic energy of the oscillator increases and so does the velocity. Consequently, to be able to follow the evolution of the oscillator correctly at high temperature, the (imaginary) time step must be made even smaller. Study the effect of increasing the value for $\Delta\tau$.

4. Study the Trotter error as a function of $\Delta\tau$ by comparing the average energy with the exact value for $\lambda = 0$, and show that it is quadratic in $\Delta\tau$. This is best done at low temperature, say $\beta = 10$.

5. The probability density for finding the oscillator between x and $x + \delta x$ is of course $|\psi(x)|^2$ where $\psi(x)$ is the wave function. Using the current program, one can measure numerically $|\psi(x)|^2 \delta x$ as follows: after every Monte Carlo sweep, bin the particle position at the various imaginary time slices, x_ℓ, into a histogram. In other words, one considers the trajectory in imaginary time as equivalent to one in real time, and the position of the particle at each time is put in the appropriate bin in the histogram. At the end of the simulation, the histogram is normalized to unity.

Measure $|\psi(x)|^2\, \delta x$ for $\lambda = 0$ at $\beta = 20$ with $L_\tau = 400$ using a bin size of 10^{-2}, 5×10^5 to thermalize and 2×10^6 to measure. The temperature is low enough so that only the ground state contributes. Is that what you observe?

Study the effect of the anharmonic interaction on the energy, average occupation number and $|\psi(x)|^2\, \delta x$ for the same parameters as above. Take, for example, $1 \leq \lambda \leq 10$.

The virial theorem in quantum mechanics takes the form,

$$\langle \mathsf{T} \rangle = \frac{1}{2}\left\langle \mathsf{X} \frac{\partial U(\mathsf{X})}{\partial \mathsf{X}} \right\rangle \tag{7.182}$$

where T is the kinetic energy operator and $U(\mathsf{X})$ the potential energy operator. Show that in terms of the classical simulation variables, this expression may be written as

$$\langle \mathsf{T} \rangle = \frac{1}{2} m\omega^2 \langle x^2 \rangle + \frac{1}{2} \lambda \langle x^4 \rangle \tag{7.183}$$

where the moments of x may be calculated from the histogram obtained above. Verify numerically the virial theorem by comparing this with the direct calculation of $\langle \mathrm{T} \rangle$ from the simulation.

6. We mentioned in passing in Section 7.4.1 that correlation information in real time may be obtained by performing a Wick rotation of the corresponding quantity calculated in imaginary time. We shall now see an explicit example.

Recall the quantum mechanical result for the time separated two-point correlation function

$$\langle 0|\mathrm{X}(0)\mathrm{X}(t)|0\rangle = \langle 0|\mathrm{X}(0)\,\mathrm{e}^{\mathrm{i}Ht}\mathrm{X}(0)\,\mathrm{e}^{-\mathrm{i}Ht}|0\rangle \tag{7.184}$$

This is appropriate for $T = 0$ since we consider only the ground state $|0\rangle$ in calculating the correlation function. Show that it leads to

$$\langle 0|\mathrm{X}(0)\mathrm{X}(t)|0\rangle = \sum_{n=0}^{\infty} \mathrm{e}^{\mathrm{i}(E_n - E_0)t}|\langle 0|\mathrm{X}(0)|n\rangle|^2 \tag{7.185}$$

Performing the Wick rotation $t \to \mathrm{i}\tau$ yields

$$\langle 0|\mathrm{X}(0)\mathrm{X}(\mathrm{i}\tau)|0\rangle = \sum_{n=0}^{\infty} \mathrm{e}^{-(E_n - E_0)\tau}|\langle 0|\mathrm{X}(0)|n\rangle|^2 \tag{7.186}$$

which we write as

$$\langle 0|\mathrm{X}(0)\mathrm{X}(\tau)|0\rangle = \sum_{n=0}^{\infty} \mathrm{e}^{-(E_n - E_0)\tau}|\langle 0|x|n\rangle|^2 \tag{7.187}$$

In the present case we have $\langle 0|x|n\rangle \neq 0$ for $n \neq 0$. Consequently, for τ large, we obtain

$$\langle 0|\mathrm{X}(0)\mathrm{X}(\tau)|0\rangle = \mathrm{e}^{-(E_1 - E_0)\tau}|\langle 0|x|1\rangle|^2 \tag{7.188}$$

since all other terms decay much more rapidly. We see that by calculating the correlation function in imaginary time, we should be able to calculate the energy of the first excited state. Verify numerically that the value of E_1 obtained in this way agrees with the exact value for the case $\lambda = 0$. Study the energy of the first excited state for $\lambda \neq 0$. For these simulations, take $\beta = 20$ with $L_\tau = 400$, 5×10^5 to thermalize and 2×10^6 to measure.

We see that the Wick rotation allowed us to extract dynamic information by studying correlation functions in imaginary time. This example works very well because we have at our disposal the exact functional form in real time, Equation (7.185), on which we performed the rotation and used the new form to fit the numerical data. In cases where the exact form is not available, the rotation becomes

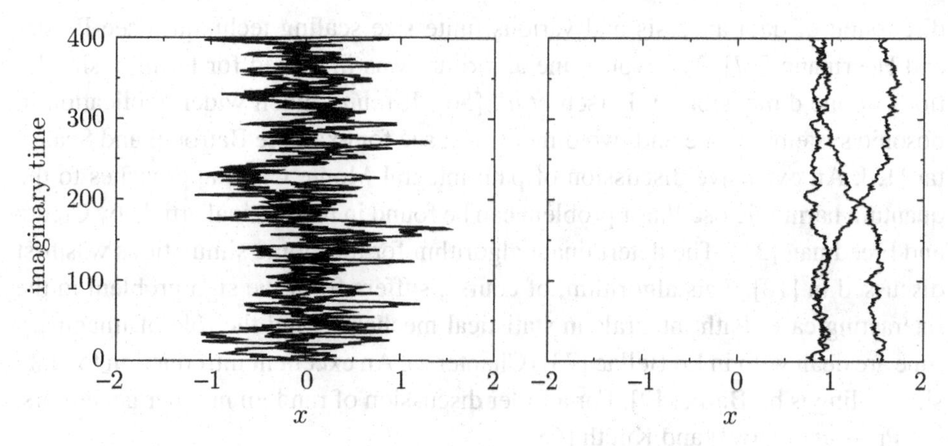

Figure 7.14 Typical world lines for two different temperatures: $\beta = 20$ left figure and $\beta = 0.1$ right figure. In both cases: $\omega = 2$, $m = 1$, $L_t = 400$. There are three configurations in each case.

much more difficult because the numerical data may be fitted by a variety of different functional forms. Very often it is not possible to choose the correct one due to the statistical errors on the numerical data.

7. In this part we consider somewhat higher temperatures. Start with the case $\lambda = 0$ and measure $|\psi(x)|^2$ for $\omega = 2, m = 1, \beta = 5, 0.4, 0.25$. Plot your results and comment on the qualitative difference in the shape. Do you observe that the $\beta = 0.4, 0.25$ cases give non-symmetric results? Explain.

Of course, this also happens when $\lambda \neq 0$. For example, Figure 7.14 shows snapshots of typical configurations for $\beta = 20$ (left) and $\beta = 0.1$ (right). Clearly, the $\beta = 20$ configurations fluctuate around $x = 0$ as they should since that is the minimum of the potential. The $\beta = 0.1$ configurations fluctuate around $x \approx 1$. Why? Can you then point out a difficulty with the world line algorithm as formulated in this problem?

7.10 Further reading

Detailed discussion of stochastic processes and Markov chains may be found in Risken [110], van Kampen [119] and Gardiner [46]. An excellent introduction to numerical simulations, with sample programs and many exercises and problems, is found in Gould and Tobochnik [50]. Another interesting reference is Newman and Barkema [94], which has many examples of equilibrium and non-equilibrium Monte Carlo methods and an interesting summary of the history of Monte Carlo simulations. The first use of the name Monte Carlo in connection with numerical simulations was by Metropolis and Ulam [91]; see also [90]. For an in-depth discussion of data analysis and various finite size scaling techniques see Binder and Heermann [17]. The world line algorithm was first used for fermion simulations in one dimension by Hirsch *et al.* [56]. It found much wider application in bosonic systems in one and two dimensions; see for example Batrouni and Scalettar [13]. An extensive discussion of path integral Monte Carlo approaches to the quantum harmonic oscillator problem can be found in the seminal article by Creutz and Freedman [31]. The determinant algorithm for fermionic simulations was first discussed in [18]. This algorithm, of course, suffers from the sign problem in the interacting case. Path integrals in statistical mechanics and the rôle of imaginary time are dealt with in Le Bellac [73] (Chapter 2). An excellent introduction to finite size scaling is by Barber [9]. For a fuller discussion of random number generators, see Press *et al.* [103] and Knuth [65].

For more on the XY model and the Kosterlitz–Thouless transition see the review article by Kogut [66]. An excellent discussion of superfluidity and its relation to the XY model and the role of vortices in its destruction may be found in Ma [85] (Chapter 30). The Ising model in a transverse field is discussed by Kogut [66]. For more on the world line algorithm applied to the quantum oscillator see Creutz and Freedman [31].

8

Irreversible processes: kinetic theory

In this chapter we shall discuss a microscopic theory of transport phenomena, namely kinetic theory. The central idea of kinetic theory is to explain the behaviour of out-of-equilibrium systems as the consequence of collisions among the particles forming the system. These collisions are described using the concept of cross section which will be introduced in Section 8.1 where we shall also demonstrate an elementary, but not rigorous, first calculation of transport coefficients. Then, in the following section, we introduce the Boltzmann–Lorentz model which describes collisions of molecules with fixed randomly distributed scattering centres. This model gives a good description of transport properties in several physically important systems such as the transport of electrons and holes in a semiconductor. In Section 8.3 we shall give a general discussion of the Boltzmann equation, with two important results: the derivation of hydrodynamics and that of the H-theorem, which gives an explicit proof of irreversibility. Finally, in the last section, we shall address the rigorous calculation of the transport coefficients, viscosity and thermal conductivity, in a dilute mono-atomic gas.

8.1 Generalities, elementary theory of transport coefficients

8.1.1 Distribution function

We adopt straight away the classical description where a point in phase space is given by its position, $\vec{r}$, and momentum, $\vec{p}$. The basic tool of kinetic theory is the *distribution function* $f(\vec{r}, \vec{p}, t)$, which is the density of particles in phase space: the number of particles at time t in the volume $\mathrm{d}^3r\mathrm{d}^3p$ at the phase space point $(\vec{r}, \vec{p})$ is, by definition, given by $f(\vec{r}, \vec{p}, t)\mathrm{d}^3r\,\mathrm{d}^3p$. The goal of kinetic theory is to calculate spatio-temporal dynamics of the distribution function. The particle

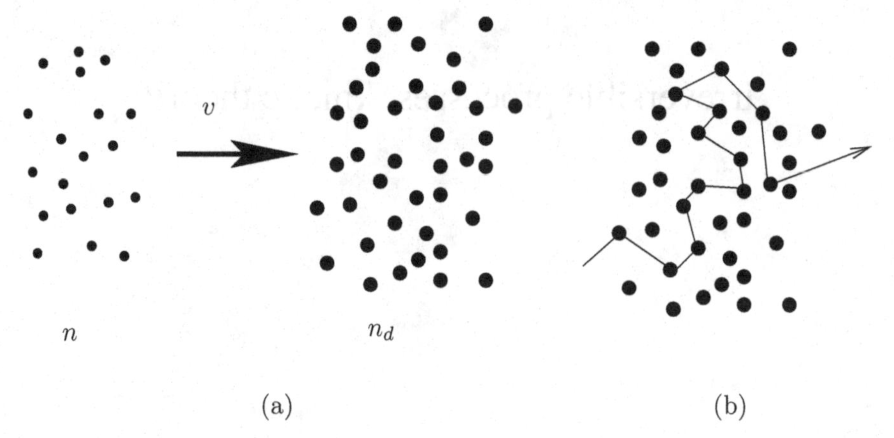

Figure 8.1 (a) Incident particles moving towards target particles. (b) Multiple collisions of an incident particle.

density in position space, $n(\vec{r}, t)$, is the momentum integral of $f(\vec{r}, \vec{p}, t)$

$$\boxed{n(\vec{r}, t) = \int \mathrm{d}^3 p \, f(\vec{r}, \vec{p}, t)} \tag{8.1}$$

Let $\mathcal{A}(\vec{r}, \vec{p})$ be a classical dynamical variable. We define its average value, $A(\vec{r}, t)$, at point $\vec{r}$ by

$$\boxed{A(\vec{r}, t) = \frac{1}{n(\vec{r}, t)} \int \mathrm{d}^3 p \, \mathcal{A}(\vec{r}, \vec{p}) f(\vec{r}, \vec{p}, t)} \tag{8.2}$$

Rather than using the momentum, it is clearly equivalent to use the velocity $\vec{v} = \vec{p}/m$, where m is the particle mass. The distribution function $f_v(\vec{r}, \vec{p}, t)$ is related to f by a simple proportionality

$$f_v(\vec{r}, \vec{v}, t) = m^3 f(\vec{r}, \vec{p}, t) \tag{8.3}$$

8.1.2 Cross section, collision time, mean free path

The central concept in the description of collisions is that of *cross section*. We start with a simple case. Consider a flux of particles of density n and momentum $\vec{p}$, incident on a target of fixed scattering centres with density n_d (Figure 8.1). The target is thin enough to ignore multiple collisions.[1] The momentum after a collision is $\vec{p}\,'$, its direction given by $\Omega' = (\theta', \varphi')$, where θ' and φ' are respectively the polar and azimuthal angles when the Oz axis is chosen parallel to $\vec{p}$. Let $\mathrm{d}\mathcal{N}/\mathrm{d}t\,\mathrm{d}V$

[1] In other words we assume that the mean free path, defined in (8.10), is large compared to the thickness of the target.

be the number of collisions per unit time and unit target volume for collisions where particles are scattered within the solid angle $\mathrm{d}\Omega'$ in the direction of $\vec{p}\,'$. This number is proportional to

(*i*) the incident flux $\mathcal{F} = nv$,
(*ii*) the solid angle $\mathrm{d}\Omega'$, which defines the cone into which the particles are scattered,
(*iii*) the density n_d of target particles as long as we ignore multiple collisions.

We can then write

$$\boxed{\frac{\mathrm{d}\mathcal{N}}{\mathrm{d}t\,\mathrm{d}V} = \mathcal{F} n_d\, \sigma(v, \Omega')\, \mathrm{d}\Omega'} \tag{8.4}$$

The coefficient of proportionality $\sigma(v, \Omega')$ in (8.4) is called the *differential scattering cross section.*[2] We often use (8.4) in the form

> $v\sigma(v, \Omega')\, d\Omega'$ *is the number of collisions per unit time and unit target volume for unit densities of target and incident particles where the scattered particle is in* $d\Omega'$ *around* $\vec{p}\,'$.

Dimensional analysis shows that σ has dimensions of area and is therefore measured in m^2.

An important concept is that of *collision time*[3] which is the elapsed time between two successive collisions of the same particle. Consider a particle in flight between the target particles (Figure 8.1(b)): to obtain the collision time from (8.4), we first divide by n and integrate over Ω'. This gives the average number of collisions per second suffered by an incident particle which is just the inverse of the collision time $\tau^*(p)$

$$\boxed{\frac{1}{\tau^*(p)} = n_d v \int \mathrm{d}\Omega'\, \sigma(v, \Omega') = n_d v \sigma_{\mathrm{tot}}(v)} \tag{8.5}$$

The *total cross section* $\sigma_{\mathrm{tot}}(v)$ is obtained by integrating the differential cross section over Ω'.

We now consider the general case. The incident particles, characterized by their density n_2 and velocity $\vec{v}_2 = \vec{p}_2/m_2$, are moving toward the target particles which in turn are characterized by their density n_1 and velocity $\vec{v}_1 = \vec{p}_1/m_1$. Clearly, in a gas containing only one kind of atom, the incident and target particles are identical but it is convenient to distinguish them at least to begin with. We may regain the previous situation by applying a Galilean transformation of velocity $\vec{v}_1$

[2] We assume that collisions are rotation invariant by not considering cases where the target or incident particles are polarized.
[3] The reader should not confuse 'collision time' with 'duration of collision'. To avoid confusion, we may use *time of flight* instead of collision time.

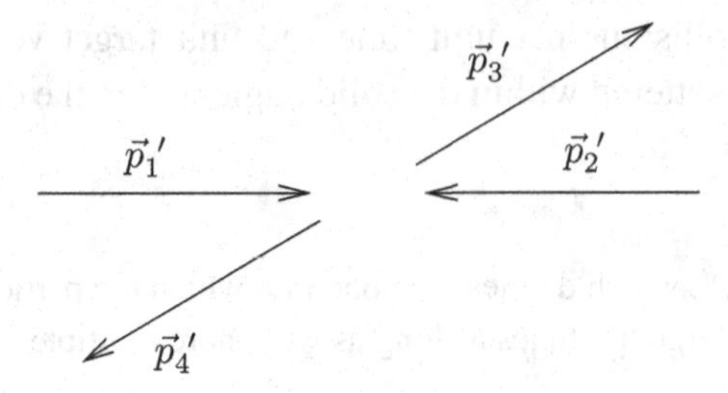

Figure 8.2 Centre-of-mass reference frame.

to place ourselves in the rest frame of the target particles i.e. the target frame.[4] In this frame, the incident particles have a velocity $(\vec{v}_2 - \vec{v}_1)$ and the incident particle flux, $\mathcal{F}_2$, i.e. the number of incident particles crossing a unit area perpendicular to $(\vec{v}_2 - \vec{v}_1)$ per unit time, is given by

$$\mathcal{F}_2 = n_2|\vec{v}_2 - \vec{v}_1| \tag{8.6}$$

In place of the target frame, it is often convenient to use the centre-of-mass reference frame, where the total momentum is zero. In this frame, two colliding particles have equal and opposite momenta (Figure 8.2) $\vec{p}_1{}' = -\vec{p}_2{}'$ and, if the masses are the same, $\vec{v}_1{}' = -\vec{v}_2{}'$. After the collision, particles (1) and (2) propagate respectively with momenta $\vec{p}_3{}'$ and $\vec{p}_4{}' = -\vec{p}_3{}'$ ($|\vec{p}_1{}'| = |\vec{p}_3{}'|$ if the collision is elastic) and $\Omega' = (\theta', \varphi')$ are the polar and azimuthal angles taken with respect to $\vec{p}_1{}'$, which defines the Oz axis. It is important to remember that by convention Ω' will always be measured in the centre-of-mass frame so that there is no limitation on the range of the polar angle θ', $0 \le \theta' < \pi$. Let $\mathrm{d}\mathcal{N}/\mathrm{d}t\,\mathrm{d}V$ be the number of collisions per unit time and unit target volume when particle $\vec{p}_1{}'$ is scattered into the solid angle $\mathrm{d}\Omega'$ around the direction $\vec{p}_3{}'$. A straightforward generalization of (8.4) gives this quantity as

$$\boxed{\frac{\mathrm{d}\mathcal{N}}{\mathrm{d}t\,\mathrm{d}V} = \mathcal{F}_2 n_1 \sigma(\vec{v}_2, \vec{v}_1, \Omega')\mathrm{d}\Omega'} \tag{8.7}$$

The coefficient of proportionality $\sigma(\vec{v}_2, \vec{v}_1, \Omega')$ in (8.7) is the differential cross section. Since all the terms in (8.7) are Galilean invariants, the differential cross section itself is also a Galilean invariant and consequently depends on $(\vec{v}_2 - \vec{v}_1)$, not on $\vec{v}_1$ or $\vec{v}_2$ separately. In addition, due to rotation invariance, the cross section depends only on the modulus of the velocity difference, $(|\vec{v}_2 - \vec{v}_1|)$. Therefore, the total cross section, obtained by integrating over the solid angle, also depends only

[4] In the previous case where the target was at rest, the target frame was the same as the laboratory frame.

on $(|\vec{v}_2 - \vec{v}_1|)$

$$\sigma_{\text{tot}}(|\vec{v}_2 - \vec{v}_1|) = \int \mathrm{d}\Omega' \, \sigma(|\vec{v}_2 - \vec{v}_1|, \Omega') \tag{8.8}$$

As before, $(|\vec{v}_2 - \vec{v}_1|)\sigma_{\text{tot}}(|\vec{v}_2 - \vec{v}_1|)$ *is the number of collisions per unit time and unit target volume for unit densities of target and incident particles.*

The concept of collision time, previously defined for fixed targets, extends to the general case: in a gas, the collision time τ^* is the average time between two successive collisions of the same particle. It is given by the generalization of Equation (8.5)

$$\boxed{\tau^* \sim \frac{1}{n\langle v\rangle\,\sigma_{\text{tot}}}} \tag{8.9}$$

where $\langle v\rangle$ is an average velocity of the particle whose definition is intentionally left imprecise. The *mean free path* ℓ is the average distance between two successive collisions: $\ell \sim \tau^*\langle v\rangle$

$$\boxed{\ell \sim \frac{1}{n\sigma_{\text{tot}}}} \tag{8.10}$$

A more precise determination of the collision time is given in Exercise 8.4.4, where we find for the Maxwell velocity distribution and a cross section independent of energy

$$\tau^* = \frac{1}{\sqrt{2}\,n\langle v\rangle\sigma_{\text{tot}}} \qquad \ell = \frac{1}{\sqrt{2}\,n\sigma_{\text{tot}}} \tag{8.11}$$

$\langle v\rangle$ being the average velocity of the Maxwell distribution (3.54b)

$$\langle v\rangle = \sqrt{\frac{8kT}{\pi m}}$$

Let us illustrate the concept of cross section with the simple example of hard spheres, radius a and velocity v_2, incident on a target of fixed point particles. During a time $\mathrm{d}t$ a sphere sweeps a volume $\pi a^2 v_2\,\mathrm{d}t$, which contains $n_1\pi a^2 v_2\,\mathrm{d}t$ target particles. The total number of collisions per unit time and unit target volume is therefore given by $n_1 n_2 \pi a^2 v_2$, which, when combined with the definition (8.5), yields $\sigma_{\text{tot}} = \pi a^2$.

Now consider spheres of radius a_1 incident on target spheres of radius a_2 (Figure 8.3). A collision takes place if the two centres pass each other at a distance $b \le (a_1 + a_2)$. The distance b is called the *impact parameter* of the collision. This situation is equivalent to the case of a sphere of radius $(a_1 + a_2)$ scattering off a

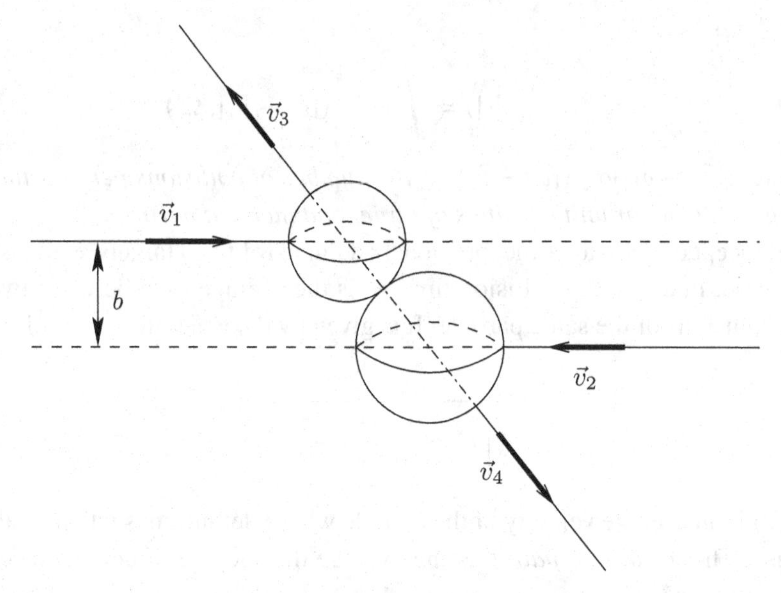

Figure 8.3 Collision of two spheres.

point particle. The total cross section is then

$$\sigma_{\text{tot}} = \pi(a_1 + a_2)^2 \tag{8.12}$$

We can show, in classical mechanics, that the differential cross section for a collision of two hard spheres depends neither on Ω' nor on the relative velocity and is therefore given by $\sigma(\Omega') = \sigma_{\text{tot}}/(4\pi)$. In general, the determination of the cross section should be done within the framework of quantum mechanics and taking into account the interaction potential $U(\vec{r})$ between the two particles. If we are given an interaction potential $U(\vec{r})$, there are standard quantum mechanical methods to calculate the cross section. Quantum mechanics shows up only during the calculation of the cross section; once this quantity is known, kinetic theory becomes a theory of classical particles.

Having defined the key ideas for describing collisions, we are now in a position to state the assumptions behind *kinetic theory*:

(*i*) The collision time τ^* is very long compared to the duration of the collision $\delta\tau$, which is the time an incident particle spends in the field of influence of a target particle: $\tau^* \gg \delta\tau$. To leading approximation, we can assume that collisions take place instantaneously and that, for the most part, the particles are free and independent. The potential energy and interactions are taken into account effectively by the collisions.

(*ii*) The probability of having three particles in close proximity is very small. Consequently, the probability of three or more particles colliding is negligible: it is sufficient to consider only binary collisions.

(*iii*) The classical gas approximation holds, in other words the thermal wavelength is very small compared to the average distance between particles, $\lambda \ll d$. This assumption may be relaxed and the theory extended to the quantum case, see Problems 8.6.2 and 8.6.7.

These conditions are satisfied by a dilute classical gas. Let us consider some orders of magnitude: For a gas such as nitrogen, under standard temperature and pressure, the density is about 2.7×10^{25} molecules/m^3 and the distance between molecules is $d \sim n^{-1/3} \sim 3 \times 10^{-9}$ m. Taking a cross section[5] of 4×10^{-19} m^2, which corresponds to a hard sphere radius $a \simeq 1.8 \times 10^{-10}$ m, we find a collision time $\tau^* \simeq 2 \times 10^{-10}$ s, and a mean free path $\ell \simeq 10^{-7}$ m. With a typical value for the average velocity $\langle v \rangle \simeq 500$ m/s, the duration of a collision is approximately $\delta\tau \sim a/\langle v \rangle = 3 \times 10^{-13}$ s. We then see a clear separation of the three length scales

$$a \ll d \ll \ell \tag{8.13}$$

and the two time scales $\delta\tau \ll \tau^*$. However, when a gas is so dilute that the mean free path is of the order of the dimension of the system, we enter a regime, called the Knudsen regime, where local equilibrium no longer exists.

8.1.3 Transport coefficients in the mean free path approximation

We shall calculate the transport coefficients in a dilute medium in the so called mean free path approximation. This elementary calculation is physically very instructive but not rigorous; for example we shall keep vague the definition of the average velocity that appears in the final equations. This calculation will allow us to identify the dependence of the transport coefficients on the relevant physical parameters, although the numerical values will be off by a factor of 2 to 3. The value of this calculation lies in the fact that a rigorous evaluation based on the Boltzmann equation is considerably more complicated as will be seen in Section 8.4.

Thermal conductivity or energy transport

Consider a stationary fluid with a temperature gradient $\mathrm{d}T/\mathrm{d}z$ in the Oz direction. Let Q be the heat flux, i.e. the amount of heat crossing the plane at z per unit area and unit time. Recall the definition (6.18) of the coefficient of thermal conductivity κ

$$Q = j_z^E = -\kappa \frac{\mathrm{d}T}{\mathrm{d}z} \tag{8.14}$$

[5] The cross section is estimated from viscosity measurements and theoretical estimates based on the Boltzmann equation.

because Q is also the z component of the energy current $\vec{j}_E$ which, in this case, is simply a heat current. Since we ignore interactions, the average energy ε of a molecule[6] is purely kinetic and therefore depends on the height z because of the temperature dependence on z.

We shall define the average velocity $\langle v \rangle$ by the following approximation. For an arbitrary function $g(\cos\theta)$, where θ is the angle between the velocity $\vec{v}$ of a molecule and the Oz axis, we have[7]

$$\int \mathrm{d}^3 p\, v_z f(\vec{p}) g(\cos\theta) \to \frac{n}{2}\langle v \rangle \int \mathrm{d}(\cos\theta)\cos\theta\, g(\cos\theta) \tag{8.15}$$

Now consider molecules whose velocity makes an angle θ with the Oz axis, $0 \le \theta \le \pi$. Such a molecule crossing the plane at z has travelled, on average, a distance ℓ and therefore its last collision took place at an altitude $z - \ell\cos\theta$. Its energy is $\varepsilon(z - \ell\cos\theta)$. The heat flux crossing the plane at z and coming from molecules whose velocities make an angle θ with the vertical is then

$$\begin{aligned}
\mathrm{d}Q(\cos\theta) &= \frac{n}{2}\mathrm{d}(\cos\theta)\langle v \rangle \cos\theta\, \varepsilon(z - \ell\cos\theta) \\
&\simeq \frac{n}{2}\mathrm{d}(\cos\theta)\langle v \rangle \cos\theta \left[\varepsilon(z) - \ell\cos\theta \frac{\mathrm{d}\varepsilon(z)}{\mathrm{d}z}\right]
\end{aligned}$$

We have made a Taylor expansion, to first order in ℓ, of the term $\varepsilon(z - \ell\cos\theta)$. The integral over $\mathrm{d}(\cos\theta)$ from -1 to $+1$ in the first term of the bracket vanishes. Using

$$\int_{-1}^{+1} \mathrm{d}(\cos\theta)\cos^2\theta = \frac{2}{3} \tag{8.16}$$

we obtain for the heat flux

$$\begin{aligned}
Q &= \int_{-1}^{1} \mathrm{d}Q(\cos\theta) = -\frac{1}{3} n\langle v \rangle \ell \frac{\mathrm{d}\varepsilon(z)}{\mathrm{d}z} \\
&= -\frac{1}{3} n\langle v \rangle \ell \frac{\mathrm{d}\varepsilon}{\mathrm{d}T}\frac{\mathrm{d}T}{\mathrm{d}z} = -\frac{1}{3} n\langle v \rangle \ell c \frac{\mathrm{d}T}{\mathrm{d}z}
\end{aligned}$$

where c is the specific heat per molecule, which is equal to $3k/2$ for mono-atomic gases, $5k/2$, etc. for poly-atomic gases (Section 3.2.4). Comparing with (8.14)

[6] Since we are now discussing gases, we talk about molecules rather than particles.

[7] The factor $n/2$ gives the correct normalization (8.2) since $\int_{-1}^{1}\mathrm{d}(\cos\theta) = 2$. Put $v_z = g = 1$ to find that both integrals give n. We have suppressed the labels $\vec{r}$ and t in f.

gives the coefficient of thermal conductivity

$$\boxed{\kappa = \frac{1}{3} n \langle v \rangle \ell c} \tag{8.17}$$

Viscosity or momentum transport

Let us now consider the fluid flow described in Section 6.3.1, Figure 6.6, and recall Equation (6.71) for the component $\mathcal{P}_{xz}$ of the pressure tensor for this situation

$$\mathcal{P}_{xz} = -\eta \frac{\mathrm{d}u_x(z)}{\mathrm{d}z} \tag{8.18}$$

The fluid flows in the Ox direction with horizontal velocity $u_x(z)$ (a function of z), η is the shear viscosity and $-\mathcal{P}_{xz}$ is the x component of the force per unit area at z applied by the fluid above z on the fluid below it. From the fundamental law of dynamics, this is also the momentum transfer per second from the fluid above z to the fluid below it. Therefore, $\mathcal{P}_{xz}$ is the momentum flux across the plane at height z with the normal to the plane pointing upward. Since the fluid is assumed to be dilute, we can ignore interactions among the molecules and the momentum flux is therefore purely convective

$$\mathcal{P}_{xz} = \int \mathrm{d}^3 p \, p_x v_z f(\vec{p}) \tag{8.19}$$

We repeat the above reasoning by considering molecules whose velocities make an angle θ with the vertical, $0 \le \theta \le \pi$. As before, a molecule crossing the plane at z has travelled, on average, a distance ℓ and therefore its last collision took place at an altitude $z - \ell \cos\theta$, and the x component of its momentum is $mu_x(z - \ell\cos\theta)$. Clearly, in this calculation we only consider the flow velocity $\vec{u}$ of the fluid and not the thermal velocity, also present, which averages to zero since its direction is random. Using the approximation (8.15), the momentum flux due to these molecules is

$$\begin{aligned}
\mathrm{d}\mathcal{P}_{xz}(\cos\theta) &= \frac{n}{2}\mathrm{d}(\cos\theta)\langle v\rangle \cos\theta \, mu_x(z - \ell\cos\theta) \\
&\simeq \frac{n}{2}\mathrm{d}(\cos\theta)\langle v\rangle \cos\theta \, m\left[u_x(z) - \ell\cos\theta \frac{\mathrm{d}u_x(z)}{\mathrm{d}z}\right]
\end{aligned}$$

We have performed a Taylor expansion of the term $u_x(z - \ell\cos\theta)$ to first order in ℓ. The integral from -1 to $+1$ over $\mathrm{d}(\cos\theta)$ of the first term in the bracket vanishes, and using (8.16) we obtain

$$\mathcal{P}_{xz} = \int_{-1}^{+1} \mathrm{d}\mathcal{P}_{xz}(\cos\theta) = -\frac{1}{3} nm\langle v\rangle \ell \frac{\mathrm{d}u_x(z)}{\mathrm{d}z} \tag{8.20}$$

which becomes, upon comparing with (8.18),

$$\boxed{\eta = \frac{1}{3} n m \langle v \rangle \ell = \frac{1}{3} \frac{m\langle v \rangle}{\sigma_{\text{tot}}}} \tag{8.21}$$

We remark that the product $n\ell = 1/\sigma_{\text{tot}}$ is constant and that, at constant temperature, the viscosity is independent of the density and the pressure. If the density is doubled, the number of molecules participating in the transport is also doubled, but the mean free path is halved and the two effects cancel out leaving the viscosity constant. In fact, the second equation in (8.21) shows that the viscosity is a function of T only through the dependence of $\langle v \rangle$ on the temperature, and that it increases with T as $\langle v \rangle \propto T^{1/2}$.[8] This should be contrasted with the case of liquids where the viscosity decreases with increasing temperature.

By comparing (8.17) and (8.21), we predict that the ratio κ/η is equal to c/m. Experimentally we have

$$1.3 \lesssim \frac{\kappa}{\eta} \frac{m}{c} \lesssim 2.5 \tag{8.22}$$

which confirms the validity of our calculation to within a factor of 2 to 3. We can understand qualitatively the origin of this factor. In the above calculation, all the molecules have the same velocity and they all travel the same distance, the mean free path, between successive collisions. However, faster molecules have a higher flux across the horizontal plane and while they transport the same horizontal momentum as the average, they transport more kinetic energy. Therefore, the ratio κ/η is underestimated in our simple calculation.

Diffusion or particle transport

As a last example, we study diffusion. Let $n(z)$ be the z dependent density of solute in a solvent. The diffusion current $\vec{j}_N$ is governed by Fick's law (6.26), which becomes in the present case

$$j_z^N = -D \frac{\mathrm{d}n}{\mathrm{d}z} \tag{8.23}$$

By using, once again, the same reasoning as above, we have

$$\mathrm{d} j_z^N(\cos\theta) = \frac{1}{2} \mathrm{d}(\cos\theta) \langle v \rangle \cos\theta \, n(z - \ell\cos\theta)$$

The Taylor expansion and integration over θ yield

$$j_z^N = -\frac{1}{3} \langle v \rangle \ell \frac{\mathrm{d}n}{\mathrm{d}z}$$

[8] In fact, the dependence of σ_{tot} on the velocity should be included. This leads to $\eta \propto T^{0.7}$.

which allows the identification of the diffusion coefficient

$$\boxed{D = \frac{1}{3}\langle v\rangle\,\ell} \tag{8.24}$$

8.2 Boltzmann–Lorentz model

8.2.1 Spatio-temporal evolution of the distribution function

The starting point will be a spatio-temporal evolution equation for the distribution function $f(\vec{r}, \vec{p}, t)$ in phase space. Consider first a collection of non-interacting particles moving under the influence of external forces. When $t \to t + \mathrm{d}t$, a particle which at time t was at point $\vec{r}(t)$ and had a momentum $\vec{p}(t)$ will at time $t + \mathrm{d}t$ find itself at $\vec{r}(t + \mathrm{d}t)$ with momentum $\vec{p}(t + \mathrm{d}t)$. Since the number of particles in a volume element in phase space is constant, we have

$$f[\vec{r}(t+\mathrm{d}t), \vec{p}(t+\mathrm{d}t), t+\mathrm{d}t]\mathrm{d}^3r'\,\mathrm{d}^3p' = f[\vec{r}(t), \vec{p}(t), t]\mathrm{d}^3r\,\mathrm{d}^3p \tag{8.25}$$

where $\mathrm{d}^3r'\,\mathrm{d}^3p'$ is the phase space volume at $t + \mathrm{d}t$ (Figure 8.4): motion in phase space takes a point in $\mathrm{d}^3r\,\mathrm{d}^3p$ to a point in $\mathrm{d}^3r'\mathrm{d}^3p'$.

Liouville's theorem (2.34) $\mathrm{d}^3r'\,\mathrm{d}^3p' = \mathrm{d}^3r\,\mathrm{d}^3p$ means that the two functions in (8.25) are equal. Expanding to first order in $\mathrm{d}t$ gives

$$\frac{\partial f}{\partial t} + \frac{\mathrm{d}\vec{r}}{\mathrm{d}t}\cdot\vec{\nabla}_{\vec{r}} f + \frac{\mathrm{d}\vec{p}}{\mathrm{d}t}\cdot\vec{\nabla}_{\vec{p}} f = \frac{\partial f}{\partial t} + \vec{v}\cdot\vec{\nabla}_{\vec{r}} f + \vec{F}\cdot\vec{\nabla}_{\vec{p}} f = 0 \tag{8.26}$$

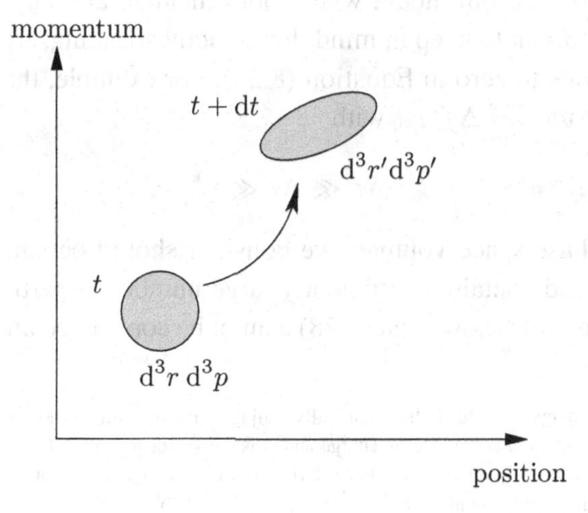

Figure 8.4 Motion in phase space.

In the presence of a magnetic field, the passage from the first to the second of equations (8.26) is not obvious (see Problem 8.5.5).[9] Comparing with (6.65), we see that the operator D

$$D = \frac{\partial}{\partial t} + \frac{\mathrm{d}\vec{r}}{\mathrm{d}t} \cdot \vec{\nabla}_{\vec{r}} + \frac{\mathrm{d}\vec{p}}{\mathrm{d}t} \cdot \vec{\nabla}_{\vec{p}} = \frac{\partial}{\partial t} + \sum_{\alpha} v_{\alpha}\, \partial_{\alpha} + \sum_{\alpha} \dot{p}_{\alpha}\, \partial_{p_{\alpha}} \tag{8.27}$$

is simply the material derivative in phase space (also see Exercise 2.7.3).[10] Equation (8.26) expresses the fact that the distribution function is constant along a trajectory in phase space.

Equation (8.26) is valid only in the absence of interactions. We now include interactions in the form of collisions among the particles. Recall that one of the assumptions of kinetic theory stipulates that the duration of collisions is very short compared with the collision time. We shall, therefore, assume instantaneous collisions. The effect of collisions is to replace the zero on the right hand side of Equation (8.26) by a *collision term*, $\mathcal{C}[f]$, which is a functional of the distribution function f. Equation (8.26) then becomes

$$\boxed{\frac{\partial f}{\partial t} + \vec{v} \cdot \vec{\nabla}_{\vec{r}} f + \vec{F} \cdot \vec{\nabla}_{\vec{p}} f = \mathcal{C}[f]} \tag{8.28}$$

The left hand side of Equation (8.28) is called the *drift* term. The collision term is in fact a balance term, and to establish its form we need to keep track of the number of particles which enter or exit the phase space volume $\mathrm{d}^3 r\, \mathrm{d}^3 p$. We note that the units of $\mathcal{C}[f]$ are inverse phase space volume per second. Equation (8.28) does not in any way specify the form of the collision term. As a result, Equation (8.28) is equally valid for the Boltzmann–Lorentz model with fixed scattering centres, as well as for the Boltzmann model where both incident and target particles are in motion. It is important to keep in mind that, strictly speaking, we cannot take time and space intervals to zero in Equation (8.28). For example, the derivative $\partial f/\partial t$ should be understood as $\Delta f/\Delta t$ with

$$\delta\tau \ll \Delta t \ll \tau^* \tag{8.29}$$

Similarly, the phase space volumes we consider should be small but not microscopic; they should contain a sufficiently large number of particles. The fact that Δt has to remain finite shows that (8.28) cannot be considered an exact equation.[11]

[9] In the presence of a magnetic field $\vec{B}$, the canonically conjugate momentum $\vec{p}$ is not the same as the momentum $m\vec{v}$: $\vec{p} = m\vec{v} + q\vec{A}$ where q is the electric charge and $\vec{A}$ is the vector potential, $\vec{B} = \vec{\nabla} \times \vec{A}$.

[10] We keep the notations of Chapter 6 for space derivatives ($\vec{\nabla}_{\vec{r}} = (\partial_x, \partial_y, \partial_z)$) and introduce the following notation for derivatives in momentum space: $\vec{\nabla}_{\vec{p}} = (\partial_{p_x}, \partial_{p_y}, \partial_{p_z})$.

[11] We do not follow the motion of each particle during infinitesimally small time intervals.

The discretization of time (8.29) cannot guarantee the reversibility of the equations of motion, and the observation of irreversible behaviour will not contradict microreversibility.

8.2.2 Basic equations of the Boltzmann–Lorentz model

The Boltzmann–Lorentz model is a model of collisions between light incident particles and randomly distributed fixed scattering centres. The gas of incident particles is assumed to be dilute enough to ignore collisions among these incident particles. For example, this model is used to describe

- electrons and holes in semiconductors,
- diffusion of impurities in a solid,
- diffusion of light solute molecules in a solvent whose molecules are heavy,
- neutron diffusion in a moderator.

This model may also be used to describe electrons in a metal if one makes the necessary changes to accommodate Fermi–Dirac statistics (Problem 8.6.2). In what follows we shall concentrate on a gas of classical particles. In order to justify the approximation of fixed scattering centres, consider a gas made of a mixture of light particles of mass m and heavy particles of mass μ. The two types of particles have kinetic energies of the order of kT, but the ratio of their momenta is $\sqrt{m/\mu}$ with the consequence that the contribution of the light particles to momentum conservation is negligible. The heavy particles can, therefore, absorb and supply momentum during collisions with the light ones, and thus they may be considered infinitely heavy and stationary. Since we assume elastic collisions, the energy of the light particles is conserved but not their momentum. As in Section 6.2.1, there will be only two conservation laws: conservation of particle number and energy along with the associated densities, n and ϵ, and their currents $\vec{j}_N$ and $\vec{j}_E$ (however, see Footnote 12).

In the absence of external forces, we can write Equation (8.28) for the distribution function $f(\vec{r}, \vec{p}, t)$ of the light particles

$$\frac{\partial f}{\partial t} + \vec{v} \cdot \vec{\nabla}_{\vec{r}} f = \mathcal{C}[f] \tag{8.30}$$

To evaluate $\mathcal{C}[f]$, we shall account for the particles which enter and leave the phase space element $\mathrm{d}^3r\,\mathrm{d}^3p$. First consider particles leaving this element. We assume this volume to be small enough so that any particle initially in it will be ejected after a collision. Before the collision the particle has momentum $\vec{p}$, after the collision its momentum is $\vec{p}\,'$, which is no longer in $\mathrm{d}^3r\,\mathrm{d}^3p$ around $\vec{p}$ (Figure 8.5).

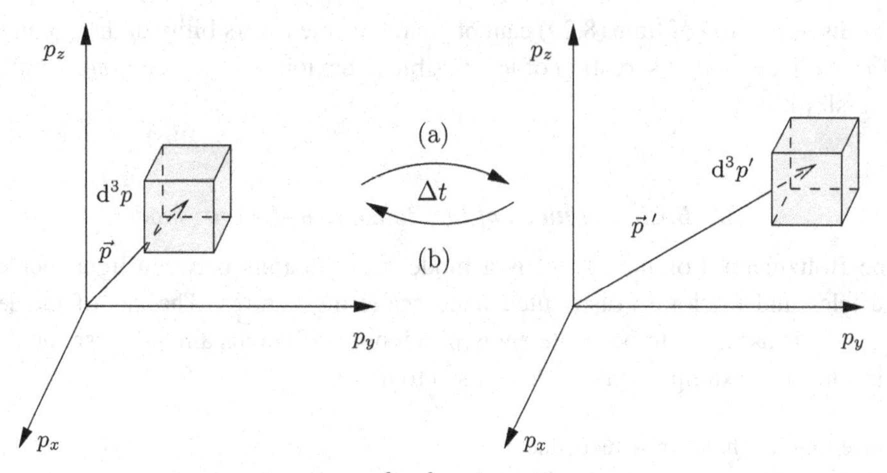

Figure 8.5 (a) A particle leaving $d^3r\, d^3p$, $\vec{p} \to \vec{p}\,'$, and (b) a particle entering it, $\vec{p}\,' \to \vec{p}$.

The collision term will be calculated in terms of the cross sections defined in (8.4). However, rather than using the variable Ω' of (8.4), it will be more convenient, for the purpose of identifying the symmetries of the problem, to use the variable $\vec{p}\,'$. This is possible because we may include in the cross section a δ-function that ensures the conservation of energy, which is purely kinetic in the present case: for a particle of momentum $\vec{p}$, the energy is $\varepsilon(\vec{p}) = \vec{p}^{\,2}/(2m)$. We note that

$$d^3p'\,\delta(\varepsilon - \varepsilon') = p'^2\,dp'\,d\Omega'\,\frac{m}{p}\,\delta(p - p') \to m^2 v\,d\Omega' \tag{8.31}$$

and we introduce the quantity $W(p, \Omega')$

$$W(p, \Omega')\,d^3p' = \frac{n_d}{m^2}\,\sigma(v, \Omega')\,d^3p'\,\delta(\varepsilon - \varepsilon') \to n_d v\,\sigma(v, \Omega')\,d\Omega' \tag{8.32}$$

$W(p, \Omega')\,d^3p'$ is the number of particles scattered in d^3p' per second and per unit target volume for unit incident particle density. The number of collisions per second in $d^3r\,d^3p$ with a final momentum in d^3p' is then

$$\frac{d\mathcal{N}}{dt} = [d^3r\,d^3p]f(\vec{r}, \vec{p}, t)W(p, \Omega')d^3p'$$

By integrating over d^3p', we obtain from the above equation the contribution to $\mathcal{C}[f]$ of the particles leaving the volume element

$$\mathcal{C}_-[f] = f(\vec{r}, \vec{p}, t)\int d^3p'\,W(p, \Omega')\quad\left(= \frac{1}{\tau^*(p)}\,f(\vec{r}, \vec{p}, t)\right) \tag{8.33}$$

Conversely, collisions with $\vec{p}\,' \to \vec{p}$ will populate the volume element $\mathrm{d}^3r\,\mathrm{d}^3p$ (Figure 8.5(b)). The number of these collisions is given by

$$[\mathrm{d}^3r\,\mathrm{d}^3p]\int \mathrm{d}^3p'\, f(\vec{r},\vec{p}\,',t)W(p',\Omega) = [\mathrm{d}^3r\,\mathrm{d}^3p]\mathcal{C}_+[f]$$

The total collision term is the difference between entering and exiting contributions

$$\mathcal{C}[f] = \mathcal{C}_+[f] - \mathcal{C}_-[f]$$

The expression for $\mathcal{C}[f]$ simplifies if we note that the collision angle is the same for the collisions $\vec{p} \to \vec{p}\,'$ and $\vec{p}\,' \to \vec{p}$ and that $p = p'$ due to energy conservation

$$W(p,\Omega') = W(p',\Omega)$$

and therefore

$$\mathcal{C}[f] = \int \mathrm{d}^3p'\,\left[f(\vec{r},\vec{p}\,',t) - f(\vec{r},\vec{p},t)\right] W(p,\Omega') \tag{8.34}$$

By using (8.32) and introducing the shorthand notation $f = f(\vec{r},\vec{p},t)$ and $f' = f(\vec{r},\vec{p}\,',t)$, we write (8.34) as

$$\boxed{\mathcal{C}[f] = \int \mathrm{d}^3p'\,\left[f' - f\right] W(p,\Omega') = v n_d \int \mathrm{d}\Omega'\,\left[f' - f\right]\sigma(v,\Omega')} \tag{8.35}$$

8.2.3 Conservation laws and continuity equations

We should be able to show that the model satisfies number and energy conservation equations. We shall obtain these equations from the following preliminary results. Let $\chi(\vec{p})$ (or $\chi(\vec{r},\vec{p})$) be an arbitrary function, and define the functional, $I[\chi]$, of χ by

$$I[\chi] = \int \mathrm{d}^3p\,\chi(\vec{p})\mathcal{C}[f] \tag{8.36}$$

First we show that $I[\chi] = 0$ if χ depends only on the modulus p of $\vec{p}$. The demonstration is simple.[12] From the definition of $I[\chi]$ and using (8.35) we may write

$$I[\chi] = \int \mathrm{d}^3p\,\mathrm{d}^3p'\,\chi(\vec{p})W(p,\Omega')[f' - f]$$

[12] The physical interpretation of this result is as follows. The collision term is a balance term in the phase space volume element $\mathrm{d}^3r\,\mathrm{d}^3p$. Like all quantities dependent only on p, which are conserved in the collisions, its integral over d^3p should vanish. Therefore, this model has an infinite number of conservation laws. A realistic model should take into account the inelasticity of collisions that is necessary for reaching thermal equilibrium.

Exchanging the integration variables p and p' gives

$$I[\chi] = -\int \mathrm{d}^3p\,\mathrm{d}^3p'\,\chi(\vec{p}\,')W(p',\Omega)[f'-f]$$

Using $W(p,\Omega') = W(p',\Omega)$ the above two equations yield

$$I[\chi] = \frac{1}{2}\int \mathrm{d}^3p\,\mathrm{d}^3p'\,\left[\chi(\vec{p}) - \chi(\vec{p}\,')\right]W(p,\Omega')[f'-f]$$

Therefore, $I[\chi] = 0$ if χ depends only on the modulus of $\vec{p}$ since $p = p'$. In the special cases $\chi = 1$ and $\chi = \varepsilon(\vec{p}) = p^2/(2m)$ we obtain

$$\int \mathrm{d}^3p\,\mathcal{C}[f] = 0 \qquad \int \mathrm{d}^3p\,\varepsilon(\vec{p})\mathcal{C}[f] = 0 \tag{8.37}$$

Integrating Equation (8.28) over d^3p and using (8.37) we have

$$\frac{\partial n}{\partial t} + \vec{\nabla}_{\vec{r}}\cdot\int \mathrm{d}^3p\,\vec{v}f = 0 \tag{8.38}$$

We thus identify the particle current and the corresponding continuity equation

$$\boxed{\vec{j}_N = \int \mathrm{d}^3p\,\vec{v}f \qquad \frac{\partial n}{\partial t} + \vec{\nabla}\cdot\vec{j}_N = 0} \tag{8.39}$$

Multiplying Equation (8.28) by $\varepsilon(\vec{p})$, integrating over d^3p and using (8.37) leads to

$$\frac{\partial \epsilon}{\partial t} + \vec{\nabla}_{\vec{r}}\cdot\int \mathrm{d}^3p\,\varepsilon(\vec{p})\vec{v}f = 0 \tag{8.40}$$

where ϵ is the energy density,[13] which is given by

$$\epsilon = \int \mathrm{d}^3p\,\varepsilon(\vec{p})f = \int \mathrm{d}^3p\,\frac{\vec{p}^{\,2}}{2m}f \tag{8.41}$$

We define, with the help of (8.40), the energy current $\vec{j}_E$

$$\boxed{\vec{j}_E = \int \mathrm{d}^3p\,\varepsilon(\vec{p})\vec{v}f = \int \mathrm{d}^3p\,\frac{\vec{p}^{\,2}}{2m}\vec{v}f \qquad \frac{\partial \epsilon}{\partial t} + \vec{\nabla}\cdot\vec{j}_E = 0} \tag{8.42}$$

8.2.4 Linearization: Chapman–Enskog approximation

The collision term $\mathcal{C}[f]$ vanishes for any distribution function which is isotropic in $\vec{p}$, $f(\vec{r},p,t)$, since in such a case $f' = f$. In particular, it vanishes for any *local*

[13] The energy density ϵ should not be confused with the dispersion law $\varepsilon(\vec{p})$.

equilibrium distribution f_0, $\mathcal{C}[f_0] = 0$, where (cf. (3.141))

$$f_0(\vec{r}, \vec{p}, t) = \frac{1}{h^3} \exp\left(\alpha(\vec{r}, t) - \beta(\vec{r}, t)\frac{\vec{p}^{\,2}}{2m}\right) \equiv f_0(\vec{r}, p, t) \tag{8.43}$$

In this equation, $\beta(\vec{r}, t)$ and $\alpha(\vec{r}, t)$ are related to the local temperature $T(\vec{r}, t)$ and chemical potential $\mu(\vec{r}, t)$ by

$$\beta(\vec{r}, t) = \frac{1}{kT(\vec{r}, t)} \qquad \alpha(\vec{r}, t) = \frac{\mu(\vec{r}, t)}{kT(\vec{r}, t)} \tag{8.44}$$

If in Equation (8.35), the difference $[f' - f]$ is not small, the collision term will be of the order of $f/\tau^*(p)$ where $\tau^*(p)$ is a microscopic time of the order of 10^{-10} to 10^{-14} s. Then the collision term leads to a rapid exponential decrease, $\exp(-t/\tau^*)$, in the distribution function. This rapid decrease evolves the distribution function toward an almost local equilibrium distribution in a time $t \gtrsim \tau^*$. The collision term of the Boltzmann equation is solely responsible for this evolution, which takes place spatially over a distance of the order of the mean free path. Subsequently, the collision term becomes small since it vanishes for a local equilibrium distribution. The subsequent evolution is hydrodynamic, and this is what we shall now examine by linearizing Equation (8.35) near a local equilibrium distribution. We write the distribution function at $t = 0$ in the form $f = f_0 + \bar{f}$. The local equilibrium distribution f_0 must obey

$$n(\vec{r}, t = 0) = \int \mathrm{d}^3 p\, f_0(\vec{r}, \vec{p}, t = 0) \tag{8.45}$$

$$\epsilon(\vec{r}, t = 0) = \int \mathrm{d}^3 p\, \frac{p^2}{2m}\, f_0(\vec{r}, \vec{p}, t = 0) \tag{8.46}$$

where $n(\vec{r}, t = 0)$ and $\epsilon(\vec{r}, t = 0)$ are the initial particle and energy densities. These densities determine the local temperature and chemical potential or, equivalently, the local parameters α and β given by

$$n(\vec{r}) = \frac{1}{h^3}\, \mathrm{e}^{\alpha(\vec{r})} \left(\frac{2\pi m}{\beta(\vec{r})}\right)^{3/2} \qquad \epsilon(\vec{r}) = \frac{3}{2\beta(\vec{r})}\, n(\vec{r}) \tag{8.47}$$

where we have suppressed the time dependence. By construction, the deviation from local equilibrium, $\bar{f}$, satisfies

$$\int \mathrm{d}^3 p\, \bar{f} = \int \mathrm{d}^3 p\, \frac{p^2}{2m}\, \bar{f} = 0 \tag{8.48}$$

Since the equilibrium distribution function f_0 is isotropic, it does not contribute to the currents which are given solely by $\bar{f}$

$$\vec{j}_N = \int \mathrm{d}^3 p\, \vec{v} \bar{f} \qquad \vec{j}_E = \int \mathrm{d}^3 p\, \vec{v}\, \frac{p^2}{2m}\, \bar{f} \tag{8.49}$$

The currents vanish to first order in f_0 and, from the continuity equations, the time derivatives of n and ϵ (or equivalently α and β) also vanish as does $(\partial f_0/\partial t)$. We also have $\mathcal{C}[f_0] = 0$ but $\vec{\nabla} f_0 \neq 0$ (unless f_0 is a global equilibrium solution in which case the problem is not interesting) and so f_0 is not a solution of (8.28). Therefore, (8.28) becomes

$$(\vec{v} \cdot \vec{\nabla}) f_0 = \mathcal{C}[\bar{f}] \tag{8.50}$$

In orders of magnitude, we have $\mathcal{C}[\bar{f}] \sim \bar{f}/\tau^*$, and $\bar{f} \sim \tau^*(\vec{v} \cdot \vec{\nabla}) f_0$. It is possible to iterate the solution by calculating the currents from $\bar{f}$, which allows us to calculate the time derivatives of n and ϵ, which in turn yields $\partial f_0/\partial t$, but instead we stay with the approximation (8.50). We therefore work at an order of approximation where the drift $(\vec{v} \cdot \vec{\nabla}) f_0$ balances the collision term $\mathcal{C}[\bar{f}]$. The expansion parameter is $\tau^*/\tau \ll 1$ where τ is a characteristic time of macroscopic (or hydrodynamic) evolution.

We now calculate the collision term $\mathcal{C}[\bar{f}]$. At point $\vec{r}$, $\vec{\nabla} f_0$ is oriented in a direction that we define as the Oz axis

$$\vec{\nabla} f_0 = \hat{z}\, \frac{\partial f_0}{\partial z}$$

Let γ be the angle between $\vec{v}$ and Oz

$$(\vec{v} \cdot \vec{\nabla}) f_0 = v \cos\gamma \frac{\partial f_0}{\partial z} \tag{8.51}$$

Also, we take γ' as the angle between $\vec{p}\,'$ and Oz, θ' the angle between $\vec{p}$ and $\vec{p}\,'$ and φ' the azimuthal angle in the plane perpendicular to $\vec{p}$ (Figure 8.6). The collision term is given by (8.35)

$$\mathcal{C}[\bar{f}] = v n_d \int \mathrm{d}\Omega'\, \sigma(v, \Omega')[\bar{f}(\vec{r}, \vec{p}\,') - \bar{f}(\vec{r}, \vec{p})] \tag{8.52}$$

We assume the solution of (8.50) takes the form

$$\bar{f}(\vec{r}, \vec{p}) = g(z, p) \cos\gamma \tag{8.53}$$

Then, using

$$\cos\gamma' = \cos\theta' \cos\gamma + \sin\theta' \sin\gamma \cos\varphi' \qquad \mathrm{d}\Omega' = \mathrm{d}(\cos\theta')\, \mathrm{d}\varphi'$$

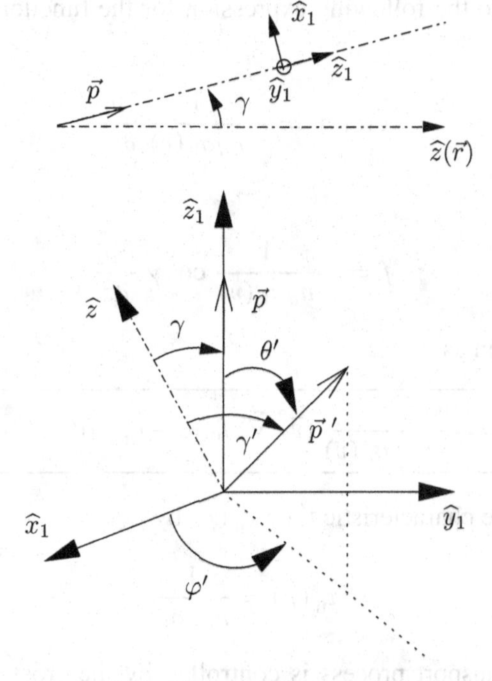

Figure 8.6 Conventions for the angles.

and the fact that $\sigma(v, \Omega') = \sigma(v, \theta')$,[14] Equation (8.52) becomes

$$\begin{aligned}\mathcal{C}[\bar{f}] &= v n_d g(z, p) \\ &\quad \times \int \mathrm{d}(\cos\theta')\, \mathrm{d}\varphi' \left[\cos\theta' \cos\gamma + \sin\theta' \sin\gamma \cos\varphi' - \cos\gamma\right] \sigma(v, \Omega') \\ &= -v n_d g(z, p) \cos\gamma \int \mathrm{d}\Omega' \,(1 - \cos\theta') \sigma(v, \Omega')\end{aligned}$$

We define the *transport cross section* σ_{tr} by[15]

$$\boxed{\sigma_{\mathrm{tr}}(v) = \int \mathrm{d}\Omega' \,(1 - \cos\theta') \sigma(v, \Omega')} \tag{8.54}$$

which when combined with (8.50) and (8.51) gives for the collision term

$$\mathcal{C}[\bar{f}] = -v n_d g(z, p) \cos\gamma\, \sigma_{\mathrm{tr}}(v) = v \cos\gamma\, \frac{\partial f_0}{\partial z}$$

[14] By rotation invariance, σ cannot depend on φ' unless the particles are polarized.

[15] This definition is specific to the process we are studying here. In other cases, the term $(1 - \cos\theta')$ may be replaced by a function $f(\cos\theta')$ satisfying $f(\pm 1) = 0$, and σ_{tr} may be different from σ_{tot} even for an isotropic differential cross section: see Section 8.4.3 where the transport cross section is defined with a $(1 - \cos^2\theta)$ factor.

This in turn leads to the following expression for the function $g(z, p)$ defined in (8.53)

$$g(z, p) = -\frac{1}{n_d \sigma_{\rm tr}(v)} \frac{\partial f_0}{\partial z} \tag{8.55}$$

and therefore

$$\bar{f} = -\frac{1}{n_d \sigma_{\rm tr}(v)} \cos\gamma \frac{\partial f_0}{\partial z}$$

which can be written as

$$\boxed{\bar{f} = -\frac{1}{n_d v \sigma_{\rm tr}(v)} (\vec{v} \cdot \vec{\nabla}) f_0 = -\tau^*_{\rm tr}(p)(\vec{v} \cdot \vec{\nabla}) f_0} \tag{8.56}$$

We have defined the characteristic time $\tau^*_{\rm tr}(p)$ by

$$\tau^*_{\rm tr}(p) = \frac{1}{n_d v \sigma_{\rm tr}} \tag{8.57}$$

We see that the transport process is controlled by the cross section $\sigma_{\rm tr}$ and not the total cross section and by $\tau^*_{\rm tr}(p)$ and not the collision time $\tau^*(p)$. When the cross section is isotropic (independent of Ω') we have $\sigma_{\rm tr}(v) = \sigma_{\rm tot}(v)$ and $\tau^*_{\rm tr}(p) = \tau^*(p)$. Since we usually consider such isotropic cases, we will take $\tau^*_{\rm tr}(p) = \tau^*(p)$ but keep in mind that these two time scales can be very different when the cross section is strongly anisotropic as in the case of Coulomb interactions.

8.2.5 Currents and transport coefficients

Equations (8.49) for the currents and (8.56) for $\bar{f}$ allow us to express the currents in the following form

$$\vec{j}_N = -\int \mathrm{d}^3 p\, \tau^*(p)\, \vec{v}(\vec{v} \cdot \vec{\nabla}) f_0 \tag{8.58}$$

$$\vec{j}_E = -\int \mathrm{d}^3 p\, \tau^*(p)\, \varepsilon(p) \vec{v}(\vec{v} \cdot \vec{\nabla}) f_0 \tag{8.59}$$

Using the result (A.34) for a function $g(|\vec{p}|)$ (see also Exercise 8.4.2)

$$\int \mathrm{d}^3 p\, v_\alpha v_\beta g(p) = \frac{4\pi}{3} \delta_{\alpha\beta} \int \mathrm{d}p\, p^2 v^2 g(p) = \frac{1}{3}\, \delta_{\alpha\beta} \int \mathrm{d}^3 p\, v^2 g(p) \tag{8.60}$$

the equations for the currents may be written as

$$\vec{j}_N = -\frac{1}{3}\int \mathrm{d}^3p\, v^2\tau^*(p)\vec{\nabla} f_0 \tag{8.61}$$

$$\vec{j}_E = -\frac{1}{3}\int \mathrm{d}^3p\, v^2\tau^*(p)\varepsilon(p)\vec{\nabla} f_0 \tag{8.62}$$

From the form of f_0 (8.43) we have

$$\vec{\nabla} f_0 \doteq \left(\vec{\nabla}\alpha - \varepsilon(p)\vec{\nabla}\beta\right) f_0$$

which leads to the final form for the currents

$$\vec{j}_N = \frac{1}{3k}\int \mathrm{d}^3p\, v^2\tau^*(p)\left[\vec{\nabla}\left(-\frac{\mu}{T}\right) + \varepsilon(p)\vec{\nabla}\left(\frac{1}{T}\right)\right] f_0 \tag{8.63}$$

$$\vec{j}_E = \frac{1}{3k}\int \mathrm{d}^3p\, v^2\tau^*(p)\varepsilon(p)\left[\vec{\nabla}\left(-\frac{\mu}{T}\right) + \varepsilon(p)\vec{\nabla}\left(\frac{1}{T}\right)\right] f_0 \tag{8.64}$$

By comparing with Equations (6.48) and (6.49) we deduce the transport coefficients

$$L_{NN} = \frac{1}{3k}\int \mathrm{d}^3p\, v^2\tau^*(p) f_0 \tag{8.65}$$

$$L_{EN} = \frac{1}{3k}\int \mathrm{d}^3p\, v^2\tau^*(p)\varepsilon(p) f_0 = L_{NE} \tag{8.66}$$

$$L_{EE} = \frac{1}{3k}\int \mathrm{d}^3p\, v^2\tau^*(p)\varepsilon^2(p) f_0 \tag{8.67}$$

The Onsager reciprocity relation $L_{EN} = L_{NE}$ is explicitly satisfied. It is also easy to verify the positivity condition (Exercise 8.4.3)

$$L_{EE}L_{NN} - L_{EN}^2 \geq 0 \tag{8.68}$$

In order to calculate explicitly these coefficients, we need $\tau^*(p)$. In the simple case where the mean free path ℓ is independent of p, we have $\tau^*(p) = m\ell/p = \ell/v$ and by defining $\tau^* = (8/3\pi)\ell/\langle v\rangle$ (Problem 8.6.2) we find

$$L_{NN} = \frac{\tau^*}{m} nT \qquad L_{EN} = \frac{2\tau^*}{m} nkT^2 \qquad L_{EE} = \frac{6\tau^*}{m} nk^2T^3 \tag{8.69}$$

Combining this with (6.27) (for a classical ideal gas), (6.50) and (6.57), we obtain the diffusion coefficient D and the electric and thermal conductivities σ_{el} and κ

$$D = \frac{\tau^*}{m} kT \qquad \sigma_{\mathrm{el}} = q^2\frac{\tau^*}{m} n \qquad \kappa = 2\frac{\tau^*}{m} nk^2T \tag{8.70}$$

which give the Franz–Wiedeman law

$$\frac{\kappa}{\sigma_{\text{el}}} = 2\frac{k^2}{q^2}T \simeq 1.5 \times 10^{-8}\,T \tag{8.71}$$

In this equation, q is the electron charge and the units are MKSA. The Franz–Wiedeman law predicts that the ratio $\kappa/\sigma_{\text{el}}$ is independent of the material and depends linearly on the temperature. Experimentally, this is well satisfied by semiconductors, but for metals we need to take into account the Fermi–Dirac statistics. The local equilibrium distribution f_0 then becomes[16]

$$f_0(\vec{r}, \vec{p}, t) = \frac{2}{h^3}\frac{1}{\exp\left[-\alpha(\vec{r}, t) + \beta(\vec{r}, t)p^2/2m\right] + 1} \tag{8.72}$$

where the factor 2 comes from the spin degree of freedom. Then the Franz–Wiedeman law becomes (Problem 8.5.2)

$$\frac{\kappa}{\sigma_{\text{el}}} = \frac{\pi^2}{3}\frac{k^2}{q^2}T \simeq 2.5 \times 10^{-8}\,T \tag{8.73}$$

which is also very well satisfied experimentally.

8.3 Boltzmann equation

8.3.1 Collision term

The Boltzmann equation governs the spatio-temporal evolution of the distribution function for a dilute gas. To the general assumptions of kinetic theory, which were discussed in Section 8.1.2, we must also add the condition of 'molecular chaos', which is crucial for writing the collision term.[17] This condition stipulates that the two-particle distribution function $f^{(2)}$, which contains the correlations, can be written as a product of one-particle distribution functions

$$f^{(2)}(\vec{r}_1, \vec{p}_1; \vec{r}_2, \vec{p}_2; t) = f(\vec{r}_1, \vec{p}_1, t)\, f(\vec{r}_2, \vec{p}_2, t) \tag{8.74}$$

In other words, the joint distribution is a product of the individual distributions: we ignore two-particle correlations (and *a fortiori* those of higher order). However, it is important to place this molecular chaos hypothesis in the general framework of

[16] The reader will correctly object that, due to the uncertainty principle, we cannot give simultaneously sharp values to the position and the momentum in a situation where quantum effects are important. However, it can be shown that (8.172) is valid provided the typical scales on which $\vec{r}$ and t vary are large enough. See Problem 8.6.7.

[17] We must also assume that the dilute gas is mono-atomic, so that the collisions are always elastic. There cannot be any transfer of kinetic energy toward the internal degrees of freedom.

the approach of Chapter 2. Among all possible dynamic variables, we focus only on the specific variable $\mathcal{A}(\vec{r}, \vec{p}, t)$

$$\mathcal{A}(\vec{r}, \vec{p}, t) = \sum_{j=1}^{N} \delta(\vec{r} - \vec{r}_j(t))\delta(\vec{p} - \vec{p}_j(t)) \tag{8.75}$$

The sum is over the total number of particles N and $\vec{r}_j(t)$ and $\vec{p}_j(t)$ are respectively the position and momentum of particle j. The average of $\mathcal{A}$ is, in fact, the distribution function f, which is given by an ensemble average[18] (Exercise 6.4.1) which generalizes (3.75)

$$f(\vec{r}, \vec{p}, t) = \langle \mathcal{A}(\vec{r}, \vec{p}, t)\rangle = \left\langle \sum_{j=1}^{N} \delta(\vec{r} - \vec{r}_j(t))\delta(\vec{p} - \vec{p}_j(t))\right\rangle \tag{8.76}$$

Knowing the distribution function is equivalent to knowing the average values of a number of dynamic variables, in fact an infinite number. The index i used in Chapter 2 to label these variables corresponds here to $(\vec{r}, \vec{p})$. As in Chapter 2, the ensemble average of the variable $\mathcal{A}(\vec{r}, \vec{p}, t)$ will be constrained to take the value $f(\vec{r}, \vec{p}, t)$ at every instant. The other dynamic variables, i.e. the correlations, are not constrained.

We start with Equation (8.28) for the function $f(\vec{r}, \vec{p}_1, t) \equiv f_1$

$$\frac{\partial f_1}{\partial t} + \vec{v}_1 \cdot \vec{\nabla}_{\vec{r}} f_1 + \vec{F}(\vec{r}) \cdot \vec{\nabla}_{\vec{p}_1} f_1 = \mathcal{C}[f_1] \tag{8.77}$$

The collision term $\mathcal{C}[f_1]$ is evaluated from the collision cross section. Let us examine, in the centre-of-mass frame, an elastic collision between two particles of equal mass, which in the laboratory frame is written as $\vec{p}_1 + \vec{p}_2 \to \vec{p}_3 + \vec{p}_4$, and let $\vec{P}$ be the total momentum. Momenta in the centre-of-mass frame will have a 'prime'. We have

$$\begin{aligned}
\vec{P} &= \vec{p}_1 + \vec{p}_2 = \vec{p}_3 + \vec{p}_4 \\
\vec{p}_3^{\,\prime} &= \vec{p}_3 - \frac{1}{2}\vec{P} = \frac{1}{2}(\vec{p}_3 - \vec{p}_4) \\
\vec{p}_4^{\,\prime} &= \vec{p}_4 - \frac{1}{2}\vec{P} = -\frac{1}{2}(\vec{p}_3 - \vec{p}_4) = -\vec{p}_3^{\,\prime}
\end{aligned}$$

[18] Ensemble averaging is effected by taking many copies of the same physical system with the same macroscopic characteristics (in this case the same distribution function) but with different microscopic configurations. The ensemble average in (8.75) takes the place of the average with respect to the Boltzmann weight in (3.75).

and similar relations for $\vec{p}_1{}'$ and $\vec{p}_2{}'$. The energies are given in terms of the momenta by

$$\varepsilon_3 = \frac{\vec{p}_3^{\,2}}{2m} = \frac{1}{2m}\left(\vec{p}_3{}' + \frac{1}{2}\vec{P}\right)^2 = \frac{1}{2m}(\vec{p}_3{}')^2 + \frac{1}{2m}\vec{p}_3{}' \cdot \vec{P} + \frac{1}{8m}\vec{P}^2$$
$$\varepsilon_4 = \frac{\vec{p}_4^{\,2}}{2m} = \frac{1}{2m}\left(-\vec{p}_3{}' + \frac{1}{2}\vec{P}\right)^2 = \frac{1}{2m}(\vec{p}_3{}')^2 - \frac{1}{2m}\vec{p}_3{}' \cdot \vec{P} + \frac{1}{8m}\vec{P}^2$$

Energy conservation is ensured by the δ function

$$\delta(\varepsilon_1 + \varepsilon_2 - \varepsilon_3 - \varepsilon_4) = \delta\left(\frac{(\vec{p}_1{}')^2}{m} - \frac{(\vec{p}_3{}')^2}{m}\right) \tag{8.78}$$

As in the Boltzmann–Lorentz model, it will be convenient to define a quantity $\overline{W}$ that is related to the cross section by

$$\overline{W}(\vec{p}_1, \vec{p}_2; \vec{p}_3, \vec{p}_4) = \frac{4}{m^2}\sigma(|\vec{v}_1 - \vec{v}_2|, \Omega') \tag{8.79}$$

Let us calculate the integral

$$\begin{aligned}
\frac{\mathrm{d}\mathcal{N}}{\mathrm{d}t} &= \int \mathrm{d}^3 p_3\, \mathrm{d}^3 p_4\, \overline{W}\delta(\vec{p}_1 + \vec{p}_2 - \vec{p}_3 - \vec{p}_4)\delta(\varepsilon_1 + \varepsilon_2 - \varepsilon_3 - \varepsilon_4) \\
&= \int \mathrm{d}^3 p_3'\, \overline{W}\,\delta\left(\frac{(\vec{p}_1{}')^2}{m} - \frac{(\vec{p}_3{}')^2}{m}\right) \\
&= \int \mathrm{d}p_3'\,(p_3')^2\,\mathrm{d}\Omega'\,\overline{W}\,\delta\left(\frac{(\vec{p}_1{}')^2}{m} - \frac{(\vec{p}_3{}')^2}{m}\right) \\
&= \frac{m}{2}p_1' \int \mathrm{d}\Omega'\,\overline{W} = \frac{2p_1'}{m}\int \mathrm{d}\Omega'\,\sigma(|\vec{v}_1 - \vec{v}_2|, \Omega')
\end{aligned}$$

But $2p_1'/m$ is just the absolute value of the relative velocity

$$|\vec{v}_1 - \vec{v}_2| = \frac{|\vec{p}_1 - \vec{p}_2|}{m} = 2\frac{p_1'}{m}$$

Consequently we have

$$\frac{\mathrm{d}\mathcal{N}}{\mathrm{d}t} = |\vec{v}_1 - \vec{v}_2| \int \mathrm{d}\Omega'\,\sigma(|\vec{v}_1 - \vec{v}_2|, \Omega') = |\vec{v}_1 - \vec{v}_2|\sigma_{\mathrm{tot}}(|\vec{v}_1 - \vec{v}_2|) \tag{8.80}$$

Recall that from (8.7) $\mathrm{d}\mathcal{N}/\mathrm{d}t$ is the number of collisions per second for unit densities of target and incident particles. In other words the quantity $W\mathrm{d}^3 p_3\, \mathrm{d}^3 p_4$, defined by

$$W(\vec{p}_1, \vec{p}_2 \to \vec{p}_3, \vec{p}_4)\, \mathrm{d}^3 p_3\, \mathrm{d}^3 p_4 = \overline{W}\, \delta(\vec{p}_1 + \vec{p}_2 - \vec{p}_3 - \vec{p}_4) \\ \delta(\varepsilon_1 + \varepsilon_2 - \varepsilon_3 - \varepsilon_4)\mathrm{d}^3 p_3\, \mathrm{d}^3 p_4 \quad (8.81)$$

is the number of collisions per second with final momenta in $\mathrm{d}^3 p_3\, \mathrm{d}^3 p_4$ for unit densities of target and incident particles. Under these conditions, the term $\mathrm{d}^3 r\, \mathrm{d}^3 p_1 \mathcal{C}_-[f_1]$, which counts the number of particles leaving the volume $\mathrm{d}^3 r\, \mathrm{d}^3 p_1$, is obtained by multiplying $W\mathrm{d}^3 p_3\, \mathrm{d}^3 p_4$ by the number of particles in $\mathrm{d}^3 r\, \mathrm{d}^3 p_1$, namely $f(\vec{r}, \vec{p}_1, t)\mathrm{d}^3 r\, \mathrm{d}^3 p_1$, and then integrating over the distribution of incident particles and over all the final configurations of $\vec{p}_3$ and $\vec{p}_4$

$$\mathcal{C}_-[f_1]\, \mathrm{d}^3 r\, \mathrm{d}^3 p_1 = f(\vec{r}, \vec{p}_1, t)\mathrm{d}^3 r\, \mathrm{d}^3 p_1 \\ \times \int \mathrm{d}^3 p_2\, \mathrm{d}^3 p_3\, \mathrm{d}^3 p_4\, f(\vec{r}, \vec{p}_2, t) W(\vec{p}_1, \vec{p}_2 \to \vec{p}_3, \vec{p}_4) \quad (8.82)$$

In the same way, the number of particles entering $\mathrm{d}^3 r\, \mathrm{d}^3 p_1$ is

$$\mathcal{C}_+[f_1]\, \mathrm{d}^3 r\, \mathrm{d}^3 p_1 = \mathrm{d}^3 r\, \mathrm{d}^3 p_1 \int \mathrm{d}^3 p_2\, \mathrm{d}^3 p_3\, \mathrm{d}^3 p_4\, f(\vec{r}, \vec{p}_3, t) \\ \times f(\vec{r}, \vec{p}_4, t) W(\vec{p}_3, \vec{p}_4 \to \vec{p}_1, \vec{p}_2) \quad (8.83)$$

We note that the molecular chaos hypothesis (8.74) was used in the two cases to decouple incident and target particles.

Before adding these two terms we are going to exploit some symmetry properties of the function W. The interactions which intervene in the collisions are electromagnetic and are known to be invariant under rotation (R), space inversion or parity (P) and time reversal (T). These invariances lead to the following symmetries of W (see Figure 8.7):

(*i*) rotation: $W(R\vec{p}_1, R\vec{p}_2 \to R\vec{p}_3, R\vec{p}_4) = W(\vec{p}_1, \vec{p}_2 \to \vec{p}_3, \vec{p}_4)$,
(*ii*) parity: $W(-\vec{p}_1, -\vec{p}_2 \to -\vec{p}_3, -\vec{p}_4) = W(\vec{p}_1, \vec{p}_2 \to \vec{p}_3, \vec{p}_4)$,
(*iii*) time reversal: $W(-\vec{p}_3, -\vec{p}_4 \to -\vec{p}_1, -\vec{p}_2) = W(\vec{p}_1, \vec{p}_2 \to \vec{p}_3, \vec{p}_4)$.

In the above, $R\vec{p}$ is the result of rotating $\vec{p}$ by R and the effect of time reversal is to change the sign of the momenta. Combining properties (*ii*) and (*iii*) yields

$$W(\vec{p}_3, \vec{p}_4 \to \vec{p}_1, \vec{p}_2) = W(\vec{p}_1, \vec{p}_2 \to \vec{p}_3, \vec{p}_4) \quad (8.84)$$

The collision $\vec{p}_3 + \vec{p}_4 \to \vec{p}_1 + \vec{p}_2$ is called the inverse collision of $\vec{p}_1 + \vec{p}_2 \to \vec{p}_3 + \vec{p}_4$. By using the symmetry relation (8.84), which relates a collision to its inverse, we can combine $\mathcal{C}_-[f_1]$ and $\mathcal{C}_+[f_1]$ to obtain $\mathcal{C}[f_1] = \mathcal{C}_+[f_1] - \mathcal{C}_-[f_1]$ in

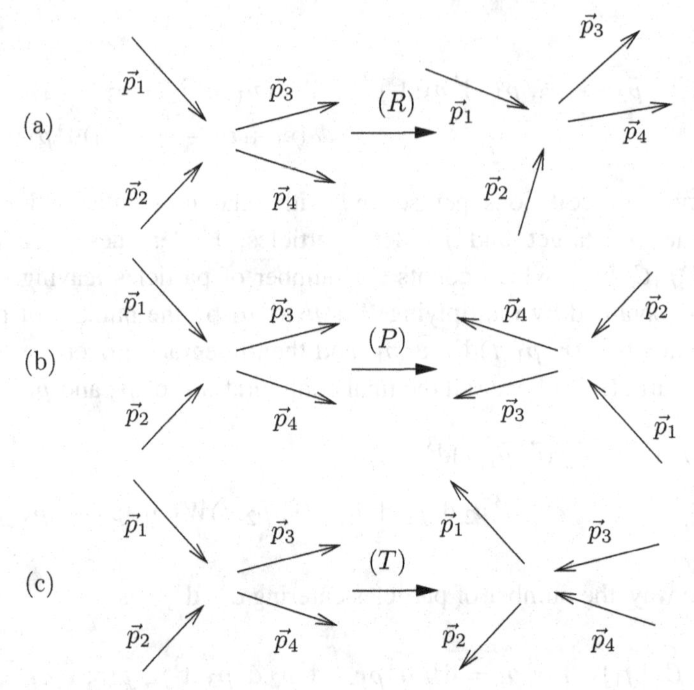

Figure 8.7 The effects of rotation, space inversion and time reversal symmetries on collisions.

the form

$$\mathcal{C}[f_1] = \int \prod_{i=2}^{4} \mathrm{d}^3 p_i \, W(\vec{p}_1, \vec{p}_2 \to \vec{p}_3, \vec{p}_4)[f_3 f_4 - f_1 f_2] \tag{8.85}$$

with the notation $f_i = f(\vec{r}, \vec{p}_i, t)$. The Boltzmann equation then takes on its final form

$$\boxed{\begin{aligned} \frac{\partial f_1}{\partial t} + \vec{v}_1 \cdot \vec{\nabla}_{\vec{r}} f_1 + \vec{F}(\vec{r}) \cdot \vec{\nabla}_{\vec{p}_1} f_1 &= \int \prod_{i=2}^{4} \mathrm{d}^3 p_i \, W(\vec{p}_1, \vec{p}_2 \to \vec{p}_3, \vec{p}_4)[f_3 f_4 - f_1 f_2] \\ &= \int \mathrm{d}\Omega' \int \mathrm{d}^3 p_2 \, \sigma(|\vec{v}_1 - \vec{v}_2|, \Omega') |\vec{v}_1 - \vec{v}_2| [f_3 f_4 - f_1 f_2] \end{aligned}} \tag{8.86}$$

One should be careful to integrate over only half the phase space to take into account the fact that the particles are identical. To make the connection with the Boltzmann–Lorentz model, it is sufficient to take $f_2 = f_4 = n_d \delta(\vec{p})$. The Boltzmann equation explicitly breaks time reversal invariance: $f(\vec{r}, -\vec{p}, -t)$ does not obey the Boltzmann equation since the drift term changes sign while the collision term does not.

8.3.2 Conservation laws

We have *a priori* five conservation laws: particle number (or mass), energy and the three components of momentum. These laws, in the form of continuity equations for the densities and currents, are the result of the conservation of mass, energy and momentum in each collision $\vec{p}_1 + \vec{p}_2 \to \vec{p}_3 + \vec{p}_4$. The argument is a simple generalization of the one given for the Boltzmann–Lorentz model. Let $\chi(\vec{p})$ be a conserved quantity in the collision $\vec{p}_1 + \vec{p}_2 \to \vec{p}_3 + \vec{p}_4$

$$\chi_1 + \chi_2 = \chi_3 + \chi_4 \tag{8.87}$$

where we use the notation $\chi_i = \chi(\vec{p}_i)$. We shall demonstrate the following preliminary result

$$I[\chi] = \int \mathrm{d}^3 p_1\, \chi(\vec{p}_1)\mathcal{C}[f_1] = 0 \tag{8.88}$$

Taking into account Equation (8.85) for the collision term,[19] we have

$$I[\chi] = \int \prod_{i=1}^{4} \mathrm{d}^3 p_i\, \chi_1 W(12 \to 34)[f_3 f_4 - f_1 f_2]$$

Since the particles 1 and 2 are identical, we have $W(12 \to 34) = W(21 \to 34)$, and changing variables $\vec{p}_1 \rightleftarrows \vec{p}_2$, we obtain a second expression for $I[\chi]$

$$I[\chi] = \int \prod_{i=1}^{4} \mathrm{d}^3 p_i\, \chi_2 W(12 \to 34)[f_3 f_4 - f_1 f_2]$$

A third expression is obtained by exchanging (12) and (34) and by using the property of the inverse collision $W(34 \to 12) = W(12 \to 34)$

$$I[\chi] = -\int \prod_{i=1}^{4} \mathrm{d}^3 p_i\, \chi_3 W(12 \to 34)[f_3 f_4 - f_1 f_2]$$

and a fourth by exchanging particles 3 and 4. Finally we obtain

$$I[\chi] = \frac{1}{4} \int \prod_{i=1}^{4} \mathrm{d}^3 p_i\, [\chi_1 + \chi_2 - \chi_3 - \chi_4] W(12 \to 34)[f_3 f_4 - f_1 f_2] \tag{8.89}$$

Consequently, from (8.87), $I[\chi] = 0$ if the quantity χ is conserved in the collision (see Footnote 12 for the physical interpretation of this result). This demonstration is unchanged if χ also depends on $\vec{r}$. We multiply the Boltzmann equation (8.86) by the conserved quantity $\chi(\vec{r}, \vec{p}_1)$, change notation $\vec{p}_1 \to \vec{p}$ and integrate over $\vec{p}$.

[19] We use the notation $W(12 \to 34) = W(\vec{p}_1, \vec{p}_2 \to \vec{p}_3, \vec{p}_4)$.

We obtain

$$\int \mathrm{d}^3p\, \chi(\vec{r},\vec{p})\left(\frac{\partial f}{\partial t} + \vec{v}\cdot\vec{\nabla}_{\vec{r}} f + \vec{F}(\vec{r})\cdot\vec{\nabla}_{\vec{p}} f\right) = 0 \tag{8.90}$$

This may be explicitly expressed in terms of physical quantities. To do this we use[20]

$$\int \mathrm{d}^3p\, \chi v_\alpha\, \partial_\alpha f = \partial_\alpha \int \mathrm{d}^3p\, (\chi v_\alpha f) - \int \mathrm{d}^3p\, f v_\alpha\, \partial_\alpha \chi \tag{8.91}$$

and

$$\begin{aligned}\int \mathrm{d}^3p\, \chi F_\alpha\, \partial_{p_\alpha} f &= \int \mathrm{d}^3p\, \partial_{p_\alpha}(\chi F_\alpha f) - \int \mathrm{d}^3p\, \left(\partial_{p_\alpha}\chi\right) F_\alpha f - \int \mathrm{d}^3p\, \chi\, \left(\partial_{p_\alpha} F_\alpha\right) f \\ &= -\int \mathrm{d}^3p\, \left(\partial_{p_\alpha}\chi\right) F_\alpha f\end{aligned} \tag{8.92}$$

The second line in (8.92) is obtained by noting that the first term in the first line is the integral of a divergence that can be written as a surface integral which vanishes since $f \to 0$ rapidly for $|\vec{p}| \to \infty$. In addition we assume, for simplicity, that the force does not depend on the velocity, which eliminates the third term.

From (8.2) the average value of χ is

$$n\langle\chi\rangle = \int \mathrm{d}^3p\, \chi f \tag{8.93}$$

and the current is given by a simple convection term since we neglect interactions

$$\vec{j}_\chi = \int \mathrm{d}^3p\, \vec{v}\, \chi f = n\langle\vec{v}\chi\rangle \tag{8.94}$$

The above results allow us to put (8.90) in the final form

$$\boxed{\frac{\partial}{\partial t}(n\langle\chi\rangle) + \partial_\alpha(n\langle v_\alpha \chi\rangle) = n\,\langle v_\alpha\, \partial_\alpha \chi\rangle + n\left\langle F_\alpha\, \partial_{p_\alpha}\chi\right\rangle} \tag{8.95}$$

In the absence of an external source, Equation (8.95) has the form of a continuity equation $\partial_t \rho_\chi + \vec{\nabla}\cdot\vec{j}_\chi = 0$.

The velocity $\vec{v}$ may be decomposed into two components $\vec{v} = \vec{u} + \vec{w}$. The first component is an average velocity $\vec{u} = \langle\vec{v}\rangle$, which is nothing more than the flow velocity of the fluid introduced in Section 6.3.1. The second component, $\vec{w}$, which has zero average, $\langle\vec{w}\rangle = 0$, is the velocity measured in the fluid rest frame and is, therefore, the velocity due to thermal fluctuations. By taking $\chi = m$, we obtain a

[20] Repeated indexes are summed.

continuity equation for the mass (see Table 6.1)

$$\frac{\partial \rho}{\partial t} + \vec{\nabla} \cdot \vec{g} = 0 \tag{8.96}$$

with $\rho = nm$ and $\vec{g} = \rho\langle\vec{v}\rangle = \rho\vec{u}$. The momentum continuity equation is obtained by taking $\chi = mv_\beta$, $\partial_{p_\alpha}\chi = \delta_{\alpha\beta}$

$$\frac{\partial}{\partial t}\left(\rho\langle v_\beta\rangle\right) + \partial_\alpha\left(\rho\langle v_\alpha v_\beta\rangle\right) = nF_\beta \tag{8.97}$$

where $nF_\beta = f_\beta$ is the force density. This equation allows us to obtain the momentum current $T_{\alpha\beta}$ (compare with (6.69) and (6.74))

$$T_{\alpha\beta} = \rho\langle v_\alpha v_\beta\rangle \tag{8.98}$$

By writing $\vec{v} = \vec{u} + \vec{w}$, we obtain

$$T_{\alpha\beta} = \rho u_\alpha u_\beta + \rho\langle w_\alpha w_\beta\rangle$$

and thus the pressure tensor $\mathcal{P}_{\alpha\beta}$ (see Table 6.1) is

$$\mathcal{P}_{\alpha\beta} = \rho\langle w_\alpha w_\beta\rangle \tag{8.99}$$

We note that the trace of the pressure tensor has the remarkable value

$$\mathcal{P}_{\alpha\alpha} = nm\langle\vec{w}^{\,2}\rangle = \rho\langle\vec{w}^{\,2}\rangle \tag{8.100}$$

Finally, let us take χ as the energy in the absence of external forces: $\chi = m\vec{v}^{\,2}/2$. The energy density is

$$\epsilon = \frac{1}{2}\rho\langle\vec{v}^{\,2}\rangle \tag{8.101}$$

and the associated current is

$$\vec{j}_E = \frac{1}{2}\rho\langle\vec{v}^{\,2}\vec{v}\rangle \tag{8.102}$$

Equation (8.95) ensures the conservation of energy in the absence of external forces

$$\frac{\partial \epsilon}{\partial t} + \vec{\nabla} \cdot \vec{j}_E = 0 \tag{8.103}$$

It is possible to relate the energy current to the heat current $\vec{j}^{\,\prime}_E = \vec{j}_Q$, which is the current measured in the fluid rest frame. Using $\vec{v} = \vec{u} + \vec{w}$, it is easy to show that (Exercise 8.4.5)

$$j^E_\alpha = \epsilon u_\alpha + \sum_\beta u_\beta \mathcal{P}_{\alpha\beta} + j^{E'}_\alpha$$

which is just Equation (6.83). The local temperature is defined in the fluid rest frame by

$$\boxed{\frac{1}{2}m\langle\vec{w}^{\,2}\rangle = \frac{3}{2}kT(\vec{r},t)} \tag{8.104}$$

which gives

$$\rho\langle\vec{w}^2\rangle = 3nkT = 3\mathcal{P}$$

and by comparing with (8.100) we have

$$\sum_\alpha \mathcal{P}_{\alpha\alpha} - 3\mathcal{P} = 0 \tag{8.105}$$

This property of the trace of the pressure tensor allows us to show, when used in (6.88), that the bulk viscosity ζ vanishes for an ideal mono-atomic gas.

8.3.3 H-theorem

We end our discussion of this section with a demonstration of the increase of the Boltzmann entropy. We adopt the framework defined in Chapter 2 by considering a set of dynamic variables $\mathcal{A}_i$ whose average values A_i are fixed.[21] This allows us to construct the corresponding Boltzmann (or relevant) entropy. In the present case, the dynamic variables are the one-particle distributions (8.75) whose average values are the distribution functions $f(\vec{r},\vec{p},t)$. The index i in Chapter 2 here represents the variables $(\vec{r},\vec{p})$ which label the dynamic variables. The Lagrange multipliers λ_i become $\lambda(\vec{r},\vec{p})$ with the corresponding notation[22]

$$i \to (\vec{r},\vec{p}) \qquad \lambda_i \to \lambda(\vec{r},\vec{p}) \qquad \sum_i \to \int \mathrm{d}^3r\,\mathrm{d}^3p$$

$$\sum_i \lambda_i A_i \to \int \mathrm{d}^3r\,\mathrm{d}^3p\,\lambda(\vec{r},\vec{p})\sum_{j=1}^{N}\delta(\vec{r}-\vec{r}_j)\,\delta(\vec{p}-\vec{p}_j) = \sum_{j=1}^{N}\lambda(\vec{r}_j,\vec{p}_j)$$

Recall the semi-classical expression (3.42) for the trace, which we generalize to the grand canonical ensemble

$$\mathrm{Tr} = \sum_N \frac{1}{N!}\int\prod_{j=1}^{N}\frac{\mathrm{d}^3r_j\,\mathrm{d}^3p_j}{h^3} \tag{8.106}$$

[21] Since our discussion is classical, we use the notation $\mathcal{A}$ for a dynamic variable and not A.
[22] We have taken in (8.75) $t=0$ and $\vec{r}_i(t=0)=\vec{r}_i$, $\vec{p}_i(t=0)=\vec{p}_i$.

The grand partition function is

$$\mathcal{Q} = \sum_N \frac{1}{N!} \int \prod_{j=1}^{N} \left(\frac{d^3 r_j \, d^3 p_j}{h^3} \exp[\lambda(\vec{r}_j, \vec{p}_j)] \right)$$
$$= \sum_N \frac{1}{N!} \left[\int \frac{d^3 r' \, d^3 p'}{h^3} \exp[\lambda(\vec{r}\,', \vec{p}\,')] \right]^N = \exp\left(\frac{1}{h^3} \int d^3 r' \, d^3 p' \exp[\lambda(\vec{r}\,', \vec{p}\,')] \right)$$

which yields

$$\ln \mathcal{Q} = \frac{1}{h^3} \int d^3 r' \, d^3 p' \, e^{\lambda(\vec{r}\,', \vec{p}\,')} \tag{8.107}$$

The distribution function is given by the functional derivative of $\ln \mathcal{Q}$ (see Appendix A.6)

$$f(\vec{r}, \vec{p}) = \frac{\delta \ln \mathcal{Q}}{\delta \lambda(\vec{r}, \vec{p})} = \frac{1}{h^3} \exp\left[\lambda(\vec{r}, \vec{p})\right] \tag{8.108}$$

which allows us to identify the Lagrange multiplier $\lambda(\vec{r}, \vec{p})$

$$\lambda(\vec{r}, \vec{p}) = \ln(h^3 f(\vec{r}, \vec{p})) \tag{8.109}$$

We finally arrive at the Boltzmann entropy by using (2.65)

$$S_B = k\left(\ln \mathcal{Q} - \sum_i \lambda_i A_i\right)$$

which gives in the present case

$$\boxed{S_B = k \int d^3 r \, d^3 p \, f(\vec{r}, \vec{p}) \left[1 - \ln(h^3 f(\vec{r}, \vec{p}))\right]} \tag{8.110}$$

Equation (8.110) permits the determination of the entropy density, s_B, and current, $\vec{j}_S$, by using once more the fact that this current is purely convective within kinetic theory

$$s_B = k \int d^3 p \, f(\vec{r}, \vec{p}) \left[1 - \ln(h^3 f(\vec{r}, \vec{p}))\right] \tag{8.111}$$

$$\vec{j}_S = k \int d^3 p \, \vec{v} f(\vec{r}, \vec{p}) \left[1 - \ln(h^3 f(\vec{r}, \vec{p}))\right] \tag{8.112}$$

We note that we are working with the classical approximation where $h^3 f \ll 1$. Therefore, the 1 in the brackets in (8.111) and (8.112) may be dropped.

This form of the Boltzmann entropy (8.110) may be obtained with a more intuitive argument by writing the density in phase space (see Section 2.2.2) as a product of one-particle distribution functions

$$D_N(\vec{r}_1, \vec{p}_1; \ldots; \vec{r}_N, \vec{p}_N) \propto \prod_{i=1}^{N} f(\vec{r}_i, \vec{p}_i)$$

which is equivalent to ignoring correlations. We thus obtain Equation (8.110) for the entropy up to factors of 1 and h. Boltzmann called this expression $-H$

$$H(t) = k \int \mathrm{d}^3 r\, \mathrm{d}^3 p\, f(\vec{r}, \vec{p}, t) \ln f(\vec{r}, \vec{p}, t)$$

hence the name 'H-theorem' for Equation (8.113) below.

Having expressed the entropy in terms of f, we use the Boltzmann equation (8.86) to write the entropy continuity equation. To this end, we multiply the two sides of (8.86) by $-k \ln(h^3 f_1)$ and integrate over $\vec{p}_1$. Using

$$\mathrm{d}\left(f\left[1 - \ln(h^3 f(\vec{r}, \vec{p}))\right]\right) = -\ln[h^3 f(\vec{r}, \vec{p})]\mathrm{d}f$$

we obtain

$$\begin{aligned}
\frac{\partial s_{\mathrm{B}}}{\partial t} + \vec{\nabla} \cdot \vec{j}_S &= -k \int \prod_{i=1}^{4} \mathrm{d}^3 p_i \, \ln(h^3 f_1)[f_3 f_4 - f_1 f_2] W(\vec{p}_1, \vec{p}_2 \to \vec{p}_3, \vec{p}_4) \\
&= \frac{k}{4} \int \prod_{i=1}^{4} \mathrm{d}^3 p_i \, \ln \frac{f_1 f_2}{f_3 f_4} [f_1 f_2 - f_3 f_4] W(\vec{p}_1, \vec{p}_2 \to \vec{p}_3, \vec{p}_4) \\
&= \frac{k}{4} \int \prod_{i=1}^{4} \mathrm{d}^3 p_i \, \ln \frac{f_1 f_2}{f_3 f_4} \left[\frac{f_1 f_2}{f_3 f_4} - 1\right] f_3 f_4 \, W(\vec{p}_1, \vec{p}_2 \to \vec{p}_3, \vec{p}_4) \geq 0
\end{aligned} \tag{8.113}$$

To obtain the second line of (8.113) we have used the same symmetry properties used to obtain (8.89). The last inequality comes from the fact that $(x-1)\ln x \geq 0$ for all x. Therefore, there is an entropy source on the right hand side of this continuity equation. This means that the total entropy increases:[23] $\mathrm{d}S_{\mathrm{B}}/\mathrm{d}t \geq 0$. The above calculation shows that the origin of the source term in the continuity equation is the collisions suffered by the particles. It is, therefore, these collisions that lead to the entropy increase.

That entropy increases becomes clear if we reason in terms of the relevant entropy. The available information is contained in the one-particle distribution functions. Collisions create correlations, but information about these correlations is lost in the picture using one-particle functions: there is a leak of information toward correlations.

[23] This statement is valid on the average. One may observe fluctuations during which $S_{\mathrm{B}}(t)$ decreases for a short period of time. Such fluctuations are found in molecular dynamics simulations.

Let f_0 be a local equilibrium distribution corresponding to local temperature $T(\vec{r}, t)$, local chemical potential $\mu(\vec{r}, t)$ and local fluid velocity $\vec{u}(\vec{r}, t)$,

$$f_0(\vec{r}, \vec{p}, t) = \frac{1}{h^3} \exp\left(\alpha(\vec{r}, t) - \beta(\vec{r}, t)\frac{(\vec{p} - m\vec{u}(\vec{r}, t))^2}{2m}\right) \tag{8.114}$$

with $\alpha = \mu/kT$ and $\beta = 1/kT$. As implied by Equation (8.118), the collision term vanishes, $C[f_0] = 0$, and $\mathrm{d}S_{\mathrm{tot}}/\mathrm{d}t = 0$. Conversely, one can show (Exercise 8.5.6) that $f(\vec{r}, \vec{p}, t)$ may be written in the form (8.114) if the collision term vanishes. *A fortiori* the entropy is constant for a global equilibrium distribution.

In order to avoid confusion, we revisit some key points in the preceding discussion. The first observation is that the increase in the Boltzmann entropy $\mathrm{d}S_{\mathrm{B}}/\mathrm{d}t \geq 0$ rests on the assumption that there are no correlations between molecules prior to collision. Of course, two molecules are correlated after suffering a collision. It is for this reason that there is an asymmetry in the temporal evolution: times 'before' and 'after' a collision are not equivalent. However, because the gas is dilute, two such molecules have negligible probability of colliding again, rather, they will collide with other molecules with which they are not correlated and the molecular chaos hypothesis continues to hold beyond the initial time. However, by waiting long enough, one would observe 'long time tails', namely a power law, and not exponential, dependence of the time-correlation functions, which behave as $t^{-3/2}$ for $t \gg \tau^*$. These long time tails are due to the long lived hydrodynamic modes of the fluid (see [108], Chapter 11). The second remark concerns the comparison between the Boltzmann and thermodynamic entropies. The quantity S_{B} is indeed the Boltzmann (or relevant) entropy in the sense of the construction in Chapter 2, but the density s_{B} in (8.111) should not be identified with a thermodynamic entropy density at local equilibrium. In fact, the distribution $f(\vec{r}, \vec{p})$ is *a priori* not a Maxwell distribution and does not allow the definition of a temperature: the kinetic description is more detailed than the thermodynamic one and cannot, in general, be reduced to it. After a time of the order of the collision time, a local equilibrium of the type (8.114) is established. A straightforward calculation allows one to check that the density of Boltzmann entropy defined in (8.111) coincides with the local thermodynamic entropy (3.17). Under these conditions the collision term vanishes. Thus, the entropy increases following the mechanisms described in Chapter 6, which are governed by the viscosity and thermal conductivity transport coefficients. The 'H-theorem' is then not related to the increase of thermodynamic entropy, which is defined only at local equilibrium. Finally, the assumption of dilute gas is crucial: the Boltzmann equation breaks down if the potential energy becomes important. In such a case, the equation satisfied by $f(\vec{r}, \vec{p}, t)$ exhibits memory effects and is no longer an autonomous equation like the Boltzmann equation. The entropy (8.110) is still defined but it loses its utility since the property of entropy increase (8.113) is no longer automatically valid except at $t = 0$.

8.4 Transport coefficients from the Boltzmann equation

In this last section, we show how the transport coefficients may be computed from the Boltzmann equation. We shall limit ourselves to the computation of the shear viscosity, which is slightly simpler than that of the thermal conductivity, for which we refer to Problem 8.6.8.

8.4.1 Linearization of the Boltzmann equation

As in the case of the Boltzmann–Lorentz model, we follow the Chapman–Enskog method by linearizing the Boltzmann equation in the vicinity of a local equilibrium distribution of the form (8.114), which we rewrite by introducing the local density $n(\vec{r}, t)$

$$f_0(\vec{r}, \vec{p}, t) = n(\vec{r}, t)\left(\frac{\beta(\vec{r}, t)}{2\pi m}\right)^{3/2} \exp\left[-\frac{1}{2} m\beta(\vec{r}, t)(\vec{v} - \vec{u}(\vec{r}, t))^2\right] \tag{8.115}$$

The local density, the local inverse temperature $\beta(\vec{r}, t)$ and the local fluid velocity $\vec{u}(\vec{r}, t)$ are defined from f_0 following (8.45)–(8.46) and (8.104)

$$\begin{aligned} n(\vec{r}, t) &= \int \mathrm{d}^3 p\, f_0(\vec{r}, \vec{p}, t) \\ \vec{u}(\vec{r}, t) &= \frac{1}{n(\vec{r}, t)} \int \mathrm{d}^3 p\, \vec{v}\, f_0(\vec{r}, \vec{p}, t) \\ kT(\vec{r}, t) = \frac{1}{\beta(\vec{r}, t)} &= \frac{m}{3n(\vec{r}, t)} \int \mathrm{d}^3 p\, f_0(\vec{r}, \vec{p}, t)(\vec{v} - \vec{u}(\vec{r}, t))^2 \end{aligned} \tag{8.116}$$

The collision term vanishes if computed with the local equilibrium distribution (8.115): $\mathcal{C}[f_0] = 0$. Indeed, because of the conservation laws in the two-body elastic collisions

$$\vec{p}_1 + \vec{p}_2 = \vec{p}_3 + \vec{p}_4 \quad \text{and} \quad \vec{p}_1^{\,2} + \vec{p}_2^{\,2} = \vec{p}_3^{\,2} + \vec{p}_4^{\,2} \tag{8.117}$$

one verifies at once the identity

$$\ln f_{01} + \ln f_{02} = \ln f_{03} + \ln f_{04}$$

so that

$$f_{01} f_{02} = f_{03} f_{04} \tag{8.118}$$

However f_0 is *not* a solution of the Boltzmann equation, as the drift term does not vanish, except in the trivial case where the local equilibrium reduces to a global one

$$Df_0 = \left(\frac{\partial}{\partial t} + \vec{v}\cdot\vec{\nabla}\right) f_0 \neq 0 \qquad (8.119)$$

In the previous equation, and in all that follows, we have assumed that there are no external forces. As in the Boltzmann–Lorentz model we write the distribution f in terms of a small deviation $\overline{f}$ from the local equilibrium distribution f_0

$$f = f_0 + \overline{f}$$

Because of (8.116), $\overline{f}$ must obey the following three conditions, which are the analogues of (8.48) in the Boltzmann–Lorentz model

$$\int \mathrm{d}^3p\,\overline{f} = \int \mathrm{d}^3p\,\vec{p}\,\overline{f} = \int \mathrm{d}^3p\,\varepsilon\,\overline{f} = 0 \qquad (8.120)$$

We now follow the reasoning of Section 8.2.4: collisions bring the gas to local equilibrium after a time of the order of the collision time τ^*, and one observes afterwards a slow relaxation toward global equilibrium during which a time-independent drift term $Df = \vec{v}\cdot\vec{\nabla}f$ is balanced by the collision term as in (8.50). It will be convenient to write f as

$$f = f_0\left(1 - \overline{\Phi}\right) \qquad \overline{f} = -f_0\overline{\Phi}, \quad |\overline{\Phi}| \ll 1 \qquad (8.121)$$

so that, with the notation $\overline{\Phi}_i = \overline{\Phi}(\vec{p}_i)$ and keeping only terms linear in $\overline{\Phi}_i$

$$f_i f_j \simeq f_{0i} f_{0j}(1 - \overline{\Phi}_i - \overline{\Phi}_j)$$

Taking into account (8.118), the collision term becomes

$$\mathcal{C}[f] = \int \prod_{i=2}^{4} \mathrm{d}^3p_i\, W f_{01} f_{02}\left[\overline{\Phi}_1 + \overline{\Phi}_2 - \overline{\Phi}_3 - \overline{\Phi}_4\right]$$

and the linearized Boltzmann equation reads

$$\begin{aligned} Df_{01} &= \frac{\beta}{m} f_{01}\mathsf{L}\left[\overline{\Phi}\right] \\ \mathsf{L}\left[\overline{\Phi}\right] &= \frac{m}{\beta}\int \prod_{i=2}^{4} \mathrm{d}^3p_i\, W f_{02}\Delta\overline{\Phi} \\ \Delta\overline{\Phi} &= \overline{\Phi}_1 + \overline{\Phi}_2 - \overline{\Phi}_3 - \overline{\Phi}_4 \end{aligned} \qquad (8.122)$$

where the factor m/β has been introduced for later purposes. The functional $\mathsf{L}\left[\overline{\Phi}\right]$ may be considered as a *linear* operator acting on a space of functions $\overline{\Phi}(\vec{p})$. Let

$\overline{\Psi}(\vec{p})$ be an arbitrary function, let us multiply both sides of (8.122) by $\overline{\Psi}(\vec{p})$ and integrate over $\vec{p}_1$

$$\int \mathrm{d}^3 p_1 \, \overline{\Psi}_1 D f_{01} = \frac{\beta}{m} \int \mathrm{d}^3 p_1 \, \overline{\Psi}_1 \, f_{01} \mathsf{L}\left[\, \overline{\Phi} \,\right]$$
$$= \frac{1}{4} \int \prod_{i=1}^{4} \mathrm{d}^3 p_i \; \Delta\overline{\Psi} \, W f_{01} \, f_{02} \Delta\overline{\Phi} \tag{8.123}$$

where we have used the symmetry properties of the collision term, as in the derivation of (8.89). The right hand side of (8.123) defines a scalar product $(\overline{\Psi}, \overline{\Phi})$, which will turn out to be most useful in the derivation of a variational method, as explained in Section 8.4.2

$$\left(\overline{\Psi}, \overline{\Phi}\right) = \frac{1}{4} \int \prod_{i=1}^{4} \mathrm{d}^3 p_i \; \Delta\overline{\Psi} \, W f_{01} \, f_{02} \Delta\overline{\Phi} \tag{8.124}$$

which is positive semi-definite

$$||\overline{\Phi}||^2 = \left(\overline{\Phi}, \overline{\Phi}\right) = \frac{1}{4} \int \prod_{i=1}^{4} \mathrm{d}^3 p_i \; W f_{01} \, f_{02} (\Delta\overline{\Phi})^2 \geq 0$$

because W and f_0 are positive functions. More precisely, the left hand side of the previous equation will be strictly positive unless $\overline{\Phi}(\vec{p})$ is one of the five conserved densities, also called 'zero modes' of the linearized Boltzmann equation, which obey $\mathsf{L}[\,\overline{\Phi}\,] = 0$

$$\overline{\Phi}^{(1)}(\vec{p}) = 1 \quad \overline{\Phi}^{(2)}(\vec{p}) = p_x \quad \overline{\Phi}^{(3)}(\vec{p}) = p_y \quad \overline{\Phi}^{(4)}(\vec{p}) = p_z \quad \overline{\Phi}^{(5)}(\vec{p}) = p^2$$

Otherwise we shall have $||\overline{\Phi}||^2 = 0 \Leftrightarrow \overline{\Phi} = 0$.

8.4.2 Variational method

We now specialize our analysis to the calculation of the shear viscosity coefficient η, by using a particular form of the local equilibrium distribution f_0 in (8.115). We assume the temperature to be uniform, while the fluid velocity is directed along the x-axis and depends on the z coordinate (see Figure 6.6)

$$\vec{u} = (u_x(z), 0, 0) \tag{8.125}$$

The time-independent drift term becomes

$$
\begin{aligned}
Df \simeq Df_0 &= n\left(\frac{\beta}{2\pi m}\right)^{3/2} v_z \frac{\partial}{\partial z}\exp\left(-\frac{1}{2}\beta m\left[(v_x - u_x(z))^2 + v_y^2 + v_z^2)\right]\right) \\
&= \beta m f_0(\vec{r}, \vec{p}) v_z (v_x - u_x)\frac{\partial u_x(z)}{\partial z} \\
&= \beta m f_0(\vec{r}, \vec{p}) w_z w_x \frac{\partial u_x(z)}{\partial z}
\end{aligned}
\tag{8.126}
$$

with $\vec{w} = \vec{v} - \vec{u}$. The linearized Boltzmann equation (8.122) becomes

$$
\beta m w_{1x} w_{1z} \frac{\partial u_x(z)}{\partial z} = \int \prod_{i=2}^{4} \mathrm{d}^3 p_i \, W f_{02} \Delta\overline{\Phi}
$$

We consider a fixed point $\vec{r}$ in ordinary space and use the Galilean frame where the fluid is locally at rest, $u_x(z) = 0$ (but $\partial u_x/\partial z \neq 0$!) so that $f_0(\vec{r}, \vec{p}) \to f_0(p)$ and $\vec{w} = \vec{v}$. Instead of $\overline{\Phi}$ in (8.121), it is more convenient to use Φ defined by

$$
f(\vec{p}) = f_0(p)\left(1 - \overline{\Phi}(\vec{p})\right) = f_0\left(1 - \Phi(\vec{p})\frac{\partial u_x}{\partial z}\right)
\tag{8.127}
$$

which corresponds to an expansion to first order in $\partial u_x/\partial z$. With this definition the linearized Boltzmann equation now reads

$$
\boxed{p_{1x} p_{1z} = \frac{m}{\beta}\int \prod_{i=2}^{4} \mathrm{d}^3 p_i \, W f_{02} \Delta\Phi = \mathsf{L}[\Phi]}
\tag{8.128}
$$

The left hand side of (8.128) is a second rank tensor T_{xz}, and since $\vec{p}$ is the only vector at our disposal, Φ must be proportional to $p_x p_z$ and its most general form is

$$
\Phi(\vec{p}) = A(p) p_x p_z
\tag{8.129}
$$

so that $\Delta\Phi$ reads

$$
\Delta\Phi = A(p_1) p_{1x} p_{1z} + A(p_2) p_{2x} p_{2z} - A(p_3) p_{3x} p_{3z} - A(p_4) p_{4x} p_{4z}
$$

It is important to remark that the three conditions (8.120) are verified by our choice (8.129) for Φ. The pressure tensor $\mathcal{P}_{xz}$ is given from (8.19) by

$$
\begin{aligned}
\mathcal{P}_{xz} &= \int \mathrm{d}^3 p \, p_x v_z \overline{f} = -\int \mathrm{d}^3 p \, p_x v_z f_0 \Phi \frac{\partial u_x}{\partial z} \\
&= -\frac{1}{m}\frac{\partial u_x}{\partial z}\int \mathrm{d}^3 p \, A(p) f_0(p) p_x^2 p_z^2 = -\eta \, \frac{\partial u_x}{\partial z}
\end{aligned}
$$

and we get the following expression for η

$$\eta = \frac{1}{m}\int \mathrm{d}^3 p\, A(p) f_0(p) p_x^2 p_z^2 \tag{8.130}$$

We now rewrite the linearized Boltzmann equation (8.128) by using a Dirac notation in a (real) Hilbert space where a function $F(\vec{p})$ is represented by a vector $|F\rangle$ and the scalar product of two functions $F(\vec{p})$ and $G(\vec{p})$ is defined by

$$\langle F|G\rangle = \frac{1}{m}\int \mathrm{d}^3 p\, F(\vec{p}) f_0(p) G(\vec{p}) \tag{8.131}$$

This scalar product is obviously positive definite: $\langle F|F\rangle \geq 0$ and $\langle F|F\rangle = 0 \Leftrightarrow F = 0$. We represent the function $p_x p_z$ in (8.128) by a vector $|X\rangle$. Note that because $f_0(p)$ decreases exponentially with p^2 at infinity, the norm $\langle X|X\rangle$ is finite and $|X\rangle$ belongs to our Hilbert space. With these notations, the linearized Boltzmann equation reads in operator form

$$p_{1x} p_{1z} = |X\rangle = \mathsf{L}|\Phi\rangle \tag{8.132}$$

and the viscosity is given from (8.16) by

$$\eta = \langle\Phi|X\rangle = \langle\Phi|\mathsf{L}|\Phi\rangle = \frac{|\langle\Phi|X\rangle|^2}{\langle\Phi|\mathsf{L}|\Phi\rangle} \tag{8.133}$$

This formula will serve as the starting point for a variational method. Indeed, Φ is an unknown function of $\vec{p}$, or, in other words, we do not know the functional form of $A(p)$, and an exact solution for $A(p)$ would require rather complicated methods. An efficient way to proceed is to use a variational method by introducing a trial function $\Psi(\vec{p})$, where we must restrict our choice to functions orthogonal to the zero modes. We define a Ψ-dependent viscosity coefficient $\eta[\Psi]$ as a functional of Ψ by

$$\eta[\Psi] \hat{=} \frac{|\langle X|\Psi\rangle|^2}{\langle\Psi|\mathsf{L}|\Psi\rangle} = \frac{|\langle\Psi|\mathsf{L}|\Phi\rangle|^2}{\langle\Psi|\mathsf{L}|\Psi\rangle} \tag{8.134}$$

The choice $\Psi = \Phi$ gives the exact value of the viscosity: $\eta \equiv \eta[\Phi]$. The quantity $\langle\Psi|\mathsf{L}|\Phi\rangle$ can be rewritten by using the (positive definite) scalar product already introduced in (8.124)[24]

$$\langle\Psi|\mathsf{L}|\Phi\rangle = (\Psi, \Phi) = \frac{1}{4\beta}\int \prod_{i=1}^{4} \mathrm{d}^3 p_i\, \Delta\Psi W f_{01} f_{02} \Delta\Phi \tag{8.135}$$

[24] Within a $1/\beta$ multiplicative factor.

The Schwartz inequality leads to an inequality on $\eta[\Psi]$

$$\eta[\Psi] = \frac{|(\Psi, \Phi)|^2}{(\Psi, \Psi)} \leq \frac{(\Psi, \Psi)(\Phi, \Phi)}{(\Psi, \Psi)} = \langle \Phi | \mathsf{L} | \Phi \rangle = \eta[\Phi]$$

We have thus derived an inequality typical of a variational method

$$\eta[\Psi] \leq \eta_{\text{exact}} = \eta[\Phi] \tag{8.136}$$

8.4.3 Calculation of the viscosity

One possible choice for a trial function would be

$$\Psi_\alpha = A p^\alpha p_x p_z$$

and the best choice for α would be obtained by minimizing $\eta[\Psi_\alpha]$ with respect to the parameter α. We shall limit ourselves to the simple case $\alpha = 0$, which already gives results that do not differ from those of an exact calculation by more than a few percent. The computation of $\langle X|\Psi\rangle$ is then straightforward, with $f_0(p)$ given by

$$f_0(p) = n \left(\frac{\beta}{2\pi m} \right)^{3/2} \exp\left(-\frac{\beta p^2}{2m} \right) \tag{8.137}$$

One obtains

$$\begin{aligned} \langle X|\Psi\rangle &= \frac{1}{m} \int \mathrm{d}^3 p \; p_x p_z f_0(p) A p_x p_z \\ &= \frac{4\pi}{15} \frac{A}{m} \int_0^\infty \mathrm{d}p \; p^6 f_0(p) \end{aligned}$$

where we have computed the angular average from (A.36)[25]

$$\langle p_x p_z p_x p_z \rangle_{\text{ang}} = \frac{1}{15} p^4$$

The p-integration is completed thanks to the Gaussian integration (A.37) in the form

$$\int_0^\infty \mathrm{d}p \; p^n \, \mathrm{e}^{-\alpha p^2} = \frac{1}{2} \Gamma\left(\frac{n+1}{2} \right) \alpha^{-(n+1)/2}$$

[25] In this elementary case, one can also use

$$\int \frac{\mathrm{d}\Omega}{4\pi} p_x^2 p_z^2 = \frac{p^4}{4\pi} \int_{-1}^{1} \mathrm{d}(\cos\alpha) \sin^2\alpha \cos^2\alpha \int_0^{2\pi} \mathrm{d}\phi \cos^2\phi = \frac{1}{15}$$

with the result

$$\langle X|\Psi\rangle = \frac{nAm}{\beta^2} \tag{8.138}$$

The calculation of $\langle\Psi|\mathsf{L}|\Psi\rangle$ is slightly more involved

$$\langle\Psi|\mathsf{L}|\Psi\rangle = \frac{1}{4\beta}\int\prod_{i=1}^{4}\mathrm{d}^3p_i\, f_{01}f_{02}W(\Delta\Psi)^2 \tag{8.139}$$

Using the centre-of-mass kinematics we get

$$\vec{p}_1 = \vec{p} + \frac{1}{2}\vec{P} \qquad \vec{p}_2 = -\vec{p} + \frac{1}{2}\vec{P}$$
$$\vec{p}_3 = \vec{p}\,' + \frac{1}{2}\vec{P} \qquad \vec{p}_4' = -\vec{p}\,' + \frac{1}{2}\vec{P}$$

where $\vec{P}$ is the centre-of-mass momentum; we recall the expression of the relative velocity, $v_{\mathrm{rel}} = 2p/m$. These equations give

$$p_{1x}p_{1z} + p_{2x}p_{2z} = 2p_xp_z + \frac{1}{4}P_xP_z$$
$$p_{3x}p_{3z} + p_{4x}p_{4z} = 2p_x'p_z' + \frac{1}{4}P_xP_z$$

so that

$$\Delta\Psi = 2A(p_xp_z - p_x'p_z')$$

The differential cross section $\sigma(\Omega, p)$ in the Boltzmann equation depends on the angle θ between $\vec{p}$ and $\vec{p}\,'$. The integration in (8.139) leads to an angular average at fixed (Ω, p), which is again computed thanks to (A.36)

$$\begin{aligned}\langle(\Delta\Psi)^2\rangle_{\mathrm{ang}} &= 4A^2\Big\langle p_x^2p_z^2 + p_x'^2p_z'^2 - 2p_xp_x'p_zp_z'\Big\rangle_{\mathrm{ang}}\\ &= 4A^2p^4\left[\frac{2}{15} - \frac{1}{15}(-1+3\cos^2\theta)\right]\\ &= \frac{4}{5}A^2p^4(1-\cos^2\theta)\end{aligned}$$

Inserting this result in (8.138) and using the relation between W and $\sigma(\Omega, p)$ (see (8.86))

$$\int\mathrm{d}^3p_3\,\mathrm{d}^3p_4\,W \to \frac{2p}{m}\int\mathrm{d}\Omega\,\sigma(\Omega, p)$$

yields

$$\langle \Psi | \mathsf{L} | \Psi \rangle = \frac{2A^2}{5\beta m} \int \mathrm{d}^3 p_1 \, \mathrm{d}^3 p_2 \, \mathrm{d}\Omega \, f_{01} f_{02} \, p^5 (1 - \cos^2\theta) \sigma(\Omega, p) \tag{8.140}$$

The result features the transport cross section $\sigma_{\mathrm{tr}}(p)$, and *not* the total cross section $\sigma_{\mathrm{tot}}(p)$

$$\sigma_{\mathrm{tr}}(p) = \int \mathrm{d}\Omega (1 - \cos^2\theta) \sigma(\Omega, p) \tag{8.141}$$

The physical reason behind the occurrence of the transport cross section is that forward scattering is very inefficient in transferring momentum, hence the suppression factor $(1 - \cos\theta)$ (and $(1 + \cos\theta)$ for the backward scattering). We also need to write the product $f_{01} f_{02}$ in terms of the centre-of-mass variables p and P

$$f_{01} f_{02} = n^2 \left(\frac{\beta}{2\pi m} \right)^3 \exp\left(-\frac{\beta p^2}{m} \right) \exp\left(-\frac{\beta P^2}{4m} \right)$$

It remains to use the change of variables with unit Jacobian

$$\mathrm{d}^3 p_1 \, \mathrm{d}^3 p_2 = \mathrm{d}^3 P \, \mathrm{d}^3 p$$

and the Gaussian integration (A.37) to compute

$$\begin{aligned} &\int \mathrm{d}^3 p_1 \, \mathrm{d}^3 p_2 \, f_{01} f_{02} \, p^5 \sigma_{\mathrm{tr}}(p) \\ &= 4\pi n^2 2^{3/2} \left(\frac{\beta}{2\pi m} \right)^{3/2} \int_0^\infty \mathrm{d}p \, p^7 \exp\left(-\frac{\beta p^2}{m} \right) \sigma_{\mathrm{tr}}(p) = \frac{12}{\sqrt{\pi}} n^2 \left(\frac{m}{\beta} \right)^{5/2} \sigma_{\mathrm{tr}} \end{aligned}$$

where, in the last line of the previous equation, we have assumed σ_{tr} to be independent of p. If this is not the case, the integral in the first line may be used to define an effective T-dependent transport cross section $\sigma_{\mathrm{tr}}^{\mathrm{eff}}(T)$ which should be used instead of a T-independent σ_{tr}. We thus get

$$\langle \Psi | \mathsf{L} | \Psi \rangle = \frac{24A^2}{5\sqrt{\pi}} n^2 m^{3/2} \beta^{-7/2} \sigma_{\mathrm{tr}} \tag{8.142}$$

Gathering (8.134), (8.138) and (8.142) we obtain the following result for η

$$\boxed{\eta = \frac{5\sqrt{\pi}}{24} \frac{\sqrt{mkT}}{\sigma_{\mathrm{tr}}}} \tag{8.143}$$

In the case of a hard sphere gas, the transport cross section is 2/3 of the total cross section $\sigma_{\rm tot}$

$$\sigma_{\rm tot} = 4\pi R^2 \qquad \sigma_{\rm tr} = \frac{8\pi}{3} R^2 = \frac{2}{3}\sigma_{\rm tot}$$

where R is the radius of the spheres, and one may rewrite (8.143)

$$\eta = \frac{5\sqrt{\pi}}{16}\frac{\sqrt{mkT}}{\sigma_{\rm tot}} \simeq 0.553\,\frac{\sqrt{mkT}}{\sigma_{\rm tot}} \tag{8.144}$$

This result is to be compared with the qualitative estimate (8.21), which may be written

$$\eta = 0.377\,\frac{\sqrt{mkT}}{\sigma_{\rm tot}} \tag{8.145}$$

where we have used the mean free path (8.11) of the Maxwell distribution and the corresponding mean value of the velocity

$$\ell = \frac{1}{\sqrt{2}\,n\sigma_{\rm tot}} \qquad v = \sqrt{\frac{8kT}{\pi m}}$$

An analogous calculation (Problem 8.6.8) gives for the coefficient of thermal conductivity, assuming a p-independent transport cross section

$$\kappa = \frac{25\sqrt{\pi}}{32\sigma_{\rm tr}}\,k\sqrt{\frac{kT}{m}} \tag{8.146}$$

and the ratio κ/η is

$$\boxed{\frac{\kappa}{\eta} = \frac{15}{4}\frac{k}{m} = \frac{5}{2}\frac{c}{m}} \tag{8.147}$$

instead of the qualitative estimate $\kappa/\eta = c/m$ derived in Section 8.1.3. The factor $5/2$ in (8.147) is in excellent quantitative agreement with the experimental results on mono-atomic gases.

8.5 Exercises

8.5.1 Time distribution of collisions

We consider the collisions of a labelled molecule starting at the initial time $t = 0$. An excellent approximation consists of considering the collisions as independent: the collision process is without memory. Let λ be the average number of collisions per unit time suffered by a molecule. What is the probability $P(n, t)$ that the molecule undergoes n collisions in the time interval $[0, t]$? What is the survival probability $P(t)$, i.e. the probability that the molecule has not suffered any collisions in the interval $[0, t]$? What is the probability $\mathcal{P}(t)\,\mathrm{d}t$ that the molecule will suffer its first collision in the interval $[t, t + \mathrm{d}t]$? Use these results to find an expression for the collision time τ^* defined as the average time from $t = 0$ for the molecule to undergo its first collision. Since the process is Markovian (without memory), τ^* is also the average time between collisions as well as the time elapsed since the last collision.

8.5.2 Symmetries of an integral

Show that if $\vec{a}$ is a fixed vector and $g(p)$ a function of $|\vec{p}| = p$, then

$$\vec{I} = \int \mathrm{d}^3 p\ (\vec{p} \cdot \vec{a})\,\vec{p}\,g(p) = \frac{1}{3}\vec{a} \int \mathrm{d}^3 p\ p^2 g(p) \tag{8.148}$$

or that equivalently

$$I_{\alpha\beta} = \int \mathrm{d}^3 p\ p_\alpha p_\beta g(p) = \frac{1}{3}\delta_{\alpha\beta} \int \mathrm{d}^3 p\ p^2 g(p) \tag{8.149}$$

Use this to obtain relation (8.60).

8.5.3 Positivity conditions

Demonstrate the positivity condition (8.68) for the transport coefficients.

$$L_{EE}L_{NN} - L_{EN}^2 \geq 0$$

8.5.4 Calculation of the collision time

1. Show that the collision time τ^* of a molecule in a gas is given by

$$\frac{1}{\tau^*} = \frac{1}{n} \int \mathrm{d}^3 p_1\, \mathrm{d}^3 p_2\, f(\vec{p}_1) f(\vec{p}_2) |\vec{v}_2 - \vec{v}_1| \sigma_{\mathrm{tot}}(|\vec{v}_2 - \vec{v}_1|) \tag{8.150}$$

Hint: First calculate the collision time of a particle with momentum $\vec{p}_1$.

2. A simple calculation of the collision time is obtained in the case where all the molecules have the same absolute value v_0 for the velocity and therefore momentum p_0. The distribution function then is

$$f(\vec{p}) = \frac{n}{4\pi p_0^2}\,\delta(p - p_0) \tag{8.151}$$

Verify that the distribution is correctly normalized. Assuming that the total cross section is independent of the velocity, show that the collision time is

$$\tau^* = \frac{3}{4n v_0 \sigma_{\rm tot}} \tag{8.152}$$

and calculate the mean free path ℓ.

3. We now assume we have a Maxwell distribution for the momenta

$$f(\vec{p}) = n \left(\frac{1}{2\pi mkT} \right)^{3/2} \exp\left(-\frac{\vec{p}^{\,2}}{2mkT} \right) \tag{8.153}$$

Again assuming that the total cross section is independent of velocity, show that the collision time is given by (8.11).

8.5.5 Derivation of the energy current

Demonstrate the expression (6.83) for the energy current starting with the Boltzmann equation.

8.5.6 Equilibrium distribution from the Boltzmann equation

1. We first assume a situation with no external forces. Show that the Boltzmann entropy $S_{\rm B}(t)$ (8.110) must tend to a constant when $t \to \infty$

$$\lim_{t\to\infty} S_{\rm B}(t) = \text{const}$$

In this limit, show that one must have

$$\ln f_1 + \ln f_2 = \ln f_3 + \ln f_4$$

2. From this condition, deduce that $\ln f$ must be of the form

$$\ln f = \chi^{(1)}(\vec{p}) + \chi^{(2)}(\vec{p}) + \cdots$$

where the $\chi^{(i)}$s are conserved quantities in the collisions

$$\chi^{(i)}(\vec{p}_1) + \chi^{(i)}(\vec{p}_2) = \chi^{(i)}(\vec{p}_3) + \chi^{(i)}(\vec{p}_4)$$

From this, deduce that $f(\vec{p})$ can be written as

$$f(\vec{p}) = -A(\vec{p} - \vec{p}_0)^2 + B$$

where A, B and $\vec{p}_0$ are constants. Compute the density, the flow velocity and the temperature as functions of A, B and $\vec{p}_0$ and show that $f(\vec{p})$ must be a Maxwell distribution centred at $\vec{p}_0$.

3. In the presence of external forces

$$\vec{F}(\vec{r}) = -\vec{\nabla}\Phi(\vec{r})$$

show that the previous result is multiplied by

$$\exp\left(-\frac{\Phi(\vec{r})}{kT}\right)$$

8.6 Problems

8.6.1 Thermal diffusion in the Boltzmann–Lorentz model

The Boltzmann–Lorentz model is particularly well suited to describe the scattering of light solute molecules (density n) on the heavy molecules of a solvent (density n_d). We consider a stationary situation, and we assume that the solution is not subject to any external force ($\vec{F} = \vec{0}$). The equilibrium distribution of the solute is that of a non-relativistic classical ideal gas

$$f_0(\vec{r}, \vec{p}) = \frac{1}{h^3} \exp\left[\beta(\vec{r})\mu(\vec{r}) - \beta(\vec{r})\frac{\vec{p}^{\,2}}{2m}\right] = \frac{n(\vec{r})}{\left[2\pi mkT(\vec{r})\right]^{3/2}} \exp\left[-\beta(\vec{r})\frac{\vec{p}^{\,2}}{2m}\right] \tag{8.154}$$

1. The density and temperature vary with space but the solution is maintained at constant and uniform pressure

$$\mathcal{P} = \bar{n}(\vec{r})kT(\vec{r})$$

where $\bar{n} = n + n_d$ is the total particle density. In this situation, we define two response coefficients: the diffusion coefficient D and the thermodiffusion coefficient λ. These coefficients appear in the phenomenological expression for the particle current

$$\vec{j}_N = -\bar{n}D\vec{\nabla}c - \frac{\bar{n}c}{T}\lambda\vec{\nabla}T \tag{8.155}$$

where c is the concentration of light particles $c = n/\bar{n}$. Calculate the particle current $\vec{j}_N$ in the framework of the Chapman–Enskog approximation and establish the microscopic expressions for D and λ

$$\begin{aligned} D &= \frac{1}{3}\frac{1}{n_d}\left\langle\frac{v}{\sigma_{\rm tot}(v)}\right\rangle = \frac{1}{3}\langle v^2\tau\rangle \\ \lambda &= \frac{1}{3}\frac{T^2}{n_d}\frac{\partial}{\partial T}\left[\frac{1}{T}\left\langle\frac{v}{\sigma_{\rm tot}(v)}\right\rangle\right] \end{aligned} \tag{8.156}$$

where $\langle(\bullet)\rangle$ is the average value defined in (8.2)

$$\langle(\bullet)\rangle = \frac{1}{n(\vec{r})}\int \mathrm{d}^3p\,(\bullet)\,f_0(\vec{r},\vec{p}) = \frac{1}{\left[2\pi mkT(\vec{r})\right]^{3/2}}\int \mathrm{d}^3p\,(\bullet)\exp\left[-\beta(\vec{r})\frac{\vec{p}^{\,2}}{2m}\right]$$

2. We now take the solute molecules to be spheres of diameter a, which have only elastic collisions with the solvent molecules assumed to be point particles. This is the hard sphere model with $\sigma_{\mathrm{tot}} = \pi a^2$. Verify that in this framework we have

$$\begin{aligned} D &= \frac{1}{3\pi a^2 n_d}\sqrt{\frac{8kT}{\pi m}} \\ \lambda &= -\frac{1}{6\pi a^2 n_d}\sqrt{\frac{8kT}{\pi m}} \end{aligned} \tag{8.157}$$

When the diffusion and thermodiffusion currents equilibrate, i.e. when the particle current vanishes, what region of the gas will contain the highest concentration of solute molecules?

3. We are now interested in the energy current. We introduce the two phenomenological coefficients, κ, coefficient of thermal conductivity, and γ, the Dufour coefficient

$$\vec{j}_E = -\bar{n}\gamma\vec{\nabla}c - \kappa\vec{\nabla}T \tag{8.158}$$

Establish the microscopic expression for $\vec{j}_E$ and show that

$$\begin{aligned} \kappa &= \frac{1}{6}mcT\frac{\partial}{\partial T}\left[\frac{1}{T}\left\langle\frac{v^3}{\sigma_{\mathrm{tot}}(v)}\right\rangle\right] \\ \gamma &= \frac{1}{6}\frac{mkT}{\mathcal{P}}\left\langle\frac{v^3}{\sigma_{\mathrm{tot}}(v)}\right\rangle \end{aligned} \tag{8.159}$$

where it should be recalled that in the Boltzmann–Lorentz model we have $n \ll n_d$.

4. Do the response coefficients in (8.155) and (8.158) obey an Onsager relation? If not, why?

8.6.2 Electron gas in the Boltzmann–Lorentz model

A Introduction

We consider a non-relativistic ideal gas of electrons, mass m and charge q ($q < 0$), obeying the Boltzmann–Lorentz equation (8.28). When the differential cross section $\sigma(v,\Omega)$ is independent of Ω, $\sigma(v,\Omega) = \sigma(v)/(4\pi)$ where $\sigma(v)$ is the total cross section and $v = p/m$, we can solve this equation using the Chapman–Enskog method by following a simpler procedure than that in Section 8.2.4. We write $f = f_0 + \bar{f}$ where f_0 is a local equilibrium distribution and we seek a solution in the stationary regime satisfying

$$\int \mathrm{d}\Omega' \, \bar{f}(\vec{r}, \vec{p}') = 0$$

Show that the collision term (8.35) becomes

$$\mathcal{C}[\bar{f}] = -\frac{1}{\tau^*(p)} \bar{f}$$

Use this to obtain f_0. In what follows we will assume that the differential cross section $\sigma(v, \Omega)$ is independent of velocity

$$\sigma(v, \Omega) = \frac{\sigma}{4\pi}$$

Show that in the absence of external forces, we can write the particle and energy densities in the form

$$\vec{j}_N = -\frac{\ell}{3} \int \mathrm{d}^3 p \, v \vec{\nabla} f_0 \qquad \vec{j}_E = -\frac{\ell}{3} \int \mathrm{d}^3 p \, \varepsilon v \vec{\nabla} f_0$$

where $\varepsilon = p^2/(2m)$. What is the physical interpretation of ℓ?

B Classical ideal gas

We assume that the electrons form a classical ideal gas, which is a good approximation for semiconductors. The local equilibrium distribution is then

$$f_0(\vec{r}, \vec{p}) = \frac{2}{h^3} \exp\left(\alpha(\vec{r}) - \beta(\vec{r}) \frac{p^2}{2m} \right) \quad \alpha(\vec{r}) = \frac{\mu(\vec{r})}{kT(\vec{r})} \quad \beta(\vec{r}) = \frac{1}{kT(\vec{r})}$$

where $\mu(\vec{r})$ and $T(\vec{r})$ are the local chemical potential and temperature and k the Boltzmann constant.

1. Express the four transport coefficients L_{ij}: $(i, j) = (E, N)$ in the form

$$L_{ij} = \frac{8\pi}{3} \frac{\ell m}{k} \int_0^\infty \mathrm{d}\varepsilon \, \varepsilon^\nu f_0$$

ν is an integer whose value, which depends on the transport coefficient, is to be determined for the four cases. We give the integral

$$I = \frac{8\pi}{3}\frac{\ell m}{k}\int_0^\infty \mathrm{d}\varepsilon\, \varepsilon^\nu f_0 = \frac{\tau^* \nu!}{m} n k^{\nu-1} T^\nu$$

where the collision time τ^* is defined in terms of the average electron velocity $\langle v\rangle = \sqrt{\dfrac{8kT}{\pi m}}$ by $\tau^* = \dfrac{8}{3\pi}\dfrac{\ell}{\langle v\rangle}$. Express the L_{ij} in terms of n, T, m and τ^*.

2. Show that the diffusion coefficient D and the thermal conductivity coefficient κ defined by

$$\vec{j}_N = -D\,\vec{\nabla} n \quad (T \text{ constant}) \qquad\qquad \vec{j}_E = -\kappa \vec{\nabla} T \quad (\vec{j}_N = 0)$$

can be expressed in terms of the L_{ij}. Give their explicit expressions in terms of n, T, m and τ^*.

3. We now subject the electrons to an electric potential $\Phi(z)$ which is time independent but varies slowly with space along the z direction. We assume the system is still at *equilibrium*: the *total* particle current vanishes and T is uniform. How does the density n vary with z? Establish a relation between the diffusion coefficient and the electrical conductivity $\sigma_{\rm el}$ defined in terms of the electric current by

$$\vec{j}_{\rm el} = \sigma_{\rm el}\vec{E} = -\sigma_{\rm el}\,\vec{\nabla}\Phi$$

and show that

$$\sigma_{\rm el} = \frac{q^2 n\tau^*}{m}$$

Obtain this expression using an elementary argument and calculate $\kappa/\sigma_{\rm el}$.

4. *We now introduce an external force* $\vec{F} = -\vec{\nabla}_{\vec{r}} V(\vec{r})$. The Boltzmann–Lorentz equation then becomes

$$\frac{\partial f}{\partial t} + \vec{v}\cdot\vec{\nabla}_{\vec{r}} f - \vec{\nabla}_{\vec{r}} V\cdot\vec{\nabla}_{\vec{p}} f = \mathcal{C}[f]$$

The local equilibrium distribution takes the form

$$f_0(\vec{r},\vec{p}) = \frac{2}{h^3}\exp\left(\alpha(\vec{r}) - \beta(\vec{r})\left[\frac{p^2}{2m} + V(\vec{r})\right]\right)$$

Give the expression for $\bar{f}$ in terms of f_0, α, V and their gradients. Indicating the response coefficients in the absence of an external force with primes, demonstrate the following transformation laws

$$\begin{aligned} L_{NN} &= L'_{NN} \\ L_{NE} &= L'_{EE} + VL'_{NN} \\ L_{EE} &= L'_{EE} + 2VL'_{NE} + V^2L'_{NN} \end{aligned}$$

5. We can calculate the electric conductivity without using L_{NN}. Consider a stationary situation with uniform density where the electrons are subjected to a constant uniform electric field $\vec{E}$. With the results of Part A and the Chapman–Enskog approximation, the Boltzmann–Lorentz equation then becomes

$$q\vec{E}\cdot\vec{\nabla}_{\vec{p}}f_0 = -\frac{v}{\ell}\bar{f}$$

Show that the electric current may be written as

$$\vec{j}_{\rm el} = -\frac{q^2\ell}{3m}\vec{E}\int \mathrm{d}^3p\, p\,\frac{\mathrm{d}f_0}{\mathrm{d}\varepsilon}$$

and find $\sigma_{\rm el}$.

C Ideal Fermi gas

We now assume that the electrons form an ideal Fermi gas whose temperature is low compared to the Fermi temperature ($T \ll T_{\rm F}$), which is an excellent approximation for electrons in a metal.

1. Derive the relation between n and μ for $T = 0$.

2. Show that, due to the Pauli principle, the distribution function $f(\vec{r}, \vec{p})$ in the collision term needs to be replaced by

$$f(\vec{r}, \vec{p})\left(1 - \frac{h^3}{2}f(\vec{r}, \vec{p}\,')\right)$$

Hint: What is the number of microscopic states in the phase space volume element $\mathrm{d}^3r\,\mathrm{d}^3p'$? What is the occupation probability of such a state? How should we modify the term $f(\vec{r}, \vec{p}\,')$ in $C[f]$? Show that the two modifications cancel out and we regain the initial $C[f]$. What would be the corresponding reasoning for a boson gas?

3. The local equilibrium distribution now is

$$f_0 = \frac{2}{h^3}\frac{1}{\exp[-\alpha(\vec{r}) + \beta(\vec{r})p^2/(2m)] + 1}$$

We assume that the local temperature $T(\vec{r})$ is small enough compared to T_F to justify the use of the Sommerfeld approximation (5.29). Show that

$$\vec{\nabla} f_0 = -c_1\Big[\vec{\nabla}\Big(-\frac{\mu}{T}\Big) + \varepsilon\,\vec{\nabla}\Big(\frac{1}{T}\Big)\Big]\Big[\delta(\varepsilon-\mu) + c_2\delta''(\varepsilon-\mu)\Big]$$

and determine the coefficients c_1 and c_2. Show that the coefficients L_{ij} may be put in the form

$$L_{ij} = \frac{16\pi}{3}\frac{m\ell T}{h^3}\int_0^\infty \mathrm{d}\varepsilon\,\varepsilon^{\nu}\left(\delta(\varepsilon-\mu) + c_2\delta''(\varepsilon-\mu)\right)$$

and determine the integer ν for all values of (i, j).

4. Show that the term δ'' does not contribute to L_{NN} and that

$$L_{NN} = \frac{\tau_F}{m} nT \qquad \tau_F = \frac{\ell}{v_F}$$

Why are the electrons with velocities $v \simeq v_F = p_F/m$ the only ones that contribute to the transport coefficients?

5. The expressions for the diffusion coefficient D and electrical conductivity σ_{el} established in Part B for a classical ideal gas are *a priori* no longer valid for a Fermi ideal gas. Give the new expression for D in terms of the L_{ij} and show that the expression for σ_{el} does not change. Express D and σ_{el} in terms of n, T, m, τ_F.

6. Calculate the numerical value of the collision time τ_F in copper with mass density $8.9 \times 10^3\,\mathrm{kg\,m^{-3}}$, atomic number $A = 63.5$ and conductivity $\sigma_{el} = 5 \times 10^7\,\Omega^{-1}\,\mathrm{m}^{-1}$. Copper has one conduction electron per atom.

7. Calculate the other transport coefficients and the ratio κ/σ_{el} (Franz–Wiedemann law)

$$\frac{\kappa}{\sigma_{el}} = \frac{\pi^2}{3}\frac{k^2}{q^2}T$$

Compare this expression with that previously obtained in B.3. Calculate the electrical conductivity using the method of Question B.5.

8.6.3 Photon diffusion and energy transport in the Sun

In this problem we study the transfer of heat from the centre of the Sun (or more generally a star) to its surface. This is what allows energy to be radiated into space. The main contribution to this heat transfer (and the only mechanism we consider) is the scattering of photons by electrons in the Sun.

Solar data

Radius	$R_\odot = 7 \times 10^8$ m
Mass	$M_\odot = 2 \times 10^{30}$ kg
Mean specific mass	$\rho = 1.4 \times 10^3$ kg m^{-3}
Specific mass at the centre	$\rho_c = 10^5$ kg m^{-3}
Surface temperature	$T_s = 6000$ K
Temperature at the centre	$T_c = 1.5 \times 10^7$ K

A Preliminaries

1. Consider a photon gas in equilibrium at temperature T. The photon momentum is $\vec{p}$, energy $\varepsilon = pc$ and velocity $\vec{v} = c\,\vec{p}/p = c\,\hat{p}$. The equilibrium distribution is given by (5.38)

$$f_{\text{eq}}(\vec{p}) = \frac{2}{h^3}\frac{1}{e^{\beta\varepsilon} - 1}$$

which satisfies

$$\int \mathrm{d}^3 p f_{\text{eq}}(\vec{p}) = \frac{N}{V} = n$$

Show that the number of photons per unit volume n and the energy density ϵ are of the form

$$n = \lambda' T^3 \qquad \epsilon = \lambda T^4$$

Determine λ and λ' in terms of $\hbar$, k and c. Calculate the numerical values of n and ϵ at the centre of the Sun as well as the pressure of the photon gas. We give $k^4/(\hbar^3 c^3) = 1.16 \times 10^{-15}$ MKSA.

2. What is, in terms of ϵ, the energy emitted per second per unit area by a black body? Assuming the solar surface is a black body, calculate in terms of λ, c, $R_\odot$ and T_s the power (energy per second) $L_\odot$ emitted by the Sun. This quantity is called the luminosity.

3. We assume that the Sun contains only protons and electrons that behave as ideal gases. We may ignore m_e compared to m_p. Calculate the Fermi momentum and energy at the centre of the Sun in terms of ρ_c, $\hbar$, m_e and m_p. Evaluate numerically the Fermi energy in eV and the Fermi temperature in K. Conclude that the electron gas is neither degenerate nor relativistic.

Show that the pressure at the centre of the Sun is given by

$$P_c = 2\frac{\rho_c}{m_p} k T_c$$

and compare this expression with that for a photon gas.

B Scattering equation

Our hypothesis is that photons are scattered only by electrons and that this scattering is elastic. The photons are described by the Boltzmann–Lorentz equation with the electrons playing the rôle of the randomly distributed heavy scattering centres. The transport cross section is equal to the total cross section σ and is independent of the photon energy. It is given by

$$\sigma = \frac{8\pi}{3}\left(\frac{e^2}{4\pi\varepsilon_0 m_e c^2}\right)^2 \approx 6.6 \times 10^{-29}\,\mathrm{m}^2$$

where $-e$ is the electron charge and ε_0 the vacuum permittivity. We solve the Boltzmann–Lorentz equation by writing the photon distribution in the stationary regime

$$f(\vec{r}, \vec{p}) = f_0(\vec{r}, \vec{p}) + \bar{f}(\vec{r}, \vec{p})$$

where $f_0(\vec{r}, \vec{p}) = f_0(\vec{r}, p)$ is a local equilibrium distribution

$$f_0(\vec{r}, \vec{p}) = \frac{2}{h^3}[\exp(\beta(\vec{r})\varepsilon - 1)]^{-1}$$

1. Express the collision time $\tau^*(\vec{r})$ in terms of m_p, σ, c and the specific mass $\rho(\vec{r})$ at point $\vec{r}$.

2. Calculate the photon and energy currents, $\vec{j}_N$ and $\vec{j}_E$, in terms of f_0 and show that they may be put in the form

$$\vec{j}_N(\vec{r}) = -D'(\vec{r})\vec{\nabla}n(\vec{r}) \qquad \vec{j}_E(\vec{r}) = -D(\vec{r})\vec{\nabla}\epsilon(\vec{r})$$

Give the expressions for D and D' in terms of c and $\tau^*(\vec{r})$.

3. Assume that the specific mass is uniform $\rho(\vec{r}) \equiv \rho$. Show that in a situation non-stationary but still at local equilibrium, $n(\vec{r}, t)$ satisfies a diffusion equation. If a photon is created at the centre of the Sun at $t = 0$, estimate the time it needs to leave the Sun.

C Model for the Sun

We are now back to a stationary situation and we assume that the problem has spherical symmetry with the origin of coordinates at the centre of the Sun. We call $q(r)$ the energy produced by the thermonuclear reactions per unit time per unit volume, and $Q(r)$ the energy produced per unit time in a sphere $S(r)$ of radius r

$$Q(r) = 4\pi \int_0^r \mathrm{d}r' q(r') r'^2$$

1. Relate $Q(r)$ to the flux of $\vec{j}_E$ across $S(r)$. Deduce the equation

$$-\frac{4\pi}{3} A \frac{r^2}{\rho(r)} \frac{\mathrm{d}}{\mathrm{d}r} T^4(r) = Q(r) \tag{8.160}$$

where A is a constant to be expressed in terms of λ, c, $\mathrm{m_p}$ and σ. In what follows we assume:

(*i*) the energy is produced uniformly in a sphere of radius $R_\mathrm{c} = 0.1\, R_\odot$ (the solar core),
(*ii*) the specific mass is uniform: $\rho(r) \equiv \rho$.

2. Calculate $T^4(r)$ in the regions $0 \le r \le R_\mathrm{c}$ and $R_\mathrm{c} \le r \le R_\odot$. Determine the integration constants for the differential equation (8.160) by examining the values $r = R_\odot$ and $r = R_\mathrm{c}$.

3. It is useful to express the temperature T_c at the centre of the Sun in terms of its luminosity $L_\odot$. Show that

$$T_\mathrm{c}^4 = T_\mathrm{s}^4 + \frac{3L_\odot \rho}{4\pi A R_\odot}\left(\frac{3R_\odot}{2R_c} - 1\right)$$

Use the expression for $L_\odot$ in terms of T_s to calculate the numerical value of T_c.

8.6.4 Momentum transfer in a shear flow

We consider a fluid in a stationary state flowing between two infinite parallel plates separated by a distance L in the z direction (see Figure 6.6). One of the plates is held fixed while the other moves at a constant speed u_0 in the increasing x direction always staying parallel to the first plate. The fluid is dragged by the moving plate. After enough time has elapsed since the motion started, a stationary situation is reached characterized by a linear dependence of the fluid speed ranging from 0 at one plate to u_0 at the other. Each layer of fluid between z and $z + \mathrm{d}z$ has a velocity $u_x(z)$. The flow thus established is called 'simple shear flow' or 'plane Couette flow'. The balance of the relative motion of the different layers of width $\mathrm{d}z$ leads to the appearance of a friction force that opposes the motion of the plate and tries to re-establish an equilibrium where the fluid moves uniformly. When the velocity gradients are small, we expect a linear relation between the velocity gradient and the force. Let $\mathcal{P}_{\alpha\beta}$ be the α component of the force applied on a unit surface perpendicular to β. $\mathcal{P}_{\alpha\beta}$ is given in terms of the shear viscosity η by

$$\mathcal{P}_{\alpha\beta} = -\eta\, \partial_\beta u_\alpha$$

The symmetries and invariances of the problem considered here are such that only $\mathcal{P}_{xz} = -\eta\, \partial_z u_x$ is relevant.

Table 8.1 *Experimental values of the coefficient of viscosity for air*

T (K)	911	1023	1083	1196	1307	1407
η ($\times 10^{-7}$ poise)	401.4	426.3	441.9	464.3	490.6	520.6

We only treat the simple case where the fluid may be considered as a classical ideal gas made of particles of mass m. The temperature T and density n of the fluid are uniform. We assume that the total scattering cross section σ_0 is constant and we limit our attention to the case where the pulling speed u_0 is very small compared to the average thermal velocity.

A Viscosity and momentum diffusion

1. Consider the balance of forces applied on a volume element between two plane parallel faces of unit area and located at z and $z + \mathrm{d}z$. Show that the x component of the momentum, $p_x = mu_x$, satisfies the diffusion equation

$$\frac{\partial p_x}{\partial t} - \frac{\eta}{nm}\frac{\partial^2 p_x}{\partial z^2} = 0 \tag{8.161}$$

2. Justify qualitatively the relation between the viscosity and momentum transport.

B Coefficient of viscosity

1. Starting with (8.21), we easily show that

$$\eta = A\sqrt{T} \tag{8.162}$$

where A is a constant. Measurement of the coefficient of viscosity for air yielded Table 8.1. Does the temperature dependence of η given by (8.162) agree with experiments?

2. Relation (8.162) predicts that η does not depend on the density of the gas (or its pressure). Although non-intuitive, this has been established experimentally over a rather wide range of density values. Give a qualitative interpretation of this result.

3. Expression (8.21) for the coefficient of viscosity neglects collisions of more than two particles. It is therefore valid only for small density, i.e. $\ell \gg \sqrt{\sigma_{\text{tot}}}$. However, the density cannot be too small. What phenomenon, so far ignored, must be taken into account if $\ell \simeq L$?

C Coefficient of viscosity in the relaxation time approximation

1. In the reference frame that follows the motion of the fluid layer between z and $z + \mathrm{d}z$ (the fluid rest frame) we have at equilibrium a Maxwell–Boltzmann distribution in the velocity $\vec{w} = \vec{v} - \vec{u}$

$$f_0'(\vec{w}) = n\left(\frac{m}{2\pi kT}\right)^{3/2} \exp\left(-\frac{m\vec{w}^{\,2}}{2kT}\right)$$

Write down the equilibrium velocity distribution $f_0(\vec{v})$ in the laboratory frame.

2. Like the Boltzmann–Lorentz equation, the Boltzmann equation may be linearized around a local equilibrium solution. This leads to the following expression for the velocity distribution law in the fluid

$$f \simeq f_0 - \tau^* \vec{v} \cdot \vec{\nabla} f_0 \tag{8.163}$$

τ^* is the characteristic time to return to equilibrium, which is of the order of the time of flight between two collisions. Show that (8.163) allows us to write

$$\mathcal{P}_{xz} = m\frac{\mathrm{d}u_x}{\mathrm{d}z}\int \mathrm{d}^3w\, \tau^* w_x w_z^2 \frac{\partial f_0'}{\partial w_x} \tag{8.164}$$

3. The relaxation time depends *a priori* on v. To calculate (8.164) explicitly, we need assumptions on the behaviour of τ^*. To begin with we assume it to be constant. Verify that this assumption leads to results in disagreement with experiments which show that the viscosity coefficient depends linearly on the temperature.

If the cross section is almost independent of the velocity, a physically more reasonable assumption is to take the mean free path $\ell = \tau^*|\vec{v} - \vec{u}|$ to be constant. Show that

$$\mathcal{P}_{xz} = -\frac{1}{15}\frac{m^2\,\ell}{kT}\frac{\mathrm{d}u_x}{\mathrm{d}z}\int \mathrm{d}^3w\, w^3 f_0' \tag{8.165}$$

and verify that the viscosity coefficient can be written as

$$\eta = \frac{4}{15} nm\langle v\rangle \ell \tag{8.166}$$

8.6.5 Electrical conductivity in a magnetic field and quantum Hall effect

We consider a non-relativistic ideal gas of electrons, mass m and charge q ($q < 0$) obeying the Boltzmann–Lorentz equation in the presence of an external force $\vec{F}(\vec{r})$. *We assume here that the densities are independent of space and time*

$$f(\vec{r}, \vec{p}, t) = f(\vec{p}) \qquad n(\vec{r}, t) = n$$

Therefore, the first two terms in (8.28) vanish. In addition we assume that the local equilibrium distribution f_0 is a function only of energy ε: $f_0(\vec{p}) = f_0(p^2/2m = \varepsilon)$.

A Electric conductivity in the presence of a magnetic field

We consider the problem of conduction in a metal where the conduction electrons form a highly degenerate ideal Fermi gas. We subject the metal to an electric field $\vec{E}$ and a magnetic field $\vec{B}$. The force in Equation (8.28) is therefore the Lorentz force

$$\vec{F} = q\left(\vec{E} + \vec{v} \times \vec{B}\right) \tag{8.167}$$

1. First we take $\vec{B} = 0$. Calculate $\bar{f} = f - f_0$ and show that the electric current density $\vec{j}_{el}$ is given by

$$\vec{j}_{el} = -q^2 \int d^3p\, \tau^*(p)\vec{v}(\vec{v} \cdot \vec{E})\, \frac{\partial f_0}{\partial \varepsilon}$$

If f_0 is the Fermi distribution at $T = 0$

$$f_0(\varepsilon) = \frac{2}{h^3}\theta(\varepsilon_F - \varepsilon) \tag{8.168}$$

where ε_F is the Fermi energy, show that the electrical conductivity σ_{el}, defined by $\vec{j}_{el} = \sigma_{el}\vec{E}$, is given by

$$\sigma_{el} = \frac{nq^2}{m}\tau_F$$

where we took $\tau_F = \tau^*(p_F)$.

2. Now we take $\vec{B} \neq 0$. How is the conductivity modified if the applied $\vec{B}$ field is parallel to $\vec{E}$? Consider the case where the $\vec{E}$ field is in the xOy plane, $\vec{E} = (E_x, E_y, 0)$, and the $\vec{B}$ field parallel to the z-axis, $\vec{B} = (0, 0, B)$, $B > 0$. Show that (8.28) becomes

$$q\vec{v} \cdot \vec{E}\, \frac{\partial f_0}{\partial \varepsilon} + q(\vec{v} \times \vec{B}) \cdot \vec{\nabla}_{\vec{p}} \bar{f} = -\frac{\bar{f}}{\tau^*(p)} \tag{8.169}$$

We seek a solution of the form

$$\bar{f} = -\vec{v} \cdot \vec{C}\, \frac{\partial f_0}{\partial \varepsilon}$$

where $\vec{C}$ is an unknown vector to be determined that is a function of $\vec{E}$ and $\vec{B}$ but independent of $\vec{v}$. What should $\vec{C}$ be when $\vec{B} = 0$? $\vec{E} = 0$? In this last case, first estimate the average magnetic force.

3. Show that $\vec{C}$ satisfies

$$q\vec{E} + \vec{\omega} \times \vec{C} = \frac{\vec{C}}{\tau^*(p)} \tag{8.170}$$

with $\vec{\omega} = (0, 0, \omega)$, where $\omega = |q|B/m$ is the Larmor frequency. Justify that $\vec{C}$ is necessarily of the form

$$\vec{C} = \alpha\vec{E} + \delta\vec{B} + \gamma(\vec{B} \times \vec{E})$$

where α, δ, γ are real numbers. Find the expression for $\vec{C}$ and show that

$$\bar{f} = -\frac{q\tau^*}{1 + \omega^2\tau^{*2}} \left[\vec{E} + \tau^*(\vec{\omega} \times \vec{E})\right] \cdot \vec{v}\, \frac{\partial f_0}{\partial \varepsilon} \tag{8.171}$$

4. Calculate the electric current and the components $\sigma_{\alpha\beta}$ of the conductivity tensor

$$\begin{aligned} j_x^{\text{el}} &= \sigma_{xx} E_x + \sigma_{xy} E_y \\ j_y^{\text{el}} &= \sigma_{yx} E_x + \sigma_{yy} E_y \end{aligned} \tag{8.172}$$

Verify that

$$\sigma_{xy} = -\sigma_{yx}$$

and comment on this relation in terms of the Onsager relation.

B Simplified model and the Hall effect

1. To represent the effect of collisions in a simple way, we write an average equation of motion for the electrons (the Drude model)

$$\frac{\mathrm{d}\langle\vec{v}\rangle}{\mathrm{d}t} = -\frac{\langle\vec{v}\rangle}{\tau^*} + \frac{q}{m}\left(\vec{E} + \langle\vec{v}\rangle \times \vec{B}\right) \tag{8.173}$$

Give a physical interpretation for this equation. Verify that in the stationary regime we have

$$\langle v_x \rangle = \frac{q\tau^*}{m} E_x - \omega\tau^* \langle v_y \rangle$$

$$\langle v_y \rangle = \frac{q\tau^*}{m} E_y + \omega\tau^* \langle v_x \rangle$$

Show that if we take $\tau^* = \tau_{\text{F}}$, this model gives the same expressions for $\sigma_{\alpha\beta}$ found above.

2. Calculate in terms of E_x the value E_{H} of E_y which cancels j_y^{el}. Verify that the transport of electrons in this situation is the same as in the case $\vec{B} = 0$, in other

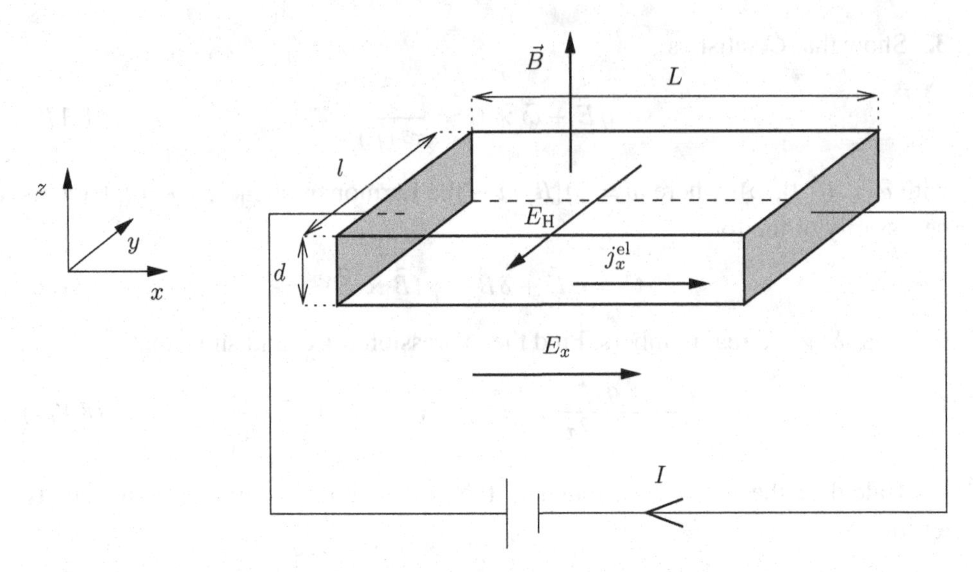

Figure 8.8 Schematic representation for an experiment on the Hall effect.

words

$$j_x = \sigma E_x$$

The field $E_{\rm H}$ is called the 'Hall field' and we define the 'Hall resistance' by

$$R_{\rm H} = \frac{V_{\rm H}}{I}$$

where $V_{\rm H}$ is the 'Hall voltage', $V_{\rm H}/l = E_{\rm H}$, and I the total current in the material (Figure 8.8). Show that $R_{\rm H}$ is given by

$$R_{\rm H} = \frac{B}{ndq}$$

By noting that $R_{\rm H}$ is independent of the relaxation time, find its expression using an elementary argument.

C Quantum Hall effect

Experiments of the type represented schematically in Figure 8.8 where electric ($\vec{E} = (E_x, 0, 0)$) and magnetic ($\vec{B} = (0, 0, B)$) fields are applied, have shown that, after transient effects have dissipated, a Hall field is established. However, in intense magnetic fields (> 1 T) and at very low temperatures, the Hall resistance is not linear in B as predicted by the Drude model. Experiments where the electrons are confined in *two-dimensional geometries* of area S and negligible thickness,

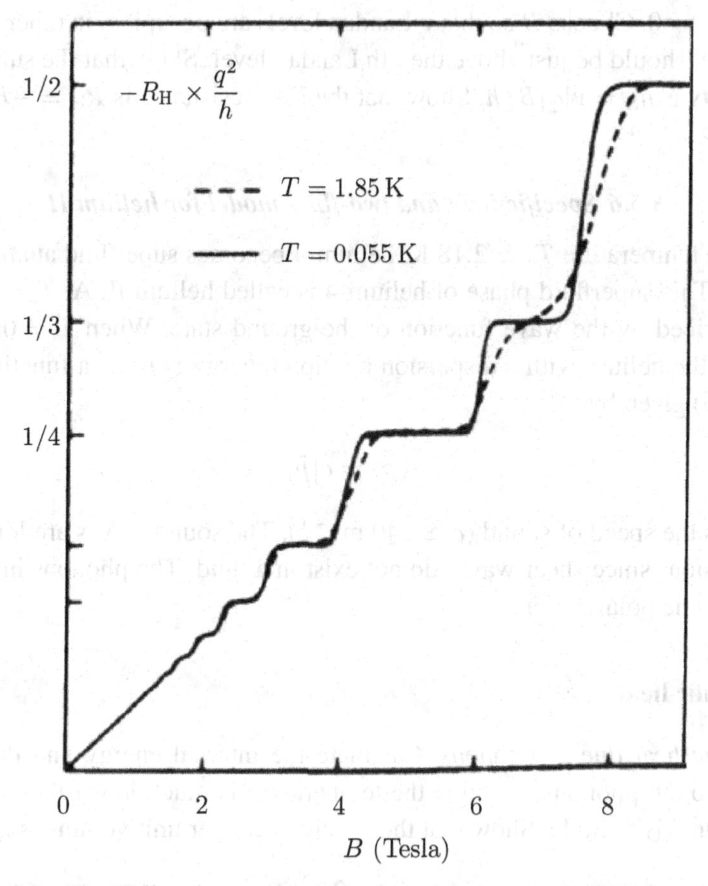

Figure 8.9 The Hall resistance as a function of the magnetic field for a heterojunction of InGaAs–InP, *Images de la Physique,* (1984).

exhibit a plateau structure for the Hall resistance as a function of magnetic field (see Figure 8.9). This suggests that the Hall resistance is quantized.

1. Calculate the energy level density $\rho(\varepsilon)$ of an electron gas in two dimensions without assuming spin degeneracy since this is lifted by the magnetic field.

2. In the presence of a magnetic field perpendicular to the surface, the energy levels (Landau levels) are labeled by an integer j: $j = 0, 1, 2, \ldots$ and have the form (see Problem 5.7.2)

$$\varepsilon_j = \hbar\omega\left(j + \frac{1}{2}\right) \tag{8.174}$$

where ω is the Larmor frequency. Calculate the degeneracy g of each level. This is equal to the number of levels, in zero field, present between $\varepsilon = \varepsilon_j$ and $\varepsilon = \varepsilon_{j+1}$.

3. Take $T = 0$. Choose B so that ν Landau levels are occupied, in other words the Fermi level should be just above the νth Landau level. Show that the surface electron density is $n_S = \nu|q|B/h$. Show that the Hall resistance is $R_H = -h/(\nu q^2)$.

8.6.6 Specific heat and two-fluid model for helium II

Below the temperature $T_\lambda \simeq 2.18$ K, helium-4 becomes superfluid at atmospheric pressure. This superfluid phase of helium-4 is called helium II. At $T = 0$, helium II is described by the wave function of the ground state. When $T \neq 0$, phonons appear in the helium with a dispersion relation (energy $\varepsilon(\vec{p})$ as a function of momentum $\vec{p}$) given by

$$\varepsilon(\vec{p}) = c|\vec{p}| \tag{8.175}$$

where c is the speed of sound ($c \simeq 240$ m s^{-1}). The sound waves are longitudinal (compression) since shear waves do not exist in a fluid. The phonons in helium II have only one polarization.

A Specific heat

1. *Specific heat due to phonons.* Calculate the internal energy and the specific heat due to the phonons. Assume the temperature is much lower than the Debye temperature $T_D \simeq 30$ K. Show that the specific heat per unit volume is given by

$$C_V^{\text{phonon}} = \frac{2\pi^2 k^4}{15\hbar^3 c^3} T^3$$

2. *Specific heat due to rotons.* In fact, the dispersion law for the phonons (which are called 'elementary excitations') is more complicated than Equation (8.175) and is given in Figure 8.10. The region around $p = p_0$ is called the 'roton' region. In the neighbourhood of $p = p_0$, we have the following approximate expression for $\varepsilon(p)$

$$\varepsilon(p) = \Delta + \frac{(p - p_0)^2}{2\mu} \tag{8.176}$$

with $\Delta/k \simeq 8.5$ K, $k_0 = p_0/\hbar \simeq 1.9$ Å^{-1} and $\mu \simeq 10^{-27}$ kg.

Show that for $T \lesssim 1$ K, we may replace the Bose distribution $n(p)$ by a Boltzmann distribution

$$n(p) \simeq e^{-\beta\varepsilon(p)}$$

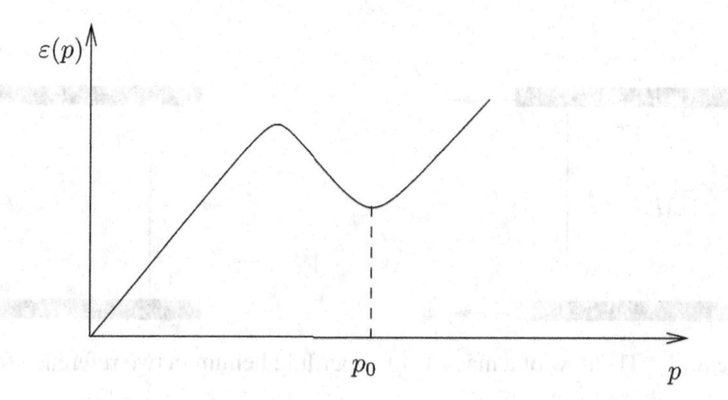

Figure 8.10 Dispersion law for elementary excitations in helium II.

Write the expression giving the contribution of the rotons to the internal energy. Compare p_0^2 and $\mu k_B T$ for $T = 1$ K and also Δ and $k_B T$. Use this to simplify the integral. Hint: Change variables $x = \sqrt{\beta/2\mu}(p - p_0)$. Then express the roton contribution to the specific heat C_V^{roton}. Compare C_V^{phonon} and C_V^{roton} for $T \to 0$ and their numerical values for $T = 1$ K.

B The two-fluid model

1. *Flow of helium II.* We consider a fluid of mass M containing N particles of mass m. The ith particle is characterized by its position $\vec{r}_i$ and momentum $\vec{q}_i$ with the total momentum $\vec{P} = \sum \vec{q}_i$. The fluid is assumed to be non-viscous and flows without friction in a tube. Consider this flow of the superfluid helium of mass M in two Galilean reference frames: in the frame R where the helium is at rest and the walls have a velocity $\vec{v}$, and the frame R' where the walls are at rest and the helium flows with velocity $-\vec{v}$. This description is possible because in the absence of viscosity the velocity of the helium does not vanish at the walls of the tube (in R'). In frame R the Hamiltonian is

$$H = \sum_{i=1}^{N} \frac{\vec{q}_i^{\,2}}{2m} + \frac{1}{2} \sum_{i \neq j} U(\vec{r}_i - \vec{r}_j) \tag{8.177}$$

Give the expression for the Hamiltonian H' in R'.

In the Landau two-fluid model, we assume that helium II is made of a superfluid with zero viscosity, and a normal viscous fluid that is identified with the phonon gas (or more generally with the gas of elementary excitations). We ignore interactions among elementary excitations and also between the normal fluid and superfluid components. The superfluid flows without friction whereas the elementary excitations are in equilibrium with the walls of the tube. The dispersion law

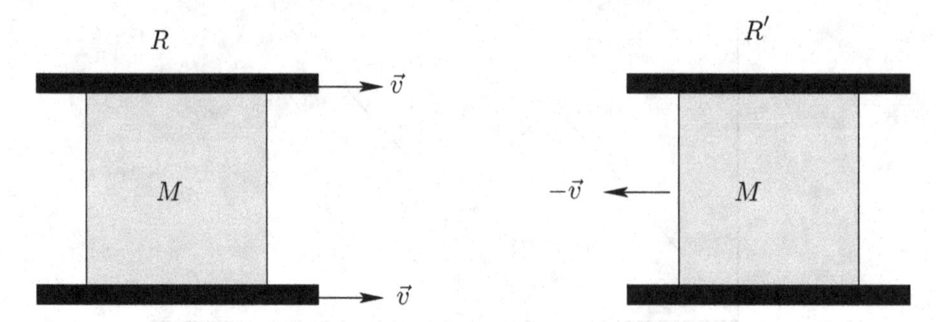

Figure 8.11 The flow of a mass M of superfluid helium in two reference frames.

$\varepsilon(\vec{p})$ of the elementary excitations is given in the frame R where the superfluid has zero velocity: the momentum $\vec{p}$ of the elementary excitations is measured in *this reference frame*.

We consider the flow of helium II in a tube in the two Galilean frames R and R' (Figure 8.11). By examining in R the creation of an elementary excitation of momentum $\vec{p}$ and energy $\varepsilon(\vec{p})$, show that in R' where the tube is at rest, the energy of this excitation is

$$\varepsilon'(\vec{p}) = \varepsilon(\vec{p}) - \vec{p} \cdot \vec{v} \tag{8.178}$$

2. *Momentum density.* The gas of elementary excitations is in equilibrium with the walls at temperature T and has a bulk velocity $\vec{v}$ in the frame R. What is the distribution $\tilde{n}(\vec{p}, \vec{v})$ of the elementary excitations in terms of the Bose distribution $n(\varepsilon) = 1/(\mathrm{e}^{\beta\varepsilon} - 1)$? Show that in the frame R the momentum density $\vec{g}$ is given to first order in v by

$$\vec{g} = -\frac{1}{3} \int \frac{\mathrm{d}^3 p}{(2\pi\hbar)^3}\, p^2 \frac{\mathrm{d}n}{\mathrm{d}\varepsilon}\, \vec{v} \tag{8.179}$$

The relation $\vec{g} = \rho_n \vec{v}$ *defines* the density ρ_n of the normal fluid. Note: In what follows it is recommended not to use the explicit expression for $n(\varepsilon)$. By integration by parts, relate ρ_n to the energy density ϵ of a phonon gas where the dispersion law is $\varepsilon = c\vec{p}$. What is the pressure $\mathcal{P}$ of the phonon gas?

3. *Energy current and momentum density.* Show that for a phonon gas, the energy current $\vec{j}_E$ calculated in the frame R is related to the momentum density by

$$\vec{j}_E = c^2 \vec{g}$$

4. *Second sound.* Use the continuity equations and Euler's equation (6.70) to show that in the absence of dissipation and for sufficiently low velocities, we have

$$\frac{\partial \epsilon}{\partial t} = -c^2 \vec{\nabla} \cdot (\rho_n \vec{v})$$

$$\frac{\partial (\rho_n \vec{v})}{\partial t} = -\frac{1}{3} \vec{\nabla} \epsilon$$

Show that ϵ obeys a wave equation and verify that the propagation velocity is $c/\sqrt{3}$. How do you interpret this wave of the so-called 'second sound'? What happens when helium II is heated locally? What is the sound velocity in a dilute ultra-relativistic gas?

8.6.7 Landau theory of Fermi liquids

Our goal in this problem is to go beyond the ideal gas approximation of the Fermi gas in the case of neutral particles, one example being liquid helium-3. We denote the dispersion law of the interacting Fermi liquid[26] by $\varepsilon(p)$: more precisely, $\varepsilon(p)$ is the extra energy due to adding *one* particle of momentum $\vec{p}$ to the system at equilibrium. In the ideal case, $\varepsilon(p) = p^2/(2m)$, but we want to consider more general dispersion laws. For simplicity, we neglect spin effects: these may be easily taken into account if one is interested, for example, in Pauli paramagnetism. Landau's first assumption is the existence of a sharp Fermi surface at zero temperature. The Fermi distribution is then the following functional of $\varepsilon(p)$ (μ was denoted by ε_0 in Section 5.2.3)

$$f_0[\varepsilon(p)] = \theta(\mu - \varepsilon(p)) \qquad \frac{\delta f_0[\varepsilon(p)]}{\delta \varepsilon(p)} = -\delta(\varepsilon(p) - \mu)$$

This distribution is the equilibrium distribution of the interacting Fermi liquid at zero temperature. The Fermi momentum p_F is still given by (5.17), but $\mu \neq p_F^2/(2m) = \varepsilon_F$.

A Static properties

1. One adds (removes) a particle of energy $\varepsilon(p)$ to (from) the Fermi sea, thus creating a quasiparticle (quasihole). Show that

$$\varepsilon(p)\Big|_{p=p_F} = \mu$$

[26] We assume an isotropic Fermi gas, which is the case for helium-3. However, electrons in metals do not obey this assumption due to the loss of rotational invariance arising from the crystal lattice.

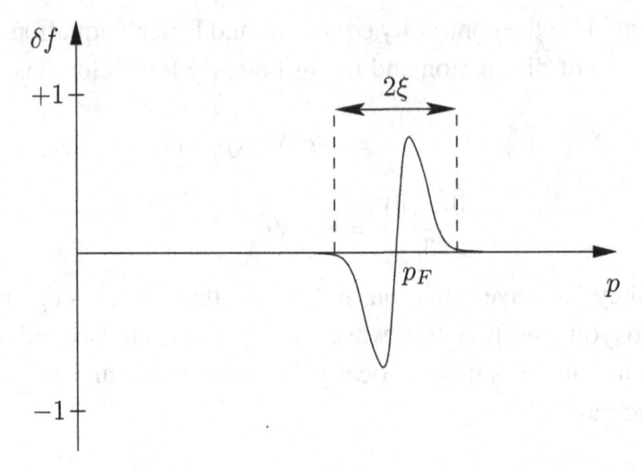

Figure 8.12 Perturbation of the equilibrium distribution. $\delta f(\vec{p})$ is non-zero only in the vicinity of the Fermi surface. The negative part of δf corresponds to the hole distribution, the positive part to that of particles.

2. One now assumes that many quasiparticles or quasiholes are added to the Fermi sea. This does not necessarily mean that the *total* number of particles varies, but the particle distribution is modified with respect to the equilibrium distribution

$$\delta f(\vec{p}) = f(\vec{p}) - f_0(\vec{p})$$

The quantity $\delta f(\vec{p})$ is a measure of the deviation of the particle distribution from the equilibrium distribution, or, in other words, it measures the 'degree of excitation' in the neighbourhood of the Fermi surface. Because it relies in fact on an expansion in powers of δf, Landau's theory will only be valid at low temperatures, and δf will differ from zero only in a neighbourhood ξ of the Fermi sea (Figure 8.12)

$$|\varepsilon(p) - \mu| \lesssim \xi \ll \mu$$

Landau's second assumption is that one may write the energy of a quasiparticle *in the environment of other quasiparticles* as

$$\tilde{\varepsilon}(\vec{p}) = \varepsilon(p) + \frac{2V}{h^3} \int \mathrm{d}^3 p' \, \lambda(\vec{p}, \vec{p}\,')\delta f(\vec{p}\,')$$

Therefore, as already mentioned, $\tilde{\varepsilon}(\vec{p}) = \varepsilon(p)$ if one adds a single particle to the system. Note that, contrary to $\varepsilon(p)$, $\tilde{\varepsilon}(\vec{p})$ is not in general an isotropic function, since $\delta f(\vec{p})$ is not isotropic. The function $\lambda(\vec{p}, \vec{p}\,')$ is defined in the vicinity of the

Fermi surface ($p = p' = p_F$) and is expanded in Legendre polynomials

$$\lambda(\vec{p}, \vec{p}\,') = \sum_{\ell=0}^{\infty} \alpha_\ell P_\ell(\cos\theta)$$

where θ is the angle between $\vec{p}$ and $\vec{p}\,'$ and the α_ℓs are real coefficients. Show that the order of magnitude of $\lambda(\vec{p}, \vec{p}\,')$ is a^3/V, where a is the range of the interaction between particles. Show that, exactly as $|\varepsilon(p) - \mu|$, the second term in the equation for $\tilde{\varepsilon}(p)$ is of the order of ξ, and may not be neglected. However, a term $\propto (\delta f(p))^2$ in the expansion of $\tilde{\varepsilon}(\vec{p})$ would be negligible. Show that if $\delta f(\vec{p})$ is isotropic, then

$$\tilde{\varepsilon}(p) = \varepsilon(p) + \alpha_0 \delta N$$

where δN is the total number of quasiparticles.

3. Instead of $\delta f(\vec{p})$, it is useful to define

$$\delta \overline{f}(\vec{p}) = f(\vec{p}) - f_0(\tilde{\varepsilon}(\vec{p}))$$

Show that

$$\delta \overline{f}(\vec{p}) = \delta f(\vec{p}) + \delta(\varepsilon - \mu)\frac{2V}{h^3}\int \mathrm{d}^3p'\, \lambda(\vec{p}, \vec{p}\,')\delta f(\vec{p}\,')$$

4. Let $f_0(\vec{p}, T)$ be the equilibrium distribution of quasiparticles at temperature T, $kT \ll \mu$

$$f_0(\vec{p}, T) = \frac{1}{1 + \exp[(\tilde{\varepsilon}(\vec{p}) - \mu)/kT]}$$

and define the 'thermal excitation' $\delta f(\vec{p}, T)$ by

$$\delta f(\vec{p}, T) = f_0(\vec{p}, T) - f_0[\varepsilon(p)]$$

Using the Sommerfeld formula (5.29), show that

$$\int \mathrm{d}^3p\, \delta f(\vec{p}, T) \propto (kT)^2 \rho'(\mu)$$

where $\rho(\varepsilon)$ is the density of states. Deduce from this result that

$$\int p^2 \mathrm{d}p\, \delta f(\vec{p}, T) \propto T^2$$

for any direction of $\vec{p}$, and that one can set $\tilde{\varepsilon}(\vec{p}) = \varepsilon(p)$ to leading order in T. Thus one can adapt the formulae of the non-interacting Fermi gas by simply replacing the free particle dispersion law by $\varepsilon(p)$. Obtain the specific heat

$$C_V = \frac{\pi^2 k^2 T}{3} \rho(\mu) = \frac{V k^2 T}{3\hbar^2} m^* p_F$$

where the *effective mass* m^* is defined by

$$\left. \frac{d\varepsilon}{dp} \right|_{p=p_F} = \frac{p_F}{m^*}$$

In the rest of the problem, we shall revert to the $T = 0$ case.

5. Compute $\delta\overline{f}(\vec{p})$ when the chemical potential varies from μ to $\mu + d\mu$. Taking into account the isotropy of $\delta f(\vec{p})$ in this case, show that

$$\frac{\partial N}{\partial \mu} = \frac{\rho(\mu)}{1 + \Lambda_0}$$

$\Lambda_0 = \alpha_0 \rho(\mu)$ is a dimensionless parameter independent of V. Show that the expression for the $T = 0$ compressibility is now given by

$$\kappa = \frac{1}{n^2} \frac{m^* p_F}{\pi^2 \hbar^3 (1 + \Lambda_0)}$$

For helium-3 at atmospheric pressure, $m^*/m \simeq 3$ and $\Lambda_0 \simeq 10$. Is the compressibility over or underestimated when it is computed in the ideal gas model?

B Boltzmann equation

1. In a local equilibrium situation, the quasiparticle energy, $\tilde{\varepsilon}(\vec{p}, \vec{r})$, is a function of $\vec{p}$ and $\vec{r}$. In order to describe transport phenomena in a Fermi liquid, Landau assumed that independent quasiparticles are described by a classical Hamiltonian $\tilde{\varepsilon}$. A kinetic theory of the ideal gas is then possible.[27] Taking into account

$$\frac{d\vec{r}}{dt} = \vec{v} = \vec{\nabla}_{\vec{p}}\, \tilde{\varepsilon} \qquad \frac{d\vec{p}}{dt} = \vec{F} = -\vec{\nabla}_{\vec{r}}\, \tilde{\varepsilon}$$

the Boltzmann equation for the distribution $f(\vec{r}, \vec{p}, t)$ of quasiparticles in the absence of collisions reads

$$\frac{\partial f}{\partial t} + \vec{\nabla}_{\vec{p}}\, \tilde{\varepsilon} \cdot \vec{\nabla}_{\vec{r}} f - (\vec{\nabla}_{\vec{r}}\, \tilde{\varepsilon}) \cdot \vec{\nabla}_{\vec{p}} f = 0$$

2. One writes for the local deviation from equilibrium

$$f(\vec{r}, \vec{p}, t) = f_0[\varepsilon(p)] + \delta f(\vec{r}, \vec{p}, t)$$

[27] Strictly speaking, the assumption of a one-particle distribution $f(\vec{r}, \vec{p}, t)$ is incompatible with the uncertainty principle. One can show that the theory is valid provided the characteristic space and time scales ℓ and τ of f obey $\hbar v_F/\ell \ll \mu$ and $\hbar/\tau \ll \mu$.

where f_0 is the equilibrium distribution. The density $n(\vec{r}, t)$ is given by

$$n(\vec{r}, t) = \frac{2}{h^3} \int \mathrm{d}^3 p\, f(\vec{r}, \vec{p}, t)$$

The local $\tilde{\varepsilon}$ is then

$$\tilde{\varepsilon}(\vec{p}, \vec{r}, t) = \varepsilon(p) + \frac{2V}{h^3} \int \mathrm{d}^3 p'\, \lambda(\vec{p}, \vec{p}\,')\delta f(\vec{r}, \vec{p}\,', t)$$

Show that $\vec{\nabla}_{\vec{r}}\, \tilde{\varepsilon}$ is first order in δf. Check that

$$\vec{\nabla}_{\vec{p}}\, f_0 = -\frac{\vec{p}}{m^*}\, \delta(\varepsilon(p) - \mu) = -\vec{v}\, \delta(\varepsilon(p) - \mu)$$

where $\vec{v}$ is the group velocity. Show that to leading order in δf, the Boltzmann equation is given by (note that the equation features the time-derivative of δf and the space-derivative of $\delta \overline{f}$!)

$$\frac{\partial\, \delta f}{\partial t} + \vec{v} \cdot \vec{\nabla}_{\vec{r}}\, \delta \overline{f} = \mathcal{C}[\delta f]$$

Do not try to give the explicit expression for the collision term $\mathcal{C}[\delta f]$.

3. The density excess of quasiparticles is given by

$$\delta n(\vec{r}, t) = \frac{2}{h^3} \int \mathrm{d}^3 p\, \delta f(\vec{r}, \vec{p}, t)$$

and the associated current $\vec{j}$ must obey the continuity equation

$$\frac{\partial \delta n}{\partial t} + \vec{\nabla}_{\vec{r}} \cdot \vec{j} = 0$$

Using the Boltzmann equation and the property of the collision term

$$\int \mathrm{d}^3 p\, \mathcal{C}[\delta f] = 0$$

show that the current

$$\vec{j} = \frac{2}{h^3} \int \mathrm{d}^3 p\, \vec{v}\, \delta \overline{f}$$

obeys the continuity equation. One could have guessed the following form of the current

$$\vec{j}\,' = \frac{2}{h^3} \int \mathrm{d}^3 p\, \vec{v}\, \delta f$$

Why is it incorrect?

4. One can nevertheless write a correct expression for the current featuring δf

$$\vec{j} = \frac{2}{h^3} \int \mathrm{d}^3 p \, \vec{j}_{\vec{p}} \, \delta f$$

with

$$\vec{j}_{\vec{p}} = \vec{v} + \frac{2V}{h^3} \int \mathrm{d}^3 p' \, \frac{\vec{p}\,'}{m^*} \lambda(\vec{p}, \vec{p}\,') \delta(\varepsilon(p') - \mu) = \vec{v} \left(1 + \frac{1}{3} \alpha_1 \rho(\mu) \right)$$

which one identifies with the current associated with a *single localized quasiparticle*. For a translation invariant system, Galilean invariance implies $\vec{j}_{\vec{p}} = \vec{p}/m$. Use this property to derive a relation between the effective mass and the mass of the free particle

$$\frac{1}{m} = \frac{1}{m^*} \left(1 + \frac{1}{3} \Lambda_1 \right) \qquad \Lambda_1 = \alpha_1 \rho(\mu)$$

8.6.8 Calculation of the coefficient of thermal conductivity

1. We assume the gas to be *at rest*, with a temperature gradient along the z direction: the temperature $T(z)$ is a function of z. The pressure $\mathcal{P}$ must be uniform, otherwise a pressure gradient would set the gas in motion. Show that the local equilibrium distribution $f_0(z)$ is given by ($\beta(z) = 1/(kT(z))$)

$$f_0(z) = \mathcal{P} \frac{[\beta(z)]^{5/2}}{(2\pi m)^{3/2}} \exp\left(- \frac{\beta(z) p^2}{2m} \right)$$

One writes, following (8.127)

$$f = f_0 \left(1 - \Phi \frac{\partial T}{\partial z} \right)$$

Show that the drift term of the Boltzmann equation is

$$\mathrm{D} f = \frac{\beta}{T} \left(\frac{5}{2} k - \varepsilon(p) \right) f_0 \frac{\partial T}{\partial z} \qquad \varepsilon(p) = \frac{p^2}{2m}$$

If the gas is not mono-atomic, one can show that

$$\Phi = \frac{\beta}{T} \left(c_P - \varepsilon(p) \right)$$

where c_P is the specific heat per particle at constant pressure, $c_P = 5k/2$ for a mono-atomic gas.

2. Following (8.128), one writes the linearized Boltzmann equation in the form

$$(\varepsilon(p) - c_P T) p_z = \frac{mT}{\beta} \int \prod_{i=2}^{4} \mathrm{d}^3 p_i \, W f_{02} \Delta \Phi = \mathsf{L}[\Phi]$$

From the symmetry of the problem, the function $\Phi(\vec{p})$ must be proportional to p_z

$$\Phi(\vec{p}) = A(p)p_z$$

One must also choose $\Phi(\vec{p})$ in such a way that (8.120) is obeyed. Show that

$$\int \mathrm{d}^3 p\, \varepsilon(p) p_z f_0(p)\Phi(\vec{p}) = \frac{Tm}{4\beta}\int \prod_{i=1}^{4} \mathrm{d}^3 p_i\, W f_{01} f_{02} (\Delta\Phi)^2$$

3. Derive the following expression for the coefficient of thermal conductivity κ

$$\kappa = \frac{1}{m}\int \mathrm{d}^3 p\, \varepsilon(p) p_z f_0(p)\Phi(\vec{p})$$

and using the scalar product defined in (8.131), show that

$$\kappa = \langle\Phi|X\rangle = \langle\Phi|\mathsf{L}|\Phi\rangle$$

where $|X\rangle$ is the Hilbert space vector representing the function $\varepsilon(p)p_z$. From these results, devise a variational method for computing κ.

4. The simplest choice one can think of as a trial function is $\Psi = Ap_z$ with a constant A. However, with this choice, the second equation in (8.120) is not satisfied. Show that the trial function

$$\Psi(\vec{p}) = A(1 - \gamma p^2)p_z$$

has all the required properties if $\gamma = \beta/(5m)$.

5. Show that

$$\langle X|\Psi\rangle = -\frac{An}{\beta^2}$$

where n is the density, and that

$$\langle\Psi|\mathsf{L}|\Psi\rangle = \frac{32}{25\sqrt{\pi}} A^2 T n^2 \frac{m}{\beta^3}\left(\frac{\beta}{m}\right)^{1/2} \sigma_{\mathrm{tr}}$$

Hint: Going to centre-of-mass variables $\vec{P}$, $\vec{p}$ and $\vec{p}\,'$, you will have to integrate an expression of the form

$$\int \mathrm{d}^3 p\, \mathrm{d}^3 P \left[(\vec{P}\cdot\vec{p})p_z - (\vec{P}\cdot\vec{p}\,')p_z'\right]^2 \cdots \to \frac{2}{9}\int \mathrm{d}^3 p\, \mathrm{d}^3 P\, P^2 p^4 (1 - \cos^2\theta)$$

First show that the angular average over the direction of $\vec{P}$ gives a factor 1/3.

6. From these results, derive a variational estimate for κ, assuming a p-independent transport cross section

$$\kappa = \frac{25\sqrt{\pi}}{32}\,\frac{k}{m}\,\frac{\sqrt{mkT}}{\sigma_{\mathrm{tr}}}$$

Compute the ratio κ/η and compare with the elementary estimate derived in Section 8.1.3.

8.7 Further reading

The classical calculation of cross sections is done in Goldstein [48] (Chapter III), Landau and Lifshitz [69] (Chapter IV) and the quantum calculation in Messiah [89] (Chapter X) and Cohen-Tannoudji *et al.* [30] (Chapter VIII). An elementary treatment of kinetic theory is in Reif [109] (Chapter 12) and Baierlein [4] (Chapter 15). At a more advanced level, the reference in the subject is the book by Lifshitz and Pitaevskii [81]. The Boltzmann–Lorentz model is treated in Balian [5] (Chapter 15) and Lifshitz and Pitaevskii [81] (Section 11). As further reading on the Boltzmann equation, we recommend Reif [109] (Chapters 13 and 14), Balian [5] (Chapter 15), Kreuzer [67] (Chapter 8), Kubo [68] (Chapter 6), McQuarrie [88] (Chapters 16 to 19) and Lifshitz and Pitaevskii [81] (Sections 1 to 10). One finds in these references a complete calculation of transport coefficients. An example of the increase of the H-function is given by Jaynes [60], but the argument is incomplete. The extension of the Boltzmann equation to the quantum case (the Landau–Uhlenbeck equation) can be found in Balian [5] (Chapter 11), Pines and Nozières [102] (Chapter 1), Baym and Pethick [15] and Lifshitz and Pitaevskii [81] (Chapter VIII).

The following references may be helpful for problems. The integer quantum Hall effect is described in the Nobel lectures by Laughlin [71] and Stormer [116]. For the fractional Hall effect see the reviews by Jain [59], Heiblum and Stern [55] and the Nobel lectures by Laughlin [71] and Stormer [116]. The two-fluid model for helium-4 is discussed in Landau and Lifshitz [70] (Volume 2), Chapter III, Goodstein [49] (Chapter 5) and Nozières and Pines [96] (Chapter 6). Landau's theory of Fermi liquids is examined in Lifshitz and Pitaevskii [81] (Section 74) to 76. See also Baym and Pethick [15] and Pines and Nozières [102] (Chapter 1).

9

Topics in non-equilibrium statistical mechanics

We have given in two previous chapters a first introduction to non-equilibrium phenomena. The present chapter is devoted to a presentation of more general approaches, in which time dependence will be made explicit, whereas in practice we had to limit ourselves to stationary situations in Chapters 6 and 8. In the first part of the chapter, we examine the relaxation toward equilibrium of a system that has been brought out of equilibrium by an external perturbation. The main result is that, for small deviations from equilibrium, this relaxation is described by *equilibrium* time correlation functions, called Kubo (or relaxation) functions: this result is also known as 'Onsager's regression law'. The Kubo functions turn out to be basic objects of non-equilibrium statistical mechanics. First they allow one to compute the dynamical susceptibilities, which describe the response of the system to an external time dependent perturbation: the dynamical susceptibilities are, within a multiplicative constant, the time derivatives of Kubo functions. A second crucial property is that transport coefficients can be expressed in terms of time integrals of Kubo functions. As we limit ourselves to small deviations from equilibrium, our theory is restricted to a linear approximation and is known as *linear response theory*.[1] The classical version of linear reponse is somewhat simpler than the quantum one, and will be described first in Section 9.1. We shall turn to the quantum theory in Section 9.2, where one of our main results will be the proof of the fluctuation-dissipation theorem. In Section 9.3 we shall describe the projection method, restricting ourselves to the simplest case, the Mori projection method. The idea that underlies this method is that one is not interested in the full dynamics of an N-body system, but only in that of slow modes, for example hydrodynamic modes. One is led to project the full dynamics on that of its slow modes. The back action of fast modes on slow modes can be described thanks to memory effects. The projection

[1] The validity of linear response theory has been challenged by van Kampen in *Physica Norvegica* **5**, 279 (1971). For an answer to van Kampen's arguments, see for example Dorfman [33] Chapter 6.

method leads in a natural way to a description of the dynamics in Section 9.4 by a stochastic differential equation, the Langevin equation. It will be shown in Section 9.5 that the probability distribution associated with the Langevin equation obeys a partial differential equation, the Fokker–Planck equation, whose solutions will be obtained by using an analogy with the Schrödinger equation in imaginary time. Finally numerical studies of the Langevin equation will be examined in Section 9.6.

9.1 Linear response: classical theory

9.1.1 Dynamical susceptibility

Our goal is to study small deviations from equilibrium driven by a (small) perturbation applied to the system, when the equilibrium situation is described by an unperturbed classical Hamiltonian $H(\boldsymbol{p}, \boldsymbol{q})$. As in Section 3.2.1, $(\boldsymbol{p}, \boldsymbol{q})$ is a shorthand notation for the full set of canonical variables $(p_1, \ldots, p_N; q_1, \ldots, q_N)$, N being the number of degrees of freedom, and the partition function is

$$Z(H) = \int \mathrm{d}\boldsymbol{p}\, \mathrm{d}\boldsymbol{q}\, \mathrm{e}^{-\beta H(\mathbf{p},\mathbf{q})} \tag{9.1}$$

The equilibrium average A_i of a classical dynamical variable $\mathcal{A}_i(\boldsymbol{p}, \boldsymbol{q})$ is given from (3.43) by

$$A_i \equiv \langle \mathcal{A}_i \rangle \equiv \langle \mathcal{A}_i \rangle_{\text{eq}} = \int \mathrm{d}\boldsymbol{p}\, \mathrm{d}\boldsymbol{q}\, D_{\text{eq}}(\boldsymbol{p}, \boldsymbol{q}) \mathcal{A}_i(\boldsymbol{p}, \boldsymbol{q}) \tag{9.2}$$

where the probability density D_{eq} is the normalized Boltzmann weight

$$D_{\text{eq}}(\mathbf{p}, \mathbf{q}) = \frac{1}{Z(H)} \mathrm{e}^{-\beta H(\mathbf{p},\mathbf{q})} \tag{9.3}$$

In the present chapter, the notation $\langle \bullet \rangle$ *will always stand for an average value computed with the equilibrium probability density (9.3), or its quantum analogue (9.55).* Let us perturb the Hamiltonian H by applying a perturbation $\mathcal{V}$ of the form

$$\mathcal{V} = -\sum_j f_j \mathcal{A}_j$$

so that

$$H \to H_1 = H + \mathcal{V} = H - \sum_j f_j \mathcal{A}_j \tag{9.4}$$

Comparing with (2.64), we see that the rôle of the Lagrange multipliers λ_j is now played by βf_j. In the present context, the f_js are often called the *external forces*, or simply the forces. This terminology is borrowed from the forced one-dimensional harmonic oscillator (Exercise 9.7.1), where the perturbation may be written in terms of the dynamical variable $x(t)$ as $\mathcal{V} = -f(t)x(t)$. We denote by $\overline{\mathcal{A}_i}$ the average of the dynamical variable $\mathcal{A}_i$ *computed with the perturbed Hamiltonian* H_1. The response of the average $\overline{\mathcal{A}_i}$ to a variation of the Lagrange multiplier f_j is given by the fluctuation-response theorem, which we may use in the form (2.70) since the classical variables $\mathcal{A}_i$ commute

$$\frac{\partial \overline{\mathcal{A}_i}}{\partial f_j} = \beta \overline{(\mathcal{A}_i - \overline{\mathcal{A}_i})(\mathcal{A}_j - \overline{\mathcal{A}_j})} \tag{9.5}$$

If we restrict ourselves to small deviations from the equilibrium Hamiltonian H, we can take the limit $f_j \to 0$ in (9.5), which becomes, since the average values are now computed with D_{eq},

$$\begin{aligned} \left.\frac{\partial \overline{\mathcal{A}_i}}{\partial f_j}\right|_{f_k=0} &= \beta\langle(\mathcal{A}_i - \langle\mathcal{A}_i\rangle)(\mathcal{A}_j - \langle\mathcal{A}_j\rangle)\rangle \\ &= \beta\langle\delta\mathcal{A}_i\,\delta\mathcal{A}_j\rangle = \langle\mathcal{A}_i\,\mathcal{A}_j\rangle_{\mathrm{c}} \end{aligned} \tag{9.6}$$

where we have defined $\delta\mathcal{A}_i$ by

$$\delta\mathcal{A}_i = \mathcal{A}_i - \langle\mathcal{A}_i\rangle = \mathcal{A}_i - A_i \tag{9.7}$$

so that its (equilibrium) average value vanishes: $\langle\delta\mathcal{A}_i\rangle = 0$. The subscript c stands for 'cumulant' or 'connected'. In the *linear approximation* (9.6) implies at once

$$\overline{\mathcal{A}_i} = \langle\mathcal{A}_i\rangle + \beta\sum_j f_j\langle\delta\mathcal{A}_i\,\delta\mathcal{A}_j\rangle = \langle\mathcal{A}_i\rangle - \beta\langle\mathcal{A}_i\mathcal{V}\rangle_{\mathrm{c}} \tag{9.8}$$

Thus, within this approximation, the deviation from equilibrium is linearly related to the perturbation.

So far we have limited ourselves to static situations: the perturbed Hamiltonian H_1 is time-independent and has been used to define a new probability density D_1

$$D_1 = \frac{1}{Z(H_1)}\,\mathrm{e}^{-\beta H_1}$$

We now assume that the Hamiltonian is equal to H_1 for $t < 0$ during which the system is at equilibrium with a probability density D_1, but we introduce an explicit time dependence by switching off suddenly the perturbation $\mathcal{V}$ at $t = 0$. The Hamiltonian $\tilde{H}_1(t)$ is now

$$\tilde{H}_1(t) = H - \sum_i f_j\mathcal{A}_j = H_1 \quad \text{if} \quad t < 0 \qquad \tilde{H}_1(t) = H \ \text{if} \ t \geq 0 \tag{9.9}$$

Another equivalent way to proceed is to switch on adiabatically[2] the perturbation at $t = -\infty$ and to switch it off suddenly at $t = 0$,[3]

$$\tilde{H}_1(t) = H - \theta(-t)\mathrm{e}^{\eta t}\sum_j f_j \mathcal{A}_j \qquad \eta \to 0^+ \tag{9.10}$$

where $\theta(-t)$ is a step function. Note that the perturbation $\mathcal{V}$ is a *mechanical perturbation* and the time evolution is Hamiltonian. Energy flows into and out of the system in the form of work only, and we assume that the system remains thermally isolated. We expect that the system will relax for $t \to +\infty$ to an equilibrium situation described by H. It is very difficult to characterize this relaxation in a general way. However, if the deviation from equilibrium is not large, we may work in the framework of the *linear* approximation, and the analysis is straightforward. For $t \le 0$, $\overline{\mathcal{A}_i}$ takes the time-independent value (9.8), and in particular at $t = 0$

$$\begin{aligned}\overline{\delta\mathcal{A}_i}(t=0) = \overline{\mathcal{A}_i}(t=0) - \langle\mathcal{A}_i\rangle &= \beta\sum_j f_j \langle\mathcal{A}_i(t=0)\mathcal{A}_j\rangle_{\mathrm{c}} \\ &= \beta\sum_j f_j \langle\delta\mathcal{A}_i(t=0)\delta\mathcal{A}_j\rangle\end{aligned}$$

The non-equilibrium ensemble average is obtained by integrating over all possible initial conditions at $t = 0$ with the weight $D_1 = \exp(-\beta H_1)/Z(H_1)$. It is convenient to write the dynamical variable $\mathcal{A}_i$ for positive t as a function of the initial conditions at $t = 0$, $\boldsymbol{p} = \boldsymbol{p}(t=0)$, $\boldsymbol{q} = \boldsymbol{q}(t=0)$, and of t: $\mathcal{A}_i(t) = \mathcal{A}_i(t; \boldsymbol{p}, \boldsymbol{q})$. Contrary to the $t \le 0$ case, $\overline{\mathcal{A}_i}$ will be time dependent for $t \ge 0$ because the time evolution of $\mathcal{A}_i$ for positive t is governed by H

$$t > 0: \qquad \partial_t \mathcal{A}_i = \{\mathcal{A}_i, H\}$$

while the probability density at $t = 0$ is determined by H_1 (9.4)

$$H_1 = H - \sum_j f_j \mathcal{A}_j = H - \sum_j f_j \mathcal{A}_j(0)$$

Then the non-equilibrium ensemble average of $\mathcal{A}_i(t)$ is

$$\overline{\mathcal{A}_i}(t) = \int \mathrm{d}\boldsymbol{p}\,\mathrm{d}\boldsymbol{q}\, D_1(\boldsymbol{p}, \boldsymbol{q})\mathcal{A}_i(t; \boldsymbol{p}, \boldsymbol{q})$$

[2] In the present context, adiabatically means infinitely slowly. The relation between this meaning of adiabatic and the thermodynamic one is complex, and is discussed for example in Balian [5], Chapter 5.

[3] In all that follows, η will denote a positive infinitesimal number.

and we get from (9.8)[4]

$$\boxed{\overline{\delta\mathcal{A}_i}(t) = \overline{\mathcal{A}_i}(t) - \langle\mathcal{A}_i\rangle = \beta\sum_j f_j\langle\mathcal{A}_i(t)\mathcal{A}_j(0)\rangle_{\mathrm{c}} = \beta\sum_j f_j\langle\delta\mathcal{A}_i(t)\delta\mathcal{A}_j(0)\rangle} \tag{9.11}$$

Equation (9.11) has introduced the *Kubo function* (or relaxation function) $C_{ij}(t)$

$$\boxed{C_{ij}(t) = \langle\mathcal{A}_i(t)\mathcal{A}_j(0)\rangle_{\mathrm{c}} = \langle\delta\mathcal{A}_i(t)\delta\mathcal{A}_j(0)\rangle} \tag{9.12}$$

which is the time correlation function of $\delta\mathcal{A}_i(t)$ and $\delta\mathcal{A}_j(0)$ *computed at equilibrium* with the probability density (9.2). Equation (9.11) is often called *Onsager's regression law*: for small deviations from equilibrium, the relaxation toward equilibrium is governed by equilibrium fluctuations. Physically, this result can be understood as follows. One may obtain deviations from equilibrium by applying an external perturbation, as described above, but such deviations may also occur as the result of spontaneous fluctuations, and in both cases the relaxation toward equilibrium should be governed by the same laws.

The Kubo function (9.12) is directly linked to the *dynamical susceptibility* $\chi_{ij}(t)$, which is defined by writing the most general formula for the dynamical linear response to an external time-dependent perturbation $\sum_j f_j(t)\mathcal{A}_j$

$$\overline{\delta\mathcal{A}_i}(t) = \sum_j \int_{-\infty}^{t} \mathrm{d}t'\, \chi_{ij}(t-t') f_j(t') \tag{9.13}$$

In Fourier space, and supposing that $\chi_{ij}(t-t')$ vanishes for $t' > t$, the convolution in (9.13) is transformed into a product

$$\overline{\delta\mathcal{A}_i}(\omega) = \sum_j \chi_{ij}(\omega) f_j(\omega) \tag{9.14}$$

In the case of a sudden switching off of a constant external perturbation at $t = 0$, as in (9.9), Equation (9.13) becomes

$$\overline{\delta\mathcal{A}_i}(t) = \sum_j f_j \int_{-\infty}^{0} \mathrm{d}t'\chi_{ij}(t-t') = \sum_j f_j \int_{t}^{\infty} \chi_{ij}(\tau)\,\mathrm{d}\tau$$

[4] One may wonder why $\delta\mathcal{A}_j$ in (9.11) is taken at $t' = 0$, while any $t' < 0$ would be *a priori* possible. Indeed, any $t' < 0$ would be fine, because $\tilde{H}_1(t')$ in (9.9) is time independent for $t' < 0$, but $H(t')$ and $\mathcal{V}(t')$ are not *separately* time independent. In (9.11) H is taken implicitly at $t' = 0$, which implies that $\mathcal{V}$ should also be taken at $t' = 0$. Note that the correct boundary condition (9.8) is ensured at $t = 0$

$$\overline{\delta\mathcal{A}_i}(t=0) = \beta\sum_j f_j\langle\mathcal{A}_i(0)\mathcal{A}_j(0)\rangle_{\mathrm{c}}$$

Differentiating this equation with respect to t and comparing with (9.11) yields

$$\frac{\mathrm{d}}{\mathrm{d}t}\overline{\delta\mathcal{A}_i}(t) = -\sum_j f_j\,\chi_{ij}(t) = \beta\sum_j f_j\theta(t)\dot{C}_{ij}(t)$$

so that the dynamical susceptibility is nothing other than minus the time derivative of the Kubo function times β

$$\boxed{\chi_{ij}(t) = -\beta\,\theta(t)\dot{C}_{ij}(t)} \tag{9.15}$$

9.1.2 Nyquist theorem

As a simple application of the preceding considerations, let us derive the Nyquist theorem, which relates the electrical conductivity σ_{el} to the equilibrium fluctuations of the electric current. Since we need consider only two dynamical variables, we simplify the notations by setting $\mathcal{A}_i = \mathcal{B}$ and $\mathcal{A}_j = \mathcal{A}$. Using time-translation invariance at equilibrium yields

$$\dot{C}_{BA}(t) = \langle\dot{\mathcal{B}}(t)\mathcal{A}(0)\rangle_{\mathrm{c}} = -\langle\mathcal{B}(t)\dot{\mathcal{A}}(0)\rangle_{\mathrm{c}} \tag{9.16}$$

From (9.13) and (9.15) we can write the Fourier transform $\overline{\delta\mathcal{B}}(\omega)$ as

$$\overline{\delta\mathcal{B}}(\omega) = \beta f_A(\omega)\int_0^\infty \mathrm{d}t\,\mathrm{e}^{\mathrm{i}\omega t}\,\langle\mathcal{B}(t)\dot{\mathcal{A}}(0)\rangle_{\mathrm{c}} \tag{9.17}$$

In case of convergence problems, ω should be understood as $\lim_{\eta\to 0^+}(\omega + \mathrm{i}\eta)$ (see Section 9.1.3). Let us use this result in the following case. We consider charge carriers with charge q and mass m in a one-dimensional conductor and take as dynamical variables the following $\mathcal{A}$ and $\mathcal{B}$

$$\mathcal{A} = q\sum_i x_i \qquad \mathcal{B} = \dot{\mathcal{A}} = q\sum_i \dot{x}_i = V j_{\mathrm{el}} \tag{9.18}$$

where x_i is the position of carrier i, j_{el} the current density and V the volume of the conductor. The external force is an external (uniform) time dependent electric field $E(t)$ and the perturbation $\mathcal{V}(t)$ is

$$\mathcal{V}(t) = -qE(t)\sum_i x_i = -E(t)\mathcal{A}$$

so that from (9.17)

$$\overline{\delta \mathcal{B}}(\omega) = V j_{\rm el}(\omega) = \beta V^2 E(\omega) \int_0^\infty \mathrm{d}t\, \mathrm{e}^{\mathrm{i}\omega t}\, \langle j_{\rm el}(t) j_{\rm el}(0)\rangle|_{E=0}$$

$$= \beta q^2 E(\omega) \int_0^\infty \mathrm{d}t\, \mathrm{e}^{\mathrm{i}\omega t} \sum_{i,k} \langle \dot{x}_i(t)\dot{x}_k(0)\rangle|_{E=0}$$

Since the average equilibrium (or $E = 0$) current density vanishes, we may write $j_{\rm el}$ instead of $\delta j_{\rm el}$. This equation is nothing other than the time dependent Ohm's law $j_{\rm el}(\omega) = \sigma_{\rm el}(\omega)E(\omega)$. We have thus shown that the electrical conductivity $\sigma_{\rm el}(\omega)$ is given by the Fourier transform of the time correlation of the current density in the absence of an external electric field

$$\boxed{\sigma_{\rm el}(\omega) = \beta V \int_0^\infty \mathrm{d}t\, \mathrm{e}^{\mathrm{i}\omega t}\, \langle j_{\rm el}(t) j_{\rm el}(0)\rangle|_{E=0}} \tag{9.19}$$

In the zero frequency limit $\omega = 0$ we get the following formula for the static conductivity $\sigma_{\rm el}$

$$\sigma_{\rm el} = \beta V \int_0^\infty \mathrm{d}t\, \langle j_{\rm el}(t) j_{\rm el}(0)\rangle|_{E=0} \tag{9.20}$$

It may be necessary to include a factor $\exp(-\eta t)$ in (9.20) in order to ensure the convergence of the integral. Equation (9.20) is one version of the *Nyquist theorem*, and is typical of a *Green–Kubo formula*, which gives a transport coefficient (in the present case the static electrical conductivity) in terms of the integral of a time correlation function. Let us give a rough estimate of (9.19). As we have seen in Section 3.3.2, velocities of different particles are uncorrelated in classical statistical mechanics. Introducing a microscopic relaxation (or collision) time $\tau^* \sim 10^{-14}$ s

$$\langle \dot{x}_i(t)\dot{x}_k(0)\rangle = \delta_{ik}\langle \dot{x}(t)\dot{x}(0)\rangle \sim \delta_{ik}\frac{kT}{m}\mathrm{e}^{-|t|/\tau^*}$$

leads to the familiar result, already obtained in (6.59) in the case $\omega = 0$

$$\boxed{\sigma_{\rm el}(\omega) = \frac{nq^2\tau^*}{m(1 - \mathrm{i}\omega\tau^*)}} \tag{9.21}$$

where n is the density of carriers. Of course, (9.21) may be obtained by much more elementary methods, but the point is that (9.19) and (9.20) are *exact* results, and at least one knows where to start from if one wishes to derive better approximations.

9.1.3 Analyticity properties

In this subsection,[5] we shall work for simplicity with a single dynamical variable $\mathcal{A}$, $\chi_{AA}(t) = \chi(t)$, but the results generalize immediately to any $\chi_{ij}(t)$, provided $\mathcal{A}_i$ and $\mathcal{A}_j$ have the same parity under time reversal (see (9.67)). One very important property of $\chi(t)$ is *causality*: $\chi(t) = 0$ if $t < 0$, which reflects the obvious requirement that the effect must follow the cause. This property allows us to define the Laplace transform of $\chi(t)$

$$\chi(z) = \int_0^\infty \mathrm{d}t\, \mathrm{e}^{\mathrm{i}zt}\chi(t) \tag{9.22}$$

for any complex value of z such that $\mathrm{Im} z > 0$. Indeed, if we write $z = z_1 + \mathrm{i}z_2$, $z_2 > 0$, causality provides in (9.22) a convergence factor $\exp(-z_2 t)$ and this equation defines an analytic function of z in the half plane $\mathrm{Im} z > 0$. Following standard notations, we define $\chi''(t)$ by

$$\chi''(t) = \frac{\mathrm{i}}{2}\beta\dot{C}(t) \tag{9.23}$$

or, equivalently, from (9.15), $\chi''(t) = -(\mathrm{i}/2)\chi(t)$ for $t > 0$; note that $\chi''(t)$ is an odd function of t, as the Kubo function $C_{AA}(t) = C(t)$ is an even function of t. Furthermore, $\chi''(t)$ is pure imaginary, so that its Fourier transform $\chi''(\omega)$ is an odd and real function of ω. Now, for $t > 0$

$$\chi(t) = 2\mathrm{i}\chi''(t) = 2\mathrm{i}\int \frac{\mathrm{d}\omega}{2\pi}\,\mathrm{e}^{-\mathrm{i}\omega t}\chi''(\omega)$$

Plugging this expression for $\chi(t)$ in (9.22) and exchanging the t and ω integrations leads to a *dispersion relation*[6] for $\chi(z)$

$$\chi(z) = \int_{-\infty}^{\infty} \frac{\mathrm{d}\omega'}{\pi}\,\frac{\chi''(\omega')}{\omega' - z} \tag{9.24}$$

[5] The reader is strongly advised to solve the example of the forced harmonic oscillator of Exercise 9.7.1, in order to get some familiarity with the results of this section in an elementary case.

[6] We have assumed in (9.24) that the ω'-integral is convergent at infinity. If this is not the case, one uses subtracted dispersion relations, for example the once-subtracted dispersion relation

$$\chi(z) - \chi(z=0) = z\int_{-\infty}^{\infty} \frac{\mathrm{d}\omega'}{\pi}\,\frac{\chi''(\omega')}{\omega'(\omega' - z)}$$

However, this subtraction is done at the expense of introducing an unknown parameter $\chi(z=0)$.

The existence of a dispersion relation is of course directly linked to causality. Since $\chi(z)$ is analytic in the upper half plane, $\chi(\omega)$ is the limit $\eta \to 0^+$ of $\chi(\omega + i\eta)$ and $\chi''(\omega)$ is the imaginary part of $\chi(\omega)$

$$\mathrm{Im}\,\chi(\omega) = \chi''(\omega) \tag{9.25}$$

where we have used in (9.24) $z = \omega + i\eta$ and

$$\frac{1}{\omega' - \omega - i\eta} = P\frac{1}{\omega' - \omega} + i\pi\delta(\omega' - \omega)$$

where P indicates the Cauchy principal value. A more elementary derivation uses a periodic external force

$$f(t) = f_\omega \cos\omega t$$

so that from (9.13), writing

$$\chi(\omega) = \chi'(\omega) + i\chi''(\omega)$$

we get

$$\overline{\delta A}(t) = f_\omega\left[\chi'(\omega)\cos\omega t + \chi''(\omega)\sin\omega t\right]$$

In this simple case, the reactive part of the response, in phase with the force, is controlled by the real part $\chi'(\omega)$ of $\chi(\omega)$, while the dissipative part, in quadrature with the force (i.e. out of phase by $\pi/2$ with the force), is controlled by its imaginary part $\chi''(\omega)$.[7]

The susceptibility $\chi(z)$ may also be expressed in terms of the Kubo function and of the static susceptibility $\chi = \lim_{\omega\to 0}\chi(\omega + i\eta)$[8]

$$\chi(z) = -\beta\int_0^\infty dt\, e^{izt}\dot{C}(t) = \beta C(t=0) + iz\beta\int_0^\infty dt\, e^{izt}C(t) = \chi + iz\beta C(z)$$

Solving for $C(z)$ and using the fact that $\overline{\delta A}(t=0) = \chi f_A$ allows us to derive an expression for $\overline{\delta A}(z)$ that depends on $\overline{\delta A}(t=0)$ (and not on f_A)

$$\overline{\delta A}(z) = \frac{1}{iz}\left(\frac{\chi(z)}{\chi} - 1\right)\overline{\delta A}(t=0) \tag{9.26}$$

[7] However remember our warning: this is only true if $\mathcal{A}_i$ and $\mathcal{A}_j$ have the same parity under time reversal.

[8] The static susceptibility χ is in general different from the isothermal susceptibility χ_T, which is related to the Kubo function by $\chi_T = \beta C(t=0)$. The two susceptibilities coincide if the integral in the second expression of $\chi(z)$ converges at infinity. If the integral does not converge, one writes

$$\chi(z) = \beta[C(0) - C(\infty)] + iz\beta\int_0^\infty dt\, e^{izt}\,[C(t) - C(\infty)]$$

and $\chi = \beta[C(0) - C(\infty)]$.

The relation which matches (9.24) for $C(z)$ is

$$C(z) = -\frac{\mathrm{i}}{\beta}\int_{-\infty}^{\infty}\frac{\mathrm{d}\omega'}{\pi}\,\frac{\chi''(\omega')}{\omega'(\omega'-z)} \tag{9.27}$$

Furthermore, from (9.23) and from our conventions (9.22) for Fourier transforms: $\partial_t \to -i\omega$

$$\chi''(\omega) = \frac{1}{2}\,\beta\omega\, C(\omega) \tag{9.28}$$

This is the classical version of the *fluctuation-dissipation theorem*. On the right hand side of (9.28), $C(\omega)$, being the Fourier transform of $\langle \mathcal{A}(t)\mathcal{A}(0)\rangle$, is clearly a measure of the equilibrium fluctuations of $\mathcal{A}$, but we have still to justify that $\chi''(\omega)$ does describe dissipation: this is shown formally in Section 9.2.4 and Exercise 9.7.6. However, from experience with the forced harmonic oscillator and other systems, we already know that dissipation is governed by that part of the response which is in quadrature of phase with the driving force. As already mentioned, and as is shown explicitly in Exercise 9.7.1, the real part of $\chi(\omega)$ gives the reactive component of the response, while its imaginary part gives the dissipative component (see, however, the comments following (9.68)).

9.1.4 Spin diffusion

As an illustration of the preceding considerations, let us consider, following Kadanoff and Martin [61], a fluid of particles carrying a spin 1/2 aligned in a fixed direction (or, equivalently, an Ising spin) with which is associated a magnetic moment μ. A practical example would be helium-3. We assume that all spin flip processes may be neglected. To the magnetization density

$$n(\vec{r},t) = \mu[n_+(\vec{r},t) - n_-(\vec{r},t)]$$

where n_+ (n_-) is the density of particles with spin up (down), corresponds a magnetization current $\vec{j}(\vec{r},t)$ such that magnetization is locally conserved

$$\partial_t n(\vec{r},t) + \vec{\nabla}\cdot\vec{j}(\vec{r},t) = 0 \tag{9.29}$$

Equation (9.29) is exact under the no spin flip assumption. It implies that the magnetization can change in a volume of space only because particles move into and out of this volume. As a consequence, a local magnetization imbalance cannot disappear locally, but only by slowly spreading over the entire system. The second relation we need is a phenomenological one and is inspired by Fick's law (6.26).

It relates the magnetization current to the density gradient

$$\vec{j}(\vec{r}, t) = -D\vec{\nabla}n(\vec{r}, t) \tag{9.30}$$

where D is the spin diffusion coefficient. Combining (9.29) and (9.30) leads, of course, to a diffusion equation (6.21) for n

$$\left(\frac{\partial}{\partial t} - D\nabla^2\right) n = 0 \tag{9.31}$$

or, in Fourier space

$$(\omega + \mathrm{i}Dk^2)n(\vec{k}, \omega) = 0 \tag{9.32}$$

Note that our convention for space-time Fourier transforms is

$$\begin{aligned} f(\vec{k}, \omega) &= \int \mathrm{d}t\, \mathrm{d}^3r\, \mathrm{e}^{-\mathrm{i}(\vec{k}\cdot\vec{r}-\omega t)} f(\vec{r}, t) \\ f(t, \vec{r}) &= \int \frac{\mathrm{d}\omega}{2\pi} \frac{\mathrm{d}^3k}{(2\pi)^3} \mathrm{e}^{\mathrm{i}(\vec{k}\cdot\vec{r}-\omega t)} f(\vec{k}, \omega) \end{aligned} \tag{9.33}$$

A mode with a dispersion law $\omega = -i\,Dk^2$ is called a *diffusive mode* and is characteristic of the relaxation of a conserved quantity. Indeed, let us consider a fluctuation of the magnetization density with wavelength λ. Since the relaxation occurs via diffusion, the characteristic time τ is linked to λ through $\lambda^2 \sim D\tau$ so that $\tau \sim \lambda^2/D$ or $\omega \sim Dk^2$.

Let us assume that we have created at $t = 0$ an off-equilibrium magnetization density $n(\vec{r}, t = 0)$, or, in Fourier space, $n(\vec{k}, t = 0)$. Then, for positive times, the evolution of $n(\vec{r}, t)$ is governed by the diffusion equation (9.31), which we write in $(\vec{k}, t)$ space

$$\partial_t n(\vec{k}, t) = -Dk^2 n(\vec{k}, t) \tag{9.34}$$

Taking the Laplace transform of both sides of (9.34) gives

$$n(\vec{k}, z) = \frac{\mathrm{i}}{z + \mathrm{i}Dk^2} n(\vec{k}, t = 0) \tag{9.35}$$

Let us now make the link with linear response. If our magnetic fluid is placed in a time-independent, but space-dependent magnetic field, the perturbed Hamiltonian reads (we set for simplicity $\mu = 1$)

$$H_1 = H - \int \mathrm{d}^3r\, n(\vec{r})B(\vec{r}) \tag{9.36}$$

Using invariance under space translations, the fluctuation-response theorem gives

$$\overline{\delta n}(\vec{r}) = \int \mathrm{d}^3 r' \, \chi(\vec{r} - \vec{r}\,') B(\vec{r}\,') \tag{9.37}$$

where the static susceptibility $\chi(\vec{r} - \vec{r}\,')$ is given from (4.28) by the space correlations of the density

$$\chi(\vec{r} - \vec{r}\,') = \beta \langle n(\vec{r}) n(\vec{r}\,') \rangle_c$$

Then, in Fourier space, (9.37) becomes

$$\overline{\delta n}(\vec{k}) = \chi(\vec{k}) B(\vec{k}) \tag{9.38}$$

In Fourier space, the Fourier components are decoupled, so that all preceding results on the static suceptibility χ or on the dynamic susceptibility $\chi(t)$ apply without modification to $\chi(\vec{k})$ and $\chi(\vec{k}, t)$ respectively. Of course, this simplicity is a consequence of space translation invariance: the equations would be much more complicated in the absence of this invariance. Comparing (9.35) and (9.26), which reads in the present case

$$\overline{\delta n}(\vec{k}, z) = \frac{1}{\mathrm{i}z} \left(\frac{\chi(\vec{k}, z)}{\chi(\vec{k})} - 1 \right) \overline{\delta n}(\vec{k}, t = 0)$$

one derives the following expression for the dynamical susceptibility

$$\chi(\vec{k}, z) = \frac{\mathrm{i}Dk^2}{z + \mathrm{i}Dk^2} \chi(\vec{k}) \tag{9.39}$$

One notes that the susceptibility depends on a thermodynamic quantity $\chi(\vec{k})$ and on a transport coefficient D. Taking the imaginary part of both sides of (9.39), we obtain $\chi''(\vec{k}, \omega)$

$$\chi''(\vec{k}, \omega) = \frac{\omega D k^2}{\omega^2 + D^2 k^4} \chi(\vec{k}) \tag{9.40}$$

This last expression leads to a Green–Kubo formula. Defining $\chi = \chi(\vec{k} = 0)$, one verifies that

$$D\chi = \lim_{\omega \to 0} \lim_{k \to 0} \frac{\omega}{k^2} \chi''(k, \omega) \tag{9.41}$$

where the order of limits is crucial. The result can be transformed into (see Exercise 9.7.3)

$$D\chi = \frac{1}{3}\beta \int_0^\infty \mathrm{d}t \int \mathrm{d}^3 r\, \mathrm{e}^{-\eta t} \left\langle \vec{j}(t,\vec{r}) \cdot \vec{j}(0,\vec{0}) \right\rangle_{\mathrm{eq}} \tag{9.42}$$

where the factor $\exp(-\eta t)$ has been added in case convergence problems are encountered for $t \to \infty$. One can show that the correct way to proceed to evaluate expressions like (9.42) is to keep η finite for a finite volume, perform the t integral, and then take the thermodynamic limit $V \to \infty$.

The fluctuation-dissipation theorem (9.28) joined to (9.40) gives the Kubo function $C(\vec{k}, \omega)$

$$C(\vec{k}, \omega) = S(\vec{k}, \omega) = \frac{2}{\beta} \frac{Dk^2}{\omega^2 + D^2 k^4} \chi(\vec{k}) \tag{9.43}$$

We have defined a new function $S(\vec{k}, \omega)$, which is called the *dynamical structure factor*. It is the space-time Fourier transform of the (connected) density–density correlation function $\langle n(\vec{r}, t)\, n(\vec{0}, 0)\rangle_c$

$$S(\vec{k}, \omega) = \int \mathrm{d}t\, \mathrm{d}^3 r\, \mathrm{e}^{-\mathrm{i}(\vec{k}\cdot\vec{r} - \omega t)} \langle n(\vec{r}, t)\, n(\vec{0}, 0)\rangle_c \tag{9.44}$$

In the classical case, the structure factor is identical to the Kubo function, but we shall see shortly that the two functions are different in the quantum case. The dynamical structure factor generalizes the static structure factor introduced in Section 3.4.2, and it may also be measured in (inelastic) neutron scattering experiments: see Problem 9.8.1 for an analogous case, that of inelastic light scattering by a suspension of particles in a fluid.

In the case of light scattering by a simple fluid, the dynamical structure factor, plotted as a function of ω, displays three peaks instead of the single peak of width Dk^2 at $\omega = 0$ found in (9.43): see Figure 9.1 and Problem 9.8.2. The central peak at $\omega = 0$ is called the *Rayleigh peak*, and its width $D_{\mathrm{T}} k^2$ is determined by the coefficient of thermal conductivity κ, because it corresponds to light scattering by heat diffusion. The other two peaks are the *Brillouin peaks*: they are centred at $\omega = \pm ck$, where c is the sound speed, and they correspond to light scattering by sound waves. Their width $\Gamma k^2/2$ depends also on the shear viscosity η and the bulk viscosity ζ defined in (6.88)

$$D_{\mathrm{T}} = \frac{\kappa}{mnc_P} \qquad \Gamma = D_T\Big(\frac{c_P}{c_V} - 1\Big) + \Big(\frac{4}{3}\eta + \zeta\Big)\frac{1}{mn} \tag{9.45}$$

where mc_P and mc_V are the specific heat per particle at constant pressure and volume, m is the mass of the particles and n the fluid density. As in the spin diffusion

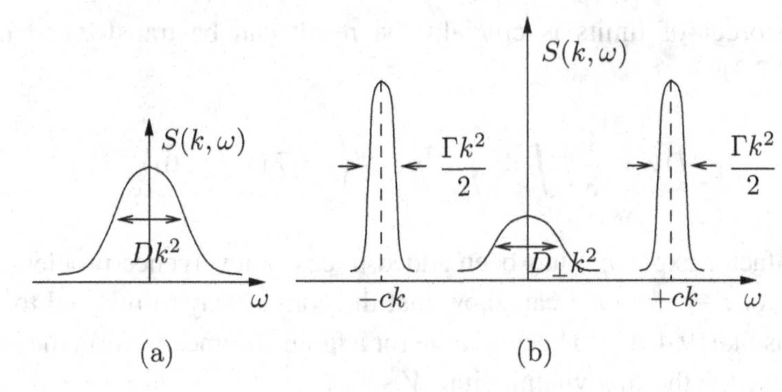

Figure 9.1 Dynamical structure factors. (a) Spin diffusion, (b) light scattering by a simple fluid.

case, the susceptibilities depend on thermodynamic quantities and on transport coefficients. It can also be shown that the transport coefficients κ, η and ζ may be written in the form of Green–Kubo formulae analogous to (9.42).

9.2 Linear response: quantum theory

9.2.1 Quantum fluctuation response theorem

Let us first recall from Chapter 2 that the equilibrium density operator $D \equiv D_{\rm B}$ is given as a function of the relevant observables A_j by (2.64)

$$D_{\rm B} \equiv D = \frac{1}{Z}\exp(\lambda_j \mathsf{A}_j) \qquad Z = \mathrm{Tr}\,\exp(\lambda_j A_j) \tag{9.46}$$

where we adopt from now on the summation convention over repeated indices. Let B be an observable that may or may not be a member of the set $\{\mathsf{A}_j\}$; its expectation value $\langle \mathsf{B}\rangle_D$ is

$$\langle \mathsf{B}\rangle_D = \mathrm{Tr}\,(\mathsf{B}D) = \frac{1}{Z}\,\mathrm{Tr}\left[\mathsf{B}\exp(\lambda_j \mathsf{A}_j)\right]$$

We now compute $\partial\langle \mathsf{B}\rangle_D/\partial\lambda_i$

$$\frac{\partial\langle \mathsf{B}\rangle_D}{\partial\lambda_i} = \frac{1}{Z}\frac{\partial}{\partial\lambda_i}\mathrm{Tr}\left[\mathsf{B}\exp(\lambda_j \mathsf{A}_j)\right] - \langle \mathsf{A}_i\rangle_D\langle \mathsf{B}\rangle_D \tag{9.47}$$

To compute the λ_i-derivative in (9.47), we use the identity (2.120) for an operator A which depends on a parameter α

$$\frac{\partial}{\partial\alpha}\,\mathrm{e}^{\mathsf{A}(\alpha)} = \int_0^1 \mathrm{d}x\,\mathrm{e}^{x\mathsf{A}(\alpha)}\,\frac{\partial \mathsf{A}(\alpha)}{\partial\alpha}\,\mathrm{e}^{(1-x)\mathsf{A}(\alpha)} \tag{9.48}$$

and obtain

$$\frac{\partial}{\partial \lambda_i}\mathrm{Tr}\left[\mathsf{B}\exp(\lambda_j \mathsf{A}_j)\right] = \int_0^1 \mathrm{d}x\,\mathrm{Tr}\Big(\mathsf{B}\,\mathrm{e}^{x\lambda_j \mathsf{A}_j}\mathsf{A}_i\,\mathrm{e}^{(1-x)\lambda_j \mathsf{A}_j}\Big)$$

so that

$$\frac{\partial \langle \mathsf{B}\rangle_D}{\partial \lambda_i} = \int_0^1 \mathrm{d}x\,\mathrm{Tr}\left[\delta\mathsf{B}\,D^x\,\delta\mathsf{A}_i\,D^{1-x}\right] \tag{9.49}$$

where $\delta\mathsf{B} = \mathsf{B} - \langle \mathsf{B}\rangle_D$ and $\delta\mathsf{A}_i = \mathsf{A}_i - \langle \mathsf{A}_i\rangle_D$. The preceding equation gives the quantum generalization of the fluctuation-response theorem (9.5), and it suggests that it is useful to introduce the following combination of two operators A and B

$$\boxed{\langle \mathsf{B};\mathsf{A}\rangle_D = \int_0^1 \mathrm{d}x\,\mathrm{Tr}\left[\mathsf{B}\,D^x\,\mathsf{A}^\dagger\,D^{1-x}\right]} \tag{9.50}$$

It is easily checked (Exercise 9.7.4) that Equation (9.50) defines on the vector space of operators a positive definite scalar product of the two operators A and B, called *Mori's scalar product*. Note that $\langle \mathsf{B};\mathsf{A}\rangle_D^* = \langle \mathsf{A};\mathsf{B}\rangle_D$ and that, in the classical limit, Mori's scalar product is simply an equilibrium average of two classical variables $\mathcal{B}$ and $\mathcal{A}$

$$\langle \mathcal{B};\mathcal{A}\rangle_D^{\text{classical}} = \langle \mathcal{B}\,\mathcal{A}^*\rangle_D$$

Using the definition (9.50), the quantum fluctuation-response theorem (9.49) may be written in the following equivalent forms

$$\boxed{\frac{\partial \langle \mathsf{B}\rangle_D}{\partial \lambda_i} = \langle \delta\mathsf{B};\delta\mathsf{A}_i^\dagger\rangle_D = \langle \delta\mathsf{B};\mathsf{A}_i^\dagger\rangle_D = \langle \mathsf{B};\delta\mathsf{A}_i^\dagger\rangle_D = \langle \mathsf{B};\mathsf{A}_i^\dagger\rangle_{D,c}} \tag{9.51}$$

From now on we restrict ourselves to Hermitian A_is, leaving the non-Hermitian case to Exercise 9.7.4. Setting $\mathsf{B} = \mathsf{A}_j$ in (9.51) we recover (2.121)

$$\frac{\partial \langle \mathsf{A}_j\rangle_D}{\partial \lambda_i} = \frac{\partial^2 \ln Z}{\partial \lambda_i \partial \lambda_j} = \langle \delta\mathsf{A}_j;\delta\mathsf{A}_i\rangle_D = C_{ji} \tag{9.52}$$

The matrix C_{ji} is symmetric and positive definite for Hermitian A_is (Exercise 2.7.8) and has thus an inverse C_{ji}^{-1}. We may write

$$\mathrm{d}\langle \mathsf{A}_j\rangle_D = C_{ji}\,\mathrm{d}\lambda_i \quad \text{or} \quad \mathrm{d}\lambda_i = C_{ij}^{-1}\mathrm{d}\langle \mathsf{A}_j\rangle_D$$

so that

$$\mathrm{d}\langle \mathsf{B}\rangle_D = \langle \delta\mathsf{B};\delta\mathsf{A}_i\rangle_D \mathrm{d}\lambda_i = C^{-1}_{ij}\langle \delta\mathsf{B};\delta\mathsf{A}_i\rangle_D\, \mathrm{d}\langle \mathsf{A}_j\rangle_D \tag{9.53}$$

It is important to understand the meaning of (9.53). The average value $\langle \mathsf{B}\rangle_D$ is a function of the $\langle \mathsf{A}_j\rangle$s, since it is the set of $\langle \mathsf{A}_j\rangle_D$s which determines the density operator D through the requirement

$$\langle \mathsf{A}_i\rangle_D = \mathrm{Tr}(D\mathsf{A}_i) = A_i$$

Then (9.53) controls the variation of the average value $\langle \mathsf{B}\rangle_D$ when the $\langle \mathsf{A}_j\rangle_D$s are modified. Although we shall not use it, the preceding structure turns out to play an important rôle because it may be generalized to non-equilibrium situations, and it underlies more general approaches to the projection method than that which will be described in this book. Note also that, for the time being, we have not used any linear approximation. The linear approximation will come shortly.

9.2.2 *Quantum Kubo function*

As in the classical case, we perturb a Hamiltonian H

$$H \to H_1 = H - f_i\mathsf{A}_i \qquad \exp(-\beta H) \to \exp(-\beta H_1) = \exp[-\beta(H - f_i\mathsf{A}_i)] \tag{9.54}$$

One of the observables of the preceding subsection, A_0, is identified with the Hamiltonian H, and the corresponding Lagrange multiplier λ_0 is identified with $-\beta$. As in the previous section, the other Lagrange multipliers $\lambda_i, i = 1, \ldots, N$ are identified with βf_i, and, as we are interested in linear response, we set $\lambda_i = f_i = 0$, $i = 1, \ldots, N$ once the derivatives have been taken. The equilibrium density matrix to be used in the computation of the average values is

$$D_{\mathrm{eq}} = \frac{1}{Z(H)}\exp(-\beta H) \qquad Z(H) = \mathrm{Tr}\,\exp(-\beta H) \tag{9.55}$$

and, as in (9.2), we shall use the convention $\langle \bullet\rangle \equiv \langle \bullet\rangle_{\mathrm{eq}} = \mathrm{Tr}\,(\bullet\, D_{\mathrm{eq}})$. As in the classical case we take external forces of the form

$$f_i(t) = \mathrm{e}^{\eta t}\theta(-t) f_i$$

and in (9.51) we choose as observable B the observable A_j in the Heisenberg picture, setting from now on $\hbar = 1$,

$$\mathsf{A}_{j\mathrm{H}}(t) \equiv \mathsf{A}_j(t) = \mathrm{e}^{\mathrm{i}Ht}\,\mathsf{A}_j \mathrm{e}^{-\mathrm{i}Ht} \tag{9.56}$$

$\mathsf{A}_i = \mathsf{A}_i(0)$ is the observable in the Schrödinger picture; we have written $\mathsf{A}_i(t)$ instead of $\mathsf{A}_{i\mathrm{H}}(t)$, as the explicit time dependence implies that the observable $\mathsf{A}_i(t)$

is taken in the Heisenberg picture. From (9.49) with $\delta \mathsf{B} = \delta \mathsf{A}_j(t)$, the ensemble average $\overline{\delta \mathsf{A}_j}(t)$ is given in the linear approximation by

$$\boxed{\overline{\delta \mathsf{A}_j}(t) = \beta f_i \langle \delta \mathsf{A}_j(t); \delta \mathsf{A}_i(0) \rangle} \tag{9.57}$$

It is customary to make in (9.50) the change of variables $\alpha = \beta x$ and to write (9.57) as

$$\overline{\delta \mathsf{A}_j}(t) = f_i \int_0^\beta \mathrm{d}\alpha \left\langle \delta \mathsf{A}_j(t)\, \mathrm{e}^{-\alpha H}\, \delta \mathsf{A}_i(0)\, \mathrm{e}^{\alpha H} \right\rangle \tag{9.58}$$

Comparing with (9.11) and (9.12), we see that the Kubo function $C_{ji}(t)$ is now

$$\boxed{C_{ji}(t) = \langle \delta \mathsf{A}_j(t); \delta \mathsf{A}_i(0) \rangle = \frac{1}{\beta} \int_0^\beta \mathrm{d}\alpha \left\langle \mathsf{A}_j(t)\, \mathrm{e}^{-\alpha H}\, \mathsf{A}_i(0)\, \mathrm{e}^{\alpha H} \right\rangle_{\mathrm{c}}} \tag{9.59}$$

Note that, at equilibrium, the matrix C_{ji} introduced in (9.52) is nothing other than the Kubo function taken at $t = 0$: $C_{ji} = C_{ji}(t = 0)$. In the classical limit, all operators commute and one recovers (9.12). The Kubo function may be written in a different form, using the property that the inverse temperature generates translations in imaginary time as already explained in Chapter 7. We may write from (1.58), with $t = \mathrm{i}\beta$

$$\mathrm{e}^{\mathrm{i}(\mathrm{i}\beta H)} \mathsf{A}_i(0) \mathrm{e}^{-\mathrm{i}(\mathrm{i}\beta H)} = \mathsf{A}_i(\mathrm{i}\beta)$$

so that

$$C_{ji}(t) = \frac{1}{\beta} \int_0^\beta \mathrm{d}\alpha \left\langle \mathsf{A}_j(t) \mathsf{A}_i(\mathrm{i}\alpha) \right\rangle_{\mathrm{c}} \tag{9.60}$$

The dynamical susceptibility is related to the Kubo function exactly as in the classical case

$$\begin{aligned} \chi_{ij}(t) &= -\beta \theta(t) \dot{C}_{ij}(t) \\ \chi''_{ij}(t) &= \frac{\mathrm{i}}{2}\, \beta\, \dot{C}_{ij}(t) \end{aligned} \tag{9.61}$$

All the analyticity properties derived in Section 9.1.3 are also valid in the quantum case, since they depend only on causality.

9.2.3 Fluctuation-dissipation theorem

The susceptibility and its imaginary part are given from (9.61) by the time derivative of the Kubo function, and it turns out that this derivative has a much simpler expression than the function itself. Using time translation invariance, we write the time derivative of $C_{ij}(t)$ as follows

$$\dot{C}_{ij}(t) = -\frac{1}{\beta}\int_0^\beta \mathrm{d}\alpha \left\langle \mathsf{A}_i(t) \mathrm{e}^{-\alpha H} \left.\frac{\mathrm{d}\mathsf{A}_j}{\mathrm{d}t}\right|_{t=0} \mathrm{e}^{\alpha H}\right\rangle_{\mathrm{c}}$$

From $\partial_t \mathsf{A} = i[H, \mathsf{A}]$

$$\dot{C}_{ij}(t) = \frac{i}{\beta}\int_0^\beta \mathrm{d}\alpha \left\langle \mathsf{A}_i(t) \frac{\mathrm{d}}{\mathrm{d}\alpha}\left(\mathrm{e}^{-\alpha H} \mathsf{A}_j \mathrm{e}^{\alpha H}\right)\right\rangle_{\mathrm{c}}$$

The α-integrand is then a total derivative, the integration is trivial and leads to the important result

$$\boxed{\chi''_{ij}(t) = \frac{1}{2}\langle [\mathsf{A}_i(t), \mathsf{A}_j(0)]\rangle} \tag{9.62}$$

The function $\chi''_{ij}(t)$ is given by the average value of a commutator.[9] Similarly, the susceptibility is given by the average value of a *retarded* commutator

$$\boxed{\chi_{ij}(t) = \mathrm{i}\theta(t)\langle [\mathsf{A}_i(t), \mathsf{A}_j(0)]\rangle} \tag{9.63}$$

As in the classical case, this equation shows that $\chi_{ij}(t)$ is a real function of t if A_i and A_j are Hermitian operators. Let us now evaluate (9.62) by inserting complete sets of states $|n\rangle$ of the Hamiltonian, $H|n\rangle = E_n|n\rangle$. The first term in the commutator on the right hand side of (9.62) is the dynamical structure factor $S_{ij}(t) = \langle \mathsf{A}_i(t)\mathsf{A}_j(0)\rangle_{\mathrm{c}}$

$$S_{ij}(t) = \frac{1}{Z}\sum_{n,m} \exp(-\beta E_n + \mathrm{i}(E_n - E_m)t)\langle n|\delta\mathsf{A}_i|m\rangle\langle m|\delta\mathsf{A}_j|n\rangle$$

and, taking the time Fourier transform

$$S_{ij}(\omega) = \frac{1}{Z}\sum_{n,m} \exp(-\beta E_n)\delta(\omega + E_n - E_m)\langle n|\delta\mathsf{A}_i|m\rangle\langle m|\delta\mathsf{A}_j|n\rangle$$

The second term in the commutator is computed by noting that it may be written as $S_{ji}(-t)$, so that its Fourier transform is $S_{ji}(-\omega)$; exchanging the indices n and

[9] Note that $\chi''(\omega)$ is the imaginary part of $\chi(\omega)$, but that $\chi''(t)$ is *not* the imaginary part of $\chi(t)$. We have kept this standard notation, which is admittedly somewhat confusing.

m in the expression of this second term, one readily finds

$$\boxed{\chi''_{ij}(\omega) = \frac{1}{2\hbar}(1 - \exp(-\beta\omega\hbar))S_{ij}(\omega)} \tag{9.64}$$

where we have for once written explicitly the Planck constant $\hbar$. This is the quantum version of the fluctuation-dissipation theorem, and one recovers its classical version (9.28) in the limit $\hbar \to 0$. As was already mentioned, the structure factor $S_{ij}(t)$ and the Kubo function $C_{ij}(t)$ are identical in the classical case. However, they are *different* in the quantum case. Adding space dependence as in Section 9.1.4, we get a structure factor $S_{ij}(\vec{k}, \omega)$, which may be measured in inelastic scattering experiments (light, X-rays, neutrons, electrons ...), while $\chi''_{ij}(\vec{k}, \omega)$ describes dissipation.

9.2.4 Symmetry properties and dissipation

Before showing explicitly that χ''_{ij} describes dissipation, we need to derive its symmetry properties. They follow from various invariances.

(*i*) From time translation invariance

$$\chi''_{ij}(t) = -\chi''_{ji}(-t) \qquad \text{or} \qquad \chi''_{ij}(\omega) = -\chi''_{ji}(-\omega) \tag{9.65}$$

(*ii*) From the Hermiticity of the A_is (see Exercise 9.7.5 for non-Hermitian operators)

$$\chi''^{*}_{ij}(t) = -\chi''_{ij}(t) \qquad \text{or} \qquad \chi''^{*}_{ij}(\omega) = -\chi''_{ij}(-\omega) \tag{9.66}$$

Then $\chi''_{ij}(t)$ is pure imaginary and $\chi_{ij}(t)$ is real from (9.61).

(*iii*) From time reversal invariance

$$\chi''_{ij}(t) = -\varepsilon_i\varepsilon_j\chi''_{ij}(-t) \qquad \text{or} \qquad \chi''_{ij}(\omega) = -\varepsilon_i\varepsilon_j\chi''_{ij}(-\omega) \tag{9.67}$$

In (9.67), ε_i is the parity under time reversal of the observable A_i

$$\Theta\,\mathsf{A}_i(t)\Theta^{-1} = \varepsilon_i\mathsf{A}_i(-t) \tag{9.68}$$

where Θ is the (antiunitary) time reversal operator. We have assumed that the observables have, as is generally the case, a definite parity ε_i under time reversal. The proof of (9.67) is immediate in the classical case, where one first derives the parity property of the Kubo function $C_{ij}(t) = \varepsilon_i\varepsilon_j C_{ij}(-t)$, from which (9.67) follows. The most common case is $\varepsilon_i\varepsilon_j = +1$, for example if $\mathsf{A}_i = \mathsf{A}_j = \mathsf{A}$, then $\chi''_{ij}(\omega)$ is a real and odd function of ω. If $\varepsilon_i\varepsilon_j = -1$, $\chi''_{ij}(\omega)$ is imaginary and even in ω. The symmetry properties under time reversal are, of course, intimately related to the symmetry properties (6.38) of Onsager's coefficients of irreversible thermodynamics.

Let us now justify that χ'' does describe dissipation. Instead of Heisenberg's picture, it is slightly more convenient to use Schrödinger's picture, labelled by a superscript S. Let $D_1^{\mathrm{S}}(t)$ denote the perturbed density operator in the presence of a time-dependent pertubation $\mathsf{V}^{\mathrm{S}}(t) = -\sum_i f_i(t)\mathsf{A}_i$, $\mathsf{A}_i = \mathsf{A}_i^{\mathrm{S}}$. The time evolution of D_1^{S} is governed from (2.14) by

$$i\partial_t D_1^{\mathrm{S}}(t) = [H_1^{\mathrm{S}}(t), D_1^{\mathrm{S}}(t)] = [H + \mathsf{V}^{\mathrm{S}}(t), D_1^{\mathrm{S}}(t)] \tag{9.69}$$

The system is driven by external forces $f_i(t)$, which we assume to be periodic. As a simple example, it is useful to recall the damped mechanical harmonic oscillator driven by an external force: energy is dissipated in the viscous medium which damps the oscillations. The rate $\mathrm{d}W/\mathrm{d}t$ at which the external forces do work on the system is equal to the variation per unit time of the energy $E(t)$ of the system

$$\frac{\mathrm{d}W}{\mathrm{d}t} = \frac{\mathrm{d}E}{\mathrm{d}t} = \frac{\mathrm{d}}{\mathrm{d}t}\mathrm{Tr}\Big(D_1^{\mathrm{S}}(t)H_1^{\mathrm{S}}(t)\Big) = \mathrm{Tr}(D_1^{\mathrm{S}}\dot{H}_1^{\mathrm{S}}) + \mathrm{Tr}(\dot{D}_1^{\mathrm{S}}H_1^{\mathrm{S}}) \tag{9.70}$$

The last term in (9.70) vanishes because of (9.69) and of $\mathrm{Tr}\Big([H_1^{\mathrm{S}}, D_1^{\mathrm{S}}]H_1^{\mathrm{S}}\Big) = 0$. Then

$$\frac{\mathrm{d}W}{\mathrm{d}t} = -\sum_i \mathrm{Tr}[D_1^{\mathrm{S}}(t)\mathsf{A}_i]\dot{f}_i(t) = -\sum_i \overline{\mathsf{A}_i}(t)\dot{f}_i(t) \tag{9.71}$$

Let us choose periodic $f_i(t)$s

$$f_i(t) = \frac{1}{2}\Big(f_i^{\omega}\mathrm{e}^{-\mathrm{i}\omega t} + f_i^{\omega *}\mathrm{e}^{\mathrm{i}\omega t}\Big) = \mathrm{Re}\Big(f_i^{\omega}\mathrm{e}^{-\mathrm{i}\omega t}\Big) \tag{9.72}$$

and take a time average of $\mathrm{d}W/\mathrm{d}t$ over a time interval $T \gg \omega^{-1}$. We may use $\overline{\delta\mathsf{A}_i}$ instead of $\overline{\mathsf{A}_i}$ in (9.71) because $\langle\mathsf{A}_i\rangle$ gives a vanishing contribution to a time average. As $\overline{\delta\mathsf{A}_i}(t)$ is given by (9.13), plugging (9.13) in (9.70) and taking the time average gives after an elementary calculation (see Exercises 9.7.1 and 9.7.6)

$$\boxed{\left\langle\frac{\mathrm{d}W}{\mathrm{d}t}\right\rangle_T = \frac{1}{2}\,\omega f_i^{\omega *}\chi_{ij}''(\omega)f_j^{\omega}} \tag{9.73}$$

It is easy to check that the right hand side of (9.73) is a real quantity, even if $\chi_{ij}''(\omega)$ is imaginary, because combining (9.65) and (9.66), which are independent of time reversal, gives $(\chi_{ij}''(\omega))^* = \chi_{ji}''(\omega)$. From the second law of thermodynamics, the right hand side of (9.73) must always be positive, otherwise one would obtain work from a single source of heat, which implies that the matrix $\omega\chi_{ij}''(\omega)$ must be positive. The proof is left to Exercise 9.7.6.

To conclude this subsection, let us write the most general form of linear response, including space-dependent terms, which we have already introduced in

Section 9.1.4. The generalization of (9.13) is

$$\overline{\delta A_i}(\vec{r}, t) = \int \mathrm{d}t'\, \mathrm{d}^3r'\, \chi_{ij}(\vec{r}, \vec{r}\,', t - t') f_j(\vec{r}\,', t') \tag{9.74}$$

In general, space translation invariance holds and χ depends only on the difference $\vec{r} - \vec{r}\,'$. Then one may take the space Fourier transform of (9.73) to cast the convolution into a product

$$\overline{\delta A_i}(\vec{k}, t) = \int \mathrm{d}t'\, \chi_{ij}(\vec{k}, t - t') f_j(\vec{k}, t') \tag{9.75}$$

so that all Fourier components are decoupled, and the preceding results can be immediately transposed to each individual Fourier component. However, in using the symmetry property (9.66), one must be careful that $\mathsf{A}_i(\vec{k}, t)$ is not Hermitian, even though $\mathsf{A}_i(\vec{r}, t)$ is Hermitian since

$$[\mathsf{A}_i(\vec{k}, t)]^\dagger = \mathsf{A}_i(-\vec{k}, t)$$

9.2.5 Sum rules

The dynamical susceptibility obeys sum rules that are very useful to constrain phenomenological expressions such as those written in Section 9.1.4. Let us start from the representation (9.24) of $\chi_{ij}(z)$

$$\chi_{ij}(\vec{k}, z) = \int_{-\infty}^{\infty} \frac{\mathrm{d}\omega'}{\pi} \frac{\chi''_{ij}(\vec{k}, \omega')}{\omega' - z} \tag{9.76}$$

and assume that $\chi''_{ij}(\vec{k}, \omega)$ is odd and real ($\varepsilon_i \varepsilon_j = +1$). The so-called *thermodynamic sum rule* is obtained in the static limit z or $\omega \to 0$

$$\chi_{ij}(\vec{k}, \omega = 0) = \int_{-\infty}^{\infty} \frac{\mathrm{d}\omega'}{\pi} \frac{\chi''_{ij}(\vec{k}, \omega')}{\omega'} \tag{9.77}$$

$\chi_{ij} = \lim_{k \to 0} \chi_{ij}(\vec{k}, \omega = 0)$ is a thermodynamic quantity, hence the terminology 'thermodynamic sum rule'.

By examining the high frequency limit, we obtain the so-called *f-sum rule* (or *Nozières–Pines sum rule*). Let us look at the behaviour $|z| \to \infty$ of (9.76)

$$\frac{1}{\omega - z} = -\frac{1}{z}\left(1 + \frac{\omega}{z} + \frac{\omega^2}{z^2} + \cdots\right)$$

Since χ''_{ij} is an odd function of ω

$$\chi_{ij}(\vec{k}, z) = -\frac{1}{z^2} \int_{-\infty}^{\infty} \frac{\mathrm{d}\omega}{\pi}\, \omega \chi''_{ij}(\vec{k}, \omega) + \mathcal{O}\Big(\frac{1}{z^4}\Big) \tag{9.78}$$

This equation gives the leading term in an expansion in $1/z$ of $\chi_{ij}(\vec{k}, z)$. We remark that $\omega\chi''_{ij}(\vec{k}, \omega)$ is the time-Fourier transform of $\mathrm{i}\partial_t \chi''(\vec{k}, \omega)$

$$\omega\chi''_{ij}(\vec{k}, \omega) = \int \mathrm{d}t\, \mathrm{e}^{\mathrm{i}\omega t}\Big[\mathrm{i}\partial_t \chi''_{ij}(\vec{k}, t)\Big]$$

so that, from (9.62), generalized to space-dependent observables $\mathsf{A}_i(\vec{r}, t)$, we get after a Fourier transformation

$$\Big[\mathrm{i}\partial_t \chi''_{ij}(\vec{k}, t)\Big]_{t=0} = \int_{-\infty}^{\infty} \frac{\mathrm{d}\omega}{\pi}\, \omega\chi''_{ij}(\vec{k}, \omega) = \frac{\mathrm{i}}{V} \langle [\dot{\mathsf{A}}_i(\vec{k}, t), \mathsf{A}_j(-\vec{k}, 0)] \rangle \Big|_{t=0}$$

where V is the total volume of the sample. By using the commutation relation

$$\dot{\mathsf{A}}_i(\vec{k}, t) = \mathrm{i}[\mathsf{A}_i(t), H]$$

we finally get

$$\int_{-\infty}^{\infty} \frac{\mathrm{d}\omega}{\pi}\, \omega\chi''_{ij}(\vec{k}, \omega) = \frac{1}{V} \langle [[\mathsf{A}_i(\vec{k}), H], \mathsf{A}_j(-\vec{k})] \rangle \tag{9.79}$$

The most important example is that of the density–density correlation function χ''_{nn}. Let us perform the calculation in the classical limit, leaving the quantum case to Exercise 9.7.7. From the classical fluctuation-dissipation theorem

$$\chi''(\vec{k}, \omega) = \frac{1}{2}\beta\omega S(\vec{k}, \omega)$$

and the integral to be computed is

$$I = \beta \int \frac{\mathrm{d}\omega}{2\pi}\, \omega^2 S_{nn}(\vec{k}, \omega)$$

Using time translation invariance leads to

$$V S_{nn}(\vec{k}, t - t') = \langle n(\vec{k}, t) n(-\vec{k}, t') \rangle_{\mathrm{c}}$$

and the integral I is

$$I = \frac{\beta}{V} \langle \dot{n}(\vec{k}, t)\, \dot{n}(-\vec{k}, t) \rangle_{\mathrm{c}}$$

Using the spatial Fourier transform of the continuity equation (9.29) $\partial_t n + \mathrm{i}k_l j_l = 0$ finally yields

$$I = \frac{\beta}{V} k_l k_m \langle j_l(\vec{k}, t)\, j_m(\vec{k}, t)\rangle$$

The current density $\vec{j}$ is given by a sum over the N particles of the system as a function of their velocities $\vec{v}^{\,\alpha}$

$$j_l(\vec{r}, t) = \sum_{\alpha=1}^{N} v_l^{\alpha} \delta(\vec{r} - \vec{r}^{\,\alpha}(t)) \tag{9.80}$$

or, in Fourier space

$$j_l(\vec{k}, t) = \sum_{\alpha=1}^{N} v_l^{\alpha} \exp[-\mathrm{i}\vec{k} \cdot \vec{r}^{\,\alpha}(t)] \tag{9.81}$$

In the classical limit, the velocities of different particles are uncorrelated

$$\langle v_l^{\alpha} v_m^{\gamma} \rangle = \frac{1}{3} \delta_{\alpha\gamma} \delta_{lm} \langle \vec{v}^{\,2} \rangle = \delta_{\alpha\gamma} \delta_{lm} \frac{1}{m\beta}$$

and we get the f-sum rule

$$\boxed{\int_{-\infty}^{\infty} \frac{\mathrm{d}\omega}{\pi}\, \omega \chi''_{nn}(\vec{k}, \omega) = \frac{n k^2}{m}} \tag{9.82}$$

where $n = N/V$ is the density.

9.3 Projection method and memory effects

In this section, we first give a phenomenological introduction to memory effects, and then we show that these effects can be accounted for by using the so-called *projection method.* The idea which underlies the projection method is that it is often possible to distinguish between macroscopic variables, which vary slowly in time, and microscopic variables, which, on the contrary, exhibit fast variation with time. The former variables will be called slow modes and the latter fast modes. In general, we are not interested in the fast modes, associated with the microscopic behaviour of the system. For example, in the case of Brownian motion, we are not interested in the fast motion of the fluid molecules, but only in the slow motion of the Brownian particle. The idea is then to project the dynamics onto a subspace spanned by the slowly varying observables, in order to keep only the modes we are interested in. The success of the method will depend on our ability to identify all the slow modes and to restrict the dynamics to that of the slow modes only, which is called the *reduced dynamics*. The presence of slow modes can usually be traced back to the following sources.

(*i*) Existence of local conservation laws: we have seen an example in Section 9.1.4. In general, the corresponding slow modes are called *hydrodynamic modes*.
(*ii*) Existence of one heavy particle: this is the example of Brownian motion, where a heavy particle of mass M is put in a fluid of particles of mass m, with $M/m \gg 1$. This example will be studied at the end of the present section and in Problem 9.8.3.
(*iii*) Existence of Goldstone modes, associated with a broken continuous symmetry, for example in magnets, superfluids and liquid crystals. These slow modes will not be examined in this book, and we refer to the literature for an account of this very interesting case.

Of course, one cannot completely eliminate the fast modes, and the reduced dynamics cannot be described by a closed set of differential (or partial differential) equations. The back action of the fast modes on the reduced dynamics will appear through a memory term and a stochastic force.

9.3.1 Phenomenological introduction to memory effects

Our hydrodynamic theory of Section 9.1.4 suffers from a major failure: the *f*-sum rule (9.82) does not converge, since from (9.40) $\chi''(\omega) \sim 1/\omega$ for $\omega \to \infty$. As the continuity equation (9.29) is exact, the weak link of the hydrodynamic description must be Fick's law (9.30). Let us try to correct it, in a heuristic manner, by allowing for memory effects in such a way that the current does not instantaneously follow the density gradient

$$\vec{j}(\vec{r}, t) = -\int_0^t \mathrm{d}t' \gamma(t - t') \vec{\nabla} n(\vec{r}, t') \tag{9.83}$$

$\gamma(t)$ is called the *memory function*. One could also introduce a spatial dependence in the memory function; this generalization is easily handled by going to Fourier space and is left as an exercise for the reader. Let us try the simple parametrization

$$\gamma(t) = \frac{D}{\tau^*} \mathrm{e}^{-|t|/\tau^*} \tag{9.84}$$

where τ^* is a *microscopic* time ($\tau^* \sim 10^{-12} - 10^{-14}$ s), characteristic of relaxation towards *local equilibrium* (let us recall that a hydrodynamic description assumes that a situation of local equilibrium has been reached). If $\vec{\nabla} n$ varies slowly

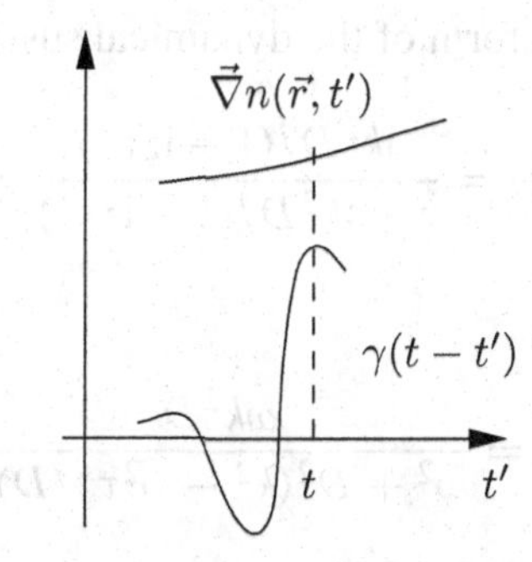

Figure 9.2 A slowly varying density distribution.

on a time scale $\sim \tau^*$, one recovers Fick's law (9.30) for $t \gg \tau^*$ (Figure 9.2)

$$\int_0^t \mathrm{d}t' \gamma(t-t')\vec{\nabla}n(\vec{r},t') \simeq \vec{\nabla}n(\vec{r},t) \int_0^\infty \mathrm{d}t' \frac{D}{\tau^*} \mathrm{e}^{-t'/\tau^*} = D\vec{\nabla}n(\vec{r},t) \tag{9.85}$$

so that (9.30) is recovered. With memory effects included, the continuity equation reads in Fourier space

$$\partial_t n(\vec{k},t) = -k^2 \int_0^t \mathrm{d}t' \gamma(t-t') n(\vec{k},t') \tag{9.86}$$

We take the Laplace transform of (9.86) and note that the Laplace transform of the convolution is transformed into a product

$$\int_0^\infty \mathrm{d}t \int_0^t \mathrm{d}t' \, \mathrm{e}^{\mathrm{i}z(t-t')} \mathrm{e}^{\mathrm{i}zt'} \gamma(t-t') \, n(\vec{k},t') = \gamma(z) \, n(\vec{k},z)$$

so that

$$n(\vec{k},t=0) + \mathrm{i}zn(\vec{k},z) = k^2\gamma(z)n(\vec{k},z)$$

Solving for $n(\vec{k},z)$ and using (9.26) allows us to identify the new expression for the susceptibility

$$\chi(\vec{k},z) = \frac{\mathrm{i}k^2\gamma(z)}{z+\mathrm{i}k^2\gamma(z)} \chi(\vec{k}) \tag{9.87}$$

while $\gamma(z)$ becomes in the approximation (9.84)

$$\gamma(z) = \frac{D}{\tau^*} \int_0^\infty \mathrm{d}t \, \mathrm{e}^{\mathrm{i}zt - t/\tau^*} = \frac{D}{1 - \mathrm{i}z\tau^*} \tag{9.88}$$

This yields the Laplace transform of the dynamical susceptibility

$$\chi(\vec{k}, z) = \frac{\mathrm{i}k^2 D/(1 - \mathrm{i}z\tau^*)}{z + \mathrm{i}k^2 D/(1 - \mathrm{i}z\tau^*)}\,\chi(\vec{k}) \tag{9.89}$$

and its imaginary part

$$\chi''(\vec{k}, \omega) = \frac{\omega k^2 D}{\omega^2 + D^2(k^2 - \omega^2\tau^*/D)^2}\,\chi(\vec{k}) \tag{9.90}$$

As $\chi''(\omega) \sim 1/\omega^3$ for $\omega \to \infty$, the *f*-sum rule is now convergent. Since the integral in the sum rule is given by the coefficient of $-1/z^2$ in an expansion of $\chi(\vec{k}, z)$ for $|z| \to \infty$, and since from (9.89)

$$\chi(\vec{k}, z) \sim -\frac{k^2 D}{z^2\tau^*}\,\chi(\vec{k}) \quad \text{for } |z| \to \infty$$

Equation (9.82) leads to an interesting relation between the diffusion coefficient and the microscopic time τ^*

$$D = \frac{n\tau^*}{m\chi(\vec{k})} \tag{9.91}$$

Incidentally, this equation shows that D (or the memory function) must depend on k. In the example of polarized helium-3 at low temperatures, one may prove that $\tau^* \sim 1/T^2$ because of the Pauli principle, and (9.91) predicts that the same behaviour should be observed for D, which is experimentally verified. However, our approximation (9.84) for the memory function $\gamma(t)$ cannot be the last word, as sum rules with factors of ω^{2n+1} instead of ω remain divergent.

9.3.2 Projectors

We now turn to a formal derivation of memory effects. We shall need the *Liouvillian*, which is an operator acting on the vector space of observables A_i. The definition of the Liouvillian $\mathcal{L}$ follows from Heisenberg's equation of motion for an observable A

$$\boxed{\partial_t \mathsf{A} = \mathrm{i}[H, \mathsf{A}] = \mathrm{i}\mathcal{L}\mathsf{A}} \tag{9.92}$$

The operators acting in the vector space of observables will be denoted by script letters: $\mathcal{L}$, $\mathcal{P}$, $\mathcal{Q}$. Equation (9.92) can be integrated formally as

$$\mathsf{A}(t) = \mathrm{e}^{\mathrm{i}\mathcal{L}t}\,\mathsf{A}(0) \tag{9.93}$$

An explicit expression for the Liouvillian can be obtained by choosing a basis in the Hilbert space of states in which (9.92) reads

$$\mathrm{i}\partial_t \mathsf{A}_{mn}(t) = \mathsf{A}_{m\nu}(t)H_{\nu n} - H_{m\nu}\mathsf{A}_{\nu n}(t)$$

As the observables may be considered as elements of a vector space with components labelled by two indices (m, n) or (μ, ν), the matrix elements of the Liouvillian are then labelled by couples of indices

$$\mathcal{L}_{mn;\mu\nu} = H_{m\mu}\delta_{n\nu} - H_{\nu n}\delta_{m\mu} \tag{9.94}$$

As explained in the introduction to the present section, we want to project the dynamics on the subspace $\mathcal{E}$ spanned by a set of slowly varying observables A_i and the identity operator $\mathcal{I}$: $\mathcal{E} \equiv \{\mathcal{I}, \mathsf{A}_i\}$. Actually the equations we are going to derive in this subsection are valid for any set of observables, but they are physically useful only if $\mathcal{E}$ is a set of slow variables. We are looking for a projector $\mathcal{P}$ ($\mathcal{P}^2 = \mathcal{P}$) on the subspace $\mathcal{E}$ such that

$$\mathcal{P}B = B \qquad \text{if} \qquad B \in \mathcal{E} \equiv \{\mathcal{I}, \mathsf{A}_i\} \tag{9.95}$$

The complementary projector will be denoted by $\mathcal{Q}$: $\mathcal{Q} = \mathcal{I} - \mathcal{P}$. One should be aware that the set $\{\mathsf{A}_i\}$ is defined at $t = 0$ and in general the observables $\mathsf{A}_i(t)$ at a time $t \neq 0$ do *not* belong to the subspace $\mathcal{E}$: only the *projected* operators $\mathcal{P}\mathsf{A}_i(t)$ belong to $\mathcal{E}$.

In order to compute $\mathcal{P}$, let us recall an elementary example. Assume that we are given in $\mathbb{R}^N$ a set of M ($M < N$) non-orthogonal, non-normalized vectors $\vec{e}_1, \ldots, \vec{e}_M$. The action on a vector $\vec{V}$ of the projector $\mathcal{P}$ that projects on the subspace spanned by these vectors is

$$\mathcal{P}\vec{V} = C^{-1}_{ij}(\vec{V}\cdot\vec{e}_i)\vec{e}_j \qquad \text{where} \qquad C_{ij} = \vec{e}_i\cdot\vec{e}_j \tag{9.96}$$

Equation (9.96) is generally valid provided one is given a scalar product C_{ij} of two vectors $\vec{e}_i$ and $\vec{e}_j$. In particular we can use (9.96) in the space of observables and choose Mori's scalar product (9.51) to define the projector $\mathcal{P}$ on the subspace $\mathcal{E}$,[10]

$$\boxed{\mathcal{P} = \delta\mathsf{A}_j C^{-1}_{jk}\delta\mathsf{A}_k \qquad C_{jk} = \langle\delta\mathsf{A}_j;\delta\mathsf{A}_k\rangle} \tag{9.97}$$

[10] Strictly speaking, there is one term missing in (9.97), because one must have $\mathcal{P}I = I$. Setting $\delta\mathsf{A}_0 = I$, we may write

$$\mathcal{P} = \sum_{j,k=0} \delta\mathsf{A}_j C^{-1}_{jk}\delta\mathsf{A}_k$$

where we have used $\langle\delta\mathsf{A}_0;\delta\mathsf{A}_0\rangle = 1$ and $\langle\delta\mathsf{A}_0;\delta\mathsf{A}_i\rangle = 0$ for $i \neq 0$.

and the average $\langle \bullet \rangle$ is computed with the *equilibrium density operator* (9.55); recall that $\delta \mathsf{A}_i = \mathsf{A}_i - A_i$. The explicit action of $\mathcal{P}$ on an observable B is

$$\mathcal{P}\mathsf{B} = \delta \mathsf{A}_j \, C^{-1}_{jk} \langle \delta \mathsf{A}_k; \mathsf{B} \rangle = \frac{\partial B}{\partial A_j} \delta \mathsf{A}_j$$

where the last equation follows from (9.53) and Hermiticity of A_k and B. It is easily checked that $\mathcal{P}^2 = \mathcal{P}$ and that $\mathcal{P}^\dagger = \mathcal{P}$. Moreover we also note that the Liouvillian is Hermitian with respect to Mori's scalar product (Exercise 9.7.4)

$$\langle \mathsf{A}; \mathcal{L}\mathsf{B} \rangle = \langle \mathcal{L}\mathsf{A}; \mathsf{B} \rangle \tag{9.98}$$

An important consequence of (9.98) is the antisymmetry property deduced from $\partial_t = \mathrm{i}\mathcal{L}$

$$\langle \mathsf{A}; \dot{\mathsf{B}} \rangle = -\langle \dot{\mathsf{A}}; \mathsf{B} \rangle \tag{9.99}$$

where $\dot{\mathsf{A}} = \partial_t \mathsf{A}$, from which $\langle \mathsf{A}; \dot{\mathsf{A}} \rangle = 0$ follows.

9.3.3 Langevin–Mori equation

We now wish to derive equations of motion for the observables of the set A_i.[11] Let us start from the trivial identity $\mathcal{P} + \mathcal{Q} = \mathcal{I}$ and write

$$\dot{\mathsf{A}}_i(t) = \mathrm{e}^{\mathrm{i}\mathcal{L}t} \mathcal{Q} \dot{\mathsf{A}}_i + \mathrm{e}^{\mathrm{i}\mathcal{L}t} \mathcal{P} \dot{\mathsf{A}}_i \tag{9.100}$$

It is easy to evaluate the second term in (9.100)

$$\begin{aligned} \mathrm{e}^{\mathrm{i}\mathcal{L}t} \mathcal{P} \dot{\mathsf{A}}_i &= \mathrm{e}^{\mathrm{i}\mathcal{L}t} \delta \mathsf{A}_j C^{-1}_{jk} \langle \delta \mathsf{A}_k; \mathrm{i}\mathcal{L}\mathsf{A}_i \rangle \\ &= \Omega_{ji} \, \mathrm{e}^{\mathrm{i}\mathcal{L}t} \, \delta \mathsf{A}_j \end{aligned} \tag{9.101}$$

the frequencies Ω_{ji} being defined by

$$\Omega_{ji} = C^{-1}_{jk} \langle \delta \mathsf{A}_k; \mathrm{i}\mathcal{L}\mathsf{A}_i \rangle \tag{9.102}$$

Note that $\Omega_{ji} = 0$ if all observables in $\mathcal{E}$ have the same parity under time reversal (see Exercise 9.7.4). In this case $\langle \dot{\mathsf{A}}_i; \mathsf{A}_j \rangle = 0 \, \forall (i, j)$ and $\dot{\mathsf{A}}_i \in \mathcal{E}_\perp$. In particular $\Omega = 0$ if there is only one slow observable.

In order to deal with the first term in (9.100) we use the operator identity (2.118), which can be cast into the form

$$\mathrm{e}^{\mathrm{i}\mathcal{L}t} = \mathrm{e}^{\mathrm{i}\mathcal{Q}\mathcal{L}t} + \mathrm{i} \int_0^t \mathrm{d}t' \, \mathrm{e}^{\mathrm{i}\mathcal{L}(t-t')} \mathcal{P}\mathcal{L} \mathrm{e}^{\mathrm{i}\mathcal{Q}\mathcal{L}t'} \tag{9.103}$$

[11] We are going to derive Mori's version of the projection method. There exist many other versions of the projection method, but Mori's turns out to be the easiest to derive and the most useful close to equilibrium.

We then transform (9.103) by introducing operators $\mathsf{f}_i(t)$, which are called *stochastic forces* for reasons that will become clear in Section 9.3.5, through

$$\boxed{\mathsf{f}_i(t) = \mathrm{e}^{\mathrm{i}\mathcal{Q}\mathcal{L}\mathcal{Q}t}\,\mathcal{Q}\dot{\mathsf{A}}_i} \tag{9.104}$$

The stochastic forces live entirely in the space $\mathcal{E}_\perp$ orthogonal to $\mathcal{E}$. First $\mathcal{Q}$ projects $\dot{\mathsf{A}}_i$ on this subspace,[12] and then the operator $\mathcal{Q}\mathcal{L}\mathcal{Q}$, which has non-zero matrix element only in $\mathcal{E}_\perp$, is the evolution operator in $\mathcal{E}_\perp$: $\exp(\mathrm{i}\mathcal{Q}\mathcal{L}\mathcal{Q}t)\mathcal{Q}\dot{\mathsf{A}}_i$ does not leave $\mathcal{E}_\perp$. The stochastic forces may thus vary with a time scale entirely different from that characteristic of $\mathcal{P}\mathsf{A}_i(t)$, which should evolve slowly with time. Note also that the stochastic forces have zero equilibrium average values, $\langle \mathsf{f}_i(t)\rangle = 0$, and that they are orthogonal by construction to the A_is: $\langle \mathsf{A}_i; \mathsf{f}_i(t)\rangle = 0$. Given the definition (9.104), the first term in (9.103) yields, when applied to $\mathcal{Q}\dot{\mathsf{A}}_i$

$$\mathrm{e}^{\mathrm{i}\mathcal{Q}\mathcal{L}t}\,\mathcal{Q}\dot{\mathsf{A}}_i = \mathrm{e}^{\mathrm{i}\mathcal{Q}\mathcal{L}\mathcal{Q}t}\,\mathcal{Q}\dot{\mathsf{A}}_i = \mathsf{f}_i(t)$$

where we have used

$$\left(I + \mathrm{i}\mathcal{Q}\mathcal{L}t + \frac{\mathrm{i}^2}{2!}\mathcal{Q}\mathcal{L}\mathcal{Q}\mathcal{L}t^2 + \cdots\right)\mathcal{Q}$$
$$= \left(I + \mathrm{i}\mathcal{Q}\mathcal{L}\mathcal{Q}t + \frac{\mathrm{i}^2}{2!}(\mathcal{Q}\mathcal{L}\mathcal{Q})(\mathcal{Q}\mathcal{L}\mathcal{Q})t^2 + \cdots\right)\mathcal{Q}$$

because $\mathcal{Q}^2 = \mathcal{Q}$. The last contribution comes from the second term in (9.103)

$$\int_0^t \mathrm{d}t'\,\mathrm{e}^{\mathrm{i}\mathcal{L}(t-t')}\,\delta\mathsf{A}_j\,C^{-1}_{jk}\big\langle\delta\mathsf{A}_k;\,\mathcal{L}\mathrm{e}^{\mathrm{i}\mathcal{Q}\mathcal{L}\mathcal{Q}t'}\,\mathcal{Q}\,\delta\dot{\mathsf{A}}_i\big\rangle$$
$$= \int_0^t \mathrm{d}t'\,\mathrm{e}^{\mathrm{i}\mathcal{L}(t-t')}\delta\mathsf{A}_j\,C^{-1}_{jk}\big\langle\delta\mathsf{A}_k;\,\mathrm{i}\mathcal{L}\,\mathsf{f}_i(t')\big\rangle$$

Using the Hermiticity (9.98) of $\mathcal{L}$, we transform the scalar product

$$\big\langle\delta\mathsf{A}_k;\,\mathrm{i}\mathcal{L}\,\mathsf{f}_i(t')\big\rangle = -\big\langle\mathrm{i}\mathcal{L}\,\delta\mathsf{A}_k;\,\mathcal{Q}\,\mathsf{f}_i(t')\rangle = -\langle\mathsf{f}_k;\,\mathsf{f}_i(t')\big\rangle$$

and cast the second term coming from (9.103) into the form

$$-\int_0^t \mathrm{d}t'\,\delta\mathsf{A}_j(t-t')C^{-1}_{jk}\big\langle\mathsf{f}_k;\,\mathsf{f}_i(t')\big\rangle = -\int_0^t \mathrm{d}t'\,\delta\mathsf{A}_j(t-t')\gamma_{ji}(t')$$

[12] If all slow observables have the same parity under time reversal, $\mathcal{Q}\dot{\mathsf{A}}_i = \dot{\mathsf{A}}_i$.

where we have defined the *memory matrix* $\gamma_{ji}(t)$

$$\boxed{\gamma_{ji}(t) = C^{-1}_{jk}\langle \mathsf{f}_k; \mathsf{f}_i(t)\rangle} \tag{9.105}$$

Gathering all terms in the preceding discussion leads to the *Langevin–Mori equations*, which are the promised equations of motion for the slow observables A_i

$$\boxed{\partial_t \mathsf{A}_i(t) = \dot{\mathsf{A}}_i(t) = \Omega_{ji}\,\delta\mathsf{A}_j(t) - \int_0^t \mathrm{d}t'\,\gamma_{ji}(t')\delta\mathsf{A}_j(t-t') + \mathrm{e}^{\mathrm{i}\mathcal{QLQ}t}\,\mathsf{f}_i} \tag{9.106}$$

The Langevin–Mori equations contain a frequency term, a memory term and a stochastic force. We emphasize that no approximations have been made in deriving (9.106), and *the Langevin–Mori equations are exact.* As already mentioned, they are valid for any set of observables $\{\mathsf{A}_i\}$, but they are useful only if this is a set of slow observables. Note that $\dot{\mathsf{A}}_i(t) = \delta\dot{\mathsf{A}}_i(t)$ as A_i is time independent. Other exact equations can be obtained if we assume a time-dependent situation given by the quantum version of (9.9). If D is the density operator at $t = 0$, we can compute the average value of (9.106) with D, and using the same notation as in (9.11), $\overline{\mathsf{A}_i} = \langle \mathsf{A}_i\rangle_D$, we obtain for $t > 0$

$$\partial_t\,\overline{\delta\mathsf{A}_i}(t) = \Omega_{ji}\,\overline{\delta\mathsf{A}_j}(t) - \int_0^t \mathrm{d}t'\,\gamma_{ji}(t')\overline{\delta\mathsf{A}_i}(t-t') + \langle \mathrm{e}^{\mathrm{i}\mathcal{QLQ}t}\,\mathsf{f}_i\rangle_D \tag{9.107}$$

In the absence of the last term in (9.107), we would have a simple linear integro-differential equation for the average values $\overline{\delta\mathsf{A}_i}(t)$. The complexity of the dynamics lies in the average value of the stochastic force, which reflects the dynamics of the fast modes.

The stochastic forces may be eliminated if we restrict ourselves to equations of motion for the Kubo functions. Indeed, taking the scalar product of the Langevin–Mori equation (9.106) with $\delta\mathsf{A}_k$ we get

$$\boxed{\dot{C}_{ki}(t) = \Omega_{ji}\,C_{kj}(t) - \int_0^t \mathrm{d}t'\,\gamma_{ji}(t')C_{kj}(t-t')} \tag{9.108}$$

In the case of a single slow mode, this equation simplifies to

$$\boxed{\dot{C}(t) = -\int_0^t \mathrm{d}t'\,\gamma(t')C(t-t')} \tag{9.109}$$

Using (9.57) and (9.108), we can derive equations of motion for the average values $\overline{\delta\mathsf{A}_i(t)}$, which are then given by (9.107), but *without the stochastic term.* We thus have a closed sytem of integro-differential equations of motion for the $\overline{\delta\mathsf{A}_i(t)}$s. However, *this system of equations is only valid close to equilibrium*, because we have used (9.57), which implies linear response, and thus small deviations from equilibrium. In the case of a single observable, this system reduces to

$$\partial_t \overline{\delta\mathsf{A}(t)} = -\int_0^t \mathrm{d}t' \gamma(t') \overline{\delta\mathsf{A}(t-t')} \tag{9.110}$$

These equations can also be derived directly, by using in (9.107) the linear approximation to the density operator D which follows from (2.119)

$$D \simeq D_{\mathrm{eq}}\Big(I - f_i \int_0^\beta \mathrm{d}\alpha\, \mathrm{e}^{\alpha H}\, \mathsf{A}_i\, \mathrm{e}^{-\alpha H}\Big)$$

and observing that $\langle \mathsf{f}_i(t)\rangle = \langle I; \mathsf{f}_i(t)\rangle = 0$ and $\langle \mathsf{A}_i; \mathsf{f}_j(t)\rangle = 0$. Comparing (9.110) with (9.83), we clearly see the connection with our previous heuristic approach to memory effects.

9.3.4 Brownian motion: qualitative description

We consider in classical mechanics a Brownian particle, namely a 'heavy' (by microscopic standards) particle of mass M in a heat bath of light molecules of mass m: $m/M \ll 1$. The first effect one can think of is viscosity. If the heavy particle has a velocity $\vec{v}$ in the positive x direction, the fluid molecules coming from the right will appear to have a larger velocity than those coming from the left, and because of its collisions with the fluid molecules and of this simple Doppler effect, the particle will be submitted to a force directed toward the left

$$\vec{\mathcal{F}} = -\alpha \vec{v} \tag{9.111}$$

where α is the friction coefficient; $\tau = M/\alpha = 1/\gamma$ defines a characteristic *macroscopic* time scale for the particle. However, there is another time scale in the problem, a *microscopic* time scale τ^*. Due to the random character of the collisions with the fluid molecules, one observes fluctuations of the force on a time scale $\sim 10^{-12} - 10^{-14}$ s on the order of the duration of a collision. The separation of time scales relies on the inequality $m/M \ll 1$: compare on Figure 9.3 the behaviour of the velocity of a Brownian particle and that of a ^{17}O molecule in a gas of ^{16}O molecules. In the latter case the velocity changes suddenly on a time

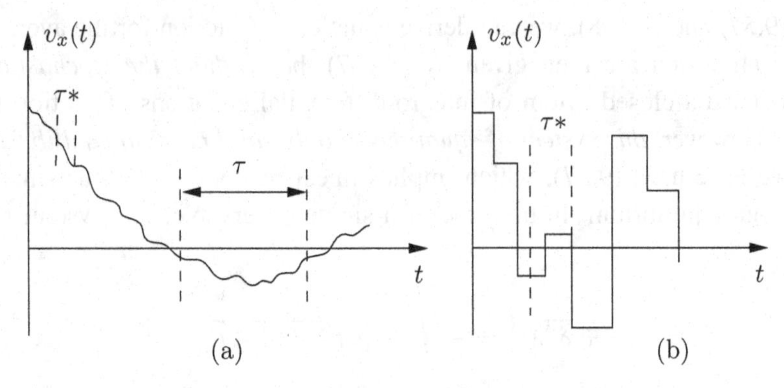

Figure 9.3 x-component of the velocity. (a) Heavy (Brownian) particle in a bath of light particles, (b) ^{17}O molecules in a gas of ^{16}O molecules.

scale τ^* (it may even change sign), while in the case of the Brownian particle, the time scale for velocity changes is τ because of its large inertia, although a short time scale τ^* is superimposed on this average motion. Although the Brownian particle may suffer large accelerations, the mean velocity varies very little on a time scale τ^*, and the average motion is a smooth one.

Let $\mathcal{A} = v$ be the velocity of a one-dimensional Brownian particle (or B-particle for short) which we identify with the (classical) observable in (9.110). From the preceding qualitative analysis, we may hope that v (but not $\dot{v}$!) is a slow variable, and we note that $\delta v = v$ since $\langle v \rangle = 0$. In the classical limit

$$\langle v(0);\, v(t) \rangle \to \langle v(0) v(t) \rangle = C_{vv}(t)$$

$C_{vv}(t)$ is the velocity autocorrelation function. Equation (9.109) becomes in this case

$$\dot{C}_{vv}(t) = -\int_0^t \mathrm{d}t'\, \gamma(t')\, C_{vv}(t - t')$$

If the characteristic time scale of $\gamma(t)$ is much shorter than that of $C_{vv}(t)$, we may use a Markovian approximation as in (9.85), and we get an ordinary differential equation

$$\dot{C}_{vv}(t) = -\gamma\, C_{vv}(t) \qquad \text{with} \qquad \gamma = \int_0^\infty \mathrm{d}t\, \gamma(t)$$

whose solution, taking into account the parity property of C, leads to an exponential decrease of the correlation function

$$C_{vv}(t) = \mathrm{e}^{-\gamma|t|} C_{vv}(0) = \mathrm{e}^{-|t|/\tau}\, C_{vv}(0) \tag{9.112}$$

9.3.5 Brownian motion: the $m/M \to 0$ limit

Since the friction force $\vec{\mathcal{F}}$ in (9.111) is due to the collisions with the fluid molecules, it should be related to the stochastic force. In this subsection, we shall derive this relation in the limit $m/M \to 0$.[13] Let us write a classical Hamiltonian for the system composed of the fluid and Brownian particle

$$H = \sum_\alpha \frac{p_\alpha^2}{2m} + \frac{1}{2}\sum_{\alpha\neq\beta} V(\vec{r}_{\alpha\beta}) + \frac{P^2}{2M} + \sum_\alpha U(\vec{R} - \vec{r}_\alpha) \tag{9.113}$$

where p_α and m denote the momenta and mass of the fluid molecules, V their potential energy, P and M the momentum and mass of the B-particle and U its potential energy in the fluid. The Liouvillian $\mathcal{L}$ corresponding to (9.113) may be written as $\mathcal{L} = \mathcal{L}_\mathrm{f} + \delta\mathcal{L}$, where $\mathcal{L}_\mathrm{f}$ is the fluid Liouvillian corresponding to the first two terms in (9.113)

$$\mathrm{i}\mathcal{L}_\mathrm{f} = \sum_\alpha \frac{\vec{p}_\alpha}{m}\cdot\frac{\partial}{\partial \vec{r}_\alpha} - \frac{1}{2}\sum_{\alpha\neq\beta}\vec{\nabla}V(\vec{r}_{\alpha\beta})\cdot\left(\frac{\partial}{\partial\vec{p}_\alpha} - \frac{\partial}{\partial\vec{p}_\beta}\right) \tag{9.114}$$

and

$$\mathrm{i}\delta\mathcal{L} = \frac{\vec{P}}{M}\cdot\frac{\partial}{\partial\vec{R}} - \sum_\alpha \vec{\nabla}U(\vec{R}-\vec{r}_\alpha)\cdot\frac{\partial}{\partial\vec{P}} + \sum_\alpha \vec{\nabla}U(\vec{R}-\vec{r}_\alpha)\cdot\frac{\partial}{\partial\vec{p}_\alpha} \tag{9.115}$$

In (9.115), the first term, denoted by $\mathcal{L}_\mathrm{B}$, is the Liouvillian of a free B-particle, the second $\mathcal{L}_{\mathrm{f}\to\mathrm{B}}$ represents the action of the fluid molecules on the B-particle and the third $\mathcal{L}_{\mathrm{B}\to\mathrm{f}}$ the action of the B-particle on the fluid molecules. Our slow variables are the components P_i of the momentum $\vec{P}$; remember that $\dot{P}_i$ is *not* a slow variable, as is clear from Figure 9.3, and does not belong to $\mathcal{E}$. Let $C_{ij}(t)$ be the equilibrium time correlation function of the components of $\vec{P}$

$$C_{ij}(t) = \langle P_i; \mathrm{e}^{\mathrm{i}\mathcal{L}t}P_j\rangle \to \langle P_i(0)P_j(t)\rangle \tag{9.116}$$

From rotational invariance (no summation over i in (9.117) and (9.118))

$$C_{ij}(t) = \delta_{ij}C(t) \qquad C_{ii}(0) = \langle P_i^2\rangle = MkT \tag{9.117}$$

The memory matrix γ_{ij} is also proportional to δ_{ij}: $\gamma_{ij} = \delta_{ij}\gamma(t)$

$$\gamma(t) = \frac{1}{MkT}\langle \dot{P}_i; \mathrm{e}^{\mathrm{i}\mathcal{QLQ}t}\dot{P}_i\rangle \tag{9.118}$$

Since $\langle\vec{P}\rangle = 0$, the projector $\mathcal{Q}$ is, from (9.97) and (9.117),

$$\mathcal{Q} = \mathcal{I} - \vec{P}\frac{1}{MkT}\vec{P} \tag{9.119}$$

[13] Our derivation follows closely Foerster [43], Section 6.1.

We now use the property that $p \sim \sqrt{mkT}$ and $P \sim \sqrt{MkT}$ to remark that in (9.114) and (9.115)

$$\frac{P/M}{p/m} \sim \left(\frac{m}{M}\right)^{1/2} \qquad \frac{\partial/\partial P}{\partial/\partial p} \sim \left(\frac{m}{M}\right)^{1/2}$$

so that

$$\frac{\mathcal{L}_{\mathrm{B}} + \mathcal{L}_{\mathrm{f}\to\mathrm{B}}}{\mathcal{L}_{\mathrm{f}} + \mathcal{L}_{\mathrm{B}\to\mathrm{f}}} \sim \left(\frac{m}{M}\right)^{1/2}$$

and $\mathcal{L} \to \mathcal{L}_0 = \mathcal{L}_{\mathrm{f}} + \mathcal{L}_{\mathrm{B}\to\mathrm{f}}$ in the limit $m/M \to 0$. But $\mathcal{L}_0$ does not act on the B-particle: $\mathcal{L}_0 \vec{P} = 0$, so that

$$\lim_{m/M\to 0} \gamma(t) = \gamma_\infty(t) = \frac{1}{MkT}\langle \dot{P}_i ; \mathrm{e}^{\mathrm{i}\mathcal{L}_0 t} \dot{P}_i \rangle$$

Now,

$$\frac{\mathrm{d}}{\mathrm{d}t}\vec{P} = \mathrm{i}\mathcal{L}\vec{P} = -\sum_\alpha \vec{\nabla} U(\vec{R} - \vec{r}_\alpha) = \vec{F}$$

where $\vec{F}$ is the *instantaneous* force that the fluid molecules exert on the B-particle, not to be confused with the friction force $\vec{\mathcal{F}}$ (9.111), which represents a mean effect. Thus we get our final result for the memory function

$$\gamma_\infty(t) = \frac{1}{3MkT}\langle \vec{F}_\infty(t) \cdot \vec{F}_\infty(0) \rangle \tag{9.120}$$

where $\vec{F}_\infty(t)$ is the force on an *infinitely heavy* B-particle, namely a B-particle *at rest* in the fluid, so that we are in an equilibrium situation. This force, which has zero average value, $\langle \vec{F}_\infty \rangle = 0$, varies randomly in time with a characteristic scale τ^*, which justifies the terminology 'stochastic force' in the definition (9.104). In a Markovian approximation, the viscosity parameter $\gamma = \alpha/M$, where α is defined in (9.111), is given by

$$\gamma = \frac{1}{6MkT}\int_{-\infty}^{\infty} \mathrm{d}t \, \langle \vec{F}_\infty(t) \cdot \vec{F}_\infty(0) \rangle \tag{9.121}$$

This is a Green–Kubo formula; another derivation of this result is proposed in Exercise 9.7.2. To summarize, we have derived a stochastic differential equation for the B-particle

$$\frac{\mathrm{d}}{\mathrm{d}t}\vec{P} = -\gamma\vec{P} + \vec{F}_\infty(t) \tag{9.122}$$

where γ is given by (9.121) and $\vec{F}_\infty(t)$ is the stochastic force acting on a B-particle *at rest* in the fluid.

9.4 Langevin equation

Equation (9.122) is the prototype of a stochastic differential equation and is called a Langevin equation. We have just seen that it is able to describe Brownian motion, but its importance lies in the fact that it has many other applications in physics, due to its versatility in the description of noise in general physical systems. One important particular case of the Langevin equation corresponds to the so-called Ornstein–Uhlenbeck process, where, in addition to the stochastic force, the Brownian particle is submitted to an harmonic force in the strong friction limit.

9.4.1 Definitions and first properties

We have derived in the case of Brownian motion a stochastic differential equation (9.122). In order to encompass more general situations, we rewrite it in one dimension as

$$\boxed{\dot{V}(t) = -\gamma V(t) + f(t) \qquad f(t) = \frac{1}{m} F_\infty(t)} \tag{9.123}$$

where, from now on, we denote by m the mass of the B-particle. In (9.123), $V(t)$ is a *random function*, which we write with an upper case letter V in order to make the distinction with the *number* $v(t)$, which is the value of the velocity for a particular realization of $f(t)$. The function $f(t)$ is a random function with zero average value, $\overline{f(t)} = 0$, and with a characteristic time scale τ^* much smaller than the characteristic time scale $\tau = 1/\gamma$ of the velocity: $\tau^* \ll \tau$. In general, $\overline{X}$ denotes an average of the random variable X taken over all the realizations of the random function $f(t)$, while $\langle X \rangle$ denotes as before an equilibrium average. In the case of Brownian motion with a B-particle at equilibrium, $\overline{X} = \langle X \rangle$ and (9.121) becomes in the one-dimensional case

$$\gamma = \frac{m}{2kT} \int_{-\infty}^{\infty} \mathrm{d}t \, \langle f(t) f(0) \rangle \qquad \langle f(t) \rangle = 0 \tag{9.124}$$

In order to write the Langevin equation in a precise form, we must define the time autocorrelation function of $f(t)$ in (9.123). Since $\tau^* \ll \tau$, we shall approximate the time autocorrelation of $f(t)$ by a δ-function

$$\boxed{\overline{f(t) f(t')} = 2A\, \delta(t - t')} \tag{9.125}$$

In the case of Brownian motion, the coefficient A is then given from (9.124) by

$$\boxed{A = \gamma \frac{kT}{m}} \tag{9.126}$$

Since $f(t)$ is the result of a large number of random collisions, it is customary to assume (but we shall not need this assumption), that in addition to $\overline{f}(t) = 0$ and (9.125), $f(t)$ is a Gaussian random function. These assumptions on $f(t)$ specify completely the stochastic differential equation (9.123).

Equation (9.123) may be solved as a function of the initial velocity $v_0 = V$ $(t = 0)$ as

$$V(t) = v_0 \mathrm{e}^{-\gamma t} + \mathrm{e}^{-\gamma t} \int_0^t \mathrm{d}t' \, \mathrm{e}^{\gamma t'} f(t') \tag{9.127}$$

from which we derive, by taking an average over all realizations of $f(t)$,[14]

$$\begin{aligned}\overline{\left(V(t) - v_0 \mathrm{e}^{-\gamma t}\right)^2} &= \mathrm{e}^{-2\gamma t} \int_0^t \mathrm{d}t' \, \mathrm{d}t'' \, \mathrm{e}^{2\gamma(t'+t'')} \overline{f(t') f(t'')} \\ &= 2A \, \mathrm{e}^{-2\gamma t} \int_0^t \mathrm{d}t' \, \mathrm{e}^{2\gamma t'} = \frac{A}{\gamma}\left(1 - \mathrm{e}^{-2\gamma t}\right)\end{aligned} \tag{9.128}$$

If $t \gg 1/\gamma$ we reach an equilibrium situation

$$\left\langle \left(V(t) - v_0 \mathrm{e}^{-\gamma t}\right)^2 \right\rangle \to \langle V^2 \rangle = \frac{A}{\gamma} = \frac{kT}{m}$$

and we recover (9.126). We can also compute the position $X(t)$ of the Brownian particle from

$$X(t) = x_0 + \int_0^t \mathrm{d}t' V(t')$$

An elementary, but somewhat tedious, calculation leads to

$$\overline{(X(t) - x_0)^2} = \left(v_0^2 - \frac{kT}{m}\right) \frac{1}{\gamma^2} \left(1 - \mathrm{e}^{-\gamma t}\right)^2 + \frac{2kT}{m\gamma}\left(t - \frac{1}{\gamma}\left[1 - \mathrm{e}^{-\gamma t}\right]\right)$$

[14] We cannot use an equilibrium average because the B-particle is not at equilibrium if it is launched in the thermal bath with a velocity $v_0 \gg \sqrt{kT/m}$.

For large values of t, $t \gg 1/\gamma$, one recovers a diffusive behaviour (see Exercise 9.7.8 for another derivation of (9.129))

$$\langle (X(t) - x_0)^2 \rangle = 2\frac{kT}{m\gamma} t = 2D\,t \tag{9.129}$$

The *diffusion coefficient* D is given by Einstein's relation (6.61)

$$\boxed{D = \frac{kT}{m\gamma} = \frac{1}{\beta m\gamma}} \tag{9.130}$$

For $t \ll 1/\gamma$, one observes a ballistic behaviour: $\overline{(X(t) - x_0)^2} = v_0^2 t^2$.

9.4.2 Ornstein–Uhlenbeck process

We are now interested in writing down a Langevin equation for the position $X(t)$ of a Brownian particle submitted to an external (deterministic) force $F(x)$. We assume that we may use a *strong friction limit*,[15] where the Brownian particle takes almost instantaneously its limit velocity v_L. Neglecting diffusion for the time being

$$\dot{v} = -\gamma v + \frac{F(x)}{m} \tag{9.131}$$

and the limit velocity is given by $\dot{v} = 0$, or $v_L(x) = F(x)/(m\gamma)$. Then one adds to $\dot{x} = v_L$ a random force $b(t)$

$$\dot{X}(t) = \frac{F(x)}{m\gamma} + b(t) \tag{9.132}$$

In the absence of $F(x)$, one should recover the diffusive behaviour (9.129), and this is satisfied if

$$\overline{b(t)b(t')} = 2D\,\delta(t - t') \tag{9.133}$$

A more rigorous derivation of (9.132) is proposed in Exercise 9.7.9 in the case of the harmonic oscillator, and in Problem 9.8.4, where one first writes an exact system of coupled equations for X and V (Kramer's equation). The strong friction limit is then obtained as a controlled approximation to this equation. The Ornstein–Uhlenbeck (O–U) process is obtained if one chooses $F(x)$ to be a harmonic force,

[15] The strong friction limit was introduced in Section 6.2.2.

$F(x) = -m\,\omega_0^2\, x$

$$\frac{F(x)}{m\gamma} = -\overline{\gamma}\, x \qquad \overline{\gamma} = \frac{\omega_0^2}{\gamma} \tag{9.134}$$

The O–U process is thus defined by the stochastic differential equation

$$\boxed{\dot{X}(t) = -\overline{\gamma}\, X(t) + b(t)} \tag{9.135}$$

where $b(t)$ is a Gaussian random function of zero mean that obeys (9.133). Since the Fourier tranform of $\overline{b(t)b(t')}$ is a constant, $b(t)$ is also called a *Gaussian white noise*.

We now wish to compute the *conditional* probability $P(x, t|x_0, t_0)$ of finding the particle at point x at time t, knowing that it was at point x_0 at time t_0. We shall take for simplicity $t_0 = 0$, and write $P(x, t|x_0, t_0) = P(x, t|x_0)$. From (9.127) we can solve (9.135) for $X(t)$

$$Y = X(t) = x_0 \mathrm{e}^{-\overline{\gamma} t} + \mathrm{e}^{-\overline{\gamma} t} \int_0^t \mathrm{d}t'\, \mathrm{e}^{\overline{\gamma} t'} b(t') \tag{9.136}$$

Equation (9.136) defines a random variable Y. We are going to show that the probability distribution of Y is Gaussian. Let us divide the $[0, t]$ interval in N small intervals of length $\varepsilon = t/N$, $N \gg 1$, with $t_i = i\varepsilon$ and define the random variable B_ε^i by

$$B_\varepsilon^i = \int_{t_i}^{t_i+\varepsilon} \mathrm{d}t'\, b(t')\, \mathrm{d}t' \tag{9.137}$$

B_ε^i is time independent due to time translation invariance. We note that $\overline{B_\varepsilon^i} = 0$ and that from (9.133)

$$\overline{B_\varepsilon^i\, B_\varepsilon^j} = \int \mathrm{d}t'\, \mathrm{d}t''\, \overline{b(t')b(t'')} = 2\varepsilon D\, \delta_{ij} \tag{9.138}$$

The definition (9.137) allows us to write a Riemann approximation to the integral in (9.136)

$$Y \simeq \mathrm{e}^{-\overline{\gamma} t} \sum_{i=0}^{N-1} \mathrm{e}^{\overline{\gamma} t_i}\, B_\varepsilon^i$$

which shows that Y is the sum of a large number of independent random variables. From the central limit theorem, the probability distribution of Y is Gaussian with a variance given by $\overline{Y^2}$

$$\overline{Y^2} = \mathrm{e}^{-2\overline{\gamma}t} \sum_{i,j=0}^{N-1} \mathrm{e}^{\overline{\gamma}t_i}\, \mathrm{e}^{\overline{\gamma}t_j}\, \overline{B^i_\varepsilon B^j_\varepsilon}$$

$$= \varepsilon \mathrm{e}^{-2\overline{\gamma}t} \sum_{i=0}^{N} \mathrm{e}^{2\overline{\gamma}t_i} \to \mathrm{e}^{-2\overline{\gamma}t} \int_0^t \mathrm{d}t'\, \mathrm{e}^{2\overline{\gamma}t'} = \frac{D}{\gamma}\left(1 - \mathrm{e}^{-2\overline{\gamma}t}\right)$$

This gives the probability distribution of Y, or equivalently of $X(t)$

$$\boxed{P(x, t|x_0) = \left[\frac{\overline{\gamma}}{2\pi D(1 - \mathrm{e}^{-2\overline{\gamma}t})}\right]^{1/2} \exp\left[-\frac{\overline{\gamma}(x - x_0\mathrm{e}^{-\overline{\gamma}t})^2}{2D(1 - \mathrm{e}^{-2\overline{\gamma}t})}\right]} \tag{9.139}$$

The final result (9.139) has a simple interpretation: as is clear from the previous derivation, it is a Gaussian distribution for the centred variable $y = (x - x_0\mathrm{e}^{\overline{\gamma}t})$ with a variance $\sigma^2(t) = (D/\overline{\gamma})(1 - \exp(-2\overline{\gamma}t))$.

It is instructive to look at the short and long time limits of (9.139). In the long time limit $t \gg 1/\overline{\gamma}$, one reaches an equilibrium situation governed by a Boltzmann distribution

$$P(x, t|x_0) \to \left(\frac{\overline{\gamma}}{2\pi D}\right)^{1/2} \exp\left(-\frac{\overline{\gamma}x^2}{2D}\right) = P_{\mathrm{eq}}(x) \propto \exp\left(-\frac{m\,\omega_0^2 x^2}{2k_\mathrm{B}T}\right) \tag{9.140}$$

Equation (9.140) gives another derivation of Einstein's relation (9.130), since it leads to

$$\frac{\overline{\gamma}}{D} = \frac{m\,\omega_0^2}{k_\mathrm{B}T} \quad \text{with} \quad \overline{\gamma} = \frac{\omega_0^2}{\gamma}$$

The limit $t \ll 1/\overline{\gamma}$

$$P(x, t|x_0) \to \frac{1}{(4\pi Dt)^{1/2}} \exp\left[\frac{(x - x_0)^2}{4Dt}\right] \tag{9.141}$$

shows that *the short time limit is dominated by diffusion*

$$\langle (x - x_0)^2 \rangle \sim 2D\,t$$

As one can write $\langle |X(t + \varepsilon) - x(t)| \rangle \propto \sqrt{t}$, one sees that the trajectory $x(t)$ is a continuous, but non-differentiable function of t (see also Exercise 9.7.12).

Of course (9.139) may also be used to obtain $P(v, t|v_0)$ from (9.132): one has only to make in (9.139) the substitutions $x \to v$, $D \to A$ and $\overline{\gamma} \to \gamma$ to get

$$P(v, t|v_0) = \left[\frac{\gamma}{2\pi A(1 - \mathrm{e}^{-2\gamma t})}\right]^{1/2} \exp\left[-\frac{\gamma(v - v_0\mathrm{e}^{-\gamma t})^2}{2A(1 - \mathrm{e}^{-2\gamma t})}\right] \tag{9.142}$$

The long time limit of (9.143) gives the Maxwell distribution $\exp[-mv^2/(2k_BT)]$, and comparison of the Maxwell distribution with the long time limit of (9.143) leads once more to (9.126).

9.5 Fokker–Planck equation

The probability distribution (9.141) derived from the Langevin equation obeys a partial differential equation, the Fokker–Planck equation. This equation displays a remarkable analogy with a Schrödinger equation in imaginary time, an analogy that we shall use to study convergence to equilibrium.

9.5.1 Derivation of Fokker–Planck from Langevin equation

We wish to derive a partial differential equation (PDE) for the conditional probability $P(x, t|x_0)$, when $x(t)$ obeys a Langevin equation of the form

$$\dot{X}(t) = a(x) + b(t) \tag{9.143}$$

where we have set $a(x) = F(x)/(m\gamma)$. This PDE is the Fokker–Planck (F–P) equation. Note that (9.143) is a generalization of the O–U equation (9.135), where $a(x) = -\overline{\gamma}\, x$; the random function $b(t)$ has the same properties as in the preceding section, and in particular it obeys (9.133). Equation (9.143) defines a Markovian process, because it is first order in time and because of the delta function in (9.133): if $b(t)$ had a finite (microscopic) autocorrelation time τ^*, or, in other words, if $b(t)$ was not strictly a white noise, then (9.133) would not define a Markovian process. Having a Markovian process allows us to write down a Chapman–Kolmogorov equation for P

$$P(x, t+\varepsilon|x_0) = \int \mathrm{d}y\, P(x, t+\varepsilon|y, t) P(y, t|x_0) \tag{9.144}$$

and integrating (9.143) over an infinitesimal time ε gives the random trajectory $X_y^{[b]}(t+\varepsilon; t)$ for a particular realization of the random force $b(t)$ and of the initial position $X(t) = y$

$$X_y^{[b]}(t+\varepsilon; t) = y + \varepsilon a(y) + \int_t^{t+\varepsilon} \mathrm{d}t'\, b(t') = y + \varepsilon a(y) + B_\varepsilon$$

where we have used the definition (9.137) of B_ε. From this equation follows, to order ε

$$P(x, t+\varepsilon|y, t) = \overline{\delta(x - y - \varepsilon a(y) - B_\varepsilon)}$$
$$\simeq (1 - \varepsilon a'(x))\,\overline{\delta(x - y - \varepsilon a(x) - B_\varepsilon)}$$

Indeed, at order ε, we may write $a(x) = a(y) + \mathcal{O}(\varepsilon)$; we have also used the standard identity

$$\delta(f(y)) = \frac{1}{|f'(y)|}\,\delta(y - y_0) \qquad f(y_0) = 0$$

We expand formally the δ-function in powers of ε,[16] noting that one must expand to order B_ε^2, because B_ε is in fact of order $\sqrt{\varepsilon}$

$$\delta(x - y - \varepsilon a(x) - B_\varepsilon) = \delta(x - y) + [\varepsilon a(x) + B_\varepsilon]\delta'(x - y)$$
$$+\frac{1}{2}[\varepsilon a(x) + B_\varepsilon]^2\delta''(x - y) + \cdots$$

and plug the result in the Chapman–Kolmogorov equation, keeping only terms of order $\sqrt{\varepsilon}$ and ε. This leads to the integral

$$\int \mathrm{d}y\, P(y, t|x_0)$$
$$\times\overline{\left[(1 - \varepsilon a'(x))\delta(y - x) + (\varepsilon a(x) + B_\varepsilon)\delta'(y - x) + (B_\varepsilon^2/2)\delta''(y - x)\right]}$$

which is evaluated thanks to $\overline{B_\varepsilon} = 0$ and $\overline{B_\varepsilon^2} = 2D\varepsilon$. Performing the now trivial integrations gives to order ε

$$P(x, t+\varepsilon|x_0) = P(x, t|x_0) + \varepsilon\,\frac{\partial P}{\partial t}$$
$$= P(x, t|x_0) + \varepsilon\left[-a'(x)P(x, t|x_0) - a(x)\frac{\partial}{\partial x}\,P(x, t|x_0)\right.$$
$$\left.+D\frac{\partial^2}{\partial x^2}\,P(x, t|x_0)\right] + \mathcal{O}(\varepsilon^2)$$

[16] This is simply a shorthand notation; for example

$$\int \mathrm{d}x f(x)\delta(x - (x_0 + \varepsilon)) = f(x_0 + \varepsilon) = f(x_0) + \varepsilon f'(x_0)$$
$$\int \mathrm{d}x f(x)[\delta(x - x_0) - \varepsilon\delta'(x - x_0)] = f(x_0) + \varepsilon f'(x_0)$$

and we obtain the Fokker–Planck equation

$$\boxed{\frac{\partial}{\partial t} P(x,t|x_0) = -\frac{\partial}{\partial x}\Big[a(x)P(x,t|x_0)\Big] + D\frac{\partial^2}{\partial x^2} P(x,t|x_0)} \tag{9.145}$$

The clearest physical interpretation of the F–P equation follows from writing it in the form of a continuity equation. Defining the current $j(x,t)$

$$j(x,t) = a(x)P(x,t) - D\frac{\partial P(x,t)}{\partial x} = \frac{F(x)}{m\gamma} P(x,t) - D\frac{\partial P(x,t)}{\partial x} \tag{9.146}$$

where we have used the shorthand notation $P(x,t) = P(x,t|x_0)$, (9.145) becomes a continuity equation

$$\frac{\partial P(x,t)}{\partial t} + \frac{\partial j(x,t)}{\partial x} = 0 \tag{9.147}$$

We find the important physical result that the current is the sum of the usual deterministic part $a(x)P = \dot{x}P$ in the absence of diffusion, and of a diffusive part $-D\partial P/\partial x$. Another useful expression of the current is obtained by introducing the potential $V(x)$: $F(x) = -\partial V/\partial x$ and using Einstein's relation (9.130)

$$j(x,t) = -D\Big(\beta P\frac{\partial V}{\partial x} + \frac{\partial P}{\partial x}\Big) \tag{9.148}$$

9.5.2 *Equilibrium and convergence to equilibrium*

There is a remarkable correspondence between the F–P equation and the Schrödinger equation in imaginary time.[17] Let us write the Schrödinger equation for a particle moving in one dimension in a potential $U(x)$ ($\hbar = 1$)

$$\mathrm{i}\frac{\partial\psi(x,t')}{\partial t'} = -\frac{1}{2m}\frac{\partial^2\psi(x,t')}{\partial x^2} + U(x)\psi(x,t') = H\psi(x,t')$$

and make the change of variables $t' = -\mathrm{i}t$

$$\frac{\partial\psi(x,t)}{\partial t} = \frac{1}{2m}\frac{\partial^2\psi(x,t)}{\partial x^2} - U(x)\psi(x,t) = -H\psi(x,t) \tag{9.149}$$

Taking $U = 0$ in (9.149), one recognizes the diffusion equation (6.21) if one identifies $D = 1/(2m)$. The F–P equation (9.145) is not yet in the form of (9.149), but it will not be difficult to find the transformation that casts (9.145) in the form

[17] In other words, the F–P equation is related to the Schrödinger equation by the Wick rotation already encountered in Chapter 7.

of (9.149). Let us first find the equilibrium distribution; we set $\beta = 1$ in order to simplify the notations and use (9.147) and (9.148)

$$\frac{\partial P}{\partial t} = D\frac{\partial}{\partial x}\left(P\frac{\partial V}{\partial x} + \frac{\partial P}{\partial x}\right) = -\frac{\partial j}{\partial x}$$

A sufficient condition for equilibrium is that $j = 0$,[18] leading to the Boltzmannn distribution (with $\beta = 1$)

$$P_{\rm eq}(x) \propto \exp(-V(x))$$

Let us define $\rho(x, t)$ by

$$P(x, t) = \exp\left(-\frac{1}{2}V(x)\right)\rho(x, t) \tag{9.150}$$

so that

$$\frac{\partial P}{\partial x} + P\frac{\partial V}{\partial x} = \exp\left(-\frac{1}{2}V(x)\right)\left[\frac{\partial \rho}{\partial x} + \frac{1}{2}\rho\frac{\partial V}{\partial x}\right]$$

A straightforward calculation shows that the unwanted terms of (9.149) cancel out, leaving us with the desired result

$$\begin{aligned}\frac{\partial \rho(x, t)}{\partial t} &= D\frac{\partial^2 \rho(x, t)}{\partial x^2} - U(x)\rho(x, t) = -H\rho(x, t)\\ H &= -D\frac{\partial^2}{\partial x^2} + U(x) \qquad U(x) = \frac{D}{4}\left(\frac{\partial V}{\partial x}\right)^2 - \frac{D}{2}\frac{\partial^2 V}{\partial x^2}\end{aligned} \tag{9.151}$$

Let us define $\psi_0(x) = \mathcal{N}\exp(-\frac{1}{2}V(x))$, where $\mathcal{N}$ is a normalization constant such that

$$\int {\rm d}x\, |\psi_0(x)|^2 = 1$$

The result $P_{\rm eq}(x) \propto \exp(-V(x))$ is equivalent to $H\psi_0 = 0$; indeed

$$\left[\frac{\partial}{\partial x} + \frac{1}{2}\frac{\partial V}{\partial x}\right]\psi_0(x) = 0$$

which corresponds to $j = 0$. Since $\psi_0(x)$ has no nodes (zeroes), one knows from a standard theorem in quantum mechanics that $\psi_0(x)$ is the ground state wave function, and it has energy $E_0 = 0$. All excited states have energies $E_n > 0$. In order to obtain the time evolution, we expand the initial condition at time $t = 0$ on a complete set of eigenfunctions $\psi_n(x) = \langle x|n\rangle$ of H

$$H\psi_n(x) = E_n\psi_n(x) \tag{9.152}$$

[18] In one dimension $j = $ const implies $j = 0$, but one may have stationary non-equilibrium currents in higher dimensions.

which can be chosen to be real. Then

$$\rho(x,0|x_0) = \sum_n c_n \psi_n(x)$$

$$c_n = \int \mathrm{d}x\, \psi_n(x)\rho(x,0|x_0) = \psi_n(x_0)\mathrm{e}^{\frac{1}{2}V(x_0)} = \langle n|x_0\rangle \mathrm{e}^{\frac{1}{2}V(x_0)}$$

since $\rho(x,0|x_0) = \exp(\frac{1}{2}V(x_0))\delta(x-x_0)$ and we have intoduced Dirac's bra and ket notation. We get $\rho(x,t|x_0)$ from the time evolution of the ψ_ns: $\psi_n(x,t) = \exp(-E_n t)\psi_n(x)$

$$\begin{aligned}\rho(x,t|x_0) &= \sum_n c_n \mathrm{e}^{-E_n t}\psi_n(x)\\ &= \sum_n \mathrm{e}^{\frac{1}{2}V(x_0)}\langle x|n\rangle \mathrm{e}^{-E_n t}\langle n|x_0\rangle\\ &= \mathrm{e}^{\frac{1}{2}V(x_0)}\langle x|\mathrm{e}^{-tH}|x_0\rangle\end{aligned}$$

Summarizing

$$\boxed{P(x,t|x_0) = \mathrm{e}^{-\frac{1}{2}(V(x)-V(x_0))}\langle x|\mathrm{e}^{-tH}|x_0\rangle} \tag{9.153}$$

For large times

$$\mathrm{e}^{-tH} \simeq |0\rangle\langle 0| + \mathrm{e}^{-E_1 t}|1\rangle\langle 1|$$

so that the approach to equilibrium is controlled by the energy E_1 of the first excited state $\psi_1(x)$: $H\psi_1(x) = E_1\psi_1(x)$.[19] One checks from (9.153) that

$$\lim_{t\to\infty} P(x,t|x_0) = \mathcal{N}^2\mathrm{e}^{-V(x)}$$

9.5.3 Space-dependent diffusion coefficient

It is possible to generalize the F–P equation (9.145) to a space-dependent diffusion coefficient $D(x)$

$$\boxed{\frac{\partial}{\partial t}P(x,t|x_0) = -\frac{\partial}{\partial x}\Big[a(x)P(x,t|x_0)\Big] + \frac{\partial^2}{\partial x^2}\Big[D(x)P(x,t|x_0)\Big]} \tag{9.154}$$

Note that we have written *a priori* $D(x)$ *inside* the *x*-derivative. Further comments on this apparently arbitrary choice will be given later on, but for the time being let us use (9.154) as it stands to compute the first and second moments of the trajectory. Let us expand $P(x,t+\varepsilon|x_0,t)$ in powers of ε and use (9.154) to express

[19] The reader will remark the analogy with Equation (7.188) of Problem 7.9.8.

$\partial P/\partial t$

$$P(x, t+\varepsilon|x_0, t) = \delta(x - x_0) + \varepsilon\frac{\partial P}{\partial t} + O(\varepsilon^2)$$
$$= \delta(x - x_0) - \varepsilon\frac{\partial}{\partial x}[a(x)P] + \varepsilon\frac{\partial^2}{\partial x^2}[D(x)P] + \mathcal{O}(\varepsilon^2)$$

The first moment is

$$\lim_{\varepsilon\to 0}\frac{1}{\varepsilon}\overline{X(t+\varepsilon) - x_0} = \int \mathrm{d}x\,(x - x_0)\Big[-\frac{\partial}{\partial x}[(a(x)P] + \frac{\partial^2}{\partial x^2}[D(x)P)]\Big] \tag{9.155}$$

where we have used $(x - x_0)\,\delta(x - x_0) = 0$. We integrate (9.155) by parts and use $\lim_{|x|\to\infty} P(x, t|x_0) = 0$, so that only the first term in the square bracket of (9.155) contributes and

$$\boxed{\lim_{\varepsilon\to 0}\frac{1}{\varepsilon}\overline{X(t+\varepsilon) - x_0} = a(x_0)} \tag{9.156}$$

The second moment $(1/\varepsilon)\overline{(X(t+\varepsilon) - x_0)^2}$ is also given by an integration by parts, but it is now the second term in the square bracket of (9.155) that contributes

$$\boxed{\lim_{\varepsilon\to 0}\frac{1}{\varepsilon}\overline{(X(t+\varepsilon) - x_0)^2} = 2D(x_0)} \tag{9.157}$$

Thus, given the first two moments of the trajectory (9.156) and (9.157), one can write the corresponding F–P equation (9.154). These results are easily generalized to multivariate F–P equations (Exercise 9.7.11).

From these results, one would be tempted to conclude that $X(t)$ obeys a Langevin equation

$$\dot{X}(t) = a(x) + \sqrt{D(x)}\,b(t) \qquad \overline{b(t)b(t')} = 2\delta(t - t') \tag{9.158}$$

However, because of the delta function in (9.158), the function $b(t)$ is singular, and the product $\sqrt{D(x)}\,b(t)$ is not defined: this leads to the famous Itô vs. Stratonovitch dilemma (see Problem 9.8.4 for more details). Let us define $C(x) = \sqrt{D(x)}$ and try to integrate Equation (9.158) over a small time interval ε

$$X(t+\varepsilon) - x(t) = \varepsilon a(x(t)) + \int_t^{t+\varepsilon} \mathrm{d}t'\, C(x(t'))b(t')$$

There are many possible prescriptions for handling the product $C(x(t'))b(t')$, and each of them leads to a different F–P equation. One can show that giving a finite width τ^* to the time autocorrelation function of $b(t)$ leads to the Stratonovitch prescription

$$\int_t^{t+\varepsilon} \mathrm{d}t'\, C(x(t'))b(t') \to C\Big[\frac{x(t)+x(t+\varepsilon)}{2}\Big] \int_t^{t+\varepsilon} \mathrm{d}t'\, b(t')$$

and a corresponding F–P equation

$$\frac{\partial}{\partial t} P(x,t|x_0) = -\frac{\partial}{\partial x}\Big[a(x)P(x,t|x_0)\Big] + \frac{\partial}{\partial x}\Big[C(x)\frac{\partial}{\partial x}C(x)P(x,t|x_0)\Big] \tag{9.159}$$

while the Itô prescription

$$\int_t^{t+\varepsilon} \mathrm{d}t'\, C(x(t'))b(t') \to C(x(t)) \int_t^{t+\varepsilon} \mathrm{d}t'\, b(t')$$

leads to the F–P equation (9.154). Other prescriptions and the corresponding F–P equations are examined in Problem 9.8.4. One may always write the F–P equation in the form (9.154); however, the various prescriptions correspond to modifying the drift velocity $a(x)$.

9.6 Numerical integration

In this section it will be convenient to work in a system of units with $D = m\gamma = 1$ (and $kT = 1$ from Equation (9.130)). Then as we saw in Section 9.5, the Langevin equation

$$\dot{X}(t) = a(x) + b(t) \tag{9.160}$$

where $a(x) = -\partial V(x)/\partial x$ and $\overline{b(t)b(t')} = 2\delta(t-t')$, admits a stationary solution of the form

$$P(x,t|x_0) \propto \mathrm{e}^{-V(x)} \tag{9.161}$$

In other words, for $t \to \infty$, the configurations are given by the time-independent probability distribution (9.161). It was also shown in the previous section that the approach to this stationary solution is controlled by the spectrum of the Fokker–Planck Hamiltonian, which is not known in general.

It is therefore evident that numerical solutions of the Langevin equation are important in the non-equilibrium case where transient effects are dominant. However, it should be evident from the discussions of Chapter 7 and Equation (9.161) that the Langevin equation can also be used as a tool to perform numerical simulations at equilibrium. In Chapter 7 we discussed how to construct dynamics (for example

Metropolis, Glauber or Wolff) which, in the long time limit, generate configurations with the correct Boltzmann distribution, $P_{\rm B} = \exp(-\beta E)$. Equations (9.160) and (9.161) give another way with $V = \beta E$. Therefore, the numerical methods that we now discuss may be applied both to equilibrium and non-equilibrium situations. We stress, however, that in the equilibrium case, the Langevin dynamics may or may not describe the actual approach to the stationary state: one may use the Langevin equation even if it does not describe the true dynamics of the system if the interest is only in the equilibrium properties.

In the case of a deterministic differential equation,

$$\frac{{\rm d}x(t)}{{\rm d}t} = -\frac{\partial V(x)}{\partial x} \tag{9.162}$$

one may use the simplest (Euler) discretization of the time derivative,

$$x(t+\varepsilon) \approx x(t) - \varepsilon\frac{\partial V(x)}{\partial x} + \mathcal{O}(\varepsilon^2) \tag{9.163}$$

The error committed in this case is of the order of ε^2. We shall now apply the same approximation to the Langevin equation (9.160)[20]

$$x(t+\varepsilon) \approx x(t) - \varepsilon\frac{\partial V(x)}{\partial x} + \varepsilon b(t) \tag{9.164}$$

However, care must be taken with the stochastic noise $b(t)$ since t is now discrete and we can no longer have $\overline{b(t)b(t')} = 2\delta(t-t')$. Recalling that the dimension of $\delta(t-t')$ is t^{-1}, we see that the discrete time form of the delta function becomes

$$\overline{b(t)b(t')} = 2\frac{\delta_{tt'}}{\varepsilon} \tag{9.165}$$

We may then rescale the noise, $b(t) \to b(t)\sqrt{2/\varepsilon}$, leading to the discrete time Langevin equation

$$x(t+\varepsilon) \approx x(t) - \varepsilon\frac{\partial V(x)}{\partial x} + \sqrt{2\varepsilon}b(t) \tag{9.166}$$

where we now have

$$\overline{b(t)b(t')} = \delta_{tt'} \tag{9.167}$$

Since the noise term in Equation (9.166) is of the order of $\sqrt{\varepsilon}$, the numerical error due to this discretization is $\mathcal{O}(\varepsilon)$ and not $\mathcal{O}(\varepsilon^2)$ as for the deterministic equation. This may also be seen clearly from the derivation of the Fokker–Planck equation where the same discretization was used for the Langevin equation and where we neglected terms of order ε in the final result.

[20] For the numerical integration of the Langevin equation we make no distinction between $X(t)$ and $x(t)$ and write all variables with lower case.

In the general case, where we may have, say, N variables (for example N B-particles) the Langevin equations are given by

$$x_i(t+\varepsilon) \approx x_i(t) - \varepsilon \frac{\partial V(\{x\})}{\partial x_i} + \sqrt{2\varepsilon} b_i(t) \tag{9.168}$$

with

$$\overline{b_i(t) b_j(t')} = \delta_{tt'} \delta_{ij} \tag{9.169}$$

The numerical integration is now simple to implement: choose an initial configuration, $\{x_i(t_0)\}$, at $t = t_0$, calculate the deterministic force $-\partial V(\{x\})/\partial x_i$, generate N random numbers, $b_i(t_0)$, and use Equation (9.168) to obtain $x_i(t_0 + \varepsilon)$ and repeat for as many time steps as is desired.

It is clear that random numbers $b_i(t)$ satisfying (9.169) may be easily generated using a Gaussian random number generator. However, while this is sufficient, it is not necessary: all one needs for (9.168) is a random number such that $\overline{b_i(t)} = 0$ and whose second moment is given by (9.169) with no conditions given for higher moments. For example, random numbers drawn from a uniform distribution between $-\sqrt{3}$ and $+\sqrt{3}$ satisfy these conditions and are much faster to generate than Gaussian ones.

To obtain higher precision with this algorithm, it is necessary to take smaller time steps. But, in order to keep the same physical time, the number of steps must be increased correspondingly, which might be costly in computer time.

It is therefore desirable to have a higher order algorithm that yields smaller errors for the same time step. A very simple such algorithm is the second order Runge–Kutta discretization. As for deterministic equations, first do a tentative update of the variables using a simple Euler step,

$$x_i'(t+\varepsilon) = x_i(t) - \varepsilon \frac{\partial V(\{x\})}{\partial x_i} + \sqrt{2\varepsilon} b_i(t) \tag{9.170}$$

and then use this $x_i'(t+\varepsilon)$ with $x_i(t)$ to get the final evolution of one time step,

$$x_i(t+\varepsilon) = x_i(t) - \frac{\varepsilon}{2} \left(\left. \frac{\partial V(\{x\})}{\partial x_i} \right|_{\{x_i(t)\}} + \left. \frac{\partial V(\{x\})}{\partial x_i} \right|_{\{x_i'(t+\varepsilon)\}} \right) + \sqrt{2\varepsilon} b_i(t) \tag{9.171}$$

For example, taking the simple case of $V(x) = x^2/2$, the tentative update (9.170) becomes

$$x'(t+\varepsilon) = x(t) - \varepsilon x(t) + \sqrt{2\varepsilon} b(t) \tag{9.172}$$

and the final update (9.171) is obtained from

$$\begin{aligned} x(t+\varepsilon) &= x(t) - \frac{\varepsilon}{2}\left(\frac{\partial V(\{x\})}{\partial x}\bigg|_{\{x(t)\}} + \frac{\partial V(\{x\})}{\partial x}\bigg|_{\{x'(t+\varepsilon)\}} \right) + \sqrt{2\varepsilon} b(t) \\ &= x(t) - \frac{\varepsilon}{2}\left(x(t) + x'(t+\varepsilon)\right) + \sqrt{2\varepsilon} b(t) \\ &= x(t) - \varepsilon\left(1 - \frac{\varepsilon}{2}\right)x(t) + \sqrt{2\varepsilon}\left(1 - \frac{\varepsilon}{2}\right)b(t) \end{aligned} \tag{9.173}$$

Although the Runge–Kutta algorithm is rather simple in general for these stochastic equations, the above particularly simple form is a special case for the quadratic case.

Three very important remarks are in order. First, it is crucial to emphasize that the *same* random number $b_i(t)$ is used in both (9.170) and (9.171): *we do not generate a number for step (9.170) and another for (9.171).* This condition is required to ensure that the error is $\mathcal{O}(\varepsilon^2)$ and has the additional advantage that it reduces the amount of work. The second remark is that it is no longer sufficient to use a random number uniformly distributed between $-\sqrt{3}$ and $+\sqrt{3}$: in the proof that the error in this algorithm is $\mathcal{O}(\varepsilon^2)$, the fourth moment of the random number is needed. The easiest way to satisfy all the conditions in this case is to use a Gaussian random number generator. The third remark is to note that this algorithm is in fact *not* equivalent to two successive Euler steps, which would still lead to an order $\mathcal{O}(\varepsilon)$ algorithm. The error of the second order Runge–Kutta algorithm just presented is $\mathcal{O}(\varepsilon^2)$. This may be shown with a tedious calculation that follows steps leading to Equation (9.145) but using (9.170) and (9.171) for the discretized Langevin equation [12, 52]. The behaviour of the errors will be studied numerically in the exercises. Higher order Runge–Kutta discretizations are also available but become very complicated.

One final comment concerns the stability of the integration. Consider the one-variable Langevin equation in the Euler discretization (9.164) with $V(x) = ax^2/2$ and with a a constant.[21] Iterating this equation a few times one observes the appearance of a term of the form $(1 - \varepsilon a)^n$ where n is the number of iterations. Clearly, for the iterations to converge, the condition $\varepsilon a < 1$ must be satisfied. In the more general case where $V(\{x_i\}) = x_i M_{ij} x_j / 2$ (i and j are summed), the stability condition becomes $\varepsilon\lambda_{\max} < 1$ where $\lambda_{\max}$ is the largest eigenvalue of the matrix M. We therefore arrive at the very important result that the time step is set by the largest eigenvalue, i.e. by the fastest mode. On the other hand, we saw at the end of Section 9.5.2 that the relaxation time is controlled by the lowest excited state, in other words the *smallest* eigenvalue, $\lambda_{\min}$. So the relaxation time

[21] While we present the argument based on the Euler discretization, the conclusions are in fact general and apply for the Runge–Kutta case too.

is $\tau \sim \lambda_{\rm min}^{-1}$ while the time step is $\varepsilon \sim \lambda_{\rm max}^{-1}$. Therefore the number of iterations required to decorrelate the configurations is

$$n_{\rm corr} \sim \lambda_{\rm max}/\lambda_{\rm min} \tag{9.174}$$

For this reason, $n_{\rm corr}$ can be very large indeed. When $\lambda_{\rm max}/\lambda_{\rm min} \gg 1$, the matrix M is said to be ill-conditioned. This is another example of critical slowing down which was discussed in Chapter 7.

If one is interested only in the stationary solution of the Langevin equation and not in how that solution is approached, one may modify the dynamics to *precondition M* and greatly accelerate the convergence. This topic is beyond our scope; see Reference [12].

9.7 Exercises

9.7.1 Linear response: forced harmonic oscillator

1. Let us consider a forced one-dimensional harmonic oscillator with mass m and damping constant γ

$$\ddot{x} + \gamma\dot{x} + \omega_0^2 x = \frac{f(t)}{m}$$

and define its dynamical susceptibility $\chi(t)$ by

$$x(t) = \int \mathrm{d}t'\, \chi(t-t') f(t')$$

Show that the Fourier transform of $\chi(t)$ is

$$\chi(\omega) = \frac{1}{m[-\omega^2 - \mathrm{i}\omega\gamma + \omega_0^2]}$$

Write the explicit expression of $\chi''(\omega)$. Find the location of the poles $\omega_\pm$ of $\chi(\omega)$ in the complex ω-plane and show that one must distinguish between the cases $\gamma < 2\omega_0$ and $\gamma > 2\omega_0$; $\gamma = 2\omega_0$ is the *critical damping*. When $\gamma \geq 2\omega_0$, one says that *the oscillator is overdamped.*

2. Find the static susceptibility χ

$$\chi = \lim_{\omega\to 0} \chi(\omega) = \frac{1}{m\omega_0^2}$$

and show that in the overdamped case one may write

$$\chi(\omega) = \frac{1}{m}\frac{1}{\omega_0^2 - \mathrm{i}\omega\gamma} = \frac{\chi}{1-\mathrm{i}\omega\tau} \qquad \tau = \frac{\gamma}{\omega_0^2} \tag{9.175}$$

Write $\chi''(\omega)$ explicitly. Note that working in the strong friction limit amounts to neglecting inertia.

3. Starting from the work per unit of time done by the external force on the oscillator

$$\frac{\mathrm{d}W}{\mathrm{d}t} = f(t)\dot{x}(t)$$

and taking a periodic $f(t)$

$$f(t) = \mathrm{Re}\Big(f_\omega\, \mathrm{e}^{-\mathrm{i}\omega t}\Big) = \frac{1}{2}\Big(f_\omega\, \mathrm{e}^{-\mathrm{i}\omega t} + f_\omega^*\, \mathrm{e}^{\mathrm{i}\omega t}\Big)$$

show that the time average of $\mathrm{d}W/\mathrm{d}t$ over a time interval $T \gg \omega^{-1}$ is

$$\left\langle\frac{\mathrm{d}W}{\mathrm{d}t}\right\rangle_T = \frac{1}{2}\,\omega|f_\omega|^2 \int_0^\infty \mathrm{d}t\, \chi(t)\sin\omega t$$

Observing that $\chi(t) = 2\mathrm{i}\,\theta(t)\,\chi''(t)$ and that $\chi''(t) = -\chi''(-t)$, deduce from this equation

$$\left\langle\frac{\mathrm{d}W}{\mathrm{d}t}\right\rangle_T = \frac{1}{2}\,\omega|f_\omega|^2\,\chi''(\omega) \tag{9.176}$$

9.7.2 Force on a Brownian particle

Let us consider a Brownian particle with mass M in a fluid. We call m the mass of the fluid molecules, $m \ll M$ and $\vec{v}$ the velocity of the Brownian particle with respect to the fluid. In the Galilean frame where the Brownian particle is at rest, the Hamiltonian reads

$$H_1 = \sum_{i=1}^{N} \frac{1}{2}m(\vec{v}_i - \vec{v})^2 + \text{potential energy}$$

where $\vec{v}_i$ is the velocity of molecule i. Define the dynamical variable $\vec{A}$ by

$$\vec{A} = m\sum_{i=1}^{N} \vec{v}_i$$

In the linear approximation, the perturbation is $\mathcal{V} = -\vec{A}\cdot\vec{v}$. Use linear response theory to compute $\mathrm{d}\delta\vec{A}/\mathrm{d}t$, and show that the viscosity coefficient α is given by

$$\alpha = \frac{1}{3kT}\int_0^\infty \mathrm{d}t\,\langle\vec{F}(t)\cdot\vec{F}(0)\rangle$$

where $\vec{F}$ is the force on the particle assumed to be *at rest* in the fluid.

9.7.3 Green–Kubo formula

Starting from

$$D\chi = \lim_{\omega\to 0}\lim_{k\to 0}\frac{\omega}{k^2}\chi''(k,\omega)$$

show that the diffusion coefficient is given by the Green–Kubo formula

$$D\chi = \frac{\beta}{3}\int_0^\infty \mathrm{d}t \int \mathrm{d}^3r\, \langle \vec{j}(t,\vec{r})\cdot\vec{j}(0,\vec{0})\rangle$$

Hints

(*i*) Using rotational invariance, show that $(l, m = x, y, z)$

$$\int \mathrm{d}^3r\, \mathrm{e}^{-i\vec{k}\cdot\vec{r}}\langle\vec{\nabla}\cdot\vec{j}(t,\vec{r})\,\vec{\nabla}\cdot\vec{j}(0,\vec{0})\rangle = \sum_{l,m} k_l k_m\Big(H(t,k^2)\delta_{lm} + K(t,k^2)k_l k_m\Big)$$

(*ii*) Study the limit

$$\lim_{\omega\to 0}\lim_{k\to 0}\frac{1}{3}\int_{-\infty}^{\infty}\mathrm{d}t\,\mathrm{e}^{\mathrm{i}\omega t}\int \mathrm{d}^3r\,\mathrm{e}^{-i\vec{k}\cdot\vec{r}}\langle\vec{j}(t,\vec{r})\cdot\vec{j}(0,\vec{0})\rangle$$

9.7.4 Mori's scalar product

1. One defines for two operators A and B and a density operator D the scalar product

$$\langle\mathsf{B};\mathsf{A}\rangle_D = \int_0^1 \mathrm{d}x\,\mathrm{Tr}\Big[\mathsf{B}\,D^x\,\mathsf{A}^\dagger\,D^{1-x}\Big]$$

Show that $\langle\mathsf{B};\mathsf{A}\rangle_D$ defines a Hermitian scalar product: it is linear in B, antilinear in A and it obeys

$$\langle\mathsf{A};\mathsf{B}\rangle_D = \langle\mathsf{B};\mathsf{A}\rangle_D^*$$

Hint: $[\mathrm{Tr}(ABC)]^* = \mathrm{Tr}(C^\dagger B^\dagger A^\dagger)$. Furthermore, show that $\langle\mathsf{A};\mathsf{A}\rangle_D \geq 0$ and that $\langle\mathsf{A};\mathsf{A}\rangle_D = 0$ implies that $\mathsf{A} = 0$. These last two results show that Mori's scalar product is positive definite.

2. Show that $\mathcal{L}$ is Hermitian with respect to Mori's scalar product

$$\langle \mathsf{A}; \mathcal{L}\mathsf{B}\rangle = \langle \mathcal{L}\mathsf{A}; \mathsf{B}\rangle$$

Hint: First derive

$$\langle \mathsf{A}; \mathcal{L}\mathsf{B}\rangle = \frac{1}{\beta}\langle [\mathsf{A}, \mathsf{B}]\rangle \tag{9.177}$$

3. Show that

$$\langle \mathsf{A}_i; \dot{\mathsf{A}}_j\rangle = -\varepsilon_i\varepsilon_j\langle \mathsf{A}_i; \dot{\mathsf{A}}_j\rangle \tag{9.178}$$

where ε_i (ε_j) is the parity of A_i (A_j) under time reversal.

9.7.5 Symmetry properties of χ''_{ij}

1. Show the following properties for Hermitian A_is:

(*i*) from time translation invariance

$$\chi''_{ij}(t) = -\chi''_{ji}(-t) \quad \text{or} \quad \chi''_{ij}(\omega) = -\chi''_{ji}(-\omega)$$

(*ii*) from the Hermiticity of the A_is

$$\chi''^{*}_{ij}(t) = -\chi''_{ij}(t) \quad \text{or} \quad \chi''^{*}_{ij}(\omega) = -\chi''_{ij}(-\omega)$$

(*iii*) from time reversal invariance

$$\chi''_{ij}(t) = -\varepsilon_i\varepsilon_j\chi''_{ij}(-t) \quad \text{or} \quad \chi''_{ij}(\omega) = -\varepsilon_i\varepsilon_j\chi''_{ij}(-\omega)$$

where ε_i (ε_j) is the parity of A_i (A_j) under time reversal. Combine (*i*) and (*ii*) to show that $(\chi''_{ij}(\omega))^* = \chi''_{ji}(\omega)$. Hint for (*iii*): If H is invariant under time reversal, $\Theta H\Theta^{-1} = H$, show that for two operators A_i and A_j

$$\langle \mathsf{A}_i(t)\,\mathsf{A}_j(0)\rangle = \varepsilon_i\varepsilon_j\langle \mathsf{A}^{\dagger}_j(0)\,\mathsf{A}^{\dagger}_i(-t)\rangle$$

by noticing that $|\tilde{n}\rangle = \Theta|n\rangle$ is an eigenvector of H

$$H|\tilde{n}\rangle = E_n|\tilde{n}\rangle$$

if $H|n\rangle = E_n|n\rangle$.

2. In the case of two operators A and B, not necessarily Hermitian, one defines

$$\chi''_{AB}(t) = \frac{1}{2}\langle [\mathsf{A}(t), B^{\dagger}(0)]\rangle \tag{9.179}$$

Show the following properties:

(*i*) from time translation invariance

$$\chi''^{*}_{AB}(t) = -\chi''_{B^\dagger A^\dagger}(-t)$$

(*ii*) from Hermitian conjugation

$$\chi''_{AB}(t) = -\chi''_{A^\dagger B^\dagger}(t)$$

(*iii*) from time reversal invariance

$$\chi''_{AB}(t) = -\varepsilon_A\varepsilon_B\chi''_{A^\dagger B^\dagger}(-t) = \varepsilon_A\varepsilon_B\chi''_{B^\dagger A^\dagger}(t)$$

9.7.6 Dissipation

Give the detailed proof of

$$\left\langle\frac{\mathrm{d}W}{\mathrm{d}t}\right\rangle_T = \frac{1}{2}\sum_{i,j} f_i^{\omega *}\omega\chi''_{ij}(\omega) f_j^{\omega}$$

Hint: Use (9.13) in the form

$$\overline{\delta \mathsf{A}_i}(t) = \int \mathrm{d}t'\ \chi_{ij}(t') f_j(t-t')$$

and integrate over t in the range $[0, T]$, $T \gg \omega^{-1}$. It is useful to remark that $\chi''_{ij}(t) = -\chi''_{ji}(-t)$. Furthermore, show that $\omega\chi''_{ij}(\omega)$ is a positive matrix. Hint: Study

$$\sum_{i,j}\int_0^T \mathrm{d}t\,\mathrm{d}t' a_i \mathrm{e}^{\mathrm{i}\omega t} a_j^* \mathrm{e}^{-\mathrm{i}\omega t'}\langle \mathsf{A}_i(t)\mathsf{A}_j(t')\rangle$$

9.7.7 Proof of the f-sum rule in quantum mechanics

The particle density operator n is defined as

$$\mathsf{n}(\vec{r}) = \sum_{\alpha=1}^{N}\delta(\vec{r} - \vec{\mathsf{r}}^{\alpha})$$

for a system of N particles of mass m, while the current density operator is $(i = x, y, z)$

$$\mathsf{j}_i(\vec{r}) = \sum_{\alpha=1}^{N}\frac{\mathsf{p}_i^{\alpha}}{m}\delta(\vec{r} - \vec{\mathsf{r}}^{\alpha})$$

More precisely, one should use a symmetrized version of the right hand side in the above definition; $\vec{\mathsf{r}}^{\alpha}$ and $\vec{\mathsf{p}}^{\alpha}$ are the position and momentum of particle α, which obey the commutation relations ($\hbar = 1$)

$$[\mathsf{x}_i^\alpha, \mathsf{p}_j^\beta] = \mathrm{i}\delta_{ij}\delta_{\alpha\beta} I$$

1. Show that the sum rule (9.79) may be written

$$\int \frac{\mathrm{d}\omega}{\pi} \omega \chi''_{nn}(\vec{r}, \vec{r}\,'; \omega) = -\mathrm{i}\nabla_x \cdot \langle [\vec{\mathsf{j}}(\vec{r}), \mathsf{n}(\vec{r}\,')] \rangle$$

2. Take the Fourier transform of both sides and remark, for example in the case of the density, that

$$\mathsf{n}(\vec{q}) = \sum_{\alpha=1}^{N} \exp(\mathrm{i}\vec{q} \cdot \vec{\mathsf{r}}^\alpha)$$

3. Compute the commutator. For a single variable

$$[\mathsf{p}, \exp(\mathrm{i}q\mathsf{x})] = -\mathrm{i}\frac{\partial}{\partial x} \exp(\mathrm{i}q\mathsf{x}) = q \exp(\mathrm{i}q\mathsf{x})$$

9.7.8 Diffusion of a Brownian particle

1. Let $I(T)$ be the integral

$$I(T) = \int_{-T/2}^{T/2} \mathrm{d}t_1 \int_{-T/2}^{T/2} \mathrm{d}t_2 \, g(t)$$

where the function $g(t)$ depends only on the difference $t = t_1 - t_2$. Show that

$$I(T) = T \int_{-T}^{+T} \mathrm{d}t \, g(t) \left(1 - \frac{|t|}{T}\right) \tag{9.180}$$

The second term may be neglected if the function $g(t)$ decreases rapidly on a time scale $\tau \ll T$. Application:

$$\int_{-T/2}^{T/2} \mathrm{d}t_1 \int_{-T/2}^{T/2} \mathrm{d}t_2 \, \mathrm{e}^{-|t|/\tau} = 2T \left[\tau \left(1 - \frac{\tau}{T}\right) + \frac{\tau^2}{T} \mathrm{e}^{-T/\tau} \right] \tag{9.181}$$

2. Starting from the *equilibrium* velocity autocorrelation function (9.112) $C_{vv}^{\mathrm{eq}}(t)$, compute

$$\langle (\Delta X(t))^2 \rangle = \langle (X(t) - x(0))^2 \rangle$$

and show that one obtains a diffusive behaviour

$$\langle (\Delta X(t))^2 \rangle = 2Dt$$

when $t \gg 1/\gamma$.

9.7.9 Strong friction limit: harmonic oscillator

We consider again the forced harmonic oscillator of Exercise 9.7.1, assuming that the external force $F(t)$ is a stationary random force

$$\ddot{X} + \gamma \dot{X} + \omega_0^2 X = \frac{F(t)}{m}$$

Let $C_{xx}(t)$ denote the position autocorrelation function

$$C_{xx}(t) = \overline{X(t'+t)X(t')}$$

and $C_{pp}(t)$ the momentum autocorrelation function

$$C_{pp}(t) = \overline{P(t'+t)P(t')}$$

τ_x and τ_p are the characteristic times of $C_{xx}(t)$ and $C_{pp}(t)$.

1. Using the Wiener–Kinchin theorem (9.189), compute the Fourier transform $C_{xx}(\omega)$ as a function of the autocorrelation of the force $C_{FF}(\omega)$. If $C_{FF}(t)$ is given by (9.125)

$$C_{FF}(t) = \overline{F(t'+t)F(t')} = 2A\delta(t)$$

show that

$$C_{xx}(\omega) = \frac{1}{m^2} \frac{2A}{(\omega^2 - \omega_0^2)^2 + \gamma^2 \omega^2} \tag{9.182}$$

2. The strong friction limit corresponds to $\gamma \gg \omega_0$. Draw qualitatively $C_{xx}(\omega)$ in this limit, show that the width of the curve is $\simeq \omega_0^2/\gamma$ and estimate τ_x.

3. What is the relation between $C_{xx}(\omega)$ and $C_{pp}(\omega)$? Draw qualitatively $C_{pp}(\omega)$ in the strong friction limit and determine its width. Deduce from this width that $\tau_p \simeq 1/\gamma$ and that $\tau_x \gg \tau_p$. Discuss the physical significance of this result.

4. Show that taking the strong friction limit amounts to neglecting the inertial term $\ddot{X}$ in the equation of motion and recover the Ornstein–Uhlenbeck equation for $\dot{X}$ as well as τ_x.

9.7.10 Green's function method

Let $G(t)$ be the retarded Green's function of the damped harmonic oscillator ($G(t) = 0$ if $t < 0$)

$$\left(\frac{\mathrm{d}^2}{\mathrm{d}t^2} + \gamma\frac{\mathrm{d}}{\mathrm{d}t} + \omega_0^2\right)G(t) = \delta(t)$$

If $\gamma < 2\omega_0$, show that

$$G(t) = \frac{\theta(t)}{\omega_1}\mathrm{e}^{-\gamma t/2}\sin\omega_1 t \qquad \omega_1 = \frac{1}{2}(4\omega_0^2 - \gamma^2)^{1/2}$$

where $\theta(t)$ is the step function. We want to solve the following stochastic differential equation for the random function $X(t)$, with initial conditions $x(0)$ and $\dot{x}(0)$

$$\left(\frac{\mathrm{d}^2}{\mathrm{d}t^2} + \gamma\frac{\mathrm{d}}{\mathrm{d}t} + \omega_0^2\right)X(t) = b(t)$$

where $b(t)$ is a stochastic force

$$\overline{b(t)b(t')} = \frac{2A}{m^2}\delta(t-t')$$

Show that

$$X(t) - x_0(t) = Y(t) = \int_0^t \mathrm{d}t'\, G(t-t')b(t')$$

where $x_0(t)$ is the solution of the homogeneous equation with initial conditions $x(0)$ and $\dot{x}(0)$. What is the characteristic damping time of $x_0(t)$? Compute $\langle Y(t)Y(t+\tau)\rangle$ for $t \gg 1/\gamma$

$$\overline{Y(t)Y(t+\tau)} \simeq \frac{A\mathrm{e}^{-\gamma\tau/2}}{\gamma\omega_0^2}\left[\cos\omega_1 t + \frac{\gamma}{2\omega_1}\sin\omega_1 t\right]$$

9.7.11 Moments of the Fokker–Planck equation

Let $P(x,t)$, $x = (x_1, \ldots, x_N)$ be a multivariate probability distribution that obeys the Fokker–Planck equation

$$\frac{\partial P}{\partial t} = -\sum_{i=1}^{N}\frac{\partial}{\partial x_i}\Big[A_i(x)P\Big] + \sum_{i,j=1}^{N}\frac{\partial^2}{\partial x_i\partial x_j}\Big[D_{ij}(x)P\Big] \tag{9.183}$$

Define $\Delta X_i = X_i(t+\varepsilon) - x_i(t)$. Show that

$$\lim_{\varepsilon \to 0} \frac{1}{\varepsilon} \overline{\Delta X_i} = A_i(x)$$
$$\lim_{\varepsilon \to 0} \frac{1}{\varepsilon} \overline{\Delta X_i \Delta X_j} = 2D_{ij}(x) \tag{9.184}$$

Hint: Integrate by parts. What is the corresponding Langevin equation if D_{ij} is x-independent?

9.7.12 Backward velocity

Let v^+ be the forward velocity

$$v^+ = \lim_{\varepsilon \to 0} \frac{1}{\varepsilon} \overline{X(t+\varepsilon) - x} = a(x)$$

for the Langevin equation

$$\frac{\mathrm{d}X}{\mathrm{d}t} = a(x) + b(t) \qquad \overline{b(t)b(t')} = 2D\delta(t-t')$$

Assume that one *knows* that the particle is at x at time t. One now wishes to determine the backward velocity

$$v^- = \lim_{\varepsilon \to 0} \frac{1}{\varepsilon} \overline{x - X(t-\varepsilon)}$$

Show that

$$v^- = v^+ - 2D \frac{\partial \ln P(x, t|x_0)}{\partial x}$$

Hint: Use $P(x, t|y, t-\varepsilon)$. This result shows clearly that the trajectory is not differentiable.

9.7.13 Numerical integration of the Langevin equation

In this exercise we shall test the Euler and Runge–Kutta integration schemes discussed in Section 9.6.

1. Write a program to implement the Euler approximation (9.166) with $V(x) = x^2/2$. Run your program doing of the order of 10^4 thermalization iterations and then about 10^6 measurement iterations. Measure $\langle x^2 \rangle$ as a function of the discrete time step ε for $0.1 \le \varepsilon \le 0.5$. Compare with the exact result, which you can calculate easily. Time your program and verify that, for the same precision and the same *physical time* $t = n\varepsilon$ where n is the number of iterations, the Runge–Kutta method is much more efficient than Euler's.

Do the same for $V(x) = x^4/4$. In this case, the exact result is unavailable but you will see that both Euler and Runge–Kutta tend to the same value as $\varepsilon \to 0$. In this case do the simulations for $5 \times 10^{-3} \leq \varepsilon \leq 0.1$.

Note: It is important to work in double precision otherwise you will quickly lose accuracy as ε gets smaller.

2. For the case $V(x) = x^2/2$ calculate the autocorrelation function $\langle x(t_0)x(t_0 + t)\rangle$ and verify that it decays exponentially. Use the Runge–Kutta method with $\varepsilon = 0.2$. What is the relaxation time? Does it agree with the prediction of the Fokker–Planck equation? Compare with the exact result. See Equations (9.151) and (9.153) but note that whereas in (9.153) we took $E_0 = 0$, in the numerical integration it is not!

Use this method to determine $E_0 - E_1$ for the case $V(x) = x^4/4$. In this case take $\varepsilon = 0.01$ or smaller.

9.7.14 Metastable states and escape times

In this exercise we shall study the tunneling time τ of a particle trapped in a metastable state (local minimum). We shall assume the particle dynamics to be described by the Langevin equation,

$$\frac{\mathrm{d}x(t)}{\mathrm{d}t} = -\frac{\partial V(x)}{\partial x} + \sqrt{T}b(t) \tag{9.185}$$

where we have introduced the temperature explicitly. The potential

$$V(x) = (ax^2 - b)^2 \tag{9.186}$$

has two degenerate minima at $x = \pm\sqrt{b/a}$ and looks like a section of the Mexican hat potential in Chapter 4. This degeneracy is lifted by the application of an external 'field' h

$$V(x) = (ax^2 - b)^2 - hx \tag{9.187}$$

where, to fix ideas, we take $h > 0$. This lifts the minimum at $-\sqrt{a/b}$ to a higher energy than the minimum at $+\sqrt{a/b}$; it also slightly shifts their positions. So now the minimum at $x < 0$ is a local minimum, and a particle trapped in it will eventually tunnel out into the global minimum.[22] Our goal is to study this tunneling time as a function of temperature.

Take $a = 1$, $b = 1$ and $h = 0.3$, and use the Runge–Kutta discretization with $\varepsilon = 0.01$ to study the escape of the particle. We consider the particle to have

[22] If one waits long enough the particle will tunnel back into the local minimum for a while and then tunnel out of it again.

escaped the local minimum as soon as $x > 0$ since then it will have passed the peak of the barrier.

To determine the average escape time, we place the particle at $x = -1$ and start the Runge–Kutta integration. We stop when $x > 0$ and record the time it took to reach this point. We repeat this, say, 5000 times and calculate the average of the escape times thus found.[23] Of course this is equivalent to taking a population of 5000 particles in the local minimum and observing it decay to the global one.

1. For several temperatures, say $T = 0.075$, 0.08, 0.09, 0.1, make a separate histogram of the escape times recorded for each temperature. Does the shape of the histogram suggest a form for the dependence of the population on time? Calculate the average escape time τ for each of these temperatures and try a fit of the form $\exp(-t/\tau)$ for the histograms.

2. Determine τ for many temperatures $0.075 \leq T \leq 1.5$ and plot τ versus T. Compare your numerical results with the theoretical calculation[24]

$$\tau = \frac{2\pi}{\sqrt{V''(x_A)|V''(x_B)|}} \mathrm{e}^{(V(x_B)-V(x_A))/T} \tag{9.188}$$

where $V''(x)$ is the second derivative of the potential, $x_A < 0$ is the position of the local minimum, $V'(x_A) = 0$, and x_B is the position of the peak separating the two minima, $V'(x_B) = 0$. Is the agreement better at low or at high temperature? At what temperature does the agreement become bad?

In fact, the very definition of escape time τ presupposes that τ is much longer than the time needed to establish local equilibrium in the local minimum. Since as the temperature of the particle increases its energy gets closer to the top of the barrier, this condition is no longer satisfied and the agreement between theory and simulation deteriorates.

9.8 Problems

9.8.1 Inelastic light scattering from a suspension of particles

In this problem we limit ourselves to the classical approximation. When light travels through a dielectric medium, the medium is polarized, and the polarization acts as the source of the electromagnetic field. We split the dielectric response into a component ε independent of space and time and a fluctuation $\delta\varepsilon(\vec{r}, t)$; the polarization is written accordingly

$$\vec{P} = \varepsilon_0(\varepsilon - 1)\vec{E}(\vec{r}, t) + \delta\vec{P}(\vec{r}, t)$$

[23] Although we are always starting the particle at the same initial position, it will not follow the same path since the noise (sequence of random numbers) is different.

[24] See for example van Kampen [119] Chapter XI.

where ε_0 is the vacuum permittivity and $\vec{E}(\vec{r}, t)$ the electric field.

1. Show that in the gauge defined by

$$\frac{\varepsilon}{c^2}\frac{\partial \varphi}{\partial t} + \vec{\nabla} \cdot \vec{A} = 0$$

where φ is the scalar potential, the vector potential $\vec{A}$ obeys

$$\nabla^2 \vec{A} - \frac{\varepsilon}{c^2}\frac{\partial^2 \vec{A}}{\partial t^2} = -\frac{1}{\varepsilon_0 c^2}\frac{\partial \delta \vec{P}}{\partial t}$$

What is the corresponding equation for the scalar potential φ? Note that $c/\sqrt{\varepsilon}$ is the light velocity in the medium.

2. It is convenient to use the Hertz vector $\vec{Z}$

$$\varphi = -\vec{\nabla} \cdot \vec{Z} \qquad \vec{A} = \frac{\varepsilon}{c^2}\frac{\partial \vec{Z}}{\partial t}$$

Write the partial differential equation obeyed by $\vec{Z}$ as a function of $\delta \vec{P}$. Give the expression of $\vec{E}$ as a function of $\vec{Z}$.

3. Show that if the wave is scattered by a fluctuation $\delta \vec{P}$ of frequency ω located around $\vec{r}\,'$, then for $r = |\vec{r}| \gg |\vec{r}\,'|$ (see Figure 9.4 for the geometry of the scattering)

$$\vec{Z}(\vec{r}, \omega) \simeq \frac{1}{4\pi \varepsilon_0 \varepsilon}\frac{e^{ikr}}{r}\int d^3r' \, e^{-i\vec{k}\cdot\vec{r}\,'} \delta \vec{P}(\vec{r}\,', \omega)$$

In the above equation, $\vec{k} = k\hat{r}$ ($\hat{r} = \vec{r}/r$) is the wave vector of the scattered light, $k = \omega\sqrt{\varepsilon}/c$. Obtain from this equation the scattered electric field

$$\vec{E}(\vec{r}, \omega) \simeq \frac{\omega^2}{4\pi \varepsilon_0 c^2}\frac{e^{ikr}}{r}\int d^3r' \, e^{-i\vec{k}\cdot\vec{r}\,'} \hat{k} \times (\delta \vec{P}(\vec{r}\,', \omega) \times \hat{k})$$

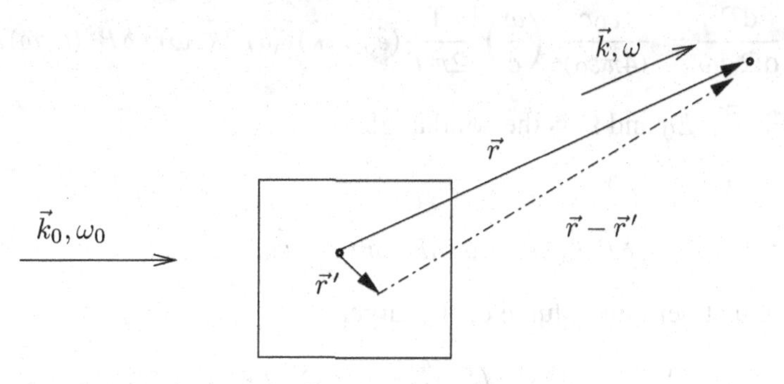

Figure 9.4 Geometry of the scattering.

4. Show the *Wiener–Kinchin theorem*: If $X(t)$ is a stationary random function and $X_T(t)$ is defined by

$$X_T(t) = X(t) \qquad -T \le t \le T, \quad \text{otherwise } X_T(t) = 0$$

where T is a time chosen to be much larger than all the characteristic times of the problem, then

$$\boxed{\langle X_T(\omega) X_T^*(\omega)\rangle \simeq T S_T(\omega)} \tag{9.189}$$

where $S_T(\omega)$ is the Fourier transform of the autocorrelation function of $X_T(t)$

$$\boxed{S_T(\omega) = \int \mathrm{d}t\, \mathrm{e}^{\mathrm{i}\omega t} \langle X_T(t) X_T^*(0)\rangle} \tag{9.190}$$

Hint: Use (9.180). In what follows, the subscript T will be suppressed.

5. We consider light scattering by small particles suspended in a fluid. The fluctuations of the polarization $\delta\vec{P}(\vec{r}, t)$ of the medium submitted to an incident electric field

$$\vec{E}(\vec{r}, t) = \vec{E}_0\, \mathrm{e}^{\mathrm{i}(\vec{k}_0\cdot\vec{r} - \omega_0 t)}$$

are given by

$$\delta\vec{P}(\vec{r}, t) = \alpha\varepsilon_0 \delta n(\vec{r}, t) \vec{E}_0 \mathrm{e}^{\mathrm{i}(\vec{k}_0\cdot\vec{r} - \omega_0 t)}$$

where ε_0 is the vacuum polarizability, α the difference between the particle dielectric constant and that of the fluid and $\delta n(\vec{r}, t)$ the density fluctuations of the particles. One notes the similarity of this problem with the spin diffusion considered in Section 9.1.4: here the rôle of the magnetic field, conjugate to the magnetization density, is played by the chemical potential, conjugate to the density.

Show that the power $\mathrm{d}\mathcal{P}$ radiated in a solid angle $\mathrm{d}\Omega$ and a frequency range $\mathrm{d}\omega$ is

$$\frac{\mathrm{d}\mathcal{P}}{\mathrm{d}\Omega\, \mathrm{d}\omega} = \frac{\varepsilon_0 c}{(4\pi\varepsilon_0)^2} \left(\frac{\omega}{c}\right)^4 \frac{1}{2\pi T} (\hat{e}_0 \times \hat{k})^2 \langle \delta\vec{P}(\vec{k}, \omega) \cdot \delta\vec{P}^*(\vec{k}, \omega)\rangle$$

where $\hat{e}_0 = \vec{E}_0 / E_0$ and Ω is the solid angle of $\vec{k}$.

6. Show that

$$\delta\vec{P}(\vec{k}, \omega) = \alpha\varepsilon_0 \vec{E}_0\, \delta n(\vec{k} - \vec{k}_0, \omega - \omega_0)$$

and deduce that per unit volume of the target

$$\frac{\mathrm{d}\mathcal{P}}{\mathrm{d}\Omega\, \mathrm{d}\omega} = \frac{\varepsilon_0 c}{32\pi^3} \left(\frac{\omega}{c}\right)^4 (\hat{e}_0 \times \hat{k})^2\, \alpha^2 \vec{E}_0^2\, S_{nn}(\vec{k}', \omega')$$

where $\vec{k}' = \vec{k} - \vec{k}_0$, $\omega' = \omega - \omega_0$ and $S_{nn}(\vec{r}, t)$ is the density autocorrelation function

$$S_{nn}(\vec{r}, t) = \langle \delta n(\vec{r}, t) \delta n(\vec{0}, 0) \rangle$$

7. One takes for $S_{nn}(\vec{k}', \omega')$ the following expression

$$S_{nn}(\vec{k}', \omega') = \frac{2}{\beta} \chi(\vec{k}') \frac{Dk'^2}{\omega'^2 + (Dk'^2)^2}$$

where D is the diffusion coefficient of the particles in the fluid and $\chi(\vec{k}')$ the Fourier transform of the static susceptibility $\chi(\vec{r} - \vec{r}') = \delta n(\vec{r})/\delta\mu(\vec{r}')$. By appealing to Section 9.1.4, give a justification of this expression.

8. Taking into account that $r^2 \langle \vec{E}(\vec{r}, \omega) \cdot \vec{E}^*(\vec{r}, \omega) \rangle$ is in fact a function of $\vec{k}'$ and ω, one defines $\vec{\mathcal{E}}(\vec{r}, t) = r\vec{E}(\vec{r}, t)$ and $S_{EE}(\vec{k}', \omega)$ as the time autocorrelation function of $\vec{\mathcal{E}}(\vec{r}, t)$

$$S_{EE}(\vec{k}', \omega) = \int \mathrm{d}t \, \mathrm{e}^{\mathrm{i}\omega t} \langle \vec{\mathcal{E}}(\vec{r}, t) \cdot \vec{\mathcal{E}}^*(\vec{r}, 0) \rangle$$

Show that

$$S_{EE}(\vec{k}', \omega) = A \frac{2Dk'^2}{(\omega - \omega_0)^2 + (Dk'^2)^2}$$

and determine the coefficient A. What is the value of $\langle \vec{\mathcal{E}}(\vec{r}, t) \cdot \vec{\mathcal{E}}^*(\vec{r}, 0) \rangle$? One recalls that

$$\int \mathrm{d}t \, \mathrm{e}^{\mathrm{i}\omega t - \gamma|t|} = \frac{2\gamma}{\omega^2 + \gamma^2}$$

Note also that $\omega' \ll \omega_0$: $A(\omega) \simeq A(\omega_0)$.

9. Instead of $\mathrm{d}\mathcal{P}/\mathrm{d}\Omega \, \mathrm{d}\omega$, one can measure experimentally the intensity correlation function $S_{II}(\vec{r}, t)$ of $r^2 I(\vec{r}, t) = \vec{\mathcal{E}}(\vec{r}, t) \cdot \vec{\mathcal{E}}^*(\vec{r}, t)$ whose Fourier transform is $\mathcal{I}(\vec{k}', t)$

$$S_{II}(\vec{k}', \omega) = \int \mathrm{d}t \, \mathrm{e}^{i\omega t} \langle \mathcal{I}(\vec{k}', t) \mathcal{I}(\vec{k}', 0) \rangle$$

Assuming that $\vec{E}(\vec{r}, t)$ is a Gaussian random function, show that $S_{II}(\vec{k}', \omega)$ is proportional to

$$2\pi\delta(\omega) + \frac{4Dk'^2}{\omega^2 + (2Dk'^2)^2}$$

Knowing that $Dk'^2 \simeq 500$ Hz, why is it more advantageous to measure $S_{II}(\vec{k}', \omega)$ rather than $S_{EE}(\vec{k}', \omega)$?

9.8.2 Light scattering by a simple fluid

We consider a simple fluid close to *static equilibrium* and call n, ε and $\vec{g} = 0$ the *equilibrium* particle, energy and momentum densities. The notations are those of Section 6.3. As in Section 9.1, we assume that a non-equilibrium situation has been created at $t = 0$ and we wish to study the relaxation toward equilibrium for $t > 0$. This study will allow us to compute the Kubo function, from which one deduces the dynamical structure factors which govern light scattering by the fluid. We denote the *deviations from equilibrium* by $\overline{n}(\vec{r}, t)$, $\overline{\varepsilon}(\vec{r}, t)$ and $\vec{g}(\vec{r}, t)$: since $\vec{g} = 0$ at equilibrium, there is no need for a bar. We assume that all these deviations are small, so that we may work in the linear approximation and use a linearized form of the hydrodynamic equations of Section 6.3.

1. *Linearized hydrodynamics.* Show that in the linear approximation, the Navier–Stoke equation (6.91) becomes

$$\partial_t \vec{g} + \vec{\nabla} P - \frac{1}{mn}\left(\zeta + \frac{\eta}{3}\right)\vec{\nabla}(\vec{\nabla}\cdot\vec{g}) - \frac{\eta}{mn}\nabla^2\vec{g} = 0 \qquad (9.191)$$

where P is the pressure and we have written the mass densiy $\rho = mn$, m being the mass of the fluid molecules. Note that the advection term $\vec{u}\cdot\vec{\nabla}\vec{u}$ in (6.91), which is quadratic in the fluid velocity $\vec{u}$, has been eliminated in the linear approximation. Show that within the same approximation the energy current $\vec{j}_E$ becomes

$$\vec{j}_E = (\varepsilon + P)\vec{u} - \kappa\vec{\nabla}T \qquad (9.192)$$

2. *Decomposition into transverse and longitudinal components.* As for any vector field, the momentum density $\vec{g}$ may be split into a transverse component $\vec{g}_{\mathrm{T}}$ and a longitudinal component $\vec{g}_{\mathrm{L}}$

$$\vec{g} = \vec{g}_{\mathrm{T}} + \vec{g}_{\mathrm{L}} \qquad \vec{\nabla}\cdot\vec{g}_{\mathrm{T}} = 0 \qquad \vec{\nabla}\times\vec{g}_{\mathrm{L}} = 0$$

This terminology reflects a property of the spatial Fourier transforms: $\vec{g}_{\mathrm{T}}(\vec{k}, t)$ is perpendicular to $\vec{k}$ (transverse), $\vec{k}\cdot\vec{g}_{\mathrm{T}}(\vec{k}, t) = 0$ and $\vec{g}_{\mathrm{L}}(\vec{k}, t)$ is parallel to $\vec{k}$ (longitudinal), $\vec{k}\times\vec{g}_{\mathrm{L}}(\vec{k}, t) = 0$. Show that the continuity and Navier–Stokes equations for $\overline{n}(\vec{r},t)$, $\vec{g}(\vec{r},t)$ and $\overline{\varepsilon}(\vec{r},t)$ take the form

$$\partial_t \overline{n} + \frac{1}{n}\vec{\nabla}\cdot\vec{g}_{\rm L} = 0 \tag{9.193}$$

$$\left(\partial_t - \frac{\eta}{mn}\nabla^2\right)\vec{g}_{\rm T} = 0 \tag{9.194}$$

$$\partial_t \vec{g}_{\rm L} + \vec{\nabla}P - \frac{\zeta'}{mn}\nabla^2\vec{g}_{\rm L} = 0 \tag{9.195}$$

$$\partial_t \overline{\varepsilon} + \frac{\varepsilon + P}{mn}\vec{\nabla}\cdot\vec{g}_{\rm L} - \kappa\nabla^2 T = 0 \tag{9.196}$$

where $\zeta' = (4\eta/3 + \zeta)$. Hint: To show (9.195), remember that for a vector field $\vec{V}(\vec{r})$

$$\vec{\nabla}\times\vec{\nabla}\times\vec{V} = \vec{\nabla}(\vec{\nabla}\cdot\vec{V}) - \vec{\nabla}^2\vec{V}$$

From (9.193) and (9.196), show that the continuity equation for the energy may be put into the following form

$$\partial_t \overline{\varepsilon} - \frac{\varepsilon + P}{n}\partial_t\overline{n} - \kappa\nabla^2 T = 0 \tag{9.197}$$

This equation suggests that it is useful to introduce the following quantity

$$q(\vec{r},t) = \overline{\varepsilon}(\vec{r},t) - \frac{\varepsilon + P}{n}\overline{n}(\vec{r},t) \tag{9.198}$$

3. *Transverse component.* Equation (9.194) shows that $\vec{g}_{\rm T}$ obeys a diffusion equation. Show that this equation implies that the Fourier–Laplace transform (9.22) $\vec{g}_{\rm T}(\vec{k},z)$ obeys

$$\vec{g}_{\rm T}(\vec{k},z) = \frac{\mathrm{i}}{z + \mathrm{i}k^2\eta/(mn)}\,\vec{g}(\vec{k},t=0)$$

Discuss the physical interpretation of this equation by comparing with Section 9.1.4 and give the expression for the dynamical structure factor $S_{\rm T}(\vec{k},\omega)$.

4. *Thermodynamic identities.* The study of the longitudinal component $\vec{g}_{\rm L}$ is unfortunately somewhat more complicated than that of $\vec{g}_{\rm T}$. The dynamical variables are $\overline{\varepsilon}$, $\overline{n}$ and $\vec{g}_{\rm L}$ and they obey a system of *coupled* partial differential equations. Furthermore, we want to use as independent thermodynamic variables n and ε, and we must express $\vec{\nabla}P$ and $\vec{\nabla}T$ as functions of these variables. We consider a subsystem of the fluid containing a fixed number N of molecules (it has of course a variable volume V) and all thermodynamic derivatives will be taken at *fixed* N; note that fixed V is then equivalent to fixed density n. Let us start from

$$\delta P = \left.\frac{\partial P}{\partial n}\right|_{\varepsilon}\delta n + \left.\frac{\partial P}{\partial \varepsilon}\right|_{n}\delta\varepsilon$$

where $(\partial P/\partial n)_\varepsilon$ and $(\partial P/\partial \varepsilon)_n$ are *equilibrium* thermodynamic derivatives. Show that

$$\left.\frac{\partial P}{\partial \varepsilon}\right|_n = \frac{V}{T}\left.\frac{\partial P}{\partial S}\right|_n$$

where S is the entropy of the subsystem. To evaluate $(\partial P/\partial n)_\varepsilon$, start from

$$V\,\mathrm{d}\varepsilon = T\,\mathrm{d}S - (\varepsilon + P)\,\mathrm{d}V$$

and show that

$$\left.\frac{\partial P}{\partial V}\right|_\varepsilon = \frac{\varepsilon + P}{T}\left.\frac{\partial P}{\partial S}\right|_n + \left.\frac{\partial P}{\partial V}\right|_S$$

It is then easy to derive

$$\vec{\nabla} P = \left.\frac{\partial P}{\partial n}\right|_S \vec{\nabla}\overline{n} + \frac{V}{T}\left.\frac{\partial P}{\partial S}\right|_n \vec{\nabla} q$$

The calculation of $\vec{\nabla} T$ follows exactly the same lines: one simply makes the substitution $P \to T$ and obtains

$$\vec{\nabla} T = \left.\frac{\partial T}{\partial n}\right|_S \vec{\nabla}\overline{n} + \frac{V}{T}\left.\frac{\partial T}{\partial S}\right|_n \vec{\nabla} q$$

5. *Equation for* $\vec{g}_\mathrm{L}$. From the results of Question 4, transform the coupled equations for q, $\overline{n}$ and $\vec{g}_\mathrm{L}$ into

$$\left[\partial_t - \frac{\zeta'}{mn}\nabla^2\right]\vec{g}_\mathrm{L} + \left.\frac{\partial P}{\partial n}\right|_S \vec{\nabla}\overline{n} + \frac{V}{T}\left.\frac{\partial P}{\partial S}\right|_n \vec{\nabla} q = 0 \qquad (9.199)$$

$$\left[\partial_t - \kappa\frac{V}{T}\left.\frac{\partial T}{\partial S}\right|_n \nabla^2\right] q - \kappa\left.\frac{\partial T}{\partial n}\right|_S \nabla^2\overline{n} = 0 \qquad (9.200)$$

Show also that δq (9.198) is related to the entropy density

$$\frac{T}{V}\,\delta S = \delta\varepsilon - \frac{\varepsilon + P}{n}\,\delta n = \delta q$$

It is convenient to define the following quantities

$$D_\mathrm{L} = \frac{\zeta'}{mn} = \frac{1}{mn}\left(\frac{4}{3}\eta + \zeta\right)$$

$$mnc_V = \frac{T}{V}\left.\frac{\partial S}{\partial T}\right|_n \qquad mnc_P = \frac{T}{V}\left.\frac{\partial S}{\partial T}\right|_P$$

$$c^2 = \frac{1}{m}\left.\frac{\partial P}{\partial n}\right|_S$$

c is the sound velocity, mc_V and mc_P are the specific heat capacities per fluid molecule.

6. *Relaxation of $\overline{\varepsilon}$, q and $\vec{g}_{\mathrm{L}}$*. Equations (9.193), (9.199) and (9.200) define a system of coupled PDE for $\overline{\varepsilon}$, q and $\vec{g}_{\mathrm{L}}$. Show that they can be solved by taking the spatial Fourier and time Laplace transforms

$$\begin{pmatrix} z & -\dfrac{k}{m} & 0 \\ -kmc^2 & z + \mathrm{i}k^2 D_{\mathrm{L}} & -\dfrac{V}{T}\left.\dfrac{\partial P}{\partial S}\right|_n k \\ \mathrm{i}k^2\kappa\left.\dfrac{\partial T}{\partial n}\right|_S & 0 & z + \mathrm{i}k^2\dfrac{\kappa}{mnc_V} \end{pmatrix} \begin{pmatrix} \overline{n}(\vec{k}, z) \\ g_{\mathrm{L}}(\vec{k}, z) \\ q(\vec{k}, z) \end{pmatrix} = \mathrm{i} \begin{pmatrix} \overline{n}(\vec{k}, t = 0) \\ g_{\mathrm{L}}(\vec{k}, t = 0) \\ q(\vec{k}, t = 0) \end{pmatrix} \tag{9.201}$$

where we have defined $g_{\mathrm{L}}(\vec{k}, z)$ by $\vec{g}_{\mathrm{L}}(\vec{k}, z) = \hat{k}\, g_{\mathrm{L}}(\vec{k}, z)$.

7. *Poles of $\vec{g}_{\mathrm{L}}(\vec{k}, z)$*. An explicit solution of (9.201) requires the inversion of the 3×3 matrix M in this equation. We shall limit ourselves to finding the poles of the functions $\overline{n}(\vec{k}, z)$, $q(\vec{k}, z)$ and $\vec{g}_L(\vec{k}, z)$, which are given by the zeroes of the determinant of M. First we use a low temperature approximation, where P is essentially a function of n and S essentially a function of T, so that

$$\left.\frac{\partial P}{\partial S}\right|_n = \left.\frac{\partial T}{\partial n}\right|_S \simeq 0$$

and $c_P \simeq c_V$. Within this approximation, show that the poles are located at

$$z \simeq -\mathrm{i}\frac{\kappa}{mnc_V}k^2$$
$$z \simeq \pm ck - \frac{\mathrm{i}}{2}D_{\mathrm{L}}k^2$$

In the general case, show that the additional term in the determinant is proportional to k^4 with a coefficient X

$$X = \mathrm{i}\frac{\kappa V}{T}\left.\frac{\partial T}{\partial P}\right|_S \left.\frac{\partial P}{\partial S}\right|_n$$

and that X may be written as

$$X = \mathrm{i}\frac{\kappa VT}{C_P C_V}\left.\frac{\partial V}{\partial T}\right|_P \left.\frac{\partial P}{\partial T}\right|_V$$

Use (1.38) to express the result in terms of $C_P - C_V$. Show that the position of the poles is now

$$z \simeq -\mathrm{i}\frac{\kappa}{mnc_P}k^2$$
$$z \simeq \pm ck - \frac{\mathrm{i}}{2}k^2\left[D_{\mathrm{L}} + \frac{\kappa}{mnc_P}\left(\frac{c_P}{c_V} - 1\right)\right] = \pm ck - \mathrm{i}\frac{\Gamma}{2}k^2$$

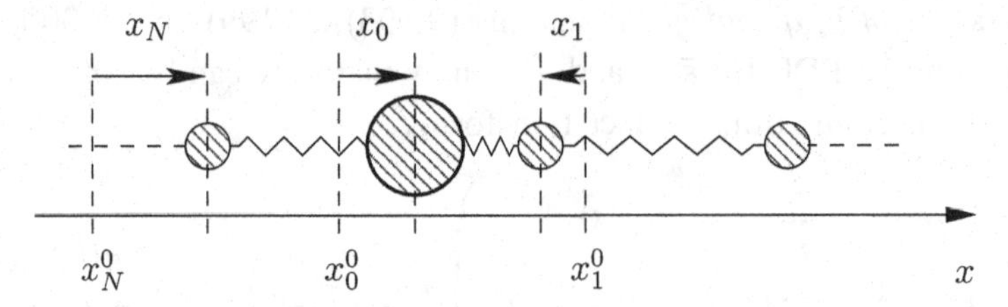

Figure 9.5 Springs and masses.

These equations give the positions of the heat pole and the two sound poles of Figure 9.1. If laser light is scattered from a simple fluid with wave vector transfer $\vec{k}$, the position of the sound poles gives the sound velocity and the width of the poles gives a combination of transport coefficients and thermodynamic quantities. It remains to compute explicitly the Kubo functions and the structure factors describing light scattering. This computation is now straightforward but cumbersome, and we refer the courageous reader who has followed us up to this point to Kadanoff and Martin [61] or Foerster [43].

9.8.3 Exactly solvable model of a Brownian particle

The model considered in this problem is an exactly solvable model, in which one can compute explicitly the memory function and the stochastic force. The model consists of a chain of coupled harmonic oscillators, with N ($N \to \infty$) light oscillators of mass m, and a heavy (Brownian) oscillator of mass M, with $M \gg m$, although the model can be solved for any ratio m/M. The goal of the problem is to compute the memory function, the memory time τ^*, the velocity relaxation time τ, and to show that $\tau^*/\tau \sim m/M$. We use a quantum mechanical treatment, as it is no more complicated than a classical one.

A Preliminary results

Let us consider N identical masses m, linked by identical springs with spring constant K (Figure 9.5). The equilibrium position of mass n is labelled by $x_n^0 = n$, where the lattice spacing is taken to be unity, and $n = 1, \ldots, N$; x_n denotes the deviation with respect to equilibrium of the position of mass n. The extremities of the first and last springs are fixed: $x_0 = x_{N+1} = 0$. Let q_k denote the (lattice) Fourier tranform of x_n

$$q_k = \sum_n C_{kn} x_n$$

with

$$C_{nk} = C_{kn} = \sqrt{\frac{2}{N+1}} \sin\left(\frac{\pi k n}{N+1}\right)$$

1. Show that C_{nk} is an orthogonal matrix

$$\sum_{k=1}^{N} C_{nk} C_{km} = \delta_{nm}$$

2. The Hamiltonian H of the chain reads

$$H = \sum_{n=1}^{N} \frac{p_n^2}{2m} + \frac{1}{2} K \sum_{n=0}^{N} (x_{n+1} - x_n)^2$$

Show that

$$H = \sum_{k=1}^{N} \frac{r_k^2}{2m} + \frac{1}{2} m \sum_{k=1}^{N} \omega_k^2 q_k^2 \qquad \omega_k = \sqrt{\frac{4K}{m}} \sin \frac{\pi k}{2(N+1)}$$

and r_k is the Fourier transform of p_n.

B The model

We use as quantum Hamiltonian of the model

$$H = \frac{\mathsf{p}_0^2}{2M} + \sum_{n=1}^{N} \frac{\mathsf{p}_n^2}{2m} + \frac{1}{2} K \sum_{n=0}^{N} (\mathsf{x}_{n+1} - \mathsf{x}_n)^2$$

Note the periodic boundary condition: $\mathsf{x}_{N+1} = \mathsf{x}_0$. The strategy will be to project the dynamics on the slow variable p_0, namely the momentum of the heavy particle.

3. Using the relation (see Exercise 9.7.4)

$$\langle \mathsf{A}; \mathcal{L}\mathsf{B} \rangle = \frac{1}{\beta} \langle [\mathsf{A}^\dagger, \mathsf{B}] \rangle$$

where $\langle \bullet; \bullet \rangle$ denotes Mori's scalar product and $\mathcal{L}$ the Liouvillian, show that

$$\mathrm{i}\mathcal{L}\mathsf{x}_0 = \frac{\mathsf{p}_0}{M} \qquad \langle \mathsf{p}_0; \mathsf{p}_0 \rangle = \frac{M}{\beta} \qquad \langle \mathsf{p}_0; \mathsf{x}_0 \rangle = 0$$

4. Let $\mathcal{P}$ denote the projector on p_0 (with respect to the Mori scalar product), and $\mathcal{Q} = \mathcal{I} - \mathcal{P}$. Show that for any dynamical variable B that is a linear combination of p_n and x_n

$$\mathsf{B} = \sum_{n=0}^{N} (\lambda_n \mathsf{p}_n + \mu_n \mathsf{x}_n)$$

if one defines $\overline{H} = H - \mathsf{p}_0^2/2M$, then

$$\mathcal{Q}\mathcal{L}\mathsf{B} = [\overline{H}, \mathsf{B}] = \overline{\mathcal{L}}\mathsf{B}$$

The potential energy can be re-expressed as a function of the variables $\mathsf{x}'_n = \mathsf{x}_n - \mathsf{x}_0$, and the projected dynamics corresponds to that of a heavy particle at rest.

5. Show that the stochastic force $\mathsf{f}(t) = \exp(\mathrm{i}\overline{\mathcal{L}}t)\mathcal{Q}\,\mathrm{i}\,\mathcal{L}\,\mathsf{p}_0$ is given by

$$\mathsf{f}(t) = K \exp(\mathrm{i}\overline{\mathcal{L}}t)\,(\mathsf{x}_1 + \mathsf{x}_N - 2\mathsf{x}_0)$$

6. By going to Fourier space, show that $\mathsf{f}(t)$ is given as a function of the operators x_n and p_n in the Schrödinger picture by

$$\mathsf{f}(t) = K \sum_{k,n=1}^{N} (C_{1k} + C_{Nk})\Big[C_{kn}(\mathsf{x}_n - \mathsf{x}_0)\cos\omega_k t + C_{kn}\frac{\mathsf{p}_n}{m\omega_k}\sin\omega_k t\Big]$$

and the memory function by

$$\gamma(t) = \frac{\beta}{M}\langle \mathsf{f}(0); \mathsf{f}(t)\rangle = \frac{K}{M}\sum_{k,n=1}^{N} C_{kn}(C_{1k} + C_{Nk})\cos\omega_k t$$

7. Compute $\gamma(t)$ in the thermodynamic limit $N \to \infty$

$$\lim_{N\to\infty} \gamma(t) = \frac{m}{M}\alpha^2 \frac{J_1(\alpha t)}{\alpha t}$$

with $\alpha = \sqrt{4K/m}$, and where J_1 is a Bessel function. One may use the following representation of the Bessel function

$$\frac{J_1(\alpha t)}{\alpha t} = \frac{1}{\pi}\int_{-\pi/2}^{\pi/2} \cos^2\varphi \cos(\alpha t \sin\varphi)\mathrm{d}\varphi$$

Compute the velocity relaxation time τ and give an estimate of the memory time τ^*. Show that $\tau^*/\tau \sim m/M$ and give a physical discussion of your results.

9.8.4 Itô versus Stratonovitch dilemma

When the diffusion coefficient is x-dependent, the Langevin equation is ambiguous. Let us write

$$\frac{\mathrm{d}X}{\mathrm{d}t} = a(x) + \sqrt{D(x_0)}\,b(t)$$

with

$$\overline{b(t)b(t')} = 2\,\delta(t - t')$$

while, for an infinitesimal time interval $[t, t+\varepsilon]$, x_0 is defined as a function of a parameter q, $0 \le q \le 1$, by

$$x_0 = x(t) + (1-q)[X(t+\varepsilon) - x(t)] = y + (1-q)[X(t+\varepsilon) - y]$$

We define as in (9.137)

$$B_\varepsilon = \int_t^{t+\varepsilon} \mathrm{d}t'\, b(t') \qquad \overline{B_\varepsilon} = 0 \qquad \overline{B_\varepsilon^2} = 2$$

1. *Itô prescription.* We first study the Itô prescription, which corresponds to $q = 1$: $x_0 = x(t)$

$$X(t+\varepsilon) = y + \varepsilon a(y) + \sqrt{D(y)} \int_t^{t+\varepsilon} \mathrm{d}t'\, b(t')$$

As in Section 9.5.1, we start from the Chapman–Kolmogorov equation (9.144) with

$$P(x, t+\varepsilon | y, t) = \overline{\delta(x - y - \varepsilon a(y) - \sqrt{D(y)}\, B_\varepsilon)}$$

Show that, to order ε, the δ-function may be written as $f'(x)\delta(y - f(x))$ where

$$f(x) = x - \varepsilon a(x) - B_\varepsilon \sqrt{D(x)} + \frac{1}{2} B_\varepsilon^2 D'(x) \tag{9.202}$$

Taking an average over all realizations of $b(t)$, derive the F–P equation (9.154)

$$\frac{\partial P}{\partial t} = -\frac{\partial}{\partial x}[a(x)P] + \frac{\partial^2}{\partial x^2}[D(x)P]$$

2. *General case.* In the general case, show that

$$\overline{X(t+\varepsilon) - y} = \varepsilon[a(y) + (1-q)D'(y)]$$

and, with respect to the preceding case, one has to make the substitution

$$a(y) \to a(y) + (1-q)D'(y)$$

Derive the F–P equation for arbitrary q

$$\frac{\partial P}{\partial t} = -\frac{\partial}{\partial x}[a(x)P] + \frac{\partial}{\partial x}\left[D^{1-q}(x)\frac{\partial}{\partial x}D^q(x)P\right]$$

The Stratonovitch prescription corresponds to $q = 1/2$. One notes that the various prescriptions differ by the drift velocities if one wants to write the F–P equation in the form (9.154).

9.8.5 Kramers equation

Consider a particle of mass m, moving in one dimension, which is subjected to a deterministic force $F(x) = -\partial V/\partial x$, a viscous force $-\gamma p$ and a random force $f(t)$

$$\overline{f(t)f(t')} = 2A\,\delta(t-t')$$

The equations of motion are

$$\dot{P} = F(x) - \gamma P + f(t) \qquad \dot{X} = \frac{P}{m}$$

Note that X and P alone are not Markovian variables, but the set (X, P) is Markovian.

1. By examining the moments $\langle \Delta X\rangle$, $\langle \Delta P\rangle$, $\langle (\Delta P)^2\rangle$, $\langle (\Delta X)^2\rangle$ and $\langle \Delta P \Delta X\rangle$ and by using (9.183) and (9.184), show that the probability distribution $P(x, p; t)$ obeys the *Kramers equation*

$$\boxed{\left[\frac{\partial}{\partial t} + \frac{p}{m}\frac{\partial}{\partial x} + F(x)\frac{\partial}{\partial p}\right] P = \gamma \left[\frac{\partial}{\partial p}(pP) + mkT\frac{\partial^2 P}{\partial p^2}\right]}$$

with $kT = A/(m\gamma)$. This equation can be simplified in the strong friction limit. Let us define the density

$$\rho(x,t) = \int \mathrm{d}p\, P(x, p; t)$$

and the current

$$j(x,t) = \int \mathrm{d}p\, \frac{p}{m}\, P(x, p; t)$$

2. From the $|p| \to \infty$ behaviour of $P(x, p; t)$, prove the (exact) continuity equation

$$\frac{\partial \rho}{\partial t} + \frac{\partial j}{\partial x} = 0$$

In the strong friction limit, show that one expects

$$P(x, p; t) \simeq \rho(x,t)\sqrt{\frac{1}{2\pi mkT}}\, \exp\left(-\frac{[p - \overline{p}(x)]^2}{2mkT}\right)$$

with $\overline{p}(x) = F(x)/\gamma$.

3. Let us finally define

$$K(x,t) = \int \mathrm{d}p \, \frac{p^2}{m} \, P(x,p;t)$$

What is the physical meaning of K? Prove the (exact) continuity equation

$$m\frac{\partial j}{\partial t} + \frac{\partial K}{\partial x} - F(x)\rho = -\gamma m j(x,t)$$

and show that in the strong friction limit

$$\left|\frac{\partial j}{\partial t}\right| \ll \gamma |j|$$

and that

$$K(x,t) \simeq \rho(x,t)\left[kT + \frac{\overline{p}^2(x)}{m}\right]$$

Using the continuity equation for $K(x,t)$, show that $\rho(x,t)$ obeys a Fokker–Planck equation when $kT \gg \overline{p}^2(x)/m$

$$\frac{\partial \rho}{\partial t} + \frac{\partial}{\partial x}\left[\frac{F(x)}{m\gamma}\rho(x,t) - D\frac{\partial}{\partial x}\rho(x,t)\right] = 0$$

Give the explicit expression of the diffusion coefficient D for x-independent γ and A. Can you generalize to x-dependent γ and A?

4. Show that for $F(x) = 0$, $X(t)$ obeys a diffusion equation with a space-dependent diffusion coefficient. Write this diffusion equation in the form of a Langevin equation and find the prescription (Itô, Stratonovitch or other, see Problem 9.8.4) that must be used in the following two cases (*i*) γ is x-independent and A is x-dependent and (*ii*) vice versa.

9.9 Further reading

An elementary introduction to the topics of this chapter may be found in Chandler [28]. The material in Sections 9.1 and 9.2 is detailed by Foerster [43] (Chapter 2) and by Fick and Sauermann [41] (Chapter 9). Spin diffusion is treated in Foerster [43] (Chapter 2) and in Kadanoff and Martin [61]. Good references on the projection method are: Balian *et al.* [7], Fick and Sauermann [41] (Chapters 14 to 18), Grabert [51], Zubarev *et al.* [126] (Chapter 2) and Rau and Müller [107]. The discussion of Section 9.3.5 follows Foerster [43] (Chapter 6); see also Lebowitz and Résibois [76]. The application of the projection method to Goldstone modes is discussed in great detail by Foerster [43] (Chapters 7 to 11); see also Chaikin and Lubensky [26] (Chapter 8). Although the articles are some fifty years old, the collection by Wax [121] is still a very useful reference on random functions; see also, e.g. Mandel and Wolff [86] (Chapters 1 and 2). There are many references on the Langevin and Fokker–Planck equations, among them: Parisi [99] (Chapter 19), van Kampen [119], Risken [110], Reichl [108] (Chapter 5).

Appendix

A.1 Legendre transform

A.1.1 Legendre transform with one variable

Let $f(x)$ be a convex function[1] of x, which we suppose, for simplicity, to be twice differentiable, $f''(x) \geq 0$. Then, the equation $f'(x) = u$ admits a unique solution $x(u)$ (Figure A.1). The function $F(x, u) \equiv ux - f(x)$, considered as a function of x for fixed u, has a unique maximum, $x(u)$, given by

$$\frac{\partial F}{\partial x} = u - f'(x) = 0 \tag{A.1}$$

This is indeed a maximum since $\partial^2 F/\partial x^2 = -f''(x) \leq 0$.

The Legendre transform, $g(u)$, of $f(x)$ is by definition

$$g(u) = \text{Max}\Big|_x F(u, x) = \text{Max}\Big|_x (ux - f(x)) \tag{A.2}$$

$$g(u) = ux(u) - f(x(u)) \tag{A.3}$$

To obtain the derivative of $g(u)$ we write

$$\mathrm{d}g = x\,\mathrm{d}u + u\,\mathrm{d}x - f'\,\mathrm{d}x = x(u)\,\mathrm{d}u$$

which gives

$$\frac{\mathrm{d}g}{\mathrm{d}u} = x(u) \tag{A.4}$$

The function $g(u)$ is a convex function of u

$$\frac{\mathrm{d}^2 g}{\mathrm{d}u^2} = x'(u) = \frac{1}{f''(x(u))} \geq 0 \tag{A.5}$$

[1] The following considerations apply equally to concave functions.

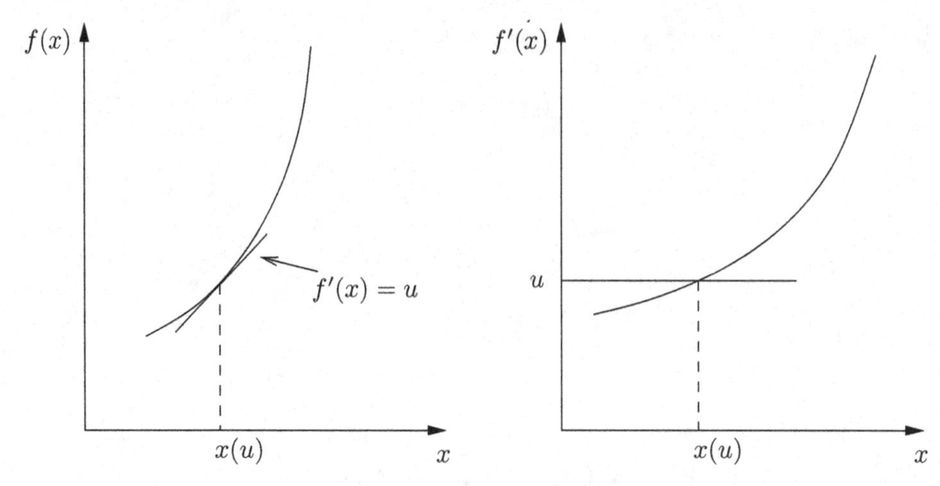

Figure A.1 Two graphical interpretations of the solution of the equation $f'(x) = u$.

Note that the Legendre transform of $g(u)$ gives back $f(x)$: the Legendre transform is involutive.

The above definition of the Legendre transform is the usual one used in mechanics. The definition used in thermodynamics differs by a sign: $g(u) = f(x(u)) - ux(u)$, with the consequence that to a convex function corresponds a concave Legendre transform and vice versa. When we deal with functions of several variables, the general statement of the convexity properties of the Legendre transform becomes complicated, whereas it remains very simple in the mechanics definition (see the following section). For this reason, in this appendix we stay with the mechanics definition of the Legendre transform.

A.1.2 Multivariate Legendre transform

Let $f(x_1, \dots, x_N)$ be a convex function of N variables. If this function is twice differentiable, the matrix C_{ij} with elements

$$C_{ij} = \frac{\partial^2 f}{\partial x_i \, \partial x_j}$$

is positive. Let $u_i = \partial f / \partial x_i$, the general Legendre transform g of the function f is then given by

$$g(u_1, \dots, u_N) = \sum_{i=1}^{N} u_i x_i - f(x_1, \dots, x_N) \tag{A.6}$$

Using the relation

$$\mathrm{d}g = \sum_{i=1}^{N} \left[u_i \,\mathrm{d}x_i + x_i \,\mathrm{d}u_i - \frac{\partial f}{\partial x_i} \mathrm{d}x_i \right] = \sum_{i=1}^{N} x_i \,\mathrm{d}u_i$$

we deduce

$$\frac{\partial g}{\partial u_i} = x_i \qquad \frac{\partial^2 g}{\partial u_i \,\partial u_j} = \frac{\partial x_i}{\partial u_j}$$

which yields the relation

$$\sum_j \frac{\partial^2 f}{\partial x_i \,\partial x_j} \frac{\partial^2 g}{\partial u_j \,\partial u_k} = \sum_j \frac{\partial u_i}{\partial x_j} \frac{\partial x_j}{\partial u_k} = \delta_{ik} \tag{A.7}$$

This means that if the matrix D_{ij} exists, where the matrix elements are given by

$$D_{ij} = \frac{\partial^2 g}{\partial u_i \,\partial u_j}$$

then D_{ij} is the inverse of C_{ij}. In order for D_{ij} to exist, C_{ij} must be positive definite. Its inverse is, therefore, positive definite, which means that g is a convex function. The convexity property still holds when the Legendre transform is taken over only some of the variables.

A.2 Lagrange multipliers

To obtain the extrema of a function of N variables $F(x_1, \ldots, x_N)$, it suffices to determine the points $x_1^* \ldots , x_N^*$ where the N partial derivatives vanish

$$\left.\frac{\partial F}{\partial x_1}\right|_{x_i = x_i^*} = \ldots = \left.\frac{\partial F}{\partial x_N}\right|_{x_i = x_i^*} = 0 \tag{A.8}$$

It is slightly more complicated to obtain the extrema if the variables are subject to constraints.

We start with the simplest example: a function, $F(x_1, x_2)$, of two variables subject to one constraint $f(x_1, x_2) = 0$. This constraint defines a curve Γ in the (x_1, x_2) plane. The vector $\vec{u} = \vec{\nabla} f$ is perpendicular to Γ at all points (Figure A.2). At a point M on Γ, an extremum must satisfy $\mathrm{d}F = 0$ for any variation $\mathrm{d}\vec{r}$ parallel to Γ, $\mathrm{d}\vec{r} \cdot \vec{\nabla} F = 0$. In other words, the vector $\vec{\nabla} F$ must be parallel to $\vec{u}$, i.e. $\vec{\nabla} F = \lambda \vec{u}$ which means

$$\frac{\partial}{\partial x_1}(F - \lambda f) = 0 \qquad \frac{\partial}{\partial x_2}(F - \lambda f) = 0 \tag{A.9}$$

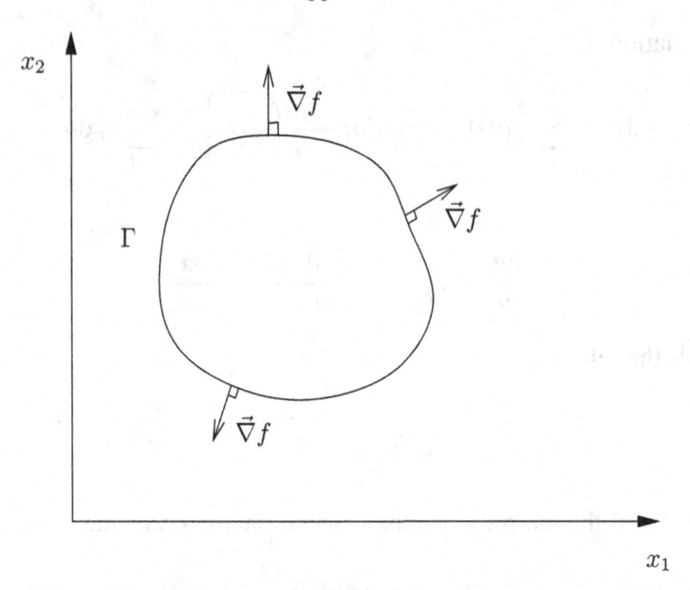

Figure A.2 The geometry for two variables.

λ is the Lagrange multiplier. If the system of equations (A.9) has solutions $(x_1^*(\lambda), x_2^*(\lambda))$, we substitute them in the constraint, $f(x_1^*(\lambda), x_2^*(\lambda)) = 0$, which then determines λ and consequently the values of x_1^* and x_2^* of the extrema.

In the general case, one has n constraints $f_p(x_1, \ldots, x_N) = 0$, $1 \leq p \leq n < N$ which define an n-dimensional hypersurface. The hyperplane tangent to this surface at a point M is perpendicular to the n vectors $\vec{u}_p$ given by

$$\vec{u}_p = \left(\frac{\partial f_p}{\partial x_1}, \ldots, \frac{\partial f_p}{\partial x_N} \right)$$

For any $\mathrm{d}\vec{r}$ in the tangent hyperplane, an extremum must satisfy $\mathrm{d}\vec{r} \cdot \vec{\nabla} F = 0$, which implies that $\vec{\nabla} F$ is a linear combination of $\vec{u}_p$. There are, then, n constants, or Lagrange multipliers, $\lambda_1, \ldots, \lambda_n$ such that

$$\vec{\nabla} F = \lambda_1 \vec{u}_1 + \cdots + \lambda_n \vec{u}_n$$

which yield the N equations

$$\frac{\partial}{\partial x_k}(F - \lambda_1 f_1 - \cdots - \lambda_n f_n) = 0 \qquad k = 1, \ldots, N \tag{A.10}$$

These equations determine the solutions $x_k^*(\lambda_1, \ldots, \lambda_n)$. Substituting these solutions in the constraints yields $\lambda_1, \ldots, \lambda_n$ from which we obtain x_k^*, the values of the extrema. In practice, we seldom need to perform these last two operations; usually it is sufficient to find the solutions to (A.10).

A.3 Traces, tensor products

A.3.1 Traces

By definition, the trace of an operator A acting in a Hilbert space $\mathcal{H}$ of dimension N is the sum of its diagonal elements

$$\mathrm{Tr}\,\mathsf{A} = \sum_n A_{nn} \tag{A.11}$$

We limit ourselves to a vector space of finite dimension. The precautions to take in the case of an infinite-dimensional space will be discussed at the end of this section. The trace is invariant under cyclical permutations

$$\mathrm{Tr}\,\mathsf{AB} = \mathrm{Tr}\,\mathsf{BA} \qquad \mathrm{Tr}\,\mathsf{ABC} = \mathrm{Tr}\,\mathsf{BCA} = \mathrm{Tr}\,\mathsf{CAB} \tag{A.12}$$

This means that the trace is independent of the basis used: changing basis means $\mathsf{A}' = S^{-1}\mathsf{A}S$, and from (A.12) we obtain $\mathrm{Tr}\,\mathsf{A}' = \mathrm{Tr}\,\mathsf{A}$. Equation (A.12) also allows us to show the following often used identities

$$\mathrm{Tr}\,\mathsf{A}[\mathsf{B},\mathsf{C}] = \mathrm{Tr}\,\mathsf{C}[\mathsf{A},\mathsf{B}] = \mathrm{Tr}\,\mathsf{B}[\mathsf{C},\mathsf{A}] \tag{A.13}$$

Let $|\varphi\rangle$ and $|\psi\rangle$ be two vectors in $\mathcal{H}$ and $|n\rangle$ an orthonormal basis. We then have,

$$\mathrm{Tr}\,|\varphi\rangle\langle\psi| = \sum_n \langle n|\varphi\rangle\langle\psi|n\rangle = \sum_n \langle\psi|n\rangle\langle n|\varphi\rangle$$

and by using the completeness relation $\sum_n |n\rangle\langle n| = 1$, the above may be written as

$$\mathrm{Tr}\,|\varphi\rangle\langle\psi| = \langle\psi|\varphi\rangle \tag{A.14}$$

In the same way we show

$$\mathrm{Tr}\,\mathsf{A}|\varphi\rangle\langle\psi| = \langle\psi|\mathsf{A}|\varphi\rangle \tag{A.15}$$

In an infinite dimensional Hilbert space, the trace of an operator, even a bounded one, may not be defined. The trace exists only for a special class of operators called trace-class operators: for example, $\exp(-\beta H)$ is such an operator. Equations (A.11) and (A.12) must be used with caution. For example, one could conclude from (A.12) that the canonical commutation relation $[\mathsf{A},\mathsf{B}] = \mathrm{i}\hbar$ never has a solution. In fact, this commutation relation does not have a solution in a finite dimensional Hilbert space because the trace of the left hand side vanishes whereas the trace of the right hand side is $\mathrm{i}\hbar N$ where N is the dimension of the vector space. Of course the canonical commutation relation has a solution, unique up to unitary transformations, with self-adjoint operators A and B in an infinite dimensional Hilbert space. In this case, the two operators A and B are necessarily unbounded.

A.3.2 Tensor products

Let $\mathcal{H}^{(a)}$ and $\mathcal{H}^{(\alpha)}$ be two Hilbert spaces with dimensions N and M respectively, and orthonormal bases $|n\rangle$, $n = 1, \ldots, N$ and $|\nu\rangle$, $\nu = 1, \ldots, M$. We form the tensor product space $\mathcal{H} = \mathcal{H}^{(a)} \otimes \mathcal{H}^{(\alpha)}$ of dimension NM by defining in $\mathcal{H}$ a basis of NM vectors $|n \otimes \nu\rangle$ and the tensor product $|a \otimes \alpha\rangle$ of vectors $|a\rangle$ and $|\alpha\rangle$

$$|a\rangle = \sum_n c_n |n\rangle \qquad\qquad |\alpha\rangle = \sum_\nu c_\nu |\nu\rangle$$

by

$$|a \otimes \alpha\rangle = \sum_{n,\nu} c_n c_\nu |n \otimes \nu\rangle \tag{A.16}$$

This is a manifestly linear operation. In general, a vector in $\mathcal{H}$ cannot be written as a tensor product (A.16). Similarly, we define the tensor product $\mathsf{A}^{(a)} \otimes \mathsf{B}^{(\alpha)}$ of two operators $\mathsf{A}^{(a)}$ and $\mathsf{B}^{(\alpha)}$ acting respectively in $\mathcal{H}^{(a)}$ and $\mathcal{H}^{(\alpha)}$ by their matrix elements

$$\left[\mathsf{A}^{(a)} \otimes \mathsf{B}^{(\alpha)}\right]_{a\alpha;b\beta} = A^{(a)}_{ab} B^{(\alpha)}_{\alpha\beta} \tag{A.17}$$

In general, an operator C acting in $\mathcal{H}$ cannot be written as a tensor product. We define the partial trace of C relative to $\mathcal{H}^{(\alpha)}$ by

$$C_{ab} = \sum_\alpha C_{a\alpha;b\alpha} = \left[\mathrm{Tr}_\alpha \mathsf{C}\right]_{ab} \tag{A.18}$$

We thus obtain an operator acting in $\mathcal{H}^{(a)}$.

Let us take as example the density operator D of a coupled system (a, α) whose state space is $\mathcal{H}$. The partial trace of D over $\mathcal{H}^{(\alpha)}$

$$D^{(a)} = \mathrm{Tr}_\alpha D \qquad\qquad D^{(a)}_{ab} = \sum_\alpha D_{a\alpha;b\alpha} \tag{A.19}$$

yields the density operator of system (a) if systems (a) and (α) are independent. In such a case, D is a tensor product

$$\sum_\alpha D_{a\alpha;b\alpha} = \sum_\alpha D^{(a)}_{ab} D^{(\alpha)}_{\alpha\alpha} = D^{(a)}_{ab} \sum_\alpha D^{(\alpha)}_{\alpha\alpha} = D^{(a)}_{ab}$$

since $\mathrm{Tr}\, D^{(\alpha)} = 1$. Let us consider the operator defined by $\mathsf{C}^{(a)} = \mathrm{Tr}_\alpha[\mathsf{A}^{(\alpha)}, D]$ and acting in the space $\mathcal{H}^{(a)}$. Since $\mathsf{A}^{(\alpha)}$ acts only in the space $\mathcal{H}^{(\alpha)}$

$$A^{(\alpha)}_{a\alpha;c\gamma} = A^{(\alpha)}_{\alpha\gamma} \delta_{ac}$$

we obtain

$$\sum_\alpha \left[\mathsf{A}^{(\alpha)} D\right]_{a\alpha;b\alpha} = \sum_{\alpha,\gamma} A^{(\alpha)}_{\alpha\gamma} D_{a\gamma;b\alpha}$$

whereas

$$\sum_{\alpha}\left[D\mathsf{A}^{(\alpha)}\right]_{a\alpha;b\alpha}=\sum_{\alpha,\gamma}D_{a\alpha;b\gamma}A^{(\alpha)}_{\gamma\alpha}$$

Exchanging the summation indexes α and γ shows that these two quantities are the same and that

$$\mathrm{Tr}_{\alpha}[\mathsf{A}^{(\alpha)},D]=0 \tag{A.20}$$

A.4 Symmetries

A.4.1 Rotations

The invariance of a physical problem under operations that form a group, e.g. a crystallographic group, imposes very important restrictions on the equations one may write to describe a physical situation, for example the equations relating the currents to the affinities in Chapter 6. These restrictions are known as the *Curie principle*. In this appendix, we restrict attention to the rotation group that allows us to apply the Curie principle to an isotropic system. Since in this appendix there is no possible confusion in the indexes, we use the letters $i, j, k, \ldots$ instead of Greek letters to denote the components of vectors. Thus, a vector $\vec{V}$ has components V_i, $i=(x,y,z)$. Under a rotation by an angle θ around a unit vector $\hat{n}$ of components n_i, a vector $\vec{V}$ is transformed into $\vec{V}\,'=\vec{V}(\theta)$

$$\vec{V}\,'=\vec{V}+(1-\cos\theta)(\vec{V}\cdot\hat{n})\hat{n}+\sin\theta\,(\hat{n}\times\vec{V}) \tag{A.21}$$

To demonstrate (A.21), we decompose $\vec{V}$ into a component parallel to $\hat{n}$ and a component $\vec{W}$ perpendicular to $\hat{n}$

$$\vec{V}=(\vec{V}\cdot\hat{n})\hat{n}+(\vec{V}-(\vec{V}\cdot\hat{n})\hat{n})=(\vec{V}\cdot\hat{n})\hat{n}+\vec{W}$$

Under this rotation, the component $(\vec{V}\cdot\hat{n})\hat{n}$, parallel to $\hat{n}$, is unchanged (Figure A.3) whereas $\vec{W}$ is rotated into $\vec{W}\,'$

$$\vec{W}\,'=\cos\theta\,\vec{W}+\sin\theta\,(\hat{n}\times\vec{W})$$

For an infinitesimal rotation, the change $\mathrm{d}\vec{V}$ in $\vec{V}$ becomes

$$\mathrm{d}\vec{V}=\vec{V}\,'-\vec{V}=(\hat{n}\times\vec{V})\,\mathrm{d}\theta \tag{A.22}$$

which we can also write in the form of a differential equation for the vector $\vec{V}(\theta)=\vec{V}\,'$

$$\frac{\mathrm{d}\vec{V}(\theta)}{\mathrm{d}\theta}=\hat{n}\times\vec{V}(\theta) \tag{A.23}$$

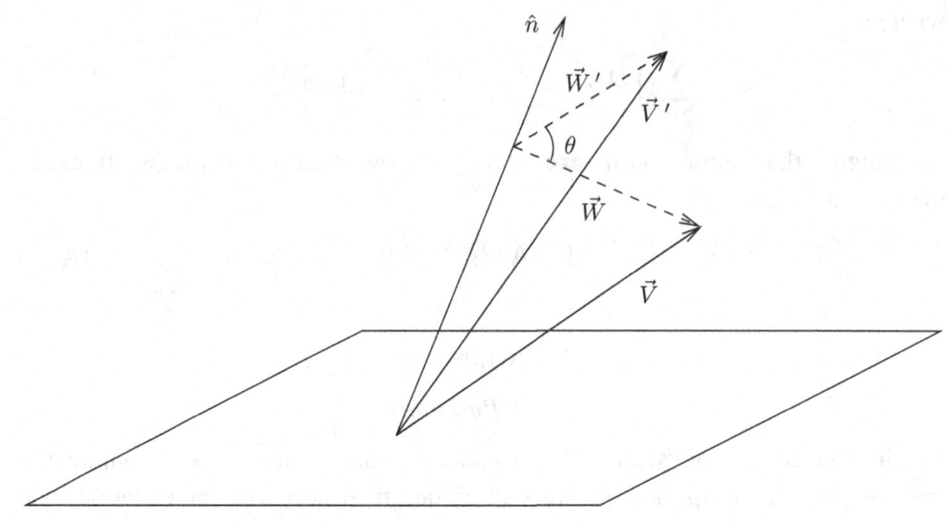

Figure A.3 Rotation of a vector $\vec{V}$ by an angle θ.

This equation is often encountered in physics. For example, it describes the motion of a magnetic moment $\vec{M}$ in a constant magnetic field $\vec{B}$

$$\frac{d\vec{M}}{dt} = -\gamma \vec{B} \times \vec{M} \tag{A.24}$$

where γ is the gyromagnetic ratio.[2] Comparing with (A.23) shows that the vector $\vec{M}$ precesses around $\vec{B}$ with an angular frequency $|\gamma B|$.

It is easy to deduce from (A.21) the transformation law for the components of $\vec{V}$:

$$V_i' = R_{ij}(\hat{n}, \theta) V_j \tag{A.25}$$

where $R_{ij}(\hat{n}, \theta)$ is an orthogonal rotation matrix, $R^T = R^{-1}$. We have adopted the convention of summation over repeated indexes. Since we do not need the explicit form of R, we leave it as an exercise for the reader. We will only give the infinitesimal form (A.22) since, according to (A.23), all finite rotations may be constructed from the infinitesimal ones.[3] An infinitesimal rotation (A.22) can be conveniently written using the completely antisymmetric tensor ε_{ijk}:[4] $\varepsilon_{ijk} = +1$ (-1) for (ijk) an even (odd) permutation of (xyz), $\varepsilon_{ijk} = 0$ if two or more indexes

[2] If the magnetic moment is due to a rotating charge q, $\gamma = q/(2m)$, where m is the mass of the charge distribution.

[3] Mathematically, this is due to the fact that the rotation group is a Lie group.

[4] The matrix T_i whose matrix elements $(T_i)_{jk} = \varepsilon_{ijk}$ is the generator of rotations around the i axis. It is an element of the Lie algebra $[T_i, T_j] = \varepsilon_{ijk} T_k$ of the rotation group.

are equal

$$\varepsilon_{123} = \varepsilon_{231} = \varepsilon_{312} = -\varepsilon_{213} = -\varepsilon_{132} = -\varepsilon_{321} = 1 \tag{A.26}$$

It is very easy to verify that the components of a vector product $\vec{Z} = \vec{V} \times \vec{W}$ can be written as

$$Z_i = (\vec{V} \times \vec{W})_i = \varepsilon_{ijk} V_j W_k \tag{A.27}$$

Equation (A.22) then becomes

$$\mathrm{d}V_i = \mathrm{d}\theta \; \varepsilon_{ijk} n_j V_k \tag{A.28}$$

We note the useful property of the tensor ε_{ijk}

$$\varepsilon_{ijk}\varepsilon_{ilm} = \delta_{jl}\delta_{km} - \delta_{jm}\delta_{kl} \tag{A.29}$$

We must have $j \neq i$ for a non-zero result, and so either $j = l$ or $j = m$. We may also show

$$\varepsilon_{ijk}\varepsilon_{ijl} = 2\,\delta_{kl} \qquad \varepsilon_{ijk}\varepsilon_{ijk} = 6 \tag{A.30}$$

Equation (A.29) allows us to demonstrate easily classic identities on vector operators. We take two examples

- $\vec{\nabla} \times (\vec{\nabla} \times \vec{V})$

$$\begin{aligned}
\left(\vec{\nabla} \times (\vec{\nabla} \times \vec{V})\right)_i &= \varepsilon_{ijk}\partial_j(\vec{\nabla} \times \vec{V})_k \\
&= \varepsilon_{ijk}\varepsilon_{klm}\partial_j\partial_l V_m \\
&= (\delta_{il}\delta_{jm} - \delta_{im}\delta_{jl})\partial_j\partial_l V_m \\
&= \partial_i(\partial_m V_m) - \partial^2 V_i
\end{aligned}$$

  which is the classic identity often used in electromagnetism

$$\vec{\nabla} \times (\vec{\nabla} \times \vec{V}) = \vec{\nabla}(\vec{\nabla} \cdot \vec{V}) - \nabla^2 \vec{V}$$

- $\vec{V} \times (\vec{\nabla} \times \vec{V})$

$$\begin{aligned}
\left(\vec{V} \times (\vec{\nabla} \times \vec{V})\right)_i &= \varepsilon_{ijk}\varepsilon_{klm} V_j\partial_l V_m \\
&= (\delta_{il}\delta_{jm} - \delta_{jl}\delta_{im}) V_j\partial_l V_m \\
&= V_j\partial_i V_j - V_j\partial_j V_i
\end{aligned}$$

  which is the classic identity often used in hydrodynamics

$$\vec{V} \times (\vec{\nabla} \times \vec{V}) = \frac{1}{2}\,\vec{\nabla}(\vec{V}^2) - (\vec{V} \cdot \vec{\nabla})\vec{V}$$

A.4.2 Tensors

By definition, a vector is an object that transforms under rotation according to (A.25). A rank two tensor, T_{ij}, is an object that transforms under rotation according to

$$T'_{ij} = R_{ik} R_{jl} T_{kl} \tag{A.31}$$

Let us find the rotation invariant rank two tensors, $T'_{ij} = T_{ij}$. To this end, consider the object $V_i T_{ij} W_j$ constructed with the help of the two vectors $\vec{V}$ and $\vec{W}$. We have

$$V_i T_{ij} W_j = V_i T'_{ij} W_j = V_i R_{ik} T_{kl} R_{jl} W_j = R^T_{ki} V_i R^T_{lj} W_j T_{kl} = V'_k T_{kl} W'_l$$

where $\vec{V}\,'$ and $\vec{W}\,'$ are the vectors obtained by rotating $\vec{V}$ and $\vec{W}$ with the rotation matrix R^T. But the only rotationally invariant quantity we can construct from two vectors is their scalar product. Therefore, $T_{ij} \propto \delta_{ij}$: the only rotation invariant rank two tensor is δ_{ij}.[5]

Similarly, the only rotation invariant quantity we can construct with three vectors is their mixed product. This entails that the only rotation invariant rank three tensor is ε_{ijk}. Finally, with the four vectors, $\vec{V}_1$, $\vec{V}_2$, $\vec{V}_3$, $\vec{V}_4$, we can form the invariant quantity $(\vec{V}_1 \cdot \vec{V}_2)(\vec{V}_3 \cdot \vec{V}_4)$ and two others obtained by permutation. Consequently, the rank four rotation invariant tensors are

$$\delta_{ij}\delta_{kl} \qquad \delta_{ik}\delta_{jl} \qquad \delta_{il}\delta_{jk} \tag{A.32}$$

The most general relation between a current $\vec{j}$ and a (scalar) affinity γ involves a rank two tensor T_{ij}

$$j_i = T_{ij} \partial_j \gamma$$

In an isotropic system, the tensor T_{ij} must be rotation invariant, i.e. $T_{ij} = C\delta_{ij}$, and so the Curie principle leads to

$$j_i = C\delta_{ij}\,\partial_j \gamma = C\,\partial_i \gamma \quad \text{or} \quad \vec{j} = C\,\vec{\nabla}\gamma$$

In the same way, for a vector affinity, $\vec{\gamma}$, the most general relation is

$$j_i = C\varepsilon_{ijk}\,\partial_j \gamma_k \quad \text{or} \quad \vec{j} = C\,\vec{\nabla} \times \vec{\gamma}$$

However, this equation is not invariant under space inversion and is thus unacceptable unless $\vec{\gamma}$ is a pseudo-vector.

[5] It is equally easy to show these results using the infinitesimal form of the rotations (A.28).

We now use rotation symmetry considerations to evaluate integrals. Consider the angular integral

$$T_{ij} = \int \mathrm{d}\Omega \; p_i p_j f(p) \tag{A.33}$$

where the function $f(p)$ depends only on the length $p = |\vec{p}|$ of the vector $\vec{p}$, and $\Omega = (\theta, \varphi)$ gives the direction of $\vec{p}$. The integral is a rank two tensor since it is constructed from $\vec{p}$, but it is also rotation invariant. Therefore $T_{ij} = C\delta_{ij}$. Multiplying both sides of (A.33) by δ_{ij} and summing over the indexes i and j gives

$$3C = \int \mathrm{d}\Omega \; p^2 f(p) = 4\pi p^2 f(p)$$

which leads to

$$\int \mathrm{d}\Omega \; p_i p_j f(p) = \frac{4\pi}{3} \, p^2 f(p) \, \delta_{ij} \tag{A.34}$$

A more complicated example is as follows: consider an integral of a function $p_i p_j p'_k p'_l f(p, p'; \cos\alpha)$, over the angles Ω and Ω' of the vectors $\vec{p}$ and $\vec{p}\,'$ and where α is the angle between $\vec{p}$ and $\vec{p}\,'$

$$T_{ijkl} = \int \mathrm{d}\Omega \, \mathrm{d}\Omega' \; p_i p_j p'_k p'_l f(p, p'; \cos\alpha) \tag{A.35}$$

By rotation invariance, and according to (A.32) we have

$$T_{ijkl} = A\delta_{ij}\delta_{kl} + B\left[\delta_{ik}\delta_{jl} + \delta_{il}\delta_{jk}\right]$$

Multiplying (A.35) by $\delta_{ij}\delta_{kl}$ and then by $\delta_{ik}\delta_{jl}$ and summing over the indexes we find

$$T_{ijkl} = 8\pi^2 \int \mathrm{d}(\cos\alpha) \left(A'\delta_{ij}\delta_{kl} + B'\left[\delta_{ik}\delta_{jl} + \delta_{il}\delta_{jk}\right]\right) f(p, p'; \cos\alpha) \tag{A.36}$$

with

$$A' = \frac{1}{15}\left(2p^2 p'^2 - (\vec{p}\cdot\vec{p}\,')^2\right)$$
$$B' = \frac{1}{30}\left(-p^2 p'^2 + 3(\vec{p}\cdot\vec{p}\,')^2\right)$$

This result, which is written for three-dimensional space, may be easily generalized to any dimensionality D.

A.5 Useful integrals

A.5.1 Gaussian integrals

The basic Gaussian integral is

$$I_n = \int_0^\infty \mathrm{d}x\, x^n \exp\left(-\frac{1}{2}Ax^2\right) = \frac{2^{(n-1)/2}\Gamma\left(\frac{n+1}{2}\right)}{A^{(n+1)/2}} \tag{A.37}$$

The constant A must be strictly positive for the integral (A.37) to exist. Γ is the 'Euler gamma function', which, for integer arguments, is the factorial function

$$\Gamma(p) = (p-1)! \tag{A.38}$$

Equation (A.37) is valid for all values of n, integer or not, as long as $n > -1$. In practice, we need in Equation (A.38) integer or half-integer values of p. In the latter case, we can calculate $\Gamma(p)$ from $\Gamma(1/2) = \sqrt{\pi}$.

When calculating average values in statistical mechanics, it is generally not necessary to calculate the integral (A.37). It is more useful to evaluate the generating function $f(u)$

$$f(u) = \mathcal{N}\int_{-\infty}^{\infty} \mathrm{d}x\, \exp\left(-\frac{1}{2}Ax^2 + ux\right) \tag{A.39}$$

The normalization constant $\mathcal{N}$ is chosen so that $f(0) = 1$. We 'complete the square' in the exponent of Equation (A.39) by changing variables $x = x' - u/A$, which gives

$$f(u) = \mathcal{N}\exp\left(\frac{u^2}{2A}\right)\int_{-\infty}^{\infty} \mathrm{d}x'\, \exp\left(-\frac{1}{2}Ax'^2\right) = \exp\left(\frac{u^2}{2A}\right) \tag{A.40}$$

We see that it was not necessary to calculate the constant $\mathcal{N}$. The average values of x^{2n} for integer n are obtained by differentiation

$$\left\langle x^2\right\rangle = \left.\frac{\mathrm{d}^2 f(u)}{\mathrm{d}u^2}\right|_{u=0} = \frac{1}{A} \tag{A.41}$$

$$\left\langle x^4\right\rangle = \left.\frac{\mathrm{d}^4 f(u)}{\mathrm{d}u^4}\right|_{u=0} = \frac{3}{A^2} \tag{A.42}$$

These results are easy to generalize to multi-dimensional Gaussian integrals where the quadratic terms in the exponent are related by a symmetric positive definite $N \times N$ matrix, A_{ij}. We use matrix notation where the N variables $(x_1, \ldots, x_N)$

form a column vector x and

$$\sum_{i,j=1}^{N} x_i A_{ij} x_j = x^T A x \qquad \sum_{i=1}^{N} x_i u_i = x^T u$$

The generating function now becomes

$$f(u_1, \ldots, u_N) = \mathcal{N} \int \prod_{i=1}^{N} \mathrm{d}x_i \, \exp\left(-\frac{1}{2} x^T A x + u^T x\right) \tag{A.43}$$

where again $\mathcal{N}$ is chosen such that $f(u_i = 0) = 1$. The variable change

$$x = x' - A^{-1} u$$

generalizes the result (A.40)

$$f(u) = \exp\left(\frac{1}{2} u^T A^{-1} u\right) \mathcal{N} \int \prod_{i=1}^{N} \mathrm{d}x'_i \, \exp\left(-\frac{1}{2} x'^T A x'\right) = \exp\left(\frac{1}{2} u^T A^{-1} u\right) \tag{A.44}$$

Again, average values are obtained by differentiation. For example, the average of two variables is given by

$$\langle x_i x_j \rangle = \left. \frac{\partial^2 f}{\partial u_i \partial u_j} \right|_{u_i = 0} = (A^{-1})_{ij} \tag{A.45}$$

and for four variables we have

$$\begin{aligned} \langle x_i x_j x_k x_l \rangle &= \left. \frac{\partial^4 f}{\partial u_i \, \partial u_j \, \partial u_k \, \partial u_l} \right|_{u_i = 0} \\ &= (A^{-1})_{ij} (A^{-1})_{kl} + (A^{-1})_{ik} (A^{-1})_{jl} + (A^{-1})_{il} (A^{-1})_{jk} \\ &= \langle x_i x_j \rangle \langle x_k x_l \rangle + \langle x_i x_k \rangle \langle x_j x_l \rangle + \langle x_i x_l \rangle \langle x_j x_k \rangle \end{aligned} \tag{A.46}$$

In general, the even order moments of a centred Gaussian distribution are expressed in terms of the second moment. The normalization constant $\mathcal{N}$ is calculated by diagonalizing the matrix A with an orthogonal transformation of unit Jacobian. We find

$$\mathcal{N} = \frac{(\det A)^{1/2}}{(2\pi)^{N/2}} \tag{A.47}$$

A.5.2 Integrals of quantum statistics

Integrals over the Bose–Einstein or Fermi–Dirac distributions can be expressed in terms of the following two integrals for integer or half-integer values of n:

$$B_n = \int_0^\infty \mathrm{d}x \, \frac{x^n}{\mathrm{e}^x - 1} = \Gamma(n+1)\zeta(n+1) \tag{A.48}$$

$$F_n = \int_0^\infty \mathrm{d}x \, \frac{x^n}{\mathrm{e}^x + 1} = \left(1 - \frac{1}{2^n}\right)\Gamma(n+1)\zeta(n+1) \tag{A.49}$$

The function $\zeta(n)$ is the 'Riemann zeta function'

$$\zeta(n) = \sum_{p=0}^{\infty} \frac{1}{p^n} = 1 + \frac{1}{2^n} + \frac{1}{3^n} + \cdots + \frac{1}{p^n} + \cdots \tag{A.50}$$

The most often used values of $\zeta(n)$ are: for integer n, $\zeta(2) = \pi^2/6$, $\zeta(3) \simeq 1.202$, $\zeta(4) = \pi^4/90$, and for half-integer n, $\zeta(3/2) \simeq 2.612$, $\zeta(5/2) \simeq 1.342$. The most common special cases of (A.48) for the Bose–Einstein distribution are

$$B_1 = \int_0^\infty \mathrm{d}x \, \frac{x}{\mathrm{e}^x - 1} = \frac{\pi^2}{6} \tag{A.51}$$

$$B_2 = \int_0^\infty \mathrm{d}x \, \frac{x^2}{\mathrm{e}^x - 1} = 2\zeta(3) \tag{A.52}$$

$$B_3 = \int_0^\infty \mathrm{d}x \, \frac{x^3}{\mathrm{e}^x - 1} = \frac{\pi^4}{15} \tag{A.53}$$

and of (A.49) for the Fermi–Dirac distribution are

$$F_1 = \int_0^\infty \mathrm{d}x \, \frac{x}{\mathrm{e}^x + 1} = \frac{\pi^2}{12} \tag{A.54}$$

$$F_2 = \int_0^\infty \mathrm{d}x \, \frac{x^2}{\mathrm{e}^x + 1} = \frac{3}{2}\zeta(3) \tag{A.55}$$

$$F_3 = \int_0^\infty \mathrm{d}x \, \frac{x^3}{\mathrm{e}^x + 1} = \frac{7\pi^4}{120} \tag{A.56}$$

A.6 Functional derivatives

Let $\mathcal{F}$ be a linear functional space and I a mapping of $\mathcal{F}$ into the space $\mathbb{R}$ of real numbers[6]

$$\mathcal{F} \longrightarrow \mathbb{R}$$

Thus, to every function $f \in \mathcal{F}$, there corresponds a real number $I(f) : f \longmapsto I[f]$ and $I[f]$ is a *functional* of f. In order to avoid any confusion, we put the argument of a functional between square brackets. An elementary example of a functional is

$$I_1[f] = \int \mathrm{d}x\, g(x) f(x) \tag{A.57}$$

where $g(x)$ is a given function. Other examples are $I_{x_0}[f] = f(x_0)$ or $I_M[f] = \mathrm{Max}\,|f(x)|$. The *functional derivative* $\delta I/\delta f(x)$, which is a *function* of x, is defined thanks to a generalization of the Taylor expansion. Let ε be a small number, $h(x) \in \mathcal{F}$ an arbitrary function of x and expand $I[f + \varepsilon h]$ to first order in ε

$$\boxed{I[f + \varepsilon h] = I[f] + \varepsilon \int \mathrm{d}x\, \frac{\delta I}{\delta f(x)} h(x) + \mathcal{O}(\varepsilon^2)} \tag{A.58}$$

Equation (A.58) *defines* the functional derivative of $I[f]$. Let us illustrate (A.58) with a few examples

(*i*) Our first example is a slight generalization of (A.57)

$$I_p[f] = \int \mathrm{d}x\, g(x) f^p(x)$$

In this case the Taylor expansion (A.58) reads

$$\begin{aligned} I_p[f + \varepsilon h] &= \int \mathrm{d}x\, g(x)\Big(f^p(x) + \varepsilon p f^{p-1}(x) h(x)\Big) + \mathcal{O}(\varepsilon^2) \\ &= I_p[f] + \varepsilon \int \mathrm{d}x\, p g(x) f^{p-1}(x) + \mathcal{O}(\varepsilon^2) \end{aligned}$$

and identification with the definition (A.58) yields

$$\frac{\delta I_p}{\delta f(x)} = p g(x) f^{p-1}(x) = g(x) \frac{\mathrm{d}}{\mathrm{d}x} f^p(x)$$

(*ii*)

$$I_V = \int \mathrm{d}x\, V[f(x)]$$

[6] The case of a mapping into $\mathbb{R}$ is taken for simplicity, but one can easily generalize to mappings into $\mathbb{R}^N$ or to spaces of complex numbers.

One may use a Taylor expansion of $V[f]$ in powers of f to show that

$$\frac{\delta I_V}{\delta f(x)} = V'[f(x)] \tag{A.59}$$

(*iii*)

$$I_D[f] = \int_{x_1}^{x_2} \mathrm{d}x \left(\frac{\mathrm{d}f}{\mathrm{d}x}\right)^2$$

The Taylor expansion gives

$$I[f + \varepsilon h] = I[f] + 2\varepsilon \int_{x_1}^{x_2} \mathrm{d}x\, f'(x)h'(x) + \mathcal{O}(\varepsilon^2)$$

In order to display $h(x)$ instead of its derivative, one integrates by parts and finds

$$\boxed{\frac{\delta I_D}{\delta f(x)} = -2f''(x) + 2f'(x)(\delta(x - x_2) - \delta(x - x_1))} \tag{A.60}$$

The coefficient of $f'(x)$ depends on the boundary conditions, which can generally be ignored.

(*iv*) By choosing $g(x) = \delta(x - x_0)$ in (A.57) one finds

$$I_{x_0}[f] = f(x_0) \qquad \frac{\delta I_{x_0}}{\delta f(x)} = \delta(x - x_0) \tag{A.61}$$

It is left to the reader to show that

$$I_M[f] = \mathrm{Max}\,|f(x)| \qquad \frac{\delta I_M}{\delta f(x)} = \delta(x - x_0)$$

if $f(x)$ has a *unique* extremum at $x = x_0$. When $f(x)$ has many extrema, the functional derivative of I_M is not defined: $I[f]$ is not 'differentiable' at 'point' f.

Let us illustrate the concept of functional derivative on a very important physical example. Consider a particle of unit mass moving on a line (just to simplify the discussion), its position $q(t)$ being a function of time. The *action* S between two times t_1 and t_2 is a functional of $q(t)$ defined by

$$S[q(t)] = \int_{t_1}^{t_2} \mathrm{d}t \left(\frac{1}{2}\dot{q}(t)^2 - \frac{1}{2}V[q(t)]\right) \tag{A.62}$$

where $V(t)$ is the potential energy. Consider a small variation $\delta q(t)$ of the trajectory with the boundary conditions

$$\delta q(t_1) = \delta q(t_2) = 0$$

Maupertuis' principle implies that the physical trajectory $\overline{q}(t)$ should be an extremum of the action

$$\left.\frac{\delta S}{\delta q(t)}\right|_{q(t)=\overline{q}(t)} = 0$$

Then, from (A.59)–(A.60) and the boundary conditions, we obtain the equation of motion (Newton's law)

$$\ddot{q}(t) + V'[(q(t)] = 0 \tag{A.63}$$

The general Taylor expansion of a functional reads

$$I[f+h] = I[f] + \sum_{N=1}^{\infty} \frac{1}{N!} \int \mathrm{d}x_1 \cdots \mathrm{d}x_N \, \frac{\delta^N I}{\delta f(x_1) \cdots \delta f(x_N)} h(x_1) \cdots h(x_N) \tag{A.64}$$

Let us give an example

$$I[f] = \exp\left(\int \mathrm{d}x \, f(x) g(x)\right) = \sum_{N=0}^{\infty} \frac{1}{N!} \left(\int \mathrm{d}x \, f(x) \, g(x)\right)^N$$
$$= \sum_{N=1}^{\infty} \frac{1}{N!} \int \mathrm{d}x_1 \cdots \mathrm{d}x_N \, g(x_1) \cdots g(x_N) \, f(x_1) \cdots f(x_N)$$

This is a Taylor expansion around $f = 0$ and we find

$$\left.\frac{\delta^N I}{\delta f(x_1) \cdots \delta f(x_N)}\right|_{f=0} = g(x_1) \cdots g(x_N)$$

Let us finally give the differentiation rule which generalizes the formula for ordinary functions $(f(g(x)))' = g'(x) f'(g(x))$. Let $\varphi(y)$ be a functional of $f(x)$, for example

$$\varphi(y) = \int \mathrm{d}x \, K(y, x) f(x)$$

and compute $\delta I[\varphi[f]]/\delta f(x)$

$$I\big[\varphi[f + \varepsilon h]\big] \simeq I\left\{\varphi[f] + \varepsilon \int \mathrm{d}x \, \frac{\delta \varphi(y)}{\delta f(x)} h(x)\right\}$$
$$\simeq I\big[\varphi[f]\big] + \varepsilon \int \mathrm{d}x \, \mathrm{d}y \frac{\delta I}{\delta \varphi(y)} \frac{\delta \varphi(y)}{\delta f(x)} h(x)$$

From the definition (A.58) of the functional derivative we get

$$\boxed{\frac{\delta I}{\delta f(x)} = \int \mathrm{d}y \, \frac{\delta I}{\delta \varphi(y)} \frac{\delta \varphi(y)}{\delta f(x)}} \tag{A.65}$$

A.7 Units and physical constants

The physical constants below are given with a relative precision of 10^{-3}, which is sufficient for the numerical applications in this book.

Speed of light in vacuum	$c = 3.00 \times 10^{8}\ \mathrm{m\,s^{-1}}$
Planck constant	$h = 6.63 \times 10^{-34}\ \mathrm{J\,s}$
Planck constant divided by 2π	$\hbar = 1.055 \times 10^{-34}\ \mathrm{J\,s}$
Electronic charge (absolute value)	$e = 1.602 \times 10^{-19}\ \mathrm{C}$
Gravitational constant	$G = 6.67 \times 10^{-11}\ \mathrm{N\,m^2 kg^{-2}}$
Electron mass	$m_{\mathrm{e}} = 9.11 \times 10^{-31}\ \mathrm{kg} = 0.511\ \mathrm{MeV\ c^{-2}}$
Proton mass	$m_{\mathrm{p}} = 1.67 \times 10^{-27}\ \mathrm{kg} = 938\ \mathrm{MeV\ c^{-2}}$
Bohr magneton	$\mu_{\mathrm{B}} = e\hbar/(2m_{\mathrm{e}}) = 5.79 \times 10^{-5}\ \mathrm{eV\ T^{-1}}$
Nuclear magneton	$\mu_{\mathrm{N}} = e\hbar/(2m_{\mathrm{p}}) = 3.15 \times 10^{-8}\ \mathrm{eV\ T^{-1}}$
Avogadro's number	$\mathcal{N} = 6.02 \times 10^{23}\ \mathrm{mol^{-1}}$
Ideal gas constant	$R = 8.31\ \mathrm{J\ K^{-1}\ mol^{-1}}$
Boltzmann constant	$k = R/\mathcal{N} = 1.38 \times 10^{-23}\ \mathrm{J\ K^{-1}}$
Density of ideal gas	$n = 2.7 \times 10^{25}\ \mathrm{m^{-3}}$ (Standard temperature and pressure)
Electronvolt and temperature	$1\ \mathrm{eV} = 1.602 \times 10^{-19}\ \mathrm{J} = k \times 11\,600\ \mathrm{K}$

References

[1] D. Amit, *Field Theory, the Renormalization Group and Critical Phenomena*, Singapore, World Scientific, 1984.

[2] M. H. Anderson, J. R. Ensher, M. R. Matthews, C. E. Wieman, and E. A. Cornell, Observation of Bose–Einstein Condensation in a Dilute Atomic Vapor, *Science*, **269** (1995), 198.

[3] N. Ashcroft and N. Mermin, *Solid State Physics*, Philadelphia, Saunders College, 1976.

[4] R. Baierlein, *Thermal Physics*, Cambridge, Cambridge University Press, 1999.

[5] R. Balian, *From Microphysics to Macrophysics*, Berlin, Springer-Verlag, 1991.

[6] R. Balian, Incomplete Descriptions and Relevant Entropies, *American Journal of Physics*, **67** (1999), 1078.

[7] R. Balian, Y. Alhassid, and H. Reinhardt, Dissipation in Many-body Theory: A Geometric Approach Based on Information Theory, *Physics Reports*, **131** (1986), 1.

[8] R. Balian and J.-P. Blaizot, Stars and Statistical Physics: a Teaching Experience, *American Journal of Physics*, **67** (1999), 1189.

[9] M. Barber, Finite Size Scaling, in *Phase Transitions and Critical Phenomena Volume 8*, C. Domb and J. Lebowitz, eds., London, Academic Press, 1983.

[10] G. Batrouni, Gauge Invariant Mean-Plaquette Method for Lattice Gauge Theories, *Nuclear Physics B*, **208** (1982), 12.

[11] G. Batrouni, E. Dagotto, and A. Moreo, Mean-Link Analysis of Lattice Spin Systems, *Physics Letters B*, **155** (1984), 263.

[12] G. Batrouni, G. Katz, A. Kronfeld, G. Lepage, B. Svetitsky, and K. Wilson, Langevin Simulations of Lattice Field Theories, *Physical Review D*, **32** (1985), 2736.

[13] G. Batrouni and R. Scalettar, World Line Simulations of the Bosonic Hubbard Model in the Ground State, *Computer Physics Communications*, **97** (1996), 63.

[14] R. Baxter, *Exactly Solved Models in Statistical Mechanics*, London, Academic Press, 1982.

[15] G. Baym and C. Pethick, *Landau-Fermi Liquid Theory*, New York, John Wiley, 1991.

[16] K. Bernardet, G. G. Batrouni, J.-L. Meunier, G. Schmid, M. Troyer, and A. Dorneich, Analytical and Numerical Study of Hardcore Bosons in Two Dimensions, *Physical Review B*, **65** (2002), 104519.

[17] K. Binder and D. Heermann, Monte Carlo Simulations in Statistical Physics, Berlin, Springer-Verlag, 1992.

[18] R. Blankenbecler, D. Scalapino, and R. Sugar, Monte Carlo Calculations of Coupled Boson-Fermion Systems, *Physical Review D*, **24** (1981), 2278.

[19] W. Brenig, Statistical Theory of Heat: Nonequilibrium Phenomena, Berlin, Springer-Verlag, 1989.

[20] J. Bricmont, Science of Chaos or Chaos in Science?, *Annals of the NY Academy of Sciences*, **79** (1996), 131.

[21] A. Bruce and D. Wallace, Critical Phenomena: Universality of Physical Laws for Large Length Scales, in *The New Physics*, Cambridge, Cambridge University Press, 1992.

[22] S. Brush, History of the Lenz-Ising Model, *Reviews of Modern Physics*, **39** (1967), 883.

[23] K. Burnett, M. Edwards, and C. Clark, The Theory of Bose-Einstein Condensation of Dilute Gases, *Physics Today*, **52** (1999), 37.

[24] H. Callen, *Thermodynamics and an Introduction to Thermostatistics*, New York, John Wiley, 1985.

[25] J. Cardy, *Scaling and Renormalization in Statistical Physics*, Cambridge, Cambridge University Press, 1996.

[26] P. Chaikin and T. Lubensky, *Principles of Condensed Matter Physics*, Cambridge, Cambridge University Press, 1995.

[27] M. Challa and D. Landau, *Critical Behavior of the Six-state Clock Model in Two Dimensions*, *Physical Review B*, **33** (1986), 437.

[28] D. Chandler, *Introduction to Modern Statistical Mechanics*, Oxford, Oxford University Press, 1987.

[29] A. I. Chumakov and W. Sturhahn, Experimental Aspects of Inelastic Nuclear Resonance Scattering, *Hyperfine Interactions*, **123/124** (1999), 781.

[30] C. Cohen-Tannoudji, B. Diu, and F. Laloë, *Quantum Mechanics*, New York, John Wiley, 1977.

[31] M. Creutz and J. Freedman, A Statistical Approach to Quantum-mechanics, *Annals of Physics*, **132** (1981), 427.

[32] F. Dalfovo, S. Giorgini, L. Pitaevskii, and S. Stringari, Theory of Bose-Einstein Condensation in Trapped Gases, *Reviews of Modern Physics*, **71** (1999), 463.

[33] J. Dorfman, *An Introduction to Chaos in Nonequilibrium Statistical Mechanics*, Cambridge, Cambridge University Press, 1999.

[34] B. Doubrovine, S. Novikov, and A. Fomenko, *Géométrie Contemporaine*, Éditions de Moscou, 1985.

[35] J. Drouffe and J. Zuber, Strong Coupling and Mean Field Methods in Lattice Gauge-theories, *Physics Reports*, **102** (1983), 1.

[36] F. Dyson and A. Lenard, Stability of Matter, *Journal of Mathematical Physics*, **8** (1967), 423.

[37] T. E. Faber, *Hydrodynamics for Physicists*, Cambridge, Cambridge University Press, 1995.

[38] P. Fazekas and P. Anderson, On the Ground State Properties of the Anisotropic Triangular Antiferromagnet, *Philosophical Magazine*, **38** (1974), 423.

[39] R. Feynman, *The Character of Physical Law*, Cambridge, MA, MIT Press, 1967.

[40] R. Feynman and A. Hibbs, *Quantum Mechanics and Path Integrals*, New York, McGraw-Hill, 1965.

[41] E. Fick and G. Sauermann, *The Quantum Statistics of Dynamic Processes*, Berlin, Springer-Verlag, 1990.

[42] K. A. Fisher and J. A. Hertz, *Spin Glasses*, Cambridge, Cambridge University Press, 1993.
[43] D. Foerster, *Hydrodynamic Fluctuations, Broken Symmetry, and Correlation Functions*, New York, W. A. Benjamin, 1975.
[44] C. M. Fortuin and P. W. Kasteleyn, Random-cluster Model 1: Introduction and Relation to other Models, *Physica*, **57** (1972), 536.
[45] D. Fried, T. C. Killian, L. Willmann, D. Landhuis, J. C. Moss, D. Kleppner, and T. J. Greytak, Bose-Einstein Condensation of Atomic Hydrogen, *Physical Review Letters*, **81** (1998), 3811.
[46] C. Gardiner, *Handbook of Stochastic Methods: For Physics, Chemistry and the Natural Sciences*, Berlin, Springer-Verlag, 1996.
[47] P. Gaspard, *Chaos, Scattering and Statistical Mechanics*, Cambridge, Cambridge University Press, 1998.
[48] H. Goldstein, *Classical Mechanics*, Reading, Addison Wesley, 1980.
[49] D. Goodstein, *States of Matter*, Englewood Cliffs, New Jersey, Prentice-Hall, 1975.
[50] H. Gould and J. Tobochnik, *An Introduction to Computer Simulation Methods*, Reading, Addison-Wesly, 1996.
[51] H. Grabert, *Projection Operator Techniques in Nonequilibrium Statistical Mechanics*, Berlin, Springer-Verlag, 1982.
[52] H. Greenside and E. Helfand, Numerical-integration of Stochastic Differential Equations, *Bell Systems Technical Journal*, **60** (1981), 1927.
[53] E. Guyon, J.-P. Hulin, L. Petit, and C. D. Mitescu, *Physical Hydrodynamics*, New York, Oxford University Press, 2001.
[54] J.-P. Hansen and I. Mc Donald, *Theory of Simple Liquids*, New York, Academic Press, 1997.
[55] M. Heiblum and A. Stern, Fractional Quantum Hall Effect, *Physics World*, **13** (2000), 37.
[56] J. Hirsch, R. Sugar, D. Scalapino, and R. Blankenbecler, Monte Carlo Simulations of One-dimensional Fermion Systems, *Physical Review B*, **26** (1982), 5033.
[57] K. Huang, *Statistical Mechanics*, New York, John Wiley, 1963.
[58] C. Itzykson and J. Drouffe, Statistical Field Theory, Cambridge, Cambridge University Press, 1989.
[59] J. K. Jain, The Composite Fermion: A Quantum Particle and its Quantum Fluids, *Physics Today*, **53** (2000), 39.
[60] E. Jaynes, Violation of Boltzmann's H-theorem in Real Gases, *Physical Review A*, **4** (1971), 747.
[61] L. Kadanoff and P. Martin, Hydrodynamic Equations and Correlation Functions, *Annals of Physics*, **24** (1963), 419.
[62] W. Ketterle, Experimental Studies of Bose-Einstein Condensation, *Physics Today*, **52** (1999), 30.
[63] C. Kittel, *Quantum Theory of Solids*, New York, John Wiley, 1987.
[64] C. Kittel, *Introduction to Solid State Physics*, New York, John Wiley, 1996.
[65] D. Knuth, *The Art of Computer Programming: Semi Numerical Algorithms*, vol. II, Reading, Addison-Wesley, 1981.
[66] J. Kogut, *An Introduction to Lattice Gauge Theory and Spin Systems*, *Reviews of Modern Physics*, **51** (1979), 659.
[67] H. J. Kreuzer, *Non Equilibrium Thermodynamics and its Statistical Foundations*, Oxford, Clarendon Press, 1981.
[68] R. Kubo, *Statistical Mechanics*, Amsterdam, North Holland, 1971.

[69] L. Landau and E. Lifschitz, *Mechanics*, Oxford, Pergamon Press, 1976.
[70] L. Landau and E. Lifschitz, *Statistical Physics*, Oxford, Pergamon Press, 1980.
[71] R. B. Laughlin, Fractional Quantization, *Reviews of Modern Physics*, **71** (1999), 863.
[72] M. Le Bellac, *Quantum and Statistical Field Theory*, Oxford, Clarendon Press, 1991.
[73] M. Le Bellac, *Thermal Field Theory*, Cambridge, Cambridge University Press, 1996.
[74] J. Lebowitz, Boltzmann's Entropy and Time's Arrow, *Physics Today*, **46** (1993), 32.
[75] J. Lebowitz, Microscopic Origins of Irreversible Macroscopic Behavior, *Physica A*, **263** (1999), 516.
[76] J. Lebowitz and P. Résibois, Microscopic Theory of Brownian Motion in an Oscillating Field; Connection with Macroscopic Theory, *Physical Review*, **139** (1963), 1101.
[77] A. Leggett, The Physics of Low Temperatures, Superconductivity and Superfluidity, *in The New Physics*, Cambridge, Cambridge University Press, 1992.
[78] A. Leggett, Superfluidity, *Reviews of Modern Physics*, **71** (1999), 318.
[79] D. Levesque and L. Verlet, Molecular Dynamics and Time Reversibility, *Journal of Statistical Physics*, **72** (1993), 519.
[80] J.-M. Lévy-Leblond and F. Balibar, *Quantics: Rudiments of Quantum Physics*, New York, North Holland, 1990.
[81] E. Lifschitz and L. Pitaevskii, *Physical Kinetics*, Oxford, Pergamon Press, 1981.
[82] O. V. Lounasmaa, *Experimental Principles and Methods below 1 K*, London, Academic Press, 1974.
[83] O. V. Lounasmaa, Towards the Absolute Zero, *Physics Today*, **32** (1979), 32.
[84] S. Ma, *Modern Theory of Critical Phenomena*, Philadelphia, Benjamin, 1976.
[85] S. Ma, *Statistical Mechanics*, New York, John Wiley, 1985.
[86] L. Mandel and E. Wolf, *Optical Coherence and Quantum Optics*, Cambridge, Cambridge University Press, 1995.
[87] F. Mandl, *Statistical Physics*, New York, John Wiley, 1988.
[88] D. McQuarrie, *Statistical Mechanics*, New York, Harper & Row, 1976.
[89] A. Messiah, *Quantum Mechanics*, Mineola, Dover Publications, 1999.
[90] N. Metropolis, A. Rosenbluth, M. Rosenbluth, A. H. Teller, and E. Teller, Equation of State Calculations by Fast Computing Machines, *Journal of Chemical Physics*, **21** (1953), 1087.
[91] N. Metropolis and S. Ulam, The Monte Carlo Method, *Journal of the American Statistical Association*, **44** (1949), 335.
[92] M.-O. Mewes, M. R. Andrews, N. J. van Druten, D. M. Kurn, D. S. Durfee, and W. Ketterle, Bose-Einstein Condensation in a Tightly Confining dc Magnetic Trap, *Physical Review Letters*, **77** (1996), 416.
[93] F. Mila, Low Energy Sector of the $s = 1/2$ Kagome Antiferromagnet, *Physical Review Letters*, **81** (1998), 2356.
[94] M. Newman and G. Barkema, *Monte Carlo Methods in Statistical Physics*, Oxford, Oxford University Press, 1999.
[95] M. P. Nightingale and H. W. J. Blöte, Dynamic Exponent of the Two-dimensional Ising Model and Monte Carlo Computation of the Subdominant Eigenvalue of the Stochastic Matrix, *Physical Review Letters*, **76** (1996), 4548.
[96] P. Nozières and D. Pines, *The Theory of Quantum Liquids*, vol. II, New York, Addison-Wesley, 1990.

[97] L. Onsager, Liquid Crystal Statistics I: A Two-dimensional Model with an Order-disorder Transition, *Physical Review*, **65** (1944), 117.
[98] L. Onsager and O. Penrose, Bose-Einstein Condensation and Liquid Helium, *Physical Review*, **104** (1956), 576.
[99] G. Parisi, *Statistical Field Theory*, New York, Addison-Wesley, 1988.
[100] R. Penrose, *The Emperor's New Mind*, Oxford, Oxford University Press, 1989.
[101] P. Pfeuty and G. Toulouse, *Introduction to the Renormalization Group and Critical Phenomena*, New York, John Wiley, 1977.
[102] D. Pines and P. Nozières, *The Theory of Quantum Liquids*, vol. I New York, Addison-Wesley, 1989.
[103] W. Press, S. Teukolsky, W. Vetterling, and B. Flannerry, *Numerical Recipes*, Cambridge, Cambridge University Press, 1992.
[104] I. Prigogine, Laws of Nature, Probability and time Symmetry Breaking, *Physica A*, **263** (1999), 528.
[105] E. Purcell and R. Pound, A Nuclear spin System at Negative Temperature, *Physical Review*, **81** (1951), 279.
[106] A. Ramirez, A. Hayashi, R. Cava, R. Siddhartan, and B. Shastry, Zero Point Entropy in "Spin Ice", *Nature*, **399** (1999), 333.
[107] J. Rau and B. Müller, From Reversible Quantum Dynamics to Irreversible Quantum Transport, *Physics Reports*, **272** (1996), 1.
[108] L. E. Reichl, *A Modern Course in Statistical Physics*, New York, John Wiley, 1998.
[109] F. Reif, *Fundamentals of Statistical and Thermal Physics*, New York, McGraw-Hill, 1965.
[110] H. Risken, *The Fokker-Planck Equation: Methods of Solution and Applications*, Berlin, Springer Verlag, 1996.
[111] W. K. Rose, *Advanced Stellar Astrophysics*, Cambridge, Cambridge University Press, 1998.
[112] L. Saminadayar, D. C. Glattli, Y. Jin, and B. Etienne, Observation of the *e*/3 Fractionally Charged Laughlin Quasiparticle, *Physical Review Letters*, **79** (1997), 2526.
[113] R. Savit, Duality in Field Theory and Statistical Systems, *Reviews of Modern Physics*, **52** (1980), 453.
[114] D. Schroeder, *Thermal Physics*, New York, Addison-Wesley, 2000.
[115] F. Schwabl, *Statistical Mechanics*, Berlin, Springer-Verlag, 2002.
[116] H. Stormer, The Fractional Quantum Hall Effect, *Reviews of Modern Physics*, **71** (1999), 875.
[117] R. Streater and A. Wightman, *PCT, Spin and Statistics and All That*, New York, Benjamin, 1964.
[118] D. R. Tilley and J. Tilley, *Superfluidity and Superconductivity*, Bristol, IOP Publishing, 1990.
[119] N. van Kampen, *Stochastic Processes in Physics and Chemistry*, Amsterdam, North-Holland, 2001.
[120] G. Wannier, Antiferromagnetism: The Triangular Ising Net, *Physical Review*, **79** (1950), 357.
[121] N. Wax, *Selected Papers on Noise and Stochastic Processes*, New York, Dover, 1954.
[122] S. Weinberg, *The First Three Minutes: A Modern View of the Origin of the Universe*, New York, Basic Books, 1993.

[123] J. Wilks, *Liquid and Solid Helium*, Oxford, Clarendon Press, 1967.
[124] M. Zemansky, *Heat and Thermodynamics*, New York, McGraw-Hill, 1957.
[125] J. Zinn-Justin, *Quantum Field Theory and Critical Phenomena*, Oxford, Oxford University Press, 2002.
[126] D. Zubarev, V. Morozov, and G. Röpke, *Statistical Mechanics of Nonequilibrium Processes*, Berlin, Akademie Verlag, 1996.

Index